MW01675291

Community and Family-Focused Public Health and Sustainable Development

Community and Family-Focused Public Health and Sustainable Development

Editors

Wioletta Zukiewicz-Sobczak
Paulina Wojtyla-Buciora
Izabela Racka
Andrzej Wojtyla

MDPI • Basel • Beijing • Wuhan • Barcelona • Belgrade • Manchester • Tokyo • Cluj • Tianjin

Editors

Wioletta Zukiewicz-Sobczak
Calisia University
Poland

Paulina Wojtyla-Buciora
Calisia University
Poland

Izabela Racka
State University of Applied Sciences
Poland

Andrzej Wojtyla
State University of Applied Sciences
Polan

Editorial Office
MDPI
St. Alban-Anlage 66
4052 Basel, Switzerland

This is a reprint of articles from the Special Issue published online in the open access journal *International Journal of Environmental Research and Public Health* (ISSN 1660-4601) (available at: https://www.mdpi.com/journal/ijerph/special_issues/family_health).

For citation purposes, cite each article independently as indicated on the article page online and as indicated below:

LastName, A.A.; LastName, B.B.; LastName, C.C. Article Title. *Journal Name* **Year**, *Volume Number*, Page Range.

ISBN 978-3-0365-3783-2 (Hbk)
ISBN 978-3-0365-3784-9 (PDF)

© 2022 by the authors. Articles in this book are Open Access and distributed under the Creative Commons Attribution (CC BY) license, which allows users to download, copy and build upon published articles, as long as the author and publisher are properly credited, which ensures maximum dissemination and a wider impact of our publications.
The book as a whole is distributed by MDPI under the terms and conditions of the Creative Commons license CC BY-NC-ND.

Contents

About the Editors

Wioletta Zukiewicz-Sobczak is an Associate Professor at Calisia University and graduated from the University of Life Sciences in Lublin. In 2009, she obtained a PhD in the field of food and nutrition technology from the same university. In 2014, she received a postdoctoral degree in health sciences at the Medical University of Poznań. Currently, she is an Associate Professor and an academic teacher at the Department of Food and Nutrition. As a researcher, she conducts research in the field of health sciences, dietetics, food, nutrition, and sustainable development. She has been the author and co-author of over 200 peer-reviewed publications and 2 patents. She has also acted as a reviewer in foreign and Polish journals. She has also been a Guest Editor for five special issues in collaboration with MDPI, two of which are finished. Within the framework of scientific cooperation, Prof. Wioletta Agnieszka Żukiewicz-Sobczak has participated in internships at institutions such as at the University of Bari (Italy) and the University in Lund (Sweden) as well as at the Institute of Sustainability in Rio de Janerio (Brasil). She is currently the manager of two research projects.

Paulina Wojtyla-Buciora is an Associate Professor at Calisia University and graduated from the Adam Mickiewicz University. In 2012, she she obtained a PhD in the field of health sciences at the Medical University of Poznań. In 2019, she received a postdoctoral degree in health sciences at the Institute of Rural Medicine in Lublin. Currently, she is an Associate Professor and an academic teacher at the Department of Public Health. As a researcher, she conducts research in the field of health sciences, epidemiology, food, lifestyle and lifestyle change, and the impact of environmental threats on the health of the population. She has been the author and co-author of over 80 peer-reviewed publications and 2 patents. She has acted as a reviewer in foreign and Polish journals. She is also the Editorial Secretary and Technical Editor of the Journal of Health Inequalities. She has served as the Guest Editor of two Special Issues in collaboration with MDPI, one of which is finished. Within the framework of scientific cooperation, Prof. Paulina Wojtył-Buciora has participated internships at institutions such as the International Alliance of Patient Organizations (IAPO) based in London and the International Prevention Research Institute, (IPRI) in Lyon, France. She is currently the manager of one research projects.

Izabela Racka is an Assistant Professor at Calisia University—Kalisz, Poland and graduated from the Poznań University of Economics and Business. In 2016, she obtained a PhD in the field of economics from the same university. Currently, she is an Assistant Professor and an academic teacher at the Department of Management. As a researcher, she conducts research covering real estate economics, the housing market, cities, urban renewal, and sustainable development. She has been the author and co-author of over 50 peer-reviewed publications. She has also served as a reviewer in foreign and Polish journals. She has been a Guest Editor for two Special Issues published in collaboration with MDPI, both of which are finished. Within the framework of scientific cooperation, Dr. Izabela Racka scientifically cooperates with the Aldo Moro University in Bari (Italy) and the University of Economics—Varna (Bulgaria). She is currently the manager of one research project.

Andrzej Wojtyla is the Rector and Associate Professor at Calisia University. He graduated from the Faculty of Medicine at the University of Medical Sciences, Poznan, Poland. In 2005, he obtained a PhD in the field of medical sciences at the Institute of Rural Medicine in Lublin. In 2014, he received a postdoctoral degree in health sciences at the Medical University of Poznan. From 1992–1994, he served as the Minister of Health and Social Welfare of Poland, and from 2005–2006, he served as the Undersecretary of State in the Polish Ministry of Health. From 2006–2010, he was the Chief Sanitary Inspector of Poland. From 2010–2012, he was the Director and Head of the Health Promotion, Food and Nutrition Department, Institute of Rural Health, Lublin, Poland. He has been the coordinator for the development of the Medical Center in Pleszew, Poland, since 2012. From 2013–2015, he worked as a lecturer at the Department of Mother and Child Development—Poznań University of Medical Sciences. Since 2018, he has been the Rector of the State University of Applied Sciences in Kalisz, Poland, and the Director of the World Institute of Family Health at the same university. As a researcher, he conducts research in the field of health sciences, medicine, epidemiology, food, nutrition, the health behaviors of pregnant women, lifestyle and lifestyle change, and the impact of environmental threats on the health of the population. He has been the author and co-author of over 140 peer-reviewed publications and 3 patents. He has served a reviewer in foreign and Polish journals. He is currently the Editor-in-Chief of the Journal of Health Inequalities. He has served as a Guest Editor for two Special Issue published in collaboration with MDPI, one of which is finished. Within the framework of scientific cooperation Prof. Andrzej Wojtyła has participated in internships at institutions such as the Institute for Health Research and Policy, Georgetown University Medical Center—Washington, DC, USA, and as a paediatric specialist in children and adolescent development at the Child Development Center at the Georgetown University Hospital in Washington, DC,USA. From 1995–1996, he completed an internship on health insurance management at the Health Maintenance Organization in Washington. From 1993–1996 he completed postgraduate foreign studies on the administration and management of public health at Georgetown University in Washington and received a Public Health Administration Diploma in 1996. From 1994–1996 he took postgraduate courses abroad on health management administration at the George Washington University, USA, and received a Diploma of Health Management Administration in 1996.

Preface to "Community and Family-Focused Public Health and Sustainable Development"

This book highlights recent research on community and family-focused public health and sustainable development. This Special Issue encompasses novel aspects of applications that are connected with sustainability issues in community and family-focused public health studies. This book gathers selected, rigorously peer-reviewed papers relating to recent trends in public health.

Wioletta Zukiewicz-Sobczak, Paulina Wojtyla-Buciora, Izabela Racka, Andrzej Wojtyla
Editors

International Journal of
Environmental Research and Public Health

MDPI

Article

Physical Fitness and Inflammatory Response to the Training Load of Wheelchair Rugby Players

Łukasz Szymczak [1], Tomasz Podgórski [2], Jacek Lewandowski [1,3], Arkadiusz Janiak [1], Edyta Michalak [2] and Katarzyna Domaszewska [1,2,*]

1 Faculty of Health Sciences, Calisia University, 62-800 Kalisz, Poland; l.szymczak@akademiakaliska.edu.pl (Ł.S.); lewandowski@awf.poznan.pl (J.L.); arjan7@wp.pl (A.J.)
2 Department of Physiology and Biochemistry, Poznan University of Physical Education, 61-871 Poznan, Poland; podgorski@awf.poznan.pl (T.P.); michalak@awf.poznan.pl (E.M.)
3 Department of Musculoskeletal Physiotherapy, Poznan University of Physical Education, 61-871 Poznan, Poland
* Correspondence: domaszewska@onet.pl; Tel.: +48-61-8355190

Abstract: The aim of the study was the evaluation of the hormonal response of wheelchair rugby participants under the half-year training cycle. The study sample included 11 members of the Polish national wheelchair rugby team with spinal cord injury at the cervical level, ranging in age from 21 to 41 years, body weight (72.2 ± 11.53 kg), and body height (182.3 ± 6.11 cm). The disabled individuals with spinal cord injury subjected to the study constitute a homogeneous group in terms of age, body height, weight, and injury level. The study was carried out at the beginning and at the end of a 6-month training period. In the first and second examination, measurements of the peak oxygen uptake (peakVO_2) and blood biochemical analysis were performed (Lactate dehydrogenase (LDH) activity and concentration of creatinine (Cr), total testosterone (TT), free testosterone (FT), and cortisol (C)). A significant change was observed in the concentration of C in the Wheelchair Rugby players' blood between two research periods ($p < 0.05$ (ES:0.76)) and a correlation between the post-training change in FT/C concentration and the change in Cr concentration ($r = -0.6014$, $p < 0.05$). The 6-month training period did not result in overloads within the group of players. However, due to the significant loss of the capacity of the spinal cord injury (SCI) and the possibility of a life-threatening trend, the anabolic/catabolic status of the players should be monitored using blood biochemical indices.

Keywords: spinal cord injury; testosterone; cortisol; creatinine; overtraining

Citation: Szymczak, Ł.; Podgórski, T.; Lewandowski, J.; Janiak, A.; Michalak, E.; Domaszewska, K. Physical Fitness and Inflammatory Response to the Training Load of Wheelchair Rugby Players. *IJERPH* **2022**, *19*, 2228. https://doi.org/10.3390/ijerph19042228

Academic Editors: Wioletta Zukiewicz-Sobczak, Paulina Wojtyla-Buciora, Izabela Rącka, Andrzej Wojtyla and José Carmelo Adsuar Sala

Received: 28 December 2021
Accepted: 14 February 2022
Published: 16 February 2022

Publisher's Note: MDPI stays neutral with regard to jurisdictional claims in published maps and institutional affiliations.

Copyright: © 2022 by the authors. Licensee MDPI, Basel, Switzerland. This article is an open access article distributed under the terms and conditions of the Creative Commons Attribution (CC BY) license (https://creativecommons.org/licenses/by/4.0/).

1. Introduction

Spinal cord injury causes impairment, or total loss of the motor and sensory functions in the areas of thorax, upper and lower limbs, depending on the severity of the damage [1]. The vast majority of such injuries stem from traffic accidents, bullet wounds, falls, jumps into the water, and sports injuries. For the elderly, the injury factors may be a disc, primary or metastatic neoplastic lesions, developing inflammatory processes in the area of the damage, or compression by abnormal blood vessels. The most common injuries occur in the cervical region or at the border of the thoracolumbar region. However, almost half are injuries between C4 and C7 [2,3]. Impairment of the skeletal muscle function, as well as a physiological adaptation to the physical effort, impacts everyday life activities. The rehabilitation process of such persons, apart from health and compensation components, aims at the mobilization of the patient to function together with healthy persons. Accounting for modern knowledge, it was the right decision to introduce sportsmanship to the rehabilitation programs of patients with SCI [4]. SCI person activation practices may apply to several sports disciplines, including wheelchair basketball, tennis, or boccia. One of the sports dedicated to persons with SCI damage is wheelchair rugby [5,6]. This sport, despite the emergence of persons with other impairments, is still dominated by persons with spinal cord injuries involving the cervical spine (tetraplegia), which is one of the most severe musculoskeletal disabilities. Similar to other Paralympics disciplines, the problem of injuries and overloads of the musculoskeletal system in wheelchair rugby is not widely and thoroughly researched. When identifying damages and overloads in handicapped sports, it is important not only to account for the sport's characteristics but also for the type of disability of the participants. Injuries that occur during wheelchair rugby training and matches include those associated with falls and wheelchair impacts (wheelchair rugby is a contact sport and such incidents are acceptable) [5].

Another aspect of damages and overloads is the everyday life in which most of the participants use wheelchairs. Sports-related damages and overloads can overlap with overloads from daily wheelchair use.

Martin et al., Burnham et al., and Castro et al. in their research on the persons with SCI observed the transformation of slow-twitch fibers into fast-twitch fibers, primarily glycolytic [7–9]. Muscle glycogen stores and muscle aerobic potential are reduced due to decreased mitochondria and muscle fiber transformation. The abolition of sympathetic nervous system function is responsible for the rapid development of muscle fatigue in a person with cervical SCI. Incorrect physiological conditions sympathetic innervation is responsible for increasing blood flow to working skeletal muscles, therefore intensifying the cell oxygen supply, and non-muscular energy substrates such as glucose and free fatty acids. Reduced muscle blood flow in an injured situation, lack of muscle pump function, impaired cardiovascular reflexes lead to metabolic changes in the muscle and accumulation of lactate. Increased glycolytic metabolism, with concomitant disturbance of homeostasis, leads to the rapid development of fatigue in working muscles [8,10].

The disturbance in maintaining proportions between the time of training and biological rejuvenation results in decreased physical capability of the participant, greater risk of injury, and decreased body immunity, which is particularly dangerous for this group of patients during the increase in COVID-19 (SARS-CoV-2) infections [11–13]. Hence, the monitoring of fatigue levels using physiological or biochemical indicators will allow for sustaining the participants' optimal health conditions and life comfort despite numerous physical conditions affecting them.

Biochemical indicators of muscle breakdown or cell membrane damage include measurement of creatinine (Cr) or creatine kinase (CK; EC 2.7.3.2) lactate dehydrogenase (LDH; EC 1.1.1.27). Most often the testosterone (T) and cortisol concentrations in the blood are used for the diagnosis of the overload state. The literature analysis indicates that the development of anabolic–catabolic balance is not only influenced by the training's intensity and duration, but also by the training volume, or atmospheric conditions [14,15]. Cortisol is considered to be a catabolic hormone because it stimulates glycogenolysis in the liver by increasing the glucose concentration in blood. It inhibits muscle uptake of amino acids and limits muscle protein synthesis [15,16]. Testosterone via its anabolic impact on the body causes the increase in the synthesis of muscle tissue, forming mass and muscle strength. At rest, only 1% of testosterone circulates in the blood in the free form (FT) and is metabolically active. The remaining part is linked to the sex hormone-binding globulin (SHBG) and with albumins [16,17]. Analysis of the testosterone/cortisol T/C ratio is not only a good indicator of the body's metabolic balance but is also found in the literature to be an indicator of peripheral and central fatigue in athletes [18–20]. Low values of the TT/C and FT/C indicators are proposed to be early biochemical indicators of exercise-induced fatigue [14]. Esspecially in extreme cases, intensified catabolic and limited anabolic processes during athletes' training cycle, TT/C and FT/C ratios seem to be appropriate indicators of the overreaching or overtraining syndrome [20]. The value of these indicators was noticed, for example, in non-disabled rugby players [21]. For athletes with quadriplegia and limited active muscle mass, these conditions can occur at even low exercise loads, and the effects of overtraining are far more dangerous to the athlete's health and even life. In this group of persons the katabolic changes increase in muscle even in resting conditions. There is evidence that the level of endogenic, anabolic hormones may be decreased. Decreasing the testosterone and growth hormones concentration and through them insulin-like growth factor-1 (IGF-1) may intensify unbeneficial body mass composition changes, decreasing the fatigue tolerance [22–24]. Wang et al. indicated in their research that in the group of 63 patients with spinal cord injury, a comparable number of individuals had both decreased and increased blood testosterone concentrations at rest [25]. Hence, the analysis of the concentration of selected hormone levels in the group of elite wheelchair rugby competitors in different stages of preparatory training for the important sporting event was proposed, which will allow for the evaluation of the direction and character of changes.

The aim of the study was the analysis of the hormonal response (TT, FT, C, and TT/C, FT/C ratio) of wheelchair rugby participants under the training.

2. Materials and Methods

The study protocol was approved by the Ethics Committee for Human Research of the Poznan University of Medical Sciences (Ethics Approval Number: 405/12). The study was conducted according to the Declaration of Helsinki and the National Statement. All participants in this study gave their written informed consent.

The study sample included 11 members of the Polish national wheelchair rugby team with spinal cord injury at the cervical level, ranging in age from 21 to 41 years, body weight (72.2 ± 11.53 kg), body height (182.3 ± 6.11 cm). The disabled with spinal cord injury subjected to the study constitute a homogeneous group in terms of age, body height, weight, and the injury level. All of them were tetraplegics with cervical spinal cord injury at levels C5–C7. Wheelchair rugby classification system, used to assess the subjects came from the ISMGF/ISMWSF (International Stoke Mandeville Games Federation/International Stoke Mandeville Wheelchair Sport Federation) and American Spinal Injury Association (ASIA) Impairment Scale, medical classification system (Table 1).

Table 1. Classification points, spinal cord injury level, American Spinal Injury Association (ASIA) Impairment Scale of the subjects with spinal cord injury [11].

No	Spinal Cord Injury Level	American Spinal Injury Association (ASIA) Impairment Scale	Classification Points
1	C6–C7	A	2
2	C5–C6	A	2
3	C6	A	2
4	C6–C7	A	2
5	C6–C7	A	2.5
6	C5–C6	A	2
7	C6–C7	A	2
8	C6–C7	A	2
9	C5	A	2
10	C5–C6	A	1
11	C6–C7	A	0.5

Training period between research dates was characterized by build-up of general capability, with a particular focus on the aerobic capacity and improving elements of game tactics. In the later period, the participants focused their training on the speed. Research in the 2nd term took place after the beginning in a target competition, near the immediate start.

2.1. Assessment of Aerobic Fitness (Graded Exercise Test Protocol)

The subjects performed a cardiopulmonary exercise test with increasing intensity on the Sweedish MONARK REHAB TRAINER 881E manual ergometer, manufactured by Monark Exercise AB (Vansbro, Sweden) and specially adjusted to functional abilities of research participants. The initial workload amounted to 10 Watts was successively increased by 10 Watts every 3 min until the subjects achieved maximum individual workloads or refused to continue the effort. Peak oxygen uptake (peakVO_2) was measured in absolute and relative values with a German Jeager Oxycon Mobile gas analyser (Viasys Healthcare, Höchberg, Germany) [26].

2.2. Methodology of Biochemical Marking

Blood samples were collected before exercise tests (two days after the last intensive training unit) to exclude the influence of training loads on biochemical variables tested. Blood samples were taken from the ulnar vein using a S-Monovette syringe tube (Sarstedt, Nümbrecht, Germany) then placed in tubes containing a clot activator and centrifuged (1500 g, 4 °C, 4 min) in order to separate the serum. The samples were frozen and stored at −75 °C until the time the analyses were performed. Lactate dehydrogenase (LDH) activity and concentration of creatinine (Cr) was determined with the use of the Accent 200S (Cormay, Poland) biochemical analyser and sets of enzymatic reagents (Cormay, Poland). The sensitivity of the sets was 6.6 U/L for LDH and creatinine 7.07 (micro) mol/L (intra-assay coefficient of variation [CV], <10%). Concentrations of total testosterone (TT), free testosterone (FT), cortisol (C), were measured using the ELISA immune-enzymatic method (AssayPro LLC, St. Charles, MO). The sensitivity of the ELISA kits was as follows: 0.083 ng/mL ([CV], 3.61%), 0.018 pg/mL ([CV], 5.96%), 2.5 ng/mL ([CV], 5.63%).

The samples were read using a Synergy 2 SIAFRT multi-detection microplate reader (BioTek, Winusky, VT, USA) at the manufacturer's recommended wavelength.

2.3. Training Program

The preparation cycle of the Polish Wheelchair Rugby Team in 2012 included 30 weeks of training (microcycles) in preparation for the start in the target event—the International Metro Cup Tournament.

- The first stage of preparations, covering the period from March to mid-May 2012 (microcycles 1–12), was focused on building aerobic capacity (raising the level of fitness of players), strengthening muscle mass and perfecting tactical elements of the game (raising the tactical level of players, shaping the reaction speed to changing tactics of the opponent during the game and gathering players from different sports clubs together).
- The second stage of preparations, covering the period from mid-May to the end of August 2012 (microcycles 13–26), was focused on maintaining aerobic capacity and strength endurance (the beginning of the stage) and speed endurance (the end of the stage), improving elements of game tactics (shaping the habits of observation and analysis during physical effort, learning team attitudes).
- The third stage of preparation, covering the period of September 2012 (micro-cycles 27–30), (direct starting preparation) is shaping the speed abilities of players and improving elements of game tactics (evaluation of observation and analysis habits during physical effort).
- Stage IV, covering September/October 2012 (microcycles 31–32) of the year, was the start of the target event.
- The fifth stage, covering the period from October to November 2012 (microcycles 33–36), was a transitional period (retrenchment).

The exact training plan realised by the players during the period covered by the research is shown in Table 2.

Table 2. Wheelchair rugby players' weekly training frequency in a macrocycle (training load description).

Training Goal	Week	Type of Training	Type of Activity	Duration [min]	Intensity [% max]	No. of Training Sessions per Week
Aerobic endurance (development and maintenance)	1–12	Constant intensity aerobic training	Wheelchair push (indoor, outdoor)	60–100	50–60	3
	13–32	Constant and variable intensity aerobic training	Wheelchair push (indoor)	60–100	60–70	2
Speed endurance (anaerobic)	27–32	Interval training	Wheelchair push (indoor)	15–20	70–90	1
Development of muscle strength	1–12	Strength training	Gym	60–80	60–70	2
Maximum strength	13–26	Strength training/functional training	Gym/fitness	40–80	80–90	1
Strength endurance	13–32	Circuit training	Gym/fitness	40–80	50–60	1
Power/Capacity	13–32	Speed and strength training	Gym/fitness	40–80	60–70	1
Stabilisation (deep muscle training)	1–26	Circuit training (repetition)	Fitness room	30	-	1
Tactics/Team play	1–12	Small tactics training	Wheelchair rugby (key, half court play)	120–150	50–75	3
	13–32	Training of small and large tactics	Wheelchair rugby (key, half court play, regular games)	120–150	50–90	3

2.4. Statistical Analysis

All data are presented as mean, standard deviation (SD). The normality of distribution was tested with the Shapiro–Wilk test. The differences between paired and normally distributed variables were analysed using the Wilcoxon test. Spearman's rank analysis was used to calculate correlation coefficients. The level of statistical significance was set at $p \leq 0.05$. The obtained results were analysed statistically using the Dell Inc. (2016) Dell Statistica 13 software (Tulsa, Oklahoma, USA). Effect sizes (ES) were calculated as the difference between means divided by the pooled standard deviation using Cohen's criteria; effect sizes <0.20 and <0.50 were considered small, <0.50 and <0.80 medium, and <0.80 large [27].

3. Results

Resting testosterone levels in the blood of the FT and TT fractions were within the physiological limits of this parameter. The blood cortisol concentration was similar. A slight increase in LDH activity and a decrease in creatinine concentration were found. Values below the lower limit of the reference values were found in 9 persons (82%) in the first research period and 8 persons (73%) in the second research period.

Our research showed a significant change in the concentration of cortisol in the Wheelchair Rugby players' blood between two research period. ($p < 0.05$ (ES:0.76)) and a correlation between the post-training change in FT/C concentration and the change in creatinine concentration ($r = -0.6014$, $p < 0.05$) There was also a post-training insignificant decrease in peakVO$_2$ ($p > 0.05$), resulting from the increase in body weight of the subjects in the second term of the study (Table 3).

Table 3. Basic characteristics (x ± SD) of physiological and biochemical parameters measured in first and the second research period.

Parameter	First Research Period	Second Research Period	p-Value
Cr [μMol/L]	57.1 ± 13.53	57.7 ± 12.14	1.0000
LDH [U/L]	310.7 ± 68.09	307.2 ± 49.02	0.8139
TT [ng/mL]	4.2 ± 0.90	4.2 ± 1.53	0.8753
TF [pg/mL]	8.2 ± 2.71	8.2 ± 3.69	0.8139
C [ng/mL]	150.7 ± 60.14	207.1 ± 52.61	0.0342 *
TT/C	0.03 ± 0.01	0.02 ± 0.01	0.0843
TF/C	0.06 ± 0.03	0.04 ± 0.02	0.0994
Body mass [kg]	72.2 ± 11.53	73.5 ± 7.69	0.5939
peakVO$_2$ [mL/kg/min]	16.8 ± 3,49	15.9 ± 3.52	0.1386

Data are presented as mean ± SD, Cr = creatinine, LDH = Lactate dehydrogenase, TT = total testosterone, FT = free testosterone, C = cortisol, peakVO$_2$ = peak oxygen uptake. * $p < 0.05$

4. Discussion

The size of exercise loads during training camps varies and is dependent on the preparatory period of participants. It needs to be stated that weekend training camp consists of 3 training units lasting 3–4 training hours, and of one theoretical training unit. In the 1st preparatory stage during the weekend camps, participants cover the biggest number of training kilometers in comparison with the speed exercises. This ratio changes with the realization of subsequent training macrocycles.

In our research, the intensive 6-month-long training impacted the selected metabolic parameters on the verification of the overextension state occurrence. Such a state in healthy sports-playing participants requires the application of effective biological regeneration treatments, and decrease of several weeks in the training severity. Healthy persons with high levels of physical capacity have developed effective mechanisms of post-exercise rejuvenation; however, among persons with SCI this mechanism is severely limited. The typical over-exercise symptoms are the loss of physical capacity, loss of muscle mass, the occurrence of the negative nitrogen balance. Such changes are visible in the nervous system functioning, and changes to the secretion of the metabolic hormone [28,29].

From the point of view of training, there are many similarities in the physical reactions to the over-training of healthy and handicapped participants. The hormonal response to the applied workload may differ. Individuals with acquired disabilities respond differently to training loads than individuals with disabilities from birth. It has been demonstrated during the observational studies that persons with acquired disabilities are more motivated to push through their limitations and have stronger stress reactions to the training stimulus, therefore they may react with greater cortisol secretion [6].

The amount of the training load of the rugby wheelchair players during training camps varies and is dependent on the preparation period of the participants. It ought to be noted that the weekend training camps consist of three practical training units of 3–4 h each, and one theoretical training unit. In the first stage of preparation during the weekend training camps the participants cover the greatest amount of training kilometers in comparison with speed training. The proportion changes with the realization of the training macro-cycles. The first date of participants' testing was at the beginning of the preparatory period, during which the general fitness of players was built, especially aerobic capacity. The second date of testing was on the period immediately after the start of the target event. In this period, it was expected to record a significant overload of the body, and damage to the muscular system was expected.

In our research, we concluded that participants with tetraplegia had normal resting testosterone and cortisol blood levels (Table 3). Castellani et al. found a statistically significant ($p < 0.05$) decrease in this hormone in the wheelchair participants in comparison with healthy ones. They found a greater increase in exercise-induced testosterone concentration in wheelchair participants in comparison with healthy ones. They also showed a higher exercise-induced increase in testosterone concentrations in subjects with disabilities. However, there were no differences in cortisol concentrations at rest as well as after exercise [30]. The physical mechanisms leading to an increase in exercise-induced testosterone levels are a result of both its production, breakdown, and excretion. Exercise-induced changes in plasma volume can also modulate its concentration in the blood. In the research of Wheeler et al., it can be concluded that some participants diagnosed with SCI deliberately increase the catecholamine (CA) output by triggering the autonomic dysreflexia mechanism to increase their performance. Therefore, it is assumed that high exercise blood testosterone concentration is a response to catecholamine output or a protective response induced by cortisol [31,32].

There is little research on the biochemical reaction of wheelchair participants to the training load. Limiting the search to the rugby participants gives very few publications useful for the critical analysis of own research results. In the research of Bizzarini et al. on the impact of a 6-month training regime on the persons diagnosed with SCI (6 persons with damaged cervical region, 15 with the damaged thoracic and lumbar regions). There were no significant changes to the testosterone and cortisol levels after the training period. General ratio TT/C, currently perceived to be the indicator of overtraining, did not decrease in the presented study. The authors explain this fact by the lack of intensification of catabolic metabolism in response to physical activity or by the limited ability of the body to metabolically adapt to the intensity of physical exercise [33].

In the research on healthy participants, Ahtiainen et al. described post-training decrease in total testosterone levels, as well as the ratio of TT/C caused by too high training loads [34]. In our research on the SCI participants, the increase in the above-mentioned markers was not recorded, which may be indicative of the high physical capability of the research participants as well as the training load appropriate for their health condition capabilities. This is also indicated by the lack of a post-workout decrease in the peakVO_2, which determines an athlete's oxidative capacity level. The importance of cortisol analysis in the diagnosis of overload conditions in athletes was mentioned by Michailidis in his study on football players [35]. Cortisol is the stress hormone secreted both as a result of physical exertion factors as well as stress factors connected to physical activity. In many research, the post-training increase is dependent on the exercise intensity and acidification levels [36–38]. Based on the Michailidis results it can be concluded that the lack of changes in the C concentration during the participant's preparation cycle is indicative of their good psychophysical preparation [35]. Barboza et al. indicate that athletes with neurological damage (SCI), unlike other disabilities (poliomyelitis, amputation), do not respond to saliva secretion of cortisol in basketball players in wheelchairs [39]. Eventually this can cause problems with the interpretation of the SCI athletes' body response to the effects of training intensity.

Lack of changes in LDH activity or creatinine concentration in our research indicates the sustained muscle cells stability and no damage to them as a result of the training regime. The increase in cortisol in the second term of research may be reflective of stress caused by the matches itself during the target event in the year of the research. A slight increase in LDH enzyme activity and a decrease in creatinine concentration; values below the lower limit of reference values were found in more than 80% of the subjects, most likely due to increased creatinine excretion in urine [40]. Kaji et al. based on the conducted research proved the decrease in urinary creatinine excretion in a group of patients with SCI based on their study. The factor limiting the precision of defining the accuracy of the renal clearance rate in our study was lack of the creatinine concentration in urine measurements [40]. The long-term research of Elmelund et al. on the creatinine levels in the blood of persons with SCI prove that the decrease in this metabolite is caused by the loss of muscle mass with age, which naturally occurs in healthy persons as well. In persons with SCI, this mechanism is accelerated due to the extensive muscle mass loss due to injury itself [41].

The lack of changes in the creatinine blood concentration in blood in participants studied in the second term indicates sustaining the stable muscle mass and lack of damage to the kidneys. We are more interested in the presented findings in terms of the occurrence of possible overload changes resulting in a decrease in muscle mass in individuals with symptoms of overtraining of a peripheral nature. Particularly since, in the case of athletes with SCI, the source of overload is a matter of everyday life, in which most athletes use wheelchairs.

Monitoring of the metabolic hormonal state (TT/C) will allow for both trainers and psychologists treating handicapped patients to fully control the degree of physical participant's preparation as well as their psychological state and motivation. It may also serve as a sensitive biochemical marker of the way to deal with physical limitations accounting for the SCL type, functional classification, or disability due to impairment (acquired or congenital) [42,43].

5. Conclusions

The 6-month training period did not overstrain the group of players. However, due to the significant loss of SCI capacity and the possibility of a life-threatening trend, the anabolic/catabolic status could be used to diagnose early states of overload.

Limitation of the Study

The study was conducted only on persons physically active with spinal cord injuries. In order to obtain a complete picture of anabolic/catabolic biochemical changes under the influence of a 6-month training period, tests would have to be carried out on physically inactive men with spinal cord injury. Unfortunately, this study did not obtain the right consent to search for this group of people.

Author Contributions: Conceptualization, Ł.S. and K.D.; methodology, K.D., Ł.S. and T.P.; software, K.D.; validation, K.D. and T.P.; resources, K.D., Ł.S. and T.P.; data curation, K.D., Ł.S., E.M. and T.P.; writing—original draft preparation, K.D., Ł.S., T.P., J.L. and A.J.; writing—review and editing, K.D., Ł.S., T.P. and J.L.; visualization, K.D.; supervision, K.D. and J.L.; project administration, K.D. and Ł.S. All authors have read and agreed to the published version of the manuscript.

Funding: This research received no external funding.

Institutional Review Board Statement: The study was conducted according to the guidelines of the Declaration of Helsinki and approved by the IRB (Institute for Research in Biomedicine) at the Poznan University of Medical Sciences (5 April 2012; Ethics Approval Number: 405/12).

Informed Consent Statement: Informed consent was obtained from all subjects involved in the study.

Data Availability Statement: The data presented in this study are available on request from the corresponding author. The data are not publicly available due to the consent provided by participants on the use of confidential data.

Conflicts of Interest: The authors declare no conflict of interest.

References

1. Fouad, K.; Popovich, P.G.; Kopp, M.A.; Schwab, J.M.L. The neuroanatomical–functional paradox in spinal cord injury. *Nat. Rev. Neurol.* **2021**, *17*, 53–62. [CrossRef] [PubMed]
2. Maynard, F.M.; Bracken, M.B.; Creasey, G.J.F.D.; Ditunno, J.F.; Donovan, W.H.; Ducker, T.B.; Garber, S.L.; Marino, R.J.; Stover, S.L.; Tator, C.H.; et al. International standards for neurological and functional classification of spinal cord injury. *Spinal Cord* **1997**, *35*, 266–274. [CrossRef]
3. Sekhon, L.H.S.; Fehlings, M.G. Epidemiology, demographics, and pathophysiology of acute spinal cord injury. *Spine* **2001**, *26*, S2–S12. [CrossRef] [PubMed]
4. Guttmann, L. Significance of sport in rehabilitation of spinal paraplegics and tetraplegics. *JAMA* **1976**, *236*, 195–197.
5. Bauerfeind, J.; Koper, M.; Wieczorek, J.; Urbański, P.; Tasiemski, T. Sports Injuries in Wheelchair Rugby—A Pilot Study. *J. Hum. Kinet.* **2015**, *48*, 123–132. [CrossRef]
6. Samulski, D.M.; Noce, F.; Costa, V.T. Mental preparation. In *Handbook of Sports Medicine and Science—The Paralympic Athlete*; Vanlandewijck, Y.C., Thompson, W.R., Eds.; Wiley-Blackwell: Hoboken, NJ, USA, 2011; pp. 198–213.
7. Martin, T.P.; Stein, R.B.; Hoeppner, P.H.; Reid, D.C. Influence of electrical stimulation on the morphological and metabolic properties of paralyzed muscle. *J. Appl. Physiol.* **1992**, *72*, 1401–1406. [CrossRef]

8. Burnham, R.; Martin, T.; Stein, R.; Bell, G.; MacLean, I.; Steadward, R. Skeletal muscle fibre type transformation following spinal cord injury. *Spinal Cord* **1997**, *35*, 86–91. [CrossRef]
9. Castro, M.J.; Apple, D.F., Jr.; Hillegass, E.A.; Dudley, G.A. Influence of complete spinal cord injury on skeletal muscle cross-sectional area within the first 6 months of injury. *Eur. J. Appl. Physiol. Occup. Physiol.* **1999**, *80*, 373–378. [CrossRef]
10. Eggers, L.; Carson, C.; Evans, K.; Swank, A.M.; Adams, K.J.; Barnard, K.L.; Berning, J. Limiting factors for increasing VO2peak for individuals with spinal cord injury. *Clin. Exerc. Physiol.* **2001**, *1*, 10–16.
11. Szymczak, Ł.; Podgórski, T.; Domaszewska, K. Comparison of the levels of hematological parameters at rest and after maximum exercise between physically active people with spinal cord injury and able-bodied people. *Int. J. Environ. Res. Public Health* **2021**, *18*, 12323. [CrossRef]
12. Jin, C.H.; Paik, I.Y.; Kwak, Y.S.; Jee, Y.S.; Kim, J.Y. Exhaustive submaximal endurance and resistance exercises induce temporary immunosuppression via physical and oxidative stress. *J. Exerc. Rehabil.* **2015**, *11*, 198–203. [CrossRef] [PubMed]
13. Gomez-Merino, D.; Drogou, C.; Guezennec, C.Y.; Burnat, P.; Bourrilhon, C.; Tomaszewski, A.; Milhau, S.; Chennaoui, M. Comparison of systemic cytokine responses after a long distance triathlon and a 100-km run: Relationship to metabolic and inflammatory processes. *Eur. Cytokine Netw.* **2006**, *17*, 117–124. [PubMed]
14. Podgórski, T.; Kryściak, J.; Pluta, B.; Adrian, J.; Marynowicz, J.; Krzykała, M.; Konefał, M.; Chmura, P.; Chmura, J.; Andrzejewski, M. A practical approach to monitoring biomarkers of inflammation and muscle damage in youth soccer players during a 6-month training cycle. *J. Hum. Kinet.* **2021**, *80*, 185–197. [CrossRef] [PubMed]
15. Brisswalter, J.; Collardeau, M.; Rene, A. Effects of acute physical exercise characteristics on cognitive performance. *Sports Med.* **2002**, *32*, 555–566. [CrossRef]
16. Chmura, P.; Podgórski, T.; Konefał, M.; Rokita, A.; Chmura, J.; Andrzejewski, M. Endocrine Responses to Various 1 × 1 Small-Sided Games in Youth Soccer Players. *Int. J. Environ. Res. Public Health* **2019**, *16*, 4974. [CrossRef]
17. Fahrner, C.L.; Hackney, A.C. Effects of endurance exercise on free testosterone concentration and the binding affinity of sex hormone binding globulin (SHBG). *Int. J. Sports Med.* **1998**, *19*, 12–15. [CrossRef]
18. Zar, A.; Ahmadi, F.; Krustrup, P.; Fernandes, R.J. Effects of morning and afternoon high-intensity interval training (HIIT) on testosterone, cortisol and testosterone/cortisol ratio response in active men. *Trends Sport Sci.* **2021**, *28*, 179–185. [CrossRef]
19. Urhausen, A.; Kindermann, W. Diagnosis of overtraining: What tools do we have? *Sports Med.* **2002**, *32*, 95–102. [CrossRef]
20. Purvis, D.; Gonsalves, S.; Deuster, P.A. Physiological and psychological fatigue in extreme conditions: Overtraining and elite athletes. *PM&R* **2010**, *2*, 442–450. [CrossRef]
21. Twist, C.; Highton, J. Monitoring fatigue and recovery in rugby league players. *Int. J. Sports Physiol. Perform.* **2013**, *8*, 467–474. [CrossRef]
22. Tsitouras, P.D.; Zhong, Y.G.; Spungen, A.M.; Bauman, W.A. Serum testosterone and growth hormone/insulin-like growth factor-1 in adults with spinal cord injury. *Horm. Metab. Res.* **1995**, *27*, 287–292. [CrossRef] [PubMed]
23. Andrzejewski, M.; Podgórski, T.; Kryściak, J.; Chmura, P.; Konefał, M.; Chmura, J.; Marynowicz, J.; Adrian, J.; Pluta, B. Anabolic-catabolic hormonal responses in youth soccer players during a half-season. *Res. Sports Med.* **2021**, *29*, 141–154. [CrossRef] [PubMed]
24. Rowbottom, D.G.; Keast, D.; Goodman, C.; Morton, A.R. The haematological, biochemical and immunological profile of athletes suffering from the overtraining syndrome. *Eur. J. Appl. Physiol. Occup. Physiol.* **1995**, *70*, 502–509. [CrossRef] [PubMed]
25. Wang, Y.H.; Huang, T.S.; Lien, I.N. Hormone changes in men with spinal cord injuries. *Am. J. Phys. Med. Rehabil.* **1992**, *71*, 328–332. [CrossRef]
26. Domaszewska, K.; Szymczak, Ł.; Kryściak, J.; Pospieszna, B.; Laurentowska, M. Spirometric and ergospirometric evaluation of wheelchair rugby players. *Trends Sport Sci.* **2013**, *2*, 89–94.
27. Cohen, J. *Statistical Power Analysis for the Behavioral Sciences*, 2nd ed.; Lawrence Erlbaum Associates: Hillsdale, NJ, USA, 1988.
28. Fry, A.C.; Kraemer, W.J. Resistance exercise overtraining and overreaching. *Sports Med.* **1997**, *23*, 106–129. [CrossRef]
29. Lehmann, M.J.; Lormes, W.; Opitz-Gress, A.; Steinacker, J.M.; Netzer, N.; Foster, C.; Gastmann, U. Training and overtraining: An overview and experimental results in endurance sports. *J. Sports Med. Phys. Fit.* **1997**, *37*, 7–17.
30. Castellani, J.W.; Armstrong, L.E.; Kenefick, R.W.; Pasqualicchio, A.A.; Riebe, D.; Gabaree, C.L.; Maresh, C.M. Cortisol and testosterone concentrations in wheelchair athletes during submaximal wheelchair ergometry. *Eur. J. Appl. Physiol.* **2001**, *84*, 42–47. [CrossRef]
31. Levin, J.; Lloyd, C.W.; Lobotsky, J.; Friedrich, E.H. The effect of epinephrine on testosterone production. *Acta Endocrinol.* **1967**, *55*, 184–192. [CrossRef]
32. Wheeler, G.; Cumming, D.; Burnham, R.; Maclean, I.; Sloley, B.D.; Bhambhani, Y.; Steadward, R.D. Testosterone, cortisol and catecholamine responses to exercise stress and autonomic dysreflexia in elite quadriplegic athletes. *Spinal Cord* **1994**, *32*, 292–299. [CrossRef]
33. Bizzarini, E.; Saccavini, M.; Lipanje, F.; Magrin, P.; Malisan, C.; Zampa, A. Exercise prescription in subjects with spinal cord injuries. *Arch. Phys. Med. Rehabil.* **2005**, *86*, 1170–1175. [CrossRef] [PubMed]
34. Ahtiainen, J.P.; Pakarinen, A.; Kraemer, W.J.; Häkkinen, K. Acute hormonal and neuromuscular responses and recovery to forced vs. maximum repetitions multiple resistance exercises. *Int. J. Sports Med.* **2003**, *24*, 410–418. [CrossRef]
35. Michailidis, Y. Stress hormonal analysis in elite soccer players during a season. *J. Sport Health Sci.* **2014**, *3*, 279–283. [CrossRef]

36. Kraemer, W.J.; Ratamess, N.A. Hormonal responses and adaptations to resistance exercise and training. *Sports Med.* **2005**, *35*, 339–361. [CrossRef] [PubMed]
37. Banfi, G.; Dolci, A. Free testosterone/cortisol ratio in soccer: Usefulness of a categorization of values. *J. Sports Med. Phys. Fit.* **2006**, *46*, 611–616.
38. Anderson, T.; Lane, A.R.; Hackney, A.C. Cortisol and testosterone dynamics following exhaustive endurance exercise. *Eur. J. Appl. Physiol.* **2016**, *116*, 1503–1509. [CrossRef]
39. Barboza, B.; Oliveira, M.; Lambert, E.; Werle, E.; Marta, S.; Greguol, M. Does the type of disability influence salivary cortisol concentrations of athletes in official wheelchair basketball games? *Int. J. Sports Sci. Coach.* **2019**, *14*, 507–513.
40. Kaji, D.; Strauss, I.; Kahn, T. Serum creatinine in patients with spinal cord injury. *Mt. Sinai J. Med.* **1990**, *57*, 160–164.
41. Elmelund, M.; Oturai, P.S.; Biering-Sørensen, F. 50 years follow-up on plasma creatinine levels after spinal cord injury. *Spinal Cord* **2014**, *52*, 368–372. [CrossRef]
42. Papacosta, E.; Nassis, G.P. Saliva as a tool for monitoring steroid, peptide and immune markers in sport and exercise science. *J. Sci. Med. Sport* **2011**, *14*, 424–434. [CrossRef]
43. Teles, L.J.L.; Aidar, F.J.; de Matos, D.G.; Marçal, A.C.; de Almeida-Neto, P.F.; Neves, E.B.; Moreira, O.C.; Neto, F.R.; Garrido, N.D.; Vilaça-Alves, J.; et al. Static and dynamic strength indicators in paralympic power-lifters with and without spinal cord injury. *Int. J. Environ. Res. Public Health* **2021**, *18*, 5907. [CrossRef] [PubMed]

International Journal of
Environmental Research and Public Health

MDPI

Article

Population Access to Hospital Emergency Departments: The Spatial Analysis in Public Health Research

Wojciech Kisiała [1], Izabela Rącka [2,*] and Katarzyna Suszyńska [3]

1 Department of Econometrics, Institute of Informatics and Quantitative Economics, Poznań University of Economics and Business, 61-875 Poznań, Poland; wojciech.kisiala@ue.poznan.pl
2 Department of Management, Faculty of Social Sciences, Calisia University—Kalisz, 62-800 Kalisz, Poland
3 Department of Investment and Real Estate, Institute of Management, Poznań University of Economics and Business, 61-875 Poznań, Poland; katarzyna.suszynska@ue.poznan.pl
* Correspondence: i.racka@akademia.kalisz.pl; Tel.: +48-60-244-9177

Abstract: The emergency medical services support the primary health care system. Hospital emergency departments (HEDs), which provide medical assistance to all patients in a state of emergency are of considerable importance to the system. When studying access to HEDs, attention should be focused on spatial relations resulting from the location of HEDs and the places of residence of the potential patients. The aim of the paper is to explain the level of spatial accessibility of HEDs and its changes as a result of organizational and spatial transformations of HEDs' networks in Poland. The research was conducted within two time series, comparing the changes in the distribution of HEDs in 2011 and 2021. GIS techniques were used to measure the distances between emergency departments and places of residence. It was observed that the transformation of the spatial organization of the hospital emergency department network in 2011–2021 resulted in the overall improvement of the spatial accessibility of these facilities, reducing the distance between them and places of residence.

Keywords: access to health care; health care services; hospital emergency departments; spatial analysis; public health management

Citation: Kisiała, W.; Rącka, I.; Suszyńska, K. Population Access to Hospital Emergency Departments: The Spatial Analysis in Public Health Research. *IJERPH* **2022**, *19*, 1437. https://doi.org/10.3390/ijerph19031437

Academic Editor: Paul B. Tchounwou

Received: 13 December 2021
Accepted: 25 January 2022
Published: 27 January 2022

Publisher's Note: MDPI stays neutral with regard to jurisdictional claims in published maps and institutional affiliations.

Copyright: © 2022 by the authors. Licensee MDPI, Basel, Switzerland. This article is an open access article distributed under the terms and conditions of the Creative Commons Attribution (CC BY) license (https://creativecommons.org/licenses/by/4.0/).

1. Introduction

One of the most basic social needs of human beings is security, satisfied by, e.g., properly designed health protection. Due to the modern lifestyles people lead, the risk of numerous civilization and social diseases, including those caused by various types of injuries, increases every year. Therefore, it is necessary to shape the health care system in such a way that it would give citizens a sense of security, especially with regard to the pathogenic effects of an accelerating pace of life in the modern world [1,2]. The importance of this type of activities is indicated and proved by their presence in the third goal of the sustainable development programme—2030 Agenda for Sustainable Development [3]. Since a person's life and health in many cases is determined by the speed of medical actions taken, the problem of spatial organization of emergency medical services is of major importance [4–7].

Until the end of the 1990s, emergency medicine in Poland was obsolete, and emergency care was in fact limited to visits of the doctors and medical transport to hospitals. The emergency rescue system operating in Poland since the post-war years complemented the inefficient primary health care system, without guaranteeing comprehensive, adequate to the needs and sufficiently quick medical service in the case of sudden health or life-threatening events. The need for reforms in the field of emergency medical services, in addition to widespread social discontent, was also reflected in the mortality rate at the scene of accidents or deaths from sudden illnesses on the spot or on the way to the hospital, which significantly exceeded the international standards [8–10].

The increasingly visible imperfections in the functioning of the Polish emergency medical services and the positive effects of implementing modern medical systems in western European countries were incentives to organize similar structures in Poland. The works undertaken at the turn of the 20th and 21st centuries led to the implementation of the Act of 8 September 2006 on the Emergency Medical Services (EMS). According to its regulations, planning and organization of the system takes place mainly at the regional level (voivodeships), and the authorities responsible for the performance of the system's tasks are voivodes (representatives of state government administration in regions). Supervision over the system is exercised by the Minister of Health.

The task of the voivodes is to prepare regional plans for the operation of the system, which defines, e.g., the number, location and method of coordinating the activities of the units. The act distinguishes between two types of units operating within the system: hospital emergency departments (HED) and emergency medical teams.

An important element of the introduced reforms was the creation of a hospital base of emergency medical services. Hospital emergency departments (HEDs), as organizational units of hospitals specializing in emergency care, are the place of first contact for people in a state of emergency. The main task of HEDs is to admit, perform initial diagnostics and undertake the treatment necessary to stabilize the vital functions of patients who require medical assistance as a result of health- or life-threatening accident or illness. The treatment within hospital emergency departments should contribute to the reduction in mortality rate, the number and scope of disabilities caused by injuries, the number of deaths to be avoided, as well as the extent of pain and suffering resulting from accidents and sudden illnesses [11]. HEDs provide medical assistance to all patients in a state of emergency—both transported in ambulances and self-reporting, regardless of their age, place of residence, financial or social status.

In the event of a risk of loss of life or health, patients should be transferred to the hospital emergency department closest to the scene of the incident. Therefore, the problem of the distribution and access to emergency departments becomes of key importance. In this context, an important question seems to be whether the changes in the number and distribution of emergency departments, that take place after the implementation of the EMS in individual regions in Poland, result in an improvement in the level of spatial accessibility of the departments. This issue has not been widely addressed in the literature so far, although the results of such research could be a valuable hint for other countries with a similar level of development, reforming health and rescue systems for residents.

The aim of the paper is to explain the spatial accessibility of hospital emergency departments and changes in its level as a result of organizational and spatial transformations of the emergency department network in Poland. On the basis of the research results, it is possible to indicate new locations of emergency rooms, the launch of which would improve the level of spatial accessibility of emergency medical services. The research was conducted within two time series, comparing the changes in the distribution of HEDs in 2011 and 2021 and the resulting consequences for the level of spatial accessibility of these medical services. In the research procedure, GIS techniques were used to measure the distances between emergency departments and places of residence. Although the study was conducted on a local scale (approximately 2500 communes in Poland), the results were presented on a regional level.

2. Literature Review

2.1. Spatial Analysis of Emergency Department Accessibility

The emergency medical services is a field where spatial analysis is broadly used.

The crux of the emergency medical system is to save lives by providing a quick response to emergencies, thus minimizing the waiting time of the patient for professional emergency medical assistance, provided both in the pre-hospital stage (ambulance care) and as part of the hospital emergency care (i.e., in emergency departments). Shortening the time of ambulance response, patient transfer and rescue operation has clinical implications for many conditions. Yet increasing the chances of saving the patient depends in particular on the proximity to emergency units. Studies confirming this assumption have been conducted in numerous countries, e.g., Sweden [12], United Kingdom [13], Spain [14], the USA [15] and Portugal [16]. Their location should be well designed and result from the distribution and demographic structure of the population, terrain conditions, the socioeconomic situation and the existing and developing infrastructure. Geographic research on emergency medical services using quantitative methods to optimize location of the institutions focuses mainly on the aspect of spatial organization of the emergency medical system and its consequences [17,18].

In the case of pre-hospital emergency medical services, spatial analyses are related to the issues of access and effectiveness, which are closely connected to the distance expressed in patients' waiting time for the ambulances to reach the places where accidents and other emergencies occur. Due to the possibility of relocating mobile emergency medical units, most of the works concern the determination of the optimal number and distribution of ambulance stations and the emergency ambulances themselves [19–24].

On the other hand, spatial studies of emergency medical services rendered at hospitals focus mainly on spatial, organizational and social aspects of the access to emergency departments. The relationships between the level of their spatial accessibility measured in travel time units and the level of availability as perceived by the local community are analyzed [25]. The authors state that time distance has an impact on the perceived level of availability, and it depends on the individual feelings of potential patients whether, when and where they will use the emergency department services.

Another important issue addressed in the context of distance to health care facilities is the changing demand for health care [26–28]. Vaz, Ramos and Santana [16] describe this phenomenon as "the distance decay effect", resulting in decreasing demand for health care as distance to health care institutions increases. Brabyn and Beere [29] use GIS tools to calculate the closest emergency department in terms of travel time. They observe that after the hospital network reform in New Zealand, there are significant spatial inequalities in the accessibility of hospital emergency services, and the number of people living more than 60 min from the emergency department in some areas has increased dramatically. Spatial differentiation in the access to emergency departments is noticed by Carr et al. [30] and Hashtarkhani et al. [31]. The authors note a particularly unfavorable situation in rural areas and point at the need of further studies examining the relationship between the accessibility to emergency medical services in the context of mortality rate and medical results of patients in emergency departments.

The access to emergency medical services, taking into account both the ambulance arrival time to the scene and the time necessary to transport the patient to an emergency department, was tested by Jones and Bentham [4]. Using regression models to analyze data concerning rescue services for road accidents, they determined that an increased probability of death occurred when the accident participant was an elderly person (according to the model, over 60 years old), a pedestrian, or when it involved a larger number of cars or in the case of an accident on the road with a high permissible speed limit. Surprisingly, the study did not prove any relationship between the emergency action duration and the loss of life or health as a result of a road accident. The authors emphasized, however, that this could be the result of a limited research area, insufficient number of interventions analyzed, and difficulties in obtaining complete and reliable data.

The functioning of emergency departments has also been a hotspot of the current research concerning health care systems during COVID-19 pandemic and much new cognition about it has been yielded in the past two years. The increasing body of literature revealed that infrastructure problems, lack of access to medical essentials, and failures to deliver appropriate technological support were the key issues faced by health care organizations during this period. Recommendations to improve emergency department efficacy were made by numerous authors [32,33], while others focused on investigating changes in the numbers of visits to emergency departments during the pandemic and the reasons and results of these changes [34,35].

2.2. The Access to Health Services—Theoretical Background

Access to services has long been the subject of interest in many geographic, sociological and economic studies. Despite this, no exhaustive and unequivocal definition of the term has been developed so far, and most authors point out that the definitions available are vague or contradicting [36,37].

Domański [38] believes that the term access should be understood as "the possibility of taking advantage of opportunities created by economic, social, cultural and political objects and institutions". The author notes that access is influenced by many factors, among which he distinguishes distance as a determinant of spatial accessibility. Similarly, Taylor [37], defines it as "opportunities or possibilities allowing the use of various types of activities, functions, some of which may be classified as services".

When describing the concept of access, many authors consider, apart from the relations resulting from the location in space, the road network and the possibilities of transport as factors significantly determining the level of access [39–41].

A significant amount of attention has been paid to health services in spatial studies concerning accessibility. It can be stated that access to health services is equal to the possibility of obtaining advice or a medical procedure in a health care facility appropriate for the type of illness, with a minimum waiting time and minimum costs incurred. Therefore, the access to health services depends on the effective use of health care resources and their rational distribution, taking into account the population density, health condition of the population and the existing communication infrastructure and means of transport.

Most authors emphasize that access to health care is a complex concept that should be analyzed in several aspects (dimensions), among which the most important from the point of view of geographical research are spatial dimension (spatial accessibility) and the socio-organizational dimension (availability of services) [42–46]. Accessibility is identified as the relationship between the distribution of health care infrastructure and human resources and the distribution of the population reporting health needs. In turn, the availability of services is defined as the relationship between the sizes and types of services and the sizes and types of consumer needs, or sometimes as the adequate supply of medical services.

As defined above, the spatial accessibility depends on patients' distance to the treatment site. Therefore, it is associated with transportation that requires devoting a certain time and incurring certain costs. Thus, spatial accessibility can be identified as the physical, time and economic distance to the place where medical services are rendered [42,44]. However, reaching a healthcare facility is not equal to getting medical help. There are a number of additional factors that determine the possibility of using health services. The utilization of medical services is often limited by such barriers as: organizational, financial, and, finally, personal [46].

In the case of hospital emergency departments responsible for providing medical services to every person in a situation of sudden health emergency, the organizational and socioeconomic dimensions of access seem to be less important than distance in determining spatial accessibility. On the one hand, there is a legal obligation to provide medical assistance to all patients admitted to emergency departments (which in principle translates into the lack of organizational and institutional barriers), and on the other hand, in the case of sudden and serious health threats, the timing of emergency assistance is of key importance. According to the so called "golden hour" principle applied in emergency medicine (the first hour after the onset of out-of-hospital traumatic injury), after the first 60 min of emergency, the patient's chances of survival dramatically decrease. Therefore, one of the determinants of the effectiveness of emergency medical services is providing access to specialized hospital services to people in a state of sudden health emergency within no more than one hour [47]. Paying additional attention to the notion conceptualized in the literature, that the distribution of the population is the basic factor determining the spatial arrangement of the demand for health care [18,44,48,49], it can be concluded that when studying access to hospital emergency departments, attention should be focused primarily on spatial relations resulting from the location of emergency departments and the places of residence of potential patients.

3. Materials and Methods

The empirical data on the network of hospital emergency departments was obtained from two sources: (1) regional plans for the operation of the EMS, and (2) information on contracts concluded by individual Regional Departments of the National Health Fund. The preparation of the regional plans for the operation of the EMS in Poland is the task of voivodes (representatives of government administration in regions). They are the basic documents defining the organization and functioning of the system in the area of each region, and include, inter alia, information on the number, distribution and method of coordinating the activities of system units. The action plans of the EMS system served as a source of addresses of individual hospitals where HEDs operated and places where the emergency departments were planned to be launched. In turn, the National Health Fund (NHF), as a medical service payer, finances medical emergency services. The contract for the provision of services through the NHF is a prerequisite for including the hospital emergency department among the units of the EMS system. Data on concluded and implemented contracts for the provision of healthcare services in the field of hospital emergency department were collected from the NHF guide, which allowed for the verification of the HEDs actually functioning within the system. The remaining data necessary to analyse the spatial accessibility of hospital emergency departments were obtained using techniques related to geographic information systems (GIS) using the ArgGIS program. The usefulness of GIS tools in this type of research is emphasized by various authors [25,29,50–54].

The research procedure started with measuring the distance between each commune and the nearest hospital emergency department. Due to the nationwide scope of the research, measurements were made using the Euclidean metric (more precisely: orthodrome), although the authors were aware that the results of such measurements are almost always understated. Measurement in a straight line does not take into account the terrain conditions resulting from the route of the road network and the difference in altitude [55]. However, as the research of, e.g., Jones et al. [56] proves, the use of the Euclidean distance in spatial-medical research, despite the lower precision of the obtained measurements in relation to the so-called urban metric (distance measured along streets) is acceptable and does not distort the final results of the analyses.

When determining the distance, it was necessary to define points representing the individual commune. In the case of urban communes (municipalities), it was the mean centre of the city. In rural communes it was the population mean centre of the commune, while in urban-rural communes, it was the mean centre of the city being the seat of the commune.

In the next step, the data were aggregated at the regional level and the weighted average distances of places of residence to emergency departments were calculated according to the formula:

$$\overline{d}_{\mathrm{HED}} = \frac{\sum\limits_{i=1}^{n} d_i P_i}{\sum\limits_{i=1}^{n} P_i}, \quad (1)$$

where d_i is the distance to the HED from a given commune, and P_i is the number of people living in particular communes. In addition, each of the communes was included in one of the five zones of distance (remoteness) from the HED. The zones were separated on the basis of previous measurements, setting limit values every 10 km. Then, the number of people living in individual zones was summed up, both at the country level and in individual regions, and the results were presented as a percentage.

The study was conducted in two time series covering 2011 and 2021 (Figure 1). On this basis, the changes in the level of access resulting from organizational changes within the hospital emergency system were compared.

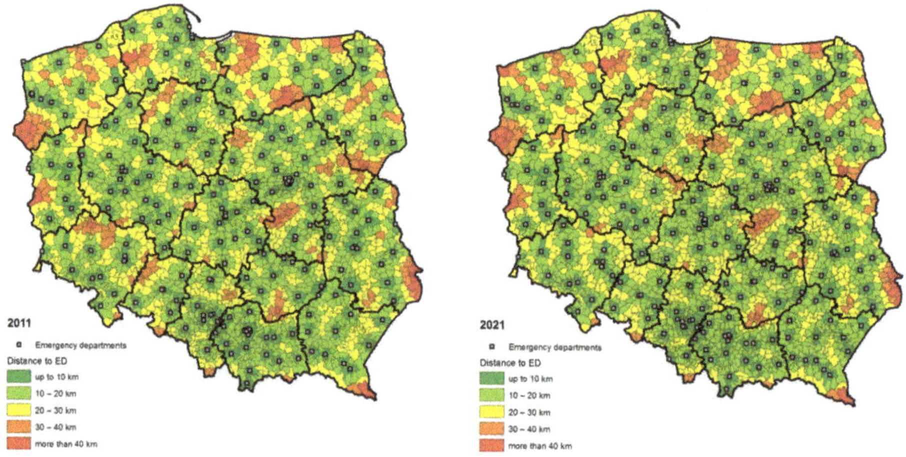

(a) (b)

Figure 1. Spatial accessibility of hospital emergency departments in the Polish communes: (**a**) in 2011; (**b**) in 2021.

4. Results

4.1. *The Hospital Emergency Department Network in Poland*

There were assumptions regarding the organization of the emergency medical service system in Poland forecasting the creation of approximately 270 hospital emergency departments, each of them covering the population from 100,000 to 300,000 with its catchment area [9,57]. Despite the relevant provisions in the regional action plans for the system, many Polish hospitals failed to meet the organizational criteria imposed by the legislator to launch the HED.

In individual regions, the organizers of the EMS system made different assumptions regarding the number and distribution of the emergency departments. The general rule was to launch HEDs in the largest population centers. In this way, emergency departments were established in administrative capitals of the regions and in all the biggest cities in the country.

Among the cities with over 100,000 residents without HED were Gliwice, Bytom, Ruda Śląska and Tychy (in 2011 also the city of Zabrze). These were the cities of the Upper Silesian urban area—the largest urban area in Poland and one of the largest in the European Union. The services of hospital emergency departments in the conurbation area were secured in ten other cities. The lack of hospitals with emergency departments was also noticeable in areas adjacent to large cities (including the surroundings of Rzeszów, Olsztyn, Szczecin, Bydgoszcz, Toruń). Patients from these areas were sent to HEDs operating in the central cities.

In some regions efforts were made to ensure that HEDs were located in almost every district (poviat) (e.g., in the Małopolskie region), in others the focus was on launching emergency departments in places of the greatest population concentration (e.g., the Kujawsko-Pomorskie and Śląskie regions). When planning the location of departments, it was also important to take into consideration the structure of the national road network and highways (e.g., the Lubuskie, Dolnośląskie, Wielkopolskie regions) (Figure 2).

Figure 2. Spatial distribution of hospital emergency departments in Poland.

In 2011, out of 256 planned emergency departments, 229 (almost 90%) were in operation. The plans for the organization of HEDs were fully implemented in several regions: Kujawsko-Pomorskie, Lubelskie, Lubuskie, Łódzkie, Małopolskie and Opolskie. Slight differences in relation to the assumptions occurred in the Pomorskie, Podlaskie, Warmińsko-Mazurskie, Wielkopolskie and Zachodniopomorskie regions (each region did not activate one of the planned HEDs). The lowest number of emergency departments completed was observed in the Śląskie region, where by 2011 only 10 out of 23 planned departments (43%) were established. A relatively low degree of implementation of the adopted assumptions was also recorded in the Dolnośląskie (81%) and Świętokrzyskie (82%) regions (Table 1).

In the following years, the hospital emergency department network was transformed. Both the number of HEDs planned by voivodes and the number of departments under operation changed. In some regions, the emergency medical system planners amended the assumptions regarding the number and distribution of emergency departments in the action plans. While shaping the network of emergency departments, on the one hand, efforts were made to maximize spatial accessibility, and on the other hand, the economic efficiency of providing emergency and medical services was taken into account.

As a result of the ongoing organizational and spatial transformation up to 2021, a network of hospital emergency departments included 241 out of the planned 267 venues. The degree of implementation of the assumptions stipulated in the plan exceeded 90%, and this value did not differ significantly from that from 10 years ago. Compared to 2011, the number of regions where all emergency departments were launched (the Dolnośląskie, Lubuskie, Małopolskie, Opolskie and Wielkopolskie regions) decreased. An almost complete network of HEDs (only one department less than planned) operated in the Podkarpackie, Pomorskie, Świętokrzyskie and Warmińsko-Mazurskie regions. As in 2011, the lowest effectiveness in launching the planned number of emergency departments could be observed in the Śląskie region, yet the degree of implementation of the HED network increased in this voivodeship from 43% to 67%. Some difficulties could also be noted in the Lubelskie, Łódzkie and Mazowieckie regions, where three emergency departments remained not launched. In the remaining regions, two departments were missing to complete the plan for the construction of the HED network (Table 1).

Table 1. The network of hospital emergency departments in Polish regions.

Region	No. of HEDs in 2011			No. of HEDs in 2021		
	Functioning	Planned	Level of Completion	Functioning	Planned	Level of Completion
Dolnośląskie	13	16	81%	16	16	100%
Kujawsko-Pomorskie	12	12	100%	10	12	83%
Lubelskie	19	19	100%	17	20	85%
Lubuskie	8	8	100%	8	8	100%
Łódzkie	17	17	100%	20	23	87%
Małopolskie	22	22	100%	21	21	100%
Mazowieckie	30	32	94%	32	35	91%
Opolskie	6	6	100%	7	7	100%
Podkarpackie	12	14	86%	14	15	93%
Podlaskie	11	12	92%	13	15	87%
Pomorskie	12	13	92%	13	14	93%
Śląskie	10	23	43%	14	21	67%
Świętokrzyskie	9	11	82%	10	11	91%
Warmińsko-Mazurskie	11	12	92%	11	12	92%
Wielkopolskie	26	27	96%	25	27	93%
Zachodniopomorskie	11	12	100%	10	10	100%
POLAND	**229**	**256**	**89%**	**241**	**267**	**90%**

Source: own study based on 16 regional action plans of the Emergency Medical Services and the information on National Health Fund contracts.

4.2. Spatial Accessibility of Hospital Emergency Departments

The analysis of spatial accessibility of hospital emergency departments in 2011 showed that over half of the Polish population (52.4%) lived within 10 km from the nearest emergency department. In municipalities included in the further zones, 28.2% of the total population of the country lived 10–20 km from HEDs, 15.2% had to travel the distance of 20–30 km, 3.5% had to cover the distance of 30–40 km, and only 0.7% of the population had to travel the distance of more than 40 km to the closest HED (Table 2). The weighted average distance between emergency departments and places of residence, calculated on the basis of assumptions made, was 11 km (Table 3). This value indicated a relatively high spatial accessibility of hospital emergency departments in Poland, especially in a situation where one of the main factors taken into account in the process of shaping the network of emergency departments was economic efficiency leading to the concentration of services and the construction of health facilities mainly in the largest cities.

The conducted analysis showed quite significant regional differentiation in terms of spatial accessibility of HEDs, measured by the distance separating the potential patients from the nearest hospital with an emergency department. The shortest distances in 2011 were noted in the regions of Łódzkie ($\bar{d}_{\mathrm{HED}}$ = 8 km) and Małopolskie ($\bar{d}_{\mathrm{HED}}$ = 8.5 km), where over 60% of the population lived in the communes located within 10 km away from the HED, and less than 1% of the population had to travel the distance greater than 30 km to reach it. The weighted average distance between emergency departments and places of residence, apart from the above-mentioned regions, did not exceed 10 km in the Mazowieckie (9.2 km) and Pomorskie (10 km) regions. The lowest spatial accessibility of emergency departments was observed in the regions of Opolskie (15.4 km), Warmińsko-Mazurskie (15.2 km) and Zachodniopomorskie (15.2 km). The first of the distinguished zones of remoteness from the HED (up to 10 km) was inhabited by only every third resident of Opolskie (33.5%) and slightly more than 40% of Warmińsko-Mazurskie. In turn, as many as 5.7% of the population in the Warmińsko-Mazurskie and over 2% in the Lubuskie, Lubelskie and Zachodniopomorskie regions lived further than 40 km from the HED (Tables 2 and 3).

Table 2. Spatial accessibility of the hospital emergency departments in Poland.

Region	Share of Population Living within a Given Distance from the Nearest HED									
	2011					2021				
	Up to 10 km	10–20 km	20–30 km	30–40 km	>40 km	up to 10 km	10–20 km	20–30 km	30–40 km	>40 km
Dolnośląskie	45.6	26.0	23.6	4.5	0.2	53.0	29.8	16.5	0.6	0.0
Kujawsko-Pomorskie	53.7	25.9	17.8	2.6	0.0	50.9	24.8	20.0	4.3	0.0
Lubelskie	51.9	24.4	18.1	3.3	2.2	51.4	23.2	20.3	3.1	2.1
Lubuskie	47.6	22.9	23.6	3.5	2.4	48.0	24.0	22.3	3.4	2.3
Łódzkie	60.7	26.8	11.6	0.8	0.0	60.9	28.1	10.1	0.9	0.0
Małopolskie	60.8	33.5	5.1	0.5	0.0	60.4	33.8	5.3	0.5	0.0
Mazowieckie	58.9	26.4	10.0	4.3	0.3	60.5	27.8	8.0	3.5	0.3
Opolskie	33.5	32.6	23.5	10.5	0.0	39.1	35.0	24.1	1.7	0.0
Podkarpackie	45.1	39.1	15.6	0.1	0.1	50.2	40.8	8.8	0.1	0.1
Podlaskie	52.8	22.1	17.1	6.3	1.8	58.7	21.0	13.4	5.2	1.7
Pomorskie	57.6	22.0	15.4	4.0	1.1	58.2	26.0	13.2	1.6	1.0
Śląskie	50.2	35.7	13.3	0.7	0.0	60.7	30.0	9.0	0.3	0.0
Świętokrzyskie	50.2	30.7	12.4	6.2	0.4	49.6	31.4	13.3	5.3	0.4
Warmińsko-Mazurskie	42.8	17.5	26.3	7.8	5.7	42.0	18.8	26.0	7.6	5.7
Wielkopolskie	52.5	31.3	14.2	2.0	0.0	51.7	32.0	13.5	2.9	0.0
Zachodniopomorskie	46.1	16.7	23.6	11.5	2.1	43.4	19.3	20.5	12.1	4.8
POLAND	**52.4**	**28.2**	**15.2**	**3.5**	**0.7**	**54.7**	**28.6**	**13.2**	**2.7**	**0.8**

Source: own calculation on the basis of data acquired from Statistics Poland.

Table 3. Weighted average distances to hospital emergency departments.

Region	2011	2021
Dolnośląskie	13.2	10.7
Kujawsko-Pomorskie	10.6	11.6
Lubelskie	11.9	12.4
Lubuskie	12.6	12.5
Łódzkie	8.0	7.9
Małopolskie	8.5	8.9
Mazowieckie	9.2	8.7
Opolskie	15.4	13
Podkarpackie	11.4	10.1
Podlaskie	11.6	10.2
Pomorskie	10.0	9.3
Śląskie	11.1	9.3
Świętokrzyskie	11.1	11.1
Warmińsko-Mazurskie	15.2	15.1
Wielkopolskie	10.2	10.7
Zachodniopomorskie	15.2	16.6
POLAND	**11.0**	**10.4**

Source: own calculation on the basis of data acquired from Statistics Poland.

The repeated research procedure regarding 2021 proved that along with the transformation of the network of hospital emergency departments in Poland, the level of spatial accessibility of medical services in Poland changed. The progressing spatial reorganization of emergency departments resulted in an increase in the percentage of the population living in the two closest zones away from emergency departments—in the up to 10 km zone by more than 2 percentage points, and in the 10–20 km zone by 0.4 percentage points. On the other hand, the share of the population living in the areas farthest from the functioning HEDs decreased—in the 20–30 km zone there was a decrease by 2 percentage points, while in the 30–40 km zone by 0.7 percentage point (Table 2). The study showed that the weighted average distance between the places of residence and the nearest hospital emergency department decreased by 0.6 km on the national scale and in 2021 amounted to 10.4 km (Table 3).

The analysis of regional differentiation in terms of spatial accessibility measured by the average weighted distance to the nearest emergency department showed that in 2021 the lowest values of the indicator were characteristic of the regions: Łódzkie (7.9 km), Mazowieckie (8.7 km), Małopolskie (8.9 km), Śląskie (9.3 km) and Pomorskie (9.3 km). In turn, the greatest average distances to hospital emergency departments were identified in the Zachodniopomorskie (16.6 km), Warmińsko-Mazurskie (15.1 km) and Opolskie (13 km) regions.

Comparing the results of the analyses carried out on the data from 2011 and 2021, it can be observed that in most regions changes in the distribution of emergency departments led to a decrease in the average distance between the places of residence and emergency departments (Table 3). The greatest reduction in weighted average distance was observed in the Dolnośląskie and Opolskie regions, where, thanks to the opening of additional emergency departments, the $\overline{d}_{\mathrm{HED}}$ indicator decreased by over 2 km. To a lesser extent, due to the changes in the spatial organization of hospital emergency departments, the average distance to this type of facilities was shortened in the following regions: Śląskie (almost 2 km), Podlaskie (1.4 km), Podkarpackie (1.3 km) and Pomorskie (approx. 0.7 km). In turn, in Zachodniopomorskie and Kujawsko-Pomorskie the weighted average distance to the nearest emergency department increased by 1.4 km and 1 km, respectively. In other regions no significant changes in the level of spatial accessibility of hospital emergency departments were recorded in 2021 compared to 2011 (changes ranged within $+/-$ 0.5 km).

In the analysed decade, particularly favorable changes took place in the Śląskie and Dolnośląskie regions, where the share of population living within 10 km from the HED increased by 10.4 and 7.4 percentage points, respectively. In the Śląskie region, it happened due to the launch of emergency departments in Cieszyn, Zawiercie and Zabrze, and further improvement is expected due to the implementation of the next seven HEDs to be launched (including Racibórz, Jastrzębie-Zdrój, Żywiec and Tychy). In turn, in the Dolnośląskie region, the improvement in the accessibility of emergency services resulted from fully implemented plans for the network of emergency departments, which in recent years were expanded by the departments in Lubin, Głogów and Oława. A different situation was observed in the Zachodniopomorskie and Kujawsko-Pomorskie regions, where the liquidation of HEDs decreased the percentage of the population dwelling the in first zone of remoteness by almost 3 percentage points.

The analysis of the spatial accessibility of HEDs in Poland proved that a decreasing share of the population lives in further zones of remoteness. There are exceptions to this rule in two regions—Warmińsko-Mazurskie and Zachodniopomorskie. In these regions, the network of HEDs is not fully adjusted to the population distribution. In Warmińsko-Mazurskie, the situation will be improved by launching the planned HED in Szczytno. It seems that an additional location in Ostróda and Braniewo would be worth considering. In the Zachodniopomorskie region decision makers should consider locating new HEDs in Świnoujście, Kołobrzeg and Myślibórz (locations not included in the plan).

Among other locations where the launch of the HED would significantly reduce the percentage of people living in the farthest zones of remoteness, and thus contribute to the reduction in the average distance to the emergency department, one should mention Krosno Odrzańskie in the Lubuskie region (not included in the plan), Grójec in the Mazowieckie region (planned to be launched), Hrubieszów in the Lubelskie region (planned) and Busko-Zdrój in the Świętokrzyskie region (planned).

5. Discussion

The presented changes in the number and distribution of emergency departments indicate that despite the dozen or so years of operation of the EMS system in Poland, the network of emergency departments has not yet reached its final state. Generally, it is difficult to identify any visible regularities in the degree of plan implementation for the organization of the network of emergency departments in individual regions. Most often, the failure to open the HED was explained by financial problems and staff shortages. It stems from the fact that there are a number of structural, space and organizational requirements necessary for the functioning of the HED. These criteria indicate the minimum number of staff of emergency departments and stipulate, inter alia, that these departments are organized in hospitals with anesthesiology and intensive care units, general surgery with trauma and internal diseases, as well as 24-h laboratory diagnostics and diagnostic imaging. The hospital should also have access to a helipad for medical helicopters [10].

Initially, a big problem was to adapt the emergency departments to the technical requirements (proper location of the emergency department in the hospital structure, appropriate usable floor area, entrance for ambulances, proper patient flow inside the HED). Currently, a major barrier to the functioning of HEDs is the provision of minimum human resources, including primarily those relating to doctors in the system [58–61]. At the same time, it seems that the number of operating HEDs was largely affected by the determination and effectiveness of the directors of individual hospitals as well as the support for the idea by decision makers in each of the region [57].

Despite the above-mentioned problems, the transformation of the spatial organization of hospital emergency department network in 2011–2021 affected the level of accessibility of these facilities, both in the scale of the entire country and in the system of individual regions.

The applied research procedure proved that the introduced changes resulted in the overall improvement of the spatial accessibility of the hospital part of the medical emergency system, reducing the weighted average distance between the places of residence and the nearest HED (on the national scale by 0.6 km, and in some regions even by a few km). These results allow for a positive assessment of the changes in the distribution of emergency departments, especially in the context of the view emphasized in the literature that the distribution of the population is the basic factor determining the spatial arrangement of the demand for health services [18,44,48,49]. Numerous studies prove that distance has a negative influence on HED utilization [16,51,62,63], and decreased emergency department access can be reflected in a substantial increase in the mortality rate [7,64–66].

The relatively small distances between the places of residence and HEDs were the result of their presence in large cities, where a significant part of the population of individual regions lives. It seems that in large cities it was easier to meet the requirements of both qualified staff and professional technical and infrastructure facilities, necessary to start the HED. A much greater problem in the largest cities is the availability of HEDs resulting from their excessive use and the excessive transfer time resulting from transport congestion [67,68]. On the other hand, in small towns and in rural areas, a greater problem is the economic efficiency of emergency departments and the shortage of physicians of EMS and specialist equipment. Yet as numerous studies emphasize, the departments in these locations are important safety nets for members of the local community [69]. Their functioning is also of key importance in case of road accident victims [4].

In the context of EMS for accident victims and patients with sudden illnesses in Poland, other hospital units cooperating with the system should also be mentioned. These include trauma centers, departments in specialized hospitals, such as invasive cardiology, neurosurgery, thoracic surgery, maxillary surgery, toxicology and stroke treatment. Some authors [70] stress that getting to the right specialist center as soon as possible, even if it is not the nearest HED, increases the patient's chances of survival. However, due to the uniqueness of services provided in these centers, they should not be included in the analysis of the availability of hospital emergency departments [71]. Moreover, trauma centers and a large number of specialist departments operate in the same hospitals where

HEDs are located, therefore even including them in the adopted research procedure would not change the obtained results.

In hospitals where, for various reasons, it was not possible to organize an emergency department, alternative services are rendered in the admission room [10]. However, as emphasized by Gaca and Witkowski [57], the scope of these services is different and is not subject to the provisions of the Act on the EMS.

The presence of an emergency department in a given area that meets the criteria imposed by the legislator significantly improves the quality and availability of medical services offered, and thus increases the local residents' sense of security, by being aware of the existence of an institution that is constantly in the mode of readiness [11,57].

6. Conclusions

The Polish system of emergency medical services, introduced by the Act on the Emergency Medical Services of 2006, was characterized by many changes in the number and arrangement of facilities with hospital emergency departments in the studied period. In the light of the research results obtained, it was found that despite the still unfinished plans to build a network of HEDs in Poland, changes in the number and distribution of departments are aimed at improving the spatial accessibility of these medical services. It is manifested by the reduction in the weighted average distance separating emergency departments from the places of residence. A further increase in accessibility should be expected after launching the planned departments, especially in peripheral cities in several regions.

In order to ensure high quality and access to hospital emergency medical services, after bringing the network of HEDs to the target state, further research should focus on the availability of services, with particular emphasis on the degree of services' utilization in the departments.

Although the level of spatial accessibility of emergency departments increased on the country level and in most regions in the analyzed period, there are also regions where reverse trends were identified. This is due to department closures or failure to complete the planned network. The situation is particularly unfavorable in the Zachodniopomorskie region, where the existing HED network proved to be insufficient, and the loss of HED in the hospital in Świnoujście further reduced the spatial accessibility of the analyzed medical services.

Although the study of spatial accessibility of hospital emergency departments was limited only to issues related to the physical distance separating the HEDs from the places of residence, it should be emphasized that in the case of sudden health threats this distance significantly determines the overall waiting time for emergency services. Therefore, the results of the presented research may be the basis for the spatial evaluation of the emergency department network. Minimizing the average distances to HEDs may be a goal function when planning the launch of subsequent departments. Such an approach will make it possible to improve the spatial accessibility of the hospital base of emergency medical services, thus contributing to better protection of the health needs of residents.

Author Contributions: Conceptualization, W.K., I.R. and K.S.; methodology, W.K.; software, W.K. and K.S.; validation, W.K., I.R. and K.S.; formal analysis, W.K.; investigation, W.K., I.R. and K.S.; resources, W.K., I.R. and K.S.; data curation, W.K. and K.S.; writing—original draft preparation, W.K., I.R. and K.S.; writing—review and editing, W.K., I.R. and K.S.; visualization, W.K.; supervision, W.K., I.R. and K.S.; project administration, W.K., I.R. and K.S.; funding acquisition, W.K., I.R. and K.S. All authors have read and agreed to the published version of the manuscript.

Funding: This research received no external funding.

Institutional Review Board Statement: Not applicable.

Informed Consent Statement: Not applicable.

Data Availability Statement: The data supporting reported results can be found at websites of Statistics Poland stat.gov.pl (accessed on 1 December 2021) and National Health Fund nfz.gov.pl (accessed on 1 December 2021).

Conflicts of Interest: The authors declare no conflict of interest.

References

1. Sygit, K.M.; Sygit, M.; Wojtyła-Buciora, P.; Lubiniec, O.; Stelmach, W.; Krakowiak, J. Physical activity as an important element in organizing and managing the lifestyle of populations in urban and rural environments. *Ann. Agric. Environ. Med.* **2019**, *26*, 8–12. [CrossRef] [PubMed]
2. Gąska, I.; Sygit, K.; Cipora, E.; Sygit, M.; Pacian, A.; Surmach, M.; Kaleta, D.; Rzeźnicki, A. Assessment of the Health Behaviours and Value-Based Health Analysis of People Aged 50+ Who Were Hospitalized Due to Cardiovascular Disease. *Int. J. Environ. Res. Public Health* **2021**, *18*, 4221. [CrossRef] [PubMed]
3. Take Action for the Sustainable Development Goals. Available online: https://www.un.org/sustainabledevelopment/sustainable-development-goals/ (accessed on 1 December 2021).
4. Jones, A.P.; Bentham, G. Emergency medical service accessibility and outcome from road traffic accidents. *Public Health* **1995**, *109*, 169–177. [CrossRef]
5. Richardson, D.B. Increase in patient mortality at 10 days associated with emergency department overcrowding. *Med. J. Aust.* **2006**, *184*, 213–216. [CrossRef] [PubMed]
6. Nicholl, J.; West, J.; Goodacre, S.; Turner, J. The relationship between distance to hospital and patient mortality in emergencies: An observational study. *Emerg. Med. J.* **2007**, *24*, 665–668. [CrossRef]
7. Brown, J.B.; Rosengart, M.R.; Billiar, T.R.; Peitzman, A.B.; Sperry, J.L. Distance matters: Effect of geographic trauma system resource organization on fatal motor vehicle collisions. *J. Trauma Acute Care Surg.* **2017**, *83*, 111. [CrossRef]
8. Jakubaszko, J. Założenia organizacyjne systemu zintegrowanego ratownictwa medycznego. In *Ratownik Medyczny*; Jakubaszko, J., Ed.; Górnicki Wydawnictwo Medyczne: Wrocław, Poland, 2003.
9. Leki, K. Pomoc doraźna w świetle ustawy o Państwowym Ratownictwie Medycznym. In *Zdrowie i Jego Ochrona: Między Teorią a Praktyką*; Korporowicz, V., Ed.; Oficyna Wydawnicza SGH: Warszawa, Poland, 2006; pp. 187–199.
10. Poździoch, S. (Ed.) *Ustawa o Państwowym Ratownictwie Medycznym. Komentarz*, 2nd ed.; Wolters Kluwer Business: Warszawa, Poland, 2013.
11. Szarpak, Ł. Organizacja szpitalnego oddziału ratunkowego. *Ogólnopolski Przegląd Med.* **2009**, *12*, 22–26.
12. Magnusson, G. The role of proximity in the use of hospital emergency departments. *Sociol. Health Illn. Cardiff.* **1980**, *2*, 202–214. [CrossRef]
13. Rudge, G.M.; Mohammed, M.A.; Fillingham, S.C.; Girling, A.; Sidhu, K.; Stevens, A.J. The combined influence of distance and neighborhood deprivation on emergency department attendance in a large English population: A retrospective database study. *PLoS ONE* **2013**, *8*, e67943. [CrossRef]
14. Lee, J.E.; Sung, J.H.; Ward, W.B.; Fos, P.J.; Lee, W.J.; Kim, J.C. Utilization of the emergency room: Impact of geographic distance. *Geospat. Health* **2007**, *1*, 243–253. [CrossRef]
15. Henneman, P.L.; Garb, J.L.; Capraro, G.A.; Li, H.; Smithline, H.A.; Wait, R.B. Geography and travel distance impact emergency department visits. *J. Emerg. Med.* **2011**, *40*, 333–339. [CrossRef]
16. Vaz, S.; Ramos, P.; Santana, P. Distance effects on the accessibility to emergency departments in Portugal. *Saúde E Soc.* **2014**, *23*, 1154–1161. [CrossRef]
17. Mayer, J.D. Seattle's paramedic program: Geographical distribution, response times, and mortality. *Soc. Sci. Med.* **1979**, *13*, 45–51. [CrossRef]
18. Mayer, J.D. Relations between two traditions of medical geography: Health systems planning and geographical epidemiology. *Prog. Hum. Geogr.* **1982**, *2*, 215–230. [CrossRef]
19. Toregas, C.; Swain, R.; ReVelle, C.; Bergman, L. The Location of Emergency Service Facilities. *Oper. Res.* **1971**, *19*, 1363–1373. [CrossRef]
20. Mayer, J.D. A Method for the Geographical Evaluation of Emergency Medical Service Performance. *Public Health* **1981**, *71*, 841–844. [CrossRef]
21. Halseth, G.; Rosenberg, M.W. Locating Emergency Medical Services in Small Town and Rural Settings. *Socio-Econ. Plan. Sci.* **1991**, *25*, 295–304. [CrossRef]
22. Brotcorne, L.; Laporte, G.; Semet, F. Ambulance location and relocation models. *Eur. J. Oper. Res.* **2003**, *147*, 451–463. [CrossRef]
23. Henderson, S.G.; Mason, A.J. Ambulance Service Planning: Simulation and Data Visualization. In *Operations Research and Health Care. A Handbook of Methods and Applications*; Brandeau, M.L., Sainfort, F., Pierskalla, W.P., Eds.; Kluwer Academic Publishers: Boston, MA, USA, 2004; pp. 78–102.
24. Schmid, V.; Doerner, K.F. Ambulance location and relocation problems with time-dependent travel times. *Eur. J. Oper. Res.* **2010**, *207*, 1293–1303. [CrossRef]

25. Fone, L.D.; Christie, S.; Lester, N. Comparison of perceived and modeled geographical access to accident and emergency departments: A cross-sectional analysis from the Caerphilly Health and Social Needs Study. *Int. J. Health Geogr.* **2006**, *5*, 16. [CrossRef]
26. Acton, J.P. Nonmonetary factors in the demand for medical services: Some empirical evidence. *J. Political Econ.* **1975**, *83*, 595–614. [CrossRef]
27. Friedman, J.M.; Hagander, L.; Hughes, C.D.; Nash, K.A.; Linden, A.F.; Blossom, J.; Meara, J.G. Distance to hospital and utilization of surgical services in Haiti: Do children, delivering mothers, and patients with emergent surgical conditions experience greater geographical barriers to surgical care? *Int. J. Health Plan. Manag.* **2013**, *28*, 248–256. [CrossRef] [PubMed]
28. Alegana, V.; Wright, J.A.; Pentrina, U.; Noor, A.M.; Snow, R.W.; Atkinson, P.M. Spatial modelling of health care utilization for treatment of fever in Namibia. *Int. J. Health Geogr.* **2012**, *11*, 6. [CrossRef] [PubMed]
29. Brabyn, L.; Beere, P. Population access to hospital emergency departments and the impact of health reform in New Zealand. *Health Inform. J.* **2006**, *12*, 227–237. [CrossRef] [PubMed]
30. Carr, B.G.; Branas, C.C.; Metlay, J.P.; Sullivan, A.F.; Camargo, C.A. Access to Emergency Care in the United States. *Ann. Emerg. Med.* **2009**, *54*, 261–269. [CrossRef]
31. Hashtarkhani, S.; Kiani, B.; Bergquist, R.; Bagheri, N.; VafaeiNejad, R.; Tara, M. An age-integrated approach to improve measurement of potential spatial accessibility to emergency medical services for urban areas. *Int. J. Health Plan. Manag.* **2020**, *35*, 788–798. [CrossRef]
32. Apornak, A. Human resources allocation in the hospital emergency department during COVID-19 pandemic. *Int. J. Healthc. Manag.* **2021**, *14*, 264–270. [CrossRef]
33. Kumar, S.; Raut, R.D.; Narkhede, B.E. A proposed collaborative framework by using artificial intelligence-internet of things (AI-IoT) in COVID-19 pandemic situation for healthcare workers. *Int. J. Health Manag.* **2020**, *13*, 337–345. [CrossRef]
34. Boserup, B.; McKenney, M.; Elkbuli, A. The impact of the COVID-19 pandemic on emergency department visits and patient safety in the United States. *Am. J. Emerg. Med.* **2020**, *38*, 1732–1736. [CrossRef]
35. Golinelli, D.; Campinoti, F.; Sanmarchi, F.; Rosa, S.; Beleffi, M.; Farina, G.; Tampieri, A.; Pia Fantini, M.; Giostra, F.; Santi, L. Patterns of Emergency Department visits for acute and chronic diseases during the two pandemic waves in Italy. *Am. J. Emerg. Med.* **2021**, *50*, 22–26. [CrossRef]
36. Maćkiewicz, A.; Ratajczak, W. Towards a new definition of topological accessibility. *Transp. Res.* **1996**, *30*, 267–284. [CrossRef]
37. Taylor, Z. Dostępność miejsc pracy, nauki, usług w obszarach wiejskich jako przedmiot badań geografii społeczno-ekonomicznej–próba analizy krytycznej. *Przegląd Geogr.* **1997**, *69*, 261–283.
38. Domański, R. Accessibility, Efficiency, and Spatial Organization. *Environ. Plan. A Econol. Space* **1979**, *11*, 1189–1206. [CrossRef]
39. Vickerman, R.W. Accessibility, attraction, and potential: A review of some concepts and their use in determining mobility. *Environ. Plan. A* **1974**, *6*, 675–691. [CrossRef]
40. Ratajczak, W. *Modelowanie Sieci Transportowych*; Wydawnictwo Naukowe UAM: Poznań, Poland, 1999.
41. Gutiérrez, J. Location, economic potential and daily accessibility: An analysis of the accessibility impact of the high-speed line Madrid-Barcelona-French border. *J. Transp. Geogr.* **2001**, *9*, 229–242. [CrossRef]
42. Penchansky, R.; Thomas, J.W. The Concept of Access: Definition and Relationship to Consumer Satisfaction. *Med. Care* **1981**, *19*, 127–140. [CrossRef]
43. Drury, P. Some spatial aspects of health service development: The British experience. *Prog. Hum. Geogr.* **1983**, *7*, 60–77. [CrossRef]
44. Joseph, A.E.; Phillips, D.R. *Accessibility and Utilization. Geographical Perspectives on Health Care Delivery*; Harper & Row Ltd.: London, UK, 1984.
45. Kowalczyk, A. Accessibility and availability of health care services in rural areas of Poland: A case study of the Sierpc medical district. *Geogr. Med.* **1987**, *17*, 47–61.
46. Gulliford, M.; Figueroa-Munoz, J.; Morgan, M.; Hughes, D.; Gibson, B.; Beech, R.; Hudson, M. What does 'access to health care' mean? *J. Health Serv. Res. Policy* **2002**, *7*, 186–188. [CrossRef]
47. Lerner, E.B.; Moscati, R.M. The Golden Hour: Scientific Fact or Medical "Urban Legend"? *Acad. Emerg. Med.* **2001**, *8*, 758–760. [CrossRef]
48. Malczewski, J. *Przestrzenna Organizacja Systemu Placówek Podstawowej Ochrony Zdrowia (na Przykładzie Dzielnicy Warszawa-Wola)*; PAN IGiPZ: Warszawa, Poland, 1989.
49. Powęska, H. *Dostępność Przestrzenna Usług Medycznych a Zachowania Medyczne Ludności*; PAN IGiPZ: Warszawa, Poland, 1990.
50. Twigg, L. Health based geographical information systems: Their potential examined in the light of existing data sources. *Soc. Sci. Med.* **1990**, *30*, 143–155. [CrossRef]
51. Parker, E.B.; Campbell, J.L. Measuring access to primary medical care: Some examples of the use of geographical information systems. *Health Place* **1998**, *4*, 183–193. [CrossRef]
52. Fortney, J.; Rost, K.; Warren, J. Comparing Alternative Methods of Measuring Geographic Access to Health Services. *Health Serv. Outcomes Res. Methodol.* **2000**, *1*, 173–184. [CrossRef]
53. Brabyn, L.; Skelly, C. Modeling population access to New Zealand public hospitals. *Int. J. Health Geogr.* **2002**, *1*, 3. [CrossRef] [PubMed]
54. Patel, A.B.; Waters, N.M.; Ghali, W.A. Determining geographic areas and populations with timely access to cardiac catheterization facilities for acute myocardial infarction care in Alberta, Canada. *Int. J. Health Geogr.* **2007**, *6*, 47. [CrossRef] [PubMed]

55. Longley, P.A.; Goodchild, M.F.; Maguire, D.J.; Rhind, D.W. *Geographic Information Systems and Science*; John Wiley & Sons: Toronto, ON, Canada, 2005.
56. Jones, S.G.; Ashby, A.J.; Momin, S.R.; Naidoo, A. Spatial Implications Associated with Using Euclidean Distance Measurements and Geographic Centroid Imputation in Health Care Research. *HSR Health Serv. Res.* **2010**, *45*, 316–327. [CrossRef] [PubMed]
57. Gaca, M.; Witkowski, A. Realizacja programu medycyny ratunkowej w perspektywie województwa wielkopolskiego. *Przew. Lek.* **2007**, *2*, 240–246.
58. Kostrzewa, D.; Sierocka, A.; Woźniak, B.; Iltchev, P.; Marczak, M. Wpływ zmian finansowania świadczeń w IP/SOR na zarządzanie strategiczne w szpitalu. *Mark. Rynek* **2015**, *9*, 253–266.
59. Hoffmann, K.; Elhag, A.E.; Bryl, W. Globalna eskalacja zapotrzebowania na świadczenia w szpitalnych oddziałach ratunkowych–analiza problemu. *Hygeia* **2018**, *53*, 241–246.
60. W SOR-ach jak w Soczewce Skupiają się Problemy, z Którymi Boryka się System. Available online: https://serwisy.gazetaprawna.pl/zdrowie/artykuly/1408767,problemy-sluzby-zdrowia-sor.html (accessed on 1 December 2021).
61. Paplicki, M. *Bezpieczeństwo Zdrowotne Obywatela w Polskim Systemie Ratownictwa Medycznego*; E-Wydawnictwo, Prawnicza i Ekonomiczna Biblioteka Cyfrowa, Wydział Prawa, Administracji i Ekonomii Uniwersytetu Wrocławskiego: Wrocław, Poland, 2020. [CrossRef]
62. Mayer, J.D. The distance behaviour of hospital patients: A disaggregated analysis. *Soc. Sci. Med.* **1983**, *17*, 819–827. [CrossRef]
63. Coucelis, H.; Getis, A. A Conceptualising and measuring accessibility within physical and virtual spaces. In *Information, Place and Cyberspace: Issues and Accessibility*; Janelle, D., Hodge, D., Eds.; Springer: Berlin/Heidelberg, Germany, 2000.
64. Sánchez-Mangas, R.; García-Ferrrer, A.; De Juan, A.; Arroyo, A.M. The probability of death in road traffic accidents. How important is a quick medical response? *Accid. Anal. Prev.* **2010**, *42*, 1048–1056. [CrossRef] [PubMed]
65. Shen, Y.C.; Hsia, R.Y. Does Decreased Access to Emergency Departments Affect Patient Outcomes? Analysis of Acute Myocardial Infarction Population 1996–2005. *Health Serv. Res.* **2012**, *47*, 188–210. [CrossRef] [PubMed]
66. Nakamura, T.; Okayama, M.; Aihara, M.; Kajii, E. Injury mortality and accessibility to emergency care in Japan: An observational epidemiological study. *Open Access Emerg. Med. OAEM* **2014**, *6*, 27–32. [CrossRef] [PubMed]
67. Mościcka, A.; Bielecka, E.; Tomala, J. Mapa dostępności czasowej szpitalnych oddziałów ratunkowych w Warszawie. *Logistyka* **2014**, *5*, 1147–1153.
68. Sitek, K.; Pokojski, W.; Gołębiowska, I. Evaluation of time availability of the selected rescue service of a large city. A case study of Warsaw. *Pol. Cartogr. Rev.* **2020**, *52*, 176–186. [CrossRef]
69. Fleet, R.; Poitras, J.; Archambault, P.; Tounkara, F.K.; Chauny, J.-M.; Ouimet, M.; Gauthier, J.; Dupuis, G.; Tanguay, A.; Levesque, J.-F.; et al. Portrait of rural emergency departments in Québec and utilization of the provincial emergency department management Guide: Cross sectional survey. *BMC Health Serv. Res.* **2015**, *15*, 572. [CrossRef]
70. Poździoch, S.; Guła, P. (Eds.) *Ustawa o Państwowym Ratownictwie Medycznym. Komentarz*; Wolters Kluwer Polska Sp. zo.o.: Warszawa, Poland, 2008.
71. Luo, W.; Wang, F. Measures of spatial accessibility to health care in a GIS environment: Synthesis and a case study in the Chicago region. *Environ. Plan. B Plan. Des.* **2003**, *30*, 865–884. [CrossRef]

International Journal of
Environmental Research and Public Health

MDPI

Article

Sleep Duration and Bedtime in the PURE Poland Cohort Study and the Link with Noncommunicable Diseases

Katarzyna Zatońska [1], Alicja Basiak-Rasała [1,*], Katarzyna Połtyn-Zaradna [1], Krystian Kinastowski [2] and Andrzej Szuba [3]

[1] Department of Population Health, Wroclaw Medical University, 50-345 Wroclaw, Poland; katarzyna.zatonska@umw.edu.pl (K.Z.); katarzyna.poltyn-zaradna@umw.edu.pl (K.P.-Z.)
[2] Calisia University, 62-800 Kalisz, Poland; krystian@kinastowski.pl
[3] Department of Angiology, Hypertension and Diabetology, Wroclaw Medical University, 50-529 Wroclaw, Poland; andrzej.szuba@umw.edu.pl
* Correspondence: alicja.basiak-rasala@umw.edu.pl; Tel./Fax: +48-71-328-21-45

Abstract: (1) Background: The objective was to investigate the association between sleep duration, bedtime, and noncommunicable diseases in the PURE Poland cohort study. (2) Methods: The baseline study was conducted in 2007–2010. The study group comprised 2023 adult inhabitants of urban and rural areas in Lower Silesia, Poland. The study protocol included questionnaires, blood pressure measurements, blood draws, and anthropometric measurements. Sleep duration and bedtime were self-reported. (3) Results: The median sleep duration of women was 30 min longer than men (8 h vs. 7.5 h; $p = 0.001$). The average time of sleep increased along with the age of the participants. A sleep duration of >8 h was more common in rural than in urban participants (40.2% vs. 27.1%; respectively; $p < 0.001$). The relative risk of diabetes, stroke, hypertension, cardiovascular diseases (CVD), and obesity was significantly higher in participants who went to bed between 6 p.m. and 10 p.m. in comparison to those who went to bed between 10 p.m. and 12 a.m. (RR 2.23, 95% CI 1.06–4.67; RR 2.52, 95% CI 1.28 to 4.97; RR 1.12, 95% CI 1.04–1.20; RR 1.36; 95% CI 1.1–1.68; RR 1.38; 95% CI 1.15–1.66, respectively). The relative risk of respiratory diseases was two-fold higher in those who went to bed after midnight in comparison to those who went to bed between 10 p.m. and 12 a.m. (RR 2.24; 95% CI 1.19–4.22). (4) Conclusions: In our study, an earlier bedtime was associated with a higher risk of diabetes, stroke, obesity, hypertension, and CVD.

Keywords: sleep duration; bedtime; noncommunicable diseases; cohort study

Citation: Zatońska, K.; Basiak-Rasała, A.; Połtyn-Zaradna, K.; Kinastowski, K.; Szuba, A. Sleep Duration and Bedtime in the PURE Poland Cohort Study and the Link with Noncommunicable Diseases. *IJERPH* **2022**, *19*, 403. https://doi.org/10.3390/ijerph19010403

Academic Editors: Wioletta Zukiewicz-Sobczak and Paul B. Tchounwou

Received: 1 November 2021
Accepted: 28 December 2021
Published: 30 December 2021

Publisher's Note: MDPI stays neutral with regard to jurisdictional claims in published maps and institutional affiliations.

Copyright: © 2021 by the authors. Licensee MDPI, Basel, Switzerland. This article is an open access article distributed under the terms and conditions of the Creative Commons Attribution (CC BY) license (https://creativecommons.org/licenses/by/4.0/).

1. Introduction

Sufficient sleep is one of the key contributors to proper functioning and wellbeing. Sleep deprivation affects our immune system, metabolism, and mental health [1]. According to the American Academy of Sleep Medicine (AASM) and the Sleep Research Society (SRS), adults should sleep approximately 7 h per night, since a shorter duration of sleep can be associated with adverse health outcomes [2]. The AASM and the SRS guidelines recognize also that sleep duration exceeding 9 h per night may be beneficial for young adults, but it is uncertain if a longer duration of sleep is associated with health outcomes in adults [2]. On the other hand, both short and long sleep durations have been associated with an increased risk of mortality, type 2 diabetes, cardiovascular diseases (CVD), stroke, coronary heart disease (CHD), and obesity [3–8]. Short sleep duration in middle age has been also associated with an increased risk of dementia [9]. A sleep duration of 6–8 h per day was considered optimal and was associated with the lowest risk of mortality and CVD events [4]. The risk of short sleep duration and sleep disturbances increases with increasing age [10]. Older age has been also associated with earlier bedtime and earlier wake up [11]. Not merely the sleep duration but also the bedtime has been associated with an increased risk of metabolic disorders. Knutson et al. [12] reported an association between later sleep timing and a higher risk of insulin resistance, especially in older participants. The association between sleep duration or bedtime and CVD events and mortality has been recently investigated in the global Prospective Urban and Rural Epidemiology Study (PURE) [4,8].

According to the results of the NHANES study, the average US citizen sleeps approximately 6.9 h per night [13]. The European average sleep duration equals 7.1 h per night [14]. In general, women tend to sleep longer than men [15,16]. According to the report of the Public Opinion Research Center (CBOS), 49% of Poles sleep fewer than 6 h per night at least once a week [17]. A total of 8% of Poles regularly sleep fewer than 6 h per night [17]. Sleeping patterns and the association between bedtime and noncommunicable diseases (NCDs) are rarely researched in Poland. The literature published so far has focused mainly on sleep disturbances, obstructive sleep apnea, and bruxism. One of the largest studies conducted by Kiejna et al. [18,19] presented the results of the National Health Interview Survey. The study focused mainly on sleep disturbances, including insomnia. According to the study, one-fourth of the population suffered from insomnia, and the risk of sleep disturbances increased with age [19]. The mean duration of sleep was 7.7 h with no difference between women and men [18]. Self-reported sleep problems were also investigated in a study focusing on the risk factors for cardiovascular diseases, NATPOL [20]. Sleep duration, sleep disturbances, and bedtime were investigated in Poland in a sample of pregnant women [21]. Overweight and obesity have been associated with sleep disturbances in a study by Lizończyk et al. [22] conducted in a group of adolescents.

Poland is one of the 21 countries enrolled in the PURE study. To our knowledge, our study is one of the first of this scale in Poland investigating the relationship between the duration of sleep, bedtime, and the risk of NCDs in adults.

2. Materials and Methods

The baseline data were collected between 2007 and 2010. A Polish cohort was enrolled into the global Prospective Urban and Rural Epidemiological Study (PURE). All participants were examined in accordance with the global PURE study protocol [23], which included a questionnaire study (individual health, household, family, food frequency questionnaire, and international physical activity questionnaire (IPAQ)), anthropometric measurements, blood pressure measurement, blood draw, ECG, and spirometry. The study design and characteristics of the baseline Polish cohort were described in detail elsewhere [24]. The baseline cohort consisted of 2035 adult participants (1281 women and 754 men) aged 30–85 years (mean: 55 years, SD $\pm$ 10). The study group comprised participants from urban (the city of Wrocław) and rural (villages surrounding Wrocław) areas in Lower Silesia, Poland.

Sleep duration and bedtime were self-reported by the participants. There were three questions in the individual health questionnaire, which directly referred to sleeping habits [23]:

(1) During your longest or nocturnal sleep period, what time do you normally go to bed?
(2) During your longest or nocturnal sleep period, what time do you normally wake up?
(3) Do you usually take naps/siestas?

In the case of (1) and (2), the participants were asked to indicate the approximate hour of bedtime and wake up (00:00–23:59). If participants reported taking naps in (3), they were asked to assess the nap duration (in minutes). Total sleep duration per day was defined as the sum of estimated nocturnal sleep time and self-reported nap duration. Considering the total duration of sleep, we divided the participants into three groups: (1) <6 h of sleep, (2) 6–8 h of sleep, and (3) >8 h of sleep. Following the approach presented by Wang et al. [8], introducing 10:00 p.m. and 00:00 a.m. (midnight) as cut-offs, the participants were divided into three groups: (1) bedtime before 10 p.m., (2) bedtime between 10 p.m. and midnight, and (3) bedtime after midnight.

The body mass of the participants was measured with the use of a Tanita Ironman Body Composition Monitor Model BC-554 with an accuracy of 0.1 kg. The Body Mass Index (BMI) was calculated as weight (kg) divided by height (m) squared. Participants were divided into four BMI categories, according to the WHO guidelines: underweight (BMI < 18.5 kg/m^2), normal weight (BMI 18.5–24.9 kg/m^2), overweight (BMI 25.0–29.9 kg/m^2), and obese (BMI $\geq$ 30.0 kg/m^2). Diabetes was ascertained on the basis of self-reported diabetes and/or self-reported anti-diabetic medication and/or fasting blood glucose measurement $\geq$ 126 mg/dL [25]. Hypertension was ascertained on the basis of self-reported hypertension and/or self-reported anti-hypertensive medication and/or an average of two blood pressure measurements $\geq$ 140/90 mmHg as previously described [26]. The occurrence of cardiovascular diseases (CVD) and respiratory diseases was self-reported by the participants. The category of CVD comprised participants who reported heart failure, coronary heart disease, and other heart diseases. Respiratory diseases comprised asthma and chronic obstructive pulmonary disease (COPD). Attitudes toward tobacco smoking and alcohol consumption were self-reported by the participants. In the case of tobacco smoking, participants could have chosen one of three possible answers: "formerly used tobacco products", "currently use tobacco products", or "never used tobacco products". Similarly, in the case of the question regarding alcohol consumption ("Which best describes your history of alcohol use?"), participants could have chosen between "formerly used alcohol products", "currently use alcohol products", and "never used alcohol products". The study has been reviewed and accepted by the Bioethics Committee of the Wrocław Medical University and have therefore been performed in accordance with the ethical standards laid down in an appropriate version of the 1964 Declaration of Helsinki (positive opinion of The Bioethics Committee of the Wrocław Medical University nr KB-443/2006).

Sleeping patterns in the PURE Poland cohort study were described using basic statistical parameters. The Kolmogorov–Smirnov test with the Lillefors correction and the Shapiro–Wilk test were used to assess the normality of the distributions of the analyzed variables. As empirical distributions significantly deviated from the normal distribution, non-parametric tests were used to assess the significance of differences in sleep duration between the studied groups: for two groups, the Mann–Whitney U test was used, and when there were more groups, the Kruskal–Wallis test was used. The Pearson Chi-square test was used to assess the independence between the two variables. The relative risk (RR) with 95% confidence intervals (CI) was calculated to take into account the risk of the occurrence of the selected NCD in groups of participants differentiated according to bedtime. The relative risk was calculated directly from the contingency tables (C $\times$ R). A bedtime between 10 p.m. and 12 a.m. was considered a reference value. In all analyses, the level of statistical significance was set at $p \leq 0.05$. Statistical analysis was performed using the Statistica 13.3 software (TIBCO. Software Inc., Palo Alto, CA, USA).

3. Results

A total of 2023 participants were enrolled in the analysis. Eleven participants were excluded due to a lack of information on bedtime. Shift workers who declared bedtime between 4 a.m. and 12 p.m. (5 participants) were also excluded.

3.1. Total Sleep Duration

The characteristics of the total sleep duration in our study are presented in Table 1. There was a significant difference in total sleep duration between men and women ($p < 0.001$). Women slept an average of 7.9 h $\pm$ 1.4, whereas men 7.7 h $\pm$ 1.3. The median sleep duration of women was approximately 30 min longer than men (8 h vs. 7.5 h; $p = 0.001$). Among women, 9.1% reported sleep duration < 6 h, 55.5% reported 6–8 h, and 35.4% reported > 8 h of sleep. Among men, 11.9% reported sleep duration < 6 h, 60.7% reported 6–8 h of sleep, and 27.4% reported > 8 h of sleep. Total sleep duration was significantly differentiated by sex, age, place of residence, education, professional activity, bedtime, attitudes toward tobacco smoking and alcohol drinking, level of physical activity, WHR, and the occurrence of chosen non-communicable diseases. In our study, the total sleep duration was longer in older participants. The average time of sleep in 30- to 44-

year-olds was significantly shorter than in 45- to 64-year-olds (7.5 h vs. 7.8 h, respectively; $p = 0.024$) and shorter than in participants > 64 years of age (7.5 h vs. 8.5 h; $p < 0.001$). Moreover, 45- to 64-year-olds also slept shorter than participants > 64 years of age (7.8 h vs. 8.5 h, respectively; $p < 0.001$). Every 10 years, the duration of sleep increased, on average, by 0.3 h. An earlier bedtime was associated with a longer sleep duration. Participants who declared a bedtime between 6 p.m. and 10 p.m. slept, on average, 8.6 h ± 1.1; participants who declared bedtime between 10 p.m. and 0 a.m. slept, on average, 7.4 h ± 1.1; and those who went to bed between midnight and 4 a.m. slept, on average, 6.5 h ± 1.4; $p < 0.001$.

Table 1. Total sleep duration in the study population.

Variable	Total Sleep Duration (Hours)			*p*-Value **
	<6 **n = 205**	**6–8** **n = 1162**	**>8** **n = 656**	
Age (years), Me (Q1, Q3)	53 (45, 57)	54 (47, 60)	60 (52, 65)	<0.001
Sex				
Female, *n* (%)	116 (9.1)	707 (55.5)	451 (35.4)	<0.001
Male, *n* (%)	89 (11.9)	455 (60.7)	205 (27.4)	
Place of residence				
Urban, *n* (%)	145 (12.1)	730 (60.8)	325 (27.1)	<0.001
Rural, *n* (%)	60 (7.3)	432 (52.5)	331 (40.2)	
Marital status				
Never married, *n* (%)	18 (12.3)	92 (63.0)	36 (24.7)	0.057
Married/living together, *n* (%)	152 (10.1)	870 (58.0)	478 (31.9)	
Separated/divorced/widowed, *n* (%)	35 (9.3)	200 (53.1)	142 (37.6)	
Education				
Primary or unknown, *n* (%)	13 (4.2)	126 (41.2)	167 (54.6)	<0.001
Vocational, *n* (%)	25 (7.7)	193 (59.6)	106 (32.7)	
Secondary, *n* (%)	97 (12.2)	444 (56.1)	251 (31.7)	
Higher, *n* (%)	70 (11.6)	399 (66.4)	132 (22.0)	
Professional activity				
Professionally active, *n* (%)	161 (14.5)	756 (67.9)	196 (17.6)	<0.001
Retired, *n* (%)	35 (5.0)	316 (45.0)	352 (50.0)	
Pensioner, *n* (%)	9 (4.3)	90 (43.5)	108 (52.2)	
Bedtime				
6 p.m.–10 p.m., *n* (%)	7 (1.0)	348 (45.4)	411 (53.6)	<0.001
10 p.m.–0 a.m., *n* (%)	147 (12.9)	760 (66.6)	234 (20.5)	
0 a.m.–4 a.m., *n* (%)	51 (43.9)	54 (46.6)	11 (9.5)	
Naps				
Taking naps, *n* (%)	74 (11.7)	346 (54.6)	214 (33.7)	0.137
Average nap time (min), M ± SD	47.5 ± 36.4	38.5 ± 24.3	42.9 ± 27.0	0.015
Attitudes toward tobacco smoking				
Never smokers, *n* (%)	89 (9.3)	520 (54.3)	349 (36.4)	<0.001
Former smokers, *n* (%)	59 (9.2)	401 (62.5)	182 (28.3)	
Current smokers, *n* (%)	57 (13.5)	241 (57.0)	125 (29.6)	

Table 1. *Cont.*

Variable	Total Sleep Duration (Hours)			p-Value **
	<6 n = 205	6–8 n = 1162	>8 n = 656	
	Attitudes toward alcohol consumption			
Never drinkers, n (%)	44 (9.7)	240 (53.1)	168 (37.2)	<0.001
Former drinkers, n (%)	18 (8.7)	99 (47.6)	91 (43.6)	
Current drinkers, n (%)	143 (10.5)	823 (60.4)	397 (29.1)	
Physical activity, MET × min/week *	N = 190	N = 1063	N = 612	
Low (<600), n (%)	3 (4.9)	33 (54.1)	25 (41)	0.157
Moderate (600–3000), n (%)	48 (9.4)	281 (54.9)	183 (35.7)	
High (>3000), n (%)	139 (10.8)	749 (58.0)	404 (31.3)	
	Blood pressure			
SBP (mm Hg), Mean ± SD	145 ± 19	144 ± 21	147 ± 23	0.015
DBP (mm Hg), Mean ± SD	87 ± 11	86 ± 12	86 ± 11	0.314
HR (bmp), Mean ± SD	74 ± 13	72 ± 10	72 ± 11	0.042
	Body Mass Index (BMI)			
BMI (kg/m^2), Mean ± SD	27.9 ± 4.8	28.0 ± 5.0	28.5 ± 5.2	0.059
Underweight, n (%)	1 (6.6)	7 (46.7)	7 (46.7)	0.271
Normal, n (%)	61 (10.7)	336 (58.9)	173 (30.4)	
Overweight, n (%)	84 (10.4)	476 (58.9)	248 (30.7)	
Obesity, n (%)	59 (9.4)	343 (54.4)	227 (36.2)	
	Waist-to-hip ratio (WHR)			
WHR (-), Mean ± SD	0.884 ± 0.094	0.880 ± 0.094	0.900 ± 0.100	0.005
Normal, n (%)	78 (11.0)	449 (63.3)	182 (25.7)	<0.001
Central obesity, n (%)	127 (9.7)	713 (54.3)	474 (36.1)	
	Noncommunicable diseases			
Diabetes, n (%)	12 (6.0)	86 (43.2)	101 (50.8)	<0.001
Hypertension, n (%)	125 (10.3)	664 (54.5)	430 (35.3)	0.002
Cardiovascular diseases, n (%)	41 (9.3)	212 (48.3)	186 (42.4)	<0.001
Respiratory diseases, n (%)	14 (12.8)	60 (55.1)	35 (32.1)	0.621

* Physical activity, unlike other variables, was assessed in the group of 1865 participants, the number of participants in categories differentiated by sleep duration in this analysis has been placed in the same verse; ** statistically significant p-value has been highlighted in bold. Abbreviations: Me—median, M—mean; SD—standard deviation; SBP—systolic blood pressure; DBP—diastolic blood pressure; HR—heart rate; BMI—Body mass index; WHR—waist to hip ratio.

Sleep duration has been differentiated by the place of residence and level of education ($p < 0.001$). Among urban participants, 12.1% slept <6 h, 60.8% slept 6–8 h, and 27.1% slept more than 8 h. In comparison, among rural participants, 7.3% slept < 6 h, 52.5% slept 6–8 h, and 40.2% slept more than 8 h. Considering the level of education, 4.2% of participants with primary or unknown education slept < 6 h, 41.2% slept 6–8 h, and 54.6% slept more than 8 h, respectively. The 11.6% of participants with higher education reported fewer than 6 h of sleep, 66.4% reported 6–8 h of sleep, and 22.0% reported more than 8 h of sleep.

Total sleep duration was not significantly differentiated by BMI (p = 0.059). On the other hand, an average BMI was slightly higher among participants who slept, on average, > 8 h per night, followed by those who slept 6–8 h (mean 28.5 kg/m^2 ± 5.2 and 28.0 kg/m^2 ± 5.0, respectively). The mean WHR was the highest among those who slept more than 8 h (0.90 ± 0.1; $p = 0.005$). The mean systolic blood pressure (SBP) was the highest in participants who slept more than 8 h (147 ± 23; $p = 0.015$). On the other hand, heart rate was higher in those who slept < 6 h (74 ± 13; $p = 0.042$).

3.2. Naps

Taking naps during the day was also more prevalent in men than women (38.7% vs. 27.1%, respectively; $p < 0.001$). The duration of the naps of those participants who slept 6–8 h was, on average, 9 min shorter than of those who slept < 6 h (38.5 min vs. 47.5 min; $p = 0.015$). Patients who took naps were older by 2 years than those who did not take naps (56.1 min ± 9.7 vs. 53.9 min ± 9.8; $p = 0.001$). Naps were significantly more common in older participants.

3.3. Total Sleep Duration and Bedtime According to Chosen Non-Communicable Diseases

A total of 13% of participants went to sleep between 6 p.m. and 10 p.m., 71% went to sleep between 10 p.m. and 12 a.m., and 17% between 12 a.m. and 4 a.m. The basic characteristics of bedtime according to chosen non-communicable diseases are presented in Figure 1. A bedtime between 6 p.m. and 10 p.m. was more common for participants with diabetes and stroke, whereas a bedtime between 10 p.m. and 12 a.m. was more prevalent in participants with hypertension, CVD, and respiratory diseases.

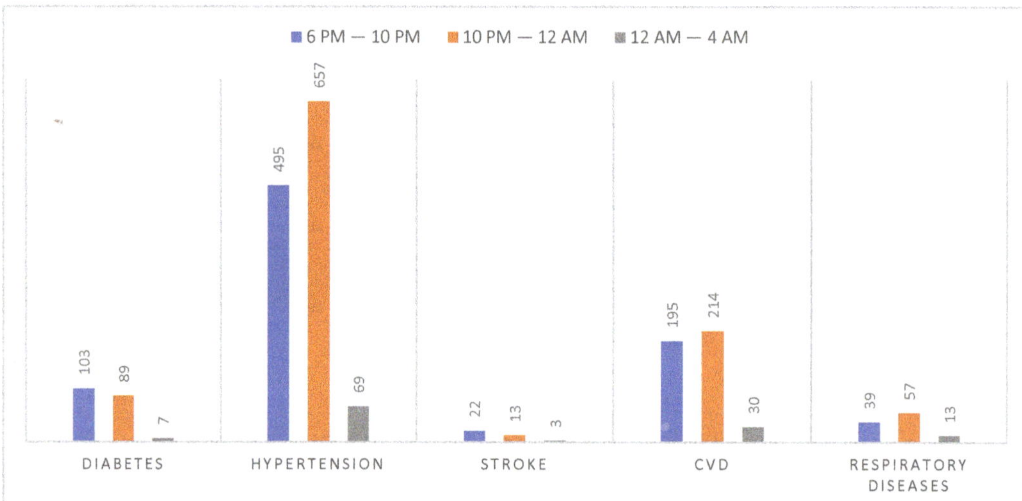

Figure 1. Bedtime in participants with chosen noncommunicable diseases. In the case of every disease, there were statistically significant differences between the categories differentiated by bedtime ($p < 0.02$). CVD—cardiovascular diseases.

Among participants with diabetes, 50.8% slept > 8 h, 43.2% slept 6–8 h and 6.0% slept fewer than 6 h, respectively ($p < 0.001$). The relative risk of diabetes was two-fold higher and of stroke 2.5-fold higher in participants who went to bed between 6 p.m. and 10 p.m., in comparison to those who went to bed between 10 p.m. and 12 a.m. [RR 2.23; 95% CI 1.06–4.67; RR 2.52; 95% CI 1.28 to 4.97].

Among participants with diabetes, 50.8% slept > 8 h, 43.2% slept 6–8 h, and 6.0% slept fewer than 6 h, respectively ($p < 0.001$). The relative risk of diabetes was two-fold higher and of stroke 2.5-fold higher in participants who went to bed between 6 p.m. and 10 p.m., in comparison to those who went to bed between 10 p.m. and 12 a.m. (RR 2.23; 95% CI 1.06–4.67; RR 2.52; 95% CI 1.28 to 4.97).

Among participants with hypertension, 35.3% slept > 8 h, 54.5% slept 6–8 h, and 10.3% slept fewer than 6 h, respectively ($p = 0.002$). The relative risk of hypertension was significantly higher in participants who went to bed between 6 p.m. and 10 p.m., in comparison to those who went to bed between 10 p.m. and 12 a.m. (RR 1.12; 95% CI 1.04–1.20).

Among participants with CVD, 42.4% slept > 8 h, 48.3% slept 6–8 h, and 9.3% slept fewer than 6 h, respectively ($p < 0.001$). Moreover, the relative risk of CVD was also significantly higher in those participants who went to bed between 6 p.m. and 10 p.m., in comparison to those who went to bed between 10 p.m. and midnight (RR 1.36; 95% CI 1.1–1.68).

The relative risk of obesity according to BMI was significantly higher in participants, who went to bed between 6 p.m. and 10 p.m., in comparison to those who went to bed between 10 p.m. and 12 a.m. (RR 1.38; 95% CI 1.15–1.66).

On the other hand, the relative risk of respiratory diseases was two-fold higher in those who went to bed after midnight, in comparison to those who went to bed between 10 p.m. and 12 a.m. (RR 2.24; 95% CI 1.19–4.22). Participants with respiratory diseases were not significantly differentiated by sleep duration. The analysis of relative risk with 95% confidence intervals of the occurrence of chosen non-communicable diseases in participants who went to sleep earlier (6 p.m.–10 p.m.) and later (12 a.m.–4 a.m.) in comparison to those who went to sleep between 10 p.m. and 12 a.m. is presented in Figure 2.

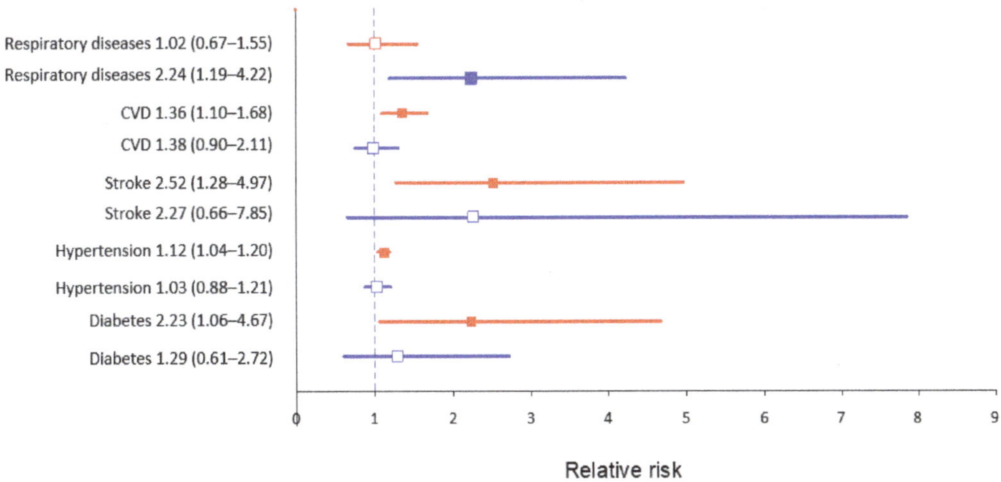

Figure 2. Relative risk with 95% confidence intervals of the occurrence of chosen non-communicable diseases in participants who went to sleep earlier (6 p.m.–10 p.m., indicated by the red color) and later (12 a.m.–4 a.m., indicated by the blue color) in comparison to those who went to sleep between 10 p.m. and 12 a.m. (reference value).

4. Discussion

Deprivation and low quality of sleep have been linked to many metabolic diseases, including obesity and type 2 diabetes. The presented article aimed to characterize sleep duration and bedtime in the PURE Poland cohort study in the view of sociodemographic characteristics and chosen NCDs. The PURE Poland study is one of the few longitudinal cohort studies in Poland. In our cohort, sleep duration was significantly differentiated by sociodemographic factors (sex, age, place of residence, level of education, professional activity status), behavioral factors (bedtime, taking naps, tobacco smoking, alcohol drinking), anthropometric factors (BMI, WHR), and some NCDs (diabetes, hypertension, stroke, CVD). The relative risk of diabetes, obesity, hypertension, stroke, and CVDs was significantly higher in participants who went to bed between 6 p.m. and 10 p.m. in comparison to the control (bedtime 10 p.m.–12 a.m.).

In our cohort, the average duration of sleep was significantly longer in women than in men. On the other hand, more men than women declared taking naps during the day. Conforming to our results, in the global PURE study analysis, a longer duration of sleep was more prevalent in women, participants > 50 years old, and in those living in rural areas [8]. In our cohort, there were more long sleepers among the participants with primary education.

In our study, a shorter sleep duration was more common in current smokers and alcohol drinkers. Similarly, in the Whitehall study [27], a shorter sleep duration was more common in participants consuming alcohol in comparison to never-drinkers.

Too-short and too-long durations of sleep have been previously associated with an increased risk of type 2 diabetes [28,29]. In a study published by Cappuccio et al. [28], a long duration of sleep (>8–9 h/night) was associated with a relative risk (RR) of type 2 diabetes of 1.48 [CI 1.13–1.96]. Both short and long durations of sleep have been associated with increased risk of CHD, stroke, and total CVD [30]. In a prospective study and meta-analysis by Leng et al. [31], a long duration of sleep has been significantly associated with an increased risk of stroke. On the other hand, in a meta-analysis of prospective cohort studies conducted by Li et al. [32], both short and long durations of sleep were associated with an increased risk of stroke, but a long sleep duration was a significant marker of stroke mortality. In our cohort, hypertensive participants had a tendency for a longer duration of sleep, which is consistent with the results from the global PURE study [8]. It is speculated that fatigue or a longer duration of sleep can be a preliminary symptom of slowly developing health problems, hence the significant association between a long duration of sleep and mortality or morbidity observed in the studies [8]. Having said that, we cannot determine the causality at this stage. In our cohort, the duration of sleep increased along with the increasing age of participants. As a general principle, aging is associated with disturbed sleep, a decreased ability to maintain sleep, and a shorter sleep duration [33]. In our study, total sleep duration combined both nocturnal sleep and the sum of daytime naps, which may have contributed to this association. It has been previously observed that the frequency of daytime napping increases with age [34]. There were also studies that indicated that retirement promoted longer sleep duration [35]. In a study by Basner et al. [36], participants who were unemployed or retired reported longer sleep duration and were less likely to be short sleepers.

In our study, an earlier bedtime was associated with a higher risk of the chosen NCDs, including diabetes and hypertension. On the contrary to our results, in a study by Yan et al. [37], a later bedtime (after midnight) was associated with an increased prevalence of diabetes (OR 1.446; 95% CI 1.107–1.888]. A U-shaped relationship between the prevalence of hypertension and bedtime was observed by Jansen et al. [38]. The risk of hypertension was almost two-fold higher in those participants who went to bed before 9 p.m., as well as after 11 p.m., in comparison to those, who went to bed between 9 p.m. and 11 p.m. (RR 1.96; 95% CI 1.27–3.01; RR1.87; 95% CI 1.09–2.21, respectively) [38]. Similarly, a U-shaped relationship between bedtime and health outcomes was observed in the global PURE study [8]. Both those who went to bed early (before 10 p.m.) and late (after midnight) had a higher risk of mortality and major CVD events [8]. In our study, the relative risk of respiratory diseases was two-fold higher in those who went to bed after midnight in comparison to those who went to bed between 10 p.m. and 12 a.m. The direct association between bedtime and respiratory diseases has been rarely described in the literature. However, the diurnal variability of lung function can influence bedtime, the duration of sleep, and the quality of sleep in asthmatic patients [39,40]. A nocturnal decline in pulmonary function has been previously associated with later bedtime and shorter sleep duration [41]. Chronic obstructive pulmonary disease (COPD) has been also associated with disturbed sleep, insomnia, and delayed initiation of sleep [42], which may partially explain the association observed in our study.

There are some limitations of our study to discuss. The current analysis focuses solely on data collected at the baseline. Our study is a cohort study, so the results should be treated with caution. Our cohort is characterized by the overrepresentation of women and elderly participants in comparison to the general Polish population. A lot of participants have already retired, which influences their sleeping patterns. Sleep disorders, including insomnia, were not investigated in the PURE study. The use of medication, which may influence sleeping patterns, was not investigated in this analysis. On the other hand, there are also several strengths to discuss. Our study is one of the few cohort studies in Poland of this size. To our knowledge, our study is one of the first of this scale in Poland investigating the relationship between the duration of sleep, bedtime, and the risk of NCDs in adults. It is planned that future research will include a longitudinal analysis of bedtime and duration of sleep in the study population.

5. Conclusions

Sleep duration and bedtime were significantly differentiated by sociodemographic and behavioral factors. In our study, an earlier bedtime was associated with a higher risk of diabetes, obesity, hypertension, and CVD.

Author Contributions: Conceptualization, A.S. and K.Z.; formal analysis, K.P.-Z. and A.B.-R.; methodology, K.Z.; visualization, K.P.-Z.; writing—original draft, K.Z. and A.B.-R.; writing—review and editing K.Z., K.K., A.S., and K.P.-Z. All authors have read and agreed to the published version of the manuscript.

Funding: The main PURE study and its components are funded by the Population Health Research Institute, the Canadian Institute of Health Research, the Heart and Stroke Foundation of Ontario, and unrestricted grants from several pharmaceutical companies. The Poland cohort study is funded by the Polish Ministry of Science and Higher Education (grant no. 290/W-PURE/2008/0). Additionally, the presented research results were conducted within the framework of the topic: "Analysis of the prevalence of risk factors of non-communicable diseases with particular emphasis on tobacco smoke exposure in the population of adult residents of the Lower Silesia Province and school-age children", as per records in SIMPLE system number SUB. C. 300.21.009. We declare that the funding body had no role in the design of the study; the collection, analysis, or interpretation of data; and no role in writing the manuscript.

Institutional Review Board Statement: The study was reviewed and accepted by the appropriate ethics committee and was therefore performed in accordance with the ethical standards laid down in an appropriate version of the 1964 Declaration of Helsinki (positive opinion of The Bioethics Committee of the Wrocław Medical University nr KB-443/2006).

Informed Consent Statement: Written and informed consent was obtained from all subjects involved in the study.

Data Availability Statement: Data available on request due to restrictions, e.g., privacy or ethical.

Conflicts of Interest: The authors declare no conflict of interest.

References

1. Medic, G.; Wille, M.; Hemels, M.E.H. Short- and long-term health consequences of sleep disruption. *Nat. Sci. Sleep* **2017**, *9*, 151–161. [CrossRef] [PubMed]
2. Watson, N.F.; Badr, M.S.; Belenky, G.; Bliwise, D.L.; Buxton, O.M.; Buysse, D.; Dinges, D.F.; Gangwisch, J.; Grandner, M.A.; Kushida, C.; et al. Recommended amount of sleep for a healthy adult: A joint consensus statement of the American Academy of Sleep Medicine and Sleep Research Society. *J. Clin. Sleep Med.* **2015**, *11*, 591–592. [CrossRef] [PubMed]
3. Jike, M.; Itani, O.; Watanabe, N.; Buysse, D.J.; Kaneita, Y. Long sleep duration and health outcomes: A systematic review, meta-analysis and meta-regression. *Sleep Med. Rev.* **2018**, *39*, 25–36.
4. Wang, C.; Bangdiwala, S.I.; Rangarajan, S.; Lear, S.A.; AlHabib, K.F.; Mohan, V.; Teo, K.; Poirier, P.; Tse, L.A.; Liu, Z.; et al. Association of estimated sleep duration and naps with mortality and cardiovascular events: A study of 116,632 people from 21 countries. *Eur. Heart J.* **2019**, *40*, 1620–1629. [CrossRef] [PubMed]
5. Itani, O.; Jike, M.; Watanabe, N.; Kaneita, Y. Short sleep duration and health outcomes: A systematic review, meta-analysis, and meta-regression. *Sleep Med.* **2017**, *32*, 246–256. [CrossRef] [PubMed]
6. Yin, J.; Jin, X.; Shan, Z.; Li, S.; Huang, H.; Li, P.; Peng, X.; Peng, Z.; Yu, K.; Bao, W.; et al. Relationship of sleep duration with all-cause mortality and cardiovascular events: A systematic review and dose-response meta-analysis of prospective cohort studies. *J. Am. Heart Assoc.* **2017**, *6*, e005947. [CrossRef] [PubMed]

7. Chaput, J.P.; McNeil, J.; Després, J.P.; Bouchard, C.; Tremblay, A. Short sleep duration as a risk factor for the development of the metabolic syndrome in adults. *Prev. Med.* **2013**, *57*, 872–877. [CrossRef]
8. Wang, C.; Hu, B.; Rangarajan, S.; Bangdiwala, S.I.; Lear, S.A.; Mohan, V.; Gupta, R.; Alhabib, K.F.; Soman, B.; Abat, M.E.M.; et al. Association of bedtime with mortality and major cardiovascular events: An analysis of 112,198 individuals from 21 countries in the PURE study. *Sleep Med.* **2021**, *80*, 265–272. [CrossRef] [PubMed]
9. Sabia, S.; Fayosse, A.; Dumurgier, J.; van Hees, V.T.; Paquet, C.; Sommerlad, A.; Kivimäki, M.; Dugravot, A.; Singh-Manoux, A. Association of sleep duration in middle and old age with incidence of dementia. *Nat. Commun.* **2021**, *12*, 1–10. [CrossRef] [PubMed]
10. Hertenstein, E.; Gabryelska, A.; Spiegelhalder, K.; Nissen, C.; Johann, A.F.; Umarova, R.; Riemann, D.; Baglioni, C.; Feige, B. Reference Data for Polysomnography-Measured and Subjective Sleep in Healthy Adults. *J. Clin. Sleep Med.* **2018**, *14*, 523–532. [CrossRef] [PubMed]
11. Evans, M.A.; Buysse, D.J.; Marsland, A.L.; Wright, A.G.C.; Foust, J.; Carroll, L.W.; Kohli, N.; Mehra, R.; Jasper, A.; Srinivasan, S.; et al. Meta-analysis of age and actigraphy-assessed sleep characteristics across the lifespan. *Sleep* **2021**, *44*, zsab088. [CrossRef] [PubMed]
12. Knutson, K.L.; Wu, D.; Patel, S.R.; Loredo, J.S.; Redline, S.; Cai, J.; Gallo, L.C.; Mossavar-Rahmani, Y.; Ramos, A.R.; Teng, Y.; et al. Association between sleep timing, obesity, diabetes: The hispanic community health study/study of latinos (hchs/sol) cohort study. *Sleep* **2017**, *40*, zsx014. [CrossRef]
13. Yang, G.; Han, Y.Y.; Sun, T.; Li, L.; Rosser, F.; Forno, E.; Patel, S.R.; Chen, W.; Celedón, J.C. Sleep duration, current asthma, and lung function in a nationwide study of U.S. Adults. *Am. J. Respir. Crit. Care Med.* **2019**, *200*, 926–929. [CrossRef] [PubMed]
14. Lakerveld, J.; Mackenbach, J.D.; Horvath, E.; Rutters, F.; Compernolle, S.; Bárdos, H.; De Bourdeaudhuij, I.; Charreire, H.; Rutter, H.; Oppert, J.M.; et al. The relation between sleep duration and sedentary behaviours in European adults. *Obes. Rev.* **2016**, *17*, 62–67. [CrossRef]
15. Jonasdottir, S.S.; Minor, K.; Lehmann, S. Gender differences in nighttime sleep patterns and variability across the adult lifespan: A global-scale wearables study. *Sleep* **2021**, *44*, zsaa169. [CrossRef]
16. Walch, O.J.; Cochran, A.; Forger, D.B. A global quantification of "normal" sleep schedules using smartphone data. *Sci. Adv.* **2016**, *2*, e1501705. [CrossRef] [PubMed]
17. Boguszewski, R. *Zdrowie i Prozdrowotne Zachowania Polaków [Health and Health-RELATED Behavior of Poles]*; Centrum Badania Opinii Społecznej: Warszawa, Poland, 2016.
18. Kiejna, A.; Rymaszewska, J.; Wojtyniak, B.; Stokwiszewski, J. Characteristics of sleep disturbances in Poland-Results of the National Health Interview Survey. *Acta Neuropsychiatr.* **2004**, *16*, 124–129. [CrossRef]
19. Kiejna, A.; Wojtyniak, B.; Rymaszewska, J.; Stokwiszewski, J. Prevalence of insomnia in Poland-Results of the National Health Interview Survey. *Acta Neuropsychiatr.* **2003**, *15*, 68–73. [CrossRef] [PubMed]
20. Nowicki, Z.; Grabowski, K.; Cubała, W.J.; Nowicka-Sauer, K.; Zdrojewski, T.; Rutkowski, M.; Bandosz, P. Prevalence of self-reported insomnia in general population of Poland. *Psychiatr. Pol.* **2016**, *50*, 165–173. [CrossRef] [PubMed]
21. Smyka, M.; Kosińska-Kaczyńska, K.; Sochacki-Wójcicka, N.; Zgliczyńska, M.; Wielgoś, M. Sleep problems in pregnancy—A cross-sectional study in over 7000 pregnant women in Poland. *Int. J. Environ. Res. Public Health* **2020**, *17*, 5306. [CrossRef] [PubMed]
22. Lizończyk, I.; Jośko-Ochojska, J. Relationship between overweight, obesity and sleep disorders in adolescents from selected cities of upper silesia, Poland. *Ann. Agric. Environ. Med.* **2021**, *28*, 193–197. [CrossRef] [PubMed]
23. Teo, K.; Chow, C.K.; Vaz, M.; Rangarajan, S.; Yusuf, S.; PURE Investigators-Writing Group. The Prospective Urban Rural Epidemiology (PURE) study: Examining the impact of societal influences on chronic noncommunicable diseases in low-, middle-, and high-income countries. *Am. Heart J.* **2009**, *158*, 1–7.e1. [CrossRef] [PubMed]
24. Zatońska, K.; Zatoński, W.A.; Szuba, A. Prospective urban and rural epidemiology Poland–study design. *J. Health Inequal.* **2016**, *2*, 136–141. [CrossRef]
25. Polskie Towarzystwo Diabetologiczne. Zalecenia kliniczne dotyczące postępowania u chorych na cukrzycę 2019. *Diabetol. Prakt.* **2019**, *5*, 7–10.
26. Szuba, A.; Martynowicz, H.; Zatońska, K.; Ilow, R.; Regulska-Ilow, B.; Różańska, D.; Wołyniec, M.; Połtyn-Zaradna, K.; Zatoński, W.A. Prevalence of hypertension in Polish population of PURE Poland study. *J. Health Inequal.* **2017**, *2*, 157–162. [CrossRef]
27. Jackson, C.L.; Gaston, S.A.; Liu, R.; Mukamal, K.; Rimm, E.B. The Relationship between Alcohol Drinking Patterns and Sleep Duration among Black and White Men and Women in the United States. *Int. J. Environ. Res. Public Health* **2018**, *15*, 557. [CrossRef]
28. Cappuccio, F.P.; D'Elia, L.; Strazzullo, P.; Miller, M.A. Quantity and quality of sleep and incidence of type 2 diabetes: A systematic review and meta-analysis. *Diabetes Care* **2010**, *33*, 414–420. [CrossRef]
29. Shan, Z.; Ma, H.; Xie, M.; Yan, P.; Guo, Y.; Bao, W.; Rong, Y.; Jackson, C.L.; Hu, F.B.; Liu, L. Sleep duration and risk of type 2 diabetes: A meta-analysis of prospective studies. *Diabetes Care* **2015**, *38*, 529–537. [CrossRef] [PubMed]
30. Cappuccio, F.P.; Cooper, D.; Delia, L.; Strazzullo, P.; Miller, M.A. Sleep duration predicts cardiovascular outcomes: A systematic review and meta-analysis of prospective studies. *Eur. Heart J.* **2011**, *32*, 1484–1492. [CrossRef] [PubMed]
31. Leng, Y.; Cappuccio, F.P.; Wainwright, N.W.J.; Surtees, P.G.; Luben, R.; Brayne, C.; Khaw, K.T. Sleep duration and risk of fatal and nonfatal stroke: A prospective study and meta-analysis. *Neurology* **2015**, *84*, 1072–1079. [CrossRef] [PubMed]

32. Li, W.; Wang, D.; Cao, S.; Yin, X.; Gong, Y.; Gan, Y.; Zhou, Y.; Lu, Z. Sleep duration and risk of stroke events and stroke mortality: A systematic review and meta-analysis of prospective cohort studies. *Int. J. Cardiol.* **2016**, *223*, 870–876. [CrossRef]
33. Li, J.; Vitiello, M.V.; Gooneratne, N.S. Sleep in Normal Aging. *Sleep Med. Clin.* **2018**, *13*, 1–11. [CrossRef]
34. Furihata, R.; Kaneita, Y.; Jike, M.; Ohida, T.; Uchiyama, M. Napping and associated factors: A Japanese nationwide general population survey. *Sleep Med.* **2016**, *20*, 72–79. [CrossRef]
35. Hagen, E.W.; Barnet, J.H.; Hale, L.; Peppard, P.E. Changes in Sleep Duration and Sleep Timing Associated with Retirement Transitions. *Sleep* **2016**, *39*, 665. [CrossRef]
36. Basner, M.; Spaeth, A.M.; Dinges, D.F. Sociodemographic characteristics and waking activities and their role in the timing and duration of sleep. *Sleep* **2014**, *37*, 1889–1906. [CrossRef] [PubMed]
37. Yan, B.; Fan, Y.; Zhao, B.; He, X.; Yang, J.; Chen, C.; Ma, X. Association between late bedtime and diabetes mellitus: A large community-based study. *J. Clin. Sleep Med.* **2019**, *15*, 1621–1627. [CrossRef] [PubMed]
38. Jansen, E.C.; Dunietz, G.L.; Matos-Moreno, A.; Solano, M.; Lazcano-Ponce, E.; Sánchez-Zamorano, L.M. Bedtimes and Blood Pressure: A Prospective Cohort Study of Mexican Adolescents. *Am. J. Hypertens.* **2020**, *33*, 269–277. [CrossRef] [PubMed]
39. Sutherland, E.R. Nocturnal asthma: Underlying mechanisms and treatment. *Curr. Allergy Asthma Rep.* **2005**, *5*, 161–167. [CrossRef] [PubMed]
40. Scheer, F.A.J.L.; Hilton, M.F.; Evoniuk, H.L.; Shiels, S.A.; Malhotra, A.; Sugarbaker, R.; Ayers, R.T.; Israel, E.; Massaro, A.F.; Shea, S.A. The endogenous circadian system worsens asthma at night independent of sleep and other daily behavioral or environmental cycles. *Proc. Natl. Acad. Sci. USA* **2021**, *118*, e2018486118. [CrossRef]
41. Krouse, H.J.; Krouse, J.H. Diurnal variability of lung function and its association with sleep among patients with asthma. *J. Asthma* **2007**, *44*, 759–763. [CrossRef]
42. Budhiraja, R.; Siddiqi, T.A.; Quan, S.F. Sleep Disorders in Chronic Obstructive Pulmonary Disease: Etiology, Impact, and Management. *J. Clin. Sleep Med.* **2015**, *11*, 259. [CrossRef] [PubMed]

International Journal of
Environmental Research and Public Health

MDPI

Article

Comparison of the Levels of Hematological Parameters at Rest and after Maximum Exercise between Physically Active People with Spinal Cord Injury and Able-Bodied People

Łukasz Szymczak [1], Tomasz Podgórski [2] and Katarzyna Domaszewska [1,2,*]

[1] Faculty of Health Sciences, Calisia University, 62-800 Kalisz, Poland; l.szymczak@akademiakaliska.edu.pl
[2] Department of Physiology and Biochemistry, Poznan University of Physical Education, 61-871 Poznan, Poland; podgorski@awf.poznan.pl
* Correspondence: domaszewska@onet.pl; Tel.: +48-501-476-716

Abstract: The aim of the study was to reveal the difference in the hematological reaction to the applied exercise-induced workload between the able-bodied and physically active people with cervical spinal cord injury. The study covered 11 males with spinal cord injury and 11 able-bodied persons. An incremental stress test was carried out until the maximum individual workloads were achieved. The peak oxygen uptake was measured with the use of the ergospirometric method. Venous blood test results at rest and after finishing the maximal exercise showed hemoglobin (Hb) concentration, hematocrit (HCT) value, erythrocytes (RBC), leukocytes (WBC) and platelets (PLT) counts as well as the relative percentage of granulocytes (GRA), lymphocytes (LYM), and monocytes (MON). RBC, HCT as well as Hb and PLT among people with the injury were statistically lower ($p < 0.001$) large effect size, than in the control group. Statistically significant difference between the test and control group, subjected to the maximal exercise stress test, was observed in the exercise induced change of the PLT [$p < 0.001$, (ES: 2.631)] WBC [$p < 0.05$, (ES: 1.429)] and the percentage of LYM and GRA [$p < 0.05$, (ES: 1.447) for LYM and (ES: 1.332) for GRA] between both groups, subjected to the maximal cardiac stress test on the manual cycloergometer. The analysis of the obtained results indicates that people with spinal cord injury are much more vulnerable to the occurrence of microcytic anemia compared to able-bodied people. The after-exercise percentage shift of selected subpopulations of leukocytes in both groups indicates a delayed post-exercise recovery among people with spinal cord injury.

Keywords: spinal cord injury; hematological parameters; peak oxygen uptake

Citation: Szymczak, Ł.; Podgórski, T.; Domaszewska, K. Comparison of the Levels of Hematological Parameters at Rest and after Maximum Exercise between Physically Active People with Spinal Cord Injury and Able-Bodied People. *IJERPH* **2021**, *18*, 12323. https://doi.org/10.3390/ijerph182312323

Academic Editors: Wioletta Zukiewicz-Sobczak, Paulina Wojtyla-Buciora, Izabela Rącka and Andrzej Wojty

Received: 1 November 2021
Accepted: 23 November 2021
Published: 24 November 2021

Publisher's Note: MDPI stays neutral with regard to jurisdictional claims in published maps and institutional affiliations.

Copyright: © 2021 by the authors. Licensee MDPI, Basel, Switzerland. This article is an open access article distributed under the terms and conditions of the Creative Commons Attribution (CC BY) license (https://creativecommons.org/licenses/by/4.0/).

1. Introduction

Hematological disorders, particularly anemia, are a very common complication of spinal cord injury in the acute phase, i.e., upon the occurrence of the injury, even when there is no blood loss (gastrointestinal bleeding, hemorrhage). The erythropoiesis rate is regulated by such factors as hormones, nervous system, and hypoxia. Hormones regulating red blood cell production rate include erythropoietin, thyroid hormones, growth hormone, and cortisol [1]. The influence of the nervous system, in turn, is related to the regulation of adrenaline and noradrenaline secretion by the adrenal medulla. These hormones work in a similar way and cause a change in the contractility of blood vessels, thereby reducing blood flow through the kidneys. It leads to hypoxia-induced increase in the erythropoietin production. Furthermore, among people with spinal cord injury, a reduction in the concentration of adrenaline circulating in the blood can be observed, which is caused by the disconnection of the adrenal sympathetic nervous system [2–4]. Understanding the etiology of anemia in this group of patients is a prerequisite to prevent its consequences. The analysis of hematological parameters of patients with spinal cord injury indicates the occurrence of different types of normocytic and microcytic anemia. A reduced level of red blood cells count does not significantly endanger patients' life but can contribute to developing secondary complications related to tissue hypoxia [5,6].

Rehabilitation of people with spinal cord injury is aimed at improving their functional capacities and increasing their injury induced decreased exercise tolerance level. Regular physical activity may sharpen their wheelchair skills and allows to maintain muscle mass. Additionally, it limits the occurrence of adverse circulatory and respiratory changes. Physical activity is also conducive to the prevention of osteoporosis, diabetes, or neoplastic diseases. Patients with cervical spinal cord injuries who undertake physical activity are less likely to suffer from bedsores, muscle contractures, muscle tension disorders, and orthostatic blood pressure decline. Moreover, regular exercises limit the occurrence of adverse hematological and biochemical changes in the blood [7–11].

The volume of plasma changes during a physiological reaction to the exercise-induced workload, which leads to hemoconcentration or hemodilution. Their occurrence depends on the intensity and duration of the exercise as well as the physical endurance capacity of the person taking the effort. Hemoconcentration increases blood viscosity and occurs during short term effort with high intensity. Long term effort leads to an increase of oncotic blood pressure and its dilution [12,13]. Along with the intensification of the physical activity, increasing blood lactate concentration and a series of hormones capable of modulation within leukocytes is released. Exercise-induced leukocytosis is not a consequence of changes in the volume of plasma but results from the increased cell traffic from bone marrow to blood, demargination from blood vessel walls, and decreased exit to tissues. Animal experiments suggest that signals from inflamed tissues are humoral, not nervous. Some of these humoral factors are adrenaline, noradrenaline, growth hormone (GH), and cortisol, as well as plasma granulocyte colony-stimulating factor (G-CSF) and interleukin-6 (IL-6). The multiple stage process of leukocytosis is characterised not only by a change in the total white blood cell count, but also in the percentage of each of their fractions. At the lymphocytic stage, during moderate effort the lymphocyte count increases to 55% and neutrophil granulocyte count decreases. The total count of leukocytes in the blood does not change. Intensification of the effort increases the neutrophil granulocyte count to 78% and decreases the lymphocyte count below the values at rest. The total leukocytes count may increase even to 12×10^9 L^{-1}. Very intense effort, though not exceeding the possibilities of the body, leads to the intoxicative phase. The white blood cell count may increase even to 20×10^9 L^{-1}. Undertaken effort that significantly exceeds one's performance capacities leads to leukopenia and a decrease in immune resistance. Most likely, this phenomenon results from metabolism disorders and insufficient capacity of protein synthesis in the changed conditions. Hormonal mechanisms responsible for the exercise-induced change in the leukocyte count also cause exercise-induced thrombocytosis [14–16].

The aim of the study was to answer the question whether there is a difference in the hematological reaction to the applied exercise-induced workload between the able-bodied and physically active people with cervical spinal cord injury.

2. Materials and Methods

The study was conducted according to the Declaration of Helsinki and the National Statement and Human Research Ethics Guidelines and approved by IRB (Institute for Research in Biomedicine) at the Poznan University of Medical Sciences (5 April 2012; Ethics Approval Number: 405/12). All participants in this study gave their written informed consent. The research covered 11 members of the Polish national wheelchair rugby team with spinal cord injury at the cervical level, ranging in age from 21 to 41 years (test group), and 11 able-bodied male students of the Poznan University of Physical Education Department (control group). These persons had an average level of physical endurance (Table 1) and had comparable training experience to the study group. The disabled with spinal cord injury subjected to the study constitute a homogeneous group in terms of age, body height, weight, and the injury level. All of them were tetraplegics with cervical spinal cord injury at levels C5–C7. Wheelchair rugby classification system, used to assess the subjects came from the ISMGF/ISMWSF (International Stoke Mandeville Games Federation/ International Stoke Mandeville Wheelchair Sport Federation) and American Spinal Injury Association (ASIA) Impairment Scale, medical classification system [17]. All the disabled research participants started training wheelchair rugby relatively early and gained considerable experience in the sport. The anthropometric characteristics of the groups as well as the degree of their spinal cord injury and training experience are presented in Tables 2 and 3.

Table 1. Comparison of peak oxygen uptake and oxygen consumption obtained at the ventilatory threshold level.

Parameter	Test Group	Control Group	p-Value
$VO_{2\,VT}$ [mL·min^{-1}]	1046.00 ± 302.860	1740.54 ± 258.058	0.0003
$VO_{2\,VT}$ [mL·kg^{-1}·min^{-1}]	14.94 ± 4.519	23.25 ± 3.245	0.0022
peakVO_2 [mL·min^{-1}]	1195.16 ± 482.197	2396.36 ± 327.269	0.0000
peakVO_2 [mL·kg^{-1}·min^{-1}]	16.82 ± 7.569	32.14 ± 4.995	0.0001
peak Load [Watt]	43.18 ± 15.706	80.00 ± 10.445	0.0000
Exercise duration [min]	13.00 ± 4.632	24.00 ± 3.133	0.0000

Data are presented as mean ± SD, peakVO_2 = peak oxygen uptake, $VO_{2\,VT}$ = oxygen uptake on VT. Bold indicates the statistically significant values.

Table 2. The anthropometric characteristics of the groups subjected to the study.

Parameter	Test Group	Control Group
Age [years]	34.18 ± 4.39	22.64 ± 2.38
Height [m]	1.81 ± 0.055	1.79 ± 0.073
Body mass [kg]	72.91 ± 11.71	75.36 ± 8.71
BMI [kg·m^{-2}]	22.40 ± 4.38	23.25 ± 1.68

Data are expressed as mean ± SD; BMI = body mass index.

Table 3. Classification points, spinal cord injury level, American Spinal Injury Association (ASIA) Impairment Scale and training experience of the subjects with spinal cord injury.

No	Spinal Cord Injury Level	American Spinal Injury Association (ASIA) Impairment Scale	Classification Points	Time after the Injury [Years]	Training Experience [Years]	Time between the Injury and the Onset of Training [Years]
1	C6–C7	A	2	13	11	2
2	C5–C6	A	2	2	1	1
3	C6	A	2	6	3	3
4	C6–C7	A	2	5	3	2
5	C6–C7	A	2.5	17	13	4
6	C5–C6	A	2	13	8	5
7	C6–C7	A	2	12	6	6
8	C6–C7	A	2	9	7	2
9	C5	A	2	15	10	5
10	C5–C6	A	1	20	12	8
11	C6–C7	A	0.5	14	13	1
			Mean	11.45	7.91	3.55
			SD	5.18	4.08	2.15

2.1. Assessment of Aerobic Fitness (Graded Exercise Test Protocol)

The exercise tests were conducted between 8:00 a.m. and noon in an air-conditioned laboratory, 2 h after consuming a light breakfast (one sandwich with butter and cheese; approx. 200 kcal). The subjects performed a cardiac stress test with increasing intensity on the REHAB TRAINER 881E arm cranking manual ergometer, manufactured by Monark Exercise AB (Vansbro, Sweden) and specially adjusted to functional abilities of research participants. The initial workload amounted to 10 Watts seconds (60 revolutions per minute RPM) was successively increased by 10 Watts every 3 min until the subjects achieved maximum in dividual workloads or refused to continue the effort. Peak oxygen uptake (peakVO_2) and oxygen consumption per minute at the level of ventilatory threshold

(VO_2VT) using the "V-slope" method were measured in absolute and relative values with the German Jeager Oxycon Mobile gas analyser (Viasys Healthcare, Höchberg, Germany).

2.2. Preparation of Blood Samples for Analysis

Venous blood was taken twice from the ulnar veins, i.e., at rest (fasting blood; 6:00 a.m.) and 5 min after finishing the exercise using a S-Monovette syringe (Sarstedt, Nümbrecht, Germany) containing K_3EDTA as an anticoagulant. Hemoglobin (Hb) concentration, hematocrit (HCT) value, total erythrocyte (RBC), leukocyte (WBC) and platelet (PLT) counts as well as the relative percentage of granulocytes (GRA), lymphocytes (LYM) and monocytes (MON) were measured immediately after blood collection and the samples were determined with the use of the MYTHIC 18® hematology analyser (PZ Cormay SA, Łomianki, Poland).

2.3. Statistical Analysis

All the data are presented as mean (standard deviation). In order to calculate the significance of changes in the studied parameters, the Wilcoxon sign test of pairs was performed. The importance of differences between the test and control group was calculated with the Mann–Whitney U test. Correlation between the variables was calculated with the Spearman's rank test. The level of statistical significance was set at $p < 0.05$. All results were statistically analysed using Dell Inc. (2016), Dell Statistica v.13. (Tulsa, OK, USA). Effect sizes [ES] were calculated as the difference between means divided by the pooled standard deviation. Using Cohen's (1988) criteria, an effect size ≥ 0.20 and <0.50 was considered small, $\geq$0.50 and <0.80 medium, and $\geq$0.80 large [18].

3. Results

Table 4 presents morphological blood parameters as mean and standard deviation, measured at rest and after exercise in both groups. RBC, HCT as well as Hb and PLT counts among people with the injury were statistically lower ($p < 0.001$) large effect size, than in the control group. The analysis of the Mann–Whitney U test did not reveal any significant differences in the exercise-induced change in the red blood cell count, hemoglobin concentration, hematocrit value and the percentage of monocytes in the blood between able-bodied and disabled research participants. However, a statistically significant difference was observed in the exercise-induced change of platelet count [$p < 0.001$, (ES: 2.631)] white blood cell count [$p < 0.05$, (ES: 1.429)] and the percentage of lymphocytes and granulocytes in the blood [$p < 0.05$, (ES: 1.447) for lymphocytes and (ES: 1.332) for granulocytes] between both groups, subjected to the maximal cardiac stress test on the manual cycloergometer.

To assess physical endurance and exercise tolerance, research participants performed an exercise with increasing intensity on the manual ergometer. Table 1 presents average values and standard deviations of oxygen consumption (in absolute and relative values) obtained at peak workload and at the level of ventilatory threshold. The difference in each aerobic capacity parameter was statistically significant between the groups ($p < 0.001$). The subjects with spinal cord injury obtained about 50% lower results both at the level of threshold and peak workloads, compared to the able-bodied subjects within the same model of physical exercise workload (Table 1).

We observed, a statistically significant positive correlation between the spinal cord injury level and the exercise-induced change in the relative percentage of neutrocytes ($r = 0.66$, $p < 0.05$), the training experience and exercise induced increase of the total leukocyte count ($r = -0.61$, $p < 0.05$), the value of peakVO_2 and the exercise induced decrease of the relative percentage of lymphocytes ($r = -0.70$, $p < 0.05$) and the change in the relative percentage of monocytes ($r = -0.69$; $p < 0.05$).

Table 4. Basic characteristics and changeability of hematological parameters measured before and after the exercise in both groups.

Parameter	Test Group			Control Group			Test Group	Control Group	
	Before Exercise	After Exercise	p-Value	Before Exercise	After Exercise	p-Value	Change After-Before	Change After-Before	p-Value
RBC $[10^{12}\cdot L^{-1}]$	4.51 ±0.220	4.74 ±0.273	0.0033	5.58 ±0.217	5.77 ±0.363	0.0262	0.23 ±0.161	0.19 ±0.147	1.0000
HCT [%]	40.97 ±2.589	43.47 ±3.229	0.0033	47.06 ±2.226	48.96 ±3.128	0.0164	2.50 ±1.327	1.90 ±0.902	0.4495
Hb $[mmol\cdot L^{-1}]$	8.69 ±0.664	9.12 ±0.758	0.0033	10.07 ±0.529	10.40 ±0.715	0.0185	0.43 ±0.266	0.33 ±0.186	0.5316
PLT $[10^{9}\cdot L^{-1}]$	185.00 ±17.378	191.82 ±18.059	0.1095	245.54 ±43.517	285.82 ±57.311	0.0033	6.82 ±11.535	40.27 ±13.794	0.0005
WBC $[10^{9}\cdot L^{-1}]$	5.68 ±1.169	6.44 ±0.905	0.0505	5.97 ±0.851	7.81 ±1.221	0.0033	0.76 ±1.002	1.84 ±0.371	0.0233
LYM [%]	38.31 ±7.103	33.72 ±5.786	0.0128	36.11 ±7.150	37.23 ±8.447	0.4498	−4.59 ±5.428	1.12 ±1.297	0.0104
MON [%]	4.34 ±0.835	4.37 ±0.921	0.4069	5.55 ±0.811	5.14 ±0.924	0.0912	0.04 ±0.384	−0.42 ±0.113	0.0659
GRA[%]	58.38 ±7.226	62.58 ±7.226	0.0262	58.34 ±7.370	57.63 ±8.579	0.5937	4.20 ±5.061	−0.70 ±1.209	0.0488

Data are presented as mean ± SD, GRA = granulocyte, Hb = hemoglobin concentration, HCT = hematocrit value, LYM = lymphocytes, MON = monocytes, PLT = thrombocyte count, RBC = red blood cell count, WBC = white blood cell count. Bold indicates the statistically significant values.

4. Discussion

The reduction of aerobic capacity of the people with cervical spinal cord injury is multifactorial. Due to the paralysis of lower limbs and partial paralysis of upper limbs, the activity of the muscle pump drops. In consequence, keeping venous blood return to the heart is insufficient. Reduced values of hemodynamic parameters and their exercise-induced course, different from the one of the able-bodied people, are caused by the lack of influence of the sympathetic system on the heart muscle, the state of blood vessels or secretory actions of endocrine glands. In the case of people with cervical spinal cord injury, the rest and exercise induced values of circulatory and respiratory parameters as well as the biochemical blood parameters differ from those of the able-bodied people or limbless amputees with maintained mechanisms of stimulation of the sympathetic nervous system [19–23]. Physical activity undertaken by people with spinal cord injury is of crucial importance in keeping the appropriate state of their physical capability and exercise tolerance. Hence, the oxygen efficiency parameters of the subjects from the test group were very high, taking into account the degree of their spinal cord injury. The results obtained by the disabled research participants in the in-house study of the peakVO_2 value averaged at 16.82 ± 7.569 $mL\cdot kg^{-1}\cdot min^{-1}$. Moreover, 10 subjects from the test group reached individual threshold workload at the level of peakVO_2 79.93%. The intended threshold oxygen consumption averaged at 1046.00 ± 302.860 $mL\cdot min^{-1}$; whereas the relative value amounted to 14.94 ± 4.519 $mL\cdot kg^{-1}\cdot min^{-1}$. Average values of this parameter observed in the in-house study were considerably higher than those obtained by other researchers of this group of people and 50% lower compared to the able-bodied subjects. This result was statistically significant ($p < 0.001$). Dreisinger et al., while studying training tetraplegics, noted the values of oxygen consumption at the level of 780 $mL\cdot min^{-1}$, whereas Wicks et al. at the level of 13.8 $mL\cdot kg^{-1}\cdot min^{-1}$ [24,25]. Burkett et al. in their quite comprehensive research concerning both training and non-training people with spinal cord injury, noted the maximal oxygen consumption for physically inactive tetraplegics oscillating between 5.31–11.89 $mL\cdot kg^{-1}\cdot min^{-1}$. Undoubtedly, the clinical state after spinal cord injury, as well as the physical capability level, influence the values of the morphological parameters at rest and after exercise [26,27]. Huang et al. examined 28 people with cervical spinal cord injury between C3 and C7 vertebrae and revealed that several weeks after the injury, hematologic parameters of 9% of the subjects from the test group were lower than reference values. The changes concerned erythrocytes and reticulocytes counts, hematocrit value, hemoglobin, and iron concentration in the blood. Huang et al. demonstrated the occurrence of normocytic anemia among 71% of the people examined by him, whereas 14% of the subjects had microcytic anemia. These disorders influence the development of other secondary complications related to tissue hypoxia [5]. Perkash et al. and Davies et al. demonstrated the occurrence of a mild type of anemia in 52.3% out of 65 male patients with spinal cord injury subjected to the study. They did not confirm the correlation between age, duration of the injury or damage degree and the occurrence and type of anemia [28–30]. The in-house study demonstrated that people with cervical spinal cord injury had low values of the examined hematologic blood parameters at rest—below the range of physiological values (Table 4). RBC, HCT as well as Hb and PLT counts among people with the injury were statistically lower ($p < 0.001$) than in the control group. In case of WBC, no significant difference between the groups was observed. Only the percentage value of monocytes at rest was statistically higher in the control group ($p < 0.05$), and the maximal physical exercise caused the occurrence of poliglobulia phenomenon in each group. The exercise-induced blood density was higher among people with spinal cord injury, but the difference was not statistically significant. The exercise-induced thrombocytosis occurred in each of the tested groups, with lower statistically significant intensity among people with spinal cord injury ($p < 0.001$). Most likely, this difference results from the level of physical performance of the subjects and the degree of homeostatic imbalance of the body. The inflammatory reaction model described in the literature is a typical physiological reaction of the body to the workload in the exercise stress test [31]. In case of the people with the injury, the change of the specific fraction of white blood cells, along with a significant increase of the total leukocyte count, is the most similar to the model of changes of a neutrophil character. Furthermore we observed, a statistically significant positive correlation between the spinal cord injury level and the exercise induced change in the relative percentage of neutrocytes ($r = 0.66$, $p < 0.05$), the training experience and exercise induced increase of the total leukocyte count ($r = -0.61$, $p < 0.05$), the value of peak oxygen consumption and the exercise induced decrease of the relative percentage of lymphocytes ($r = -0.70$, $p < 0.05$) and the change in the relative percentage of monocytes ($r = -0.69$; $p < 0.05$). A statistically significant higher increase of the WBC count in the blood ($p < 0.05$) in reaction to the physical exercise of the group of able-bodied people was noted compared to the people with spinal cord injury, along with a statistically significant difference in the change in the percentage of lymphocytes between the groups ($p < 0.05$). The lack of exercise induced reduction in the percentage of lymphocytes among the able-bodied people was probably related to the lower exercise induced secretion of cortisol and growth hormone during the test effort [32]. One of the reasons responsible for the mechanisms of myogenic leukocytosis is, among others, a change in the concentration of neurohormonal factors (cortisol, catecholamines, growth hormone, endorphins, sex-steroids) and cytokines (TNF-α, IL-1, IL-6). A very similar model of reaction to exertion was described by Kouda et al., who revealed that the concentration of IL-6 and blood hematologic elements in plasma among able-bodied people increased significantly after finishing the exercise. However, they did not demonstrate that physical exercise causes such significant changes in case of the people with cervical spinal cord injury. They assumed that these differences might have been caused by muscular atrophy and sympathetic nervous system disorders among people with spinal cord injury [33]. The analysis of the obtained results revealed that people with spinal cord injury, despite high physical activity, run a much greater risk of microcytic anemias compared to able-bodied people. Furthermore, the after-exercise percentage shift of specific sub-populations of leukocytes in both groups (lymphocytes and granulocytes) indicates a later after-exercise regeneration of people with spinal cord injury.

5. Conclusions

The analysis of the obtained results indicates that people with spinal cord injury are much more vulnerable to the occurrence of microcytic anemia compared to able-bodied people. The after-exercise percentage shift of selected subpopulations of leukocytes in both groups indicates a delayed post-exercise recovery among people with spinal cord injury.

Limitation of the Study

The study was conducted on physically active people with spinal cord injuries and healthy students. For a complete view of hematological changes under the influence of exercise, it would be necessary to perform tests on non-physically active males with damaged spinal cord. Unfortunately, for this study, the appropriate consent to conduct research on this group of people was not obtained.

Author Contributions: Conceptualization, Ł.S. and K.D.; methodology, K.D., Ł.S.,T.P.; software, K.D.; validation, K.D., and T.P.; resources, K.D., Ł.S. and T.P.; data curation, K.D. and Ł.S.; writing—original draft preparation, K.D., Ł.S. and T.P.; writing—review and editing, K.D.; visualization, K.D; supervision, K.D.; project administration, K.D. and Ł.S. All authors have read and agreed to the published version of the manuscript.

Funding: This research received no external funding.

Institutional Review Board Statement: The study was conducted according to the guidelines of the Declaration of Helsinki and approved by the IRB (Institute for Research in Biomedicine) at the Poznan University of Medical Sciences (5 April 2012; Ethics Approval Number: 405/12).

Informed Consent Statement: Informed consent was obtained from all subjects involved in the study.

Data Availability Statement: The data presented in this study are available on request from the corresponding author. The data are not publicly available due to the consent provided by participants on the use of confidential data.

Conflicts of Interest: The authors declare no conflict of interest.

References

1. Tsitouras, P.D.; Zhong, Y.G.; Spungen, A.M.; Bauman, W.A. Serum Testosterone and Growth Hormone Insulin-Like Growth Factor-I in Adults with Spinal Cord Injury. *Horm. Metab. Res.* **1995**, *27*, 287–292. [CrossRef] [PubMed]
2. Frankel, H.; Mathias, C.; Spalding, J. Mechanisms of reflex cardiac arrest in tetraplegic patients. *Lancet* **1975**, *306*, 1183–1185. [CrossRef]
3. Collins, H.L.; Rodenbaugh, D.W.; DiCarlo, S.E. Spinal cord injury alters cardiac electrophysiology and increases the susceptibility to ventricular arrhythmias. *Prog. Brain Res.* **2005**, *152*, 275–288. [CrossRef]
4. Manogue, M.; Hirsh, D.S.; Lloyd, M. Cardiac electrophysiology of patients with spinal cord injury. *Hear. Rhythm* **2017**, *14*, 920–927. [CrossRef] [PubMed]
5. Huang, C.T.; DeVivo, M.J.; Stover, S.L. Anemia in acute phase of spinal cord injury. *Arch. Phys. Med. Rehabil.* **1990**, *71*, 3–7. [PubMed]
6. Furlan, J.C.; Krassioukov, A.V.; Fehlings, M.G. Hematologic abnormalities within the first week after acute isolated traumatic cervical spinal cord injury: A case-control cohort study. *Spine* **2006**, *31*, 2674–2683. [CrossRef]
7. Dallmeijer, A.J.; Hopman, M.T.; Angenot, E.L.; Van Der Woude, L.H. Effect of training on physical capacity and physical strain in persons with tetraplegia. *Scand. J. Rehabil. Med.* **1997**, *29*, 181–186. [PubMed]
8. Janssen, T.W.J.; Dallmeijer, A.J.; Veeger, D.J.H.E.J.; Van Der Woude, L.H.V. Normative values and determinants of physical capacity in individuals with spinal cord injury. *J. Rehabil. Res. Dev.* **2002**, *39*, 29–39.
9. Ginis, K.A.M.; Latimer, A.; Arbour-Nicitopoulos, K.P.; Buchholz, A.C.; Bray, S.; Craven, B.; Hayes, K.C.; Hicks, A.L.; McColl, M.A.; Potter, P.J.; et al. Leisure Time Physical Activity in a Population-Based Sample of People With Spinal Cord Injury Part I: Demographic and Injury-Related Correlates. *Arch. Phys. Med. Rehabil.* **2010**, *91*, 722–728. [CrossRef]
10. Morgulec-Adamowicz, N.; Kosmol, A.; Molik, B.; Yilla, A.B.; Laskin, J.J. Aerobic, Anaerobic, and Skill Performance With Regard to Classification in Wheelchair Rugby Athletes. *Res. Q. Exerc. Sport* **2011**, *82*, 61–69. [CrossRef] [PubMed]
11. Molik, B.; Laskin, J.J.; Kosmol, A.; Marszałek, J.; Morgulec-Adamowicz, N.; Frick, T. Relationships between anaerobic performance, field tests, and functional level of elite female wheelchair basketball athletes. *Hum. Mov.* **2013**, *14*, 366–371. [CrossRef]
12. Van Beaumont, W.; Underkofler, S.; Van Beaumont, S. Erythrocyte volume, plasma volume, and acid-base changes in exercise and heat dehydration. *J. Appl. Physiol.* **1981**, *50*, 1255–1262. [CrossRef] [PubMed]
13. Balogh, N.; Gaál, T.; Ribiczeyné, P.S.; Petri, A. Biochemical and Antioxidant Changes in Plasma and Erythrocytes of Pentathlon Horses Before and After Exercise. *Veter. Clin. Pathol.* **2001**, *30*, 214–218. [CrossRef]
14. Foster, N.K.; Martyn, J.B.; Rangno, R.E.; Hogg, J.C.; Pardy, R.L. Leukocytosis of exercise: Role of cardiac output and catecholamines. *J. Appl. Physiol.* **1986**, *61*, 2218–2223. [CrossRef] [PubMed]
15. Chamberlain, K.G.; Tong, M.; Penington, D.G. Properties of the exchangeable splenic platelets released into the circulation during exercise-induced thrombocytosis. *Am. J. Hematol.* **1990**, *34*, 161–168. [CrossRef]
16. Risøy, B.; Raastad, T.; Hallén, J.; Lappegård, K.T.; Bæverfjord, K.; Kravdal, A.; Siebke, E.; Benestad, H.B. Delayed leukocytosis after hard strength and endurance exercise: Aspects of regulatory mechanisms. *BMC Physiol.* **2003**, *3*, 14. [CrossRef] [PubMed]
17. Kirshblum, S.C.; Burns, S.P.; Biering-Sørensen, F.; Donovan, W.; Graves, D.E.; Jha, A.; Johansen, M.; Jones, L.; Krassioukov, A.; Mulcahey, M.J.; et al. International standards for neurological classification of spinal cord injury (Revised 2011). *J. Spinal. Cord. Med.* **2011**, *34*, 535–546. [CrossRef] [PubMed]

18. Cohen, J. *Statistical Power Analysis for the Behavioral Sciences*, 2nd ed.; Lawrence Erlbaum Associates: Hillsdale, NJ, USA, 1988.
19. Westgren, N.; Levi, R. Quality of life and traumatic spinal cord injury. *Arch. Phys. Med. Rehabil.* **1998**, *79*, 1433–1439. [CrossRef]
20. Dijkers, M.P. Correlates of life satisfaction among persons with spinal cord injury. *Arch. Phys. Med. Rehabil.* **1999**, *80*, 867–876. [CrossRef]
21. Simpson, L.A.; Eng, J.J.; Hsieh, J.T.; Dalton, L. Wolfe And The Spinal Cord Injury Spinal Cord Injury Rehabilitation Evidence Scire Research Team The Health and Life Priorities of Individuals with Spinal Cord Injury: A Systematic Review. *J. Neurotrauma* **2012**, *29*, 1548–1555. [CrossRef] [PubMed]
22. Baydur, A.; Adkins, R.H.; Milic-Emili, J. Lung mechanics in individuals with spinal cord injury: Effects of injury level and posture. *J. Appl. Physiol.* **2001**, *90*, 405–411. [CrossRef] [PubMed]
23. DeVivo, M.J.; Black, K.J.; Stover, S.L. Causes of death during the first 12 years after spinal cord injury. *Arch. Phys. Med. Rehabil.* **1993**, *74*, 248–258. [PubMed]
24. Dreisinger, T.E.; Dalton, R.B.; Whiting, R.B. Exercise evaluation in the disabled San Diego Maximal wheelchair exercise: Comparison of able bodied and wheelchair bound. *Med. Sci. Sports Exerc.* **1984**, *16*, 147. [CrossRef]
25. Wicks, J.R.; Oldridge, N.B.; Cameron, B.J.; Jones, N.L. Arm cranking and wheelchair ergometry in elite spinal cord-injured athletes. *Med. Sci. Sports Exerc.* **1983**, *15*, 224–231. [CrossRef]
26. Baumgart, J.K.; Brurok, B.; Sandbakk, Ø. Peak oxygen uptake in Paralympic sitting sports: A systematic literature review, meta- and pooled-data analysis. *PLoS ONE* **2018**, *13*, e0192903. [CrossRef] [PubMed]
27. Farkas, G.J.; Gorgey, A.S.; Dolbow, D.R.; Berg, A.S.; Gater Jr, D.R. Energy expenditure, cardiorespiratory fitness, and body composition following arm cycling or functional electrical stimulation exercises in spinal cord injury: A 16-week randomized controlled trial. *Top. Spinal. Cord. Inj. Rehabil.* **2021**, *27*, 121–134. [CrossRef] [PubMed]
28. Perkash, A.; Brown, M. Anemia in Patients With Traumatic Spinal Cord Injury. *J. Am. Paraplegia Soc.* **1986**, *9*, 10–15. [CrossRef]
29. Davies, A.L.; Hayes, K.C.; Dekaban, G.A. Clinical Correlates of Elevated Serum Concentrations of Cytokines and Autoantibodies in Patients With Spinal Cord Injury. *Arch. Phys. Med. Rehabil.* **2007**, *88*, 1384–1393. [CrossRef]
30. Harrington, G.M.B.; Cool, P.; Hulme, C.; Osman, A.; Chowdhury, J.R.; Kumar, N.; Budithi, S.; Wright, K. Routinely Measured Hematological Markers Can Help to Predict American Spinal Injury Association Impairment Scale Scores after Spinal Cord Injury. *J. Neurotrauma* **2021**, *38*, 301–308. [CrossRef]
31. Ostrowski, K.; Rohde, T.; Zacho, M.; Asp, S.; Pedersen, B.K. Evidence that interleukin-6 is produced in human skeletal muscle during prolonged running. *J. Physiol.* **1998**, *508*, 949–953. [CrossRef] [PubMed]
32. Robertson, A.J.; Ramesar, K.C.; Potts, R.C.; Gibbs, J.H.; Browning, M.C.; Brown, R.A.; Hayes, P.C.; Beck, J.S. The effect of strenuous physical exercise on circulating blond lymphocytes and serum cortisol levels. *J. Clin. Lab. Immunol.* **1981**, *5*, 53–57. [PubMed]
33. Kouda, K.; Furusawa, K.; Sugiyama, H.; Sumiya, T.; Ito, T.; Tajima, F.; Shimizu, K. Does 20-min arm crank ergometer exercise increase plasma interleukin-6 in individuals with cervical spinal cord injury? *Eur. J. Appl. Physiol.* **2012**, *112*, 597–604. [CrossRef] [PubMed]

International Journal of
Environmental Research and Public Health

MDPI

Article

Mortality and Alcohol as Its Cause—Comparative Characteristics of the Two Neighboring Countries: Ukraine and Poland

Oleh Lyubinets [1], Marta Kachmarska [2], Katarzyna Maria Sygit [3,*], Elżbieta Cipora [4] and Jaroslaw Grshybowskyj [1]

1 Department of Public Health, Danylo Halytsky Lviv National Medical University, 79010 Lviv, Ukraine; g.pulmo@gmail.com (O.L.); grshybowskyj@yahoo.de (J.G.)
2 Department of Psychiatry and Psychotherapy, Danylo Halytsky Lviv National Medical University, 79010 Lviv, Ukraine; kachmarskamarta@gmail.com
3 Faculty of Health Science, Calisia University, 62-800 Kalisz, Poland
4 Medical Institute, Jan Grodek State University in Sanok, 38-500 Sanok, Poland; elacipora@interia.pl
* Correspondence: ksygit@poczta.onet.pl

Abstract: This paper presents a comparative assessment of mortality in Poland and Ukraine, including due to alcohol consumption, by sex, place of residence, and age groups. Mortality from alcohol consumption is and remains one of the health problems of the state's population. The aim of this study was to establish the difference in mortality, including due to alcohol consumption, in the two neighboring countries. The analysis was conducted in 2008 and 2018 according to statistical institutions in Poland and Ukraine. Data from the codes of the International Statistical Classification of Diseases of the 10th edition: F10, G31.2, G62.1, I42.6, K70, K86.0, and X45 were used to calculate mortality due to alcohol consumption. The share of mortality caused by alcohol consumption in Ukraine in 2008 was 3.52%, and 1.83% in 2018. At the same time, in Poland, there is an increase in this cause of death from 1.72% to 2.36%. Mortality caused by alcohol consumption is the main share of mortality in the section "Mental and behavioral disorders" in both Ukraine, at 73–74%, and Poland, at 82–92%. Changes in the mortality rate in the cities and villages of Ukraine and Poland showed different trends: Poland nated, a significant increase in mortality, while in Ukraine it has halved on average. Overall and alcohol mortality rates in both countries were higher among the male population. The analysis of mortality among people of working age showed that the highest proportion of deaths from alcohol consumption in both countries was among people aged 25–44. Despite the geographical proximity, and similarity of natural and climatic characteristics and population, mortality rates in each country reflect the difference in the medical and demographic situation, and the effectiveness of state social approaches to public health.

Keywords: mortality; alcohol; working population; Poland; Ukraine

Citation: Lyubinets, O.; Kachmarska, M.; Sygit, K.M.; Cipora, E.; Grshybowskyj, J. Mortality and Alcohol as Its Cause—Comparative Characteristics of the Two Neighboring Countries: Ukraine and Poland. *IJERPH* **2021**, *18*, 10810. https://doi.org/10.3390/ijerph182010810

Academic Editors: Wioletta Zukiewicz-Sobczak, Paulina Wojtyla-Buciora, Izabela Rącka and Andrzej Wojtyla

Received: 31 August 2021
Accepted: 11 October 2021
Published: 14 October 2021

Publisher's Note: MDPI stays neutral with regard to jurisdictional claims in published maps and institutional affiliations.

Copyright: © 2021 by the authors. Licensee MDPI, Basel, Switzerland. This article is an open access article distributed under the terms and conditions of the Creative Commons Attribution (CC BY) license (https://creativecommons.org/licenses/by/4.0/).

1. Introduction

The medical and demographic situation in Ukraine is negative, and is characterized by a decrease in the birth rate for the period 2014–2019 by 19.4% and an increase in the mortality rate by 6.16 per 100 thousand population [1,2]. Against the background of negative natural growth, there has also been a decrease in the average age of women giving birth and increased morbidity and perinatal mortality [3].

Stably high mortality rates in Ukraine compared to those in European countries require constant monitoring and identification of the causes of this phenomenon. In Ukraine, as in other countries, noncommunicable diseases remain the leading cause of morbidity, disability, and premature mortality. The most effective way to reduce them is to prevent the development of non-communicable diseases by addressing the behavioral risk factors underlying them at the population and individual levels: smoking, alcohol consumption, excessive salt intake, low physical activity, overweight and obesity, and unhealthy diet [4].

Alcohol is one of the most important health risk factors and causes of premature mortality. According to Jürgen Rehm and co-authors, alcohol consumption caused 14.6% of all premature adult mortality in eight European countries, 17.3% in men and 8.0% in women [5]. The study of medical and social areas of alcohol consumption is valuable. Thus, 83.54 ± 0.98% of the 1422 students surveyed in Ukraine consume alcoholic beverages, three-quarters of them choose mostly alcoholic beverages with low alcohol content. The main motive for drinking low-alcohol beverages is hedonism. The majority of respondents who consume low-alcohol beverages do not consider the occurrence of alcohol dependence when consuming alcohol (56.85 ± 1.44%), which indicates an insufficient level of preventive measures [6].

The issue of the impact of morbidity and mortality on alcohol consumption on the economic sphere of the state is important. Thus, Błażej Łyszczarz notes that the total production losses associated with alcohol deaths in the EU in 2016 amounted to 32.1 billion euros [7]. Expenditure per capita (share of gross domestic product expenditure) was EUR 62.88 (0.215%) for the EU as a whole and ranged from EUR 17.29 (0.062%) in Malta to EUR 192.93 (0.875%) in Lithuania.

Macroeconomic fluctuations can affect alcohol problems in different ways in socio-demographic groups. The Spanish crisis after 2008 was a turning point for such conclusions: the annual changes in the percentage of deaths directly from alcohol mortality were 6.9% in 2002–2007 and 3.7% in 2008–2011 among the employed, and −4.3% and −0.4%, respectively, among the unemployed [8].

Despite the WHO's emphasis on the importance of restricting alcohol consumption through fiscal measures, the governments of Lithuania (in 1999), Poland (in 2002), and Finland (in 2004) reduced the excise tax on alcoholic beverages by, respectively 44%, 30%, and 44%. These decisions led to immediate and impressive health consequences. In Poland, the decline in vodka prices wasaccompaniedby a sharp increase in registered alcohol sales from 7 L per capita in 2002 to almost 10 L per capita in 2008 and an annual increase in alcohol-related mortality, especially for menaged 45–64 years [9].

Given the geographical proximity of Poland and Ukraine—sharing a common border of 535 km—similarity of natural and climatic characteristics, similarity of population (as of 1 January 2019, at 38.4 and 42.3 million people, respectively), different political and economic situation, and insignificant number of publications in international scientific journals comparing mortality rates in neighboring countries, including Ukraine, and the impact of alcohol on the health of the population, the present study examined the mortality rates of the population of Ukraine and Poland and its level due to alcohol consumption.

2. Objective

This study aims to identify trends and features of changes in the overall mortality rate, as well as mortality due to alcohol consumption, in two countries with different economic development, Ukraine and Poland, for a ten-year period—in 2008 and 2018.

3. Materials and Method

Regarding the analytical strategy of the study, we assessed whether there are any basic differences between the two geographically neighboring countries—Poland and Ukraine, in terms of overall mortality and alcohol-related mortality.

Data on population mortality were taken from the database of the State Statistics Service of Ukraine and the Central Statistical Office of the Republic of Poland.

Statistical analysis of the obtained data was performed by measuring the ratio and interval measurement (2008–2018) with zero data for 2008 in each of the countries, and in comparison between countries, in the respective year.

To calculate the mortality rate due to alcohol use, data from the codes of the International Statistical Classification of Diseases of the 10th revision were used:

Chapter V Mental and behavioral disorders (F00-F99):

F10Mental and behavioral disorders due to use of alcohol.

Chapter VI Diseases of the nervous system (G00-G99):
G31.2Degeneration of nervous system due to alcohol;
G62.1Alcoholic polyneuropathy.
Chapter IX Diseases of the circulatory system (I00-I99):
I42.6Alcoholic cardiomyopathy.
Chapter XI Diseases of the digestive system (K00-K93):
K70Alcoholic liver disease;
K86.0Alcohol-induced chronic pancreatitis.
Chapter XX External causes of morbidity and mortality (V01-Y98):
X45Accidental poisoning by and exposure to alcohol.

To be able to compare the obtained data on the mortality rate of the working-age population, we determined its range of 16–65 years, as according to the Organization for Economic Cooperation and Development, in 2016, the average normal retirement age was 64.3 years for men and 63.7 years for women in all schemes and countries [10]. In Ukraine, this age is 60 years.

4. Results

The basis for comparing mortality rates in 2008 and 2018 in Poland and Ukraine was the geographical proximity of these countries, a common border, the similarity of natural and climatic characteristics, and similarity of population. At the same time, the political and economic situation of these countries differs sharply. Poland has been a part of the European Union since 2004, which played a significant role in the development of the state and contributed to obtaining significant investments for the economy and social sphere, including health systems. Its GDP per capita in 2008 was USD 13,996; in 2018 this was already equal to USD 15,468.4 (according to the World Bank). The growth over these 10 years was 10.5%.

Ukraine, as an independent state, has been taking steps since 1991 to deepen cooperation with European countries in the economic and social spheres. The Government and the President of Ukraine have set the task of gaining membership in the European Union. This became especially relevant after the beginning of Russia's aggression in 2014. Ukraine's GDP from 2008 to 2018 decreased by 20.3%, from 3887.2 to 3096.8 USD. In comparison with the data for 2018, Poland's GDP was 5 times higher than in Ukraine.

Thus, the overall mortality rate in Ukraine for 10 years decreased by 14.76% (according to the State Statistics Service of Ukraine) [11]. In Poland, the opposite changes took place—there was an increase in overall mortality by 8.33% (according to Statistics Poland) [12]. At the same time, the overall mortality rate in Ukraine was, in 2008 and 2018, higher than that of Poland.

The overall mortality rate of Ukraine in 2015 was one of the highest (after Moldova, Kazakhstan, Kyrgyzstan, and Georgia) among the countries that are members of the Word Health Organization (WHO) Regional Office for Europe [13]. Comparisons of mortality rates in Ukraine and Poland per 1000 people by main chapters of causes of death are presented in Table 1.

Alcohol is one of the most important determinants of health and the cause of premature mortality. We analyzed the mortality rates in Ukraine and Poland according to the codes specified in the section "Materials and methods of research". Comparison of the share of mortality caused by alcohol consumption, in the structure of total mortality in certain chapters of causes of death by chapters and codes of the International Statistical Classification of Diseases and changes in mortality in 2008 and 2018, are shown in Table 2.

Table 1. Mortality rates in Ukraine and Poland per 1000 people by main classes of causes of death.

The Main Classes of Causes of Mortality	Ukraine			Poland		
	Rate 2008	Rate 2018	% Change	Rate 2008	Rate 2018	% Change
The overall mortality rate	16.31	13.90	−14.76	9.95	10.78	+8.33
Chapter I. Certain infectious and parasitic diseases	0.37	0.21	−43.16	0.07	0.05	−32.39
Chapter II. Neoplasms	1.92	1.86	−3.38	2.51	2.85	+13.53
Chapter IV. Endocrine, nutritional, and metabolic diseases	0.06	0.05	−13.11	0.18	0.25	+36.41
Chapter V: Mental and behavioral disorders	0.06	0.03	−51.61	0.06	0.10	+78.18
Chapter VI. Diseases of the nervous system	0.16	0.11	−29.11	0.13	0.17	+30.77
Chapter IX. Diseases of the circulatory system	10.38	9.28	−10.64	4.54	4.37	−3.64
Chapter X. Diseases of the respiratory system	0.50	0.31	−38.77	0.51	0.72	+41.90
Chapter XI. Diseases of the digestive system	0.76	0.58	−23.92	0.45	0.45	+0.67
Chapter XX. External causes of morbidity and mortality	1.33	0.73	−44.91	0.67	0.52	−21.35

% Change—decrease or increase rate from 2008 to 2018 in percent (of observed values).

Table 2. Change in the proportion of mortality caused by alcohol consumption in certain classes of causes of death.

Mortality Due to Alcohol Consumption		Ukraine		Poland	
		% in a Certain Class of Causes of Death	Change 2008–2018	% in a Certain Class of Causes of Death	Change 2008–2018
All causes	2008	3.52	−1.69	1.72	+0.64
	2018	1.83		2.36	
Chapter V: F10 Mental and behavioral disorders due to use of alcohol	2008	72.99	+1.11	91.91	−9.08
	2018	74.10		82.83	
Chapter VI: G31.2 Degeneration of nervous system due to alcohol	2008	22.78	−8.94	1.07	−0.92
	2018	13.84		0.15	
Chapter VI: G62.1 Alcoholic polyneuropathy	2008	0.33	−0.22	0.08	−0.05
	2018	0.11		0.03	
Chapter IX: I42.6 Alcoholic cardiomyopathy	2008	1.88	−0.88	0.15	−0.08
	2018	1.00		0.07	
Chapter XI: K70 Alcoholic liver disease	2008	14.52	−5.81	17.55	+13.57
	2018	8.71		31.12	
Chapter XI: K86.0 Alcohol-induced chronic pancreatitis	2008	0.20	−0.10	0.20	+0.16
	2018	0.10		0.36	
Chapter XX: X45Accidental poisoning by and exposure to alcohol	2008	14.00	−4.00	5.07	+0.39
	2018	10.00		5.46	

Place of residence and analysis of causes of death by sex are important criteria in analyzing the level of mortality caused by alcohol consumption in a particular chapter of causes of death (Tables 3 and 4).

Table 3. Mortality caused by alcohol consumption by certain classes of causes of death depending on the place of residence.

Mortality Rate	Place of Residence	UKRAINE			POLAND			% Change* 2008	% Change* 2018
		Rate 2008	Rate 2018	% Change	Rate 2008	Rate 2018	% Change		
Overall mortality rate	urban	14.75	12.51	−15.21	9.81	11.06	+12.68	−33.50	−11.62
	rural	19.83	17.19	−13.29	10.17	10.36	+1.91	−48.71	−39.72
The mortality rate due to alcohol	urban	0.50	0.20	−59.72	0.19	0.27	+41.16	−62.17	+32.57
	rural	0.73	0.37	−49.08	0.14	0.23	+64.32	−80.66	−37.60
Chapter V. Mental and behavioral disorders	urban	0.05	0.02	−57.33	0.06	0.10	+70.74	+25.21	+401.07
	rural	0.10	0.05	−46.02	0.05	0.10	96.23	−48.50	+87.22
F10 Mental and behavioral disorders due to use of alcohol	urban	0.03	0.01	−56.59	0.05	0.08	+47.07	+62.74	+451.39
	rural	0.07	0.04	−45.39	0.04	0.08	+90.80	−38.40	115.22
Chapter VI. Diseases of the nervous system	urban	0.15	0.10	−34.54	0.14	0.19	+40.02	−7.98	+96.82
	rural	0.18	0.15	−19.20	0.12	0.14	+15.45	−33.71	−5.29
G31.2 Degeneration of nervous system due to alcohol	urban	0.032	0.012	−63.82	0.002	0.0003	−83.18	−95.19	−97.76
	rural	0.045	0.024	−45.70	0.001	0.0003	−77.25	−97.43	−98.92
G62.1 Alcoholic polyneuropathy	urban	0.0004	0.0001	−70.57	0.0001	0.003	−32.71	−63.28	−16.05
	rural	0.0009	0.0002	−82.63	0.0001	0.000	-	−92.37	-
Chapter IX. Diseases of the circulatory system	urban	9.14	8.06	−11.81	4.30	4.32	+0.68	−53.01	−46.35
	rural	13.15	12.10	−7.92	4.91	4.44	−9.64	−62.62	−63.32
I42.6 Alcoholic cardiomyopathy	urban	0.18	0.08	−56.30	0.01	0.003	−55.14	−96.03	−95.93
	rural	0.24	0.13	−45.54	0.01	0.002	−61.74	−97.31	−98.11
Chapter XI. Diseases of the digestive system	urban	0.79	0.57	−27.42	0.50	0.49	−0.59	−36.62	−13.18
	rural	0.72	0.61	−15.34	0.37	0.38	+3.79	−48.51	−36.88
K70 Alcoholic liver disease	urban	0.10	0.04	−62.40	0.09	0.16	+68.74	−10.01	+303.80
	rural	0.13	0.08	−39.33	0.06	0.11	+105.41	−56.02	+48.90
K86.0 Alcohol-induced chronic pancreatitis	urban	0.0014	0.0005	−65.67	0.001	0.002	+61.48	−23.51	+259.80
	rural	0.0018	0.0008	−52.23	0.001	0.001	+112.67	−61.84	+69.88
Chapter XX. External causes of morbidity and mortality	urban	1.20	0.65	−46.77	0.61	0.48	−20.79	−49.96	−25.54
	rural	1.59	0.93	−41.28	0.76	0.59	−22.34	−52.31	−36.93
X45Accidental poisoning by and exposure to alcohol	urban	0.16	0.06	−61.50	0.03	0.03	−19.93	−78.79	−55.88
	rural	0.25	0.10	−59.00	0.03	0.03	−8.49	−86.24	−69.29

% Change—decrease or increase rate from 2008 to 2018 in percent (of observed values). % Change*—decrease or increase rate from Ukraine to Poland in percent (of observed values).

Table 4. Mortality rate of men and women due to alcohol consumption.

Mortality Rate	Sex	Ukraine			Poland			% Change* 2008	% Change* 2018
		Rate 2008	Rate 2018	%Change	Rate 2008	Rate 2018	%Change		
Overall mortality rate	Men	18.19	14.89	−18.14	10.99	11.49	+4.60	−39.60	−22.83
	Women	14.82	13.15	−11.24	8.98	10.11	+12.57	−39.38	−23.12
Overall mortality rate caused by alcohol consumption	Men	0.98	0.44	−54.98	0.29	0.39	+33.91	−69.86	−10.35
	Women	0.24	0.10	−58.82	0.06	0.10	+76.50	−75.76	+3.91

Table 4. *Cont.*

Mortality Rate	Sex	Ukraine			Poland			% Change* 2008	% Change* 2018
		Rate 2008	Rate 2018	%Change	Rate 2008	Rate 2018	%Change		
Chapter V. Mental and behavioral disorders	Men	0.10	0.05	−52.24	0.09	0.16	+66.45	−4.55	+232.65
	Women	0.03	0.01	−52.74	0.02	0.04	+148.46	−44.15	+193.61
F10 Mental and behavioral disorders due to use of alcohol	Men	0.08	0.04	−51.79	0.09	0.15	+59.66	+10.93	+267.38
	Women	0.014	0.006	−51.18	0.01	0.02	+77.16	−14.77	+209.29
Chapter VI. Diseases of the nervous system	Men	0.23	0.15	−34.80	0.13	0.16	+21.48	−43.42	+5.41
	Women	0.10	0.08	−18.11	0.13	0.18	+39.59	+34.07	+128.52
G31.2 Degeneration of nervous system due to alcohol	Men	0.06	0.03	−55.78	0.0023	0.0004	−83.87	−96.21	−98.62
	Women	0.014	0.005	−62.22	0.0005	0.0002	−70.18	−96.45	−97.20
G62.1 Alcoholic polyneuropathy	Men	0.0010	0.0002	−84.44	0.0002	0.0001	−50.47	−78.03	−30.04
	Women	0.0001	0.0001	−26.71	0.00	0.00	-	-	-
Chapter IX. Diseases of the circulatory system	Men	10.04	8.79	−12.39	4.43	4.13	−6.74	−55.88	−53.04
	Women	10.75	9.76	−9.18	4.64	4.60	−0.86	−56.85	−52.90
I42.6 Alcoholic cardiomyopathy	Men	0.33	0.16	−51.32	0.013	0.006	−56.95	−96.15	−96.60
	Women	0.08	0.03	−56.93	0.001	0.000	−66.87	−98.63	−98.95
Chapter XI. Diseases of the digestive system	Men	1.07	0.78	−26.65	0.54	0.55	+0.92	−49.31	−30.25
	Women	0.51	0.41	−19.11	0.36	0.36	−0.02	−28.45	−11.56
K70 Alcoholic liver disease	Men	0.16	0.07	−54.91	0.12	0.21	+70.43	−23.98	+187.38
	Women	0.07	0.03	−53.37	0.04	0.07	+103.47	−45.78	+136.59
K86.0 Alcohol-induced chronic pancreatitis	Men	0.003	0.001	−52.48	0.002	0.003	+77.63	−39.17	+127.38
	Women	0.0006	0.0000	−92.67	0.0003	0.0005	+65.64	−49.60	+1039.08
Chapter XX. External causes of morbidity and mortality	Men	2.29	1.27	−44.74	1.05	0.81	−23.13	−54.17	−36.25
	Women	0.51	0.27	−46.32	0.31	0.26	−15.74	−40.33	−6.34
X45Accidental poisoning by and exposure to alcohol	Men	0.33	0.13	−59.32	0.06	0.03	−56.70	−81.50	−80.31
	Women	0.06	0.02	−67.25	0.008	0.007	−11.52	−87.09	−65.14

% Change—decrease or increase rate from 2008 to 2018 in percent (of observed values). % Change*—decrease or increase rate from Ukraine to Poland in percent (of observed values).

The analysis of the causes of death caused by alcohol consumption by sex and among residents of cities and villages of Ukraine and Poland had a clear difference.

Given the importance of the human factor as the main labor force of the country, we calculated the total mortality rate of the working-age population (Table 5), as well as analyzed the mortality rates for three age groups, such as 16–24 years, 25–44 years, and 45–64 years (Tables 6–8).

Table 5. Mortality working age population.

Mortality Rate	Ukraine			Poland			% Change* 2008	% Change* 2018
	Rate 2008	Rate 2018	% Change	Rate 2008	Rate 2018	% Change		
Overall mortality rate	7.97	5.97	−25.10	4.24	3.84	−9.46	−46.81	−35.70
Overall mortality rate caused by alcohol consumption	0.78	0.35	−55.35	0.22	0.29	+35.70	−72.16	−15.40

Table 5. *Cont.*

Mortality Rate	Ukraine			Poland			% Change* 2008	% Change* 2018
	Rate 2008	Rate 2018	% Change	Rate 2008	Rate 2018	% Change		
% of deaths due to alcohol consumption in the total number of deaths	9.77	5.82	-	5.11	7.66	-	−47.66	+31.57
Chapter V. Mental and behavioral disorders								
Mortality rate	0.07	0.03	−52.19	0.06	0.09	+45.08	−11.05	+169.90
Deaths on mental and behavioral disorders due to use of alcohol Mortality due to mental and behavioral disorders due to alcohol consumption	0.06	0.03	−51.28	0.06	0.09	+43.94	+3.44	+205.62
% of deaths due to alcohol use in this class of causes of death	84.22	85.82	-	97.94	97.17	-	+16.29	+13.23
Chapter VI. Diseases of the nervous system								
Mortality rate	0.18	0.11	−39.31	0.07	0.06	−2.72	−63.39	−41.31
Mortality due to degeneration of the nervous system caused by alcohol consumption	0.048	0.020	−58.32	0.002	0.000	−83.82	−96.50	−98.64
Mortality due to alcohol field neuropathy	0.0007	0.0002	−75.76	0.0002	0.0000	−73.99	−79.35	−77.85
% of deaths due to alcohol use in this class of causes of death	27.37	18.68	-	2.81	0.49	-	−89.74	−97.37
Chapter IX. Diseases of the circulatory system								
Mortality rate	2.86	2.35	−17.71	1.16	0.88	−23.84	−59.57	−62.58
Mortality due to alcoholic cardiomyopathy	0.264	0.126	−52.39	0.008	0.003	−60.26	−96.99	−97.49
% of deaths due to alcohol use in this class of causes of death	9.24	5.35	-	0.69	0.36	-	−92.55	−93.28
Chapter XI. Diseases of the digestive system								
Mortality rate	0.85	0.60	−29.29	0.32	0.31	−1.24	−62.66	−47.85
Mortality due to alcoholic liver disease	0.15	0.07	−54.00	0.10	0.16	+66.35	−34.61	+136.45
Mortality due to chronic alcoholic pancreatitis	0.002	0.001	−63.93	0.001	0.002	+74.49	−45.88	+161.80
% of deaths due to alcohol use in this class of causes of death	17.86	11.58	-	31.20	52.58	-	+74.69	+353.95
Chapter XX. External causes of morbidity and mortality								
Mortality rate	1.59	0.85	−46.78	0.69	0.49	−28.20	−56.88	−41.82
Mortality due to accidental poisoning and alcohol exposure	0.25	0.10	−59.52	0.05	0.04	−21.70	−82.16	−65.49
% of deaths due to alcohol use in this class of causes of death	15.85	12.06	-	6.56	7.15	-	−58.62	−40.68

% Change—decrease or increase rate from 2008 to 2018 in percent (of observed values). % Change*—decrease or increase rate from Ukraine to Poland in percent (of observed values).

Table 6. Mortality of the working age population by age groups.

Age Groups	Mortality Rate	Ukraine			Poland			% Change* 2008	% Change* 2018
		Rate 2008	Rate 2018	% Change	Rate 2008	Rate 2018	% Change		
16–24	Mortality in this age group	1.18	0.68	−42.58	0.66	0.53	−19.85	1.80	1.29
	Mortality of the population in this age group due to alcohol consumption	0.04	0.01	−79.55	0.01	0.01	5.23	8.80	1.50
	Proportion of deaths due to alcohol consumption in this age group	3.73	1.39	-	0.83	1.10	-	4.46	1.27

Table 6. *Cont.*

Age Groups	Mortality Rate	Ukraine			Poland			% Change* 2008	% Change* 2018
		Rate 2008	Rate 2018	% Change	Rate 2008	Rate 2018	% Change		
25–44	Mortality in this age group	4.87	2.89	−40.53	1.49	1.27	−15.14	3.26	2.29
	Mortality of the population in this age group due to alcohol consumption	0.63	0.27	−57.10	0.12	0.16	28.04	5.19	1.74
	Proportion of deaths due to alcohol consumption in this age group	12.89	9.28	-	8.10	12.22	-	1.59	0.76
45–64	Mortality in this age group	15.03	11.18	−25.60	9.04	8.05	−10.85	1.66	1.39
	Mortality of the population in this age group due to alcohol consumption	1.34	0.55	−59.00	0.43	0.56	31.11	3.12	0.98
	Proportion of deaths due to alcohol consumption in this age group	8.91	4.91	-	4.75	6.98	-	1.88	0.70

%Change—decrease or increase rate from 2008 to 2018 in percent (of observed values). %Change*—decrease or increase rate from Ukraine to Poland (of observed values).

Table 7. Mortality of the working age population in Poland by causes of death and by age groups.

Mortality Rate\Age Groups	16–24 Years			25–44 Years			45–64 Years		
	Rate 2008	Rate 2018	%Change	Rate 2008	Rate 2018	%Change	Rate 2008	Rate 2018	%Change
F10–F99. Chapter V. Mental and behavioral disorders	0.002	0.001	−63.67	0.033	0.043	30.36	0.129	0.185	42.99
F 10. Mortality due to mental and behavioral disorders due to alcohol consumption	0.001	0.001	−58.48	0.032	0.041	30.58	0.127	0.180	41.07
Proportion of deaths due to alcohol consumption in this age group (%)	58.33	66.67	-	96.67	96.83	-	98.64	97.32	-
G00–G99. Chapter VI. Diseases of the nervous system	0.023	0.025	8.69	0.034	0.028	−16.10	0.121	0.119	−1.80
G31.2 Mortality due to degeneration of the nervous system caused by alcohol consumption	0.000	0.000	-	0.001	0.000	-	0.003	0.001	−77.72
G62.1 Mortality due to alcohol field neuropathy	0.000	0.000	-	0.000	0.000	-	0.000	0.000	−74.53
Proportion of deaths due to alcohol consumption in this age group (%)	0.00	0.00	-	3.48	0.00	-	2.89	0.67	-
I00-I99. Chapter IX. Diseases of the circulatory system	0.035	0.018	−49.49	0.249	0.166	−33.15	2.708	2.031	−25.00
I42.6 Mortality due to alcoholic cardiomyopathy	0.000	0.000	-	0.002	0.001	−67.77	0.018	0.007	−60.57
Proportion of deaths due to alcohol consumption in this age group (%)	0.00	0.00	-	0.95	0.46	-	0.67	0.35	-
K00-K93. Chapter XI. Diseases of the digestive system	0.009	0.008	−7.25	0.150	0.150	0.56	0.659	0.618	−6.29
K70 Mortality due to alcoholic liver disease	0.002	0.003	117.97	0.050	0.088	73.36	0.199	0.310	55.69
K86.0 Mortality due to chronic alcoholic pancreatitis	0.000	0.000	-	0.001	0.001	−30.16	0.001	0.004	171.66
Proportion of deaths due to alcohol consumption in this age group (%)	17.02	40.00	-	34.73	58.87	-	30.43	50.84	-

Table 7. *Cont.*

Mortality Rate\Age Groups	16–24 Years			25–44 Years			45–64 Years		
	Rate 2008	Rate 2018	%Change	Rate 2008	Rate 2018	%Change	Rate 2008	Rate 2018	%Change
V01-Y98. Chapter XX. External causes of morbidity and mortality	0.470	0.336	−28.62	0.549	0.397	−27.69	0.949	0.664	−29.96
X45 Mortality due to accidental poisoning and alcohol exposure	0.003	0.002	−27.34	0.034	0.024	−28.35	0.079	0.060	−23.66
Proportion of deaths due to alcohol consumption in this age group (%)	0.56	0.57	-	6.16	6.11	-	8.34	9.09	-

Table 8. Mortality of the working age population of Ukraine by causes of death and by age groups.

Mortality Rate\Age Groups	16–24 Years			25–44 Years			45–64 Years		
	Rate 2008	Rate 2018	% Change	Rate 2008	Rate 2018	% Change	Rate 2008	Rate 2018	% Change
F10–F99. Chapter V. Mental and behavioral disorders	0.009	0.002	−82.16	0.066	0.027	−59.36	0.111	0.054	−51.97
F 10. Mortality due to mental and behavioral disorders due to alcohol consumption	0.005	0.001	−82.16	0.050	0.022	−56.64	0.102	0.048	−53.18
Proportion of deaths due to alcohol consumption in this age group (%)	50.00	50.00	-	75.68	80.75	-	91.31	89.01	-
G00–G99. Chapter VI. Diseases of the nervous system	0.051	0.034	−33.11	0.163	0.085	−48.02	0.264	0.160	−39.37
G31.2 Mortality due to degeneration of the nervous system caused by alcohol consumption	0.002	0.000	−89.22	0.039	0.015	−62.35	0.083	0.033	−60.45
G62.1 Mortality due to alcohol field neuropathy	0.000	0.000	-	0.000	0.000	−79.54	0.001	0.000	−75.63
Proportion of deaths due to alcohol consumption in this age group (%)	5.15	0.78	-	24.25	17.49	-	31.83	20.63	-
I00-I99. Chapter IX. Diseases of the circulatory system	0.071	0.054	−23.94	0.897	0.652	−27.33	6.519	5.023	−22.94
I42.6 Mortality due to alcoholic cardiomyopathy	0.006	0.002	−61.68	0.191	0.087	−54.31	0.483	0.210	−56.53
Proportion of deaths due to alcohol consumption in this age group (%)	7.86	3.96	-	21.30	13.39	-	7.42	4.18	-
K00-K93. Chapter XI. Diseases of the digestive system	0.052	0.023	−56.63	0.684	0.401	−41.41	1.468	1.024	−30.28
K70 Mortality due to alcoholic liver disease	0.010	0.001	−94.35	0.130	0.053	−59.20	0.248	0.110	−55.63
K86.0 Mortality due to chronic alcoholic pancreatitis	0.000	0.000	-	0.002	0.001	−48.85	0.003	0.001	−79.28
Proportion of deaths due to alcohol consumption in this age group (%)	18.26	2.38	-	19.33	13.51	-	17.09	10.80	-
V01-Y98. Chapter XX. External causes of morbidity and mortality	0.746	0.427	−42.76	1.526	0.784	−48.63	2.122	0.784	−63.07
X45 Mortality due to accidental poisoning and alcohol exposure	0.022	0.006	−73.95	0.214	0.091	−57.69	0.419	0.106	−74.78
Proportion of deaths due to alcohol consumption in this age group (%)	2.90	1.32	-	14.03	11.56	-	19.74	13.48	-

5. Discussion

Comparison of medical and demographic indicators in different countries or regions of the world is one of the elements of an analytical approach to solving public health problems. In addition to WHO data, i.e.,Global Health Observatory Data Repository and Framework for alcohol policy in the WHO European Region, this kind of analysis is presented in many works, including M. Marmot et al., F. Baum, J. P. Mackenbach, C. Bambra et al., and further Polish researchers [14–20]. Comparison of data from some geographically close regions of different countries was carried out in the study of J.Grshybowskyj and co-authors, who found a significant difference in the development processes in terms of medical and demographic indicators [21].

Our analysis of the data showed that the main causes of death in Ukraine in 2008 and in 2018 were (in % of all causes of death) (Figure 1):

Diseases of the circulatory system—63.64 and 66.98;
Neoplasms—11.80 and 13.43;
External causes of morbidity and mortality—8.14 and 5.28;
Diseases of the digestive system—4.67 and 4.18;
Diseases of the respiratory system—3.09 and 2.22.

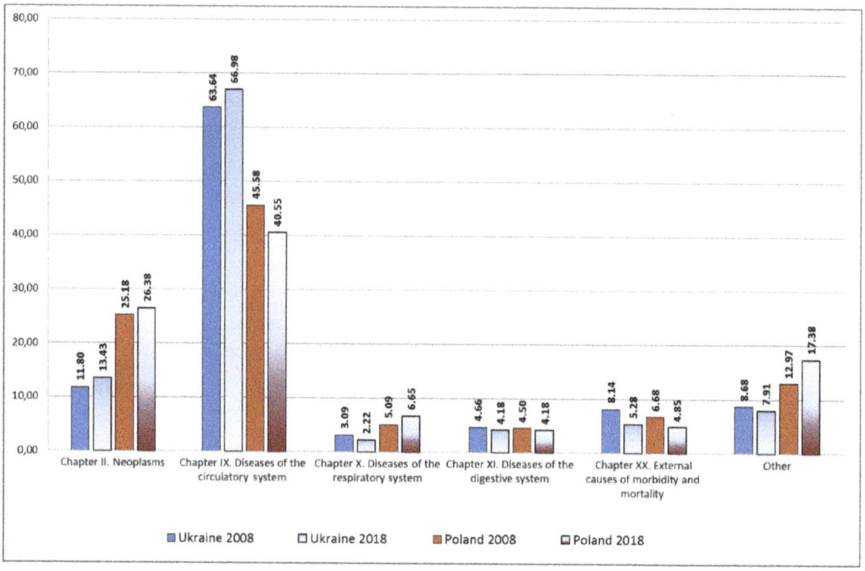

Figure 1. The main causes of mortality in Ukraine and Poland in 2008 and 2018 (in %).

It is established that, for these 10 years in Ukraine, there was an increase in the share of deaths from diseases of the circulatory system (by 3.34%) and tumors (by 1.63%).

In Poland over the years, the main causes of death were the same pathology, but in contrast in Ukraine, mortality from tumors and respiratory diseases was higher (in %% of all causes of death):

Diseases of the circulatory system—45.84 and 40.55;
Neoplasms—25.18 and 26.38;
External causes of morbidity and mortality—6.68 and 4.85;
Diseases of the respiratory system—5.09 and 6.65;
Diseases of the digestive system—4.50 and 4.18.

It should also be noted that, in ten years, Poland saw a decrease in the mortality rate from diseases of the circulatory system and external causes, while the share of deaths due to respiratory diseases (+1.57%), tumors (+1.21%), endocrine diseases, eating disorders, and metabolic disorders (+0.48%), mental and behavioral disorders (+0.36%) and diseases of the nervous system (+0.27%) all increased.

When comparing mortality rates by main classes of causes in Ukraine and Poland (Table 1), it was found that the total mortality in Ukraine exceeded the figures in Poland, both in 2008 and in 2018, by 1.64 and 1.29 times, respectively (16.31 and 13.90 in Ukraine per 1000 people and 9.95 and 10.78 in Poland), including excess mortality observed in some infectious and parasitic diseases (5.28 and 4.20 times, respectively), diseases of the circulatory system (2.29 and 2.12 times), and external causes (1.99 times in 2008). In contrast, the mortality rate in Poland was higher in endocrine diseases, eating disorders and metabolic disorders (3 and 5 times, respectively) and neoplasms (1.31 and 1.53 times, respectively). From 2008 to 2018, there was a significant increase in the difference in mortality from respiratory diseases—with almost the same figures in 2008 to 2.32 times more in Poland in 2018.It is necessary to note separately the changes that have occurred in the mortality rates of mental and behavioral disorders—in 2008, the mortality rate in Ukraine was slightly higher than in Poland (0.062 in Ukraine against 0.054 in Poland), and in 2018 the death rate in Poland from mental and behavioral disorders exceeded that in Ukraine 2.9 times (0.034 in Ukraine and 0.098 in Poland). Similar, but less pronounced changes were observed in mortality from diseases of the nervous system.

Mortality from pathology caused by alcohol consumption occupies an ambiguous share in the structure of population mortality [22,23]. In the period we are comparing, the share of mortality caused by alcohol consumption in Ukraine decreased from 3.52 to 1.83%. At the same time, in Poland, there was an increase in this cause of death from 1.72 to 2.36%. When comparing the share of such causes of death in our selected chapters, it was found that growth was observed in mental and behavioral disorders due to alcohol use (in Ukraine), andalcoholic liver disease, accidental poisoning by and exposure to alcohol, and alcohol-induced chronic pancreatitis (in Poland). Moreover, the increase in mortality from alcoholic liver disease exceeded 13% and, in Ukraine, there was a decrease in such indicatorsfrom 14.52% to 8.71%.

It should be noted that the mortality caused by alcohol consumption was the main share of mortality in the class of mental and behavioral disorders: 73% in Ukraine and 92% in Poland 2008, and 74% and 83%, respectively, in 2018 (Table 2).

Place of residence is an important criterion in analyzing the level of mortality caused by alcohol consumption in a particular chapter of causes of death, as it allows for appropriate organizational and preventive measures for medical care in the area, whether in the city or in the countryside. Studies of changes in mortality rates in cities and villages of Ukraine and Poland have shown different trends in mental and behavioral disorders (F10) and alcoholic liver disease (K70).

In 2008, mortality from mental and behavioral disorders due to alcohol consumption among peasants was higher in Ukraine than in Poland (0.071 vs. 0.044) (Table 3). However, in 2018 it became much higherin Poland: among urban residents 5.7 times (0.014 in Ukraine, 0.080 in Poland) and among rural residentsmore than 2 times (0.039 in Ukraine, 0.083 in Poland). For this reason, mortality in both urban and rural areas in Ukraine has halved on average, while in Poland the mortality rate has increased by the same amount. The mortality rate due to alcohol and behavioral disorders due to alcohol consumption in Ukraine is predominant among rural residents; in Poland it is slightly more common among urban residents.

Mortality due to alcoholic liver disease among urban residents of Poland from 2008 to 2018 increased from 0.093 to 0.157 per 1000 inhabitants, and in 2018 was 4 times higher than that among residents of Ukrainian cities (0.104 and 0.039, respectively). The same dynamics are observed among rural residents.

One of the reasons for this difference in the mortality rates of rural and urban populations of Ukraine is the problem of unemployment, especially in rural areas, and it acquires new features [24]. Despite the fact that many unemployed rural residents have their own farm as a source of livelihood, this way of employment, as well as work in agricultural production in general, which is mostly low-paid and uncreative, can provide neither material wealth nor the need for self-realization in accordance with the demands of modern young villagers. Therefore, living in the countryside today is characterized not only by physical overload but often by life's frustrations and the associated alcoholism, which shortens life. Therefore, the mortality of the rural population ofworking age is higher than in the urban [25,26]. Similar problems are observed in other countries [27,28].

Analysis of the causes of death by sex showed that the overall mortality rate and mortality caused by alcohol consumption in both Ukraine and Poland was higher among the male population in 2008 and 2018 (Table 4). The comparison of indicators makes it possible to state that, against the background of declining mortality in Ukraine and its growth in Poland, the most pronounced change was the increase in mortality caused by alcohol consumption among women in Poland, which exceeded that among women in Ukraine (in 2018,0.097 in Ukraine, 0.101 in Poland).

J. Mackenbach, N. Ryngach, and V. Kruchanytsia [29–31] indicate a high death rate from alcohol-related causes, especially in men. The effects of alcohol consumption are largely associated with smoking [32–36].

Among the male population of Ukraine by certain chapters of causes of death due to alcohol consumption, the highest mortality rate in 2008 and 2018 was observed for alcoholic cardiomyopathy (0.334 and 0.163 per 1000 cases, respectively) and for accidental poisoning and alcohol exposure (0.329 and 0.134 per 1000 cases, respectively). In Poland, the highest mortality rates for men were observed from alcoholic liver disease (0.125 and 0.213 per 1000 cases, respectively, in 2008 and 2018) and from mental and behavioral disorders (0.091 and 0.146, respectively).

However, the mortality of the male population of Ukraine caused by alcohol consumption, in the chapter "Mental and behavioral disorders" was actually the most common, and there was a tendency to increase the share of mortality in the chapter: 82.79% in 2008 and 83.57% in 2018. In Poland, the same negative dynamics of mortality weredue to alcoholic liver disease (23.12% in 2008 and 39.05% in 2018), and the highest proportion of deaths caused by alcohol consumption was in the chapter "Mental and behavioral disorders" (96.22% in 2008 and 92.29% in 2018). For other chapters of causes of death, the share in the mortality of the population caused by alcohol consumption was much lower.

Among the female population with a similar ratio of mortality to mortality in the corresponding chapter in Poland, the highest mortality rate with a tendency to increase is determined in subchapter K.86.0 "Alcoholic liver disease": 0.035 per 1000 cases in 2008 and 0.072 in 2018. When comparing these indicators with the data in Ukraine, they were twice lower in 2008 and already exceeded them in 2018 (0.065 per 1000 cases in 2008 and 0.030 in 2018).

As among the male population, among women in Ukraine and Poland in the chapter "Mental and behavioral disorders", the highest percentage of deaths was caused by alcohol consumption. However, it was twice lower than among men (Ukraine: 45.39% in 2008 and 46.89% in 2018; Poland: 69.28% in 2008 and 49.40% in 2018).

Given the importance of the human factor as the main labor resource of the country, we calculated the indicators of total mortality and mortality caused by alcohol consumption in working age (Table 5). It is established that with the overall decrease in the mortality of the working population from 2008 to 2018 in both Ukraine and Poland, mortality due to alcohol consumption in Ukraine decreased by 55.35%, and in Poland, it increased by 35.70%. Accordingly, the share of deaths due to alcohol consumption in the total number of deaths changed: in Ukraine, their share decreased from 9.77% to 5.82%, and in Poland it increased from 5.11% to 7.66%.

The same trends were observed for the mortality chapter of the working population due to mental and behavioral disorders: a significant decrease in both total and alcohol-related mortality in Ukraine (by 52.2% and 51.3%, respectively) and its increase in Poland (by 45.1% and 43.9%). In general, mortality rates in this chapter of causes in Poland have become higher than in Ukraine.

Another chapter where the same mortality trends are observed was chapter XI: Diseases of the digestive system. With the prevalence of the overall mortality rate in Ukraine over the mortality rate in Poland for this chapter of causes of death (2.9 and 1.9 times in 2008 and 2108, respectively), there were significant changes in the mortality rate due to alcoholic liver disease: a 54% decrease in mortality in Ukraine, and growth in Poland by 66.4%. In general, the share of deaths due to alcohol use in this chapter of causes of death in

Ukraine decreased from 17.86% to 11.58%, and in Poland increased from 31.20% to 52.58%, exceeding the mortality rate in Ukraine in 2018.

According to the other classes of causes of death, it is necessary to state a significant predominance of the mortality rate in Ukraine against that in Poland due to alcoholic cardiomyopathy and due to degeneration of the nervous system caused by alcohol consumption.

The analysis of mortality rates by three age groups, 16–24 years, 25–44 years, and 45–64 years, showed that mortality was predominant among people aged 45–64 (Table 6). With the general decrease in the mortality rate of the working-age population both in Poland and in Ukraine, it was most pronounced among persons aged 16–24 (19.9% in Poland and 42.6% in Ukraine). Regarding mortality due to alcohol consumption, its level in all age groups was higher in Ukraine than in Poland, except in 2018 in the age group 45–64 years, when the figures were almost the same. The highest share of deaths due to alcohol consumption in those of working-age was among persons aged 25–44, with tendenciesin Poland to increase (from 8.10% to 12.22%) and in Ukraine to decrease (from 12.89% to 9.28%).

When comparing the data on mortality rates by age groups, and by the defined chapters of causes of death, it was found that the main cause of death in the age groups 16–24 and 25–44 years in both countries were external causes of morbidity and mortality. However, in these groups in Ukraine, the mortality due to alcohol consumption was mainly due to "Accidental poisoning and exposure to alcohol" (0.022 per 1000 cases in 2008 and 0.006 in 2018). In Poland, this was observed only in 2008 among the working population aged 16–24 (0.003 per 1000 cases). In 2018, in this age group, and in 2008 and 2018 among residents aged 25–44, the highest mortality due to alcohol consumption was due to alcoholic liver disease (2008: 25–44 years—0.050 per 1000 cases; 2018: 16–24 years—0.003 and 25–44 years—0.088).

In both countries, in the age group 45–64 years, the main cause of death was "Diseases of the circulatory system"in 2008 and 2018.

The age group 45–64 years, as a non-mobile productive group of the population, accounted for a significant share of the population of Poland in 2018—43.8%, which indicates an aging population [37]. In Ukraine, it was equal to 27.4%. In this group of the population in Ukraine, the main cause of death in 2008 and 2018 was "Diseases of the circulatory system"; in Poland this was "Neoplasms".

In Poland, the main causes of death from alcohol use in this age group were alcoholic liver disease and mental and behavioral disorders (respectively: 2008: 0.199 and 0.127; 2018: 0.310 and 0.180) (Table 7). When comparing the indicators for 2008 and 2018, there is a 55.69%increase in mortality due to alcoholic liver disease and a41.07% increase in mortality due to mental and behavioral disorders due to alcohol consumption.

In all age groups, the proportion of deaths due to alcohol use in the total number of deaths by chapters of causes of death was highest in the chapter "Mental and behavioral disorders". This was higher than 95% among people aged 25–44 and 45–64.

In Ukraine, alcohol consumption has become the most common cause of death due to alcoholic cardiomyopathy (Table 8). In 2018, the mortality rate from alcoholic cardiomyopathy decreased by more than 2 times, but continued to be the leading cause of death due to alcohol consumption, and exceeded this figure among the working population of Poland by 29.5 times. As in Poland, the share of deaths due to alcohol use in the total number of deaths by chapters of causes of death was the highest in the chapter "Mental and behavioral disorders" in all age groups, and reached the maximum among persons aged 45–64 in 2008—91.3%, and in 2018—89.0%.

6. Conclusions

Despite the geographical proximity, the similarity of natural and climatic characteristics, and similarity of population, mortality rates in each country reflect the difference in the medical and demographic situation, and the effectiveness of state social approaches to public health. Alcohol plays and continues to play an important role in the mortality rate: the share of mortality caused by alcohol consumption in Ukraine in 2008 was 3.52%, and

1.83% in 2018. At the same time, in Poland, there was an increase in such a cause of death from 1.72% (2008) to 2.36%. Mortality caused by alcohol consumption is the main share of mortality in the chapter on mental and behavioral disorders. Changes in the mortality rate in cities and villages of Ukraine and Poland showed different trends: in Poland, during the study period there was a significant increase in mortality due to alcohol consumption to a greater extent among the rural population. At the same time, mortality in Ukraine has halved on average among both urban and rural residents. The overall mortality and alcohol-related mortality rates in both Ukraine and Poland were higher among the male population. Analysis of mortality among people of working age showed that the highest proportion of deaths due to alcohol consumption in both countries was among people aged 25–44. The main cause of death in this and the group of 16–24 years was due to accidental poisoning and alcohol exposure. In the age group 45–64 years in Poland, the main causes of death with a tendency to increase were alcoholic liver disease and mental and behavioral disorders due to alcohol consumption. In Ukraine, the most common cause of death was alcoholic cardiomyopathy.

Author Contributions: Conceptualization, O.L. and M.K.; methodology, J.G.; software, K.M.S.; validation, E.C.; formal analysis, J.G.; investigation, K.M.S.; resources, O.L. and E.C.; data curation, M.K.; and Writing—Original draft preparation, O.L. and M.K. All authors have read and agreed to the published version of the manuscript.

Funding: This research received no external funding.

Institutional Review Board Statement: The study was conducted according to the guidelines of the Declaration of Helsinki, and approved by the Ethics Committee of Medical University in Lviv Ukraine.

Informed Consent Statement: Not applicable.

Data Availability Statement: Not applicable.

Conflicts of Interest: The authors declare no conflict of interest.

References

1. Slabkiy, G.O.; Mironyuk, I.S.; Koshelia, I.I.; Dudnyk, S.V. МЕДИКО-ДЕМОГРАфІЧНА СИТУАЦІЯ ЯКГЛОБАЛЬНА ПРОБЛЕМА ГРОМАДСЬКОГО ЗДОРОВ'Я УКРАЇНИ. *ВісникСоціальноїГігієни Та Організації Охорони Здоров'я України* **2019**, 62–72. [CrossRef]
2. Stovban, I.V.; Marusyn, O.V. Features of Organization of Regional Monitoring of Reproductive Health in Ukraine and Its Results. In *Challenges of Medical Science and Education: An Experience of EU Countries and Practical Introduction in Ukraine, Collective Monograph*; Baltija Publishing: Riga, Latvia, 2020.
3. Rohach, I.; Keretzman, A.; Hadzheha, I. Overview of the demographic situation dynamics in Ukraine and its regions compared to EU and the world: Aspests and problems. *Probl. Clin. Pediatr.* **2019**, *2*, 49–56. [CrossRef]
4. World Health Organization. *STEPS: Prevalence of Noncommunicable Disease Risk Factors in Ukraine 2019*; WHO Regional Office for Europe: Copenhagen, Denmark, 2020.
5. Rehm, J.; Sulkowska, U.; Manczuk, M.; Boffetta, P.; Powles, J.; Popova, S.; Zatoński, W. Alcohol accounts for a high proportion of premature mortality in central and eastern Europe. *Int. J. Epidemiol.* **2007**, *36*, 458–467. [CrossRef]
6. Chemerys, N.M.; Lyubinets, O.V. *Consumption of Alcoholic Drinks among University Youth: Medical and Social Perspectives*; Nation's Health; American Public Health Association: Washington, DC, USA, 2018; Volume 47, pp. 42–47. Available online: http://nbuv.gov.ua/UJRN/Uzn_2018_1_7 (accessed on 27 January 2021).
7. Łyszczarz, B. Production Losses Associated with Alcohol-Attributable Mortality in the European Union. *Int. J. Environ. Res. Public Health* **2019**, *16*, 3536. [CrossRef]
8. Alonso, I.; Vallejo, F.; Regidor, E.; Belza, M.J.; Sordo, L.; Otero-García, L.; Barrio, G. Changes in directly alcohol-attributable mortality during the great recession by employment status in Spain: A population cohort of 22 million people. *J. Epidemiol. Commun. Health* **2017**, *71*, 736–744. [CrossRef] [PubMed]
9. Zatoński, A.W.; Sulkowska, U.; Zatoński, M.Z.; Herbeć, A.A.; Muszyńska, M.M. Alcohol taxation and premature mortality in Europe. *Lancet* **2015**, *385*, 1181. [CrossRef]
10. OECD. Pensions at a Glance 2017. Available online: https://www.oecd-ilibrary.org/docserver/pension_glance-2017-9-en.pdf?expires=1613643582&id=id&accname=guest&checksum=876E20AEC7006FEDCE89A6D4D0720D0F/ (accessed on 27 January 2021).
11. State Statistics Service of Ukraine. 2020. Available online: http://www.ukrstat.gov.ua/ (accessed on 26 March 2021).

12. Baza Demografia. Available online: http://demografia.stat.gov.pl/bazademografia/Tables.aspx (accessed on 26 January 2021).
13. Available online: https://dw.euro.who.int/api/v3/export?code=HFAMDB_113 (accessed on 19 April 2021).
14. World Health Organization. The Global Health Observatory. Explore a World of Health Data. Available online: https://www.who.int/data/gho (accessed on 15 April 2021).
15. Framework for Alcohol Policy in the WHO European Region/WHO Regional Office for Europe. Copenhagen, Denmark. Available online: https://www.euro.who.int/_data/assets/pdf_file/0007/79396/E88335.pdf (accessed on 26 January 2021).
16. Marmot, M.; Bloomer, E.; Goldblatt, P. The Role of Social Determinants in Tackling Health Objectives in a Context of Economic Crisis. *Public Health Rev.* **2013**, *35*, 1–24. [CrossRef]
17. Baum, F. *The New Public Health*, 4th ed.; Oxford University Press: Oxford, UK, 2016; p. 720.
18. MacKenbach, J.P. Nordic paradox, Southern miracle, Eastern disaster: Persistence of inequalities in mortality in Europe. *Eur. J. Public Health* **2017**, *27*, 14–17. [CrossRef]
19. Bambra, C.; Gibson, M.; Sowden, A.; Wright, K.; Whitehead, M.; Petticrew, M. Tackling the wider social determinants of health and health inequalities: Evidence from systematic reviews. *J. Epidemiol. Commun. Health* **2009**, *64*, 284–291. [CrossRef] [PubMed]
20. Banski, J.; Pantylej, W.; Janicki, W. The impact of socio-economic transformation on population potential in the eastern Poland. *Geogr. Polonica.* **2014**, *36*, 155.
21. Grshybowskyj, J.L.; Smiianov, V.A.; Myronyuk, I.M.; Lyubinets, O.V. TEN INDICATORS WHICH CHARACTERIZE MEDICAL-DEMOGRAPHIC PROCESSES IN ADJACENT REGIONS OF UKRAINE AND POLAND. *Wiad. Lek.* **2019**, *72*, 868–876. [CrossRef] [PubMed]
22. MacKenbach, J.P.; Kulhánová, I.; Bopp, M.; Borrell, C.; Deboosere, P.; Kovács, K.; Looman, C.W.N.; Leinsalu, M.; Mäkelä, P.; Martikainen, P.; et al. Inequalities in Alcohol-Related Mortality in 17 European Countries: A Retrospective Analysis of Mortality Registers. *PLoS Med.* **2015**, *12*, e1001909. [CrossRef]
23. Moskalewicz, J.; Razvodovsky, Y.; Wieczorek, Ł. East–west disparities in alcohol-related harm. *Alcohol. Drug Addict.* **2016**, *29*, 209–222. [CrossRef]
24. Gruzieva, T.S. Social gradient in the availability of health services. *East. Eur. J. Public Health* **2011**, *1*, 104–105.
25. Lysak, V.P. *Mortality of the Adult Population of Poltava Region due to Diseases of the Endocrine System*; Ukraine Nation's Health "Експерт"; Ministry of Health of Ukraine: Kyiv, Ukraine, 2016; Volume 4, pp. 146–149. Available online: http://nbuv.gov.ua/UJRN/Uzn_2016_4%281%29__27 (accessed on 26 January 2021).
26. Lyubinets, O.; Slabkyj, G. State of Mortality from Cancer Pathology (Comparative Analysis of Ukraine and Poland). In Proceedings of the VII International Scientific Oncology Conference "Prevention, Treatment and Care in Cancer Diseases", Streszcz, Sanok, Poland, 29–30 May 2014; Available online: https://up-sanok.edu.pl/en/1234/dzialalnosc-naukowa/wydawnictwa/materialypokonferencyjne (accessed on 26 January 2021).
27. Chan, G.; Leung, J.; Quinn, C.; Kelly, A.; Connor, J.; Weier, M.; Hall, W. Rural and Urban Differences in Adolescent Alcohol Use, Alcohol Supply, and Parental Drinking. *J. Rural. Health* **2015**, *32*, 280–286. [CrossRef] [PubMed]
28. Spencer, M.R.; Curtin, S.C.; Hedegaard, H. Rates of Alcohol-Induced Deaths among Adults Aged 25 and over in Urban and Rural Areas: United States, 2000–2018. 2020. Available online: https://www.cdc.gov/nchs/products/databriefs/db383.htm (accessed on 26 January 2020).
29. Janssen, F.; El Gewily, S.; Bardoutsos, A.; Trias-Llimós, S. Past and Future Alcohol-Attributable Mortality in Europe. *Int. J. Environ. Res. Public Health* **2020**, *17*, 9024. [CrossRef] [PubMed]
30. Ryngach, N. Statistical Estimation of the Awareness of the Working Age Population about Harmful Consumption of Alcohol as a Risk Factor of Cardiovascular Diseases. *Stat. Ukr.* **2019**, *84*, 70–77. [CrossRef]
31. Whitfield, J.B.; Heath, A.C.; Madden, P.A.F.; Landers, J.G.; Martin, N.G. Effects of high alcohol intake, alcohol-related symptoms and smoking on mortality. *Addiction* **2017**, *113*, 158–166. [CrossRef]
32. Kruchanytsia, V.V. ДЕТЕРМІНАНТИ ЗЛОВЖИВАННЯАЛКОГОЛЕМ НАСЕЛЕННЯМ ЗАКАРПАТТЯ. *Вісник СоціальноїГігієни Та Організації Охорони Здоров'я України* **2017**. [CrossRef]
33. Zatonski, W.A.; Zatonski, M.; Janik-Koncewicz, K.; Wojtyla, A. Alcohol-Related Deaths in Poland During a Period of Weakening Alcohol Control Measures. *JAMA* **2021**, *325*, 1108–1109. [CrossRef]
34. Zatoński, A.W.; Zatoński, M.; Janik-Koncewicz, K.; McKee, M. Alcohol-related liver cirrhosis in Poland: The reservoir effect. *Lancet Gastroenterol. Hepatol.* **2020**, *5*, 1035–1040. [CrossRef]
35. Cholankeril, G.; Dennis, B.B.; Kim, D.; Ahmed, A. Regional Trends in Mortality from Alcohol-Induced Causes in the United States, 2000–2017. *J. Gen. Intern. Med.* **2019**, *35*, 2495–2498. [CrossRef] [PubMed]
36. Batra, A.; Müller, C.A.; Mann, K.; Heinz, A. Alcohol Dependence and Harmful Use of Alcohol: Diagnosis and Treatment Options. *Dtsch. Aerzteblatt Online* **2016**, *113*, 301–310. [CrossRef] [PubMed]
37. Population. *Condition and Structure as well as Natural Movement in the Territorial Profile in 2018 as of 31 XII*; Statistics Poland Warsaw Central Statistical Office: Warsaw, Poland, 2019.

International Journal of
Environmental Research and Public Health

MDPI

Article

The Level of Zinc, Copper and Antioxidant Status in the Blood Serum of Women with Hashimoto's Thyroiditis

Joanna Szczepanik [1], Tomasz Podgórski [2] and Katarzyna Domaszewska [3,*]

1 Faculty of Health Sciences, School of Socio-Economics in Sroda Wielkopolska, 63-000 Sroda Wielkopolska, Poland; tomczewskaj@gmail.com
2 Department of Physiology and Biochemistry, Poznan University of Physical Education, 61-871 Poznan, Poland; podgorski@awf.poznan.pl
3 Faculty of Health Sciences, Calisia University, 62-800 Kalisz, Poland
* Correspondence: domaszewska@onet.pl; Tel.: +48-501-476-716

Abstract: The aim of this study was to analyze selected indicators of oxidative stress. The study subjects consisted of 42 women with Hashimoto's disease and a control group of 30 healthy women. The concentration of zinc (Zn) and copper (Cu) in the serum was determined by Atomic Absorption Spectrometry (AAS) and the total antioxidative potential by the Ferric Reducing Ability of Plasma (FRAP) method. In addition, an assessment of concentrations of thiobarbituric acid reactive substances (TBARS) and total phenolics was carried out. Our research showed a significant difference in TBARS concentration ($p < 0.0001$ (ES: 0.92)) without significant differences in Zn, Cu, FRAP and total phenolics concentrations. Analysis of the correlation of the obtained results of biochemical tests for both groups showed a highly significant dependence of FRAP and total phenolics concentration in the blood of the examined women ($r = 0.5283$, $p = 0.0003$). The obtained results indicate no differences in Cu, Zn, and FRAP concentrations in the blood between two analyzed groups and a significantly higher concentration of TBARS in Hashimoto's thyroiditis women. The concentration of total phenolics significantly influences the value of the FRAP.

Keywords: Hashimoto's disease; oxidative stress; TBARS; zinc; copper

Citation: Szczepanik, J.; Podgórski, T.; Domaszewska, K. The Level of Zinc, Copper and Antioxidant Status in the Blood Serum of Women with Hashimoto's Thyroiditis. *IJERPH* **2021**, *18*, 7805. https://doi.org/10.3390/ijerph18157805

Academic Editors: Wioletta Zukiewicz-Sobczak, Paulina Wojtyla-Buciora, Izabela Rącka and Andrzej Wojtyla

Received: 15 June 2021
Accepted: 20 July 2021
Published: 23 July 2021

Publisher's Note: MDPI stays neutral with regard to jurisdictional claims in published maps and institutional affiliations.

Copyright: © 2021 by the authors. Licensee MDPI, Basel, Switzerland. This article is an open access article distributed under the terms and conditions of the Creative Commons Attribution (CC BY) license (https://creativecommons.org/licenses/by/4.0/).

1. Introduction

Reactive oxygen and nitrogen species (RONS) are used by human organisms for such tasks as signaling, the neutralization of infectious microorganisms, the induction of apoptosis, the stimulation of antioxidants and repair processes [1]. Additionally, their excessive accumulation can impair the main molecules, including proteins, lipids and even nucleic acids, and inhibit their physiological function. In addition, oxidative damage incubated by RONS may play a role in thyroid disease [2]. There is a balance between the formation of RONS and their detoxification, known as redox homeostasis (intracellular reduction–oxidation). The most commonly generated ones are singlet oxygen, superoxide anion radical, hydroxyl radical, hydrogen peroxide, and nitric oxide [3]. Under physiological conditions, there are numerous enzymatic (superoxide dismutase (SOD), catalase (CAT), glutathione peroxidase (GPx) and glutathione reductase (GP)) and non-enzymatic (vitamin C, E, A, glutathione, bilirubin, uric acid, flavonoids, carotenoids, zinc, copper, and selenium) defense systems of the so-called antioxidants (present in erythrocytes of human blood serum, as well as in other biological fluids and tissues) aimed at preventing damage caused by RONS accumulation [3–5].

The inflammatory process within the thyroid, caused by Hashimoto's disease, is at the same time the promoter of increasing the amount of RONS and causing disturbances in the oxidative–antioxidative balance [6]. Imbalance may increase inflammation and cell damage by stimulating the release of proinflammatory cytokines and changing enzyme functions [3,4]. Thyroid hormones play a key role in the regulation of oxidative metabolism.

They increase mitochondrial respiration and produce free radicals. Furthermore, the presence of antioxidant defense system enzymes has been documented in the thyroid (such as SOD, GPX, CAT and non-enzymatic antioxidants) [7]. SOD catalyzes the conversion of superoxide molecules to H_2O_2 and oxygen. Depending on its form, it contains in the active center manganese (MnSOD) or copper together with zinc (Cu/Zn SOD) [8]. Thus, the availability of zinc and copper (SOD) may affect the body's ability to defend against RONS [9]. As demonstrated in thyroid diseases, the level of these elements may be lowered in people with thyroid diseases [10,11]. The total antioxidant status (TAS)—defined as the total body's ability to neutralize RONS—according to many authors, is lower in people with Hashimoto's disease [12–14]. Polyphenols are other phytochemicals that provide protection against oxidative stress and free radicals [15]. Many authors suggest a significant correlation between elevated levels of compounds formed as a result of lipid peroxidation and thyroid diseases [16–18].

Substances reactive with thiobarbituric acid (TBARS) are formed as bioproducts of the oxidation of primarily polyunsaturated fatty acids, which are an integral component of biological membranes. Peroxidation reactions intensify during infection, inflammation, aging, and neurodegenerative and cancer diseases [19–22]. TBARS (including malondialdehyde (MDA)) can be an indicator of the body's exposure to free radicals, and so a measure of oxidative stress.

This study is aimed at evaluating selected indicators of oxidative stress and the concentration of zinc and copper in women with Hashimoto's thyroiditis in comparison to a healthy group.

2. Materials and Methods

2.1. Study Population

The study was conducted according to the Declaration of Helsinki and the National Statement and Human Research Ethics Guidelines and approved by IRB (Institute for Research in Biomedicine) at the Poznan University of Medical Sciences (10 May 2013; Ethics Approval Number: 302/13). All subjects gave their prior consent to take part in it. Participation in the research was voluntary, explained by full and reliable information about the nature, purpose and course of the study, as well as benefits and risks associated with participation.

Hashimoto's disease is one of the most common causes of hypothyroidism, mainly among women aged 45–60 [23]. This is why the study subjects consisted of 42 women with Hashimoto's disease (based on medical examinations, elevated antibody titers). The control group of 30 was selected from among healthy women (selected in terms of sex, age and body weight). Patient exclusion criteria (presence of at least one of the factors listed below): obesity, staying on a vegetarian or any other alternative diet, patients with active or post cancerous disease (ongoing radiation and/or chemotherapy treatment), patients with liver diseases (alanine transaminase (ALT) > 3x border line) except for patients with fatty liver disease, chronic kidney disease eGFR < 30 mL/1.73 m^2/min, acute inflammation c-reactive protein (CRP) > 5 mg/dL, unstable ischemic heart disease, patients after an ischemic or hemorrhagic stroke (<6 months), post STEMI (ST Elevation Myocardial Infarction) patients with a drug-eluting stent implantation, nSTEMI (No ST Elevation Myocardial Infarction) below 12 months, inherited metabolic disorders: phenylketonuria, galactosemia, autoimmune diseases (acute thyroiditis, celiac disease, systemic connective tissue disease, hemolytic anemia, vitiligo, Addison's disease, hyperbilirubinemia), non-specific enteritis (Crohn's disease, ulcerative colitis), pregnancy, psychological disorders, eating disorders such as anorexia and bulimia, antibiotic therapy, steroid therapy (ongoing), drug or alcohol addiction (a daily consumption of more than 1 portion of alcohol). All of the patients were asked not to take minerals (especially zinc and copper) or diet supplements, which could affect the measured biochemical blood parameters in the period before and during the study.

The baseline characteristics of both groups are shown in Table 1. Hashimoto's disease was diagnosed based on value >35 U/L for anti-thyroid peroxidase (anti-TPO) antibodies and >20 U/L for thyroglobulin (anti-TG) antibodies [23,24]. In the group of women with Hashimoto's disease, 45.2% managed to balance the level of hormones regulating thyroid function. A total of 76% of patients took synthetic thyroxine. After substitution treatment, the elevated level of thyroid-stimulating hormone (TSH) was still present in 11.9% of cases (normal range for TSH: 0.35–2.80 mIU/L), elevated anti-TPO titer in 50% and elevated anti-TG titer in 37.5%.

Table 1. Anthropometric characteristics and comparison of zinc and copper levels and antioxidant potential in women's serum.

Median (Interquartile Range).			
Parameters	**Study Group (n = 42)**	**Control Group (n = 30)**	**p Value**
Age (years)	40.0 ± 32.0–49.0	41.0 ± 34.0–35.0	0.3391
Body weight (kg)	61.0 ± 55.0–70.0	64.5 ± 58.0–72.0	0.1914
Body height (cm)	168.5 ± 163.0–172.0	168.0 ± 163.0–170.0	0.9905
BMI (kg/m^2)	23.2 ± 19.3–24.7	22.1 ± 20.4–26.3	0.2881
Zn (μg/dL)	88.0 ± 82.0–95.0	86.5 ± 76.0–104.0	0.7217
Cu (μg/dL)	110.5 ± 100.0–122.0	110.0 ± 99.0–136.0	0.6908
FRAP (μmol/L)	694.1 ± 559.5–835.2	733.6 ± 617.4–855.4	0.4949
TBARS (μmol/L)	26.3 ± 20.1–35.6	3.2 ± 2.6–4.4	0.0000
Total phenolics (g GAE/L)	2.9 ± 2.6–3.2	3.1 ± 2.8–3.2	0.1857

Abbreviations: FRAP = Ferric Reducing Ability of Plasma, TBARS = thiobarbituric acid reactive substances, GAE = gallic acid.

The average length of the illness is 7.7 years, and 19% are people diagnosed with the disease in less than one year.

The most frequently recorded health problems among women with Hashimoto's disease are: weakness, fatigue, drowsiness, concentration disorders, mood swings, insomnia, lower libido, dry skin, headaches, problems with getting pregnant, intestinal problems.

2.2. Preparation of Blood Samples for Analysis

The subjects were asked to stop supplementation that could potentially affect the result. In the day before the examination, they were asked not to perform physical activity. Blood samples were taken from the ulnar vein using a S-Monovette syringe (Sarstedt, Nümbrecht, Germany), then placed in tubes containing a clot activator and centrifuged (1500× g, 4 °C, 4 min) to separate the serum (Universal 320R; Hettich Lab Technology, Tuttlingen, Germany). Samples were frozen and stored at −80 °C until analysis (U410, Ultra Low Temperature Freezer, New Brunswick Scientific, Enfield, CT, USA).

2.3. Determination of Copper and Zinc

The concentration of zinc and copper in the serum was determined by atomic absorption spectrometry (Atomic Absorption Spectrometry, AAS).

For this purpose, a Perkin-Elmer, Model 3030 Zeeman flameless absorption spectrometer equipped with a HGA-600 graphite furnace and an AS-40 (Perkin-Elmer, Norwalk, CT, USA) automated sample dispenser was used.

Determination of copper and zinc was made in accordance with generally accepted standards: at the wavelength for zinc (213.9 nm), and for copper (324.8 nm). The accuracy of the method was determined in relation to material certified for Seron Trace Elements Serum (Nycomed Pharma As, Oslo, Norway) and for a measurement error of 3–5%. The measurement was repeated three times. The volume of the sample fed into the graphite cuvette of the sample was 20 μL. Reference values: zinc 70–120 μg/dL, copper 85–155 μg/dL [25].

2.4. Determination of Antioxidant Status

In order to determine the concentrations of total antioxidative capacity of plasma, the following methods were used: colorimetry (ability to reduce plasma iron concentration, FRAP, reference values: 600–1600 µmol/L) [26], concentration of reactive substances with thiobarbituric acid (TBARS, reference values: 1–6 µmol/L) [27] and total phenolic compounds (reference values: 2.8–4.0 (g GAE, gallic acid/L)) [28]. The samples were read using a multi-detection microplate reader (Synergy 2 SIAFRT BioTek, Winooski, VT, USA).

2.5. Statistical Analysis

All data are presented as median (interquartile range). Distribution was tested with the Shapiro–Wilk test for normality. Differences between variables were examined using the Mann–Whitney test. The relationship between the variables was tested while using Spearman's rank correlation. The significance level for all statistical analysis was set at $p \leq 0.05$. All results were statistically analyzed using Dell Inc. (2016), Dell Statistica v.13.—soft-ware.dell.com (Statistica 13, Statsoft, Dell, Tulsa, OK, USA). Effect sizes (ES) were calculated as the difference between means divided by the pooled standard deviation. Using Cohen's (1988) criteria, an effect size $\geq$0.20 and <0.50 was considered small, $\geq$0.50 and <0.80 medium, and $\geq$0.80 large [29].

3. Results

Table 1 presents the anthropometric characteristics of the studied women, the average concentrations of the examined microelements and the concentration of indicators of reductive–oxidative status in women's serum.

Comparative analysis of the level of biochemical markers determined in the blood at rest between two groups showed a significant difference in TBARS concentration (Mann–Whitney U-test, $p < 0.0001$, (ES: 0.92)), with no significant differences in Zn, Cu, FRAP and total phenolics concentrations. The TBARS level was significantly higher for women with Hashimoto's disease (26.3 µmol/L compared to 3.2 µmol/L among healthy women). Analysis of the correlation of the obtained results of biochemical tests for both groups showed a highly significant dependence (positive correlation) of FRAP and total phenolics concentration in the blood of the examined persons ($r = 0.5283$, $p = 0.0003$). Additionally, a significant correlation was demonstrated between TBARS concentration and the presence of anti-TG in the studied group ($r = 0.03312$, $p = 0.0321$). There was no relationship between Zn and Cu concentration and oxidative stress indicators and serum anti-TPO and anti-TG levels.

4. Discussion

In our study, the most significant differences between people with Hashimoto's disease and healthy people were related to the level of substances reactive with thiobarbituric acid, i.e., those resulting from lipid peroxidation (damage). We showed statistically significant differences in the TBARS level in both groups. For women with Hashimoto's disease, the level of compounds formed as a result of lipid damage was significantly higher. Malondialdehyde (MDA) is the most significant among TBARS. In tissues, an increase in MDA concentration is observed depending on the increased production of RONS, and the resulting aldehyde has cytotoxic, mutagenic and carcinogenic effects [8]. Chakrabarti et al. also used MDA concentration assessment as a marker of oxidative stress [17]. These authors found that MDA levels were higher in patients with hypothyroidism prior to levothyroxine treatment and/or selenium supplementation than in the control group. MDA concentration has also been found to decrease after treatment and/or supplementation in patients with hypothyroidism. In addition, they obtained a significant positive correlation between the MDA level and baseline TSH values [17]. Other authors also obtained results indicating a higher level of MDA in patients with hypothyroidism [16,18].

In our study, serum zinc and copper levels were lower in women with Hashimoto's disease compared to the control group, but the differences were not statistically significant.

Borawska et al. suggest that in people with Hashimoto's thyroiditis, the level of zinc in blood serum is reduced and it may be related to the ongoing inflammation of the thyroid gland and may result from insufficient intake of this element with the diet. In addition, they pointed out that the increase in anti-TPO titers was inversely correlated with the level of zinc in the blood serum of the studied women [30]. Therefore, with the decrease in serum zinc concentration, the titer of antithyroid antibodies increases, which may confirm the role of zinc in the functioning of the body's immune defense [30,31]. Free triiodothyronine (FT3) and thyroxine (FT4) need zinc to fulfill their biological activity (as selenium and iodine), and a deficiency of this element negatively affects the metabolic activity of these hormones [32]. It is also possible that changes in the pool of stored elements, such as zinc, selenium and iodine, in the thyroid gland may affect the function of this gland depending on the secretion of TSH by the pituitary gland, responsible for the regulation of T3 and T4 hormones [33]. The studies showed that high levels of copper were associated with higher levels of both thyroid hormones—T3 and T4 [34]. Kucharzewski et al. proved that persons with thyroid disease (thyroid cancer, Graves-Basedow disease and nodular goiter) have significantly higher levels of copper in the blood compared to healthy people [35]. Sinha et al. obtained results that also indicate an elevated level of copper in people with hyperthyroidism [36]. Rasic-Milutinovic et al. showed that the concentration of copper in people with Hashimoto's disease was significantly higher compared to healthy people. At the same time, they suggest that the ratio of copper and selenium may affect the level of thyroid hormones, and a higher selenium level and reduced copper may promote a reduction in L-thyroxine or cause euthyreosis at lower FT4 values [11]. Mittag et al. concluded that the association between copper and selenium in the blood serum may be a marker of thyroid hormone resistance (RTH) in adults [37]. Al-Juboori et al., however, did not find differences in the level of copper in the blood between patients with hypothyroidism and healthy people. They suggest that more research is required to determine whether the level of copper can affect the level of thyroid hormones [38,39].

At the same time, there is no coherence in the results of studies assessing the level of oxidative stress and the concentration of thyroid hormones in the blood [40]. Indicators of oxidative stress in hypothyroidism may be increased [41–44], reduced [45] or unchanged [46], whereas in subclinical hypothyroidism (defined as a high level of TSH at the normal values of FT3 and FT4, which is usually the initial stage of Hashimoto's disease), the knowledge of oxidative stress is limited [2].

The FRAP assay includes many possible antioxidants present in blood: albumin, uric acid, bilirubin, vitamins C and E and phenolics. We observed correlation between FRAP and total phenolics for both groups. Reddy et al. found that FRAP was significantly reduced in patients with overt and subclinical hypothyroidism [47]. It was found that both hyperthyroidism and hypothyroidism are associated with increased oxidative stress, increased production of free radicals and oxidants has been shown [40,42,48–50], and at the same time, reduced resistance to oxidation has also been shown [42,48,51]. The most important effects of oxidative stress are: damage to mitochondria, a decrease in ATP (adenosine triphosphate) and glutathione, the breakdown of red blood cells, intracellular calcium homeostasis, the inactivation of some proteins, increased adenine nucleotide catabolism, an increase in the lipid peroxidation rate, depolarization, an increase in cell membrane permeability, DNA damage, the acceleration of cell apoptosis and changes in their functioning [8].

Recent research presented by Ruggeri et al. shows that in people with Hashimoto's disease the biological antioxidant potential is lowered and the level of reactive oxygen metabolites is elevated compared to healthy people. They proved that the oxidative stress index (estimated as the ratio of the total amount of oxidants and antioxidants) in people with Hashimoto's disease was statistically higher compared to the control group [12]. The results from 2019 confirm those obtained by the above-mentioned authors three years earlier [13], which suggests a clear imbalance between the endogenous production of free radicals and antioxidant defense, or the occurrence of oxidative stress in patients with thyroid disease, especially autoimmune inflammation of this gland [12]. An increased amount of RONS in the system of people with Hashimoto's disease may be due to a decrease in the synthesis of enzymes with antioxidant activity (superoxide dismutase and glutathione), which is the result of lowering thyroid hormone levels. Furthermore, it is also known that hyperlipidemia, which develops with decreasing thyroid hormone levels, leads to an increase in RONS [14]. Total antioxidant status (TAS) and total oxidant status (TOS) reflect the general state of redox balance in the system [14]. Ates et al. compared the levels of TAS, TOS and the oxidative stress index (OSI) in the group of people with Hashimoto's disease with overt and subclinical hypothyroidism, with euthyroidism and in healthy people. They obtained results in which TOS and OSI increased significantly in each phase of the disease and TAS decreased. In addition, there was a negative correlation between the level of antithyroid antibodies and the overall oxidative level [14,52,53]. The authors suggest that this is the first study that assesses oxidative stress at various levels of Hashimoto's disease [14]. Similar results were obtained by other researchers. Baser et al. concluded that people with Hashimoto's euthyroid disease had lower TAS and higher levels of TOS compared to healthy subjects, suggesting the role of oxidative stress in thyroid autoimmunity [51]. In the Wang et al. study, designed to determine the level of oxidative stress in patients with thyroid cancer, Graves' disease, Hashimoto's disease and the control group, TAS was lower, while TOS and OSI were higher in hypothyroidism than in a healthy control group [54]. Researchers suggest that the role of inflammation in oxidative stress can be explained in two ways: first, inflammation directly increases the level of hydrogen peroxide in thyroid epithelial cells, and second, it activates enzymes from the NADPH oxidase family in T and B lymphocytes that increase RONS production [54]. Another possible cause is that hypothyroidism is associated with a lower production of hormones secreted by this gland, and it is known that these hormones affect the synthesis and biological activity of antioxidative enzymes [55,56]. Increased levels of hormones as a result of a synthetic levothyroxine may reduce oxidative stress [55]. In a mentioned study conducted by Reddy et al., antioxidant defense in subclinical and overt hypothyroidism was evaluated and the reduction in antioxidant defense in overt hypothyroidism was found to be due to both reduced antioxidative enzyme synthesis and the low activity of these enzymes [47]. Thus, the possibility of preventing the development of overt hypothyroidism among people with Hashimoto's disease by providing exogenous antioxidants should be considered. Additionally, future research projects in this direction should be carried out [14]. It is also required to investigate whether oxidative stress is the cause or the result of Hashimoto's disease. Research results suggested that a lower value of TBARS is not dependent on a decrease in Zn and Cu values.

5. Conclusions

The results from this study indicate that:

1. There are no differences in the concentration of Cu and Zn in the blood of people with Hashimoto's disease compared to the control group.
2. In the blood of patients with Hashimoto's disease, a significantly higher TBARS concentration was found, with no difference in FRAP concentration.
3. The concentration of total phenolics has a significant positive effect on the value of the FRAP indicator.

6. Limitation of the Study

The study was performed on a small group of women. The study compared selected markers of oxidative stress in people with Hashimoto's disease and healthy people; however, the relationship between the level of thyroid hormones and the assessed parameters was not analyzed in the study group. Future research is required to assess whether specific thyroid hormone values may modulate the level of the studied parameters and how this substitution affects treatment.

Author Contributions: Conceptualization, J.S. and K.D.; methodology, K.D., J.S., T.P.; software, K.D.; validation, K.D. and J.S.; resources, K.D., J.S.; data curation, T.P., J.S., K.D.; writing—original draft preparation, J.S., K.D.; writing—review and editing, K.D.; visualization, J.S., K.D.; supervision, K.D.; project administration, K.D., J.S. All authors have read and agreed to the published version of the manuscript.

Funding: This research received no external funding.

Institutional Review Board Statement: The study was conducted according to the guidelines of the Declaration of Helsinki and approved by the IRB (Institute for Research in Biomedicine) at the Poznan University of Medical Sciences (10 May 2013; Ethics Approval Number: 302/13).

Informed Consent Statement: Informed consent was obtained from all subjects involved in the study.

Data Availability Statement: The data presented in this study are available on request from the corresponding author. The data are not publicly available due to the consent provided by participants on the use of confidential data.

Conflicts of Interest: The authors declare no conflict of interest.

References

1. Pani, G.; Calavitti, R.; Bedogni, B.; Anzevino, S.; Borrello, S.; Galotti, T. A redox signaling mechanism for density dependent inhibition of cell growth. *JBC* **2000**, *275*, 38891–38899. [CrossRef] [PubMed]
2. Öztürk, Ü.; Vural, P.; Özderya, A.; Karadag, B. Oxidative stress parameters in serum and low density lipoproteins of Hashimoto's thyroiditis patients with subclinical and overt hypothyroidism. *Int. Immunopharmacol.* **2012**, *14*, 349–352. [CrossRef]
3. Valko, M.; Leibfritz, D.; Moncol, J.; Cronin, M.T.; Mazur, M.; Telser, J. Free radicals and antioxidants in normal physiological functions and human disease. *Int. J. Biochem. Cell. Biol.* **2007**, *39*, 44–84. [CrossRef] [PubMed]
4. McCord, J.M. Human disease, free radicals, and the oxidant/antioxidant balance. *Clin. Biochem.* **1993**, *26*, 351–357. [CrossRef]
5. Brown, N.S.; Bicknell, R. Hypoxia and oxidative stress in breast cancer. Oxidative stress-its effects on the growth, metastatic potential and response to therapy of breast cancer. *Breast Cancer Res.* **2001**, *3*, 323–327. [CrossRef]
6. Lassoued, S.; Mseddi, M.; Mnif, F.; Abid, M.; Guermazi, F.; Masmoudi, H.; El Feki, A.; Attia, H. A comparative study of the oxidative profile in Graves' disease, Hashimoto's thyroiditis, and papillary thyroid cancer. *Biol. Trace Elem. Res.* **2010**, *138*, 107–115. [CrossRef]
7. Mancini, A.; Raimondo, S.; Di Segni, C.; Persano, M.; Gadotti, G.; Silvestrini, A.; Festa, R.; Tiano, L.; Pontecorvi, A.; Meucci, E. Thyroid hormones and antioxidant systems: Focus on oxidative stress in cardiovascular and pulmonary diseases. *Int. J. Mol. Sci.* **2013**, *14*, 23893–23909. [CrossRef] [PubMed]
8. Sies, H.; Cadenas, E. Oxidative stress: Damage to intact cells and organs. *Philos. Trans. R. Soc. Lond. B. Biol. Sci.* **1985**, *311*, 617–631.
9. Włochal, M.; Kucharski, M.A.; Grzymisławski, M. The effects of vitamins and trace minerals on chronic autoimmune thyroiditis. *JMS* **2014**, *2*, 167–172. [CrossRef]
10. Kandhro, G.A.; Kazi, T.G.; Afridi, H.I.; Kazi, N.; Baig, J.A.; Arain, M.B.; Sirajuddin; Shah, A.Q.; Sarfraz, R.A.; Jamali, M.K.; et al. Effect of zinc supplementation on the zinc level in serum and urine and their relation to thyroid hormone profile in male and female goitrous patients. *Clin. Nutr.* **2009**, *28*, 162–168. [CrossRef]
11. Rasic-Milutinovic, Z.; Jovanovic, D.; Bogdanovic, G.; Trifunovic, J.; Mutic, J. Potential Influence of Selenium, Copper, Zinc and Cadmium on L-Thyroxine Substitution in Patients with Hashimoto Thyroiditis and Hypothyroidism. *Exp. Clin. Endocrinol. Diabetes.* **2017**, *125*, 79–85. [CrossRef] [PubMed]
12. Ruggeri, R.M.; Jovanovic, D.; Bogdanovic, G.; Trifunovic, J.; Mutic, J. Increased serum interleukin α37 (IL α37) levels correlate with oxidative stress parameters in Hashimoto's thyroiditis. *J. Endocrinol. Invest.* **2019**, *42*, 199–205. [CrossRef]
13. Ruggeri, R.M.; Vicchio, T.M.; Cristani, M.; Certo, R.; Caccamo, D.; Alibrandi, A.; Giovinazzo, S.; Saija, A.; Campennì, A.; Trimarchi, F.; et al. Oxidative stress and advanced Glycation End Products (AGES) in Hashimoto's thyroiditis. *Thyroid.* **2016**, *26*, 504–511. [CrossRef] [PubMed]
14. Ates, I.; Yilmaz, F.M.; Altay, M.; Yilmaz, N.; Berker, D.; Güler, S. The Relationship between Oxidative Stress and Autoimmunity in Hashimoto's Thyroiditis. *Eur. J. Endocrinol.* **2015**, *173*, 791–799.
15. Cory, H.; Passarelli, S.; Szeto, J.; Tamez, M.; Mattei, J. The Role of Polyphenols in Human Health and Food Systems: A Mini-Review. *Front Nutr.* **2018**, *21*, 87. [CrossRef]
16. Baskol, G.; Atmaca, H.; Tanriverdi, F.; Baskol, M.; Kocer, D.; Bayram, F. Oxidative stress and enzymatic antioxidant status in patients with hypothyroidism and after treatment. *Exp. Clin. Endocrinol. Diabetes.* **2007**, *115*, 522–526. [CrossRef]
17. Chakrabarti, S.K.; Ghosh, S.; Banerjee, S.; Mukherjee, S.; Chowdhury, S. Oxidative stress in hypothyroid patients and the role of antioxidant supplementation. *Indian J. Endocrinol. Metab.* **2016**, *20*, 674–678. [CrossRef]
18. Lakshmi, L.J.; Mohapatra, E.; Doddigarla, Z.; Kumari, S. Serum lipids and oxidative stress in hypothyroidism. *Int. J. Adv. Res. Biol. Sci.* **2013**, *5*, 63–66.

19. Agha, A.M.; Gad, M.Z. Lipid peroxidation and lysosomal integrity in different inflammatory models in rats: The effects of indomethacin and naftazone. *Pharmacol. Res.* **1995**, *32*, 279–285. [CrossRef]
20. Banerjee, B.D.; Seth, V.; Bhattacharya, A.; Pasha, S.T.; Chakraborty, A.K. Biochemical effects of some pesticides on lipid peroxidation and free-radical scavengers. *Toxicol. Lett.* **1999**, *107*, 33–47. [CrossRef]
21. Bartsch, H. Keynote address: Exocyclic adducts as new risk markers for DNA damage in man. *IARC Sci. Publ.* **1999**, *150*, 1–16.
22. Uchida, K. 4-Hydroxy-2-nonenal: A product and mediator of oxidative stress. *Prog. Lipid Res.* **2003**, *42*, 318–343. [CrossRef]
23. Ajjan, A.R.; Weetman, A.P. The pathogenesis of Hashimoto's thyroiditis: Further developments in our understanding. *Horm. Metab. Res.* **2015**, *47*, 702–710. [CrossRef]
24. Pop, V.; Krabbe, J.; Maret, W.; Rayman, M. Plasma mineral (selenium, zinc or copper) concentrations in the general pregnant population, adjusted for supplement intake, in relation to thyroid function. *Br. J. Nutr.* **2021**, *125*, 71–78. [CrossRef]
25. Forrer, R.; Gautschi, K.; Lutz, H. Simultaneous measurement of the trace elements Al, As, B, Be, Cd, Co, Cu, Fe, Li, Mn, Mo, Ni, Rb, Se, Sr, and Zn in human serum and their reference ranges by ICP-MS. *Biol. Trace Elem. Res.* **2001**, *80*, 77–93. [CrossRef]
26. Benzie, I.; Strain, J. The Ferric Reducing Ability of Plasma (FRAP) as a Measure of "Antioxidant Power: The FRAP Assay". *Anal. Biochem.* **1996**, *239*, 70–76. [CrossRef]
27. Ohkawa, H.; Ohishi, N.; Yagi, K. Assay for lipid peroxides in animal tissues by thiobarbituric acid reaction. *Anal. Biochem.* **1979**, *95*, 351–358. [CrossRef]
28. Singleton, V.; Rossi, J. Colorimetry of Total Phenolic Compounds with Phosphomolybdic-Phosphotungstic Acid Reagents. *Am. J. Enol. Vitic.* **1965**, *16*, 144–158.
29. Cohen, J. *Statistical Power Analysis for the Behavioral Sciences*, 2nd ed.; Lawrence Erlbaum Associates: Hillsdale, NJ, USA, 1988.
30. Borawska, M.; Markiewicz-Żukowska, R.; Dziemianowicz, R.; Socha, K.; Soroczyńska, J. Wpływ nawyków żywieniowych i palenia papierosów na stężenie cynku w surowicy krwi kobiet z chorobą Hashimoto. *Bromat. Chem. Toksykol.* **2012**, *3*, 759–765.
31. Turnlund, J.R.; Jacob, R.A.; Keen, C.J.; Stain, J.J.; Darshan, S.K.; Domek, J.M.; Keyes, W.R.; Ensunsa, J.L.; Lykkesfeldt, J.; Coulter, J. Long-term high copper intake: Effects on indexes of copper status, antioxidant status, and immune function in young men. *Am. J. Clin. Nutr.* **2004**, *79*, 1037–1044. [CrossRef]
32. Freake, H.C.; Govoni, K.E.; Guda, K.; Huang, C.; Zinn, S.A. Actions and interactions of thyroid hormone and zinc status in growing rats. *JN* **2001**, *131*, 1135–1141. [CrossRef]
33. Bellisola, G.; Brätter, P.; Cinque, G.; Francia, G.; Galassini, S.; Gawlik, D.; de Brätter, V.E.N.; Azzolina, L. The TSH dependent variation of the essential elements iodine, selenium and zinc within human thyroid tissues. *J. Trace Elem. Med. Biol.* **1998**, *12*, 177–182. [CrossRef]
34. Jain, R.B. Thyroid function and serum copper, selenium, and zinc in general US population. *Biol. Trace Elem. Res.* **2014**, *159*, 87–98. [CrossRef]
35. Kucharzewski, M.; Braziewicz, J.; Majewska, U.; Góźdz, S. Copper, Zinc, and Selenium in Whole Blood and Thyroid Tissue of People with Various Thyroid Diseases. *Biol. Trace Elem. Res.* **2003**, *93*, 9–18. [CrossRef]
36. Sinha, S.; Kar, K.; Disgupta, A.; Basu, S.; Sen, S. Correlation of Serum zinc with TSH in hyperthyroidism. *AJMS.* **2015**, *7*, 66–69. [CrossRef]
37. Mittag, J.; Behrends, T.; Nordström, K.; Anselmo, J.; Vennström, B.; Schomburg, L. Serum copper as a novel biomarker for resistance to thyroid hormone. *Biochem. J.* **2012**, *443*, 103–109. [CrossRef] [PubMed]
38. Al-Juboori, I.A.; Al-Ravi, R.; A-Hakeim, H.K. Estimation of Serum Copper, Manganese, Selenium, and Zinc in Hypothyroidism Patients. *IUFS J. Biol.* **2009**, *68*, 121–126.
39. Khan, S.; Lungba, R.M.; Ajibawo-Aganbi, U.; Veliginti, S.; Perez Bastidas, M.V.; Saleem, S.; Cancarevic, I. Minerals: An Untapped Remedy for Autoimmune Hypothyroidism? *Cureus* **2020**, *12*, e11008. [CrossRef]
40. Rostami, R.; Aghasi, M.R.; Mohammadi, A.; Nourooz-Zadeh, J. Enhanced oxidative stress in Hashimoto's thyroiditis: Inter-relationships to biomarkers of thyroid function. *Clin. Biochem.* **2013**, *46*, 308–312. [CrossRef] [PubMed]
41. Lampka, M.; Junik, R.; Nowicka, A.; Kardymowicz, H.; Kaczorowski, P.; Tyrakowski, T. Evaluation of low density lipoprotein oxidation in a course of hypothyroidism. *Pol. J. Endokrynol.* **2006**, *57*, 116–121.
42. Erdamar, H.; Demirci, H.; Yaman, H.; Erbil, M.K.; Yakar, T.; Sancak, B.; Elbeg, S.; Biberoğlu, G.; Yetkin, I.; Yaman, H. The effect of hypothyroidism, hyperthyroidism, and their treatment on parameters of oxidative stress and antioxidant status. *Clin. Chem. Lab. Med.* **2008**, *46*, 1004–1010. [CrossRef] [PubMed]
43. Nanda, N.; Bobby, Z.; Hamide, A.; Koner, B.C.; Sridhar, M.G. Association between oxidative stress and coronary risk factors in hypothyroid women is independent of body mass index. *Metabolism* **2007**, *56*, 1350–1355. [CrossRef]
44. Sarandöl, E.; Taş, S.; Dirican, M.; Serdar, Z. Oxidative stress and serum paraoxonase activity in experimental hypothyroidism: Effect of vitamin E supplementation. *Cell Biochem. Funct.* **2005**, *23*, 1–8. [CrossRef]
45. Venditti, P.; Balestrieri, M.; Di Meo, S.; De Leo, T. Effect of thyroid state on lipid peroxidation, antioxidant defences, and susceptibility to oxidative stress in rat tissues. *J. Endocrinol.* **1997**, *155*, 151–157. [CrossRef]
46. Coria, M.J.; Pastran, A.I.; Gimenez, M.S. Serum oxidative stress parameters of women with hypothyroidism. *Acta Biomed.* **2009**, *80*, 135–139.
47. Resch, U.; Helsel, G.; Tatzber, F.; Sinzinger, H. Antioxidant status in thyroid dysfunction. *Clin. Chem. Lab. Med.* **2002**, *40*, 1132–1134. [CrossRef]

48. Reddy, V.S.; Gouroju, S.; Suchitra, M.M.; Suresh, V.; Sachan, A.; Srinivasa Rao, P.V.; Bitla, A.R. Antioxidant defense in overt and subclinical hypothyroidism. *Horm. Metab. Res.* **2013**, *45*, 754–758. [CrossRef]
49. Aslan, M.; Cosar, N.; Celik, H.; Aksoy, N.; Dulger, A.C.; Begenik, H.; Soyoral, Y.U.; Kucukoglu, M.E.; Selek, S. Evaluation of oxidative status in patients with hyperthyroidism. *Endocrine* **2011**, *40*, 285–289. [CrossRef]
50. Marcocci, C.; Bartalena, L. Role of oxidative stress and selenium in Graves' hyperthyroidism and orbitopathy. *J. Endocrinol. Investig.* **2013**, *36*, 15–20.
51. Torun, A.N.; Kulaksizoglu, S.; Kulaksizoglu, M.; Pamuk, B.O.; Isbilen, E.; Tutuncu, N.B. Serum total antioxidant status and lipid peroxidation marker malondialdehyde levels in overt and subclinical hypothyroidism. *Clin. Endocrinol.* **2009**, *270*, 469–474. [CrossRef] [PubMed]
52. Baser, H.; Can, U.; Baser, S.; Yerlikaya, F.H.; Aslan, U.; Hidayetoglu, B.T. Assessment of oxidative status and its association with thyroid autoantibodies in patients with euthyroid autoimmune thyroiditis. *Endocrine.* **2014**, *48*, 916–923. [CrossRef] [PubMed]
53. Nanda, N.; Bobby, Z.; Hamide, A.A. Oxidative stress and protein glycation in primary hypothyroidism. Male/female difference. *Clin. Exp. Med.* **2008**, *8*, 101–108. [CrossRef]
54. Wang, D.; Feng, J.F.; Zeng, P.; Yang, Y.H.; Luo, J.; Yang, Y.W. Total oxidant/antioxidant status in sera of patients with thyroid cancers. *Endocr. Relat. Cancer.* **2011**, *18*, 773–782. [CrossRef]
55. Pereira, B.; Rosa, L.C.; Safi, D.A.; Bechara, E.J.H.; Curi, R. Control of superoxide dismutase, catalase and glutathione peroxidase activities in rat lymphoid organs by thyroid hormones. *J. Endocrinol.* **1994**, *140*, 73–77. [CrossRef] [PubMed]
56. Fernandez, V.; Videla, L.A. Hepatic glutathione biosynthetic capacity in hyperthyroid rats. *Toxicol. Lett.* **1996**, *89*, 85–89. [CrossRef]

International Journal of
Environmental Research and Public Health

MDPI

Article

Application of Blue Filters Increases the Usefulness of Moreland Test in Anomaloscopic Color Vision Assessment for Blue–Green Color Range

Krzysztof Piotr Michalak [1,*], Jacek Zabel [1,2], Jan Olszewski [2], Paulina Wojtyła-Buciora [3] and Anna Przekoracka-Krawczyk [1]

1 Laboratory of Vision Science and Optometry, Faculty of Physics, Adam Mickiewicz University in Poznan, Uniwersytetu Poznanskiego Street 2, 61-614 Poznan, Poland; jacekz@post.pl (J.Z.); ania_pk@amu.edu.pl (A.P.-K.)
2 Department of Bionics and Experimental Medical Biology, Karol Marcinkowski University of Medical Sciences in Poznan, Parkowa Street 2, 60-775 Poznań, Poland; jolszewski@ump.edu.pl
3 Department of Health Sciences, Calisia University-Kalisz, Nowy Swiat Street 4, 62-800 Kalisz, Poland; paulinawojtyla@gmail.com
* Correspondence: kmichalak@amu.edu.pl

Abstract: The effect of blue light filters on the anomaloscopic examination was analyzed. Thirty subjects (18–43 y, 20 female, 10 male) without color vision disorders were examined in 4 filter conditions: no filter (F-0), Blue Control Hoya (F-BC), Med-1 JZO (F-Med1) and 450 Eschenbach (F-450). Both Rayleigh test (red–green axis) and Moreland test (blue–green axis) were performed. Application of F-BC filter shows negligible effect on color vision perception in both tests. Contrary to this, the application of strong F-450 filter causes significant shift in Moreland test towards tritanopy and the decrease in correlations of Moreland parameters with Rayleigh test parameters. The application of medium strong F-Med1 filter causes the slight shift in Moreland test towards the center of the Moreland scale and increases the Spearman correlations between Moreland and Rayleigh test parameters. This observation suggests that the about 15–40% reduction of blue diode intensity in the Moreland test may be beneficial in detecting mild changes in color vision perception in the blue-green axis and may improve its usefulness in evaluating the color vision perception disorders accompanying different illnesses, such as diabetes, glaucoma, neuritis optica, or cataract. The discussion concerning the modifications of Moreland test construction is also presented.

Keywords: color vision; anomaloscope; Moreland; Rayleigh

Citation: Michalak, K.P.; Zabel, J.; Olszewski, J.; Wojtyła-Buciora, P.; Przekoracka-Krawczyk, A. Application of Blue Filters Increases the Usefulness of Moreland Test in Anomaloscopic Color Vision Assessment for Blue–Green Color Range. *IJERPH* **2021**, *18*, 7654. https://doi.org/10.3390/ijerph18147654

Academic Editor: Paul B. Tchounwou

Received: 16 June 2021
Accepted: 14 July 2021
Published: 19 July 2021

Publisher's Note: MDPI stays neutral with regard to jurisdictional claims in published maps and institutional affiliations.

Copyright: © 2021 by the authors. Licensee MDPI, Basel, Switzerland. This article is an open access article distributed under the terms and conditions of the Creative Commons Attribution (CC BY) license (https://creativecommons.org/licenses/by/4.0/).

1. Introduction

Color vision disorders occur among the symptoms of numerous diseases of the human eye, including glaucoma [1,2], cataract [3,4], optic neuritis [5], and certain systemic diseases, such as diabetes mellitus [6,7]. Changes in color perception may affect the areas of red, green, and blue related to three types of cones (L, M, and S) in the retina, which are responsible for perception of three primary colors. Although color vision disorders are observed in the above diseases, they are not used to assess their severity. Such approach may be explained by several reasons. The first and the most important one is the scarcity of scientific research describing color vision deficits in particular diseases in a quantitative manner, the second is the scarcity of standardized diagnostic methods that would quantify color vision deficits, in particular in the green/blue range. Another issue is the inter-subject variability of color perception, which makes defining norms for healthy individuals ambiguous in certain cases.

Such variability might be related to:

(a) variability of primary structure of receptor proteins and minor peak shifts or other changes in their absorption spectra [8],

(b) variable number of cones of each type in the retina and the macula [9],
(c) differences in the structure or functioning of the retina's neuronal network, which codes information concerning the activity of each cone type and transfers the color information to the central nervous system [10],
(d) differences in color perception at the brain cortex level.

Particularly, aspects (c) and (d) above have not been well-researched so far in the literature.

Despite these limitations, developing a methodology for quantifying color perception and analyzing changes in color perception in individual diseases is an important direction of research. In the studies carried out to date in our laboratory, the researchers have assessed the anomaloscopic test, which enables quantitative assessment of color perception in the red/green (Rayleigh test) and green/blue (Moreland test) [11] axes. The Rayleigh test is well standardized and the results are comparable. This test is primarily used to quantify deficiencies of red–green perception. In contrast, the Moreland test has been poorly studied and is practically not used for diagnostic purposes in clinical practice. The Rayleigh test is based on a comparison of a subjective impression of the color yellow (produced by mixing green and red) with a standardized yellow color. As a result of the measurement procedure, border values of light intensity ratio are defined for the red and the green diodes (R_1 and R_2). The R value can range between 0 (100% green diode) and 73 (100% red diode) anomaloscope units [12]. The diodes allow the subject to perceive the test color as being identical to the reference color. An additional parameter is the brightness of the yellow reference color at the color adjustment borders (Y_1, Y_2). Similarly, the Moreland test is based on comparing the subjective impression of the cyan in the test field (resulting from mixing the light emitted by green and blue diodes) with the standard cyan diode. The resulting measurement values are the border values of intensity for the blue and the green diode, which give make the subject perceive the test color as being identical to the standard cyan color (M_1 and M_2). Another test outcome is the brightness of the reference cyan color at the time of shade equalization (C_1, C_2). The M value can range between 0 (100% green diode) and 100 (100% blue diode) anomaloscope units.

Recently, Zabel et al. [11] presented the results of anomaloscopic measurements of Polish subjects aged below 45 years without color vision disorders, as determined by the HRR method. The subjects were free from eye diseases that might affect changes in color perception. The Rayleigh test results obtained by the research team, as compared to the standards recommended by the anomaloscope manufacturer, were slightly shifted towards protanomaly. The recommended standard according to the anomaloscope user's manual is 34–46 units [12], while the results obtained in the study (± 2 std) were 35.4–51.3 for females and 36.8–54.8 for males. Additionally, it was observed that the distribution of the R_2 parameter was characterized by two maxima with a minimum between the peaks located around the value of 50. Values above 50 were observed mainly in male subjects. Based on the above observations, it was suggested that approximately 20% of males should be classified as a subtle protanomaly group. The result may be related to the fact that opsin (protein) is encoded on the X chromosome and men have only one copy of this gene. However, such an assumption poses a problem in determining a norm for the male population. The norms suggested above, however, include 2 std around the mean (95% of the population) and therefore the majority of the subtle protanomaly peak is included in the norm.

The results obtained for the Moreland test differed significantly from the norm proposed by the anomaloscope manufacturer. It was observed that the fitting range was significantly wider ($M_C = M_2 - M_1$) and considerable inter-subject variability for the border parameters M_1 and M_2 was also noticeable. In particular, such variability was observed in case of the M_1 parameter. For this reason, the Moreland test in its current version seems to have little value for quantification of color vision disorders in the green/blue axis.

In the majority of subjects, the impression of blue was dominant in a wide range of the proportion of green and blue diode intensities. Thus, one of the possible explanations of the large variability of the Moreland test results obtained in this study was too high intensity of the blue diode, which dominated over the green. In order to investigate this hypothesis in the present work, the Rayleigh and Moreland test were administered using additional lenses which partially filtered blue light. Among the available filters, the researchers selected three: Blue Control (F-BC) by Hoya, Med. 1 filter (F-Med1) manufactured by JZO and F-450 by Eschenbach. The results obtained using each of the three filters were compared against test results without any filter (F-0). It was expected that reducing the intensity of the blue diode should result in a reduction of the extremely high value of the M_2 parameter and a reduction of the variability of M_1 values. Consequently, a reduced width of the matching range (M_C) could contribute to a greater diagnostic value of the test with the use of a specific filter.

Anomaloscopic examination enables the assessment of three aspects of color vision:

1. The first aspect is the assessment of the relative subjective brightness of the tested primary colors red, green, and blue, which correspond to the maxima of absorption spectra of subsequent cones L, M, and S.
2. The second one is the ability to perceive changes in the different shades and tones, i.e., neighboring RGB colors may be perceived either as the same or slightly different in quality.
3. The third aspect is quite similar to the first one but, nevertheless, it seems reasonable to distinguish between them. It is the general ability to perceive the given color being understood as the sensitivity of the given retinal cones to the given wavelength.

The assessment of relative subjective brightness may be assessed based on the Y = fn(R) relation in the Rayleigh test and the C = fn(M) relation in the Moreland test, where Y is the luminance of the yellow reference diode as a function of R (the proportion of red to green), and C is the luminance of the cyan reference diode as a function of M (the proportion of blue to green). In the case of the Moreland test, the sensitivity to blue light is generally lower than sensitivity to green light and therefore the slope ($a_M = (C_2 - C_1)/(M_2 - M_1)$ is positive and approximately equal to 1. For example, if the evaluated individual perceives 436 nm blue light as relatively darker than it is perceived by others, then the a_M slope for the C = fn(M) relation is higher. As the M setting is increased (i.e., more green is added and blue is reduced), the perceived brightness increases more rapidly. In the Rayleigh test, red is perceived as a relatively darker color than green, therefore, a_R is negative. If a given subject perceives red as being darker (in comparison to other individuals' perception), then, as the proportion of red is increased, the perceived brightness decreases at a higher rate and the a_R ratio becomes more negative.

The assessment of the ability to perceive subjective changes in the color tones between blue and green (sea color/cyan/aquamarine/turquoise/olive) may be defined as the width of the matching range ($M_W = M_2 - M_1$, $R_W = R_2 - R_1$). The more narrow the color matching range, the more efficiently the adjacent shades and tones are perceived.

As for the third aspect, i.e., the overall ability to perceive colors, a reduced ability to perceive a particular primary color does not necessarily mean that the color will be perceived as darker. For example, a reduced ability to perceive red will be manifested as a shift of the $R_C = \frac{R_1+R_2}{2}$ parameter to the right. It means that a more red diode must be added for the resulting color (i.e., red mixed with green) to be perceived as identical to the reference yellow.

However, this does not necessarily mean that exemplary red will be perceived by the subject as darker (a higher slope in the Y = fn(R) relation). The overall ability to perceive a particular color will be manifested mainly when it is mixed with one of other primary colors. The differentiation between the perceived color brightness and the ability to perceive a given color has not been well researched to date and has not been discussed in the literature. In general, it is more complex to analyze than the two previously mentioned aspects.

It should be emphasized that if anomaloscope is to be used to detect subtle changes in color perception (due to disease processes) rather than to detect major deficits (related to genetic defects), the three aspects must be understood and analyzed separately. In addition, a perception disorder related to a particular disease may affect only one of the above aspects. So far, anomaloscopic examinations have been mainly based on the Rayleigh test administered in order to assess vision of the red color, i.e., the need to increase the amount of red light during the test in order to achieve an impression of yellow in the test field that matches the color in the reference field. It is manifested by a shift of R_2 and R_C to the right and widening of the matching range (R_W). The Y parameter of the Y = fn(R) function has not been used to date to assess color vision disorders resulting from particular disease processes.

In this article, apart from the assessment of blue light filters usability, an analytical proposal for the application of the above parameters is described. The analyzed parameters may be obtained from anomaloscopic examinations using Rayleigh and Moreland tests.

2. Methods

2.1. Group

Inclusion criteria were: no general or ocular diseases, and no regular medication used, normal visual acuity (VA 0.8 or better, 0.1 logMAR or better), normal color vision as determined by HRR test. The subjects were first screened by ophthalmologists, who checked their VA, ocular health (anterior and posterior segments), and then subsequently evaluated during optometric examination, including eye dominance test and confirmation of normal color vision using the HRR test.

Thirty subjects were selected to the experiment, aged 18–43 years, including 20 females and 10 males. HMC MR Oculus type 47700 anomaloscope was used for measurement of color perception with filters. This part of the research was conducted in a dimly lit room and each subject adapted to the test conditions for at least 10 min prior to the assessment. The test was performed in the dominant eye and the order of using the filters was randomized. The Rayleigh and Moreland tests were conducted in the manual mode. The detailed algorithm of the study is presented in Appendix A.

All tests were administered in the '*Zabel sc.*' Ophthalmology Consultation Center (*Poradnia Okulistyczna Zabel sc.*) in Piła, Poland. The study was carried out in accordance with the Helsinki Declaration following approval from the UMP bioethics committee at the Karol Marcinkowski Medical University of Poznan.

2.2. Procedure

Each of the subjects underwent the Rayleigh and Moreland tests four times, i.e., using three types of filters (F-BC, F-Med1, F-450) plus no filter (F-0). The border matching values have been recorded (R_1, R_2, M_1, M_2) together with the corresponding brightness levels of the reference field (Y_1, Y_2, C_1, and C_2). Based on the above values, the following parameters were calculated:

$$R_C = \frac{R_1 + R_2}{2}$$

$$R_W = R_2 - R_1$$

$$a_R = \frac{Y_2 - Y_1}{R_2 - R_1}$$

$$M_C = \frac{M_1 + M_2}{2}$$

$$M_W = M_2 - M_1$$

$$a_M = \frac{C_2 - C_1}{M_2 - M_1}$$

Initially, it was investigated which of the above parameters changed significantly following the application of a particular type of filter. The vast majority of the analyzed variables distributions were not normal in the Shapiro–Wilk test and, therefore, statistical analyses were performed using the Wilcoxon test and the Spearman's rank-order correlation.

Subsequently, the correlations between the Rayleigh and Moreland test results were examined in order to research the impact of red color perception on the ability to perceive colors in the green–blue axis and to assess the impact of blue perception on color perception in the red–green axis.

The first of the filters used in the current study was Hoya's Blue Control which reduces the 436 nm wavelength by 4%, 480 nm by 3% and 490 nm by 3% (Table 1, Figure 1). It is dedicated to users of digital devices in order to provide protection against adverse consequences of excessive blue light exposure. The filter provides a relatively low reduction of blue light and does not significantly modify color perception.

Table 1. Transmittance of filters used in the study in relation to wavelengths emitted by anomaloscope diodes.

Filter's Light Transmittance	436 nm (Blue Diode)	480 nm (Cyan Diode)	490 nm (Green Diode, Moreland)	546 nm (Green Diode, Rayleigh)
F-BC	96%	97%	97%	99.7%
F-Med1	16%	56%	64%	84%
F-450	0.3%	29%	47%	87%

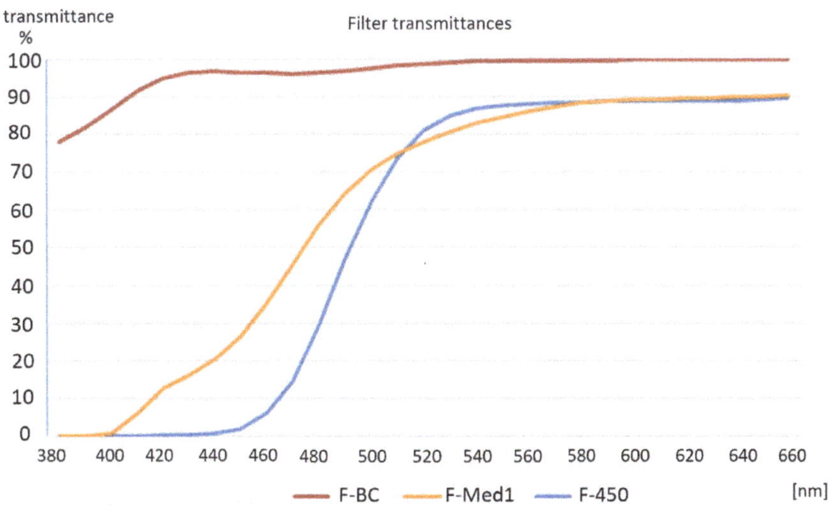

Figure 1. Characteristics of the filters used in this study: F-450 (Eschenbach), F-BC Blue Control (Hoya) and F-Med1 (JZO).

The second filter (Figure 1) was JZO's Med. 1 which reduces the 436 nm wavelength by 84%. The filter's characteristics are presented in Table 1. Its transmittance for 480 nm and 490 nm wavelengths is 56% and 64%, respectively.

The last filter (Figure 1) used was Eschenbach's F-450 edge filter which virtually eliminates the 436 nm wavelength and simulates the symptoms of severe tritanopia during the test.

Comparison of transmittance levels for each filter type, as per individual light wavelengths, is presented in Table 1.

3. Results

3.1. Basic Results of the Moreland Test

As expected, reduced intensity of blue light (436 nm) decreased the values of M_1 and M_2 matching parameters (see Figure 2). The efficacy of the Blue Control filter was too low to induce any statistically significant changes. The F-Med1 filter, which reduced the amount of 436 nm light down to 16%, caused a decrease in M_1 (from 30 to 13) and M_2 (from 85 to 62). The M_C parameter has also decreased accordingly. M_W, which describes the width of the matching range, was significantly reduced only following the application of the F-450 filter, which almost completely stops blue light (see Table 2).

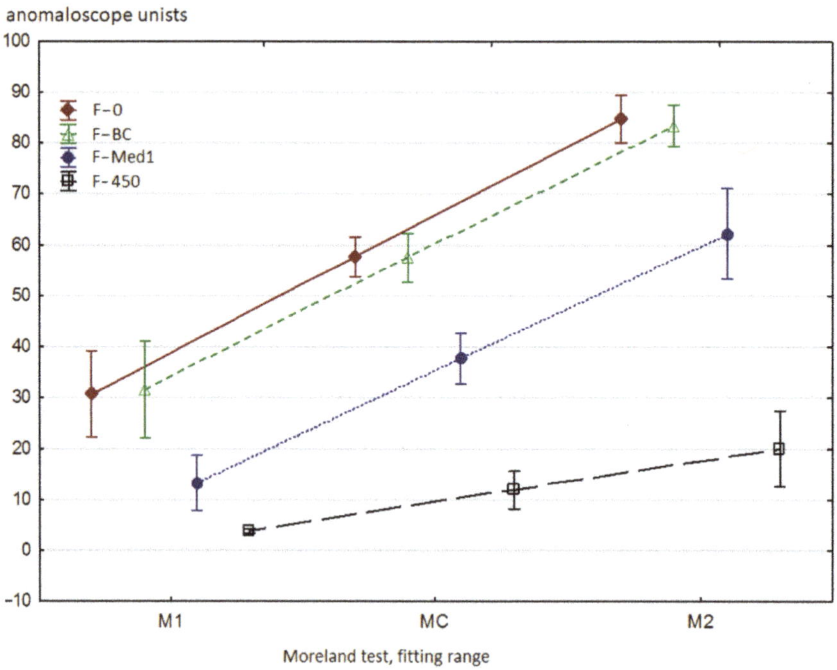

Figure 2. The results for M_1, M_C and M_2 in the Moreland test for each type of applied filters (mean/mean standard error). The anomaloscope units refer to the blue–green diode proportion in the test field: 0 au: 100% blue and 0% green; 100 au: 100% blue and 0% green. The Blue Control filter did not significantly affect the results due to its minimum absorption of blue light. Application of the F-Med1 filter significantly decreased M_1 and M_2 but did not considerably reduce the width of the matching range. The oscillation range of the M_1 variable was reduced but, conversely, the variability of M_2 increased. The use of the F-450 filter reduced M_1 and M_2 values to very low levels. Matching the test field color to the reference field color was possible only provided that a high intensity of blue diode light was set. Thus, the test results were as expected for patients with tritanopia. All the differences between the filter types were statistically significant for M_1, M_2 and M_C, except for the F-0 vs. F-BC comparison.

Table 2. The results for the analyzed parameters (mean ± standard deviation) and statistical significance of the differences in the Wilcoxon test as compared to the results in the preceding columns; no—no significance, +—$p < 0.05$, *—$p < 0.01$, **—$p < 0.001$.

Parameter	No Filter (F-0)	F-BC (F-0)	F-Med1 (F-0, F-BC)	F-450 (F-0, F-BC, F-Med1)
R_1	41.3 ± 2.3	41.6 ± 2.6 (no)	40.1 ± 2.7 (+, **)	40.9 ± 2.4 (no, +, +)
R_2	47.7 ± 2.5	47.6 ± 2.4 (no)	46.8 ± 3.4 (no, +)	46.6 ± 2.8 (*, *, no)
R_W	6.5 ± 3.3	5.9 ± 3.5 (no)	6.7 ± 4.3 (no, no)	5.7 ± 3.5 (no, no, no)
R_C	44.5 ± 1.8	44.6 ± 1.8 (no)	43.5 ± 2.1 (*, *)	43.8 ± 1.9 (+, *, no)
a_R	−0.12 ± 0.79	−0.38 ± 0.34 (no)	−0.37 ± 0.38 (no. no)	−0.39 ± 0.51 (no. no. no)
M_1	30.7 ± 22.6	31.6 ± 25.5 (no)	13.3 ± 14.1 (*, *)	3.9 ± 1.2 (*, *, *)
M_2	84.7 ± 12.5	83.4 ± 11.0 (no)	62.2 ± 23.3 (*, **)	20.1 ± 19.9 (*, *, **)
M_W	54.0 ± 30	51.8 ± 29.6 (no)	48.9 ± 28.6 (no, no)	16.2 ± 20.1 (*, *, **)
M_C	57.7 ± 10.4	57.5 ± 12.9 (no)	37.7 ± 13.0 (*, **)	12.0 ± 9.8 (*, *, **)
a_M	0.88 ± 0.23	0.89 ± 0.34 (no)	1.1 ± 0.17 (*, **)	1.39 ± 0.22 (*, *, **)

3.2. Color Matching C = fn(M) Slope

The slope (a_M) of the C = fn(M) color matching function reflects the relative difference of brightness between blue and green. The steeper the slope of the above relationship (i.e., the higher the value of a_M), the more the green color will be perceived as relatively brighter, as compared to blue. When the green to blue proportion is increased, the perception of brightness will increase more rapidly, which is manifested as an increase of luminance of the cyan field being matched (480 nm) to the reference field. The obtained a_M values for each of the filter types are presented in Figure 3.

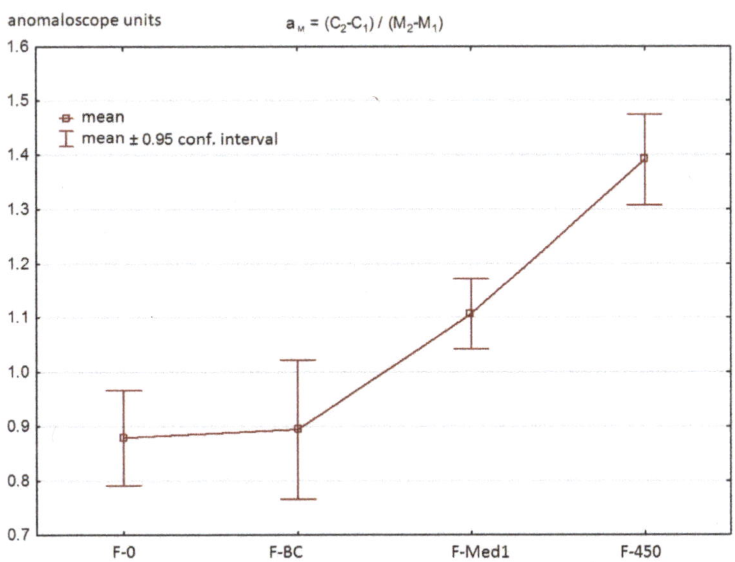

Figure 3. Values of the slope (a_M) for C = fn(M) matching (mean/mean standard error). As the amount of blue light is reduced by a filter, the value of the slope ratio increases. This reflects the reduced impression of perceived brightness of blue when a filter is introduced. Except for the F-0 vs. F-BC comparison, other differences had high statistical significance ($p < 0.001$) in the Wilcoxon test.

Subsequently, the correlations between the perceived brightness (a_M, aspect 1, see Introduction) and other parameters of the Moreland test were studied. The main goal of the research team was to identify to what extent the parameter was correlated to the values of M_1, M_2, and M_C (i.e., the ability to perceive blue and green, aspect 3). The results are presented collectively in Table 3.

Table 3. Comparison of Spearman's rank correlation coefficient (r) for a_M vs. other Moreland test parameters. The correlations were weak or none, which means that the aspect of perceived color brightness is essentially different from the general ability to perceive the blue and green colors. *—$p < 0.05$.

Moreland Test Parameter	Filter-0	F-BC	F-Med. 1	F-450
M_1	r = 0.244	r = 0.071	r = 0.171	r = −0.013
M_C	r = 0.089	r = −0.012	r = −0.187	r = 0.049
M_2	r = −0.374 *	r = −0.292	r = −0.369 *	r = 0.054
M_W	r = −0.315	r = −0.112	r = −0.406 *	r = 0.042

In the F-0 test, there was a weak negative correlation for M_2 and M_W. This means that the individuals who perceive blue as being relatively darker (high a_M) exhibit slightly lower values of M_W and M_2, which also means that the general perception of blue is relatively poorer as compared to green. However, the correlation is poor, which indicates that the relative perception of blue as being darker may only partially be explained by its lower perception. Similar results were obtained for the F-BC and F-Med1 filters, whereas the F-Med1 filter improved statistical significance of the correlation for M_W. Introduction of the F-Med1 filter shifted the Moreland test results closer to the central area of the test range (see Figure 2). Thus, the results reflect the differences between green and blue color perception somewhat better, which might have been the reason for a slightly stronger correlation between M_W and a_M. Meanwhile, the F-450 filter cut out so much light coming from the blue diode that the comparison of blue perception vs. green became problematic and all correlations in this regard disappeared.

3.3. *The Impact of Filters on the Rayleigh Test Results*

The use of the tested blue light filters could have had a minor impact on the obtained values of the Rayleigh test parameters due to the slight reduction of the 546 nm green light used in the Rayleigh test (see Table 1). However, the effect of filtering is also possible due to the fact that the L (red) and M (green) cones are also, to some extent, sensitive to blue light.

The results obtained by the researchers are presented in Figure 4. Table 4 presents the statistical significances of the differences between the various filters for parameters R_1 and R_2. A slight decrease in the values of the R_1 and R_2 parameters can be observed when using the F-Med1 and F-450 filters. The obtained results indicate a slightly weaker perception of green, which is consistent with the characteristics of the filters used as they slightly absorb green light (546 nm) used in the Rayleigh test. Interestingly, the use filters did not significantly affect the values of the R_W and a_R parameters. Cutting out a small portion of the green light used in the Rayleigh test did not affect the quality of distinguishing similar color tones in the lemon-yellow–orange area. Regarding the a_R parameter, there is a noticeable tendency for a_R to decrease (i.e., become more negative) with all filter types when compared to the test without any filter (NF). This is consistent with the fact that green light intensity is slightly lower. When analyzing the a_R parameter, it should be noted that the width of the color matching area (R_W) is narrower in the Rayleigh test than in the Moreland test. Therefore, the accuracy of estimating the values of R_1, Y_1, R_2, and Y_2 may significantly affect the accuracy of calculating the value of a_R.

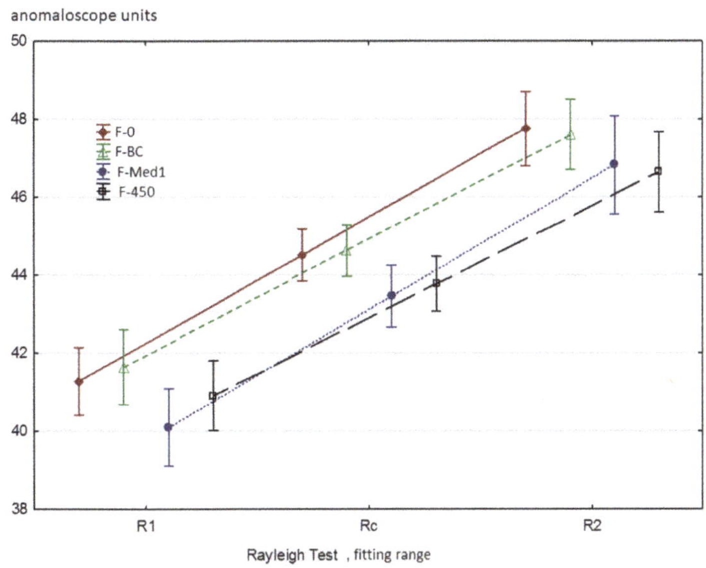

Figure 4. The impact of filters on the Rayleigh test results. The anomaloscope units refer to the red-green diode proportion in the test field: 0 au: 100% red and 0% green; 73 au: 100% red and 0% green. A certain reduction in R_1 and R_2 values is noticeable following the application of the F-Med1 and the F-450 filters. Statistical significance of the differences observed in the Wilcoxon test is presented in Table 4. R_1 results are presented top-right while R_2 results are to be found bottom-left.

Table 4. Statistical significance of the differences in the Wilcoxon test for R_1 and R_2, as per filter types (see Table 2 and Figure 4). R_1 results are presented top-right while R_2 results are to be found bottom-left. The effect of filtering on both parameters is noticeable, while there is no impact on R_W (i.e., the ability to perceive color tones between lemon and orange).

R_1 (Top/Right) R_2 (Bottom/Left)	F-0	F-BC	F-Med1	F-450
F-0	—	$p = 0.304$	$p = 0.010$ **	$p = 0.242$
F-BC	$p = 0.590$	—	$p < 0.001$ ***	$p = 0.012$ *
F-Med. 1	$p = 0.062$	$p = 0.044$ *	—	$p = 0.038$ *
F-450	$p = 0.007$ **	$p = 0.004$ **	$p = 0.959$	—

*—$p < 0.05$, **—$p < 0.01$, ***—$p < 0.001$.

3.4. Correlations between the Rayleigh and the Moreland Tests

Perception of colors, their tones and shades by the human eye and the central nervous system is a complex process involving a number of aspects which have not been thoroughly researched to date. Perception of the entire scale of color tones depends on the relative proportion of activity of only three types of cones. Therefore, it seems justified to claim that perception of various color tones and differentiation of shades depends on the mutual proportion of the extent to which all three cone types are stimulated. Therefore, it may be expected that there is a certain correlation between the results of the Rayleigh and Moreland tests. For that reason, the results of both tests were compared and Spearman's correlation factors were defined for the four blue light filtering scenarios. The results are presented collectively in Table 5.

Table 5. Spearman's rank correlations between the parameters of the Rayleigh and Moreland tests. It should particularly be noted that there is a strong correlation between R_W and most of the Moreland test parameters for each of the filter types. A narrow matching range (R_W), i.e., good perception of shades in the lemon–orange range, was correlated to high M_1, low M_2 and low M_W, which means that efficient perception of color shades in both tests was correlated. The strongest correlations ($|r| > 0.5$) were observed in case of the F-Med1 filter and were presented in detail in Figure 5. *—$p < 0.05$, **—$p < 0.01$ (marked with bold).

			F-0		
r	R_1	R_2	R_C	R_W	a_R
M_1	0.458 *	−0.099	0.14	−0.459 *	0.109
M_2	−0.094	0.134	0.108	0.246	0.125
M_C	0.415 *	−0.096	0.146	−0.404 *	0.133
M_W	−0.355	0.12	−0.05	0.420 *	0.071
a_M	0.121	0.205	0.229	0.134	0.125
			F-BC		
r	R_1	R_2	R_C	R_W	a_R
M_1	0.149	−0.261	−0.001	−0.406 *	−0.116
M_2	−0.057	0.460 *	0.205	0.445 *	0.252
M_C	0.162	−0.068	0.099	−0.266	−0.044
M_W	−0.156	0.359	0.064	0.480 *	0.113
a_M	0.039	−0.344	−0.111	−0.285	−0.077
			F-Med. 1		
r	R_1	R_2	R_C	R_W	a_R
M_1	0.347	−0.34	−0.069	−0.477 *	−0.264
M_2	−0.255	**0.583** **	0.295	**0.712** **	0.294
M_C	−0.11	0.24	0.117	0.3	0.125
M_W	−0.303	**0.640** **	0.294	**0.793** **	0.338
a_M	−0.271	−0.431 *	−0.480 *	−0.261	−0.184
			F-450		
r	R_1	R_2	R_C	R_W	a_R
M_1	0.105	−0.348	−0.131	−0.363 *	−0.103
M_2	−0.183	0.29	0.093	0.468 *	0.032
M_C	−0.165	0.255	0.11	0.437 *	−0.013
M_W	−0.173	0.307	0.091	0.480 *	0.066
a_M	0.102	0.0147	−0.02	−0.152	−0.147

4. Discussion

The data presented in Table 5 and in Figure 5 indicates that the results of the Moreland test are significantly correlated mainly with the values of R_W and, to a lesser extent, with R_1 and R_2. R_W is related to the ability to distinguish color tones in the lemon-yellow-orange axis. It is noticeable that a good ability to differentiate color tones in the Rayleigh test also allowed the subjects to better differentiate between color tones in the green–blue axis. In the case of F-0 tests, such effect was noticeable primarily as an increase in M_1 values i.e., the parameter, which shows significant inter-subject variability. M_1 was also correlated to R_1, which leads to the conclusion that the correlation mainly depends on the quality of green perception. Efficient perception of green (i.e., high R_1 value) causes statistically higher values of M_1 and more narrow matching ranges (M_C) in the Moreland test.

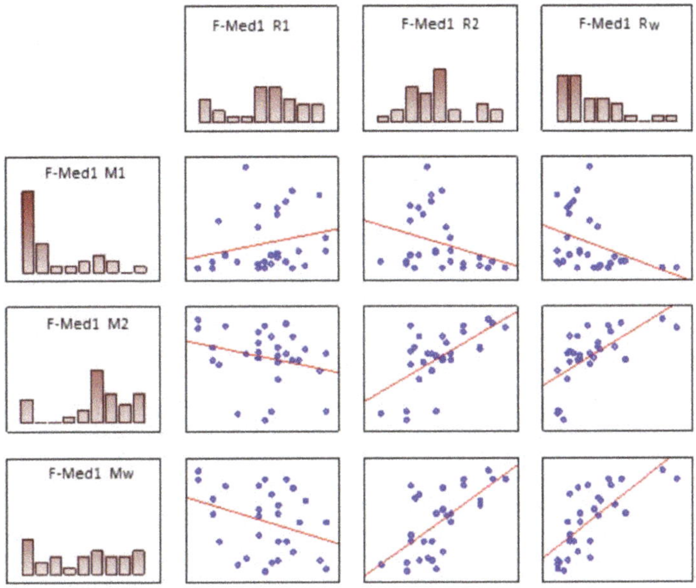

Figure 5. Selected correlations between the Moreland and the Rayleigh tests for the F-Med1 filter. Detailed correlations are listed in Table 5. Notice the particularly strong correlation between R_W and M_1 (negative), as well as M_2 and M_W (positive). Efficient perception of color shades in the lemon-orange range in the Rayleigh test was strongly correlated with efficient perception of shades in the blue-cyan-green axis. Such effect is best visible following the reduction of the amount of blue light with the F-Med1 filter (down to approx. 16%).

When blue filters were applied, the relation was somewhat different. The correlation between R_W and other Moreland test parameters was still observable while the relationship with R_1/R_2 was altered. The correlation to R_1 diminished in case of the F-BC filter, a strong correlation to R_2 occurred for the F-Med1 filter, while the correlations both to R_1 and R_2 were no longer noticeable in case of the F-450 filter.

It should be noted that, in case of the F-Med1 filter, the correlations reached values between 0.6 and 0.8, which makes the results of both tests significantly aligned.

Since R_2 is mainly related to the quality of red perception, it is reasonable to conclude that a reduction of blue light and, to a lesser extent, the amount of green light, when the F-Med1 filter is applied, increases and emphasizes the role of L cones and the L-M information in the perception of blue and green light as well as color tones in the green-cyan-blue axis. Proficiency in differentiating color tones in the lemon-yellow–orange axis increases the competence in differentiating between the color tones in the blue-cyan-green axis.

In case of F-Med1, there is also a clear negative correlation of a_M to R_2 and to R_C. This means that low values of R_2 and R_C (i.e., better red perception) is associated with higher a_M values (i.e., blue is relatively darker or green is brighter). It can be carefully concluded that higher individual number of red cones in the retina is correlated with lower number of blue cones rather than with the green ones and blue light reduction magnifies this effect.

It is arguable that a decrease in luminance of the blue diode in the Moreland test enhances the correlations between the tests in terms of green perception which is common for both test types. Such result is another indication that the intensity of light emitted by the blue diode in the Moreland test is too high and reducing it by 15–40% percent may be beneficial in the context of detecting certain abnormalities. This may be applicable especially in case of decreased green perception ability.

An important observation was also that the correlations between R_C and M_C were low while the correlations between R_1, R_2, M_1, and M_2 were high, which indicates that the limit parameters (R_1, R_2, M_1, and M_2) should be considered independently from one another and that calculation of mean values makes the related information less convincing. R_C and M_C are significant in cases of more pronounced color vision deficits but they are less meaningful when more subtle differences are considered.

The results discussed above indicate that there is a significant and unused potential of anomaloscopic examination in terms of quantitative analysis of subtle inter-subject differences in terms of color perception. The same is also true for this examination in the context of quantitative assessment of deterioration in the perception of certain colors in the course of various disease processes affecting the visual and nervous systems. The application of blue light filters presented herein suggests that there is a significant potential for further analyses and new evaluation parameters in cases of color vison disorders due to an underlying disease process.

One of the main difficulties, which hinders the acquisition of valuable results related to disease severity in the context of color vision, is inter-subject variability. The application of filters, including those with different parameters such as those reducing red or green light, may provide additional parameters which may be less related to the inter-subject variability and may serve as better indicators of the characteristics of certain disease processes. Further analyses may also shed some light on the theory of opposing colors, the characteristics of color coding in the retina as well as color perception by the central nervous system.

It also seems justified to modify the Moreland test so that the difference between the wavelengths of the green diode (490 nm) and the cyan reference diode (480 nm) becomes more significant. Thus, the color matching area in the Moreland test could be narrowed down, which may facilitate more precise analysis. However, this shall require redesigning of the anomaloscope and a revision of the reference ranges for a new set of diodes used in the green–blue axis test. It might also be interesting to design a test able to compare blue directly to red by means of comparison of the red–blue color mix in the test field with a reference violet diode. Valuable results may also be expected from the application of a filter characterized by significantly different transmittance for 480 nm and 490 nm wavelengths, as this would allow test subjects to better differentiate between the green and cyan diodes (with the design of anomaloscope remaining as is). The filter suggested here could possibly have transmittance levels of 15–40% for 436 nm, 50–80% for 480 nm, and 95–100% for 490 nm. Application of such filter might significantly reduce the excessive width of the matching range (M_C) in the Moreland test and could facilitate more efficient diagnosis of blue color vision deficits.

5. Conclusions

Reduction of blue diode intensity in the Moreland anomaloscope test by about 15–40% may be beneficial in detecting mild changes in color vision perception in the blue-green axis and may improve its usefulness in evaluating the color vision perception disorders accompanying different illnesses.

Author Contributions: Conceptualization, K.P.M., J.O., P.W.-B. and A.P.-K.; data curation, J.Z.; formal analysis, K.P.M., J.Z. and A.P.-K.; investigation, J.Z.; methodology, K.P.M. and A.P.-K.; project administration, A.P.-K.; supervision, K.P.M. and J.O.; validation, K.P.M., J.O., P.W.-B. and A.P.-K.; visualization, J.Z.; writing—original draft, K.P.M., J.Z. and A.P.-K.; writing—review & editing, K.P.M. All authors have read and agreed to the published version of the manuscript.

Funding: This research received no external funding.

Institutional Review Board Statement: The study was conducted according to the guidelines of the Declaration of Helsinki, and approved by the Ethics Committee of Poznan University of Medical Sciences (1111/18, 7 Nov 2018).

Informed Consent Statement: Informed consent was obtained from all subjects involved in the study.

Conflicts of Interest: The authors declare no conflict of interest.

Abbreviations

$R_1/R_C/R_2$—left border/center/right border of the fitting range in the Rayleigh anomaloscope test (red–green axis); $R_W = R_2 - R_1$—width of the fitting range in the Rayleigh anomaloscope test; $M_1/M_C/M_2$—left border/center/right border of the fitting range in the Moreland anomaloscope test (blue–green axis); $M_W = M_2 - M_1$—width of the fitting range in the Moreland anomaloscope test; a_R—the slope of the relation Yellow Brightness vs. Red–Green proportion in the Rayleigh test; a_M—the slope of the relation Cyan Brightness vs. Blue–Green proportion in the Moreland test.

Appendix A

HRR test: The HRR test was performed monocularly in the room with intensive artificial light. If the person had problem at anyone level of the test, he/she was excluded from the further experiment. Next, the subject who fulfilled correctly HRR test, was send for optometric examination (visual acuity and refractive error measurements).

Anomaloscope test: This test was performed after the HRR and optometric examinations. HMC Anomaloscope MR type 47700 (Oculus) was used. The non-tested eye was covered with black patch. The order of eyes was counterbalanced. The ilumination in the room was low. The persons were adopted to the room ilumination by 10 min. Before the main test, the demo test was performed in order to explain the measurement procedure.

Both Rayleigh and Moreland tests were performed manually. Firstly, the Rayleigh test was carried out. The upper and lower limit has been found using the staircase method. The initial value of the reference field was 40 units. The measurement of higher limit (R_2) preceeded lower limit test (R_1). For R_2, the reference field was increased by 5 up to reaching the lack of possibility of adjusting the reference and test field. Next, the actual range for R_2 was divided by half in order to tighten the range of R_2. This procedure was repeated up to reaching the measurement limit of the device equal to 0.5. Next, R_1 was measured in simiar way. The R_1 and R_2 values were measured with precision 0.5. After the first eye examination, 2–3 min break was performed and the second eye was measured with Rayleigh test.

After the next 2–3 min break, the Moreland test started. The initial reference value was 50 units. The measurement of M_2 preceeded M_1. For M_2, the reference value increased by 10 up to reaching the lack of possibility of adjusting the reference and test field. Next, the actual range for M_2 was divided by half in order to tighten its range. The procedure was repeated up to reaching the precision of M_2 equal to 1 unit. After one eye was measured, 2–3 min break was performed and the second eye was measured.

References

1. Papaconstantinou, D.; Georgalas, I.; Kalantzis, G.; Karmiris, E.; Koutsandrea, C.; Diagourtas, A.; Ladas, I.; Georgopoulos, G. Acquired color vision and visual field defects in patients with ocular hypertension and early glaucoma. *Clin. Ophthalmol.* **2009**, *3*, 251–257. [PubMed]
2. Zhao, J.; Dave, S.B.; Wang, J.; Subramanian, P.S. Clinical color vision testing and correlation with visual function. *Am. J. Ophthalmol.* **2015**, *160*, 547–552.e541. [CrossRef] [PubMed]
3. Langina-Jansone, Z.; Truksa, R.; Ozolinsh, M. Visual acuity and color discrimination in patients with cataracts. *J. Opt. Soc. Am. A Opt. Image Sci. Vis.* **2020**, *37*, A212–A216. [CrossRef] [PubMed]
4. Kohnen, T.; Koch, D.D. *Cataract and Refractive Surgery*; Springer Science & Business Media: Berlin/Heidelberg, Germany; New York, NY, USA, 2006.
5. Sanchez-Dalmau, B.; Martinez-Lapiscina, E.H.; Pulido-Valdeolivas, I.; Zubizarreta, I.; Llufriu, S.; Blanco, Y.; Sola-Valls, N.; Sepulveda, M.; Guerrero, A.; Alba, S.; et al. Predictors of vision impairment in Multiple Sclerosis. *PLoS ONE* **2018**, *13*, e0195856. [CrossRef]
6. Pohlen, B.; Hawlina, M.; Pompe, M.T.; Kopac, I. Do Type 1 Diabetes Mellitus and Color-Vision Deficiencies Influence Shade-Matching Ability? *Int. J. Prosthodont.* **2018**, *31*, 239–247. [CrossRef]
7. Raman, R.; Verma, A.; Srinivasan, S.; Bhojwani, D. Parietal reversal of color vision imapirment in type 2 diabetes associated with obctructive sleep apnea. *GMS Opjthalmol. Cases* **2018**, *8*, 1–4.

8. Sharpe, L.T.; Stockman, A.; Jagle, H.; Knau, H.; Klausen, G.; Reitner, A.; Nathans, J. Red, green, and red-green hybrid pigments in the human retina: Correlations between deduced protein sequences and psychophysically measured spectral sensitivities. *J. Neurosci.* **1998**, *18*, 10053–10069. [CrossRef] [PubMed]
9. Mehri, A. Non-extensive distribution of human eye photoreceptors. *J. Theor. Biol.* **2017**, *419*, 305–309. [CrossRef] [PubMed]
10. Hurvich, L.M.; Jameson, D. An opponent-process theory of color vision. *Psychol. Rev.* **1957**, *64*, 384–404. [CrossRef] [PubMed]
11. Zabel, J.; Przekoracka-Krawczyk, A.; Olszewski, J.; Michalak, K.P. Variability of Rayleigh and Moreland test results using anomaloscope in young adults without color vision disorders. *PLoS ONE* **2021**, *16*, e0251903. [CrossRef] [PubMed]
12. GmbH OO. *HMS Anomaloskop MR-Instruction Manual*; GmbH OO: Berlin, Germany, 2011; pp. 1–31.

International Journal of
Environmental Research and Public Health

MDPI

Article

The Influence of Women's Empowerment on Poverty Reduction in the Rural Areas of Bangladesh: Focus on Health, Education and Living Standard

Wei Wei [1,†], Tanwne Sarker [1,*,†], Wioletta Żukiewicz-Sobczak [2,*], Rana Roy [3], G. M. Monirul Alam [4], Md. Ghulam Rabbany [5,6], Mohammad Shakhawat Hossain [5,7] and Noshaba Aziz [8]

1 School of Economics and Finance, Xi'an Jiaotong University, Xi'an 710049, China; wei_wei@xjtu.edu.cn
2 Department of Food and Nutrition, Calisia University, 62-800 Kalisz, Poland
3 Department of Agroforestry & Environmental Science, Sylhet Agricultural University, Sylhet 3100, Bangladesh; ranaroy.aes@sau.ac.bd
4 Department of Agribusiness, Bangabandhu Sheikh Mujibur Rahman Agricultural University, Gazipur 1706, Bangladesh; gmmonirul79@gmail.com
5 College of Economics and Management, Northwest A&F University, Yangling 712100, China; tonoy_mkt@nwafu.edu.cn (M.G.R.); shossain7490@gmail.com (M.S.H.)
6 Department of Agribusiness and Marketing, Sher-e-Bangla Agricultural University, Dhaka 1207, Bangladesh
7 Economic Development and Value Chain Specialist, World Vision Bangladesh, BleNGS Project, Jamalpur 2000, Bangladesh
8 College of Economics and Management, Nanjing Agricultural University, Nanjing 210095, China; noshabaaziz@yahoo.com
* Correspondence: sarker@stu.xjtu.edu.cn or tanwnesarker@gmail.com (T.S.); wiola.zukiewiczsobczak@gmail.com (W.Ż.-S.)
† These authors have contributed equally to this work and are joint co-first authors.

Citation: Wei, W.; Sarker, T.; Żukiewicz-Sobczak, W.; Roy, R.; Alam, G.M.M.; Rabbany, M.G.; Hossain, M.S.; Aziz, N. The Influence of Women's Empowerment on Poverty Reduction in the Rural Areas of Bangladesh: Focus on Health, Education and Living Standard. *IJERPH* **2021**, *18*, 6909. https://doi.org/10.3390/ijerph18136909

Academic Editor: Paul B. Tchounwou

Received: 19 May 2021
Accepted: 24 June 2021
Published: 27 June 2021

Publisher's Note: MDPI stays neutral with regard to jurisdictional claims in published maps and institutional affiliations.

Copyright: © 2021 by the authors. Licensee MDPI, Basel, Switzerland. This article is an open access article distributed under the terms and conditions of the Creative Commons Attribution (CC BY) license (https://creativecommons.org/licenses/by/4.0/).

Abstract: Women's empowerment has a great influence on health, nutrition, education, and the overall well-being of societies as well as of the children and households. This study investigates the effect of women's empowerment on poverty reduction and focuses on household deprivation, in terms of education, health, and standard of living. Primary data was collected from 914 married women from rural areas of Bangladesh using a well-structured questionnaire and a random sampling technique. Descriptive statistics, logistic regression, and ordinary least squares models were used in this study. The results indicate that increased women's access to education, asset ownership, decision-making power on children's health and education, and access to medical facilities, have caused a significant decline in income poverty and multidimensional poverty. However, gender violence, taking resources against women's will, and preventing women from working outside, have caused a considerable decline in per capita income and an increase in income poverty and multidimensional poverty. Overall, it is found that women's empowerment has a great impact on the reduction of income poverty and multidimensional poverty in society. The findings of the study can assist and guide policymakers to initiate appropriate strategies for women's empowerment to reducing poverty in Bangladesh while making progress towards other social and developmental goals.

Keywords: Bangladesh; gender violence; multidimensional poverty; poverty reduction; women's empowerment

1. Introduction

Achieving gender equality is regarded as one of the key goals of the 2030 Agenda for Sustainable Development, as women and girls continue to suffer serious discrimination problems in every part of the world [1]. The vital aspect of the goal of women's empowerment is connected to improving health and nutrition status, ensuring food security, eliminating hunger, and reducing poverty [2,3]. "Poverty is a pronounced deprivation of well-being", whereas well-being is measured by education, assets, housing, health, nutrition, and certain human rights in society [4]. Empowering women and enhancing women's status can play a significant role in the achievement of many development programs and help to bring positive societal transformation [5]. Globally, women constitute about 43% of the total agricultural labor force [6] and also contribute to 50% of the total food production [7]. It has been found that when more power is given to women to use household income, the proportion of money spent on healthy food increases because women favor spending more money on nutritive and high-quality foods than on unhealthy foods, recreation, and alcohol [8,9]. Women's empowerment is frequently cited as a goal of rural development aimed primarily at reducing household vulnerability to poverty and food insecurity [10,11]. Empowering women is also a vital instrument in the fight against poverty [12,13].

Bangladesh has been struggling to reduce the prevalence of poverty and to improve the socio-economic conditions of poor citizens. In the early 1970s, the percentage of people living below the poverty line was 80%, and in 2016, this percentage decreased by 24.3%. Moreover, the employment and literacy rates of women have increased from 36% and 57.8% in 2011 to 88.5% and 69.5% in 2016, respectively [14]. The changes and transformations in women's lives are broadly seen as the cause and consequence of considerable human development over the past 25 years in Bangladesh [15]. Although women constitute about half of the Bangladeshi population, their social status, especially in rural areas, remains very low. Rural women are among the most disadvantaged members of society, suffering from social oppression and economic inequality; the vast majority of them are impoverished [16], and their empowerment, therefore, is critical to bringing about a positive change in their lives.

A great number of empirical studies have investigated the impact of women's empowerment on education, child health, food security, and nutrition [17–20]. However, our study specifically evaluated the impact of women's empowerment on household income poverty and multidimensional poverty, by focusing on the rural areas of Bangladesh. This study extends previous work by providing new empirical evidence on the impact of women's empowerment on their overall livelihood status. This study is a contribution not solely to descriptive literature on the present condition of women in Bangladesh, but also offers critical insights into the effects of women's empowerment on their households.

2. Background of the Study

Considerable diversity exists in the foci, agenda, and terminologies used to define women's empowerment [21]. The terms, "women's empowerment", "gender equality", "female autonomy", or "women's status", are widely used in literature, and all these terms focus on women's power and control in making life choices [21]. Women's empowerment is the process whereby women learn knowledge and skills, overcome difficulties, and benefit from useful resources [22]. It is not only an outcome but also an intermediate variable to further observe other developmental outcomes, especially poverty [23]. The idea of empowerment is often linked to the notion of power relations, which specifically refers to gaining either more or less power among actors [24,25]. However, women's empowerment is different from the empowerment of other disadvantaged groups due to intra-household dynamics [26,27]. Women's empowerment and gender equality are often considered two sides of the same coin: progress toward gender equality requires women's empowerment, on the other hand, women's empowerment contributes to increasing gender equality. In the case of gender inequality or discrimination, it is generally believed that women are excluded or disadvantaged in terms of decision-making and access to economic and social resources [28]. All definitions of women's empowerment make some reference to women's ability in certain areas, specifically in controlling their own lives, having the freedom to make decisions, and having the input to change life choices [27]. Women's empowerment is typically conflated with the status that women have, which is often represented by the acquisition of and control over resources [29], which not only include material and financial resources, but also the social and human resources that can increase a woman's ability to exercise her choices [30]. Examples of such resources include age, education, employment, social capital, networking, and ownership of property. Sharaunga et al. [31] reported that empowerment is multidimensional, and women empowered in one dimension do not necessarily empower in another. It has also been established that women require both resources and a sense of agency to achieve their live outcomes on their own. The multidimensional concept of the agency includes different domains, namely, decision-making authority, control over finances, and freedom of movement. Having the authority over making decisions and overusing money, is a more apt and precise description of women's power. Although there is no single definition of women's empowerment in the literature, it is variously conceptualized as a process or outcome, an end state or a means to an end, a capacity [32–36], a matter of gaining power, and as a matter of agency. Given the broad scope of concerns surrounding empowerment, scholars have examined many topics across a wide range of contexts [33]. Studies have examined many topics, including educational attainment [37], political participation [38], gender-based domestic violence [39], resource control [40–42], entrepreneurialism [43], well-being [44], household decision-making [45–47], time poverty [48], and health [49,50]. While women's empowerment is intrinsically important, studies in developing countries have shown that empowering women can also improve children's health and education [51], decrease child mortality [52], improve the organizational effectiveness of businesses [53], increase agricultural productivity [54], and increase economic growth and reduce poverty [55]. The empowerment of women and poverty reduction are intertwined under the framework of empowerment. A growing amount of evidence has shown that empowerment is instrumental toward reducing income and consumption poverty [34]. Empowering woman as a member of a poor household, in turn, contribute to empower the whole household and thus the issue is a vigorous researchable area in a developing country such as Bangladesh.

3. Research Design

3.1. The Survey Area, Selection of Participants, and Data Collection

The quantitative research method was used for the present study. Survey data were collected from the northern part of Bangladesh, as the poverty rate was reported the highest in this part of the country [56]. In the first stage, four divisions out of nine divisions of Bangladesh were selected purposively, i.e., Rajshahi, Rangpur, Mymensingh, and Sylhet divisions. Then, from each division, three districts were randomly selected. After that from each district, one Upazila was selected. In the next stage, two villages were randomly selected from each Upazila from the complete list of villages as recorded by Upazila authorities. Finally, from each village, forty households were selected based on the random sampling technique where women were the respondent for the study. The aggregate sample was 960 women. However, some observations were excluded from the analysis, owing to missing data. For this reason, the analysis comprised 914 samples for which complete information was available. Face-to-face interviews of respondents were conducted from May to August 2019 with the help of a peer review and a well-structured interview schedule. We restricted our investigation to married women only as conventional indicators and measures of empowerment characteristically focus on the circumstances of marriage [30], and empowerment within marriage is resulting in the economic wellbeing and health safety of women and their families [57].

3.2. Indicators of Women's Empowerment

Indicators of women's empowerment were identified based on extensive literature reviews (see Table S1) as follows: household decision-making power, gender attitude, and beliefs, physical mobility, control over resources, and relative freedom from domination by the family.

3.2.1. Household Decision-Making Power

To assess women's decision-making power within a household, a set of questions were asked relating to (1) large household purchases; (2) purchases for daily needs; (3) visits to family, friends, and relatives; (4) decisions on spending husband's earnings, and (5) decision-making on their own health care. For each question, if women by themselves or jointly (together with their husband) participate in the decision-making process, the responses were coded as one. If women do not participate in the decision-making process, the responses were coded as zero.

3.2.2. Gender Attitude and Beliefs

Women were asked whether or not a husband is justified to physically abuse his wife under the following circumstances: (1) wife goes out without her husband's permission; (2) wife argues with her husband; (3) wife refuses to have sexual relations with her husband; (4) the food is burnt when the wife is cooking, and (5) wife neglects the children. For each question, the responses were reverse-coded. If women responded negatively, it was scored one; and otherwise, zero.

3.2.3. Physical Mobility

The physical mobility of respondents was derived by asking the respondents about their going out alone to the following five places: (1) market; (2) friends' and relatives' houses; (3) children's school; (4) health care center or hospital; and (5) outside of the village. If a respondent answered positively, her response was coded as one; and otherwise, as zero.

3.2.4. Control over Resources

Control over resources by women was measured by questioning them about the following conditions: (1) ownership of land and house; (2) ownership of assets; (3) decision-making on the sale or purchase of land, house, and assets; (4) having access to loans, microcredit, and insurance; and (5) being engaged in paid work. If a respondent answered positively, her response was coded as one; and otherwise, as zero.

3.2.5. Relative Freedom from Domination by the Family

To assess relative freedom from domination by the family, the following questions were asked relating to (1) money is taken from them against their will; (2) land or jewelry or livestock is taken from them against their will, and (3) they are prevented from working outside their home. If the women responded that none of the above situations has happened in their lives, they scored one; and otherwise, zero.

3.3. Measuring the Poverty Status of Households

This part of the study focused on the poverty level of surveyed women and their households with the help of (1) the classical income poverty approach; and (2) the multi-dimensional poverty index (MPI). The MPI contains only non-monetary indicators of the three dimensions of poverty, i.e., education, health, and living standards.

3.3.1. Measuring Income Poverty

We examined the impact of women's empowerment on the income poverty of women and their households, anchored in the poverty line approach, using Foster, Greer, and Thorbecke's poverty indicators [58]. We converted per capita income in Bangladesh currency (Taka) into US dollars using purchasing power parity. In 2019, the purchasing power parity exchange rate was 1US dollar = 79.12 Taka [56]. If the household's income was positioned above the poverty line of USD1.90 per day, income poverty was scored zero, and otherwise, one. We evaluated the income poverty gap using the following Equation (1):

$$xi = \frac{p - vi}{p} \quad (1)$$

The poverty line is denoted as p in the above equation, and the per capita income of household i is denoted as vi. Households with income above the poverty line are automatically assigned a value of zero. The income poverty gap is a continuous value, ranging from zero to one [59].

3.3.2. Measuring Multidimensional Poverty

To investigate the households' deprivation in the poverty dimensions and their ability to meet basic needs, the MPI approach was employed. In association with the United Nations Development Programme, Alkire and Santos [60] developed MPI approach [61]. Among different procedures of multidimensional poverty measurement, namely cluster analysis, factor analysis, weighting procedure, and ordinal approaches [60,62,63], this study used the weighting procedure proposed by Alkire and Santos [60]. The weighting procedure of multidimensional poverty identifies a household's level of deprivation and the people who face poverty in a community. Finally, the MPI provides absolute poverty levels (in intensity terms) that allow measuring poverty across different settings.

Detailed information about all the included indicators of the poverty dimensions and weighted deprivation as illustrated by Alkire and Santos [60] are presented in Table 1. The MPI includes 10 indicators in three dimensions, which are connected to global standards [64]. For the education dimension, we included two indicators, i.e., school attendance of school-aged children and completion of five years of schooling of all household members. For the health dimension, we used nutrition and child mortality as indicators. Nutrition included the Body Mass Index (ratio of weight in kilograms and square of height in meters) for adults and weight-for-age for children. If any child under the age of five had died in a household, it was treated as a child mortality indicator. The standard of living was measured by focusing on electricity, sanitation facilities, drinking water sources, floor type, cooking fuel, and asset ownership [60].

Table 1. Dimensions, indicators, deprivation cut-off, and weights of the multidimensional poverty index.

Dimension	Indicator	Description and Deprivation Cut-Off	Relative Weight
Education	Years of schooling	No household member has completed five years of schooling	1/6
	Child's school attendance	Any school-aged child is not attending school in years 1 to 8	1/6
Health	Mortality	Any child who has died in the family	1/6
	Nutrition	Any adult to child for whom there is nutritional information is malnourished *	1/6
Living standard	Electricity	The household has no electricity	1/18
	Sanitation	The household's sanitation facility is not improved (according to MDG guidelines), or it is improved but shared with another household **	1/18
	Drinking water	The household does not have access to safe drinking water (according to MDG guidelines) or safe drinking water is more than a 30-min walk from home (round-trip) ***	1/18
	Floor	The household has dirt, sand, or dung floor	1/18
	Cooking fuel	The household cooks with dung, wood, or carbon	1/18
	Asset ownership	The household does not own one of the following assets: radio, TV, telephone, bicycle, motorbike, refrigerator, and does not own a car or truck.	1/18

* Adults are considered malnourished if their BMI is below 18.5. Children are considered malnourished if their z-score of weight-for-age is below minus two standard deviations from the median of the reference population. This was estimated following the algorithm provided by the WHO Child Growth Standards (WHO, 2006). http://www.who.int/childgrowth/software/en/ (accessed on 27 January 2020). ** A household is considered to have access to improved sanitation if it has some type of flush toilet or latrine, or ventilated improved pit or composting toilet, provided that they are not shared. *** A household has access to safe drinking water if the water source is any of the following types: piped water, public tap, borehole or pump, protected well, protected spring or rainwater, and it is within a distance of a 30-min walk (round-trip).

The weighting procedure of MPI normally requires an equal-weighted dimension and assures an equal-weighted indicator across different dimensions. The poverty cut-off confines the poor by selecting a lower level of deprivation cut-off of 33.33% (1/3). This poverty cut-off reveals a minimum value of weighted indicators to capture multidimensional poor whose deprivation score is equal to or greater than 33.33% [60].

In this study, we calculated different MPI measures for each sample household using one and zero values for each of the 10 indicators. First, we measured the "total deprivation score of each household" by the summation of weighted values for each of the 10 indicators, using weights ranging from zero to one (presented in Table 1). Second, we constructed a "multidimensional poverty dummy," taking a value of one of the total deprivation score of a household is equal to or greater than a definite threshold of 0.33, and zero otherwise [60]. Finally, we created "multidimensional poverty intensity", equal to the deprivation score if the sample household is multidimensionally poor (MPI dummy equals 1), and zero, otherwise. The interpretation of the multidimensional poverty intensity is similar to the poverty gap [59].

3.4. Estimation Strategy

Data was assembled, coded, tabulated, and analyzed using the STATA 13.1 statistical package to justify the objectives of the study. Descriptive statistics included mean, percentage distribution, and standard deviation for MPI and empowerment indicators. To estimate the impact of women's empowerment on income poverty and multidimensional poverty, we used the following regression Equation (2):

$$y_i = ß_0 + ß_1 E_i + ß_2 X_i + {}^{e}{}_{i} \quad (2)$$

In the above equation, y_i is the income and poverty status for the respondent's household i, E_i is the total empowerment score, X_i is the control variables, and ${}^{e}{}_{i}$ is the random error term. We evaluated separate models for each poverty indicator by controlling relevant household and socioeconomic characteristics that may influence poverty. Ordinary least squares (OLS) estimators were used for continuous dependent variables, including income, poverty gap, and multidimensional poverty intensity. Logit estimators were used for binary dependent variables, including income poverty dummy and MPI dummy.

The key coefficient of interest is $ß_1$, which captures the effect of women's empowerment on household income and poverty. We expected a positive coefficient or income-increasing empowerment effect when we used household income as the dependent variable and a negative coefficient or poverty-reducing empowerment effect when using poverty indicators. For these analyses, we utilized the SVY command of STATA 13.1 to adjust the village effects.

4. Results

4.1. Socio-Demographic Analysis

A total of 914 married women participated in this study. The dominant (45.3%) age group was 39–46 years of age, followed by the age group of 29–38 years (38.9%) and 18–28 years (15.8%) (Table 2). As for education level, the women and their husbands were categorized into six groups: (a) illiterate (6.3 and 0.1%, respectively); (b) primary (28.0 and 3.8%, respectively); (c) secondary (37.7 and 7.2%, respectively); (d) higher secondary (13.7 and 32.7%, respectively); (e) graduate (10.6 and 42.7%, respectively); and (f) postgraduate (3.6 and 13.5%, respectively). The average education level was relatively higher for husbands than for women. The majority of the respondents are engaged in paid work (52.4%). On average, four members live in a household and only one member earns money (Table 2).

Table 2. Socio-demographic analysis.

Variables		Percentage	Mean	Standard Deviation
Age	18–28	15.8	2.295	0.724
	29–38	38.9		
	39–46	45.3		
Education	Illiterate	6.3	3.05	1.191
	Primary	28		
	Secondary	37.7		
	Higher secondary	13.7		
	Graduate	10.6		
	Postgraduate	3.6		
Husband's education	Illiterate	0.1	4.54	0.952
	Primary	3.8		
	Secondary	7.2		
	Higher secondary	32.7		
	Graduate	42.7		
	Postgraduate	13.5		
Employment status	Engaged in paid work	52.4	0.524	0.499
	Unpaid work	47.6		
Household size		914	3.906	0.859
Earning member		914	1.41	0.492

4.2. Evidence of Women's Empowerment

In this study, we used descriptive statistics to express the respondents' empowerment status, which also provides an outline of the empowerment variables that are shown in Table 3. In terms of household decision-making, women exhibited impressive evidence of empowerment in their abilities to take decisions solely or jointly with their husband to visit people (80%), to purchase large household products (62%) and daily needs (77%), to spend their husband's earnings (75%), and their own health care (43%). When a wife disagrees with her husband, physical abuse occurs at a rate of 5%, according to gender attitude and beliefs. Our respondents are less likely to justify physical abuse if women go out without their husbands' permission (1%), burn food (2%), neglect children (3%), and refuse sex (4%). Our findings are also consistent with a previous study [65], indicating that intimate partner violence has begun to decline in Bangladesh. In terms of freedom of physical mobility, the respondents exhibited substantial empowerment in their abilities to visit hospitals alone (96%), visit relatives (70%), go outside of the village (66%), visit their children's school (64%), and go to the market (59%). Concerning control over resources, 88% of women have assets, either solely or jointly with their husband and/or others, whereas 28% have access to loans, micro-credit, and insurance facilities. With regards to freedom from family domination, 68% of the respondents do not have permission to work outside.

4.3. Evidence of Multidimensional Poverty

Our results show that the majority of households are non-deprived in terms of MPI indicators, such as years of schooling, child's school attendance, child mortality, nutrition, sanitation, drinking water, and floor, with the exception of cooking fuel facilities. Deprivation scores of households based on the multidimensional poverty index are presented in Table 4.

Table 3. Descriptive statistics of women's empowerment.

Indicators of Women's Empowerment	Yes (%)	Mean	Standard Deviation
Women's participation in household decision-making			
Making decisions on large household purchases	62.0	0.623	0.485
Making decisions on purchases for daily needs	77.0	0.774	0.419
Making decisions on visits to family, relatives, or friends	80.0	0.799	0.401
Making decisions on spending husband's earnings	75.0	0.753	0.432
Making decisions on own health care	43.0	0.434	0.496
Attitude towards wife-beating	8.0	0.923	0.266
Wife beating is justified if she goes out without telling her husband	1.0	0.987	0.114
Wife beating is justified if she argues with husband	5.0	0.950	0.219
Wife beating is justified if she refuses to have sex with her husband	4.0	0.964	0.187
Wife beating is justified if she neglects the children	3.0	0.967	0.178
Wife beating is justified if she burns the food	2.0	0.983	0.131
Physical mobility			
Going alone to the market	59.0	0.591	0.492
Going alone to visit friends, family, and relatives	70.0	0.697	0.460
Going alone to visit a health care center or hospital	96.0	0.956	0.204
Going alone outside of the village	66.0	0.656	0.475
Going alone to visit children school	64.0	0.635	0.482
Control over resources			
Ownership of land	53.0	0.526	0.500
Ownership of assets	88.0	0.875	0.331
Decision on sale and purchase of house, land, and assets	57.0	0.567	0.496
Have access to loans, micro-credit, and insurance	28.0	0.277	0.448
Engaged in paid work	52.0	0.524	0.500
Relative freedom from domination by the family			
Money and jewelry taken against her will	2.0	0.984	0.127
Land taken against her will	2.0	0.976	0.153
Prevented from working outside the home	68.0	0.323	0.468

Notes: Answer "Yes" means positive outcome for indicators of women's participation in making household decisions, indicators of physical mobility, and indicators of control of resources; while answer "Yes" means negative outcome for indicators of attitude towards wife-beating and relative freedom from domination by the family.

Table 4. Descriptive statistics of MPI elements.

MPI Indicator	Deprived (%)	Non-Deprived (%)	Mean	Standard Deviation
Years of schooling	21.0	79.0	0.208	0.406
Child school attendance	22.0	78.0	0.218	0.413
Mortality	5.0	95.0	0.046	0.209
Nutrition	30.0	70.0	0.299	0.458
Electricity	0.0	100.0	0.000	0.000
Sanitation	10.0	90.0	0.103	0.304
Drinking water	4.0	96.0	0.036	0.187
Floor	34.0	66.0	0.342	0.475
Cooking fuel	59.0	41.0	0.589	0.492
Asset ownership	0.0	100.0	0.000	0.000

Surprisingly, no household is deprived of electricity and asset ownership. In line with our findings, DHS [66] reported that 91% of households have access to electricity, and nearly every household owns a single set of assets. Concerning education, 21% of households are deprived of both years of schooling and a child's school attendance. In terms of the health dimension, only 5% of households are deprived of child mortality indicators. However, the nutrition indicator is quite high (29%). For living conditions, results show that less than 5% of households are deprived in terms of drinking water. There are various sources of drinking water, such as tube well water, pump water, public taps, and so on, and most of the participants are aware of the importance of drinking clean water. Most respondents are accustomed to hygienic sanitation as a result of increased awareness through newspaper reading, media exposure, and promotion of worthwhile and low maintenance cost of hygienic sanitation by government departments. For cooking fuel indicators, more than half of all households (60%) are considered deprived. Because of the availability of local resources, agricultural residue, and animal excreta in Bangladesh, most rural people use these resources for fuel, which makes many rural households to be deprived of cooking fuel indicators.

4.4. Impact Assessment on Poverty

Table 5 represents the effects of women's empowerment on the per capita income of households and estimates of log-transformation of per capita income using the OLS model.

Table 5. Effect of women's empowerment on per capita income.

Variables	Per Capita Income	Log of per Capita Income
	OLS	OLS
Empowerment score	1.483 * (0.786)	0.008 ** (0.003)
Education of respondents (years)	5.840 *** (2.011)	0.032 *** (0.009)
Education of husband (years)	6.046 *** (2.245)	0.038 *** (0.010)
Asset ownership of women (dummy)	2.321 (6.809)	0.002 (0.029)
Engaged in paid work (dummy)	−0.493 (3.920)	0.002 (0.017)
Experiencing gender violence (dummy)	−13.330 (15.099)	−0.121 * (0.065)
Political knowledge (dummy)	10.759 (10.415)	0.058 (0.045)
Control over use of household income (dummy)	12.168 (7.740)	0.058 * (0.033)
Prevented from working outside (dummy)	−12.303 ** (4.797)	−0.044 ** (0.021)
Money and jewelry taken against their will (dummy)	−14.761 (14.024)	−0.044 (0.060)
Access to loans, micro-credit, and insurance (dummy)	1.105 (4.692)	−0.002 (0.020)
Decision on children's health and education (dummy)	2.612 (4.338)	0.039 ** (0.019)
Access to healthcare facilities (dummy)	5.948 *** (2.169)	0.020 ** (0.009)
Child mortality (dummy)	2.559 (8.270)	−0.014 (0.035)
Number of dependent persons in the household (number)	−7.764 * (4.565)	−0.033 * (0.020)
Constant	9.912 (19.247)	1.422 *** (0.082)
(Pseudo) R-squared	0.090	0.151
Observations	914	914

Notes: Coefficient estimates are reported. Standard errors are shown in parentheses. * Significant at 10% level; ** Significant at 5% level; *** Significant at 1% level.

Results are shown as coefficients and their 90% confidence intervals for the variables. Women's empowerment score has a significantly synergistic (positive) effect on household income. Table 5 reports that a 0.1 increase (10 percentage points) in women's empowerment would raise per capita income by 14%. Moreover, women's education, husband's education, and access to healthcare facilities have significant synergistic effects on both per capita income and log of per capita income. However, preventing women from going outside to work and the number of dependent people in the household has significantly antagonistic (negative) effects on both per capita income and log of per capita income.

Table 6 represents that women's empowerment score, women's education, husband's education, and decision on children's health and education, all have significantly antagonistic effects on both income poverty and income poverty gap; while experiencing gender violence has a significant synergistic effect on both income poverty and the income poverty gap.

Table 6. Effect of women's empowerment on income poverty.

Variables	Income Poverty (Dummy)	Income Poverty Gap (0–1)
	Logit	OLS
Empowerment score	−0.099 *** (0.037)	−0.005 ** (0.002)
Education of respondents (years)	−0.465 *** (0.107)	−0.015 *** (0.005)
Education of husband (years)	−0.617 *** (0.111)	−0.027 *** (0.006)
Asset ownership of women (dummy)	−0.169 (0.301)	0.014 (0.018)
Engaged in paid work (dummy)	−0.342 * (0.195)	−0.016 (0.010)
Experiencing gender violence (dummy)	1.791 ** (0.858)	0.105 *** (0.040)
Political knowledge (dummy)	−0.344 (0.591)	−0.023 (0.027)
Control over use of household income (dummy)	0.022 (0.331)	−0.041** (0.020)
Prevented from working outside (dummy)	−0.031 (0.243)	0.007 (0.013)
Money and jewelry taken against their will (dummy)	0.146 (0.624)	−0.011 (0.037)
Access to loans, micro-credit, and insurance (dummy)	0.050 (0.233)	0.005 (0.012)
Decision on children's health and education (dummy)	−0.818 *** (0.205)	−0.054 *** (0.011)
Access to healthcare facilities (dummy)	−0.246 ** (0.104)	−0.002 (0.006)
Child mortality (dummy)	0.609 (0.383)	0.043 ** (0.022)
Number of dependent persons in household (number)	0.506 ** (0.228)	0.020 (0.012)
Constant	5.402 *** (0.910)	0.377 *** (0.051)
(Pseudo) R-squared	0.212	0.188
Observations	914	914

Notes: Coefficient estimates are reported. Standard errors are shown in parentheses. * Significant at 10% level; ** Significant at 5% level; *** Significant at 1% level.

Table 7 represents that empowerment score, women's education, husbands' education, women's asset ownership, engaged in paid work, decision on children's health and education, and access to healthcare facilities, have significantly antagonistic effects on both multidimensional poverty and multidimensional poverty intensity. However, experiencing gender violence and child mortality have significantly synergistic effects on both multidimensional poverty and multidimensional poverty intensity.

Table 7. Effect of women's empowerment on multidimensional poverty.

Variables	Multidimensional Poverty (Dummy)	Multidimensional Poverty Intensity (0–1)
	Logit	OLS
Empowerment score	−0.101 ** (0.046)	−0.567 ** (0.224)
Education of respondents (years)	−0.306 ** (0.124)	−2.019 *** (0.574)
Education of husband (years)	−0.481 *** (0.133)	−2.136 *** (0.640)
Asset ownership of women (dummy)	−0.809 ** (0.365)	−6.764 *** (1.943)
Engaged in paid work (dummy)	−0.543 ** (0.244)	−2.148 * (1.119)
Experiencing gender violence (dummy)	2.511 *** (0.818)	15.683 *** (4.309)
Political knowledge (dummy)	−0.861 (0.866)	−2.339 (2.972)
Control over use of household income (dummy)	−0.295 (0.404)	0.166 (2.209)
Prevented from working outside (dummy)	−0.382 (0.321)	0.267 (1.369)
Money and jewelry taken against their will (dummy)	0.822 (0.713)	7.634 * (4.003)
Access to loans, micro-credit, and insurance (dummy)	0.323 (0.294)	1.499 (1.339)
Decision on children's health and education (dummy)	−2.703 *** (0.240)	−21.482 ** (1.238)
Access to healthcare facilities (dummy)	−0.376 *** (0.128)	−1.591 ** (0.619)
Child mortality (dummy)	2.709 *** (0.418)	23.409 *** (2.360)
Number of dependent persons in the household (number)	0.270 (0.298)	1.329 (1.303)
Constant	5.516 *** (1.089)	51.766 *** (5.493)
(Pseudo) R-squared	0.344	0.426
Observations	914	914

Notes: Coefficient estimates are reported. Standard errors are shown in parentheses. * Significant at 10% level; ** Significant at 5% level; *** Significant at 1% level.

5. Discussion

This study has examined the impact of women's empowerment on poverty reduction in the rural areas of Bangladesh. Different indicators that were used to measure the empowerment status of women are given in Table 3. Our findings show that a 10% increase in women's empowerment scores will increases per capita income by 14%, while also lowering the prevalence of income poverty by 0.99% and multidimensional poverty by 1.01%, as shown in Tables 5–7. It is also found that the education of women and their husbands has a significantly positive relationship with the increase of per capita income and decrease of poverty. The level of education improves people's participation in the capital and labor markets and is widely recognized as an authentic tool for eliminating poverty and improving the well-being of people [67]. Moreover, qualified women employ themselves in different kinds of prestigious jobs [68], hence reducing their economic dependence and household poverty [69,70]. Generally, highly educated women have more economic and decision-making power and also social freedom, which ultimately reduce their physical, sexual, and emotional vulnerability within the household [71]. Highly educated women can make their own healthcare decisions, ensure better health status for their children, spend their husband's earnings well, and reject views of physical abuse [72,73]. In addition to this, highly educated household members know a better way to tackle the effect of poverty and chronic poverty [74,75]. The education of women and their husbands acts as a significant catalyst to reduce income poverty, and households with educated couples have a lower risk of being poor [76]. On the other hand, households with a higher number of dependents are multidimensionally poorer than households with a lower number of dependents [60]. Overall, our analysis demonstrates that women and their husbands' education have a significant effect on reducing income poverty and multidimensional poverty in households. Similar to our study, Sell and Minot [77] stated that education empowers women to hold a better position within the household and to reduce their household poverty.

The measure of household resources (such as the value of a household's assets) is a long-run measure of economic status rather than a measure of income [78]. Assets, such as mobile phones, televisions, refrigerators, fans, air conditioners, motorcycles, etc., could improve an individual's health, peace of mind, and mental development, as well as make life more comfortable after a day's work. Women's having business equipment (like sewing and embroidery machines, electric irons, incubators, etc.) and skills could improve their income and living conditions as well as decrease household poverty [18]. In line with this result, our study also found that with the increase of assets in the household, multidimensional poverty declines by 0.809%, as shown in Table 7. Our findings support DHS [66], who reported that asset ownership has increased dramatically in both urban and rural areas of Bangladesh, with nearly every household owning a number of assets.

Education is found to be a very important determinant of women's favorable attitude towards gender equality. The majority of educated women in rural areas work as teachers, are engaged in trade or moneylending, do small businesses, or do sewing and embroidery. We can observe from Tables 6 and 7 that women engaged in paid work, have a negative correlation with income poverty and multidimensional poverty. Khatun and Kabir [79] studied women's empowerment in Bangladesh through entrepreneurship and found that women who are not permitted to go outside for work but earn income through running a business such as tailoring, making cakes, gardening, etc. Furthermore, rural women earn income by rearing livestock, money saved from their husband's family, money received from their natal family, or microcredit from various non-government organizations (NGOs) ultimately increase their economic empowerment [68]. These kinds of income-generating activities empower women not only within the household but also to contribute to reducing income poverty and multidimensional poverty [80].

Intimate partner violence is a crime encompassing physical, sexual, psychological, and emotional abuse by a husband or partner [81]. Frequent abuse is the main cause of deterioration in mental health and happiness [82]. Gender-based violence violates women's fundamental rights, laws, and regulations, and also limits the potential sources of women's empowerment. The environment in which women are victims of violence can weaken women's empowerment [5]. The experience of poverty exacerbates the risk of violence for women [83,84]. However, when men rely on their wives' financial contributions and maintain a certain standard of living, violence against women decreases [68]. Economic advancements in rural areas have reduced poverty, and households with more assets have a lower rate of domestic violence [78].

Women's economic independence is an important factor for improving their empowerment. Women have gained freedom of mobility, decision-making power, awareness of their rights as women, and self-confidence from earning and managing their own income [85]. Women in Bangladesh do household work that is not paid work, like working on their husband's farm and caring for their children. This makes them financially dependent on their husbands and other male relatives [86]. However, it has been found that women's income significantly increases overall household income and the income of the husband and wife is pooled as family income [17]. Empowered women contribute not only by providing their personal income, but also by assisting their husbands with various income-generating activities, such as working in farms and shops, giving advice, taking loans from various NGOs, and supervising family resources [68]. World Vision Ghana states that if women have the opportunity to get credit for employment, their contribution to the well-being of their family would help to reduce family poverty and women's economic dependence on their husbands [80]. As a result, when restrictions are imposed on women from working outside the home, the overall income is significantly reduced. Our study has similar findings that prohibiting women from working outside significantly reduces per capita income, as shown in Table 5.

Women's capability to participate in and make household decisions is one of the key elements of empowerment. Women have the power to make decisions on household expenses (groceries, clothes, and expenditure on children's health and school fees) in Indonesia, Myanmar, the Philippines, and Thailand, while men occasionally participate in major expenditure decisions [17]. Moreover, credit decisions are also made by mutual agreement of women and men in all the mentioned countries [17]. The results of Wouterse [9] show that when more power is given to women over household expenditure, income spent on food quality increases as opposed to income spent on unhealthy food, recreation, and alcohol. In this study, we find that women's participation in household decision-making and control over resources, considerably increases per capita income, while at the same time, significantly decreases income poverty and multidimensional poverty, as presented in Tables 5–7. Similar findings have been reported by previous studies that women play a significant role in household food security, children's education, and the healthy lives of household members [17,68]. However, contradictory trends are observed in poor Ghanaian communities, where men have very strong domination over women within the household [80].

In this study, we also observe a significant relationship between child mortality and income poverty and multidimensional poverty in the rural areas of Bangladesh. Women's relative unequal access to education, employment, finance, decision-making power, basic health care facilities, and other productive resources, is considered as the prime reason for their ill-health and that of their children. Relatively economically poor and less empowered women are less likely to receive proper health care facilities than wealthy and empowered women [87]. According to Lachaud [88], in rural areas, lower living standards of households, in terms of assets, are associated with high child mortality. Furthermore, it has been discovered that living below or slightly above the national poverty line, as well as a lack of insurance coverage, are risk factors for children's health.

A wealth of research has reported that access to loans and micro-credit has played a remarkable role in reducing poverty [80,89,90]. However, in this study, we find no significant relationship between women's access to loans and microcredit and per capita income or reducing income poverty, which can be verified from Tables 5 and 6. Freedom of mobility indicator determines not only the extent to which women can go outside the home but also their personal autonomy in terms of not having to seek permission from their husbands or any other household member [91]. In Bangladesh, women who do not have any freedom to go outside on their own, are possibly deprived of empowerment programs, such as microfinance, and therefore, microfinance oftentimes does not reach the poor [92]. This is a common scenario in rural areas compared to urban areas because, in rural areas of Bangladesh, the purdah practice confines women to the home and compound; they mostly have to obtain permission either from their husband or any other responsible person or at least, tell them when they are going outside. Mahmud et al. [93] reported that women who work outside the home do not usually ask for permission when they go outside for work, but do so when they go out for other purposes.

The different selected indicators considerably influence women's empowerment status, showing an increment in per capita income and a decrease in different poverty indices. Educated women have more opportunities to participate in income-generating activities, household decision-making, and resource control, which help to empower them. Empowered women are more conscious of their own rights as well as of the wellbeing of their children and family members, which eventually can reduce household income poverty and multidimensional poverty. If women become conscious about their rights and supportive policies that empower women socioeconomically are in place, that would generate a righteous succession of women's overall empowerment in both their household and societal spheres.

6. Conclusions

Women's empowerment is a critical human rights issue with implications for the well-being of women, their families, and society as well as socio-economic development and poverty reduction, especially in developing countries like Bangladesh. In this paper, we have examined the impact of women's empowerment on income, income poverty, and multidimensional poverty in the rural area of Bangladesh. Data were collected from 914 respondents. To measure women's empowerment, this study used different indicators which were selected based on an extensive literature review. The results reveal that women's empowerment score contributes to increase per capita income and decrease income poverty and multidimensional poverty. Women's education significantly reduces multidimensional poverty and income poverty. Whereas gender violence, taking resources against women's will and preventing women from working outside, caused a considerable decline in per capita income and, increase in income poverty and multidimensional poverty. Working women who earn cash have high household autonomy, high freedom of movement, no gender preference and they face lower levels of domestic violence. Violence against women remains one of the most widespread and persistent human rights abuses in the world, stemming from deep-rooted notions of women's unequal status. It is a major obstacle to the fulfillment of women's and girls' human rights and the achievement of the 2030 Agenda for Sustainable Development. Violence against women is no longer viewed as an inevitable part of family life, of social relations, of the workplace, or war; indeed, violence against women cannot be justified under any circumstances. However, the disempowerment of women in Bangladesh is not solely linked to gender, because working women who earn cash have high household autonomy, high freedom of movement, no gender preference and they do not rationalize domestic violence, all of which indicate that women's empowerment is positively associated with lower levels of domestic violence.

The concept of empowerment is closely related to agency and thereby to human development. Any policy aiming at human development needs to be informed about the factors that enhance agency and contribute to empowerment as both agency and empowerment are not only intrinsically valuable but also instrumentally important for poverty reduction. The government and non-government organizations' programs directed to repositioning family planning services should be geared up and centered around various dimensions of women's empowerment, especially focusing on their economic existence in society and decision-making power. Educational programs should prioritize achieving gender equity in schooling outcomes. Although this study examines a number of indicators accompanied by women's empowerment, such as asset ownership, employment status, household decision-making power, educational status, physical mobility, and so on. Future research could add other different indicators, like assets brought to the marriage, proximity to other family members, characteristics of the respondent's parents and upbringing, and the influence of community norms, which could explain empowered women's roles in household poverty reduction. The findings may be useful as a policy tool for planners, administrators, and development workers to initiate appropriate strategies for women's empowerment to reduce poverty in the rural areas of Bangladesh, which can also be applied in other similar contexts.

Supplementary Materials: The following are available online at https://www.mdpi.com/article/10.3390/ijerph18136909/s1, Table S1: Indicators of women's empowerment in previous studies.

Author Contributions: Conceptualization, W.W. and T.S.; data curation, T.S. and N.A.; formal analysis, T.S.; funding acquisition, W.Ż.-S.; investigation, W.W., T.S., R.R., M.G.R. and M.S.H.; methodology, W.W. and T.S.; project administration, W.W.; resources, W.W. and T.S.; software, T.S.; supervision, W.W. and T.S.; validation, T.S.; visualization, T.S., R.R., M.G.R. and M.S.H.; writing—original draft, T.S.; writing—review and editing, W.W., T.S., W.Ż.-S., R.R. and G.M.M.A. All authors have read and agreed to the published version of the manuscript.

Funding: This research received no external funding.

Institutional Review Board Statement: This study was approved by the School of Economics and Finance, Xi'an Jiaotong University, Xi'an, China.

Informed Consent Statement: Not applicable.

Data Availability Statement: The data that support the findings of the study are available from the corresponding author upon reasonable request.

Conflicts of Interest: The authors declare no conflict of interest.

References

1. United Nations. Transforming our world: The 2030 Agenda for Sustainable Development. *United Nations Sustainable knowledge platform.* 2015. Available online: https://sustainabledevelopment.un.org/post2015/transformingourworld (accessed on 27 January 2020).
2. Wei, W.; Sarker, T.; Roy, R.; Sarkar, A.; Ghulam Rabbany, M. Women's empowerment and their experience to food security in rural Bangladesh. *Sociol. Health Illn.* **2021**, 1–24. [CrossRef]
3. Malapit, H.J.L.; Quisumbing, A.R. What dimensions of women's empowerment in agriculture matter for nutrition in Ghana? *Food Policy* **2015**, *52*, 54–63. [CrossRef]
4. World Bank. *World Development Report 2000/2001: Attacking Poverty*; World Bank: Washington, DC, USA, 2000.
5. Gupta, K.; Yesudian, P.P. Evidence of women's empowerment in India: A study of socio-spatial disparities. *GeoJournal* **2006**, *65*, 365–380. [CrossRef]
6. Doss, C. If women hold up half the sky, how much of the world's food do they produce? In *Gender in Agriculture: Closing the Knowledge Gap*; Springer: Dordrecht, The Netherlands, 2014; ISBN 9789401786164.
7. Food and Agriculture Organization. *The State of Food and Agriculture: Women in Agriculture: Closing the Gap for Development*; Food and Agriculture Organization: Rome, Italy, 2011.
8. Sraboni, E.; Malapit, H.J.; Quisumbing, A.R.; Ahmed, A.U. Women's empowerment in agriculture: What role for food security in Bangladesh? *World Dev.* **2014**, *61*, 11–52. [CrossRef]
9. Wouterse, F.S. *The Distribution of Power and Household Behavior: Evidence from Niger*; IFPRI: Washington, DC, USA, 2016.
10. Action Aid International. *What Women Farmers Need: A blueprint for Action*; Action Aid International: Johannesburg, South Africa, 2011.
11. Sharaunga, S.; Mudhara, M.; Bogale, A. The impact of "women's empowerment in agriculture" on household vulnerability to food insecurity in the KwaZulu-Natal province. *Forum Dev. Stud.* **2015**, *42*, 195–223. [CrossRef]
12. Sharma, V.; Seth, P.; Niyazi, S. Capability Enhancement of Women through Information Technology to Reduce Poverty, Inequality and Radical Feminism. *Int. J. Asian Bus. Inf. Manag.* **2012**, *3*, 1–10. [CrossRef]
13. Faborode, H.; Alao, T.O. The battle against rural poverty and other challenges of development: Empirical analysis of women empowerment programme of Justice, Development and Peace Movement (JDPM) in Osun State, Nigeria. *Acta Agron.* **2016**, *65*, 149–155. [CrossRef]
14. BBS. *Bangladesh Statistics 2016*; BBS: Dhaka, Bangladesh, 2017.
15. Chowdhury, A.M.R.; Bhuiya, A.; Chowdhury, M.E.; Rasheed, S.; Hussain, Z.; Chen, L.C. The Bangladesh paradox: Exceptional health achievement despite economic poverty. *Lancet* **2013**, *382*, 1734–1745. [CrossRef]
16. Parveen, S.; Leonhäuser, I. Empowerment of Rural Women in Bangladesh: A Household Level Analysis. In Proceedings of the Conference on Rural Poverty Reduction through Research for Development and Transformation, Berlin, Germany, 5–7 October 2004.
17. Akter, S.; Rutsaert, P.; Luis, J.; Htwe, N.M.; San, S.S.; Raharjo, B.; Pustika, A. Women's empowerment and gender equity in agriculture: A different perspective from Southeast Asia. *Food Policy* **2017**, *69*, 270–279. [CrossRef]
18. Oladokun, Y.O.M.; Adenegan, K.O.; Salman, K.K.; Alawode, O.O. Level of asset ownership by women in rural North-East and South-East Nigeria. *Womens. Stud. Int. Forum* **2018**, *70*, 68–78. [CrossRef]
19. Sinharoy, S.S.; Waid, J.L.; Ali, M.; Yount, K.M.; Thilsted, S.H.; Girard, A.W. Resources for women's agency, household food security, and women's dietary diversity in urban Bangladesh. *Glob. Food Sec.* **2019**, *23*, 1–8. [CrossRef]
20. Sraboni, E.; Quisumbing, A. Women's empowerment in agriculture and dietary quality across the life course: Evidence from Bangladesh. *Food Policy* **2018**, *81*, 21–36. [CrossRef]

21. Narayan, D. Conceptual framework and methodological challenges. In *Measuring Empowerment: Cross-Disciplinary Perspectives*; The World Bank: Washington, DC, USA, 2005; Volume 6, pp. 363–365, ISBN 9786468600.
22. Cornwall, A. Women's empowerment: What works. *J. Int. Dev.* **2016**, *28*, 342–359. [CrossRef]
23. Moyo, C.; Francis, J.; Ndlovu, P. Community-Perceived State of Women Empowerment in Some Rural Areas of Limpopo Province, South Africa. *Gend. Behav.* **2012**, *10*, 4418–4432.
24. Rowlands, J. *Empowerment Examined*; Taylor & Francis: Milton Park, UK, 1995; Volume 5, ISBN 0961452951.
25. Sen, G. *Empowerment as an Approach to Poverty*; Human Development Report Office: New York, NY, USA, 1997.
26. Doepke, M.; Tertilt, M. Does Female Empowerment Promote Economic Development? The World Bank: Washington, DC, USA, 2014.
27. Malhotra, A.; Schuler, S.R. Women's empowerment as a variable in International Development. In *Measuring Empowerment: Cross-Disciplinary Perspectives*; The World Bank: Washington, DC, USA, 2005; pp. 71–88, ISBN 0821360574.
28. Sundar, I. Gender equality and women empowerment. *Int. J. Humanit. Soc. Sci.* **2017**, *7*, 7–21.
29. Mosedale, S. Assessing women's empowerment: Towards a conceptual framework. *J. Int. Dev.* **2005**, *17*, 243–257. [CrossRef]
30. Kabeer, N. Resources, agency, achievements: Reflections on the measurement of women's empowerment. *Dev. Chang.* **1999**, *30*, 435–464. [CrossRef]
31. Sharaunga, S.; Mudhara, M.; Bogale, A. Conceptualisation and Measurement of Women's Empowerment Revisited. *J. Hum. Dev. Capab.* **2019**, *20*, 1–25. [CrossRef]
32. Kabeer, N. Conflicts over credit: Re-evaluating the empowerment potential of loans to women in rural Bangladesh. *World Dev.* **2001**, *29*, 63–84. [CrossRef]
33. Malhotra, A.; Schuler, S.R.; Boender, C. Conceptualizing and measuring women's empowerment as a variable in international development. In Proceedings of the Paper Prepared for the World Bank Workshop on Poverty and Gender: New Perspectives; 2003. Final Version: 28 June 2002. Available online: http://sitesresources.worldbank.org/INTGENDER/Resources/MalhotraSchulerBoender.pdf (accessed on 30 May 2013).
34. Alsop, R.; Bertelsen, M.F.; Holland, J. *Empowerment in Practice: From Analysis to Implementation*; World Bank: Washington, DC, USA, 2006; ISBN 9780821364505.
35. Martinez, E. *The Courage to Change: Confronting the Limits and Unleashing the Potential of CARE's Programming for Women*; CARE: Atlanta, GA, USA, 2006.
36. Njoh, A.J.; Ananga, E. The Development Hypothesis of Women Empowerment in the Millennium Development Goals Tested in the Context Women's Access to Land in Africa. *Soc. Indic. Res.* **2016**, *128*, 89–104. [CrossRef]
37. Takayanagi, T. Rethinking women's learning and empowerment in Kenya: Maasai village women take initiative. *Int. Rev. Educ.* **2016**, *62*, 671–688. [CrossRef]
38. Grabe, S. Participation: Structural and relational power and Maasai women's political subjectivity in Tanzania. *Fem. Psychol.* **2015**, *25*, 528–548. [CrossRef]
39. Wekwete, N.; Sanhokwe, H.; Murenjekwa, W.; Takavarasha, F.; Madzingira, N. The Association between Spousal Gender Based Violence and Women's Empowerment among Currently Married Women aged 15–49 in Zimba-bwe: Evidence from the 2010–2011 Zimbabwe Demographic and Health Survey. *African Popul. Stud.* **2014**, *28*, 1413–1431. [CrossRef]
40. Rao, N. Assets, Agency and Legitimacy: Towards a Relational Understanding of Gender Equality Policy and Practice. *World Dev.* **2017**, *95*, 43–54. [CrossRef]
41. Solanke, B.L.; Kupoluyi, J.A.; Abe, J.O.; Bankole, O.T. Polygyny and Resources for Empowerment and Equality in Anglo-Phone West Africa: Implications for Childbearing and Women's Well-Being. *Eur. Sci. J.* **2018**, *14*, 174–194. [CrossRef]
42. Ragsdale, K.; Read-Wahidi, M.R.; Wei, T.; Martey, E.; Goldsmith, P. Using the WEAI+ to explore gender equity and agricultural empowerment: Baseline evidence among men and women smallholder farmers in Ghana's Northern Region. *J. Rural Stud.* **2018**, *64*, 123–134. [CrossRef]
43. Kapinga, A.F.; Montero, C.S. Exploring the socio-cultural challenges of food processing women entrepreneurs in IRINGA, TANZANIA and strategies used to tackle them. *J. Glob. Entrep. Res.* **2017**, *7*, 17. [CrossRef]
44. Fielding, D.; Lepine, A. Women's Empowerment and Wellbeing: Evidence from Africa. *J. Dev. Stud.* **2017**, *53*, 826–840. [CrossRef]
45. De Brauw, A.; Gilligan, D.O.; Hoddinott, J.; Roy, S. The Impact of Bolsa Família on Women's Decision-Making Power. *World Dev.* **2014**, *59*, 487–504. [CrossRef]
46. Sumner, D.; Christie, M.E.; Boulakia, S. Conservation agriculture and gendered livelihoods in Northwestern Cambodia: Decision-making, space and access. *Agric. Human Values* **2017**, *34*, 347–362. [CrossRef]
47. Maligalig, R.; Demont, M.; Umberger, W.J.; Peralta, A. Off-farm employment increases women's empowerment: Evidence from rice farms in the Philippines. *J. Rural Stud.* **2019**, *71*, 62–72. [CrossRef]
48. Bain, C.; Ransom, E.; Halimatusa'diyah, I. 'Weak winners' of Women's empowerment: The gendered effects of dairy livestock assets on time poverty in Uganda. *J. Rural Stud.* **2018**, *61*, 100–109. [CrossRef]
49. Hindin, M.J. Women's power and anthropometric status in Zimbabwe. *Soc. Sci. Med.* **2000**, *51*, 1517–1528. [CrossRef]
50. Badejo, A.F.; Majekodunmi, A.O.; Kingsley, P.; Smith, J.; Welburn, S.C. The impact of self-help groups on pastoral women's empowerment and agency: A study in Nigeria. *Pastoralism* **2017**, *7*, 28. [CrossRef]
51. World Bank. *Gender equality in development*; World Bank: Washington, DC, USA, 2012; ISBN 9780821388105.

52. Gakidou, E.; Cowling, K.; Lozano, R.; Murray, C.J. Increased educational attainment and its effect on child mortality in 175 countries between 1970 and 2009: A systematic analysis. *Lancet* **2010**, *376*, 959–974. [CrossRef]
53. McKinsey & Company. Gender Equality. In *Global Research on the Gender Gap and the Case for Greater Diversity in the Workplace*; McKinsey & Company: New York, NY, USA, 2017.
54. FAO. *Food and Agriculture Organization. Tanzania Mainland Country Profile: Gender Inequalities in Rural Employment in Tanzania Mainland, an Overview*; FAO: Rome, Italy, 2014.
55. OECD. *Organization for Economic Co-operation and Development. OECD Week 2012—Gender Equality in Education, Employment and Entrepreneurship—Final Report to the MCM 2012*; OECD: Paris, France, 2012.
56. BBS. *Bangladesh Bureau of Statistics. Bangladesh Statistics 2019*; Ministry of Planning: Dhaka, Bangladesh, 2019.
57. Yount, K.M. Resources, family organization, and domestic violence against married women in Minya, Egypt. *J. Marriage Fam.* **2005**, *67*, 579–596. [CrossRef]
58. Foster, J.; Greer, J.; Thorbecke, E. A Class of Decomposable Poverty Measures. *Econometrica* **1984**, *52*, 761–766. [CrossRef]
59. Ogutu, S.O.; Qaim, M. Commercialization of the small farm sector and multidimensional poverty. *World Dev.* **2019**, *114*, 281–293. [CrossRef]
60. Alkire, S.; Santos, M.E. Measuring Acute Poverty in the Developing World: Robustness and Scope of the Multidimensional Poverty Index. *World Dev.* **2014**, *59*, 251–274. [CrossRef]
61. UNDP. *United Nations Development Program. Human Development Report 2013*; Oxford University Press: New York, NY, USA, 2013.
62. Arndt, C.; Distante, R.; Hussain, M.A.; Østerdal, L.P.; Huong, P.L.; Ibraimo, M. Ordinal Welfare Comparisons with Multiple Discrete Indicators: A First Order Dominance Approach and Application to Child Poverty. *World Dev.* **2012**, *40*, 2290–2301. [CrossRef]
63. Kakwani, N.; Silber, J. *Quantitative Approaches to Multidimensional Poverty Measurement*; Palgrave Macmillan: London, UK, 2008; ISBN 9780230582354.
64. UNESCO United Nations Educational, Scientific and Cultural Organization, Institute for Statistics database, Table 1. Education systems (UIS, 2010). Available online: http://stats.uis.unesco.org/unesco/TableViewer/%0AtableView.aspx?ReportId=163 (accessed on 27 January 2020).
65. Schuler, S.R.; Lenzi, R.; Nazneen, S.; Bates, L.M. Perceived decline in intimate partner violence against women in Bangladesh: Qualitative evidence. *Stud. Fam. Plann.* **2013**, *44*, 243–257. [CrossRef]
66. DHS. *Bangladesh Demographic and Health Survey*; Ministry of Health and Family Welfare: Dhaka, Bangladesh, 2018.
67. Akerele, D.; Adewuyi, S.A. Analysis of poverty profiles and socio economic determinants of welfare among urban households of Ekiti state, Nigeria. *Curr. Res. J. Soc. Sci.* **2011**, *3*, 1–7.
68. Schuler, S.R.; Nazneen, S. Does Intimate Partner Violence Decline as Women's Empowerment Becomes Normative? Perspectives of Bangladeshi Women. *World Dev.* **2018**, *101*, 284–292. [CrossRef]
69. Schuler, S.R.; Lenzi, R.; Badal, S.H.; Bates, L.M. Women's Empowerment as a Protective Factor Against Intimate Partner Violence in Bangladesh: A Qualitative Exploration of the Process and Limitations of Its Influence. *Violence Against Women* **2017**, *23*, 1100–1121. [CrossRef] [PubMed]
70. Conroy, A.A. Gender, Power, and Intimate Partner Violence: A Study on Couples from Rural Malawi. *J. Interpers. Violence* **2014**, *29*, 866–888. [CrossRef]
71. Kreager, D.A.; Felson, R.B.; Warner, C.; Wenger, M.R. Women's Education, Marital Violence, and Divorce: A Social Exchange Perspective. *J. Marriage Fam.* **2013**, *75*, 565–581. [CrossRef] [PubMed]
72. Bizzego, A.; Gabrieli, G.; Bornstein, M.H.; Deater-Deckard, K.; Lansford, J.E.; Bradley, R.H.; Costa, M.; Esposito, G. Predictors of contemporary under-5 child mortality in low-and middle-income countries: A machine learning approach. *Int. J. Environ. Res. Public Health* **2021**, *18*, 1315. [CrossRef]
73. Habibov, N.; Barrett, B.J.; Chernyak, E. Understanding women's empowerment and its determinants in post-communist countries: Results of Azerbaijan national survey. *Womens. Stud. Int. Forum* **2017**, *62*, 125–135. [CrossRef]
74. Awotide, B.A.; Diagne, A.; Awoyemi, T.T.; Ebiohomon, V.; Ojehomon, T. Household Endowments and Poverty Reduction in Rural Nigeria: Evidence from Rice Farming Households. *Agric. J.* **2011**, *6*, 274–284. [CrossRef]
75. Glauben, T.; Herzfeld, T.; Wang, X. The persistence of poverty I rural China: Applying an ordered Probit and a hazard approach. In Proceedings of the Contributed paper for presentation at the International Association of Agricultural Economics Conference, Gold Coast, Australia, 12–18 August 2006.
76. Alkire, S.; Fang, Y. Dynamics of Multidimensional Poverty and Uni-dimensional Income Poverty: An Evidence of Stability Analysis from China. *Soc. Indic. Res.* **2019**, *142*, 25–64. [CrossRef]
77. Sell, M.; Minot, N. What factors explain women's empowerment? Decision-making among small-scale farmers in Uganda. *Womens. Stud. Int. Forum* **2018**, *71*, 46–55. [CrossRef]
78. Heath, R. Women's Access to Labor Market Opportunities, Control of Household Resources, and Domestic Violence: Evidence from Bangladesh. *World Dev.* **2014**, *57*, 32–46. [CrossRef]
79. Khatun, F.; Fardaus Kabir, A.T.M. Women SME Entrepreneurs in ensuring women empowerment in Bangladesh: A Study on Women SME Entrepreneurs in Bangladesh. *ISSN ISSN (CD-ROM Am. Int. J. Res. Humanit. Arts Soc. Sci. AIJRHASS* **2014**, *5*, 61–68.
80. Ganle, J.K.; Afriyie, K.; Segbefia, A.Y. Microcredit: Empowerment and disempowerment of rural women in Ghana. *World Dev.* **2015**, *66*, 335–345. [CrossRef]

81. Pocock, M.; Jackson, D.; Bradbury-Jones, C. Intimate partner violence and the power of love: A qualitative systematic review. *Health Care Women Int.* **2020**, *41*, 621–646. [CrossRef] [PubMed]
82. Bonilla-Algovia, E.; Rivas-Rivero, E.; Vázquez, J.J. Impact of gender-based violence on psychological distress and happiness in León (Nicaragua). *Health Care Women Int.* **2020**, *41*, 673–689. [CrossRef]
83. Muluneh, M.D.; Francis, L.; Agho, K.; Stulz, V. A systematic review and meta-analysis of associated factors of gender-based violence against women in sub-Saharan Africa. *Int. J. Environ. Res. Public Health* **2021**, *18*, 4407. [CrossRef]
84. Gillum, T.L. The intersection of intimate partner violence and poverty in Black communities. *Aggress. Violent Behav.* **2019**, *46*, 37–44. [CrossRef]
85. Gurman, T.A.; Ballard, A.; Kerr, S.; Walsh, J.; Petrocy, A. Waking Up the Mind: Qualitative Study Findings About the Process Through Which Programs Combining Income Generation and Health Education Can Empower Indigenous Guatemalan Women. *Health Care Women Int.* **2016**, *37*, 325–342. [CrossRef]
86. Hashemi, S.M.; Schuler, S.R.; Riley, A.P. Rural Credit Programs and Women'S Empowerment in Bangladesh. *World Dev.* **1996**, *24*, 635–653. [CrossRef]
87. Zaky, H.H.M.; Armanious, D.M.; Hussein, M.A. Testing for the endogenous nature between women's empowerment and antenatal health care utilization: Evidence from a cross-sectional study in Egypt. *Biomed Res. Int.* **2014**, *2014*, 1–7. [CrossRef] [PubMed]
88. Lachaud, J. Modelling determinants of child mortality and poverty in the Comoros. *Health Place* **2004**, *10*, 13–42. [CrossRef]
89. Agbola, F.W.; Acupan, A.; Mahmood, A. Does microfinance reduce poverty? New evidence from Northeastern Mindanao, the Philippines. *J. Rural Stud.* **2017**, *50*, 159–171. [CrossRef]
90. Yasmine, F.N. Microcredit and the socio-economic wellbeing of women and their families in Cairo. *J. Socio. Econ.* **2008**, *37*, 644–656. [CrossRef]
91. Mahmud, S.; Shah, N.M.; Becker, S. Measurement of Women's Empowerment in Rural Bangladesh. *World Dev.* **2012**, *40*, 610–619. [CrossRef] [PubMed]
92. Roodman, D.; Morduch, J. The Impact of Microcredit on the Poor in Bangladesh: Revisiting the Evidence. *J. Dev. Stud.* **2014**, *50*, 583–604. [CrossRef]
93. Mahmud, S.; Maheen, S. *Community Health Workers in Bangladesh: Continuity and Change*; BRAC Development Institute, BRAC University: Dhaka, Bangladesh, 2010.

International Journal of
Environmental Research and Public Health

MDPI

Article

Physical Activity and the Quality of Life of Female Students of Universities in Poland

Katarzyna Kotarska [1], Małgorzata Paczyńska-Jędrycka [1], Katarzyna Sygit [2,*], Kamila Kmieć [2], Aleksandra Czerw [3] and Maria Alicja Nowak [1]

1 Institute of Physical Culture Sciences, University of Szczecin, 71-065 Szczecin, Poland; katarzyna.kotarska@usz.edu.pl (K.K.); malgorzata.paczynska-jedrycka@usz.edu.pl (M.P.-J.); maria.nowak@usz.edu.pl (M.A.N.)
2 Faculty of Health Sciences, Calisia University, 62-800 Kalisz, Poland; k.kmiec@akademia.kalisz.pl
3 Department of Health Economics and Medical Law, Faculty of Health Sciences, Medical University of Warsaw, 02-091 Warszawa, Poland; aleksandra.czerw@wum.edu.pl
* Correspondence: k.sygit@akademiakaliska.edu.pl

Citation: Kotarska, K.; Paczyńska-Jędrycka, M.; Sygit, K.; Kmieć, K.; Czerw, A.; Nowak, M.A. Physical Activity and the Quality of Life of Female Students of Universities in Poland. *IJERPH* **2021**, *18*, 5194. https://doi.org/10.3390/ijerph18105194

Academic Editors: Wioletta Zukiewicz-Sobczak, Paulina Wojtyla-Buciora, Izabela Rącka and Andrzej Wojtyla

Received: 19 April 2021
Accepted: 11 May 2021
Published: 13 May 2021

Publisher's Note: MDPI stays neutral with regard to jurisdictional claims in published maps and institutional affiliations.

Copyright: © 2021 by the authors. Licensee MDPI, Basel, Switzerland. This article is an open access article distributed under the terms and conditions of the Creative Commons Attribution (CC BY) license (https://creativecommons.org/licenses/by/4.0/).

Abstract: Physical activity increases human health potential and has an impact on achieving a higher quality of life in society. The aim of our research was to determine the relationship between a physically active lifestyle and the quality of life of female students in the context of demographic and social factors (major, age, marital status, professional activity). The research was conducted among a group of 285 women studying physical culture and social sciences in Poznań and Szczecin (Poland). Average age: 22.7 ± 4.90. The standardized World Health Organization Quality of Life—BREF (WHQOL-BREF) questionnaire was used to assess the quality of life of female students, and the original survey technique was used to study the lifestyle of people undertaking physical activity in the context of socio-demographic factors. Nonparametric statistics were applied in the analyses of the results. The effect size was calculated for each test: E2R for the Kruskal–Wallis H test, Glass rank biserial correlation (rg) for the Mann–Whitney U test, and Cramér's V for the $\chi 2$ test. The value of $p \leq 0.05$ was assumed to be a significant difference. In the study, it was shown that a higher overall quality of life and health satisfaction, as well as better results in the physical, psychological, and environmental domains, were achieved by female students who assessed their lifestyle as physically active in comparison to those physically inactive. Higher scores of overall quality of life and satisfaction with health were found among female students of physical education and people participating in physical recreation, who also achieved better results in the environmental domain. Female students aged 23–25 had a higher quality of life in the physical, psychological, and social domains. Having a partner or spouse had a positive effect on the quality of life of female students defined by the social domain. A higher overall quality of life and satisfaction with health were characteristic of people who were employed. In the search of factors positively influencing the quality of life of society, it seems necessary to promote a physically active lifestyle among students. The observed differences in the quality of life and health satisfaction of female students of selected majors require targeted programs and interventions that improve the quality of their lives at various stages of their studies. Such activities increase the health potential of the individual and society, not only in the biological, but also psychosocial dimension.

Keywords: physical activity; quality of life; health training; women; WHOQOL-BREF questionnaire

1. Introduction

The World Health Organization (WHO, Geneva, Switzerland) defines quality of life as an individual's perception of their position in life in the context of the culture and value systems in which they live and in relation to their goals, expectations, standards, and concerns (The World Health Organization Quality of Life–WHOQOL) [1]. Quality of life is perceived in the context of self-assessment on many levels (including physical, psychological, social, and spiritual), which is related to the individual level of health, satisfaction with life and health, and in relation to the physical, psychological, social, and environmental domains [2–4]. Quality of life has an effect on the wellbeing of an individual and his/her satisfaction with life [5]. The sense of quality of life and the factors influencing it evolve over the course of human life. Analyses conducted on the basis of the Social Diagnosis in Poland prove that the best predictors of the overall quality of life of Poles were the level of education, marriage, being an entrepreneur, or working in the public sector [6].

The quality of human life is related to the preferred lifestyle, which is also subject to constant changes. It includes everyday behavior, the characteristic ways of living of an individual or a social group. A feature that characterizes lifestyle is the possibility to make choices in certain spheres of life (e.g., education, work, family life, leisure time) above the breadline [7].

The health and physical activity of society are treated as a condition of basic professional and social competences, as well as the possibility of achieving a better quality of life. Physical activity is especially perceived as one of the most desirable forms of recreation and ways of spending leisure time due to its fundamental importance for health and the prevention of diseases of affluence. The use of leisure time by young people and representatives of various social groups, especially taking up its active forms (recreation, sport, tourism), is an important research subject. The diagnosis of real participation in these forms and the assessment of the general level of physical activity in society may constitute a measure of the consumption of leisure time as an essential element of modern human life [8].

Physical activity falls into the category of leisure time, which should be considered in the context of family and professional life, as well as in relation to the age of the respondents. Regular physical activity (recreational, sport) undertaken by women is a specific distinguishing featureof the lifestyle of an individual, as well as of a specific social group—a certain category of women. In the category of a physically active lifestyle, we observe the phenomenon of the interpenetration of a conscious choice of this behavior, leisure time, and life circumstances [9]. On the one hand, physical activity satisfies biological needs, and on the other, it supports social development [10]. It is also a tool for the health education of society; it shapes a physically active/sport lifestyle that is often associated with the choice of other pro-health behaviors (proper nutrition, avoiding smoking and consuming alcoholic beverages, systematic monitoring of one′s health) [11].

Lifestyle covers all behaviors of an individual or population, depending on his/her own interests, worldview, and hierarchy of values, as well as demographic factors [12]. A survey conducted at the beginning of 2019 on a representative random sample of 1858 Poles aged 15 and over showed that 64% of them declared physical activity at least once a month. This means that over one third of Poles did not undertake any physical activity, not even once a month. The most active were representatives of the young generation aged 15 to 24 (80%), persons in education (90%) or with higher education (78%), as well as Poles whose monthly income exceeded PLN 5000 (83%), above the national average [13].

Scientific research has repeatedly indicated the positive relationship between physical activity and people′s quality of life, health, and wellbeing [14]. The inhabitants of Wrocław aged 18–64 whose level of activity was assessed as high (at least 1500 METs/min/week) had a higher overall quality of life, as well as a higher perception of health and quality of life in the physical, psychological, social, and environmental domains. These relationships were weaker in the case of low physical activity. The chances of having a high overall perceived quality of life increase with growing levels of physical activity.

Quality of life and lifestyle (including physical activity) are determined by demographic, social, and economic factors. Some works emphasize the combined influence of many factors [15]. The research among Brazilian medical students indicates positive (male sex, self-efficacy, psychological and physical factors, including physical activity in leisure time, satisfaction with one′s body) and negative (female sex, economic status, place of residence, year of studies, chronic diseases, BMI, sleep problems, headaches) predictors of quality of life [16]. Similar results were also found among medical students in Pakistan and Chile [17,18].

The influence of peer groups, social norms, and environmental factors on the choice of specific behaviors is confirmed by research conducted in South Asia [19]. The level of physical activity, sleep intensity, and interpersonal, peer, and family relationships determined low scores of the quality of life of Chinese medical students [20]. High levels of anxiety and depression determined low scores of quality of life among Malay students of medical faculties [21]. Due to the unsatisfactory quality of life and health problems of students, it seems necessary to pay attention to their lifestyle and quality of life by applying a broad spectrum of research on health-promoting behaviors (including physical activity) of this social group. University students who are not related to physical culture and health show little interest in physical activity. There are concerns that groups of young, educated people not related to physical and health culture may avoid physical activity in the future and choose a physically inactive lifestyle that may be perceived as a role model, consequently lowering the quality of life of society. Universities of physical education, sport, tourism and recreation, and public health should intensify their efforts to ensure health for the individual and society. In the future, male and female students should promote a modern, physically active lifestyle that can contribute to improving the quality of life in society. In order to optimize our existence, it becomes necessary to monitor and diagnose conscious choices regarding physically active life in the context of quality of life, especially by women who undertake physical activity less frequently [17,18,22,23]. The aim of this work was to determine the relationship between a physically active lifestyle (assessment of lifestyle, physical activity, doing sports) and the quality of life of female students of selected majors in the context of demographic and social factors. A hypothesis regarding the relationship between a physically active lifestyle and the quality of life of female students was adopted. It was also assumed that major, age, marital status, and professional activity of the female respondents may modify this relationship.

2. Participants and Ethical Considerations

The research was conducted in the 2018/2019 academic year (October–December) among a group of 285 women studying physical education sciences (tourism and recreation–T&R, physical education–PE, public health–PH) and social sciences–SS in Poznań and Szczecin (Poland). Female first-cycle students accounted for 59.7%, and second-cycle students40.3%. According to research, 35.8% studied SS, 25.3% PH, 20.7% PE, and 18.2% T&R. The largest number of female students were aged 20–22 (48.1%), while the lowest number was aged $\geq$26 (11.6%). Most of the respondents were employed (62.5%).The female respondents most often lived in cities (68.1%).More than half (56.5%) had partners, and 7.7% were married. The female students declared having a physically active lifestyle and being rather physically active (75.4%). According to the study, 11.6% of the respondents practiced competitive sports at that time, and 68.4% of the women participated in physical recreation.

The inclusion criteria for the study group were as follows: age 18–35 years; female students of Polish universities; and fields of studies: physical education, public health, tourism and recreation, and social sciences. The criteria for exclusion from the study group were: age over 35, male gender, and women who were non-students or studying any other field of study than physical education, public health, tourism and recreation, and social sciences.

This study is a fragment of a largerstudy conducted as part of a research project on the health and quality of life of female students of universities in Poland. The first stage of the study (presented in the article) was carried out in two cities in Poland: Szczecin and Poznań. The authors of the study obtained the appropriate approvals to conduct research at the University of Szczecin and the Academy of Physical Education in Poznań, departments of physical education, public health, tourism and recreation, and social sciences (Bioethics Committee at the District Chamber of Physicians in Szczecin, Poland No. 15/KB/V/2015).

All authors of the article were involved in collecting the data, while the data analysis was carried out by two authors of this paper.

The analyzed study programs envisage the implementation of classes in the field of physical activity, varying in terms of content and hours. First-cycle PE (physical education) students had 460 movement classes, while second-cycle students had 200 such classes. These included, among others, team sports games, gymnastics, athletics, and swimming. The T&R program included 200 h of classes during first-cycle studies and 60 h during second-cycle studies (including movement games, animation forms, activity in the aquatic environment, and health training). The PH study program included 60 h during first-cycle studies and 120 h during second-cycle studies (30 h are spent on physical education, and the rest for recreation in public health and dance classes). Female SS students completed 30 h of physical education during their first-cycle studies. There were no physical activities for them during second-cycle studies.

The research was approved by the Bioethics Committee at the District Chamber of Physicians in Szczecin No. 15/KB/V/2015.The research was conducted in accordance with the Helsinki Declaration of 1975. Written informed consent was obtained from each subject included in the study.

3. Data Collection

The standardized WHOQOL-BREF (World Health Organization Quality of Life—BREF) questionnaire was used to assess the quality of life of female students. The shortened version, adapted to Polish conditions in terms of language, culture, and psychometry, contained 26 questions [4]. The first two questions about the degree of life satisfaction and health were analyzed separately. The remaining questions covered four areas of quality of life: physical (pain and discomfort, drug and treatment dependence, energy and fatigue, mobility, rest and sleep, everyday tasks, ability to work);psychological (joy of life, meaning in life, concentration, body image, self-esteem, feeling of sadness);social (personal relationships, sexual activity, social support); and environmental elements (safety, neighborhood, financial resources, access to information, recreation and leisure time, housing, access to medical care, transport). Scoring for the areas was determined by taking the arithmetic mean for each of them. It reflects the individual perception of the quality of life in individual domains. The higher the number of points, the better the quality of life.

In order to demonstrate the relationship between a physically active lifestyle and quality of life in the context of socio-demographic variables, the diagnostic poll method was used. The original survey technique was applied to study the lifestyle of people undertaking physical activity (assessment of their lifestyle as being physically active, confirmed by participation in physical recreation and/or doing sports). The information was confirmed with an analysis of the study programs, especially in terms of planned and ongoing physical activities. For characteristics of the respondents and their assessment of their quality of life, socio-demographic variables (major, age, marital status, professional activity) were used.

4. Data Analysis

Before selecting the methods of statistical inference, the distribution normality of the variables was examined. In all cases, an abnormal distribution was found, and therefore, non-parametric statistics were used. Nonparametric statistics were applied in the analyses of the results. The Kruskal–Wallis test (H) was used to compare several independent samples. In the case of determining the statistical significance of differences for the comparison of two independent samples, the Mann–Whitney (U) test was employed. In qualitative analyses, the trait frequency and the independence chi-squared test were used.

The effect size was calculated for each test: E^2_R for the Kruskal–Wallis H test, Glass rank biserial correlation (rg) for the Mann–Whitney U test, and Cramér's V for the χ^2 test. The value of $p \leq 0.05$ was assumed to be significantly different. Statistical calculations were made with Statistica 13.1 for Windows (StatSoft Sp. zo.o., Krakow, Poland) and Microsoft Office Excel 2007 (Microsoft Sp. z o.o., Warsaw, Poland).

5. Results

According to the research, 285 women (mean age: 22.7 ± 4.90) participated in the study. Most female students were in the age group of 20–22 (48.1%). The greatest number of people aged ≤19 were female students of physical education (40.7%). The oldest (≥26 years) were female students of social sciences (24.5%) ($p < 0.001$ for the $\chi 2$ test, $p = 0.3$ for Cramér's V). Female first-cycle students accounted for 59.7% ($p < 0.001$ for the $\chi 2$ test, $p = 0.3$ for Cramér's V). Most of the respondents were employed (62.5%), lived in cities (68.1%), and achieved secondary education (66%) ($p < 0.05$ for the $\chi 2$ test, $p = 0.2$ for Cramér's V). More than half of the surveyed women had partners without being in formal relationships (56.5%). Only 7.7% declared being married ($p < 0.001$ for the $\chi 2$ test, $p = 0.2$ for Cramér's V) (Table 1).

Table 1. Socio-demographic characteristics of students (independence χ^2 test and Cramér's V.

Variables	Major				Total (285)		p for χ^2	Cramér's V
	T&R (n = 52)	PE (n = 59)	PH (n = 72)	SS (n = 102)	n	%		
Age								
≤19	11.6	40.7	15.3	1.0	42	14.7	$p < 0.001$	0.3
20–22	44.2	44.1	52.8	49.0	137	48.1		
23–25	42.3	13.6	22.2	25.5	73	25.6		
≥26	1.9	1.7	9.7	24.5	33	11.6		
Degree of Study								
First-cycle	55.8	84.8	65.3	43.1	170	59.7	$p < 0.001$	0.3
Second-cycle	44.2	15.2	34.7	56.9	115	40.3		
Gainful Employment								
Yes	61.5	64.4	54.2	67.7	178	62.5	n.s.	-
No	38.5	35.6	45.8	32.2	107	37.5		
Place of Residence								
City <100 thousand	17.7	34.5	20.8	28.7	73	25.9	n.s.	-
City >100 thousand	58.8	41.4	45.8	31.7	119	42.2		
Village	23.5	24.1	33.4	39.6	90	31.9		
Marital Status								
I am not in a relationship	50.0	39.0	41.7	22.6	102	35.8	$p < 0.001$	0.2
I have a partner	48.1	55.9	55.7	61.8	161	56.5		
Married	1.9	5.1	2.8	15.7	22	7.7		

* n.s., not significant; T&R, tourism and recreation; PE, physical education; PH, public health; SS, social sciences.

The majority of female students declared having a physically active lifestyle and being rather physically active (75.4%) (Table 2). Female PE students most often assessed their lifestyle as physically active (66.1%), while SS students were the least likely to do so (14.7%). According to the study, 11.6% of the respondents practiced competitive sports at that time, most often studying PE (30.5%). Research analysis showed that 68.4% of women participated in physical recreation; most often, they werePE students (88.1%), and least often were SS students (52.0%). All female students who assessed their lifestyle as physically active or rather physically active undertook physical activity at a similar level (participation in physical recreation and doing sports—in total 80.0%). Female PE students more often participated in recreation and practiced competitive sports, as well as more often assessed their lifestyle as physically active ($p < 0.001$ for the $\chi 2$ test), but the effect size for these relationships was average ($p = 0.3$ for Cramér's V).

Table 2. Physically active lifestyle as assessed by the female respondents, participation in physical recreation, and doing sports by major (independence χ^2 test and Cramér's V).

Variables	Major				Total (285)		p for χ^2	Cramér's V
	T&R * (n =52)	PE * (n =59)	PH * (n =72)	SS * (n =102)	n	%		
Assessment of One's Lifestyle								
As physical active	25.0	66.1	20.8	14.7	82	28.8	$p < 0.001$	0.3
As rather physically active	53.9	30.5	52.8	49.0	134	47.0		
As not physically active	21.1	3.4	26.4	36.3	69	24.2		
Competitive Sport								
Yes	9.6	30.5	4.2	6.9	33	11.6	$p < 0.001$	0.3
No	90.4	69.5	95.8	93.1	252	88.4		
Participation in Physical Recreation								
Yes	80.8	88.1	66.7	52.0	195	68.4	$p < 0.001$	0.3
No	19.2	11.9	33.3	48.0	90	31.6		

* T&R, tourism and recreation; PE, physical education; PH, public health; SS, social sciences.

General diversity was found between satisfaction with life and health and the assessment of one's lifestyle ($p < 0.001$, $E^2_R = 0.06$; $p < 0.001$, $E^2_R = 0.07$, respectively) (Table 3). People who evaluated their lifestyle as physically active or rather physically active assessed their satisfaction with life and satisfaction with health more highly than those evaluating theirlifestyle as inactive ($p < 0.001$; $p < 0.01$ for the U test; $p < 0.001$; $p < 0.05$ for the U test, respectively). These differences were confirmed by the above-average effect strength (rg = 0.4). Moreover, female students who evaluated their lifestyle as physically active assessed their satisfaction with life and health more highly compared to those who evaluated their lifestyle as rather physically active (effect strength below average). Overall diversity was observed in the physical, psychological, and environmental domains. In the physical domain, higher scores were achieved by female students who assessed their lifestyle as rather physically active compared to perceiving it as inactive ($p < 0.05$ for the U test). In the psychological domain, female students who perceived their lifestyle as physically or rather physically active achieved higher scores compared to inactive ones ($p < 0.01$ for the U test U, rg = 0.3; $p < 0.01$ for the U test, respectively). In the environmental domain, the lowest results were achieved by female students who perceived their lifestyle as inactivecompared to physically active and rather active ($p < 0.01$; $p < 0.01$ for the U test, respectively). There was no such diversity in the social domain.

Table 3. Satisfaction with life, health, and particular domains of quality of life (WHOQOL-REF) of female students by a physically active or not physically active lifestyle (U test, rg).

Specification	Assessment of One's Lifestyle	Value of p for U Statistics		Glass Rank Biserial Correlation (rg)		Rank Means
		As Rather Physically Active	As Not Physically Active	As Rather Physically Active	As Not Physically Active	
Satisfaction with LifeH(2279) = 17.15E^2_R = 0.06 $p < 0.001$	As Physically Active As Rather Physically Active As Not Physically Active	0.034	0.001 0.006	0.2	0.4 0.2	161.94 140.93 111.96
Satisfaction with HealthH(2279) = 20.81 $E^2_R = 0.07$ $p < 0.001$	As Physically Active As Rather Physically Active As not Physically Active	0.002	0.000 0.022	0.2	0.4 0.2	167.30 137.28 112.76
Physical DomainH(2285) = 5.75 $E^2_R = 0.02$ $p < 0.05$	As Physically Active As Rather Physically Active As Not Physically Active	0.714	0.059 0.021	−0.1	0.2 0.2	147.13 150.99 122.57
Psychological Domain H(2285) = 11.0 $E^2_R = 0.03$ $p < 0.001$	As Physically Active As Rather Physically Active As Not Physically Active	0.397	0.002 0.004	0.1	0.3 0.2	157.38 148.49 115.25
Social Domain H(2285) = 0.34 $E^2_R = 0.01$ n.s.	As Physically Active As Rather Physically Active As Not Physically Active	0.567	0.874 0.733	−0.1	−0.1 0.1	139.40 145.89 141.67
Environmental DomainH(2285) = 9.29 $E^2_R = 0.01$ $p < 0.05$	As Physically Active As Rather Physically Active As Not Physically Active	0.869	0.010 0.004	0.1	0.2 0.2	152.21 150.86 116.78

The female respondents participating in physical recreation had a higher assessment of quality of life and health, and better results in the environmental domain compared to physically passive women ($p < 0.01$; $p < 0.05$; $p < 0.05$ for the U test, respectively). Similarly, women who declared doing sports had higher satisfaction with their health compared to those not doing sports ($p < 0.05$ for the U test), but the effect strength was small (rg = 0.1–0.2) (Table 4).

Table 4. Satisfaction with life, health, and particular domains of quality of life (WHOQOL-BREF) of female students by participation in physical recreation and sports (U test, rg).

Specification		Answers	Value of p for U Statistics	Glass Rank Biserial Correlation (rg)	Rank Means
Physical Activity	Satisfaction with life	Yes No	0.010	0.2	147.44 123.83
	Satisfaction with health	Yes No	0.028	0.1	146.49 125.90
	Environmental domain	Yes No	0.025	0.2	150.36 127.03
Doing Competitive Sport	Satisfaction with health	Yes	0.016	0.2	168.96

There was no general difference in the assessment of overall quality of life between female students of selected majors. These differences were observed between the assessment of quality of life by female PE and SS students ($p < 0.01$ for the U test; rg = 0.4) (Table 5). Female PE students rated their quality of life higher (the strength of the effect was above average). A general differentiation was found in the assessment of satisfaction with health by the respondents ($p < 0.05$ for the H test). Female PE students were also more satisfied with health than female SS students ($p<0.01$ for the U test). This was confirmed by the high strength of the effect (rg = 0.5). The general differentiation between women of selected majors in the physical domain was confirmed ($p < 0.05$ for the H test). Differences were observed between women studying PH and PE and PH and SS ($p < 0.05$, $p < 0.05$ for the U test; rg = 0.4, respectively). Female PH students, compared to female PE and SS students, achieved higher results in the physical domain (above average effect strength). Pairwise comparisons indicated differences in the physical domain between female T&R

and SS students ($p < 0.05$ for the U test; rg = 0.4). Female T&R students achieved higher scores in the environmental domain. There were no differences in the psychological and social domains.

Table 5. Satisfaction with life, health, and particular domains of quality of life (WHOQOL-BREF) of female students by major (H test, E^2_R, U test, rg).

Specification	Majors	PH *	PE *	SS *	PH *	PE *	SS *	Rank Means
		Value of p for U Statistics			Glass Rank Biserial Correlation (rg)			
Satisfaction with Life H(3279) = 6.65 $E^2_R = 0.02$ n.s.	T&R	0.890	0.134	0.356	0.1	−0.4	0.2	140.37
	PH		0.086	0.399		−0.3	0.1	138.57
	PE			0.011			0.4	160.29
	SS							129.04
Satisfaction with Health H(3, 279) = 9.61 $E^2_R = 0.03$ $p < 0.05$	T&R	0.860	0.228	0.107	0.1	−0.3	0.3	144.42
	PH		0.141	0.114		−0.3	0.2	142.12
	PE			0.002			0.5	160.75
	SS							124.2
Physical Domain H(3285) = 8.61 $E^2_R = 0.03$ $p < 0.05$	T&R	0.657	0.105	0.067	−0.1	0.4	0.3	155.74
	PH		0.035	0.014		0.4	0.4	161.52
	PE			0.780			0.1	132.05
	SS							129.75
Environmental Domain H(3, 285) = 6.45 $E^2_R = 0.02$ n.s.	T&R	0.580	0.216	0.016	0.2	0.3	0.4	161.10
	PH		0.539	0.092		0.1	0.3	150.95
	PE			0.258			0.2	142.67
	SS							128.33

* T&R, tourism and recreation; PE, physical education; PH, public health; SS, social sciences.

The overall quality of life and health satisfaction of the female students were not related to their age; however, the general differentiation in the physical, psychological, and social domains was confirmed ($p<0.001$; $p<0.01$; $p<0.05$ for the H-test, respectively) (Table 6). Women aged 23–25 years had better results in the physical domain compared to those aged ≤19, 20–22, and ≥26 ($p<0.001$, $p<0.01$, $p<0.05$ for the U test, respectively). In the psychological domain, women aged 23–25 had higher scores compared to the group of ≤19 year olds ($p<0.001$ for the U test; rg =−0.4) and 20–22 year olds ($p<0.01$ for the U test; rg = −0.2). Female students aged 23–25 achieved higher results compared to female respondents aged ≤19 in the social domain ($p<0.01$ for the U test; rg = −0.3).

Table 6. Satisfaction with life, health, and particular domains of quality of life (WHOQOL-BREF) of female students by age (H test, E^2_R, U test, rg).

Specification	Age	20–22	23–25	≥26	20–22	23–25	≥26	Rank Means
		Value of p for U Statistics			Glass Rank Biserial Correlation (rg)			
Physical Domain H(3285) = 16.3432 $E^2_R = 0.06$ $p < 0.001$	≤19	0.068	0.000	0.499	−0.2	−0.4	−0.1	112.58
	20–22		0.008	0.416		−0.2	0.1	140.39
	23–25			0.013			0.3	172.30
	≥26							127.69
Psychological Domain H(3, 285) = 13.91 $E^2_R = 0.05$ $p < 0.001$	≤19	0.108	0.000	0.154	−0.2	−0.4	−0.2	113.98
	20–22		0.005	0.789		−0.2	−0.1	137.54
	23–25			0.119			0.2	170.28
	≥26							142.22
Social Domain H(3, 285) = 7.95 $E^2_R = 0.03$ $p < 0.05$	≤19	0.161	0.009	0.605	−0.1	−0.3	−0.1	121.59
	20–22		0.080	0.496		−0.2	0.1	141.93
	23–25			0.053			0.2	162.85
	≥26							130.75

There was a general differentiation between the social domain of quality of life of female students by marital status ($p<0.001$ for the H test; $E^2_R=0.2$). Unmarried female students achieved lower results compared to married women and women with partners ($p<0.001$ for the U test, rg = −0.5; $p<0.01$ for the U test, rg =−0.4) (Table 7). These differences were confirmed by the effect strength.

Table 7. Satisfaction with life, health, and particular domains of quality of life (WHOQOL-BREF) of female students by marital status (H test, E^2_R, U test, rg).

Specification	Marital Status	Married Woman	Having a Partner	Married Woman	Having a Partner	Rank Means
		Value of p for U Statistics		Glass Rank Biserial Correlation (rg)		
Social Domain H(3, 285) = 49.31E^2_R=0.2 p<0.001	Single	0.000	0.005	−0.5	−0.4	97.82
	Married Woman		0.430		0.1	170.13
	Having a Partner					153.88

Female students undertaking work had higher scores of overall quality of life and health satisfaction compared to those non-working ($p<0.05$; $p<0.001$ for the U test, respectively), however, the effect size was small (Table 8). In terms of individual areas of quality of life, these differences were not found.

Table 8. Satisfaction with life, health, and particular domains of quality of life (WHOQOL-BREF) of female students by gainful employment/professional activity (H test, E^2_R, U test, rg).

Specification	Gainful Employment	No	No	Rank Means
		Value of p for U Statistics	Glass Rank Biserial Correlation (rg)	
Satisfaction with Life	Yes	0.025	0.1	147.71
	No			127.59
Satisfaction with Health	Yes	0.000	0.2	151.72
	No			121.15

6. Discussion

When undertaking the study, ahypothesis regarding the relationship between a physically active lifestyle and the quality of life of female students was adopted. It was also assumed that major, age, marital status, and professional activity of the female respondents may modify this relationship. Relationships between the assessment of one's lifestyle and the quality of life of female students of selected majors (physical education, tourism and recreation, public health, social sciences) was observed. The assessment by female students of their lifestyle as physically active or rather physically active was correlated with a better overall quality of life in particular domains (except for the social domain) and greater satisfaction with health. There were no differences in terms of personal relationships, sexual activity, and social support (social domain). The majority of female students assessing their lifestyle as physically active participated in physical recreation, and over 1/10 of them practiced competitive sports. Earlier studies of physical activity of female students of tourism and recreation and physical education confirmed greater physical activity and more frequent practicing of sports by female students of physical education [24].

In nationwide diagnostic studies, it was found that people who undertook any kind of physical activity achieved higher health scores [6]. It was also noticed that students who regularly practiced aerobic exercise and strength training had a higher quality of life than those physically inactive. It was also demonstrated that a significant modifier of quality of life is high self-esteem resulting from physical activity [25–27]. Studies among female students of various majors (including social sciences) indicated that a healthy lifestyle (including activity and physical fitness) had an effect on a higher evaluation of quality of life [28]. Students who were engaged in physical activity more typical for their lifestyle experienced greater satisfaction with life [29]. A high level of quality of life was observed among students of various majors from Lviv who had a higher level of physical activity [30]. Studies among Croatian youth show a higher level of physical activity among physiotherapy students and their higher quality of life (in the physical and social fields) compared to social sciences students who had lower levels of physical activity, but better results in terms of the mentality domain [26]. At the same time, there are studies regarding students of physiotherapy that do not confirm the relationship between a very high level of physical activity of respondents and their self-esteem and quality of life, regardless of gender [31]. However, a significant positive correlation between high levels of physical activity and quality of life (in all domains) was found among Brazilian medical students (both sexes). For students with low levels of physical activity, the correlation was only significant in the case of the physical and social domains [32].

Differences in the quality of life of female students of selected majors were observed. Physical education female students had a higher overall evaluation of quality of life and expressed greater satisfaction with health compared to those studying social sciences. Public health female students achieved higher results in the physical domain than those studying physical education and social sciences. In the environmental domain, female T&R students scored higher than female students in social sciences. Female social science students had the lowest evaluation for overall quality of life, as well as in the physical and environmental domains, and were the least satisfied with their health. Higher results of public health female students in the physical domain (pain and discomfort, drug and treatment dependency, energy and fatigue, mobility, rest and sleep, everyday activities, ability to work) may result from better coping with various situations by these female students, who, as a result of their preparation, are more aware of how their bodies work and are more prone to pro-health behavioral choices. Similarly, higher scores of female T&R students in the environmental domain (safety, neighborhood, financial resources, access to information, recreation and leisure time, housing, access to medical care, transport) are probably the result of studies in this field.

Comparisons of students of physical education and other faculties of the T.C.Hitit University (Corum, Turkey) showed that quality of life, along with the physical, mental, social, and environmental domains, was higher among people studying physical education [33]. It has also been shown that the probability of a high level of perceived overall quality of life rises with the increase in the level of physical activity of people aged 18–64 [14]. Physical activity and sleep duration had a positive effect on the quality of life of Chinese medical students [34]. Studies of Polish students of various majors (public health, physiotherapy, tourism and recreation, psychology, pedagogy, and theology) indicated that only some types of physical activity show a positive relationship with quality of life. Physical activity in the household was most positively and significantly correlated with quality of life [35]. It should be noted that women are more active in housework, and men are more active in their leisure time [36]. Housework comprises activities that result from daily duties. Despite the physical exertion associated with them, this activity cannot be classified as a lifestyle, the essence of which is having a choice. Despite some inconsistencies of the results of various studies resulting from the comparison of female students of many majorsin different periods of time, often using different methods, there is a relationship between the choices regarding a physically active lifestyle and quality of life.

It was confirmed that socio-demographic factors (age, marital status, and professional activity) modify the relationship between the choice of a physically active lifestyle and quality of life of female students of selected majors. In the study, it was observed that women aged 23–25 assessed their quality of life the highest in the physical, psychological, and social domains, while the youngest students aged ≤19 assessed them the lowest. Additionally, a tendency towards lower quality of life scores among women aged ≤26 was noticed. These results have been partially confirmed by other studies [5]. Senior students perceived quality of life in the physical and environmental domains as lower. The results in the physical domain were significantly higher for the fourth year respondents compared to the first year (younger) ones. International studies of nursing students (Chile, Egypt,

Greece, Hong Kong, India, Kenya, Oman, Saudi Arabia, and the USA) showed that they achieved the highest level of quality of life in the physical domain and the lowest in the social domain. The main determinants of quality of life include, among others, age [37].

It was observed that the quality of life of female students in the social domain increased with being in a formal or informal relationship. A study of dentistry students in the United States showed that single students (especially men) perceived quality of life lower than married students [5].The quality of life of medical students improved among married people [38]. Research conducted among Chinese medical students (without a gender breakdown) shows that, among others, family satisfaction had an effect on quality of life [20].

It was noticed that female students undertaking paid work were characterized by a higher quality of life and health satisfaction. However, there were no differences between working and non-working women in the particular domains of quality of life. Research has shown that increasing working hours has a negative impact on academic performance. Students who work full-time while studying are less likely to complete their studies than those who work part-time ordo not work at all [39].

The obtained research results show the importance of physical activity for the quality of life of the respondents. The study confirmed other studies conducted by researchers in various countries around the world, as outlined above. However, a comparison of the obtained results with the research conducted in the United States, among others, deserves attention. The results of a study conducted at US universities showed that women taking up sports (regardless of the chosen field of study) displayed higher health-related satisfaction. Particular attention was paid to the consistency of undertaken physical exercises and the choice of various forms of physical activity [40]. Research conducted in France also proved the impact of various forms of physical activity on overall quality of life, as well as on health satisfaction [41]. The impact of physical activity on health is a widely studied topic among Chinese students. Our study confirmed that socio-demographic factors have a significant impact on quality of life. The research carried out by Chinese researchers showed (and confirmed the results of this study) that female students who were not in a relationship and did not take up paid work were characterized by a lower quality of life [42].

In search of factors positively influencing the quality of life of society, it seems necessary to promote a physically active lifestyle (regular physical activity, participation in physical recreation, doing sports) not only among students of physical culture sciences and public health, but also other majors. The observed differences in the quality of life and satisfaction with health of female students of particular majors require targeted programs improving the quality of life of female students at various stages of their studies. Such activities increase the health potential of the individual and society, not only in the biological, but also psychosocial dimension.

The use of standardized questionnaires in further research will allow for a wider observation of the quality of life of women. In addition, increasing the number of respondents would help in further determining the quality of life of the respondents, as well as in assessing the impact of an active lifestyle on quality of life, which the respondents considered crucial for strengthening and improving health. After extensive research in the group of university students, more constructive conclusions can be drawn that would allow specific preventive measures to be taken from an early age in order to change incorrect health behaviors (especially increasing physical activity), and thus improve the health and quality of women's lives.

7. Conclusions

A higher overall quality of life and satisfaction with health of female students evaluating their lifestyle as physically active compared to those who were inactive was indicated in the research. Better results in the physical, psychological, and environmental domains of women with a physically active lifestyle were also confirmed. Higher scores for overall quality of life and in the environmental domain, as well as greater satisfaction with health, were characteristic of female students participating in physical recreation.

The impact of the selected descriptive variables on overall quality of life was observed. Higher scores for overall quality of life and health satisfaction were seen among female PE students. PH students scored better in the physical sphere, while T&R students were better in the environmental sphere. Students aged 23–25 performed better in physical, psychological, and social domains. The influence of respondents' marital status on quality of life was also observed: having a partner or a spouse had a positive effect on the quality of life of female students, defined by the social domain. Students who took up gainful employment were characterized by a high overall quality of life and health satisfaction.

Author Contributions: Conceptualization, M.A.N. and K.S.; methodology, M.A.N. and K.K. (Katarzyna Kotarska); software, K.K. (Katarzyna Kotarska); validation, K.K. (Kamila Kmieć.); formal analysis, M.P.-J.; investigation, K.S.; resources, K.S. and M.P.-J.; data curation, K.K. (Kamila Kmieć) and A.C.; writing—original draft preparation, M.A.N. and A.C. All authors have read and agreed to the published version of the manuscript.

Funding: This research received no external funding.

Institutional Review Board Statement: The study was conducted according to the guidelines of the Declaration of Helsinki and approved by the Ethics Committee of Pomeranian Medical University of Szczecin Poland (protocol code: 15/KB/V/2015, date: 10 December 2015).

Informed Consent Statement: Informed consent was obtained from all subjects involved in the study.

Data Availability Statement: Data are not publicly available and data sharing is not applicable to this article.

Conflicts of Interest: The authors declare no conflict of interest.

References

1. WHO. *WHOQOL Measuring Quality of Life*; World Health Organization: Geneva, Switzerland, 1997; Available online: https://www.who.int/toolkits/whoqol (accessed on 8 January 2020).
2. Dziurowicz-Kozłowska, A. Around the Concept of Quality of Life. *Psychol. Qual. Life* **2000**, *2*, 77–99.
3. Panzini, R.G.; Mosqueiro, B.P.; Zimpel, R.R.; Bandeira, D.R.; Rocha, N.S.; Fleck, M.P. Quality-of-life and spirituality. *Int. Rev. Psychiatry* **2017**, *29*, 263–282. [CrossRef] [PubMed]
4. Harmin, E.K.; Gustafsson, G.; Jaracz, K. Quality of life among the elderly with locomotor disabilities in Sweden and Poland in the 1990s. *Qual. Life Res.* **2012**, *21*, 281–289.
5. Andre, A.; Pierre, G.C.; McAndrew, M. Quality of Life Among Dental Students: A Survey Study. *J. Dent. Educ.* **2017**, *81*, 1164–1170. [CrossRef]
6. Włodarczyk, K. Quality of life perceived by Poles in the 21st century. *Konsumpcja Rozw.* **2015**, *1*, 3–16.
7. Siciński, A. *Lifestyle. Culture. Choice*; IFiS PAN: Warszawa, Poland, 2002.
8. Biernat, E. Sport and other motor activities of Warsaw students. *Biom. Hum. Kinet.* **2011**, *3*, 10–13. [CrossRef]
9. Nowak, M. Physical Activity in a Healthy Lifestyle of Women. In *Monografie*; AWF: Poznań, Poland, 2008.
10. Osiński, W. The Theory of Physical Education. In *Monografie*; AWF: Poznań, Poland, 2011.
11. Nowak, M.A.; Kotarska, K.; Nowak, L. Physical Activity, Health and Physical Fitness of Students, Their Parents and Grandparents. *Coll. Antropol.* **2019**, *43*, 1–10.
12. Iftikhar, A. Life style and health education. *Gomal J. Med. Sci.* **2019**, *17*, 63–64.
13. MultiSport Index. Survey of Physical Activity Among Poles. 2019. Available online: https://www.benefitsystems.pl/en/about-us/press-centre/press-release/multisport-index-2019-survey-of-physical-activity-among-poles/ (accessed on 27 April 2020).
14. Puciato, D.; Rozpara, M.; Borysiuk, Z. Physical Activity as a Determinant of Quality of Life in Working-Age People in Wrocław, Poland. *Int. J. Environ. Res. Public Health* **2018**, *15*, 623. [CrossRef]
15. Al-Nagar, R.A.; Osman, M.T.; Musa, R. Quality of Life among University Students in a Single Malaysian Institute. *Pensee J.* **2013**, *75*, 165–179.
16. Solis, A.C.; Lotufo-Neto, F. Predictors of quality of life in Brazilian medical students: A systematic review and meta-analysis. *Braz. J. Psychiatry* **2019**, *41*, 556–567. [CrossRef]
17. Aziz, Y.; Khan, A.Y.; Shahid, I.; Khan, M.A. Quality of life of students of a private medical college. *Pak. J. Med. Sci.* **2020**, *36*, 255–259. [CrossRef]
18. Irribarra, T.L.; Mery, I.P.; Lira, S.M.J.; Campos, D.M.; González, L.F.; Irarrázaval, D.S. Quality of life scores among 411 medical students. *Rev. Med. Chil.* **2018**, *146*, 1294–1303. [CrossRef]

19. Kabir, A.; Miah, S.; Islam, A. Factors influencing eating behavior and dietary intake among resident students in a public university in Bangladesh: A qualitative study. *PLoS ONE* **2018**, *13*, 17. [CrossRef]
20. Qiu, Y.; Yao, M.; Guo, Y.; Zhang, X.; Zhang, S.; Zhang, Y.; Huang, Y.; Zhang, L. Health-Related Quality of Life of Medical Students in a Chinese University: A Cross-Sectional Study. *Int. J. Environ. Res. Public Health* **2019**, *16*, 5165. [CrossRef]
21. Gan, G.G.; Yuen-Ling, H. Anxiety, depression and quality of life of medical students in Malaysia. *Med. J. Malaysia* **2019**, *74*, 57–61.
22. Griban, G.P.; Yavorska, T.Y.; Tkachenko, P.P.; Kuvaldina, O.V.; Dikhtiarenko, Z.M.; Yeromenko, E.A.; Lytvynenko, A.M.; Hresa, N.V.; Okhrimenko, I.M.; Ovcharuk, I.S.; et al. Motor activity as the basis of a healthy lifestyle of student youth. *Med. News* **2020**, *73*, 1199–1206. [CrossRef]
23. Gruzieva, T.S.; Galiienko, L.I.; Pelo, I.M.; Omelchuk, S.T.; Antonuk, O.Y. Health and lifestyle of students' youth: Status, problems and ways of solution. *Med. News* **2018**, *71*, 1753–1758.
24. Lic, Ł.; Frankiewicz, M. Health-promoting Elements of the Lifestyle of Female Students of The University School of Physical Education in Krakow. *Med. Nov.* **2007**, *76*, 233–236.
25. Dyrbye, L.N.; Satele, D.; Shanafelt, T.D. Healthy Exercise Habits Are Associated With Lower Risk of Burnout and Higher Quality of Life Among, U.S. Medical Students. *Acad. Med.* **2017**, *92*, 1006–1011. [CrossRef]
26. Sklempe-Kokic, I.; Znika, M.; Brumnic, V. Physical activity, health-related quality of life and musculoskeletal pain among students of physiotherapy and social sciences in Eastern Croatia—Cross-sectional survey. *Ann. Agric. Environ. Med.* **2019**, *26*, 182–190. [CrossRef]
27. Joseph, R.P.; Royse, K.E.; Benitez, T.J.; Pekmezi, D.W. Physical activity and quality of life among university students: Exploringself-efficacy, self-esteem, and affect as potentialmediators. *Qual. Life Res.* **2014**, *23*, 659–667. [CrossRef]
28. Rogo, B. Female Student Lifestyle and Their Health Beliefs and Preferred Values. In *Pedagogical Dimension of Physical and Health Culture in the Life of Modern Man*; Kaźmierczak, A., Kowalska, J.E., Maszorek-Szymala, A., Makarczuk, A., Eds.; University of Łódź: Łódź, Poland, 2016; pp. 229–246.
29. Maher, J.P.; Doerksen, S.E.; Elavsky, S.; Conroy, D.E. Daily satisfaction with life is regulated by both physical activity and sedentary behavior. *J. Sport Exerc. Psychol.* **2014**, *36*, 166–178. [CrossRef]
30. Pavlova, J.; Vynogradskyi, B.; Kurchaba, T.; Zikrach, D. Influence of leisure-time physical activity on quality of life of Ukrainian students. *J. Phys. Educ. Sport* **2017**, *17*, 1037–1042.
31. Gruszczyńska, M.; Skorupa, P. Physical activity, quality of life and self-esteem. *Probl. Hig. Epidemiol.* **2018**, *99*, 69–73.
32. Peleias, M.; Tempski, P.; Paro, H.B.; Perotta, B.; Mayer, F.B.; Enns, S.C.; Gannam, S.; Pereira, M.A.D.; Silveira, P.S.; Santos, I.S.; et al. Leisure time physical activity and quality of life in medical students: Results from a multicentre study. *BMJ Open Sport Exerc. Med.* **2017**, *3*, e000213. [CrossRef]
33. Ciçek, G. Quality of Life and Physical Activity among University students. *Univ. J. Educ. Res.* **2018**, *6*, 1141–1148. [CrossRef]
34. Ge, Y.; Xin, S.; Luan, D.; Zou, Z.; Liu, M.; Bai, X.; Gao, Q. Association of physical activity, sedentary time, and sleep duration on the health-related quality of life of college students in Northeast China. *Health Qual. Life Outcomes* **2019**, *17*, 124. [CrossRef]
35. Nowak, P.F.; Bożek, A.; Blukacz, M. Physical Activity, Sedentary Behavior, and Quality of Life among University Students. *Biomed. Res. Int.* **2019**, *18*, 1324–1329. [CrossRef]
36. Raposa, B.; Melczer, C.; Rozmann, N.; Károly, T.; Takacs, G.; Doma, V.; Premusz, V.; Pintér, M.; Macharia, J.; Ács, P. Investigation of physical activity and sports consumption habits in university students. *Health Prob. Civil.* **2020**, *14*, 211–220. [CrossRef]
37. Cruz, J.P.; Felicilda-Reynaldo, R.F.D.; Lam, S.C.; Machuca Contreras, F.A.; Cecily, H.S.J.; Papathanasiou, I.V.; Fouly, H.A.; MachariaKamau, S.; Valdez, G.F.D.; Adams, K.A.; et al. Quality of life of nursing students from nine countries: A cross-sectional study. *Nurse Educ. Today* **2018**, *66*, 135–142. [CrossRef] [PubMed]
38. Heidari, M.; Majdzadeh, R.; Pasalar, P.; Nedjat, S. Quality of life of medical students in Tehran University of Medical Sciences. *Acta Med. Iran.* **2014**, *52*, 390–399. [PubMed]
39. Hovdhaugen, E. Working While Studying: The Impact of Term-Time Employment on Dropout Rates. *J. Educ. Work* **2015**, *28*, 631–651. [CrossRef]
40. Snedden, T.; Scerpella, J.; Kliethermes, S.; Norman, R.; Blyholder, L.; Sanfilippo, J.; McGuine, T.; Heiderscheit, B. Sport and Physical Activity Level Impacts Health-Related Quality of Life Among Collegiate Students. *Am. J. Health Promot.* **2019**, *33*, 675–682. [CrossRef]
41. Diot, P.; Arnault, É. Health students quality of life: A challenge for universities. *Presse Med.* **2019**, *48*, 97–99. [CrossRef]
42. Mak, Y.W.; Kao, A.H.F.; Tam, L.W.Y.; Tse, V.W.C.; Tse, D.T.H.; Leung, D.Y.P. Health-promoting lifestyle and quality of life among Chinese nursing students. *Prim. Health Care Res. Dev.* **2018**, *19*, 629–636. [CrossRef]

International Journal of
Environmental Research and Public Health

MDPI

Article

Morphological and Motor Fitness Determinants of Shotokan Karate Performance

Paweł Przybylski [1], Arkadiusz Janiak [2], Piotr Szewczyk [2], Dariusz Wieliński [3] and Katarzyna Domaszewska [4,*]

1 Department of Various Sports and Camp Organisation, Poznan University of Physical Education, 61-871 Poznań, Poland; przybylski@awf.poznan.pl
2 Faculty of Health Sciences, Calisia University, 62-800 Kalisz, Poland; a.janiak@akademiakaliska.edu.pl (A.J.); p.szewczyk@akademiakaliska.edu.pl (P.S.)
3 Department of Anthropology and Biometry, Poznan University of Physical Education, 61-871 Poznań, Poland; wielinski@awf.poznan.pl
4 Department of Physiology and Biochemistry, Poznan University of Physical Education, 61-871 Poznań, Poland
* Correspondence: domaszewska@awf.poznan.pl; Tel.: +48-(61)-835-51-90

Citation: Przybylski, P.; Janiak, A.; Szewczyk, P.; Wieliński, D.; Domaszewska, K. Morphological and Motor Fitness Determinants of Shotokan Karate Performance. *IJERPH* **2021**, *18*, 4423. https://doi.org/10.3390/ijerph18094423

Academic Editors: Wioletta Zukiewicz-Sobczak, Paulina Wojtyla-Buciora, Izabela Rącka, Andrzej Wojtyla and Paul B. Tchounwou

Received: 15 February 2021
Accepted: 20 April 2021
Published: 21 April 2021

Publisher's Note: MDPI stays neutral with regard to jurisdictional claims in published maps and institutional affiliations.

Copyright: © 2021 by the authors. Licensee MDPI, Basel, Switzerland. This article is an open access article distributed under the terms and conditions of the Creative Commons Attribution (CC BY) license (https://creativecommons.org/licenses/by/4.0/).

Abstract: The achievement of high performance levels in a complex structured sport such as karate is determined by the competitor's physical fitness, fighting technique, tactics and mental state. This study aimed to identify the most important determinants of top-level performance in karate. Methods: The participants were 32 karate competitors (12 women and 20 men) aged 18–25 years. A series of tests measuring 11 anthropometric features was undertaken twice during a year, separated by a 6-week interval during a training camp at the Olympic Preparation Center in Walcz, Poland. Motor skills were measured with strength, speed, endurance, flexibility and reaction time tests. Special motor fitness was assessed with tests of karate technical skills. The results were subject to statistical analysis using multiple stepwise regression of the Polish Karate Federation ranking points as the dependent variable. Results: The multiple regression analysis revealed two main determinants of high scores in female and male karate competitors. In women, these were thigh circumference and the speed of the mawashi-geri-kick roundhouse technique (i.e., the maximum number of delivered kicks in 30 s), whilst for men they were the extent of the sideway leg swing to the highest possible height (yoko-geri) and general endurance assessed with the bent arm hang test. Conclusion: Karate training should account for the determinants of high-level competitive karate performance identified in this study. Strengthening the lower limbs, exercises increasing hip joint mobility, low position movements, performing leg techniques in various planes and applying external loads undoubtedly increase a karate athlete's strength and lead to the development of a more extensive repertoire of karate leg techniques, especially at the highest-scoring head level (jodan).

Keywords: special motor fitness; speed; flexibility; endurance

1. Introduction

The martial art of karate requires highly developed technical skills, which, amongst others, include control of static and dynamic movements. Karate is a highly dynamic tournament sport. Both kumite and kata are karate techniques performed in an extremely dynamic, precise and coordinated fashion. Karate has a complex structure, and the competitors' physical fitness, technique, tactics and mental state all affect successful performance [1,2]. During a kumite fight, attacks need to be executed at maximum speed to ensure that the attacker's hand or the foot reaches its target without the opponent being able to respond [3]. Such fights are highly dynamic and intense [4]. Karate competitors' actions are based on acyclic movements, and numerous muscle groups and all four limbs are engaged during combat. Karate fighters need to possess a great deal of strength to ensure that rapid movements are deployed with high precision. Attacking actions are often executed at very high speeds in karate training and competition, during which the fighters must strike before their opponents have any chance to defend themselves from an attack or counterattack. During combat, a karate contestant's effectiveness is determined by multiple factors, such as on their adopted technique, tactics and motor skills [5,6]. Achieving championship-level performance requires undertaking long-term training to ensure the body is properly, morphologically and physiologically, conditioned. Successful performance in karate is known to depend on many factors, such as the competitor's somatic build, fitness, level of technical preparation and mental state [4,7,8].

Becoming a sport champion is a long-term process that begins with the involvement of children and adolescents in sport. The first stage of a sports career is a functional component of competitive sport and it always lays the foundations for future championships. Maintaining an appropriate training system is key in championship building. Training should be optimized for achieving the best results and aimed at rationalizing the improvement of a contestant's functions and skills at every stage of their biological and sporting development. Solutions are still being sought to develop the best selection criteria and to overcome limitations. Performance outcomes have always constituted such criteria, whilst limitations have been the principal determinants of the body's biological development and functional capabilities. High-level sport performance can be only achieved by contestants in the right conditions. Selecting the appropriate morphological criteria for a given sport is vital because the body's somatic structure is strongly determined by genetics, which can be modified by training only to a limited extent [9–11].

The long-term process of rationally targeted athlete training is determined by a model use of specific requirements for any given sport defined as target characteristics. Obviously, such requirements vary from sport to sport, including the rate at which a contestant is prepared, and are set against the functional capabilities of the developing body. From a holistic perspective, the optimal development of physical fitness is required for any athlete [12,13].

The correct assessment of the effects of movement stimulation depends on carrying out accurate and reliable diagnoses of an individual's physical and psychomotor development at various stages of ontogenesis and training. Such diagnostic outcomes allow the development of optimal programming of various elements of sports training and the coordination of athlete's motor skills, according to the goals set for each stage of sports advancement.

The methods of assessment of physical fitness have been developed for broadly understood karate and other combat sports. The most comprehensive approach is Specific Physical Fitness Tests (SPFT), consisting of the hip speed test, "jodan-tsuki" punch test, "chudan-tsuki", "mawashi-geri" flexibility test, "mawashi-geri" kick test for the "jodan" level and flexibility test [14]. The SPFT was subsequently modified by Sterkowicz and Franchini [15], and Adamczyk et al. [14] proposed a modification of Story's speed rating by assessing the strength measured with the bench press test in various body positions and endurance based on the number of circular kicks at the chest level in 90 s per punching bag. Koropanovski et al. [16] used the Sideward Leg Splits Test (SdLS) [17] for flexibility, acceleration and speed; the strength Countermovement Jump (CMJ), also used in judo; Standing Triple Jump (STJ); as well as the shuttle run for aerobic capacity assessment [18]. The same battery of tests was used by Koropanovski et al. [19] in their assessment of advanced athletes. The treadmill has also been used to assess aerobic capacity in trials following R.A. Bruce's modified protocol. Anaerobic capacity has been assessed with the Wingate cycloergometer test [20]. Simple Reaction Time (SRT), Choice Reaction Time (CRT) and Decision Time (DT) ratings are also included in the test pool [21].

The aim of the present study was to determine the morphological and motor determinants of successful Shotokan karate performance in male and female practitioners.

2. Materials and Methods

The initial study pool consisted of 35 karate competitors; however, 3 were excluded due to injury. The number and classification of karate competitors are determined at the beginning of each year by the Board of the Polish Karate Federation. The participants were members of the World Karate Federation, who took part in major international competitions, e.g., the World Championships, European Championships and Olympic Games.

The study group comprised 32 karate competitors (12 women and 20 men) aged 18–25 years ranked from 1 kyu to 3 Dan. A series of tests was performed twice (i.e., at two stages) during one year, at an interval of 6 weeks during training camp events at the Olympic Preparation Center in Walcz, Poland. Eleven anthropometric features were measured describing the basic characteristics of the contestants' somatic build. Physical fitness was measured with the use of 16 motor skills tests of strength, speed, endurance, flexibility and reaction time. Special motor fitness was assessed with 5 tests of karate-specific technical skills, measuring special speed, endurance, and flexibility. The study was conducted in accordance with the Declaration of Helsinki and the National Statement of Intent as well as the Human Research Ethics Guidelines, and was approved by the Institute for Research in Biomedicine of the Poznan University of Medical Sciences (2008-11-13; Ethics Approval Number: 1101/08). All participants provided their written consent to take part in the tests after familiarizing themselves with the study protocol.

The study was conducted during the first 3 days of the athletes' stay at the Olympic Preparation Centre, in the same sequence, on each study day. On the first day, anthropometric characteristics of athletes' somatic build were measured, including the times of simple and complex reaction to optical and acoustic stimuli. On the second day, the remaining tests of physical fitness were carried out, with the exception of the athletes' endurance assessment. On the third day, karate-specific tests of special motor fitness, special speed, special endurance and special flexibility were performed. The Beep-Test was carried out as the last measurement. On each day, the tests were performed 2 h after a standard breakfast.

2.1. Anthropometrics

Body mass and body height were measured with the WPT 60/150 OW medical scales (Radwag®, Radom, Poland) to the nearest 100 g. The first author of the study carried the anthropometric measurements of the circumferences of the arm, forearm, thigh and lower leg as well as the length of the upper and lower limbs using a tape measure graduated in 0.1 cm increments, following the procedure in Norton [22]. The thickness of three skinfolds was also measured at the right side of the body using a Harpenden Skinfold Caliper to the nearest 0.1 mm (graduation 0.2 mm, range 80 mm) at the following sites: suprailiac (immediately superior to the iliac crest in midaxillary line), subscapular (the undermost tip of the inferior angle of the scapula) and the calf (a vertical pinch parallel to the long axis of the leg). The observed differences between the three measurements of body weight, body height, and limb length were <1%. The differences in skinfold measurements amounted to <5%.

2.2. Physical Fitness

The karate athletes' physical fitness was assessed with tests measuring the main motor skills, such as strength, speed, endurance, flexibility and reaction time. There were nine assessments of strength (static, dynamic, functional and maximal) as well as a hand movement speed test, flexibility test (torso forward), endurance test (Beep-Test) and criterion-related 20-m shuttle run test for estimating cardio-respiratory fitness [23]. There were also five tests measuring the time of simple and complex reactions of the limbs (responses to optical and acoustic stimuli). Sixteen tests assessing all round fitness were carried out in total. An in-house time counter with an electronic auxiliary MC-2 device (VOLTCRAFT®, Lindenweg, Hirschau/Germany) as well as an electronic device with an auxiliary MCZR/ATB 1.0 light (INFO-ELEKTRO, Ruda Śląska, Poland) set were used. Both devices were designed at the Research and Teaching Aid Department of the University of Physical Education in Poznan. For tests involving more than 1 repetition, each successive trial was performed in resting conditions, i.e., after the heart rate returned to baseline, so that each trial was performed under the same conditions.

2.2.1. Simple Reaction Time of the Feet (Optical Stimulation)

Each participant had to react as quickly as possible to 10 consecutive light impulses by pressing a foot against the pedal attached to a special platform on the ground. The arithmetic mean of the times was used for analysis after cutting off the three top and the three bottom times [21].

2.2.2. Simple Reaction Time of the Hands (Optical Stimulation)

The measurements were performed to the nearest 0.001 s. Each participant had to react as quickly as possible to 10 consecutive light impulses by pressing a button on a special hand-held handle with a thumb. The arithmetic mean of the times was used for analysis after cutting off the three top and three bottom times [21].

2.2.3. Complex Reaction Time of the Upper Limbs (Optical Stimulation)

The measurements were performed with an accuracy of 0.01 s. Each participant had to react as quickly as possible to 15 consecutive light impulses by pressing the buttons on hand-held handles with the thumbs. Each hand corresponded to a respective color. If an orange bulb was lit, no reaction was to be produced. The arithmetic mean of the times was used for analysis after cutting off the three top and the three bottom times [21].

2.2.4. Complex Reaction Time of the Upper and Lower Limbs (Visual and Acoustic Stimulation)

The measurements were performed with an accuracy of 0.01 s. Each participant had to react as quickly as possible to 10 consecutive light and sound impulses, pressing the thumbs on the buttons on hand-held handles, and the feet on the pedals mounted on a special platform on the ground. Each limb was assigned only one type of pulse. The arithmetic mean of the times was used for analysis after cutting off the three top and three bottom times [21].

2.2.5. Hand Movement Speed (Disc Tapping)

Each participant was to touch alternately two spaced disks with the more efficient hand as fast as possible. A participant, standing in front of the table, put his or her less efficient hand on a rectangle in the middle between the disks. At the command "Start" he or she made 25 movements back and forth (a total of 50 touches). Every touch was loudly counted, and the command "Stop" ended the test. The test was performed twice and the better score was recorded. The results were measured to the nearest 0.1 s.

2.2.6. Strength Measurement

Participants performed a forward throw with a swing of a 2 kg medicine ball while standing in front of a line. The ball was held behind the head with both hands. If, after the throw, the participant crossed the throwing line, the test was not scored. Each participant was allowed three attempts and the measurements were taken with an accuracy of 0.1 m. The better result was used for analysis.

2.2.7. Grip Strength

Hand grip strength was measured using a manual dynamometer (Lafayette 78010, Lafayette Instrument Company, Lafayette, IN, USA). Each participant took an upright posture with the arms at the sides, not touching the body, slightly bent at the elbows during the measurement [24]. The test was repeated three times with a 1-min break between the attempts to avoid the effects of muscle fatigue. Both the right and the left hands were tested, and the highest score by the hand declared as dominant was used for further analysis. The accuracy of the measurement was 0.01 kg.

2.2.8. Long Jump

Participants performed a standing broad jump on a mattress. Standing with their legs astride in front of the starting line each participants bent their knees and swung their arms backwards. After take-off, each participant attempted to make the longest possible jump possible and land on both feet. The test was performed 3 times and the best result was used for analysis.

2.2.9. Bent Arm Hang

With the examiner's assistance, each participant performed a pull up on a 2.5 mm diameter bar, with the arms bent in such a way that the chin was at the level of the bar. In this position, each participant tried to hold on the bar as long as possible. The test was completed when the participant's eyes were at the same level as the bar. The hang time was measured to the nearest 0.1 s.

2.2.10. One Repetition Maximum Bench Press Test (1 RM Test (One Repetition Maximum Bench Press Test))

The athletes lay down on a horizontal bench with their legs bent at a right angle and the feet resting on the ground. Each participant gripped the barbell placed at a distance of shoulder width using both hands. Next he or she lifted the barbell from the rack, lowered it on the chest, and lifted it up with the upper limbs fully extended. After the test the participant put the barbell back on the rack, with assistance if necessary. The starting weight was determined individually by each athlete, and the progression of the barbell load was 2.5 kg. Two failed attempts excluded the athlete from the test.

2.2.11. Attainable flexure

With the hands clasped together at the nape of the neck, in a supine position on the mattress, the participant touched the mat with their back and then returned to the supine position, with their elbows touching the knees. The feet rested on the mat and the test consisted of a maximum number of repetitions in 30 s.

2.2.12. High Jump

Each participant stood on a box equipped with a centimeter scale on one of its walls, embracing the box edge with the toes. The standing surface level was marked as 0. With the legs straightened in the knee joints and arms straightened down, the participants performed a maximum downward bend. The score was marked on the front wall of the box when the lowest position was reached. If the fingers of both hands were not on the same line, the mean distance for both hands was considered for analysis. The test was performed 3 times and the best result was used for analysis.

2.2.13. Endurance Testing: Beep Test

The test took place in a sports hall. The participants completed a number of 20 m runs from one line to another. The test started on a sound signal, and the running pace was determined by time intervals defined by successive sound signals. After each beep, the runner had to cover a distance of 20 m. The starting pace was very slow but kept increasing with the changing attempt levels. The runner had to maintain the pace, and the test ended when the participants exceeded the time limit necessary to complete the given distance twice [23].

2.3. *Special Motor Fitness*

The test aimed to assess the participants' karate-specific skills. Each participant delivered punches that were counted by the instructor and recorded with a video camera placed on a fixed tripod at the participant's side at a right angle. Special speed was measured with the choku-tsuki test (maximum number of straight hand punches performed in 10 s); special endurance test with the maximum number of uraken-uchi + gyaku-zuki combined attacks executed in 30 s and the mawashi-geri test (maximum series of kicks performed in 30 s); and special flexibility with the yoko-geri test (a side kick up to a maximum height, so-called 'Turkish twine'). In total, 5 tests were used to assess the karate competitors' special motor fitness [25–27].

2.4. Statistical Analysis

Summary statistics consisted of means and standard deviations (with 95% confidence limits). The Shapiro–Wilk test was used to test for normality of the distribution. Data were subjected to multiple stepwise regression analysis in which the best set of independent variables was selected and the coefficients of determination (R squared) were estimated for the variables in the model. Lambda ridge regression was also performed, which is a more stringent test that takes into account the variance around the coordinates' origin. It often allows those model determinants present in excess to be eliminated as well as any variables of minimal significance. All in all, the set of independent variables thus contained 33 items. The dependent variable was the number of ranking points awarded to each karate competitor by the Polish Karate Federation (PKF) according to official regulations. The PKF introduced a ranking points system in 2006. The points are used to determine the best athletes in particular competitions and weight categories and are the basis for appointing competitors to the national karate team. All karate athletes participating in the tests took part in the same competitions. It should be noted that the Poland national team members are obliged to participate in competitions in which ranking points are awarded, except for cases when a competitor is injured.

The degrees of correlation were set according to the following ranges: <0.10 trivial, 0.10 to 0.30 small, 0.30 to 0.50 moderate, 0.50 to 0.70 large, 0.70 to 0.90 very large and 0.90 to 1.00 almost perfect [28]. The level of statistical significance was set at $p < 0.05$ in all cases. The obtained results were analyzed statistically using the Dell Statistica data analysis software system (version 13, software.dell.com, Dell Inc., Round Rock, TX, USA).

3. Results

Table 1 presents the anthropometric and physical fitness characteristics and special motor fitness test results of the karate competitors under study.

Table 1. Mean values of the anthropometric and physical fitness characteristics and special motor fitness test results of male and female karate competitors.

Variable	n (20) Men		n (12) Women	
	M ± SD	95% CI	M ± SD	95% CI
Age (years)	20.40 ± 4.16	(18.45–22.35)	20.92 ± 3.00	(19.01–22.82)
Height (cm)	176.25 ± 6.31	(173.30–179.20)	165.08 ± 6.11	(161.20–168.97)
Weight (kg)	70.35 ± 8.14	(66.54–74.16)	58.58 ± 5.52	(55.08–62.09)
Lower limb length (cm)	91.75 ± 4.14	(89.81–93.69)	87.00 ± 3.28	(84.92–89.08)
Upper limb length (cm)	77.30 ± 3.73	(75.55–79.05)	73.08 ± 3.23	(71.03–75.14)
Thigh circumference (cm)	53.85 ± 4.33	(51.82–55.88)	55.58 ± 3.42	(53.41–57.76)
Shin circumference (cm)	38.15 ± 2.68	(36.90–39.40)	36.33 ± 1.92	(35.11–37.55)
Arm circumference (cm)	32.15 ± 4.06	(30.25–34.05)	28.33 ± 2.39	(26.83–29.85)
Forearm circumference (cm)	28.40 ± 2.90	(27.38–29.42)	25.00 ± 1.60	(23.99–26.01)
Shoulder skinfold (mm)	11.36 ± 2.64	(10.12–12.59)	11.52 ± 1.63	(10.48–12.55)
Stomach skinfold (mm)	10.07 ± 1.72	(9.27–10.87)	10.52 ± 2.04	(9.22–11.82)
Shin skinfold (mm)	8.54 ± 1.31	(7.93–9.15)	8.83 ± 1.04	(8.17–9.50)

Table 1. *Cont.*

Variable	n (20) Men		n (12) Women	
	M ± SD	95% CI	M ± SD	95% CI
Simple reaction time: right leg (optical stimulation) (s)	0.28 ± 0.04	(0.26–0.29)	0.29 ± 0.03	(0.27–0.31)
Simple reaction time: left leg (optical stimulation) (s)	0.27 ± 0.02	(0.25–0.28)	0.28 ± 0.03	(0.26–0.30)
Simple reaction time: hands (optical stimulation) (s)	0.23 ± 0.01	(0.22–0.23)	0.23 ± 0.02	(0.22–0.24)
Simple reaction time: hands (acoustic stimulation) (s)	0.15 ± 0.02	(0.14–0.16)	0.16 ± 0.02	(0.14–0.17)
Complex reaction time: hands (s)	0.31 ± 0.04	(0.29–0.33)	0.30 ± 0.03	(0.28–0.32)
Complex reaction time: hands + legs (s)	0.39 ± 0.05	(0.37–0.42)	0.41 ± 0.09	(0.35–0.47)
Disk tapping (number)	93.30 ± 6.22	(90.39–96.21)	95.00 ± 11.78	(87.52–102.48)
Medicine ball throw (cm)	1147.00 ± 201.50	(1052.70–1241.30)	845.00 ± 115.17	(771.83–918.17)
Grip strength (kg)	44.30 ± 9.55	(39.83–48.77)	29.00 ± 3.62	(26.70–31.30)
Long jump (cm)	254.85 ± 20.18	(245.40–264.30)	215.08 ± 15.01	(205.55–224.62)
Bent arm hang (s)	46.60 ± 9.35	(42.22–50.98)	30.42 ± 14.43	(21.25–39.58)
Bench press 1 RM test (kg)	65.50 ± 15.89	(58.07–72.93)	40.00 ±10.66	(33.23–46.77)
Flexure to sit down (amount)	41.70 ± 4.86	(39.43–43.97)	37.83 ± 6.04	(33.99–41.67)
Attainable flexure (cm)	21.20 ± 4.80	(18.95–23.45)	17.08 ± 6.47	(12.97–21.20)
High jump (cm)	55.75 ± 8.83	(51.62–59.88)	42.17 ± 6.39	(38.10–46.23)
Beep Test	57.28 ± 6.25	(54.36–60.20)	49.90 ± 9.53	(43.85–55.95)
Straight punch-choku-zuki (number of punches)	77.10 ± 10.15	(72.35–81.85)	71.00 ± 12.55	(63.03–78.97)
Hand punching combinations: uraken-uchi + gyaku-zuki, i.e., reverse fist +reverse punch (number of punches)	74.60 ± 8.95	(70.41–78.79)	64.00 ± 9.15	(58.19–69.81)
Mawashi-geri, i.e., spin kick (number of kicks)	55.90 ± 5.26	(53.44–58.36)	46.25 ± 7.14	(41.72–50.78)
Leg upswing (side kick) (cm)	175.52 ± 11.50	(170.52–181.28)	159.17 ± 12.45	(151.25–167.09)
Cross-sectional stride (cm)	22.60 ± 11.45	(17.24–24.00)	13.92 ± 7.59	(9.10–18.74)

Results expressed as M = mean, SD = standard deviation (confidence intervals 95%).

The independent variables comprised the determinants of sport performance that generally described the contestants' morphological features, physical fitness and special motor fitness.

Most of the variables considered 'independent' were, in fact, logically speaking, dependent on many factors; however, this is beyond the scope of this paper. The multi-stepwise regression allowed the best set of independent variables to be chosen and to precisely establish the sets of factors responsible for sporting outcomes in karate as well as their relative importance as determinants. In addition, a similarly acting lambda ridge regression was performed, which is, however, a more stringent procedure, which explained the variance around the coordinates' origin. It also allowed for rejecting those determinants that excessively or insignificantly contributed to the general model.

Six features were found to fit the model in the first part of the study for female participants, using stepwise multiple regression that, jointly in effect, determined the dependent variable at 100%. Each subsequent variable added was decreasingly less and less significant, but when taken together they practically accounted for the dependent variable in its entirety. Thigh circumference was found to be the most significant variable at 76% (R-squared), followed by the foot kicking rate (mawashi-geri) at 19% (R-squared change); the remaining independent variables taken together accounted for only 4% (Table 2).

Table 2. Summary regression details from the 1st study stage for female participants; dependent variable = ranking points.

Determinant/Feature	Ensuing Stage	Multiple Correlation	Cumulative Coefficient of Determination	Partial Coefficient of Determination	p
		R	R- Squared	R- Squared	
Thigh circumference (cm)	1	0.8746	0.7649	0.7649	0.1416
Mawashi-geri, i.e., spin kicks (maximum number)	2	0.9802	0.9609	0.1960	0.1254
Choku-zuki, i.e., straight punches (maximum number)	3	0.9907	0.9815	0.0206	0.2811
Long jump (cm)	4	0.9981	0.9962	0.0147	0.1800
Flexure to sit down (number)	5	0.9998	0.9996	0.0033	0.1377
Bench press 1 RM test (kg)	6	1	1	0.0003	0.0189

Likewise, the model also included six features in the second part of the study that when taken together practically determined the dependent variable in its entirety (100%). It was found that three features fit the model in both the first and second stages of the study (i.e., thigh circumference, mawashi-geri and long jump), while the other three did not (Tables 3 and 4).

Table 3. Summary regression details from the 2nd study stage for female participants; dependent variable = ranking points.

Determinant/Feature	Ensuing Stage	Multiple Correlation	Cumulative Correlation	Partial Correlation	
		R	R- Squared	R- Squared	p
Thigh circumference (cm)	1	0.8705	0.7579	0.7579	0.1443
Mawashi-geri, i.e., spin kicks (maximum number)	2	0.9713	0.9435	0.1856	0.1538
Long jump (cm)	3	0.9853	0.9709	0.0273	0.3025
Leg upswing (cm)	4	0.9950	0.9901	0.0192	0.2489
Shoulder skinfold (mm)	5	0.9990	0.9980	0.0078	0.2133
High jump (cm)	6	1	1	0.0019	0.0073

Table 4. Summary details of lambda ridge regression from the 1st study stage for female participants; dependent variable = ranking points.

L = 0.10000, R = 0.94747702, R^2 = 0.89771271
F(2.5) = 21.941 $p < 0.00335$, Estimated Standard Error: 106.04

	Standardized Regression Coefficient	Standard Error	Unstandardized Regression Coefficient	Standard Error	
	BETA	BETA	B	B	p
Free W		1760.59		843.884	0.0913
Thigh circumference (cm)	−0.6100	0.1485	−47.7082	11.615	0.0092
Mawashi-geri, i.e., spin kicks (number of kicks)	0.4670	0.1485	23.4520	7.4577	0.0255

As expected, the ridge regression left two determining variables in the model. The multiple regression analysis and the consistency of outcomes from the first and second parts of the study on female karate competitors revealed two basic variables with a decisive influence on high karate performance: thigh circumference and mawashi-geri (i.e., the maximum number of kicks performed in 30 s) (Table 5).

Table 5. Summary details of lambda ridge regression from the 2nd study stage for female participants; dependent variable = ranking points.

L = 0.10000, R = 0.96733337, R^2 = 0.93573385 **F(4.3) = 10.920 p < 0.03916, Estimated Standard Error: 108.51**					
	Standardized Regression Coefficient	**Standard Error**	**Unstandardized Regression Coefficient**	**Standard Error**	
	BETA	**BETA**	**B**	**B**	p
Free W.			1984.509	914.2573	0.1183
Thigh circumference (cm)	−0.5172	0.1888	−41.352	15.0940	0.0713
Mawashi-geri, i.e., spin kicks (maximum number)	0.3119	0.1754	13.2209	7.4355	0.1734
Hand punching combinations; uraken-uchi + gyaku-zuki, i.e., reverse fist + reverse punch (number of punches)	0.3215	0.2011	9.7345	6.0887	0.2081
Forearm circumference (cm)	−0.2041	0.1692	−32.274	26.7466	0.3140

Table 6 presents determinants of karate performance as defined by the number of ranking points. The model includes 10 characteristics that together determine the dependent variable as a whole (100%). Table 7 lists the determinants of karate performance in the order in which the variables enter the multiple regression model in consecutive "steps". The most significant determinant was yoko-geri, which explained the dependent variable in 64% (R-square), followed by lower limb length at 13% (R-square change) and lower leg circumference at 20%. The remaining independent variables collectively accounted for about 2% of the determining factors; thus, like in the case of female karate competitors, they can be ignored. Table 8 presents the results of the multiple stepwise regression analysis of the second series test results determining karate performance. The model includes 9 characteristics determining the dependent variable in almost 100%. Three variables in the model were the same as for the first series tests (yoko-geri, bent arm hang and arm circumference).

The multiple regression analysis results and the consistency of the results of the first and second series of tests revealed two variables with a decisive impact on the sport performance of male karate competitors: yoko-geri (side kicks up to the highest possible height) and bent arm hang determining overall endurance (Table 9). It can be concluded that male karate competitors with significantly high endurance levels, i.e., ability to fight without fatigue and accurately hit the opponent in upper body areas (head and neck), will have a greater chance of achieving top results in karate.

Table 6. Summary regression details from the 1st study stage for male participants; dependent variable = ranking points.

Determinant/Feature	**Ensuing Stage**	**Multiple Correlation** **R**	**Cumulative Coefficient of Determination** **R- Squared**	**Partial Coefficient of Determination** **R- Squared**	p
Leg upswing (cm)	1	0.7997	0.6395	0.6395	0.1483
Lower limb length (cm)	2	0.8787	0.7721	0.1326	0.2622
Shin circumference (cm)	3	0.9874	0.9750	0.2028	0.0785
Flexure to sit down (number)	4	0.9931	0.9864	0.0113	0.2492
Bench press 1 RM test (kg)	5	0.9966	0.9932	0.0068	0.2462
Shinfold thickness (mm)	6	0.9979	0.9959	0.0027	0.3169
Complex reaction time: hands (s)	7	0.9994	0.9989	0.0029	0.1863
Forward flexure (cm)	8	0.9999	0.9998	0.0009	0.1348
Bent arm hang (s)	9	0.9999	0.9999	0.0001	0.1009
Arm circumference (cm)	10	1	1	6.38×10^{-6}	0.0058

Table 7. Summary regression details from the 2nd study stage for male participants; dependent variable = ranking points.

Determinant/Feature	Ensuing stage	Multiple Correlation R	Cumulative Coefficient of Determination R- Squared	Partial Coefficient of Determination R- Squared	*p*
Leg upswing (cm)	1	0.7387	0.5457	0.5457	0.1879
Complex reaction time hands + legs (s)	2	0.8680	0.7534	0.2077	0.2340
Bent arm hang (s)	3	0.9276	0.8605	0.1071	0.2591
Arm circumference (cm)	4	0.9609	0.9233	0.0628	0.2696
Forearm circumference (cm)	5	0.9873	0.9748	0.0514	0.1927
Simple reaction time (optical stimulation): right leg (s)	6	0.9978	0.9956	0.0207	0.1437
Straight punch-choku-zuki (number)	7	0.9999	0.9999	0.004	0.0556
Cross-sectional stride (cm)	8	0.9999	0.9999	9.19×10^{-5}	0.1207
Forward flexure (cm)	9	1	1	6.67×10^{-6}	0.0795

Table 8. Summary details of lambda ridge regression from the 1st study stage for male participants; dependent variable = ranking points.

L = 0.10000, R = 0.92555474, R^2 = 0.85665159
F(3.8) = 15.936 $p < 0.00098$, Estimated Standard Error: 29.385

	Standardized Regression Coefficient	Standard Error	Unstandardized Regression Coefficient	Standard Error	
	BETA	BETA	B	B	*p*
Free W			−907.794	291.4988	0.0143
Leg upswing (cm)	0.8274	0.1384	7.7734	1.3006	0.0003
Shin circumference (cm)	0.4158	0.1312	10.8388	3.4223	0.0132
Lower limb length (cm)	−0.4059	0.1409	−8.8995	3.0902	0.0205

Table 9. Summary details of the lambda ridge regression from the 2nd study stage for male participants; dependent variable = ranking points.

L = 0.10000, R = 0.95387189, R^2 = 0.90987158
F(6.4) = 0.7302 $p < 0.04308$, Estimated Standard Error: 32.695

	Standardized Regression Coefficient	Standard Error	Unstandardized Regression Coefficient	Standard Error	
	BETA	BETA	B	B	*p*
Free W			−475.884	302.5304	0.1908
Leg upswing (cm)	0.3450	0.2251	3.3022	2.1549	0.2002
Complex reaction time: hands (s)	−0.3622	0.1589	−397.151	174.3357	0.0849
Bent arm hang (s)	−0.2708	0.1482	−2.0694	1.1325	0.1416
Bench press 1 RM test (kg)	−0.2183	0.1508	−1.4598	1.0085	0.2213
Hand punching combinations; uraken-uchi + gyaku-zuki, i.e., reverse fist + reverse punch (s)	0.3144	0.1990	2.5406	1.6078	0.1892
Simple reaction time (acoustic stimulation): hand (s)	0.2116	0.1924	585.9133	532.7161	0.3331

4. Discussion

The aim of the study was to identify and assess the morphological and motor fitness determinants of successful performance in karate. Most combat sports require a combination of technique, strength, aerobic fitness, power, and speed. In general, no single trait dominates in combat sports.

The multiple regression analysis revealed that two variables had the greatest impact on the karate performance of male and female competitors. For women, these were thigh circumference and the speed of the mawashi-geri-kick roundhouse technique (i.e., the maximum number of delivered kicks in 30 s), whilst for men they were the extent of the sideway swing to the highest possible height (yoko-geri) and overall endurance assessed with the bent arm hang test.

Because the actions of the upper limbs are more precise and faster than the actions of the lower limbs during karate tournament fights, the advantages of the former are clear, as observed by Chaabene et al. [5]. Nevertheless, the high-level-jodan foot techniques were scored the highest as having the greatest assigned levels of difficulty [29,30]. Chaabene et al. also pointed out that multivariate studies are needed in modern sport to assess the impact of various factors on karate performance; they are also extremely important in furthering the knowledge of the interrelationships between these methods [5].

The results of studies assessing simple or choice reaction times between successful and less successful karate athletes are also inconclusive. Fontani et al. [31] found that athletes with 3rd and 4th Dan black belts had shorter reaction times to a stimulus than athletes with 1st and 2nd Dan black belts. In contrast, Mori et al. [32] in their study four years earlier showed no significant difference in simple reaction time between athletes of varying levels of proficiency. In the present study, reaction time was not a significant determinant of high-level performance. Modern karate performance requires high levels of physical fitness, particularly flexibility, which involve a high degree of joint movement and ensures that the fighting technique is properly executed.

There are different determinants of ranking positions in women and men practicing karate [33]; however, the literature on karate performance determinants in female athletes is rather limited. Women's success in karate is strongly determined by morphological traits (i.e., strong thighs) together with the rapidity and strength of the kicks; whereas for men, it is endurance in combat and execution of accurate hits on the opponent's body in high-scoring areas. Our results demonstrate that sports training is indeed mainly focused on those important variables. Combining variables allows a broader consideration of determinants in Shotokan karate. Undoubtedly, the most significant determinant factor of successful karate performance for both sexes appears to be ideally developed strength based on the morphology of the limbs. These elements are strongly related to each other and are commonly termed the as 'the arm and leg strength factor'. This is essentially the one and only determinant of karate success in women, whilst for men it is one of the main determinants, followed by the range of mobility in the joints of the upper and lower limbs when correlated with the complex responses of the limbs. A study by Katić on karate learners showed that successful karate performance was mainly achieved thanks to their level of knowledge and ability to efficiently execute karate technique(s) as conditioned by general and specific motor skills [34,35]. The present study also identified a group of morphological, positive and negative, determinants of effective karate performance in senior contestants in reference to the used karate techniques [36]. Other authors examined the effect of cardiorespiratory performance and maximal power output (Wingate test) on the performance level. Both factors did not significantly differentiate between karate athletes at the international and national levels, despite the fact that VO_{2max} plays an important role in preventing fatigue during training and accelerates recovery between combats.

Many studies have confirmed that high-level karate performance is generally related to the explosive strength of the muscles. Furthermore, highly competitive karate athletes require both rapid acceleration and deceleration of body segments and muscular explosive power training [37–39]. Koropanovski et al. [16] in their study of morpho-functional differences between kata and kumite athletes also showed that a higher explosive power could be beneficial for kumite, while both smaller stature and higher flexibility (particularly of the lower extremity) could be important for the exceptionally low postures of the kata competitors.

Highly developed strength determines the successful performance in Shotokan karate based on the morphological properties of the limbs. Indeed, in women, this constitutes the basic and only determinant of success in karate, whereas in men it is one of a number of leading factors, closely followed by the range of mobility in the joints of the upper and lower extremities. Women's success in karate depends largely on their thigh strength as well as the speed and strength of strikes/blows inflicted with the lower limbs, albeit in men, successful karate performance is largely determined by their endurance in combat and accurate hits to the opponent's high-scoring body areas. Karate training should thus focus mainly on these factors to enable novice athletes to achieve success in competition. Undoubtedly, strength can be increased by the strengthening of the lower limbs, hip joint mobility exercises, low position movements, execution of leg techniques in various planes, and the use of external loads. All this will thereby permit a wider repertoire of leg techniques to be acquired, especially at the highest-scoring jodan-head level.

It can be concluded that these dimorphic differences in determinants of successful performance in Shotokan karate may have their evolutionary foundations. Strength is an evolutionary motor trait assigned to the male sex, and flexibility (or greater mobility in the joints) to the female sex. Karate as a sport paradoxically combines both these elements. Thus, each sex during karate training refines these evolutionarily unassigned properties, and their development guarantees an advantage in one's gender category and thus competitive success. A strong woman is likely to win over a weaker woman while both are naturally flexible; a more "flexible" man will win against a "stiffer" man, while both are naturally equally strong.

Limitation of the study: The limitation of this study was the unequal number of female and male participants. It was also impossible to carry out physiological tests such as VO_{2max} or Wingate test to assess participants' anaerobic capacity. Blood indicators of the body's response to the applied training loads were not analyzed either.

5. Conclusions

Systematic analysis of the athlete's motor level during training affects the individualization and optimization of the training process. Multiple regression analyses have shown that, in the case of both men and women, two significant variables determine high-level Shotokan karate performance to the greatest extent. In the studied group of women, they were the thigh circumference and the speed of the mawashi-geri. In men, they were the yoko-geri range and overall endurance measured with the bent arm hang test. In the case of women, successful karate performance has a strong morphological basis (strong thighs) and depends on the speed and strength of the kicks. In the studied group of men, it was endurance in combat and the precision of hitting the opponent in the highest-scoring body areas. The study results indicate that it is beneficial to focus on the above determinants in the process of training of karate athletes. These results can be directly used in the selection of athletes with appropriate motor skills, predispositions, or anthropometric features. This will allow the proper qualification of young athletes to the respective training groups and guide their sport development towards the attainment of the championship level.

The research is cross-sectional. We realize that not all variables can be controlled in our study. It seems logical to think that people with more practice time had more opportunities to be exposed to the competitive environment. In fact, practice is directly related to performance and learning. Williams and Elliot [40] showed that expert martial artists have greater anticipation skills, suggesting that perceptual skill in karate depends on task-specific knowledge structures acquired through experience.

Author Contributions: Conceptualization, D.W., P.P. and K.D.; methodology, P.P., A.J., P.S. and K.D.; validation, P.P. and K.D.; formal analysis, K.D. and P.P.; investigation, A.J., P.P. and P.S.; resources, P.S., A.J. and K.D.; data curation, P.P. and D.W.; writing—original draft preparation, K.D. and P.P.; visualization, K.D.; supervision, P.P. and K.D.; project administration, D.W. and K.D. All authors have read and agreed to the published version of the manuscript.

Funding: This research received no external funding.

Institutional Review Board Statement: The study was conducted according to the guidelines of the Declaration of Helsinki and approved by the IRB (Institute for Research in Biomedicine) at the Poznan University of Medical Sciences (13 November 2008; Ethics Approval Number: 1101/08).

Informed Consent Statement: Informed consent was obtained from all study participants.

Data Availability Statement: The data presented in this study are available on request from the corresponding author. The data are not publicly available due to the consent provided by participants on the use of confidential data.

Conflicts of Interest: The authors declare no conflict of interest.

References

1. Johnston, R.J.; Watsford, M.L.; Kelly, S.J.; Pine, M.J.; Spurrs, R.W. Validity and interunit reliability of 10 Hz and 15 Hz GPS units for assessing athlete movement demands. *J. Strength Cond. Res.* **2014**, *28*, 1649–1655. [CrossRef] [PubMed]
2. Imamura, H.; Yoshimura, Y.; Uchida, K.; Nishimura, S.; Nakazawa, A.T. Maximal oxygen uptake, body composition and strength of highly competitive and novice karate practitioners. *Appl. Human Sci.* **1998**, *17*, 215–218. [CrossRef] [PubMed]
3. Domaszewska, K.; Laurentowska, M.; Michalak, E.; Kryściak, J.; Rakowski, P. The metabolism of purines and circulation-respiratory reaction to physical effort in karate. *Pol. J. Sports Med.* **2008**, *24*, 73–80.
4. Sforza, C.; Torci, M.; Grassi, G.; Fragnito, N.; Pizzini, G.; Ferrario, V.F. The repeatability of choku-tsuki and oi-tsuki in traditional Shotokan karate: A morphological three-dimensional analysis. *Percept. Mot. Ski.* **2000**, *90*, 947–960. [CrossRef]
5. Chaabene, H.; Hachana, Y.; Franchini, E.; Mkaouer, B.; Chamari, K. Physical and physiological profile of elite karate athletes. *Sports Med.* **2012**, *42*, 829–843.
6. Sánchez-Puccini, M.B.; Argothy-Bucheli, R.E.; Meneses-Echávez, J.F.; López-Albán, C.A.; Ramírez-Vélez, R. Anthropometric and physical fitness characterization of male elite karate athletes. *Int. J. Morpho.* **2014**, *32*, 1026–1031. [CrossRef]
7. De Quel, Ó.M.; Ara, I.; Izquierdo, M.; Ayán, C. Does Physical Fitness Predict Future Karate Success? A Study in Young Female Karatekas. *Int. J. Sports Physiol. Perform.* **2020**, *15*, 868–873. [CrossRef]
8. Tabben, M.; Chaouachi, A.; Mahfoudhi, M.; Aloui, A.; Habacha, H.; Tourny, C.; Franchini, E. Physical and physiological characteristics of high-level combat sport athletes. *JCSMA.* **2014**, *5*, 1–5. [CrossRef]
9. Andreato, L.V.; Branco, B.H.M. Different sports, but the same physical and physiological profiles? *Sports Med.* **2016**, *46*, 1963–1965. [CrossRef]
10. Shariat, A.; Shaw, B.S.; Kargarfard, M.; Shaw, I.; Lam, E.T.C. Kinanthropometric attributes of elite male judo, karate and taekwondo athletes. *Rev. Bras. Med. Esporte.* **2017**, *23*, 260–263. [CrossRef]
11. Chaabène, H.; Franchini, E.; Miarka, B.; Selmi, M.A.; Mkaouer, B.; Chamari, K. Time–motion analysis and physiological responses to karate official combat sessions: Is there a difference between winners and defeated karatekas? *Int. J. Sports Physiol. Perform.* **2014**, *9*, 302–308. [CrossRef]
12. Zombra, Ž. Differences in body composition between karate athletes and non-athletes. *Sport Sci. Pract. Asp.* **2018**, *15*, 2476.
13. Rakita, D.; Rakonjac, D.; Jurisic, M.; Obradovic, J. The influence of morphological characteristics on the specific motor skills of junior-age karate athletes. *EQOL J.* **2018**, *10*, 43–49. [CrossRef]
14. Story, G. Fitness testing for karate. *Sports Coach* **1989**, 35–38.
15. Sterkowicz, S.; Franchini, E. Testing motor fitness in karate. *Arch. Budo.* **2009**, *5*, 29–34.
16. Koropanowski, N.; Berjan, B.; Bozic, P.; Pazin, N.; Sanader, A.; Jovanovic, S.; Jaric, S. Anthropometric and physical performance profiles of elite Karate kumite and kata competitors. *J. Hum. Kinet.* **2011**, *30*, 107–114. [CrossRef]
17. Bozić, P.; Pazin, N.; Berjan, B.; Planic, N.; Cuk, I. Evaluation of the field tests of flexibility of the lower extremity: Reliability and the concurrent and factorial validity. *J. Strength Cond. Res.* **2010**, *24*, 2523–2531. [CrossRef]
18. Léger, L.A.; Mercier, D.; Gadoury, C.; Lambert, J. The multistage 20 metre shuttle run test for aerobic fitness. *J. Sport Sci.* **1988**, *6*, 93–101. [CrossRef]
19. Nunan, D. Development of a sports specific aerobic capacity test for karate–A pilot study. *J. Sports Sci. Med.* **2006**, *5*, 47–53.
20. Keshishian, H. Motor Fitness Profiling of Elite and Novice Karate Practitioners. Master Thesis, Australian Catholic University, Sydney, Australia, 2013.
21. Brito, A.V.; Silva, C. Reaction time in karate athletes. IDO–Movement for Culture. *J. Martial Arts. Anthrop.* **2011**, *11*, 35–39.
22. Norton, K.; Whittingham, N.; Carter, L.; Kerr, D.; Gore, C.; Marfell-Jones, M. *Measurement Techniques in Anthropometry*; Norton, K., Olds, T., Eds.; University of New South Wales Press: Sydney, Australia, 1996; pp. 25–75.
23. Mayorga-Vega, D.; Aguilar-Soto, P.; Viciana, J. Criterion-related validity of the 20-m shuttle run test for estimating cardiorespiratory fitness: A meta-analysis. *J. Sci. Med. Sport.* **2015**, *14*, 536.
24. Pluta, B.; Galas, S.; Krzykała, M.; Andrzejewski, M. The Motor and Leisure Time Conditioning of Young Table Tennis Players' Physical Fitness. *Int. J. Environ. Res. Public Health.* **2020**, *17*, 5733. [CrossRef] [PubMed]
25. Sforza, C.; Turci, M.; Grassi, G.P.; Fragnito, N.; Serrao, G.; Ferrario, V.F. Repeatability of choku-tsuki and oi-tsuki in shotokan karate: A 3-dimensional analysis with thirteen black-belt karateka. *Percept. Mot. Ski.* **2001**, *92*, 1230–1232. [CrossRef]

26. Martin, A.P. *The Shotokan Karate Bible, In Beginner to Black Belt*, 2nd ed.; Bloomsbury Publishing: New York, NY, USA, 2016; p. 224.
27. Layton, C. Speed of technique and age in Shotokan karateka. *Percept. Mot. Ski.* **1993**, *76*, 1001–1002. [CrossRef]
28. Cohen, J. *Statistical Power Analysis for the Behavioral Sciences*, 2nd ed.; Lawrence Erlbaum Associates Inc.: Hillsdale, NJ, USA, 1988.
29. Emmermacher, P.; Witte, K.; Bystryzycki, S.; Potenberg, J. Different variations of the Karate technique Mawashi-Geri. In Proceedings of the 25th International Symposium on Biomechanics in Sports, Ouro Preto, Brazil, 23–27 August 2007; pp. 289–292.
30. Alp, M.; Gorur, B. Comparison of Explosive Strength and Anaerobic Power Performance of Taekwondo and Karate Athletes. *EduLearn* **2020**, *9*, 149–155. [CrossRef]
31. Fontani, G.; Lodi, L.; Felici, A.; Migliorini, S.; Corradeschi, F. Attention in athletes of high and low experienced engaged in different open skill sport. *Percept. Mot. Ski.* **2006**, *102*, 791–805. [CrossRef]
32. Mori, S.; Ohtani, Y.; Imanaka, K. Reaction time and anticipatory skills of karate athletes. *Hum. Mov. Science.* **2002**, *21*, 213–230. [CrossRef]
33. Jukić, J. Gender Differentiations of Some Anthropological Characteristics of Karate Players–Cadets. *Coll. Antropol.* **2015**, *39*, 83–94.
34. Katić, R.; Blažević, S.; Krstulović, S.; Mulić, R. Morphological structures of elite karateka and their impact on technical and fighting efficiency. *Coll. Antropol.* **2005**, *29*, 79–84.
35. Loturco, I.; Artioli, G.G.; Kobal, R.; Gil, S.; Franchini, E. Predicting punching acceleration from selected strength and power variables in elite karate athletes: A multiple regression analysis. *J. Strength Cond. Res.* **2014**, *28*, 1826–1832. [CrossRef]
36. Katić, R.; Jukić, J.; Čavala, M.; Vučić, D.; Blažević, S. Motor determinants of fighting efficacy in Croatian youth karateka. *Coll. Antropol.* **2013**, *37*, 1–8.
37. Ravier, G.; Grappe, F.; Rouillon, J.D. Comparison between the maximal variables of velocity, force and power from two analysis methods in the functional assessment of karate. *Sci. Sports.* **2003**, *18*, 134–140. [CrossRef]
38. Blažević, S.; Katić, R.; Popović, D. The effect of motor abilities on karate performance. *Coll. Antropol.* **2006**, *30*, 327–333.
39. Katić, R.; Blažević, S.; Zagorac, N. The impact of basic motor abilities on the specific motoricity performance in elite karateka. *Coll. Antropol.* **2010**, *34*, 1341–1345.
40. Williams, A.M.; Elliott, D. Anxiety, expertise, and visual search strategy in karate. *J. Sport Exerc. Psychol.* **1999**, *21*, 362–375. [CrossRef]

International Journal of
Environmental Research and Public Health

MDPI

Article

Assessment of the Health Behaviours and Value-Based Health Analysis of People Aged 50+ Who Were Hospitalized Due to Cardiovascular Disease

Izabela Gąska [1], Katarzyna Sygit [2,*], Elżbieta Cipora [1], Marian Sygit [2], Anna Pacian [3], Maryna Surmach [4], Dorota Kaleta [5] and Adam Rzeźnicki [6]

1 Medical Institute, Jan Godek State University in Sanok, 38-500 Sanok, Poland; izagaska@op.pl (I.G.); elacipora@interia.pl (E.C.)
2 Faculty of Health Sciences, Calisia University, 62-800 Kalisz, Poland; msygit@onet.pl
3 Department of Public Health, Faculty of Health Sciences, Medical University of Lublin, 20-093 Lublin, Poland; annapacian@umlub.pl
4 Head of the Department of Public Health and Health Services, Grodno State Medical University, 230009 Grodno, Belarus; marina_surmach@mail.ru
5 Department of Hygiene and Epidemiology, Faculty of Health Sciences, Medical University of Lodz, 90-647 Lodz, Poland; dorota.kaleta@umed.lodz.pl
6 Department of Social Medicine, Faculty of Health Sciences, Medical University of Lodz, 90-647 Lodz, Poland; adam.rzeznicki@umed.lodz.pl
* Correspondence: k.sygit@akademiakaliska.edu.pl

Citation: Gąska, I.; Sygit, K.; Cipora, E.; Sygit, M.; Pacian, A.; Surmach, M.; Kaleta, D.; Rzeźnicki, A. Assessment of the Health Behaviours and Value-Based Health Analysis of People Aged 50+ Who Were Hospitalized Due to Cardiovascular Disease. *IJERPH* **2021**, *18*, 4221. https://doi.org/10.3390/ijerph18084221

Academic Editors: Wioletta Zukiewicz-Sobczak, Paulina Wojtyla-Buciora, Izabela Rącka, Andrzej Wojtyla and Nante Nicola

Received: 22 February 2021
Accepted: 15 April 2021
Published: 16 April 2021

Publisher's Note: MDPI stays neutral with regard to jurisdictional claims in published maps and institutional affiliations.

Copyright: © 2021 by the authors. Licensee MDPI, Basel, Switzerland. This article is an open access article distributed under the terms and conditions of the Creative Commons Attribution (CC BY) license (https://creativecommons.org/licenses/by/4.0/).

Abstract: Introduction: The basic determinant of healthy behaviour—among other human behaviours—is the fact that it consistently affects health. Nowadays, health behaviour studies are considered to be an important method of measuring the health of a population. Objective: To assess the health behaviours and value-based health analysis of people aged 50+ who were hospitalized due to cardiovascular disease, depending on the selected descriptive variables. Materials and methods: The study was conducted between April 2018 and December 2018 among 411 subjects aged 50+ who were hospitalized due to cardiovascular disease at the Independent Public Health Care Unit in Sanok (Podkarpackie voivodship in Poland). The method used in the study was a diagnostic survey. The study used the authors' survey questionnaire and two standardized tests: Inventory of Health-Related Behaviour (IHB) and List of Health Criteria (LHC). A statistical analysis was carried out in the R program, version 3.5.1. The obtained results were subjected to thorough statistical analysis using the following tests: Student's *t*, Mann–Whitney U, ANOVA, Kruskal–Wallis, Fisher's Least Significant Difference (LSD), Pearson, and Spearman. Results: The strongest correlation between health status and health behaviours (according to the IHB questionnaire) was in the area of 'health practices', while the lowest correlation was found in the areas of 'correct eating habits' and 'preventive behaviours'. Based on the LHC questionnaire, the most important health criteria according to the subjects were 'not feeling any physical ailments'; 'having all body parts functional'; 'feeling well'; 'eating properly'; and 'infrequent need of going to the doctor'. A positive correlation was found in the group of respondents where the 'preventive health behaviours' were more intense; herein, the more important criterion for the respondents was 'eating properly'. Conclusions: Respondents aged 50+ and hospitalized for cardiovascular diseases indicated (based on the IHB questionnaire) that health behaviours in the area of 'health practices' had the strongest correlation with their health, while the lowest correlation was found in the areas of 'correct eating habits' and 'preventive behaviours'. According to the respondents, the most important criteria determining health (according to the LHC questionnaire) included 'not feeling any physical ailments'; 'having all body parts functional'; 'feeling well'; 'eating properly'; and 'infrequent need of going to the doctor'. Based on the information collected from the respondents, it was found that the most important criteria determining health depended on selected descriptive variables, such as age, gender, place of residence, education, and marital status.

Keywords: health; pro-health behaviours; anti-health behaviour; prevention; patient; 50+ population; chronic diseases

1. Introduction

Due to the development of the demographic situation in Poland, the issues of reducing morbidity and mortality due to cardiovascular diseases occupy an important place in scientific publications. Cardiovascular diseases have been the greatest threat to the lives of Poles for over 50 years, although the share of these diseases in the total number of deaths has decreased by over 7% over the last 20 years. Undoubtedly, the observed changes have been influenced primarily by the change in health behaviours, especially eating habits and regular preventive examinations [1–3].

However, despite the improvement in the epidemiological situation, in the last two decades the mortality rate due to cardiovascular diseases, especially the mortality rate (in people aged below 60), still has been high in Poland. Given the current disease trends and the pace of ageing of the Polish population, it is estimated that the number of deaths due to cardiovascular diseases in 2022 will exceed 200,000 [4–6].

The abovementioned epidemiological data justify the need to conduct research on the health behaviours of people of different ages in terms of assessing their health behaviours, and subsequently to take appropriate steps to eliminate incorrect health behaviours that may lead to, among others, cardiovascular diseases.

Nowadays, health behaviour studies are considered to be an important method of measuring the health of a population [1–6]. The choice of specific behaviours depends on the importance of health in one's system of values [1–5,7]. It should be emphasized, however, that the health status is conditioned by many factors, such as behavioural factors, genetic conditions, and health care [4].

Health behaviours are defined as activities aimed at maintaining or regaining health. Health behaviours are a form of a person's activity aimed at achieving health goals. The aim of the activities undertaken by healthy people is to maintain or increase health resources, while in patients it is to improve their health [2,3,8,9]. These behaviours include behaviours aimed at maintaining well-being; behaviours protecting health; behaviours reducing the risk of losing health; activities of those who noticed disease symptoms in themselves—in order to accurately assess their own health condition and take preventive measures; and activities of a person describing themselves as sick—to improve their state of health and to improve their state of health through a treatment process [8–10]. Health behaviours are reactions to all health-related situations, as well as habits and deliberate activities. Each person permanently, intentionally, consciously, and independently makes choices about behaviours that have a positive or negative impact on health. An important determinant of the decisions made is life experience and knowledge in the field of health and disease [1–5,8]. Determinants of choosing health behaviours include age, gender, life goals, social situation (education, social roles, origin, material situation, and place of residence), and cultural situation (world view, family and national traditions, and customs) [1–3]. Health behaviours that positively influence human health—apart from one's self-control, responsibility for one's own health, and a positive attitude—are mainly a healthy diet, regular physical activity, and an adequate amount of sleep per day. Disease risk factors and thus negative health behaviours include smoking, incorrect diet, low or no physical activity, and alcohol abuse [4,5,11–13]. The unsatisfactory health situation of the Polish population, with a tendency to deteriorate in some spheres, requires specific health and preventive measures. The study and analysis of health behaviours create an opportunity to diagnose the areas of ignorance in the field of health care, to determine the risks and harmfulness, as well as to determine the scale of the problem in order to prevent irregularities or pathologies [4,12].

It should be noted that the contemporary society is ageing [10–14]. Life expectancy in Poland increases but it remains significantly shorter than the average in the European Union (EU) by approximately 4 years. The development of civilization and an increasing life expectancy have become the main factors that contribute to the unsatisfactory health condition of the society, especially in the 50+ group [2,15].

The ageing period is characterized by the intensification of physical changes, which are manifested mainly by burdensome ailments and multiple diseases. The biggest threat to life related to incorrect health behaviours are cardiovascular diseases, which are a serious problem in terms of medicine, social life, and economy. Cardiovascular diseases have an epidemiological nature and are therefore referred to as 'civilization diseases'. They are the most common causes of death in Poland, Europe, and worldwide. They often cause physical and mental disability and require huge financial funding, both in terms of public finances and individual patient spending [16–22].

Hence, strengthening the correct and changing the incorrect actions that condition one's health has a significant impact on the increased quality of life, despite the unstoppable changes in the physical functioning of the ageing people [23–26]. Assessment of the level and the quality of health behaviours allows for undertaking educational activities, care, and treatment towards the senior population [24–26]

However, it should be emphasized that the key to maintaining and improving health is prevention.

Incorrect lifestyle is a factor underlying many chronic diseases. A study by Italian researchers has shown that more than half of the Italian population does not meet the WHO thresholds for at least moderate physical activity. The study assessed the impact on health and healthcare expenditure of seven public health policies aimed at promoting exercise and physical activity against a normal business scenario. The assessed policies included promotion of active transport, interventions at sedentary workplaces, investments in sports and recreation, mass-media campaigns, prescription of physical activity in primary care, school-based interventions, and mobile apps. The researchers concluded that public policies promoting physical activity could improve the health of the population and save healthcare expenditure, which would help avoid hundreds of cases of cardiovascular disease and diabetes per year, and dozens of cancer cases [27].

A similar situation in terms of a low level of physical activity can be observed in other countries, e.g., in Poland, Germany, the UK, and Belgium. Hence, appropriate steps should be taken to change the incorrect lifestyle of the society and thus prevent the occurrence of many dangerous chronic diseases in the future [28]

The aim of this study is to assess the health behaviours and value-based health analysis of people aged 50+ who were hospitalized due to cardiovascular disease, depending on the selected descriptive variables. The main hypothesis is as follows: Among the variables of the Inventory of Health-Related Behaviour (IHB), health practices had the strongest correlation with health, while understanding health, according to the List of Health Criteria (LHC), significantly affected the health behaviours of people aged 50+ who were hospitalized for cardiovascular diseases.

2. Materials and Methods

2.1. Study Design

This publication is another analysis of the results of studies conducted within a research project carried out on the territory of Poland and concerning the quality of life and health behaviours of people aged over 50 years old who were hospitalized due to cardiovascular diseases in Poland.

This study was triggered by the deteriorating health situation of people aged 50+ in Poland in the last two decades. The highest percentage of morbidity and mortality is caused by cardiovascular diseases. The risk factors for cardiovascular diseases include, among others, age, gender, genetic predisposition, as well as incorrect health behaviour, which became a starting point of this study. Standardized questionnaires—the Inventory of Health-Related Behaviour (IHB) and List of Health Criteria (LHC)—were used in the study; they are designed to study the health situation of people of different ages. These tools offer a detailed picture of health behaviours and an analysis of the respondents' health.

2.2. Studied Population

The study was conducted between April 2018 and December 2018 among 411 subjects aged 50+ who were hospitalized due to cardiovascular disease at the Independent Public Health Care Unit in Sanok (Podkarpackie voivodship in Poland).

The simple random selection method was used, and the study was carried out using 3 survey questionnaires, while the remaining data were obtained from the so-called patients' 'health cards'. The research tools allowed the authors to analyse the health situation of the respondents in the last 10 years.

Based on those records, the authors verified the existing cardiovascular diseases and obtained the necessary data for the calculation of indicators such as BMI. The (anonymous) questionnaires, along with patients' informed consent to participate in the study, were completed by the patients in accordance with the study instructions presented to them earlier.

The criteria for the study group were the following: age—50+; and gender—women and men diagnosed with cardiovascular disease and who were hospitalized due to the diseases in the Independent Public Health Care Unit in Sanok, Podkarpackie voivodship, Poland. The criteria for exclusion from the study group were age under 50, no diagnosed cardiovascular diseases, and gender—not hospitalized men and women.

2.2.1. Description of Research Tools

The method used in the study was a diagnostic survey. The study employed 3 research tools:

1. Authors' 3-part survey questionnaire. The first part focused on social information; the second part—health of the subjects; and the third part—their lifestyle.
2. The second tool was the standardized questionnaire 'List of Health Criteria' (LHC) by Zygmunt Juczyński. LHC contained 24 statements describing the positive elements of three dimensions of health: physical, mental, and social; its results help determine respondents' preferences in terms of health determinants, discover what the respondents understand by the concept of health, what it means for a given individual to 'be healthy', and discover the extent to which health is equated to state, process, or property. The respondents reported their preferences, indicating which of the given statements were important in their health assessment, and which among the selected statements were the most crucial. It resulted in a ranking of all 24 health criteria which characterized the study group. To interpret the results, one needed to consider the distribution of the number of ranks of the individual health criteria. LHC may be useful in activities aimed at modifying health behaviours, as well as in therapy and rehabilitation.
3. The third research tool was the standardized 'Inventory of Health-Related Behaviour' (IHB) developed by Zygmunt Juczyński. It consists of 25 statements describing various types of health-related behaviours. It enabled an overall assessment of the intensity of healthy behaviours, as well as 4 categories of health behaviours, namely, correct eating habits, preventive behaviour, positive mental attitude, and health practices. The value of the general indicator of health behaviours ranged from 24 to 120 points. The higher the indicator, the greater the intensity of the declared healthy behaviours. After conversion into standardized units, the indicator was subjected to interpretation as sten scores. Scores 1–4 were considered low, 5–6 average, and 7–10 high. For this scale, normative values for various age and social groups as well as for the healthy and the ill were determined. The author of the questionnaire provided sufficient reliability for the overall score of the IHB ($\alpha = 0.85$) and for the individual scales (Cronbach's alpha index ranged from 0.60 to 0.65) [29].

2.2.2. Consent of the Bioethical Committee to Conduct the Study

The study was approved by the Bioethics Committee of the Medical University of Lodz, under number RNN/156/18/KE.

2.2.3. Statistical Analysis

A statistical analysis was carried out in the R program, version 3.5.1. (The R Consortium, Vienna, Austria).

The analysis of the quantitative variables (i.e., expressed in numbers) was performed by calculating the mean, standard deviation, median, quartiles, minimum, and maximum values. The analysis of the qualitative variables (i.e., not expressed in numbers) was conducted by calculating the number and percentage of occurrences of each value. '0' means zero points, i.e., a situation in which a given criteria was not selected by the respondent.

The comparison of the values of quantitative variables in two groups was made using Student's *t*-test (when the variable had a normal distribution in these groups) or the Mann–Whitney U test (for a non-normal distribution).

A comparison of the values of the quantitative variables in three or more groups was made using ANOVA analysis (when the variable had a normal distribution in these groups) or the Kruskal–Wallis test (for a non-normal distribution). After detecting statistically significant differences, post-hoc analysis was carried out with Fisher's Least Significant Difference (LSD) test (normal distribution) or Dunn's test (non-normal distribution) to identify the statistically significant differences between groups.

Correlations between the quantitative variables were analysed using the Pearson correlation coefficient (when both variables had a normal distribution) or Spearman correlation coefficient (otherwise). The strength of the correlation was interpreted according to the following scheme:

$|r| \geq 0.9$—very strong correlation
$0.7 \leq |r| < 0.9$—strong correlation
$0.5 \leq |r| < 0.7$—medium correlation
$0.3 \leq |r| < 0.5$—weak correlation
$|r| < 0.3$—very week correlation (negligible).

Interpretation based on Hinkle D.E., Wiersma W., Jurs S.G. Applied Statistics for the Behavioural Sciences. 5th ed. Boston: Houghton Mifflin, 2003 [30].

The normality of the variable distribution was tested using the Shapiro–Wilk test. The analysis adopted the significance level of 0.05 (thus, all *p* values below 0.05 were interpreted as significant correlations) [31].

3. Results

The characteristics of the study group are presented in Table 1. The group was dominated by men—there were 223 men (54.26%) and 186 women (45.26%). The age structure of the study group: the average age of the subjects was 69.2 (SD = 9.45 and ranged from 50 to 93); the median was 69.

Table 1. Characteristics of the respondents.

Characteristics of the Respondents		N	%
Gender	Women	186	45.53
	Men	223	54.47
	No answer	2	0.49
Age (in years)	N—411	-	-
	Mean—69.2	-	-
	SD—9.45	-	-
	Median—69	-	-
	Min—50	-	-
	Max—93	-	-
	Q1—62	-	-
	Q3—76	-	-

Table 1. *Cont.*

Characteristics of the Respondents		N	%
BMI	Underweight [17–18.5]	2	0.49
	Correct weight [18.5–25]	100	24.33
	Overweight [25–30]	157	38.20
	Obesity [30–35]	125	30.41
	Class 2 obesity [35–40]	15	3.65
	Class 3 obesity [>40]	8	1.95
	No data available (weight and/or height)	4	0.97
Place of residence	City < 100,000 inhabitants	223	54.26
	City > 100,000 inhabitants	3	0.73
	Rural area	183	44.53
	No answer	2	0.49
Education	Primary	50	12.17
	Vocational	191	46.47
	Secondary	145	35.28
	Higher	22	5.35
	No answer	3	0.73
Professional activity	Professionally active	84	20.44
	Professionally inactive	313	76.16
	No answer	14	3.41
Marital status	Single	12	2.92
	Married	311	75.67
	In separation	2	0.49
	Divorced	7	1.70
	Widow/widower	73	17.76
	In a partnership	5	1.22
	No answer	1	0.24
Treatment of cardiovascular disease	Hypertension	162	39.42
	Atherosclerosis	240	58.39
	Ischemic heart disease	262	63.75
	Rhythm and cardiac conduction disorders	105	25.55
	Heart defects (congenital and acquired)	29	7.06
	Varicose veins of the lower extremities	109	26.52
	Venous thrombosis	28	6.81
	Other diseases	43	10.46

N—number of respondents; %—percentage of respondents; SD—average age of respondents; Q—quartile.

Body weight of the subject group: 157 subjects (38.20%) were overweight and 125 (30.41%) suffered from class 1 obesity.

Place of residence of the study group: the most numerous group was the inhabitants of cities below 100,000—223 people (54.26%)—while there were 183 people in rural areas (44.53%).

Education of study group: the proportion of people with vocational education was the highest—191 people (46.47%)—while 145 people (35.28%) reported secondary education.

In the study group, the majority, i.e., 313 people (76.16%), were professionally inactive, while 84 people (20.44%) were still active.

The structure of marital status: the largest group were married people—311 subjects (75.67%)—followed by widows and widowers—73 individuals (17.76%).

Analysing the incidence of cardiovascular diseases amongst subjects, it was found that 262 people (63.75%) had coronary heart disease, which was followed by atherosclerosis, 240 people (58.39%), and hypertension, 162 people (39.42%) (Table 1).

Analysis of the health behaviour results according to the Inventory of Health-Related Behaviour (IHB) and the analysis of health assessment criteria according to the List of Health Criteria (LHC) showed that people aged 50+ who were hospitalized due to cardiovascular diseases indicated the following as the important health criteria:

- not feeling any physical ailments;
- having all body parts functional;
- feeling well;
- eating properly;
- infrequent need of going to the doctor (Table 2).

Table 2. List of Health Criteria of the studied group.

Criterion	N	Mean	SD	Median	Min	Max	Q1	Q3
reaching very old age	411	0.01	0.12	0	0	1	0	0
feeling happy most of the time	411	0.17	0.75	0	0	5	0	0
getting along well with other people	411	0.55	1.09	0	0	5	0	1
knowing how to solve one's own problems	411	0.45	1.08	0	0	5	0	0
eating properly	411	1.08	1.31	0	0	5	0	2
adequate amount of rest, sleep	411	0.62	0.91	0	0	4	0	1
drinking little or no alcohol	411	0.22	0.75	0	0	5	0	0
not smoking tobacco	411	0.88	1.16	1	0	5	0	1
having a correct body weight	411	0.09	0.48	0	0	4	0	0
taking medication rarely	411	0.27	0.84	0	0	4	0	0
having a good mood	411	0.24	0.85	0	0	5	0	0
not feeling any physical ailments	411	3.11	1.77	4	0	5	2	4
ability to work without tension and stress	411	0.25	0.93	0	0	5	0	0
not getting sick, at most with flu, indigestion (rarely)	411	0.25	0.87	0	0	5	0	0
having healthy eyes, hair and skin	411	0.03	0.29	0	0	4	0	0
ability to adapt to life changes	411	0.48	1.23	0	0	5	0	0
ability to enjoy life	411	0.18	0.77	0	0	5	0	0
being responsible	411	0.05	0.37	0	0	4	0	0
ability to control one's own feelings and desires	411	0.02	0.19	0	0	3	0	0
having all body parts functional	411	2.45	2.34	3	0	6	0	5
accepting oneself, knowing one's possibilities and limitations	411	0.4	1.12	0	0	5	0	0
having a job, various interests	411	0.13	0.6	0	0	5	0	0
feeling well	411	1.7	2.01	0	0	6	0	4
infrequent need of going to the doctor	411	1	1.69	0	0	5	0	2

N—number of respondents; SD—average age of respondents; Q—quartile.

As a result of the comparison of the most important statements regarding health, taking into account the age of the respondents, it was found that age significantly and positively correlated with the following criteria ($p < 0.05$):

- drinking little or no alcohol;
- not getting sick, at most with flu, indigestion (rarely);
- having healthy eyes, hair, and skin.

It should be noted that the abovementioned health criteria were more important for the elderly respondents.

On the other hand, age correlated significantly and negatively with the following criterion ($p < 0.05$):

- ability to work without tension and stress.

For elderly patients, this criterion was less important than for the rest of the study group (Table 3).

Table 3. Correlations of health statements with the age of the respondents.

Criterion	Correlation with Age			
	Correlation Coefficient	p *	Correlation Direction	Correlation Strength
reaching very old age	−0.021	$p = 0.673$	—	—
feeling happy most of the time	0.051	$p = 0.304$	—	—
getting along well with other people	0.059	$p = 0.234$	—	—
knowing how to solve one's own problems	−0.067	$p = 0.174$	—	—
eating properly	0.009	$p = 0.858$	—	—
adequate amount of rest, sleep	−0.032	$p = 0.523$	—	—
drinking little or no alcohol	0.115	$p = 0.02$	positive	very weak
not smoking tobacco	0.026	$p = 0.593$	—	—
having a correct body weight	−0.037	$p = 0.457$	—	—
taking medication rarely	−0.002	$p = 0.976$	—	—
having a good mood	−0.038	$p = 0.442$	—	—
not feeling any physical ailments	0.021	$p = 0.671$	—	—
ability to work without tension and stress	−0.174	$p < 0.001$	negative	very weak
not getting sick, at most with flu, indigestion (rarely)	0.159	$p = 0.001$	positive	very weak
having healthy eyes, hair and skin	0.1	$p = 0.044$	positive	very weak
ability to adapt to life changes	−0.054	$p = 0.271$	—	—
ability to enjoy life	−0.064	$p = 0.198$	—	—
being responsible	−0.055	$p = 0.268$	—	—
ability to control one's own feelings and desires	−0.068	$p = 0.166$	—	—
having all body parts functional	0.001	$p = 0.978$	—	—
accepting oneself, knowing one's possibilities and limitations	−0.069	$p = 0.163$	—	—
having a job, various interests	−0.019	$p = 0.703$	—	—
feeling well	0.09	$p = 0.067$	—	—
infrequent need of going to the doctor	0.037	$p = 0.46$	—	—

* p = Non-normal distribution of both correlated variables, Pearson correlation coefficient; NP = Non-normal distribution of at least one of the correlated variables, Spearman's correlation coefficient.

After analysing the most important statements in the respondents' opinion, it was found that the perception of 9 out of 24 health criteria significantly depended on gender ($p < 0.05$).

The studied group of women perceived the following health criteria as more important than men:

- feeling happy most of the time;
- eating properly;
- drinking little or no alcohol;
- not smoking tobacco;
- ability to work without tension and stress.

Less important criteria for the studied group of women were:

- not feeling any physical ailments;

- not getting sick, at most with flu, indigestion (rarely);
- having all body parts functional;
- infrequent need of going to the doctor (Table 4).

Table 4. Correlations of health statements with the gender of the respondents.

Criterion		Women	Men	p *
reaching very old age	Median	0	0	0.806
	Quartile	0–0	0–0	
feeling happy most of the time	Median	0	0	0.048
	Quartile	0–0	0–0	
getting along well with other people	Median	0	0	0.325
	Quartile	0–1	0–0	
knowing how to solve one's own problems	Median	0	0	0.537
	Quartile	0–0	0–0	
eating properly	Median	0.5	0	0.012
	Quartile	0–3	0–2	
adequate amount of rest, sleep	Median	0	0	0.581
	Quartile	0–1	0–1	
drinking little or no alcohol	Median	0	0	0.002
	Quartile	0–0	0–0	
not smoking tobacco	Median	1	0	0.001
	Quartile	0–1	0–1	
having a correct body weight	Median	0	0	0.76
	Quartile	0–0	0–0	
taking medication rarely	Median	0	0	0.065
	Quartile	0–0	0–0	
having a good mood	Median	0	0	0.995
	Quartile	0–0	0–0	
not feeling any physical ailments	Median	4	4	0.002
	Quartile	0–4	3–5	
ability to work without tension and stress	Median	0	0	0.01
	Quartile	0–0	0–0	
not getting sick, at most with flu, indigestion (rarely)	Median	0	0	0.028
	Quartile	0–0	0–0	
having healthy eyes, hair and skin	Median	0	0	0.857
	Quartile	0–0	0–0	
ability to adapt to life changes	Median	0	0	0.14
	Quartile	0–0	0–0	
ability to enjoy life	Median	0	0	0.261
	Quartile	0–0	0–0	
being responsible	Median	0	0	0.322
	Quartile	0–0	0–0	
ability to control one's own feelings and desires	Median	0	0	0.855
	Quartile	0–0	0–0	
having all body parts functional	Median	0	4	0.023
	Quartile	0–5	0–5	

Table 4. *Cont.*

Criterion		Women	Men	p *
accepting oneself, knowing one's possibilities and limitations	Median	0	0	0.301
	Quartile	0–0	0–0	
having a job, various interests	Median	0	0	0.37
	Quartile	0–0	0–0	
feeling well	Median	0	0	0.283
	Quartile	0–4	0–4	
infrequent need of going to the doctor	Median	0	0	0.001
	Quartile	0–0	0–3	

* p = Normal distribution in groups, Student's t-test; NP = Non-normal distribution in groups, Mann–Whitney U test.

After analysing the study results, it was found that the perception of 3 out of 24 health criteria significantly depended on BMI ($p < 0.05$). A post-hoc analysis was performed, which helped discover significant relationships. The analysis showed that:

- obese people perceived the criterion 'eating properly' as less important than other respondents and obese people perceived the criterion 'having all body parts functional' as more important than those with underweight and regular body weight. (Table 5).

Table 5. Correlations of health statements with the BMI of the respondents.

Criterion		Underweight, Correct Weight	Overweight	Obesity	p *
reaching very old age	Median	0	0	0	0.249
	Quartile	0–0	0–0	0–0	
feeling happy most of the time	Median	0	0	0	0.203
	Quartile	0–0	0–0	0–0	
getting along well with other people	Median	0	0	0	0.116
	Quartile	0–1	0–1	0–0	
knowing how to solve one's own problems	Median	0	0	0	0.221
	Quartile	0–0	0–0	0–0	
eating properly	Median	0	1	0	0.025
	Quartile	0–2.75	0–2	0–2	Under.Corr.Over > O
adequate amount of rest, sleep	Median	0	0	0	0.845
	Quartile	0–1	0–1	0–1	
drinking little or no alcohol	Median	0	0	0	0.248
	Quartile	0–0	0–0	0–0	
not smoking tobacco	Median	1	1	0.5	0.544
	Quartile	0–2	0–1	0–1	
having a correct body weight	Median	0	0	0	0.652
	Quartile	0–0	0–0	0–0	
taking medication rarely	Median	0	0	0	0.565
	Quartile	0–0	0–0	0–0	
having a good mood	Median	0	0	0	0.349
	Quartile	0–0	0–0	0–0	
not feeling any physical ailments	Median	4	4	4	0.129
	Quartile	0–4	3–4	2–4	
ability to work without tension and stress	Median	0	0	0	0.236
	Quartile	0–0	0–0	0–0	

Table 5. *Cont.*

Criterion		Underweight, Correct Weight	Overweight	Obesity	p *
not getting sick, at most with flu, indigestion (rarely)	Median	0	0	0	0.769
	Quartile	0–0	0–0	0–0	
having healthy eyes, hair and skin	Median	0	0	0	0.272
	Quartile	0–0	0–0	0–0	
ability to adapt to life changes	Median	0	0	0	0.504
	Quartile	0–0	0–0	0–0	
ability to enjoy life	Median	0	0	0	0.835
	Quartile	0–0	0–0	0–0	
being responsible	Median	0	0	0	0.564
	Quartile	0–0	0–0	0–0	
ability to control one's own feelings and desires	Median	0	0	0	0.239
	Quartile	0–0	0–0	0–0	
having all body parts functional	Median	0	3	4	0.038
	Quartile	0–5	0–5	0–5	O > Under-Corr
accepting oneself, knowing one's possibilities and limitations	Median	0	0	0	0.795
	Quartile	0–0	0–0	0–0	
having a job, various interests	Median	0	0	0	0.053
	Quartile	0–0	0–0	0–0	
feeling well	Median	1	0	0	0.235
	Quartile	0–4	0–3	0–4	
infrequent need of going to the doctor	Median	0	0	0	0.01
	Quartile	0–0.75	0–0	0–3	O > Under-Corr.Over

* p = Normal distribution in groups, ANOVA + results of post-hoc analysis (Fisher's LSD test); NP = Non-normal distribution in groups, Kruskal–Wallis test + post-hoc analysis results (Dunn's test).

After analysing the study results, it was found that the perception of 9 out of 24 health criteria significantly depended on place of residence ($p < 0.05$).

The respondents who lived in cities perceived the following criteria as more important than those living in a rural environment:

- eating properly and adequate amount of rest, sleep.

On the other hand, the respondents who lived in cities perceived the following criteria as less important: drinking little or no alcohol and not getting sick, at most with flu, indigestion (rarely) (Table 6).

The statistical analysis showed that the perception of 16 out of 24 health criteria significantly depended on education ($p < 0.05$). A post-hoc analysis was performed, which helped discover that people with vocational and secondary education perceived the criterion 'getting along well with other people' as less important than people with primary and higher education and people with higher education perceived the criteria of 'knowing how to solve one's problems', 'ability to work without tension and stress', 'ability to adapt to life changes', 'ability to control one's own feelings and desires', 'accepting oneself, knowing one's possibilities and limitations', and 'having a job, various interests' as more important than other respondents. (Table 7).

Table 6. Correlations of health statements with the place of residence of the respondents.

Criterion		Urban	Rural	p *
reaching very old age	Median	0	0	0.796
	Quartile	0–0	0–0	
feeling happy most of the time	Median	0	0	0.316
	Quartile	0–0	0–0	
getting along well with other people	Median	0	0	0.229
	Quartile	0–0	0–1	
knowing how to solve one's own problems	Median	0	0	0.49
	Quartile	0–0	0–0	
eating properly	Median	1	0	0.003
	Quartile	0–2	0–2	
adequate amount of rest, sleep	Median	0	0	0.011
	Quartile	0–2	0–1	
drinking little or no alcohol	Median	0	0	0.01
	Quartile	0–0	0–0	
not smoking tobacco	Median	1	0	0.551
	Quartile	0–1	0–1	
having a correct body weight	Median	0	0	0.019
	Quartile	0–0	0–0	
taking medication rarely	Median	0	0	0.428
	Quartile	0–0	0–0	
having a good mood	Median	0	0	0.543
	Quartile	0–0	0–0	
not feeling any physical ailments	Median	4	4	0.832
	Quartile	2.25–4	2–4	
ability to work without tension and stress	Median	0	0	0.014
	Quartile	0–0	0–0	
not getting sick, at most with flu, indigestion (rarely)	Median	0	0	0.002
	Quartile	0–0	0–0	
having healthy eyes, hair and skin	Median	0	0	0.223
	Quartile	0–0	0–0	
ability to adapt to life changes	Median	0	0	0.323
	Quartile	0–0	0–0	
ability to enjoy life	Median	0	0	0.591
	Quartile	0–0	0–0	
being responsible	Median	0	0	0.764
	Quartile	0–0	0–0	
ability to control one's own feelings and desires	Median	0	0	0.071
	Quartile	0–0	0–0	
having all body parts functional	Median	3	3	0.369
	Quartile	0–5	0–5	
accepting oneself, knowing one's possibilities and limitations	Median	0	0	0.006
	Quartile	0–0	0–0	
having a job, various interests	Median	0	0	0.869
	Quartile	0–0	0–0	
feeling well	Median	0	2	0.01
	Quartile	0–3	0–4	
infrequent need of going to the doctor	Median	0	0	0.001
	Quartile	0–0	0–3	

* p = Normal distribution in groups, Student's t-test; NP = Non-normal distribution in groups, Mann–Whitney U test.

Table 7. Correlations of health statements with the education of the respondents.

Criterion		Primary	Vocational	Secondary	Higher	p *
reaching very old age	Median	0	0	0	0	0.23
	Quartile	0–0	0–0	0–0	0–0	
feeling happy most of the time	Median	0	0	0	0	0.147
	Quartile	0–0	0–0	0–0	0–0	
getting along well with other people	Median	1	0	0	1	0.001
	Quartile	0–2	0–0	0–1	0–2	H,P > S,V
knowing how to solve one's own problems	Median	0	0	0	0	0.001
	Quartile	0–1	0–0	0–0	0–2	H > P,V,S
eating properly	Median	0	1	1	0	0.001
	Quartile	0–0	0–2	0–3	0–2	S,V,H > P
adequate amount of rest, sleep	Median	0	0	0	0	0.001
	Quartile	0–0	0–1	0–2	0–0	S,V > H,P
drinking little or no alcohol	Median	0	0	0	0	0.001
	Quartile	0–1	0–0	0–0	0–0	P > S,V,H
not smoking tobacco	Median	0	1	1	0	0.689
	Quartile	0–2	0–1	0–1	0–1.75	
having a correct body weight	Median	0	0	0	0	0.007
	Quartile	0–0	0–0	0–0	0–0	S > V,P,H
taking medication rarely	Median	0	0	0	0	0.334
	Quartile	0–0	0–0	0–0	0–0	
having a good mood	Median	0	0	0	0	0.178
	Quartile	0–0	0–0	0–0	0–0	
not feeling any physical ailments	Median	4.5	4	4	0	0.001
	Quartile	3–5	3–4	2–4	0–2.5	P > V,S,H V,S > H
ability to work without tension and stress	Median	0	0	0	2.5	0.001
	Quartile	0–0	0–0	0–0	0–4	H > S,V,P
not getting sick, at most with flu, indigestion (rarely)	Median	0	0	0	0	0.001
	Quartile	0–0.75	0–0	0–0	0–0	P > V,S,H V > S, H
having healthy eyes, hair and skin	Median	0	0	0	0	0.146
	Quartile	0–0	0–0	0–0	0–0	
ability to adapt to life changes	Median	0	0	0	2	0.001
	Quartile	0–0	0–0	0–0	0–3.75	H >P, S,V
ability to enjoy life	Median	0	0	0	0	0.084
	Quartile	0–0	0–0	0–0	0–0	
being responsible	Median	0	0	0	0	0.747
	Quartile	0–0	0–0	0–0	0–0	
ability to control one's own feelings and desires	Median	0	0	0	0	0.001
	Quartile	0–0	0–0	0–0	0–0	H > S,P,V
having all body parts functional	Median	0	5	3	0	0.001
	Quartile	0–2	0–5	0–5	0–0	V > S,P,H S > P,H
accepting oneself, knowing one's possibilities and limitations	Median	0	0	0	0	0.012
	Quartile	0–0	0–0	0–0	0–0	H,S,V > P
having a job, various interests	Median	0	0	0	0	0.047
	Quartile	0–0	0–0	0–0	0–0	H > S,V,P
feeling well	Median	2	0	0	0	0.042
	Quartile	0–4	0–3.5	0–4	0–2.75	P > V,H
infrequent need of going to the doctor	Median	1	0	0	0	0.001
	Quartile	0–2	0–3	0–0	0–0	P > S,H V,S > H

* p = Normal distribution in groups, ANOVA + results of post-hoc analysis (Fisher's LSD test); NP = Non-normal distribution in groups, Kruskal–Wallis test + post-hoc analysis results (Dunn's test).

After analysing the study results, it was found that the perception of 3 out of 24 health criteria significantly depended on the professional activity ($p < 0.05$).

The respondents who were professionally active perceived the following criterion as more important than inactive ones: ability to work without tension and stress.

Meanwhile, the criteria perceived by the professionally active respondents as less important were:

- getting along well with other people;
- not getting sick, at most with flu, indigestion (rarely) (Table 8).

Table 8. Correlations of health statements with the professional activity of the respondents.

Criterion		Professionally Active	Professionally Inactive	p *
reaching very old age	Median	0	0	0.464
	Quartile	0–0	0–0	
feeling happy most of the time	Median	0	0	0.389
	Quartile	0–0	0–0	
getting along well with other people	Median	0	0	0.041
	Quartile	0–0	0–1	
knowing how to solve one's own problems	Median	0	0	0.589
	Quartile	0–0	0–0	
eating properly	Median	1	0	0.052
	Quartile	0–2.25	0–2	
adequate amount of rest, sleep	Median	0	0	0.245
	Quartile	0–1.25	0–1	
drinking little or no alcohol	Median	0	0	0.258
	Quartile	0–0	0–0	
not smoking tobacco	Median	1	0	0.566
	Quartile	0–1	0–1	
having a correct body weight	Median	0	0	0.491
	Quartile	0–0	0–0	
taking medication rarely	Median	0	0	0.879
	Quartile	0–0	0–0	
having a good mood	Median	0	0	0.812
	Quartile	0–0	0–0	
not feeling any physical ailments	Median	4	4	0.571
	Quartile	2–4	2–4	
ability to work without tension and stress	Median	0	0	0.012
	Quartile	0–0	0–0	
not getting sick, at most with flu, indigestion (rarely)	Median	0	0	0.001
	Quartile	0–0	0–0	
having healthy eyes, hair and skin	Median	0	0	0.3
	Quartile	0–0	0–0	
ability to adapt to life changes	Median	0	0	0.914
	Quartile	0–0	0–0	
ability to enjoy life	Median	0	0	0.65
	Quartile	0–0	0–0	
being responsible	Median	0	0	0.14
	Quartile	0–0	0–0	

Table 8. *Cont.*

Criterion		Professionally Active	Professionally Inactive	p *
ability to control one's own feelings and desires	Median	0	0	0.156
	Quartile	0–0	0–0	
having all body parts functional	Median	3.5	3	0.372
	Quartile	0–5	0–5	
accepting oneself, knowing one's possibilities and limitations	Median	0	0	0.153
	Quartile	0–0	0–0	
having a job, various interests	Median	0	0	0.435
	Quartile	0–0	0–0	
feeling well	Median	0	0	0.105
	Quartile	0–3	0–4	
infrequent need of going to the doctor	Median	0	0	0.109
	Quartile	0–0	0–2	

* p = Normal distribution in groups, Student's t-test; NP = Non-normal distribution in groups, Mann–Whitney U test.

The analysis of the comparison of health statements with the marital status of the respondents showed that the perception of 12 out of 24 health criteria significantly depended on marital status ($p < 0.05$).

Married respondents perceived the following criteria as more important than other respondents: eating properly; adequate amount of rest, sleep; and not smoking tobacco. (Table 9).

Table 9. Correlations of health statements with the marital status of the respondents.

Criterion		Married	Other	p *
reaching very old age	Median	0	0	0.668
	Quartile	0–0	0–0	
feeling happy most of the time	Median	0	0	0.024
	Quartile	0–0	0–0	
getting along well with other people	Median	0	0	0.007
	Quartile	0–0	0–1	
knowing how to solve one's own problems	Median	0	0	0.776
	Quartile	0–0	0–0	
eating properly	Median	1	0	0.001
	Quartile	0–2	0–1	
adequate amount of rest, sleep	Median	0	0	0.002
	Quartile	0–1	0–0	
drinking little or no alcohol	Median	0	0	0.006
	Quartile	0–0	0–0	
not smoking tobacco	Median	1	0	0.027
	Quartile	0–1	0–1	
having a correct body weight	Median	0	0	0.326
	Quartile	0–0	0–0	
taking medication rarely	Median	0	0	0.856
	Quartile	0–0	0–0	

Table 9. *Cont.*

Criterion		Married	Other	p *
having a good mood	Median	0	0	0.787
	Quartile	0–0	0–0	
not feeling any physical ailments	Median	4	3	0.036
	Quartile	2.5–4	0–4	
ability to work without tension and stress	Median	0	0	0.036
	Quartile	0–0	0–0	
not getting sick, at most with flu, indigestion (rarely)	Median	0	0	0.036
	Quartile	0–0	0–0	
having healthy eyes, hair and skin	Median	0	0	0.226
	Quartile	0–0	0–0	
ability to adapt to life changes	Median	0	0	0.167
	Quartile	0–0	0–0	
ability to enjoy life	Median	0	0	0.611
	Quartile	0–0	0–0	
being responsible	Median	0	0	0.769
	Quartile	0–0	0–0	
ability to control one's own feelings and desires	Median	0	0	0.259
	Quartile	0–0	0–0	
having all body parts functional	Median	4	0	0.002
	Quartile	0–5	0–4	
accepting oneself, knowing one's possibilities and limitations	Median	0	0	0.044
	Quartile	0–0	0–0	
having a job, various interests	Median	0	0	0.326
	Quartile	0–0	0–0	
feeling well	Median	0	2	0.001
	Quartile	0–3	0–4	
infrequent need of going to the doctor	Median	0	0	0.25
	Quartile	0–1	0–2	

* p = Normal distribution in groups, Student's t-test; NP = Non-normal distribution in groups, Mann–Whitney U test.

The analysis of the study results based on the IHB questionnaire showed that the most common health behaviours were related to 'health practices', slightly less common in the area of 'positive mental attitude', and the least common in the area of 'proper eating habits' and 'preventive behaviours' (Table 10).

By correlating the level of health behaviours according to the IHB with individual categories from the LHC, a total of 59 statistically significant relationships were observed ($p < 0.05$). A list of these dependencies, sorted from the strongest to the weakest, is presented in the table below (Table 11).

Table 10. Degree of intensity of specific categories of health behaviours according to the Inventory of Health-Related Behaviour (IHB) amongst respondents.

IHB Sub-Scales	N	Mean	SD	Median	Min	Max	Q1	Q3
Correct eating habits	411	3.28	0.75	3.17	1	5	2.83	4
Preventive behaviours	411	3	0.62	3	1.17	4.5	2.67	3.67
Positive mental attitude	411	3.59	0.52	3.83	1.33	4.67	3.33	4
Health practices	411	3.61	0.53	3.67	1.67	4.83	3.33	4

Table 11. The level of health behaviours according to the Inventory of Health-Related Behaviour (IHB) versus number of selections of individual categories of the List of Health Criteria (LHC).

LHC	IHB	Correlation Coefficient	p	Correlation Direction	Correlation Strength
Eating properly	Preventive behaviours	0.521	$p < 0.001$	positive	Average
Eating properly	Overall IHB result	0.492	$p < 0.001$	positive	Weak
Eating properly	Correct eating habits	0.414	$p < 0.001$	positive	Weak
Eating properly	Positive mental attitude	0.414	$p < 0.001$	positive	Weak
Adequate amount of rest, sleep	Preventive behaviours	0.407	$p < 0.001$	positive	Weak
Infrequent need of going to the doctor	Correct eating habits	−0.392	$p < 0.001$	negative	Weak
Infrequent need of going to the doctor	Overall IHB result	−0.354	$p < 0.001$	negative	Weak
Adequate amount of rest, sleep	Overall IHB result	0.332	$p < 0.001$	positive	Weak
Not smoking tobacco	Positive mental attitude	0.315	$p < 0.001$	positive	Weak
Adequate amount of rest, sleep	Positive mental attitude	0.313	$p < 0.001$	positive	Weak
Not smoking tobacco	Overall IHB result	0.302	$p < 0.001$	positive	Weak
Eating properly	Health practices	0.296	$p < 0.001$	positive	very weak
Feeling well	Preventive behaviours	−0.295	$p < 0.001$	negative	very weak
Feeling well	Overall IHB result	−0.291	$p < 0.001$	negative	very weak
Infrequent need of going to the doctor	Preventive behaviours	−0.285	$p < 0.001$	negative	very weak
Not getting sick, at most with flu, indigestion (rarely)	Overall IHB result	−0.279	$p < 0.001$	negative	very weak
Feeling well	Positive mental attitude	−0.276	$p < 0.001$	negative	very weak
Infrequent need of going to the doctor	Health practices	−0.274	$p < 0.001$	negative	very weak
Not smoking tobacco	Preventive behaviours	0.266	$p < 0.001$	positive	very weak
Not getting sick, at most with flu, indigestion (rarely)	Preventive behaviours	−0.263	$p < 0.001$	negative	very weak
Having all body parts functional	Preventive behaviours	0.263	$p < 0.001$	positive	very weak
Not getting sick, at most with flu, indigestion (rarely)	Correct eating habits	−0.262	$p < 0.001$	negative	very weak
Feeling well	Correct eating habits	−0.262	$p < 0.001$	negative	very weak
Adequate amount of rest, sleep	Correct eating habits	0.251	$p < 0.001$	positive	very weak
Not getting sick, at most with flu, indigestion (rarely)	Positive mental attitude	−0.239	$p < 0.001$	negative	very weak
Infrequent need of going to the doctor	Positive mental attitude	−0.238	$p < 0.001$	negative	very weak
Knowing how to solve one's own problems	Positive mental attitude	−0.232	$p < 0.001$	negative	very weak
Not smoking tobacco	Health practices	0.229	$p < 0.001$	positive	very weak
Getting along well with other people	Preventive behaviours	−0.224	$p < 0.001$	negative	very weak
Having all body parts functional	Positive mental attitude	0.22	$p < 0.001$	positive	very weak
Not smoking tobacco	Correct eating habits	0.217	$p < 0.001$	positive	very weak
Ability to adapt to life changes	Positive mental attitude	−0.196	$p < 0.001$	negative	very weak
Having all body parts functional	Overall IHB result	0.186	$p < 0.001$	positive	very weak
Knowing how to solve one's own problems	Preventive behaviours	−0.181	$p < 0.001$	negative	very weak
Not getting sick, at most with flu, indigestion (rarely)	Health practices	−0.167	$p = 0.001$	negative	very weak
Taking medication rarely	Correct eating habits	−0.158	$p = 0.001$	negative	very weak
Taking medication rarely	Health practices	−0.157	$p = 0.001$	negative	very weak
Getting along well with other people	Positive mental attitude	−0.148	$p = 0.003$	negative	very weak
Ability to work without tension and stress	Correct eating habits	0.142	$p = 0.004$	positive	very weak
Ability to adapt to life changes	Preventive behaviours	−0.139	$p = 0.005$	negative	very weak
Taking medication rarely	Overall IHB result	−0.136	$p = 0.006$	negative	very weak
Feeling well	Health practices	−0.133	$p = 0.007$	negative	very weak
Ability to enjoy life	Positive mental attitude	0.13	$p = 0.008$	positive	very weak

Table 11. *Cont.*

LHC	IHB	Correlation Coefficient	p	Correlation Direction	Correlation Strength
Being responsible	Overall IHB result	−0.129	$p = 0.009$	negative	very weak
Taking medication rarely	Preventive behaviours	−0.123	$p = 0.013$	negative	very weak
Being responsible	Preventive behaviours	−0.123	$p = 0.013$	negative	very weak
Knowing how to solve one's own problems	Overall IHB result	−0.122	$p = 0.013$	negative	very weak
Adequate amount of rest, sleep	Health practices	0.122	$p = 0.013$	positive	very weak
Ability to control one's own feelings and desires	Correct eating habits	0.121	$p = 0.014$	positive	very weak
Being responsible	Positive mental attitude	−0.114	$p = 0.021$	negative	very weak
Being responsible	Correct eating habits	−0.113	$p = 0.021$	negative	very weak
Ability to enjoy life	Overall IHB result	0.109	$p = 0.027$	positive	very weak
Getting along well with other people	Overall IHB result	−0.104	$p = 0.035$	negative	very weak
Reaching very old age	Health practices	−0.103	$p = 0.036$	negative	very weak
Feeling happy most of the time	Correct eating habits	0.102	$p = 0.039$	positive	very weak
Ability to work without tension and stress	Preventive behaviours	0.102	$p = 0.04$	positive	very weak
Ability to enjoy life	Correct eating habits	0.1	$p = 0.044$	positive	very weak
Ability to enjoy life	Health practices	0.099	$p = 0.045$	positive	very weak
Reaching very old age	Correct eating habits	−0.098	$p = 0.046$	negative	very weak

* p = Normal distribution of both correlated variables, Pearson correlation coefficient; NP = Non-normal distribution of at least one of the correlated variables, Spearman's correlation coefficient.

4. Discussion and Limitations

The results of the conducted study painted the picture of the health behaviours and health analysis of people aged 50+ who were hospitalized for cardiovascular diseases, depending on the selected descriptive variables.

The study (according to the LHC) showed that the following factors were of great importance in the selection of appropriate criteria determining the health of the respondents: gender, age, education, place of residence, and marital status.

The conducted study has shown that, for the older respondents, the most important factor determining their health is the avoidance of stimulants, while the examined group of women considered the following health criteria as more important than the studied group of men: feeling of happiness and proper nutrition. The respondents from the cities indicated that important criteria for health were proper nutrition, sleep, and rest—more often than their rural counterparts. The education of the respondents had a significant impact on the results obtained: people with higher education chose solving difficult life situations and working without stress as the most important criteria of health. The marital status of the respondents also turned out to be extremely important in choosing the criteria determining health: married respondents considered proper nutrition, as well as sleep and rest as the most important criteria of health.

The study (based on the IHB) has shown that the respondents rated the highest health practices related to proper daily health habits, such as sleep, rest, and regular meals.

Based on the obtained results, the authors initiated a discussion with other authors conducting studies with the same research tools, which helped make a precise summary.

'Having all body parts functional' was the most important criterion for people aged 50+ in the study conducted by Nowicki and Ślusarska [8]. It was followed by 'reaching very old age'. The authors' own study showed that health was understood by the respondents as *property* and *state,* similarly to the study by Bąk-Sosnowska et al. [32]. In the study by G. Nowicki and B. Ślusarska, the respondents gave the greatest importance to health as a *property* and as an *objective* [8]. The study by Cybulski et al. also showed that the elderly attached the highest importance to health understood as *property*, then as an *objective, state, process,* and a *result* [14]. Each of the studied groups understood health primary as a property. According to Juczyński, this is an instrumental approach [27]. This fact should not come as a surprise, as the respondents were a sick and hospitalized group, for whom the disease was often an unpleasant experience and a source of stress [8,33–36].

The analysis of the study results based on the Inventory of Health-Related Behaviour (IHB) showed that amongst subjects the most common health behaviours were related to 'health practices', slightly less common in the area of 'positive mental attitude' and the least common in the area of 'proper eating habits' and 'preventive behaviours'. Similar observations were made in the studies by Nowicki et al. and Cybulski et al. [8,14]. However, somewhat different results were obtained in studies of elderly people by Prakash et al. and Zanjani et al. There, health behaviours in the area of 'correct eating habits' were the most intense [23,24].

When correlating the level of health behaviours according to the IHB with individual categories of the LHC, a total of 59 statistically significant relationships were observed. A positive correlation was found in the group of respondents where the 'preventive health behaviours' were more intense; from these, the more important criterion for the respondents was 'eating properly'. Negative relationships were also found for the respondents for whom 'correct eating habits' were important; here, respondents placed more importance on 'infrequent need of going to the doctor's'.

The obtained results of the authors' own study, carried out with the use of the LHC, differed slightly from the results of other cited authors [8,14,15,37–42]. According to our respondents aged 50+, 'being health' meant above all 'not feeling any physical ailments'; 'having all body parts functional'; 'feeling well'; 'eating properly'; and 'infrequent need of going to the doctor'. The differences between the evaluation of health by the elderly presented in the available literature may result from the socio-cultural differences of individual regions of Poland, personal beliefs, general economy, politics, and organization of the society structure [2,8,14,15,35,43]. Cultural determinants—the system of norms and beliefs, patterns of behaviour, as well as all material and non-material products generated by a given social group—play an important role in shaping and choosing healthy behaviours. The determinant of social position is education, which reflects the level of knowledge, including medical knowledge. There was a clear relationship between social status and choice of health behaviours. The higher the education, the higher the requirements for the healthcare, compared to people with lower education [44,45]. Lifestyle plays a fundamental role in maintaining health and prevention of diseases, also in the elderly. Lifestyle elements that have beneficial health effects include correct diet and eating habits, optimal level of physical activity, adequate amount of sleep, satisfactory social relations, skilful use of free time, and knowledge of health prevention [2,8,35,41,46–48].

A properly balanced diet is particularly important for maintaining health. Based on the conducted study, the principles of proper nutrition were not perceived by the respondents as a significant factor for preventing or combating already existing cardiovascular diseases. Irrespectively, dietary risk is the most important behavioural health factor in the world, and appears to be the best target in the fight against cardiovascular disease. In the research by Raver A. et al., it has been proven that a proper diet, especially the Mediterranean diet, brings effective results in reducing cardiovascular risk worldwide [49].

However, an incorrect diet and low physical activity are linked to overweight and obesity, which contribute to the occurrence of cardiovascular diseases. Obesity was found in 34.87% of the respondents. Seravalle G. et al. found that obesity, and especially the excessive distribution of visceral fat, was accompanied by several changes at the hormonal, inflammatory, and endothelial levels. These changes induce the stimulation of several other mechanisms that contribute to hypertension and increase cardiovascular morbidity [50].

Given the above, the evidence clearly shows the importance of lifestyle factors (e.g., diet, exercise, and drug use) in the development of cardiovascular disease. Interventions targeted at these behaviours can greatly improve results, and therefore the health and well-being of patients with cardiovascular disease [51].

Summarizing the conducted study, it may be stated that it is important to try and promote health in the hierarchy of values of people aged 50+ and to strengthen their responsibility for their own health and life. Broadly understood education should help raise public health awareness.

As a result of the conducted study, it was found that the most important criteria (according to LHC) that determine health according to the respondents were proper nutrition, adequate sleep, rest, work without stress and tension, and the ability to solve difficult life situations. A particularly high score was given to one of four areas of health behaviours in IHB, namely, 'health practices'. It proved that the respondents had proper daily health habits, which is very valuable and may determine good health in the future.

The use of standardized questionnaires in further research will allow for a wider observation of health behaviours in people of different ages. Additionally, increasing the number of subjects would help further determine health behaviours, as well as assess the health criteria that the subjects considered as crucial. After more extensive research in the group of people aged 50+ with cardiovascular diseases, more constructive conclusions might be drawn, which would allow specific preventive measures to be taken from the earliest years in order to prevent cardiovascular diseases.

5. Conclusions

Respondents aged 50+ and hospitalized for cardiovascular diseases indicated (based on IHB questionnaire) that health behaviours in the area of 'health practices' had the strongest correlation with their health, while the lowest correlation was found in the areas of 'correct eating habits' and 'preventive behaviours'. According to the respondents, the most important criteria determining health (according to the LHC Questionnaire) included 'not feeling any physical ailments'; 'having all body parts functional'; 'feeling well'; 'eating properly'; and 'infrequent need of going to the doctor'.

Based on the information collected from the respondents, it was found that the most important criteria determining health depended on selected descriptive variables, such as age—it was found that for older respondents, avoidance of stimulants is one of the most important factors determining health; gender—the majority of the studied group of women perceived health in many aspects, including feeling happy and wellbeing; place of residence—respondents from urban environment put emphasis on the role of healthy eating in maintaining good health, as opposed to their counterparts from a rural environment; education—respondents with education higher education paid attention to many factors determining health, as opposed to respondents with lower education, e.g., solving life problems and work without stress; and lastly, marital status—married respondents attached great importance to the principles of proper nutrition in order to maintain health compared to the group of respondents who are not married.

Practical implications: Due to the differences in respondents' preferences in terms of health behaviours, as well as their approach to their health that depended on gender, age, education, and place of residence, it is particularly important to take measures aimed at providing health education to society, which can effectively overcome the problem of anti-health behaviours, existing health imbalance, and help fight cardiovascular disease.

Author Contributions: Conceptualization, I.G. and K.S.; methodology, E.C. and M.S. (Marian Sygit); software, A.P. and M.S. (Maryna Surmach); validation, D.K. and A.R.; formal analysis, I.G. and K.S.; investigation, M.S. (Marian Sygit) and A.P.; resources, M.S. (Maryna Surmach). and A.R.; data curation, D.K. and A.R.; writing—original draft preparation, I.G. and E.C. All authors have read and agreed to the published version of the manuscript.

Funding: This research received no external funding.

Institutional Review Board Statement: The study was conducted according to the guidelines of the Declaration of Helsinki, and approved by the Ethics Committee of Medical University in Lodz Poland (protocol code: RNN/156/18/KE).

Informed Consent Statement: Informed consent was obtained from all subjects involved in the study.

Data Availability Statement: Data are not publicly available and data sharing is not applicable to this article.

Conflicts of Interest: The authors declare no conflict of interest.

References

1. Kino, S.; Bernabé, E.; Sabbah, W. Socioeconomic inequality in clusters of health-related behaviours in Europe: Latent class analysis of a cross-sectional European survey. *BMC Public Health* **2017**, *17*, 451–497. [CrossRef] [PubMed]
2. Wojtyniak, B.; Goryński, P.; Moskalewicz, B. (Eds.) *The Health Situation of the Polish Population and Its Determinants*; The National Institute of Public Health-National Institute of Hygiene: Warsaw, Poland, 2012.
3. Younus, A.; Aneni, E.C.; Spatz, E.S.; Osondu, C.U.; Roberson, L.; Ogunmoroti, O.; Malik, R.; Ali, S.S.; Aziz, M.; Feldman, T.; et al. A systematic review of the prevalence and outcomes of ideal cardiovascular health in US and Non-US populations. *Mayo Clin. Proc.* **2016**, *91*, 649–670. [CrossRef] [PubMed]
4. Moor, I.; Spallek, J.; Richter, M. Explaining socioeconomic inequalities in self-rated health: A systematic review of the relative contribution of material, psychosocial and behavioural factors. *J. Epidemiol. Community Health* **2017**, *71*, 565–575. [CrossRef]
5. Maclagan, L.C.; Park, J.; Sanmartin, C.; Mathur, K.R.; Roth, D.; Manuel, U.G.; Gershon, A.; Booth, G.L.; Bhatia, S.; Atzema, C.L.; et al. The CANHEART health index: A tool for monitoring the cardiovascular health of the Canadian population. *CMAJ* **2013**, *186*, 180–187. [CrossRef] [PubMed]
6. Jankovic, S.; Stojisavljević, D.; Janković, J.; Eric, M.; Marinkovic, J. Status of cardiovascular health in a transition European country: Findings from a population-based cross-sectional study. *Int. J. Public Health* **2014**, *59*, 769–778. [CrossRef] [PubMed]
7. Zatoński, W. One hundred years of health in Poland. *J. Health Inequal.* **2019**, *5*, 11–19. [CrossRef]
8. Nowicki, G.J.; Ślusarska, B.; Piasecka, H.; Bartoszek, A.; Kocka, K.; Deluga, A. The Status of Cardiovascular Health in Rural and Urban Areas of Janów Lubelski District in Eastern Poland: A Population-Based Study. *Int. J. Environ. Res. Public Health* **2018**, *15*, 2388. [CrossRef]
9. Cabellos-García, A.C.; Castro-Sánchez, E.; Martínez-Sabater, A.; Díaz-Herrera, M.Á.; Ocaña-Ortiz, A.; Juárez-Vela, R.; Gea-Caballero, V. Relationship between Determinants of Health, Equity, and Dimensions of Health Literacy in Patients with Cardiovascular Disease. *Int. J. Environ. Res. Public Health* **2020**, *17*, 2082. [CrossRef]
10. Stefler, D.; Pikhart, H.; Jankovic, N.; Kubinova, R.; Pajak, A.; Malyutina, S.; Simonova, G.; Feskens, E.J.M.; Peasey, A.; Bobak, M. Healthy diet indicator and mortality in Eastern European populations: Prospective evidence from the HAPIEE cohort. *Eur. J. Clin. Nutr.* **2014**, *68*, 1346–1352. [CrossRef] [PubMed]
11. D'Addario, M.; Cappelletti, E.R.; Sarini, M.; Greco, A.; Steca, P. The TTCYB Study Protocol: A Tailored Print Message Intervention to Improve Cardiovascular Patients' Lifestyles. *Int. J. Environ. Res. Public Health* **2020**, *17*, 2919. [CrossRef]
12. Fernández, D.; Brotons, C.; Moral, I.; Bulc, M.; Afonso, M.; Akan, H.; Pinto, S.; Vucak, J.; Martins, C.M.D.S. Lifestyle behaviours in patients with established cardiovascular diseases: A European observational study. *BMC Fam. Pract.* **2019**, *20*, 1–6. [CrossRef] [PubMed]
13. Strzelecki, Z.; Szymborski, J. Morbidity and Mortality of Cardiovascular Diseases and the Demographic Situation of Poland Government Population Council. Available online: https://bip.stat.gov.pl/files/gfx/bip/pl/zamowieniapubliczne/426/248/1/81_gp_rrl_2015_monografia_kardiologiczna.pdf (accessed on 19 December 2020).
14. Cybulski, M.; Krajewska-Kulak, E.; Jamiolkowski, J. Preferred health behaviors and quality of life of the elderly people in Poland. *Clin. Interv. Aging* **2015**, *10*, 1555–1564. [CrossRef]
15. Deluga, A.; Kosicka, B.; Dobrowolska, B.; Chrzan-Rodak, A.; Jurek, K.; Wrońska, I.; Ksykiewicz-Dorota, A.; Jedrych, M.; Drop, B. Lifestyle of the elderly living in rural and urban areas measured by the FANTASTIC Life Inventory. *Ann. Agric. Environ. Med.* **2018**, *25*, 562–567. [CrossRef]
16. Nyberg, S.T.; Singh-Manoux, A.; Pentti, J.; Madsen, I.E.H.; Sabia, S.; Alfredsson, L.; Bjorner, J.B.; Borritz, M.; Burr, H.; Goldberg, M.; et al. Association of Healthy Lifestyle With Years Lived without Major Chronic Diseases. *JAMA Intern. Med.* **2020**, *180*, 760–768. [CrossRef] [PubMed]
17. Tang, K.L.; Rashid, R.; Godley, J.; A Ghali, W. Association between subjective social status and cardiovascular disease and cardiovascular risk factors: A systematic review and meta-analysis. *BMJ Open* **2016**, *6*, e010137. [CrossRef] [PubMed]
18. Havranek, E.P.; Mujahid, M.S.; Barr, D.A.; Blair, I.V.; Cohen, M.S.; Cruz-Flores, S.; Davey-Smith, G.; Dennison-Himmelfarb, C.R.; Lauer, M.S.; Lockwood, D.W.; et al. Social determinants of risk and outcomes for cardiovascular disease. A Scientific Statement from the American Heart Association. *Circulation* **2015**, *132*, 873–898. [CrossRef]
19. De La Cruz, N.; Shabaneh, O.; Appiah, D. The Association of Ideal Cardiovascular Health and Ocular Diseases Among US Adults. *Am. J. Med.* **2021**, *134*, 252–259.e1. [CrossRef]
20. Townsend, N.; Wilson, L.; Bhatnagar, P.; Wickramasinghe, K.; Rayner, M.; Nichols, M. Cardiovascular disease in Europe: Epidemiological update 2016. *Eur. Hear. J.* **2016**, *37*, 3232–3245. [CrossRef]
21. Benjamin, E.J.; Blaha, M.J.; Chiuve, S.E.; Cushman, M.; Das, S.R.; de Ferranti, S.D.; Floyd, J.; Fornage, M.; Gillespie, C.; Isasi, C.R.; et al. Heart disease and stroke statistics—2017 Update: A report from the American Heart Association. *Circulation* **2017**, *135*, e146–e603. [CrossRef]

22. Folsom, A.R.; Yatsuya, H.; Nettleton, J.A.; Lutsey, P.L.; Cushman, M.; Rosamond, W.D. ARIC Study Investigators Community prevalence of ideal cardiovascular health, by the American Heart Association definition, and relation to cardiovascular disease incidence. *J. Am. Coll. Cardiol.* **2011**, *57*, 1690–1696. [CrossRef]
23. Prakash, S.J.; Bihari, G.S.; Kumar, S.A. Lifestyle habits and diseases amongst rural geriatrics population. *Int. J. Community Med. Public Health* **2016**, *3*, 957–961. [CrossRef]
24. Zanjani, S.; Tol, A.; Mohebbi, B.; Sadeghi, R.; Jalyani, K.N.; Moradi, A. Determinants of healthy lifestyle and its related factors among elderly people. *J. Educ. Health Promot.* **2015**, *4*, 103–115.
25. He, Z.; Cheng, Z.; Shao, T.; Liu, C.; Shao, P.; Bishwajit, G.; Feng, D.; Feng, Z. Factors influencing health knowledge and behaviors among the elderly in rural China. *Int. J. Environ. Res. Public Health* **2016**, *13*, 975. [CrossRef]
26. Selivanova, A.; Cramm, J.M. The relationship between healthy behaviors and health outcomes among older adults in Russia. *BMC Public Health* **2014**, *14*, 1183. [CrossRef]
27. Goryakin, Y.; Aldea, A.; Lerouge, A.; Spica, V.R.; Nante, N.; Vuik, S.; Devaux, M.; Cecchini, M. Promoting sport and physical activity in Italy: A cost-effectiveness analysis of seven innovative public health policies. *Ann. Ig.* **2019**, *31*, 614–625.
28. Biernat, E.; Piątkowska, M. Recreational Physical Activity of Poles in Europe. Report for the Department of Sports Analysis and Policy of the Ministry of Sport and Tourism. Warsaw, Poland. Available online: https://www.msit.gov.pl/ (accessed on 12 December 2020).
29. Juczyński, Z. *Narzędzia Pomiaru w Promocji I Psychologii Zdrowia [Tools for Health Promotion and Psychology of Health]*, 2nd ed.; Pracownia Testów Psychologicznych: Warszawa, Poland, 2009; pp. 110–116.
30. Witz, K.; Hinkle, D.E.; Wiersma, W.; Jurs, S.G. Applied Statistics for the Behavioral Sciences. *J. Educ. Stat.* **1990**, *15*, 84. [CrossRef]
31. R Core Team. *R: A Language and Environment for Statistical Computing*; R Foundation for Statistical Computing: Vienna, Austria, 2018; Available online: https://www.R-project.org (accessed on 6 November 2020).
32. Bąk-Sosnowska, M.; Skrzypulec-Plinta, V. Health behaviors, health definitions, sense of coherence, and general practitioners' attitudes towards obesity and diagnosing obesity in patients. *Arch. Med Sci.* **2017**, *13*, 433–440. [CrossRef]
33. Stehr, P.; Weber, W.; Rossmann, C. Health information behavior of 65+: Reaching older target groups. *Bundesgesundheitsblatt Gesundh. Gesundh.* **2021**, *64*, 37–44. [CrossRef] [PubMed]
34. Benziger, C.P.; Zavala-Loayza, J.A.; Bernabe-Ortiz, A.; Gilman, R.H.; Checkley, W.; Smeeth, L.; Malaga, G.; Miranda, J.J. Low prevalence of ideal cardiovascular health in Peru. *Heart* **2018**, *104*, 1251–1256. [CrossRef] [PubMed]
35. Vandenheede, H.; Vikhireva, O.; Pikhart, H.; Kubinova, R.; Malyutina, S.; Pajak, A.; Tamosiunas, A.; Peasey, A.; Simonova, G.; Topor-Madry, R.; et al. Socioeconomic inequalities in all-cause mortality in the Czech Republic, Russia, Poland and Lithuania in the 2000s: Findings from the HAPIEE Study. *J. Epidemiol. Community Health* **2013**, *68*, 297–303. [CrossRef] [PubMed]
36. Piepoli, M.F.; Hoes, A.W.; Agewall, S.; Albus, C.; Brotons, C.; Catapano, A.L.; Cooney, M.T.; Corrà, U.; Cosyns, B.; Deaton, C.; et al. 2016 European Guidelines on cardiovascular disease prevention in clinical practice: The Sixth Joint Task Force of the European Society of Cardiology and Other Societies on Cardiovascular Disease Prevention in Clinical Practice (constituted by representatives of 10 societies and by invited experts): Developed with the special contribution of the European Association for Cardiovascular Prevention& Rehabilitation (EACPR). *Eur. J. Prev. Cardiol.* **2016**, *23*, NP1–NP96.
37. Doryńska, A.; Polak, M.; Kozel, M.; Szafraniec, K.; Piotrowski, W.; Bielecki, W.; Drygas, W.; Kozakiewicz, K.; Piwoński, J.; Tykarski, A.; et al. Cardiovascular disease (CVD) risk factors in Krakow and in the whole Poland adult population. Results from the WOBASZ study and Polish arm of the HAPPIE project. *Przegl. Epidemiol.* **2015**, *69*, 79–86. [PubMed]
38. Unger, E.; Diez-Roux, A.V.; Lloyd-Jones, D.M.; Mujahid, M.S.; Nettleton, J.A.; Bertoni, A.; Badon, S.E.; Ning, H.; Allen, N.B. Association of Neighborhood Characteristics With Cardiovascular Health in the Multi-Ethnic Study of Atherosclerosis. *Circ. Cardiovasc. Qual. Outcomes* **2014**, *7*, 524–531. [CrossRef]
39. Lloyd-Jones, D.M.; Hong, Y.; Labarthe, D.; Mozaffarian, D.; Appel, L.J.; Van Horn, L.; Greenlund, K.; Daniels, S.; Nichol, G.; Tomaselli, G.F.; et al. American heart association strategic planning task force and statistics committee defining and setting national goals for cardiovascular health promotion and disease reduction: The American heart association's strategic impact goal through 2020 and beyond. *Circulation* **2010**, *121*, 586–613. [CrossRef]
40. Roger, V.L.; Go, A.S.; Lloyd-Jones, D.M.; Benjamin, E.J.; Berry, J.D.; Borden, W.B.; Bravata, D.M.; Dai, S.; Ford, E.S.; Fox, C.S.; et al. Executive summary: Heart disease and stroke statistics-2012 update: A report from the American Heart Association. *Circulation* **2012**, *34*, 125–188.
41. Rywik, S.; Kupść, W.; Piotrowski, W.; Broda, G.; Piwoński, J.; Kurjata, P.; Waśkiewicz, A.; Gaździk, D. A multicenter na-tionwide study of the Polish population's health—WOBASZ project. Methodical assumptions and logistics. *Kardiol. Pol.* **2005**, *63* (Suppl. S4), 605–613.
42. Manczuk, M.; Vaidean, G.; Dehghan, M.; Vedanthan, R.; Boffetta, P.; Zatonski, W.A. Ideal cardiovascular health is associated with self-rated health status. The Polish Norwegian Study (PONS). *Int. J. Cardiol.* **2017**, *230*, 549–555. [CrossRef] [PubMed]
43. Jaźwiec, P.; Gać, P.; Poręba, M.; Sobieszczańska, M.; Mazur, G.; Poręba, R. The cardiovascular health score and the volume of carotid body in computed tomography angiography in patients with arterial hypertension. *J. Am. Soc. Hypertens.* **2018**, *12*, 551–560. [CrossRef] [PubMed]
44. Samet, J. The environment and health inequalities: Problems and solutions. *J. Health Inequal.* **2019**, *5*, 21–27. [CrossRef]
45. Janković, S.; Stojisavljević, D.; Erić, M.; Marinković, J. Association of socioeconomic status measured by education, and cardiovascular health: A population-based cross-sectional study. *BMJ Open* **2014**, *4*, e005222. [CrossRef]

46. Tam, H.L.; Wong, E.M.L.; Cheung, K. Effectiveness of Educational Interventions on Adherence to Lifestyle Modifications Among Hypertensive Patients: An Integrative Review. *Int. J. Environ. Res. Public Health* **2020**, *17*, 2513. [CrossRef]
47. Kim, J.I.; Sillah, A.; Boucher, J.L.; Sidebottom, A.C.; Knickelbine, T. Prevalence of the American Heart Association's "Ideal Cardiovascular Health" Metrics in a Rural, Cross-sectional, Community-Based Study: The Heart of New Ulm Project. *J. Am. Hear. Assoc.* **2013**, *2*, e000058. [CrossRef]
48. Moghaddam, M.M.; Mohebi, R.; Hosseini, F.; Lotfaliany, M.; Azizi, F.; Saadat, N.; Hadaegh, F. Distribution of ideal cardiovascular health in a community-based cohort of Middle East population. *Ann. Saudi Med.* **2014**, *34*, 134–142. [CrossRef]
49. Ravera, A.; Carubelli, V.; Sciatti, E.; Bonadei, I.; Gorga, E.; Cani, D.; Vizzardi, E.; Metra, M.; Lombardi, C. Nutrition and Cardiovascular Disease: Finding the Perfect Recipe for Cardiovascular Health. *Nutrients* **2016**, *8*, 363. [CrossRef]
50. Seravalle, G.; Grassi, G. Obesity and hypertension. *Pharmacol. Res.* **2017**, *122*, 1–7. [CrossRef]
51. Doughty, K.N.; Del Pilar, N.X.; Audette, A.; Katz, D.L. Lifestyle Medicine and the Management of Cardiovascular Disease. *Curr. Cardiol. Rep.* **2017**, *19*, 116. [CrossRef]

International Journal of
Environmental Research and Public Health

MDPI

Article

Patterns of Alcohol Consumption in the PURE Poland Cohort Study and Their Relationship with Health Problems

Katarzyna Zatońska [1,2], Piotr Psikus [2], Alicja Basiak-Rasała [1,*], Zuzanna Stępnicka [1], Maria Wołyniec [1], Andrzej Wojtyła [2], Andrzej Szuba [3] and Katarzyna Połtyn-Zaradna [1]

1 Department of Social Medicine, Wroclaw Medical University, 50-345 Wrocław, Poland; katarzyna.zatonska@umed.wroc.pl (K.Z.); zuzanna.stepnicka@gmail.com (Z.S.); maria.wolyniec@umed.wroc.pl (M.W.); Katarzyna.poltyn-zaradna@umed.wroc.pl (K.P.-Z.)

2 Calisia University, 62-800 Kalisz, Poland; burmistrz@um.kepno.pl (P.P.); a.wojtyla55@gmail.com (A.W.)

3 Department of Angiology, Hypertension and Diabetology, Wroclaw Medical University, 50-556 Wroclaw, Poland; andrzej.szuba@umed.wroc.pl

* Correspondence: alicja.basiak-rasala@umed.wroc.pl; Tel./Fax: +48-71-328-21-45

Abstract: (1) Background: Alcohol is a leading risk factor of premature morbidity and mortality. The objective of this study was to investigate the patterns of alcohol consumption in the PURE Poland cohort study baseline. (2) Methods: A Polish cohort was enrolled in the baseline study in 2007–2010. The study group consisted of 2021 adult participants of urban and rural areas from the Lower Silesia voivodeship in Poland (747 men and 1274 women). (3) Results: In the overall study population, 67.3% were current drinkers, 10.3% were former drinkers, and 22.4% were abstainers. Current use of alcohol products was more prevalent in men (77.2%), people living in urban areas (73.0%), and people with a higher level of education (78.0%). The percentage of current drinkers decreased with increasing age (from 73.4% in 30- to 44-year-olds to 48.8% in participants aged 64 and more). The majority of participants (89.2%) declared a low level of alcohol intake. The chance of high level of intake of alcohol was four times higher in men than in women (OR 4.17; CI 1.64–10.6). The majority of participants (54.6%) declared most frequent consumption of low-alcohol drinks (beer, wine) and 21% declared most frequent consumption of spirits. Current drinkers had almost 1.5-fold higher odds of diabetes and cardiovascular diseases (CVD) than never drinkers (OR 1.49, CI 1.03–2.17; OR 1.66, CI 1.27–2.18, respectively). Former drinkers had higher odds for hypertension and CVD than never drinkers (1.73, CI 1.05–2.85; OR 1.76, CI 1.22–2.53, respectively). (4) Conclusions: In our cohort study, we observed several socio-demographic factors differentiating the patterns of alcohol consumption. The preventive programs should focus predominantly on men, people aged <45 years, and those with a higher level of education.

Keywords: alcohol; cohort study; PURE

Citation: Zatońska, K.; Psikus, P.; Basiak-Rasała, A.; Stępnicka, Z.; Wołyniec, M.; Wojtyła, A.; Szuba, A.; Połtyn-Zaradna, K. Patterns of Alcohol Consumption in the PURE Poland Cohort Study and Their Relationship with Health Problems. *IJERPH* **2021**, *18*, 4185. https://doi.org/10.3390/ijerph18084185

Academic Editors: Wioletta Zukiewicz-Sobczak, Paulina Wojtyla-Buciora, Izabela Rącka and Andrzej Wojtyla

Received: 14 March 2021
Accepted: 13 April 2021
Published: 15 April 2021

Publisher's Note: MDPI stays neutral with regard to jurisdictional claims in published maps and institutional affiliations.

Copyright: © 2021 by the authors. Licensee MDPI, Basel, Switzerland. This article is an open access article distributed under the terms and conditions of the Creative Commons Attribution (CC BY) license (https://creativecommons.org/licenses/by/4.0/).

1. Introduction

Alcohol is included in many social, cultural, and religious practices. Widespread use of alcoholic beverages can have deleterious effects on health. Alcohol is a leading risk factor of premature morbidity and mortality. According to the World Health Organization (WHO), there were 2.3 billion current drinkers in 2016 worldwide [1]. It is estimated that 5.3% of overall global mortality can be attributed to excessive consumption of alcohol [1]. Based on to-date studies, there is a linear association between the level of alcohol consumption and all-cause mortality, especially cancer-related [2]. Aforementioned findings contradict previous observations, which suggested the protective impact of consumption of low to moderate amounts of alcohol [3,4]. The previous observations can be partially explained by an inadequate control for confounders and an association between light drinking and a healthier overall lifestyle [2]. Harmful use of alcohol increases the risk of premature mortality, as well as the risk of injuries, disability, and mental and psychological disorders.

It is estimated that 5.5% of cancer cases worldwide can be attributed to the consumption of alcohol [5]. Moderate to high alcohol intake has been previously associated with cancers of the larynx, esophagus, pharynx, oral cavity, colorectum, and breast, but even light alcohol intake has been linked to cancers of the pharynx, oral cavity, esophagus, and breast [6,7]. The eastern European region has one of the highest burdens of disease attributed to alcohol with the contribution of alcohol use disorders, cardiovascular diseases, and injuries [8]. According to the European Health Interview Survey (EHIS) conducted in 2014, alcohol consumption in Poland was the most frequent in 30–49-year-olds [9]. In 2014 in Poland, 63% of women and 83% of men declared at least occasional consumption of alcoholic beverages. One in three adult men and one in seven adult women reported that they consumed alcohol at least once a week [9]. One of the most comprehensive reports on alcohol epidemiology across the European Union was prepared within the framework of the Joint Action on Reducing Alcohol Related Harm (RARHA) project [10]. According to the report, 86.4% of respondents in Poland declared consumption of alcohol in the previous 12 months.

It is observed that consumption of alcohol might be differentiated by many sociodemographic factors, including sex, place of residence, and level of education [11–16]. According to the Global Burden of Disease (GBD) study [2], the attributable health burden caused by the excessive consumption of alcohol was three times higher in men than in women. In Poland in 2016, the average annual consumption of pure alcohol per capita in men was almost 3-fold higher than in women (23.8 L vs. 8.3 L, respectively) [12]. Moreover, the prevalence of heavy episodic drinking was much higher in men than in women. Recently published studies indicated a causal association between education attainment and alcohol drinking patterns [13]. A lower level of education in Poland has been associated with increased mortality caused by alcoholic liver diseases [17].

In 2016 worldwide, the most common types of alcoholic beverages were spirits (44.8%), beer (34.2%), and wine (11.7%) [1]. In contrast, in Poland, the most popular type of alcohol was beer. Based on the EHIS report, beer contributed to 70% of consumption of alcoholic beverages, whereas spirits to 20% and wine to 10% of overall consumption [9]. According to reports by The State Agency for the Prevention of Alcohol-Related Problems (PARPA), the average annual consumption of spirits, wine, and beer in 2019 per capita added up to 3.7 L, 6.2 L, and 97.1 L, respectively [18]. The average annual consumption of pure alcohol per capita in Poland in 2019 was 9.78 L. There has been a steady increase of average alcohol consumption in Poland since the 1990s. In 1993, the average annual consumption of pure alcohol per capita was approximately 6.52 L [18]. Between 1990 and 2007, the average alcohol consumption in Poland was significantly lower in comparison to other European Union (EU) countries [12]. The trend shifted after 2007, when the average alcohol consumption in Poland began to steadily increase, reaching the EU average, and then exceeding it about 2013–2014, and it has remained higher ever since [12]. The increasing trend of alcohol consumption in Poland poses a challenge for health policy makers. It is speculated that the slower rate of health improvement and slower decrease in premature mortality since 2000s in Poland can be attributed, among other factors, to increasing alcohol consumption [19–21]. A significant increase in mortality caused by alcohol-related liver cirrhosis was observed in this period of time [22]. In Poland, 14% of disability-adjusted life years (DALY) in men can be attributed to consumption of alcohol, which is a high value relative to the scale of the European Union [23]. Concomitantly, alcoholic beverages are very affordable and accessible in Poland [24].

A Polish cohort participates in the global Prospective Urban and Rural Epidemiological (PURE) study. The primary aim of the global PURE study was a longitudinal observation of modifiable and non-modifiable risk factors of non-communicable diseases. The urban–rural setting of the study enabled a comparison of risk factors between regions and tracking health inequalities. The objective of the here-presented study was to perform a sociodemographic analysis of the patterns of alcohol consumption in the PURE Poland cohort study baseline. Additionally, we assessed the prevalence of diabetes, hypertension, cardiovascular diseases (CVD), and select liver diseases in groups of participants differentiated by the attitude toward alcohol consumption. Poland is one of the most interesting countries in which to investigate alcohol consumption patterns, due to rapid changes in attitudes toward alcohol over the years. Our study is one of the few ongoing cohort studies in Poland, conducted with a consistent protocol and analyzing the socio-demographic characteristics of presented data with special emphasis on the place of residence (urban–rural). The number of prospective studies, similar in size in Poland, is limited.

2. Materials and Methods

The Polish cohort was enrolled in the baseline study between 2007 and 2010. The enrollment to the Polish cohort of the PURE study was announced in mass media (local newspapers, television, radio). The volunteers who applied had to meet the criteria of age and place of residence (urban–rural). The participants were chosen to achieve a broadly representative sample of the community [25]. The baseline cohort consisted of 2036 inhabitants aged 30–85 years of urban and rural areas from Lower Silesia voivodeship in Poland. Data are collected every three years. The full protocol of the PURE study was previously described [25,26]. Every visit in the study center included a questionnaire study (individual health, family, household, food frequency, and international physical activity questionnaires), a blood draw, blood pressure measurement, a spirometry, and anthropometric measurements [26].

The presented article includes an analysis of the baseline data of the PURE Poland cohort study. Out of 2036 baseline participants, 15 were excluded from analysis of alcohol consumption patterns due to a lack of full data regarding alcohol consumption. A total of 2021 eligible participants (747 men and 1274 women) were included for those analyses. The analysis of the occurrence of diabetes/hypertension/liver diseases/CVD and the attitudes toward alcohol consumption was conducted in 2030 participants (6 participants were excluded due to lack of data regarding the diseases or attitudes toward alcohol consumption).

The individual health questionnaire contained following questions regarding alcohol consumption:

(1) Which best describes your history of alcohol use?
(2) At what age did you start drinking alcohol?
(3) What forms of alcohol have you regularly used?
(4) At least once a month, do you consume >5 alcoholic drinks/day?

In the first question the respondent could choose between "formerly used alcohol products", "currently use alcohol products", and "never used alcohol products". In the third question, "regular use" was defined as at least once a month [25]. In the same question, the respondents were asked to assess the frequency of consumption ("daily", "weekly", "monthly"), the average number of drinks, duration of alcohol use (in years) of the following types and servings of alcohol: (a) spirits (rum, whiskey, gin, vodka, country liquor)—30 mL; (b) wine—125 mL; and beer—375 mL. In Poland, a country liquor is usually prepared by the method of maceration (with vodka or neutral spirits) of different ingredients (spices, herbs, fruit) with the addition of sugar. Typical polish country liquor contains 40–45% alcohol by volume, but some of them can be stronger. Due to the high content of alcohol, the country liquor was included in the category of spirits. The number of "drinks" that we refer to later was based on the aforementioned serving sizes of different types of alcohol. If the participants confirmed that they consume >5 alcoholic drinks per day at least once a month, they were asked two additional questions: (a) "How many times per month do you consume >5 alcoholic drinks in a day?" and (b) "What is the average number of drinks that you consume each time?".

Following the methodology adopted in a study by Smyth et al. [27], participants who self-reported alcohol abstinence were defined as "never drinkers". Participants who have ceased consumption of alcohol for at least 1 year prior to the interview were defined as "former drinkers". The level of consumption of alcohol in current drinkers was defined as "low", "moderate", or "high". "Low intake" was defined as drinking up to 7 drinks per week. "Moderate intake" was defined as drinking 7–14 drinks per week for women and 7–21 drinks for men. "High intake" was defined as drinking more than 14 drinks per week for women and more than 21 drinks per week for men. Heavy episodic drinking was defined as one episode of consumption of more than five drinks at least once per month. The number of drinks of each alcohol type was analyzed (beer, spirits, wine). The current drinker was included in the group for the type of drink according to which type they consumed most frequently. If participants declared the same frequency of consumption of more than one type of alcohol, they were defined as drinking "more than one type of alcohol" [27]. We also analyzed the association between attitudes toward alcohol consumption and the occurrence of diabetes, hypertension, CVD, and liver diseases. Diabetes was ascertained on the basis of (1) self-reported diabetes and/or (2) self-reported anti-diabetic medication and/or (3) fasting blood glucose measurement $\geq$ 126 mg/dL. Hypertension was ascertained on the basis of (1) self-reported hypertension and/or (2) self-reported anti-hypertensive medication and/or (3) an average of two blood pressure measurements $\geq$ 140/90 mmHg as previously described [28]. The category of "CVD" included coronary heart disease, stroke, heart infarction, heart failure, and other heart diseases. Liver diseases included hepatitis and jaundice. The occurrence of CVD and liver diseases were self-reported by the participants.

2.1. Statistical Analysis

Association between sociodemographic factors and consumption of alcohol was assessed with the use of logistic regression models after adjusting for age and/or gender. The strength of the association was measured by the odds ratio (OR) with 95% confidence intervals. The differences in the age of alcohol initiation between sexes and age groups were analyzed with the use of Mann–Whitney U and Kruskal–Wallis tests. Statistical analysis was done with the use of the program Statistica 13.1. We assumed a significance level of $p < 0.05$.

2.2. Ethics

The study was reviewed and accepted by the appropriate ethics committee and was therefore performed in accordance with the ethical standards laid down in an appropriate version of the 1964 Declaration of Helsinki (positive opinion of The Bioethics Committee of the Wrocław Medical University nr KB- 443/2006).

3. Results

3.1. Baseline Characteristics

At the baseline, a total of 2021 participants (747 men and 1274 women) were analyzed. The study group consisted of 17.5% of 30- to 44-year-olds, 66.4% of 45- to 64-year-olds, and 16.1% of participants aged 64 years and older. There were 59.3% of urban and 40.7% of rural participants. A total of 14.9% of participants had primary education, 16.0% had vocational education, 39.4% had secondary education, and 29.7% had higher education. At the baseline, 74.3% of participants were married/lived in a relationship, 18.5% of participants were separated/widowed/divorced, and 7.2% were never married.

3.2. Alcohol Consumption

In the overall study population, 67.3% were current drinkers, 10.3% were former drinkers, and 22.4% declared that they had never drunk alcohol. Men were more likely than women to be current drinkers (77.2% vs. 61.5%, respectively) and former drinkers (11.6% vs. 9.5%, respectively). Almost one third of women (29.0%) and 11.1% of men declared that they had never drunk alcohol. The percentage of current drinkers decreased with increasing age (from 73.4% with age 30–44 years to 48.8% with age 64 and more). Current use of alcohol products was more prevalent in people living in an urban area (73.0%) and people with a higher level of education (78.0%). The chance of being a current drinker was nearly two times higher in people living in urban vs. rural areas (OR 1.88; CI 1.55–2.29). Moreover, the chance of being a current drinker was three times higher in participants with higher education than in people with primary education (OR 3.07; CI 2.25–4.20). We observed a lower proportion of current drinkers in divorced, separated, and widowed participants than in participants in a relationship (56.7% vs. 68.3% never married and 69.8% married) (Table 1).

Table 1. Sociodemographic characteristics of attitudes toward alcohol drinking in the PURE Poland cohort study.

Characteristics	Currently Use Alcohol Products % (n)	Formerly Used Alcohol Products % (n)	Never Used Alcohol Products % (n)	p ***	OR (95% CI)
Total	67.3 (1360)	10.3 (208)	22.4 (453)		
Sex					
Men	77.2 (577)	11.6 (87)	11.1 (83)	<0.001	2.10 (1.71 to 2.58) [a]
Women	61.5 (783)	9.5 (121)	29.0 (370)		Ref.
Age					
30–44	73.4 (260)	8.2 (29)	18.4 (65)	<0.001	Ref.
45–64	70.2 (941)	8.6 (115)	21.3 (285)		0.45 (0.35 to 0.57) [b]
>64	48.8 (159)	19.6 (64)	31.6 (103)		0.39 (0.27 to 0.55) [b]
Place of residence					
Urban	73.0 (874)	7.2 (86)	19.8 (238)	<0.001	1.88 (1.55 to 2.29) [c]
Rural	59.1 (486)	14.8 (122)	26.1 (215)		Ref.
Level of education *					
Primary	52.2 (157)	20.6 (62)	27.2 (82)	<0.001	Ref.
Vocational	65.2 (210)	12.1 (39)	22.7 (73)		1.57 (1.12 to 2.22) [c]
Secondary	66.1 (525)	9.6 (76)	24.3 (193)		1.83 (1.38 to 2.41) [c]
Higher	78.0 (467)	5.0 (30)	17.0 (102)		3.07 (2.25 to 4.20) [c]
Marital status **					
Married/living together	69.8 (1048)	9.3 (139)	20.9 (314)	<0.001	Ref.
Separated/divorced/widowed	56.7 (212)	14.4 (54)	28.9 (108)		0.68 (0.53 to 0.88) [c]
Never married	68.3 (99)	10.3 (15)	21.4 (31)		0.93 (0.64 to 1.35) [c]

* 5 participants were excluded due to a lack of information about their level of education; ** 1 participant was excluded due to a lack of information about their marital status; *** Chi-square test; OR—being a current drinker, [a]—$OR_{adj.}$—odds ratio adjusted for age, [b]—$OR_{adj.}$—odds ratio adjusted for sex, [c]—$OR_{adj.}$—odds ratio adjusted for sex and age.

Table 2 characterizes the current drinkers by the level of intake of alcohol products. The majority of participants (89.2%) declared a low level of alcohol intake, whereas only 1.8% declared a high level of intake. The chance of high level of intake of alcohol was four times higher in men than in women (OR 4.17; CI 1.64–10.6). The three-times greater chance of higher level of alcohol intake was also observed in participants with a higher level of education in comparison to participants with secondary education (OR 3.21; CI 1.15–8.99). Participants not living in a relationship also had a greater chance of a high level of alcohol consumption. Separated/divorced/widowed participants had a 3-fold higher chance and never married a 4.5-fold higher chance of higher level of alcohol intake than married participants (OR 2.95; CI 1.15–7.58 and OR 4.59; CI 1.58–13.3, respectively) (Table 2).

Table 2. Baseline characteristics in current drinkers by level of intake alcohol.

Characteristics	Number of Drinks Me (IQR)	*p* ***	Alcohol Consumption Intensity % (*n*)			*p* ****	OR (95% CI)
			High	Moderate	Low		
Total	0.8 (0.2–2.0)		1.8 (24)	9.0 (122)	89.2 (1214)		
Sex							
Men	1.5 (0.5–4.8)	0.001	3.1 (18)	18.0 (104)	78.9 (455)	<0.001	4.17 (1.64–10.6) [a]
Women	0.5 (0.2–1.0)		0.8 (6)	2.3 (18)	96.9 (759)		Ref.
Age							
30–44	1.3 (0.5–3.5)		2.3 (6)	10.0 (26)	87.7 (228)		1.00 (ref.)
45–64	1.0 (0.5–2.3)	<0.001	1.6 (15)	8.8 (83)	89.6 (843)	0.888	0.69 (0.26–1.78) [b]
>64	0.8 (0.3–2.0)		1.9 (3)	8.2 (13)	89.9 (143)		0.81 (0.20–3.30) [b]
Place of residence							
Urban	1.0 (0.5–2.5)	>0.05	1.8 (16)	8.6 (75)	89.6 (783)	0.230	0.90 (0.38–2.11) [c]
Rural	1.0 (0.5–2.3)		1.6 (8)	9.7 (47)	88.7 (431)		Ref.
Level of education *							
Primary	0.8 (0.5–2.0)		0.6 (1)	8.3 (13)	91.1 (143)		0.67 (0.08–5.75) [c]
Vocational	1.0 (0.5–3.0)	<0.001	1.9 (4)	7.1 (15)	91.0 (191)	0.084	2.02 (0.54–7.60) [c]
Secondary	1.0 (0.5–2.0)		1.0 (5)	8.2 (43)	90.9 (477)		Ref.
Higher	1.3 (0.5–3.0)		3.0 (14)	10.9 (51)	86.1 (402)		3.21 (1.15–8.99) [c]
Marital status **							
Married/living together	1.0 (0.5–2.5)		1.0 (12)	9.8 (103)	89.2 (933)		Ref.
Separated/divorced/widowed	0.8 (0.5–1.5)	0.014	3.3 (7)	4.7 (10)	92.0 (195)	0.002	2.95 (1.15–7.58) [c]
Never married	1.0 (0.5–3.0)		5.1 (5)	9.1 (9)	85.9 (85)		4.59 (1.58–13.3) [c]

* 5 participants were excluded due to a lack of information about their level of education; ** 1 participant was excluded due to a lack of information about their marital status; *** ANOVA—analysis of variance; **** Chi-square test; OR—higher alcohol consumption, [a]—$OR_{adj.}$—odds ratio adjusted for age, [b]—$OR_{adj.}$—odds ratio adjusted for sex, [c]—$OR_{adj.}$—odds ratio adjusted for sex and age.

Table 3 presents the prevalence of heavy episodic drinking in our cohort. Heavy episodic drinking occurred in 1 in 10 participants (9.9%). The chance of occurrence of heavy episodic drinking was two-fold higher in men than women (OR 2.41; CI 1.67–3.48). Heavy episodic drinking was also more prevalent in participants with vocational (OR 2.22; CI 1.01–4.91) and higher education (OR 2.29; CI 1.10–4.73) in comparison to participants with primary education.

Table 3. The baseline prevalence of heavy episodic drinking among participants who declared current use of alcohol products.

Characteristics	Prevalence		p ***	OR
	Yes	No		
Total	9.9 (134)	90.1 (1226)		
Sex				
Men	14.4 (83)	85.6 (494)	<0.001	2.41 (1.67–3.48) [a]
Women	6.5 (51)	93.5 (732)		Ref.
Age				
30–44	11.5 (30)	88.5 (230)	0.409	Ref.
45–64	9.8 (92)	90.2 (849)		0.83 (0.54–1.29) [b]
>64	7.5 (12)	92.5 (147)		0.63 (0.31–1.26) [b]
Place of residence				
Urban	10.2 (89)	89.8 (785)	0.584	1.11 (0.76–1.62) [c]
Rural	9.3 (45)	90.7 (441)		Ref.
Level of education *				
Primary	5.7 (9)	94.3 (148)	0.036	1.00 (ref.)
Vocational	11.9 (25)	88.1 (185)		2.22 (1.01–4.91) [c]
Secondary	8.2 (43)	91.8 (482)		1.47 (0.70–3.08) [c]
Higher	12.2 (57)	87.8 (410)		2.29 (1.10–4.73) [c]
Marital status **				
Married/living together	10.3 (108)	89.7 (940)	0.596	1.00 (ref.)
Separated/ divorced/widowed	8.5 (18)	91.5 (194)		0.81 (0.48–1.36) [c]
Never married	8.1 (8)	91.9 (91)		0.77 (0.36–1.62) [c]

* 5 participants were excluded due to a lack of information about their level of education; ** 1 participant was excluded due to a lack of information about their marital status; *** Chi-square test, OR—heavy episodic drinking, [a]—$OR_{adj.}$—odds ratio adjusted for age, [b]—$OR_{adj.}$—odds ratio adjusted for sex, [c]—$OR_{adj.}$—odds ratio adjusted for sex and age.

Table 4 presents the types of alcoholic beverages preferred by current drinkers in our cohort. The majority of participants (54.7%) declared most frequent consumption of low-alcohol drinks (wine, beer), a further 25.3% declared most frequent consumption of spirits, and the rest of the participants declared consumption of more than one type of alcohol. Spirits were chosen more frequently by women than men (25.8% vs. 24.7%, respectively), by participants aged 44 years or older (25.7% of participants aged 45–64 years; 39.2% of participants aged >64 years vs. 15.2% of participants aged 30–44 years). Spirits were also preferred by participants living in rural areas compared to in urban areas (30.5% vs. 22.4%, respectively) and participants with primary education (40.1%). Spirits were least frequently consumed by participants with a higher level of education (17.6%).

Table 4. The baseline characteristics of the types of alcoholic beverages most frequently consumed by participants who declared current use of alcohol products.

Characteristics	Spirits % (n)	Wine % (n)	Beer % (n)	More Than One Type of Alcohol % (n)	p ***
Total	25.3 (343)	25.2 (342)	29.5 (399)	20.0 (271)	
Sex					
Men	24.7 (142)	12.9 (74)	49.4 (284)	13.0 (75)	<0.001
Women	25.8 (201)	34.4 (268)	14.7 (115)	25.1 (196)	
Age					
30–44	15.2 (39)	28.4 (73)	41.2 (106)	15.2 (39)	<0.001
45–64	25.7 (242)	24.7 (232)	28.3 (266)	21.3 (200)	
>64	39.2 (62)	23.4 (37)	17.1 (27)	20.3 (32)	
Place of residence					
Urban	22.4 (195)	28.6 (249)	28.3 (246)	20.7 (180)	<0.001
Rural	30.5 (148)	19.2 (93)	31.5 (153)	18.8 (91)	
Level of education *					
Primary	40.1 (63)	12.7 (20)	28.7 (45)	18.5 (29)	<0.001
Vocational	27.6 (58)	18.1 (38)	38.6 (81)	15.7 (33)	
Secondary	26.8 (140)	24.5 (128)	27.6 (144)	21.1 (110)	
Higher	17.6 (82)	33.5 (156)	27.6 (128)	21.3 (99)	
Marital status **					
Married/living together	24.5 (256)	24.5 (256)	31.7 (331)	19.3 (201)	0.03
Separated/divorced/widowed	28.0 (59)	28.4 (60)	19.4 (41)	24.2 (51)	
Never married	28.3 (28)	26.3 (26)	27.3 (27)	18.2 (18)	

* 5 participants were excluded due to a lack of information about their level of education; ** 1 participant was excluded due to a lack of information about their marital status; *** Chi-square test.

The average age of initiation of alcohol consumption in our cohort was 20 $\pm$ 4 years (min 10, max 58). Sex and age were factors that statistically significantly differentiated the age of alcohol initiation. The initiation of alcohol occurred later in life in women than in men (Figure 1). The age of the initiation of alcohol decreased along with the decreasing age of participants (Figure 2).

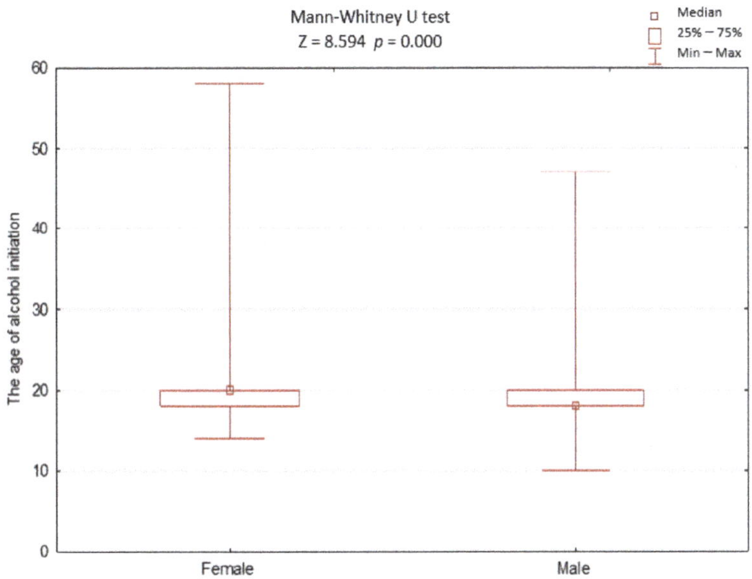

Figure 1. The age of alcohol initiation in women and men.

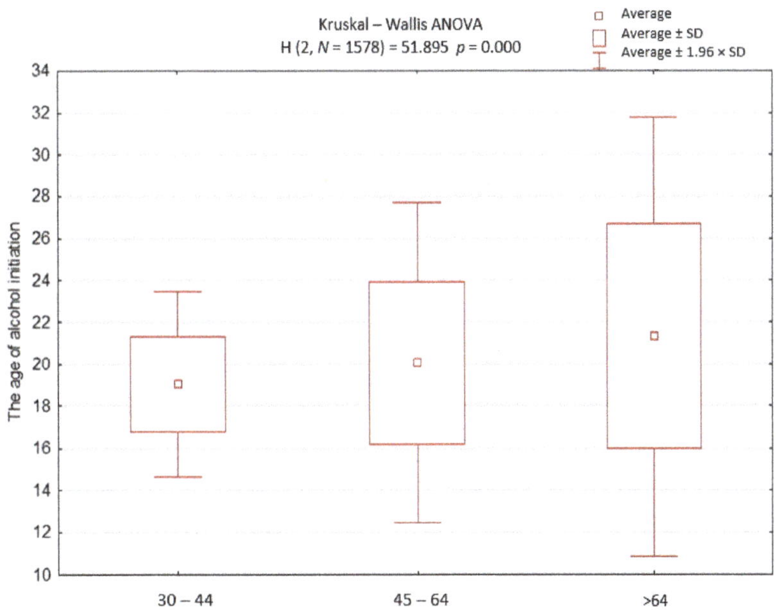

Figure 2. The age of the initiation of alcohol in different age groups.

The odds ratio (with 95% CI) for occurrence of diabetes (a), hypertension (b), CVD (c), and hepatitis or jaundice (d) in groups of participants differentiated by attitudes toward alcohol consumption are presented in Figure 3. In the model, attitudes toward alcohol consumption have been adjusted to age. Current drinkers had almost 1.5-fold higher odds of diabetes than never drinkers (OR 1.49; CI 1.03–2.17). Former drinkers had higher odds for hypertension than never drinkers (1.73; CI 1.05–2.85). Both current and former drinkers had higher odds for CVD than never drinkers (OR 1.66, CI 1.27–2.18; OR 1.76, CI 1.22–2.53, respectively). We found no significantly higher odds for the occurrence of hepatitis/jaundice between current, former, and never drinkers. The odds of occurrence of diabetes, CVD, and hypertension were not significantly associated with sex. On the other hand, the odds of occurrence of hepatitis/jaundice were lower in men than women (OR 0.67; CI 0.46–0.99).

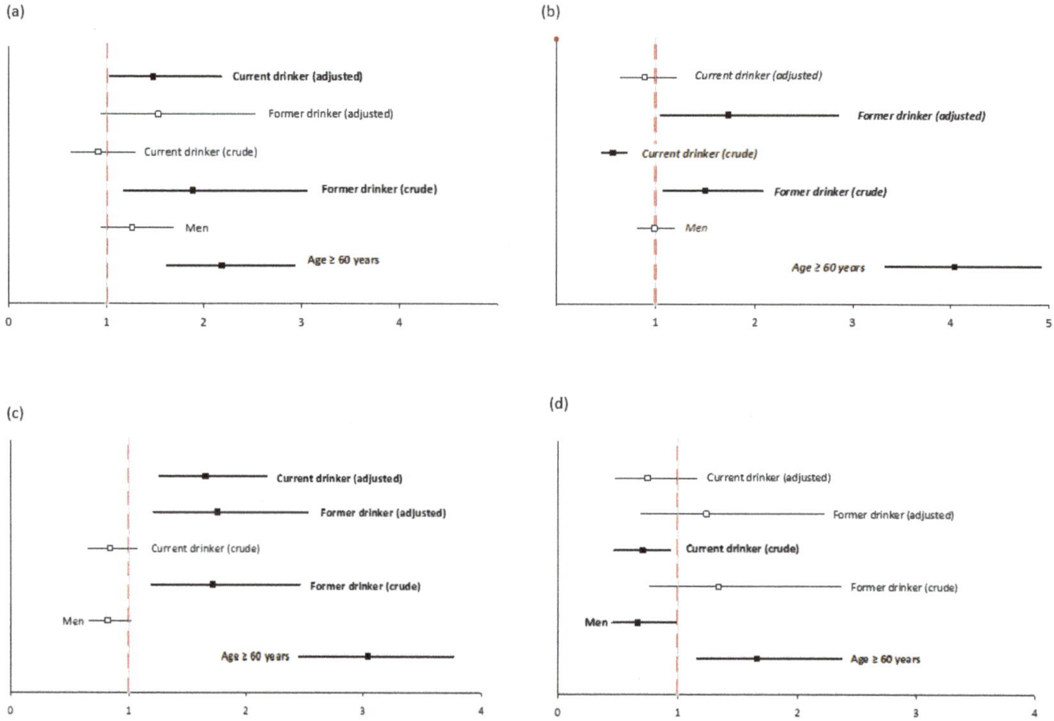

Figure 3. The odds ratio (with 95% CI) for occurrence of diabetes (**a**), hypertension (**b**), CVD (**c**), and hepatitis or jaundice (**d**) in groups of participants differentiated by attitudes toward alcohol consumption ((crude)—unadjusted OR; (adjusted)—attitudes toward alcohol consumption have been adjusted to age).

4. Discussion

This paper presents data about alcohol consumption patterns from one of the few cohort studies in Poland. Poland is one of the most interesting countries in which to investigate alcohol consumption patterns, due to rapid changes in attitudes toward alcohol over the years. It has been observed that since the 2000s, the level of alcohol consumption in Poland has steadily increased, exceeding the European average [12]. It is speculated that the increase in alcohol consumption might be contributing to a slower rate of health improvement in Poland [19,20]. The data regarding alcohol consumption coming from epidemiological studies are limited in Poland. In the National Multicenter Health Survey I (WOBASZ), conducted between 2003 and 2005, the most common type of alcohol consumed by men was beer, followed by vodka [29]. A total of 72% of men consumed 0.01–15.0 of pure ethanol per day [29]. The prevalence of current drinkers in our study was comparable to the percentage observed in the Polish–Norwegian Study (PONS) conducted between 2010 and 2011 (67.3 vs. 57.7%, respectively) [16]. Some discrepancies in a reported prevalence of attitudes towards alcohol consumption may be partially attributed to different definitions of "current alcohol drinkers" adopted in mentioned studies. According to the Public Opinion

Research Center (CBOS), in 2019, a total of 56% of Poles drank alcoholic beverages at least occasionally and a further 8% declared drinking alcohol frequently [30]. In comparison to data obtained in 2010, there was a visible increase in occasional consumption of alcoholic beverages with a concomitant decrease in the percentage of both frequent drinkers and abstainers [30]. In the European region during the time period 2000–2016, there was a stable increase in the percentage of former drinkers (9.9% in 2000 vs. 16.6% in 2016) and a decrease in percentage of current drinkers (70.1% in 2000 vs. 59.9% in 2016) [1].

Our study revealed some sociodemographic tendencies in alcohol consumption patterns. We observed more current drinkers among men than women. Moreover, men had a tendency for a higher level of alcohol consumption and more frequent heavy episodic drinking than women. Similar observations were made worldwide [1] and in Europe [12,31]. Conforming to our results, in a cross-sectional study by Mierzecki et al. [15], women drank less frequently, declared significantly lower levels of alcohol consumption, and declared that they never consumed alcohol much more frequently than men. Similarly, in the PONS study [16], men reported more frequent drinking of alcoholic beverages than women. In an analysis of alcohol consumption patterns of global PURE study participants, people with high or moderate levels of alcohol consumption were predominantly older, male, and less educated [27]. According to the current status of research on alcohol consumption in Poland prepared by the CBOS, women drank much fewer alcoholic beverages than men and were abstainers more often, but there was a significant increase in the percentage of drinking women in Poland over the years [30]. In our study, the percentage of current drinkers decreased with increasing age of participants. A similar observation was made within the framework of the RARHA project, where the highest consumption of alcohol in Poland was noted among younger age groups and that the consumption decreased with the age of the participants [10]. Concomitantly, in our study, in the older age groups, the initiation of alcohol drinking occurred several years later than among the 30- to 44-year-olds and the proportion of abstainers and former drinkers was the highest in the oldest age group. The latter observation can be partially attributed to health-related problems occurring later in life which result in decreased alcohol consumption.

Another sociodemographic factor differentiating the patterns of alcohol consumption was the level of education. In our cohort, participants with higher education consumed more alcohol than those with a lower level of education. On the other hand, participants with a higher level of education chose predominantly low-alcohol beverages. In a study by Wojtyniak et al. [14], the association between level of education and consumption of alcohol was gender-specific. Urban residency and a higher level of education were associated with lower consumption of alcohol in men, but with higher consumption of alcohol in women [14]. A higher level of education was previously associated with a lower risk of heavy episodic drinking, a lower amount of alcohol consumed, and frequency of memory loss caused by drinking [13].

In our study, there were more current drinkers in the urban areas than in rural areas. The same observation was made in the PONS study [16]. Having said that, in our study, current drinkers living in urban areas declared more frequent consumption of low-alcohol beverages (wine, beer), whereas spirits were more prevalent in the rural areas. Contrary to our results, in a study by Mierzecki et al. [15], participants living in urban areas drank less often, consumed fewer alcoholic beverages, and had a lower frequency of heavy drinking than rural inhabitants. We observed no significant differences in the level of alcohol consumption and the prevalence of heavy episodic drinking between urban and rural inhabitants participating in our study.

It was estimated in 2018 that approximately one billion current drinkers worldwide were heavy episodic drinkers [1]. The prevalence of heavy episodic drinking among current drinkers in our cohort was similar to the prevalence observed in the global PURE study (9.9% vs. 13.1%, respectively) [27]. In contrast, the estimates made by the WHO for the European region were much higher. It was estimated that in Europe, heavy episodic drinking occurred in approximately 26.4% of population above 15 years of age (42.6%

among drinkers) [1]. Europe was also the region with the highest prevalence of heavy episodic drinking [1]. The lower prevalence of heavy episodic drinking in our cohort can be partially explained by overrepresentation of participants with higher education in comparison to overall Polish population. Heavy episodic drinking is more common among people with a low socio-economic status.

There are some geographic differences when it comes to the type of preferred alcohol. Worldwide in 2018, the most commonly chosen type of alcohol was spirits (44.8%), but in the European region, the most consumed type of alcohol was beer (40.0%) [1,10]. In our cohort, beer was the most preferred type of alcohol, whereas spirits and wine were chosen equally frequently. According to data gathered by the State Agency for the Prevention of Alcohol-Related Problems (PARPA), beer was the most common type of alcohol beverage chosen by Poles in 2019 (54.6%), followed by spirits (37.8%) and wine (7.6%) [18]. Similarly, beer was the dominant type of alcohol consumed by Poles in the RARHA project [10]. In contrast, in the PONS study, the most frequently chosen type of alcohol was vodka and other spirits (75.4%), followed by wine (61.5%) and beer (54.0%) [16]. In a self-reported preference of type of alcoholic beverages prepared by CBOS in 2019, Poles chose predominantly beer (39%), then wine (25%) and spirits (16%) [30]. Conforming to our results, men preferred mostly beer (56%), whereas women preferred wine (45%) [30]. We noted differences in the preferred types of alcohol between the age groups. Participants in the youngest age group (30- to 44-year-olds) preferred beer, whereas spirits were the least popular type of alcohol in this age group. In contrast, the oldest participants (>64 years of age) preferred spirits over other types of alcohol. Similar observations can be found in the report of the RARHA project, where beer was the most common type of alcohol among younger participants [10]. The shift in the preferences is worth noting, but it is also expected. Several decades ago in Poland, spirits were the most accessible type of alcohol. The increasing consumption of beer and other low-alcohol beverages over spirits has been observed in Poland since the 1990s [32]. According to the report prepared by the Nielsen Company on the condition of the alcohol market in Poland, there was a similar increase in the sales of beer and spirits over the last years, but beer and low-alcohol beverages were still a leading category on the market [33]. It was also observed that there was an increase in interest in high quality alcoholic beverages, like cognac, brandy, or whiskey [30,33].

In our study, current drinkers had higher odds for diabetes and CVD, whereas former drinkers had higher odds for CVD and hypertension than never drinkers. This observation might suggest that giving up alcohol consumption might be related to deteriorating health. Former drinkers were considered less healthy than never drinkers and have a higher mortality rate [34]. The possible bias of considering former drinkers as abstainers and the decline in alcohol consumption along with progressing age and deterioration of health has been previously reported [35–40]. We are reporting the attitudes toward alcohol consumption and occurrence of diseases at the same point in time, so we are unable to determine causality between those events. In a meta-analysis performed by Baliunas et al. [41], a U-shape association between alcohol consumption and the risk of type 2 diabetes was observed. A moderate consumption of alcohol has been associated with a lower risk of type 2 diabetes (the most protective effect observed in case of consumption of 22 g of alcohol per day in men and 24 g of alcohol per day in women) [41]. It has been proposed however, that the favorable effect of alcohol on the diabetes risk in this meta-analysis could have resulted from some bias, e.g., including in the reference group both never drinkers and less healthy former drinkers [42]. In a more u p-to-date meta-analysis conducted by Knott et al., no reduction in diabetes risk was observed in men, regardless of the level of alcohol consumption [42]. On the other hand, a reduction in the risk of type 2 diabetes was observed in women (peak reduction at the level of consumption of 31–37 g of alcohol per day) [42]. Having said that, it was observed that among the patients with diagnosed diabetes, limiting alcohol consumption by at least 2 units/week decreased a 10-year risk of CVD (HR: 0.56, 95% CI 0.36, 0.87) [43]. Alcohol consumption is also a well-known risk factor for hypertension [44–46]. Heavy drinking has been associated with an increased risk of hemorrhagic stroke [47,48]. In the analysis of the modifiable risk factors of CVD and mortality in the global PURE study, high alcohol drinkers showed a moderate increase in CVD incidence, but a high increase in overall mortality [49]. Low-alcohol drinkers also had a lower risk of CVD incidence than abstainers [49]. In a study by Bell et al. [50], heavy drinkers and former drinkers had a higher risk of coronary death, heart failure, cardiac arrest, transient ischemic attack, ischemic stroke, intracerebral hemorrhage, and peripheral arterial disease. In the same study, abstaining from alcohol was associated with an increased risk of unstable angina, myocardial infarction, heart failure, stroke, and peripheral arterial disease in comparison to moderate drinking [50]. In a current analysis of 599,912 participants from 83 prospective studies, the lowest risk of mortality in current drinkers was observed with the consumption of 100 g of alcohol per week [51]. In the same study, when considering CVD subtypes, increased alcohol consumption was linearly associated with a higher risk of stroke and coronary artery disease [51]. For CVDs other than myocardial infarction, there was no threshold below which lower alcohol consumption ceased to be associated with a lower risk [51]. Taken together, although the impact of alcohol consumption on CVD is controversial, the position of the European Society of Cardiology is that the current limits of alcohol consumption should be lowered because the harm outweighs the possible benefits of moderate drinking [52].

There are some limitations of our study to consider. Firstly, this is a cohort study and although the population sample is quite large, the results should be treated with caution as they cannot be exactly extrapolated to the whole Polish population. The study group was not randomized, which can introduce a possible bias. Our cohort is characterized by an overrepresentation of women, elderly people, and better educated people in comparison to the Polish population overall. It is important to note that self-reported alcohol consumption can be underestimated and that there is a risk of bias. On the other hand, as stated previously in the literature, self-reported alcohol consumption, although not without some issues, is a reliable and acceptable method, especially in large-scale studies [53]. The risk of bias in our study is lowered by the fact that the questionnaires were conducted by a trained researcher. We present only the data collected at the baseline. On the other hand, there are currently only a few large cohort studies that have been conducted in Poland. Directions for the future research include a longitudinal prospective analysis of alcohol consumption patterns in the PURE Poland cohort study.

5. Conclusions

In our cohort study, we observed several socio-demographic factors differentiating the patterns of alcohol consumption. Participants living in urban areas and with a higher level of education drank more alcohol, but in contrast to rural inhabitants and participants with a lower level of education, they chose mostly low-alcohol beverages. Current drinkers had higher odds for diabetes and CVD, whereas former drinkers had higher odds for CVD and hypertension than never drinkers.

Author Contributions: Conceptualization, A.S. and K.P.-Z.; formal analysis, K.Z. and Z.S.; investigation, M.W.; methodology, K.Z.; visualization, K.P.-Z.; writing—original draft, K.Z., P.P. and A.B.-R.; writing—review and editing K.Z., A.W., A.S. and K.P.-Z. All authors have read and agreed to the published version of the manuscript.

Funding: The main PURE study and its components are funded by the Population Health Research Institute, the Canadian Institutes of Health Research, Heart and Stroke Foundation of Ontario, and through unrestricted grants from several pharmaceutical companies. The Poland cohort study is funded by the Polish Ministry of Science and Higher Education (grant no. 290/W-PURE/2008/0). Additionally, presented research results were conducted within the framework of the topic: "Analiza rozpowszechnienia czynników ryzyka chorób cywilizacyjnych w populacji osób dorosłych oraz dzieci w wieku szkolnym z uwzględnieniem miejsca zamieszkania (miasto-wieś)" (in English, "Study of prevalence of risk factors of noncommunicable diseases in adults and school-aged children considering the place of residence (urban-rural)") as per records in SIMPLE system number SUB.

C. 300.20.033. We declare that the funding body had no role in the design of the study; collection, analysis, or interpretation of data and no role in writing the manuscript.

Institutional Review Board Statement: The study was reviewed and accepted by the appropriate ethics committee and was therefore performed in accordance with the ethical standards laid down in an appropriate version of the 1964 Declaration of Helsinki (positive opinion of The Bioethics Committee of the Wrocław Medical University nr KB- 443/2006).

Informed Consent Statement: Written and informed consent was obtained from all subjects involved in the study.

Data Availability Statement: Data available on request due to restrictions, e.g., privacy or ethical.

Conflicts of Interest: The authors have no conflict of interest.

References

1. World Health Organization. *Global Status Report on Alcohol and Health 2018*; World Health Organization: Geneva, Switzerland, 2018.
2. Griswold, M.G.; Fullman, N.; Hawley, C.; Arian, N.; Zimsen, S.R.M.; Tymeson, H.D.; Venkateswaran, V.; Tapp, A.D.; Forouzanfar, M.H.; Salama, J.S.; et al. Alcohol use and burden for 195 countries and territories, 1990–2016: A systematic analysis for the Global Burden of Disease Study 2016. *Lancet* **2018**, *392*, 1015–1035. [CrossRef]
3. Gmel, G.; Gutjahr, E.; Rehm, J. How stable is the risk curve between alcohol and all-cause mortality and what factors influence the shape? A precision-weighted hierarchical meta-analysis. *Eur. J. Epidemiol.* **2003**, *18*, 631–642. [CrossRef] [PubMed]
4. Jayasekara, H.; English, D.R.; Room, R.; MacInnis, R.J. Alcohol consumption over time and risk of death: A systematic review and meta-analysis. *Am. J. Epidemiol.* **2014**, *179*, 1049–1059. [CrossRef] [PubMed]
5. Praud, D.; Rota, M.; Rehm, J.; Shield, K.; Zatoński, W.; Hashibe, M.; La Vecchia, C.; Boffetta, P. Cancer incidence and mortality attributable to alcohol consumption. *Int. J. Cancer* **2016**, *138*, 1380–1387. [CrossRef]
6. Bagnardi, V.; Rota, M.; Botteri, E.; Tramacere, I.; Islami, F.; Fedirko, V.; Scotti, L.; Jenab, M.; Turati, F.; Pasquali, E.; et al. Light alcohol drinking and cancer: A meta-analysis. *Ann. Oncol.* **2013**, *24*, 301–308. [CrossRef]
7. Secretan, B.; Straif, K.; Baan, R.; Grosse, Y.; El Ghissassi, F.; Bouvard, V.; Benbrahim-Tallaa, L.; Guha, N.; Freeman, C.; Galichet, L.; et al. A review of human carcinogens–Part E: Tobacco, areca nut, alcohol, coal smoke, and salted fish. *Lancet Oncol.* **2009**, *10*, 1033–1034. [CrossRef]
8. Shield, K.; Manthey, J.; Rylett, M.; Probst, C.; Wettlaufer, A.; Parry, C.D.H.; Rehm, J. National, regional, and global burdens of disease from 2000 to 2016 attributable to alcohol use: A comparative risk assessment study. *Lancet Public Health* **2020**, *5*, e51–e61. [CrossRef]
9. Główny Urząd Statystyczny. *Zdrowie i Zachowanie Zdrowotne Mieszkańców Polski w Świetle Europejskiego Ankietowego Badania Zdrowia (EHIS) 2014 r.*; Główny Urząd Statystyczny: Warsaw, Poland, 2015.
10. Moskalewicz, J.; Room, R.; Thom, B. *Comparative Monitoring of Alcohol Epidemiology Across The EU. Baseline Assessment and Suggestions for Future Action*; Synthesis Report; RARHA: Warsaw, Poland, 2016.
11. Wojtyła-Buciora, P.; Wojtyła, C.; Urbaniak, M.; Kapka-Skrzypczak, L.; Wojtyła, A.; Diatczyk, J.; Klimberg, A.; Piotrowski, K.; Marcinkowski, J.T. Alcohol consumption in Polish middle and high school pupils—Has this rapidly increased during 2009–11? *Ann. Agric. Environ. Med.* **2014**, *21*, 552–556. [CrossRef]
12. World Health Organization Regional Office for Europe. *Trends in Alcohol Consumption in Poland—Fact Sheets*; World Health Organization Regional Office for Europe: Copenhagen, Denmark, 2016.
13. Rosoff, D.B.; Clarke, T.K.; Adams, M.J.; McIntosh, A.M.; Davey Smith, G.; Jung, J.; Lohoff, F.W. Educational attainment impacts drinking behaviors and risk for alcohol dependence: Results from a two-sample Mendelian randomization study with ~780,000 participants. *Mol. Psychiatry* **2019**. [CrossRef]
14. Wojtyniak, B.; Moskalewicz, J.; Stokwiszewski, J.; Rabczenko, D. Gender-specific mortality associated with alcohol consumption in Poland in transition. *Addiction* **2005**, *100*, 1779–1789. [CrossRef]
15. Mierzecki, A.; Kłoda, K.; Guźmiński, B.; Morżak-Bożedajek, A.; Chełstowski, K. Alcohol drinking pattern is associated with demographic features of primary health care patients in poland: A cross-sectional study. *Med. Sci. Monit.* **2018**, *24*, 2083–2090. [CrossRef] [PubMed]
16. Przewoźniak, K.; Łobaszewski, J.; Wojtyła, A.; Bylina, J.; Mańczuk, M.; Zatoński, W.A. Alcohol drinking patterns and habits among a sample of PONS study subjects: Preliminary assessment. *Ann. Agric. Environ. Med.* **2011**, *18*, 221–228.
17. Pikala, M.; Janik-Koncewicz, K.; Zatoński, W. Educational inequalities in mortality due to alcoholic liver disease in Poland. *J. Heath Inequal.* **2020**, *6*, 134–138. [CrossRef]
18. The State Agency for the Prevention of Alcohol-Related Problems (PARPA) Alcohol Consumption in Poland—Statistics. Available online: https://www.parpa.pl/index.php/badania-i-informacje-statystyczne/statystyki (accessed on 16 December 2020).
19. Zatoński, W.A.; Sulkowska, U.; Zatoński, M.Z.; Herbeć, A.A.; Muszyńska, M.M. Alcohol taxation and premature mortality in Europe. *Lancet* **2015**, *385*, 1181. [CrossRef]
20. Zatoński, W. The alcohol crisis in Polish public health. *J. Heath Inequalities* **2019**, *5*, 122–123. [CrossRef]

21. Zatoński, W.; Janik-Koncewicz, K.; Zatoński, M.; Wojtyła, A. Pogorszenie stanu Zdrowia Polaków po 2002 r. Epidemia zgonów alkoholowych [Deterioration of the health of Poles after 2002. An epidemic of alcohol-attributable deaths]. *Menedżer Zdrowia* **2020**, *7–8*, 118–124.
22. Zatoński, W.A.; Zatoński, M.; Janik-Koncewicz, K.; McKee, M. Alcohol-related liver cirrhosis in Poland: The reservoir effect. *Lancet Gastroenterol. Hepatol.* **2020**, *5*, 1035. [CrossRef]
23. Wojtyniak, B.; Goryński, P. *Health Status of Polish Population and Its Determinants—In Short*; National Institute of Public Health-National Institute of Hygiene: Warszawa, Poland, 2018.
24. Stokłosa, M.; Drope, J.; Zatoński, M.; Zatoński, W.A. Towards improved public health: Affecting alcohol and tobacco affordability and consumption in Poland through taxation. *J. Heath Inequalities* **2016**, *2*, 101–104. [CrossRef]
25. Teo, K.; Chow, C.K.; Vaz, M.; Rangarajan, S.; Yusuf, S.; PURE Investigators-Writing Group. The Prospective Urban Rural Epidemiology (PURE) study: Examining the impact of societal influences on chronic noncommunicable diseases in low-, middle-, and high-income countries. *Am. Heart J.* **2009**, *158*, 1–7.e1. [CrossRef] [PubMed]
26. Zatońska, K.; Zatoński, W.A.; Szuba, A. Prospective urban and rural epidemiology Poland—Study design. *J. Heath Inequal.* **2016**, *2*, 136–141. [CrossRef]
27. Smyth, A.; Teo, K.K.; Rangarajan, S.; O'Donnell, M.; Zhang, X.; Rana, P.; Leong, D.P.; Dagenais, G.; Seron, P.; Rosengren, A.; et al. Alcohol consumption and cardiovascular disease, cancer, injury, admission to hospital, and mortality: A prospective cohort study. *Lancet* **2015**, *386*, 1945–1954. [CrossRef]
28. Szuba, A.; Martynowicz, H.; Zatońska, K.; Ilow, R.; Regulska-Ilow, B.; Różańska, D.; Wołyniec, M.; Połtyn-Zaradna, K.; Zatoński, W.A. Prevalence of hypertension in Polish population of PURE Poland study. *J. Heath Inequalities* **2017**, *2*, 157–162. [CrossRef]
29. Waśkiewicz, A.; Sygnowska, E. Alcohol intake and cardiovascular risk factor profile in men participating in the WOBASZ study. *Kardiol. Pol.* **2013**, *71*, 359–365. [CrossRef] [PubMed]
30. Centrum Badania Opinii Społecznej. *Konsumpcja Alkoholu w Polsce;* Komunikat z badań [Alcohol Consumption in Poland. Current Status of Research]; Centrum Badania Opinii Społecznej: Warszawa, Poland, 2019.
31. Bobak, M.; Malyutina, S.; Horvat, P.; Pajak, A.; Tamosiunas, A.; Kubinova, R.; Simonova, G.; Topor-Madry, R.; Peasey, A.; Pikhart, H.; et al. Alcohol, drinking pattern and all-cause, cardiovascular and alcohol-related mortality in Eastern Europe. *Eur. J. Epidemiol.* **2016**, *31*, 21–30. [CrossRef] [PubMed]
32. Świątkiewicz, G.; Wieczorek, Ł.; Allamani, A. What Influences Changes in Alcoholic Beverage Consumption Over Time? Poland in the Light of the European Union Amphora Study. *Subst. Use Misuse* **2014**, *49*, 1601–1610. [CrossRef]
33. Lorenc, A.; Żukowska-Socha, K.; Prokopiuk, E.; Skorupa, B.; Dużyński, A.; Grzywacz, J.; Kowal, M. *Napoje Alkoholowe w Polsce. Aktualny Obraz Rynku [Alcoholic Beverages in Poland. Current Situation of the Market];* The Nielsen Company: Warszawa, Poland, 2018.
34. Tsubono, Y.; Yamada, S.; Nishino, Y.; Tsuji, I.; Hisamichi, S. Choice of comparison group in assessing the health effects of moderate alcohol consumption. *J. Am. Med. Assoc.* **2001**, *286*, 1177–1178. [CrossRef]
35. Fillmore, K.M.; Stockwell, T.; Chikritzhs, T.; Bostrom, A.; Kerr, W. Moderate Alcohol Use and Reduced Mortality Risk: Systematic Error in Prospective Studies and New Hypotheses. *Ann. Epidemiol.* **2007**, *17*, S16–S23. [CrossRef]
36. Dent, O.F.; Grayson, D.A.; Waite, L.M.; Cullen, J.S.; Creasey, H.; Bennett, H.P.; Casey, B.J.; Broe, G.A. A Longitudinal Study of Alcohol Consumption and Functional Disability in a Community Sample of Older People. *Australas. J. Ageing* **2000**, *19*, 185–189. [CrossRef]
37. Graham, K.; Schmidt, G. The effects of drinking on health of older adults. *Am. J. Drug Alcohol Abuse* **1998**, *24*, 465–481. [CrossRef]
38. Larsson, S.C.; Wallin, A.; Wolk, A. Contrasting association between alcohol consumption and risk of myocardial infarction and heart failure: Two prospective cohorts. *Int. J. Cardiol.* **2017**, *231*, 207–210. [CrossRef]
39. Roerecke, M.; Rehm, J. Ischemic heart disease mortality and morbidity rates in former drinkers: A meta-analysis. *Am. J. Epidemiol.* **2011**, *173*, 245–258. [CrossRef]
40. Liang, W.; Chikritzhs, T. The Association between Alcohol Exposure and Self-Reported Health Status: The Effect of Separating Former and Current Drinkers. *PLoS ONE* **2013**, *8*, e55881. [CrossRef] [PubMed]
41. Baliunas, D.O.; Taylor, B.J.; Irving, H.; Roerecke, M.; Patra, J.; Mohapatra, S.; Rehm, J. Alcohol as a risk factor for type 2 diabetes: A systematic review and meta-analysis. *Diabetes Care* **2009**, *32*, 2123–2132. [CrossRef]
42. Knott, C.; Bell, S.; Britton, A. Alcohol consumption and the risk of type 2 diabetes: A systematic review and Dose-Response Meta-analysis of more than 1.9 million individuals from 38 observational studies. *Diabetes Care* **2015**, *38*, 1804–1812. [CrossRef] [PubMed]
43. Strelitz, J.; Ahern, A.L.; Long, G.H.; Boothby, C.E.; Wareham, N.J.; Griffin, S.J. Changes in behaviors after diagnosis of type 2 diabetes and 10-year incidence of cardiovascular disease and mortality. *Cardiovasc. Diabetol.* **2019**, *18*. [CrossRef]
44. Chiva-blanch, G.; Badimon, L. Benefits and risks of moderate alcohol consumption on cardiovascular disease: Current findings and controversies. *Nutrients* **2020**, *12*, 108. [CrossRef] [PubMed]
45. Lawlor, D.A.; Nordestgaard, B.G.; Benn, M.; Zuccolo, L.; Tybjaerg-Hansen, A.; Davey Smith, G. Exploring causal associations between alcohol and coronary heart disease risk factors: Findings from a mendelian randomization study in the copenhagen general population study. *Eur. Heart J.* **2013**, *34*, 2519–2528. [CrossRef] [PubMed]

46. Zhao, P.P.; Xu, L.W.; Sun, T.; Wu, Y.Y.; Zhu, X.W.; Zhang, B.; Cheng, Z.; Cai, X.; Liu, Y.C.; Zhao, T.T.; et al. Relationship between alcohol use, blood pressure and hypertension: An association study and a Mendelian randomisation study. *J. Epidemiol. Community Health* **2019**, *73*, 796–801. [CrossRef] [PubMed]
47. Zhang, C.; Qin, Y.Y.; Chen, Q.; Jiang, H.; Chen, X.Z.; Xu, C.L.; Mao, P.J.; He, J.; Zhou, Y.H. Alcohol intake and risk of stroke: A dose-response meta-analysis of prospective studies. *Int. J. Cardiol.* **2014**, *174*, 669–677. [CrossRef] [PubMed]
48. Costa, P.; Grassi, M.; Iacoviello, L.; Zedde, M.; Marcheselli, S.; Silvestrelli, G.; DeLodovici, M.L.; Sessa, M.; Zini, A.; Paciaroni, M.; et al. Alcohol intake and the risk of intracerebral hemorrhage in the elderly: The MUCH-Italy. *Neurology* **2018**, *91*, 1–9. [CrossRef]
49. Yusuf, S.; Joseph, P.; Rangarajan, S.; Islam, S.; Mente, A.; Hystad, P.; Brauer, M.; Kutty, V.R.; Gupta, R.; Wielgosz, A.; et al. Modifiable risk factors, cardiovascular disease, and mortality in 155 722 individuals from 21 high-income, middle-income, and low-income countries (PURE): A prospective cohort study. *Lancet* **2020**, *395*, 795–808. [CrossRef]
50. Bell, S.; Daskalopoulou, M.; Rapsomaniki, E.; George, J.; Britton, A.; Bobak, M.; Casas, J.P.; Dale, C.E.; Denaxas, S.; Shah, A.D.; et al. Association between clinically recorded alcohol consumption and initial presentation of 12 cardiovascular diseases: Population based cohort study using linked health records. *BMJ* **2017**, *356*. [CrossRef] [PubMed]
51. Wood, A.M.; Kaptoge, S.; Butterworth, A.; Nietert, P.J.; Warnakula, S.; Bolton, T.; Paige, E.; Paul, D.S.; Sweeting, M.; Burgess, S.; et al. Risk thresholds for alcohol consumption: Combined analysis of individual-participant data for 599 912 current drinkers in 83 prospective studies. *Lancet* **2018**, *391*, 1513–1523. [CrossRef]
52. Piepoli, M.F.; Abreu, A.; Albus, C.; Ambrosetti, M.; Brotons, C.; Catapano, A.L.; Corra, U.; Cosyns, B.; Deaton, C.; Graham, I.; et al. Update on cardiovascular prevention in clinical practice: A position paper of the European Association of Preventive Cardiology of the European Society of Cardiology. *Eur. J. Prev. Cardiol.* **2020**, *27*, 181–205. [CrossRef]
53. Del Boca, F.K.; Darkes, J. The validity of self-reports of alcohol consumption: State of the science and challenges for research. *Addiction* **2003**, *98*, 1–12. [CrossRef]

International Journal of
Environmental Research and Public Health

MDPI

Article

Physical Activity and Quality of Life of University Students, Their Parents, and Grandparents in Poland—Selected Determinants

Katarzyna Kotarska [1], Maria Alicja Nowak [1], Leonard Nowak [1], Paweł Król [2], Artur Sochacki [2], Katarzyna Sygit [3,*] and Marian Sygit [3]

[1] Institute of Physical Culture Sciences, Faculty of Physical Culture and Health, University of Szczecin, 71-065 Szczecin, Poland; katarzyna.kotarska@usz.edu.pl (K.K.); maria.nowak@usz.edu.pl (M.A.N.); leonard.nowak@usz.edu.pl (L.N.)
[2] Institute of Physical and Cultural Studies, Medical College of Rzeszów University, 35-959 Rzeszów, Poland; krolpawel1@poczta.onet.pl (P.K.); artisocha@wp.pl (A.S.)
[3] Faculty of Health Sciences, Calisia University, 62-800 Kalisz, Poland; msygit@onet.pl
* Correspondence: k.sygit@akademiakaliska.edu.pl

Abstract: Physical activity is one of the factors conditioning human health. Research shows a positive impact of regular physical activity on the quality of human life. The aim of the study was to determine the relationship between the physical activity of university students, their parents, and their grandparents, and the overall quality of their lives in individual domains (physical, mental, social, environmental), as well as the perceived state of health in relation to selected determinants. The research included 1001 participants, including 253 students related to physical culture and health promotion studying at the University of Szczecin (faculties: physical education, tourism and recreation, public health, sports diagnostics), and their 336 parents and 412 grandparents. Purposive sampling was used to outline the determinants of quality of life and family factors in physical activity. The diagnostic survey was carried out based on the standardized WHOQOL-BREF (World Health Organization Quality of Life) questionnaire. Statistically significant differences were shown in the studied generations regarding the assessment of quality of life and satisfaction with health in the physical, psychological, social, and environmental domains. The oldest generation gave the lowest assessment of quality of life and was the least satisfied with their health regarding particular domains. Female students were more satisfied with their health compared to grandmothers and grandfathers, whereas male students compared to mothers and fathers. Fathers achieved the highest scores in the psychological and social domains, but, in case of the latter, differences were found between mothers' and fathers' assessments. Intergenerational differences were found in quality of life and the assessment of health status. Current participation in broadly understood physical culture was often a result of positive attitudes towards physical education and doing sport in the past, which meant higher scores in the physical domain each time. The study demonstrated that taking up physical activity impacted the quality of life and assessment of health in the past and currently.

Keywords: lifestyle; training; exercises; intergenerational relations; academic youth; standardized questionnaire WHOQOL-BREF

Citation: Kotarska, K.; Nowak, M.A.; Nowak, L.; Król, P.; Sochacki, A.; Sygit, K.; Sygit, M. Physical Activity and Quality of Life of University Students, Their Parents, and Grandparents in Poland—Selected Determinants. *IJERPH* **2021**, *18*, 3871. https://doi.org/10.3390/ijerph18083871

Academic Editor: Wioletta Zukiewicz-Sobczak

Received: 2 March 2021
Accepted: 31 March 2021
Published: 7 April 2021

Publisher's Note: MDPI stays neutral with regard to jurisdictional claims in published maps and institutional affiliations.

Copyright: © 2021 by the authors. Licensee MDPI, Basel, Switzerland. This article is an open access article distributed under the terms and conditions of the Creative Commons Attribution (CC BY) license (https://creativecommons.org/licenses/by/4.0/).

1. Introduction

The concept of quality of life is wide, and there is no single definition of quality of life in either social sciences or medical sciences. Quality of life is studied in relation to a particular area of life and the factors that affect it. Usually, factors such as health, life satisfaction, happiness, psychosocial adjustment, well-being, or physical activity are taken into account [1–3].

According to the WHO (World Health Organization) definition, quality of life (QOL) is "an individual's perception of their position in life in the context of the culture and value systems in which they live and in relation to their goals, expectations, standards and concerns" [1,3,4]. In medicine, health-related quality of life (HRQOL), introduced by Schipper et al., is used more often [1,2,5]. It is based on the WHO definition of health. The WHO defines health as follows: "Health is a state of complete physical, mental and social well-being and not merely the absence of disease or infirmity" (the WHO definition of Health, Preamble to the Constitution of the World Health Organization as adopted by the International Health Conference) [1–5].

Over the years, however, there have been various changes to the term HRQQL. Currently, most definitions address five dimensions of health-related quality of life. These are: physical, psychological, social, and cognitive functioning, as well as a general sense of well-being [1–3,6–9]. The medical view of the quality of life emphasizes the great influence of such factors as sex, age, education, or the level of physical activity [1–3,7,10].

The factors that influence quality of life undoubtedly include physical activity, which, among others, includes various kinds of games, exercises, or amateur practices of various sports disciplines as a part of active leisure. Physical activities of this type are carried out for various purposes: for pleasure, leisure, and to maintain health, improve exercise capacity, or acquire special physical abilities and skills. Such healthy behaviors prevent the occurrence of various civilization diseases [7–9,11]. It should be emphasized that physical activity needs to accompany a person throughout their life, which is why it is so important to promote physical culture not only among children, but also among adolescents, because it provides the basis for the continuation of healthy behaviors in adulthood [4–6,12]. When examining the quality of life of healthy adults, it may be concluded that physical activity is not only strictly related to physical condition, but also to mental condition [7,8,13–17].

The assessment of quality of life depends not only on the state of physical and mental health, level of independence, social relations, and environmental factors, but also on the individual attitude of a given person. Although the abovementioned definition of health according to the WHO clearly refers to quality of life, research on this subject was initiated relatively recently; in the 1990s, systematic studies in this field started to be undertaken [4–6,18–22]. It is known from today that these studies should take into account, among others, the degree of satisfaction with health and family life, social relations, professional life, education, and specialization or universal standards affecting the quality of life in the local community, which the authors have pointed out in this study [16–18,23–29].

In recent years, there has been a lot of research on quality of life in the population of healthy and sick people, which analyzed various factors, such as lifestyle, forms of leisure, or the socioeconomic situation. Adults with frequent foot problems should be given special attention. Foot pain affects one's overall health, and often causes a person to fall. Studies conducted by scientists in Spain confirm that this is one of the reasons for the reduction of daily physical activity [30].

Taking into account the existing knowledge on quality of life and physical activity, we tried to determine how these relationships work in relation to university students (faculties of physical education, sports diagnostics, public health, tourism, and recreation), their parents, and their grandparents. The aim of the study was to determine the relationship between the physical activity of university students, their parents, and their grandparents, as well as the overall quality of their lives in individual domains (physical, mental, social, environmental) and the perceived state of health in relation to selected determinants.

A hypothesis was formulated that there is a relationship between current physical activity and quality of life within the individual generations. It was also assumed that the age, gender, marital status, education level, and employment status of respondents may modify this relationship.

2. Methodology

Participants and Ethical Considerations

The research was carried out on 1001 participants, including 253 students related to physical culture and health promotion studying at the University of Szczecin (faculties:

physical education, tourism and recreation, public health, sports diagnostics), and their 336 parents and 412 grandparents. Female students (171) were aged 18–34 years, with an average of 22.4 ± 3.44. The average age of the 82 male students, aged 19–28 years, was 21.5 ± 2.0. The mothers of the studied (205) persons were aged 35–68 years, the average being 47.7 ± 7.4. For the 131 fathers, in the 35–77 age range, the average age was 49.6 ± 8.5. The average age for the 262 grandmothers, 51–99 years old, was 73.6 ± 8.9, and for the 150 grandfathers, 55–97 years old, it was 73.8 ± 8.9. Persons over 74 constituted 14.6%, including 3.3% over 90. The majority of the respondents were women (63.8%), and their predominance was significant in every age group. The unmarried respondents constituted 48.8% (including those never married 26.5%, widowed 14.8%, and divorced 7.5%). The respondents (including students) were most likely to have secondary education (46.7%). Grandfathers and grandmothers mostly had primary education (60.4% and 51%). Mothers of the students were better educated than their fathers. Above-secondary education was reported by 45.6% of mothers and 32.8% of fathers, and below-secondary education was reported by 37.4% of fathers and 19.1% of mothers. Employment was reported by over 70% of the surveyed students and their parents (over 72%), and by about 10% of the grandparents. Respondents described their financial situation as very good (39.2%), moderate (46.4%), sufficient (10.7%), or bad (3.7%). The majority of the respondents lived in medium and large cities; only a minority lived in rural areas (21.5%).

The study protocol was approved by the appropriate ethics committee of the Local Medical Chamber in Szczecin, permit no 15/KB/V.2015, and conformed to the ethical guidelines of the 1975 Declaration of Helsinki. Written informed consent was obtained from each subject included in the study.

3. Data Collection

The diagnostic survey method was based on the standardized WHOQOL-BREF (World Health Organization Quality of Life) questionnaire. An abbreviated version, adapted to Polish language, culture, and psychometric conditions, contained 26 questions [30]. The first two questions, relating to the individual's general perception of the quality of life and health, were analyzed separately. The remaining 24 questions concerned four domains of the perceived quality of life. The physical domain included an assessment of daily life activities, drug dependence and treatment, energy and fatigue, mobility, pain and discomfort, rest and sleep, and the ability to work. The psychological domain concerned body image, joy, sense of life, mood, negative and positive feelings, self-esteem, and concentration/memory/thinking/learning. The social domain concerned personal relationships, sexual activity, and social support. The environmental domain concerned financial resources, safety, home, physical environment, acquiring new information and skills, recreation and leisure, housing conditions, and access to medical care and transportation. The scores for the aforementioned domains were determined by calculating the arithmetic mean of their constituent items. The higher the score, the higher the quality of life; the maximum value was 120 points. Our research also included our survey technique to study the lifestyles of women and men, as well as an interview. Information obtained from the interviews conducted among the students was used to supplement and verify the results of the research. Our study used personal data on physical activity, age, gender, place of residence, education, employment, and financial situation. The focus was on establishing a relationship between the quality of life and currently undertaking physical activity over the previous week and in the past (participation in physical education classes or involvement in professional sport) of the students of the Faculty of Physical Culture and Health Promotion of the University of Szczecin, their parents, and their grandparents. The quality of life of respondents were also analyzed in relation to socio-demographic variables.

4. Data Analysis

The one-way analysis of variance (F-test) for independent groups was used (ANOVA), as well as Student's t-test. The effect size in ANOVA was expressed using ω^2. For the

Student's *t*-test, we used Hedges' g—a measure of effect size. In qualitative analyses, the trait frequency and the chi-square test of independence were used, as well as Cramér's V for the χ^2 test. Differences were deemed statistically significant at $p \leq 0.05$. Statistical calculations were made with Statistica 12 for Windows, Microsoft Excel 2007, and JASP 0.10.0.0 [31,32].

5. Results

The attitudes of the respondents towards compulsory participation in physical education classes, practicing professional sport in the past, and current recreational physical activity are presented in Figure 1.

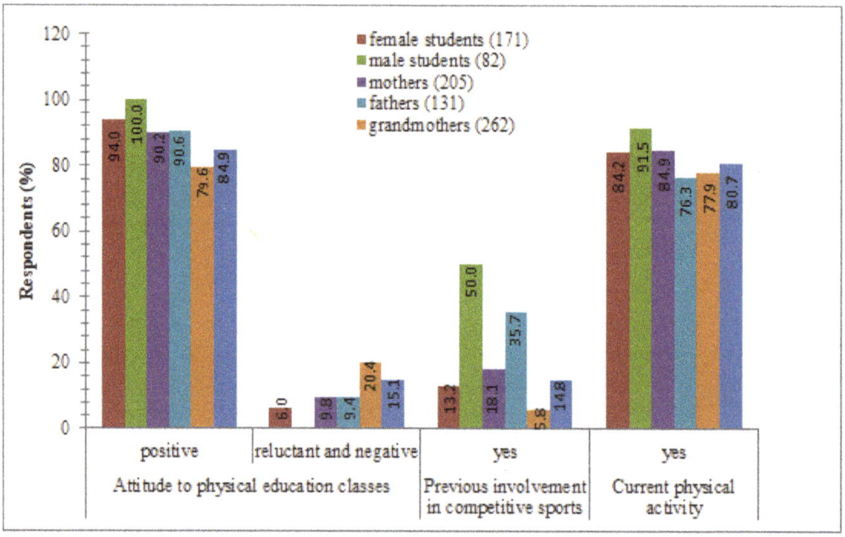

Figure 1. Physical activity characteristics of university students, their parents, and their grandparents (independence $\chi 2$ test and Cramér's V).

The majority of the respondents had willingly participated in physical education classes in the past (88.1%). A reluctant and negative attitude towards those lessons was more often shown by the oldest generation (grandmothers 20.4%; grandparents 15.1%). Few respondents (2%) had been exempted from these classes. All of the university students willingly participated in compulsory physical education classes (94% of women and 100% of men). According to research, 18.4% of respondents declared involvement in competitive sport in the past, men more often than women (students 50%; fathers 35.7%; grandparents 14.8%) ($p < 0.001$ for the χ^2 test, Cramér's V = 0.3). As a result of the analysis, it was found that 81.7% of respondents declared current physical activity. The percentages of physically active female students and their mothers were similar. Grandmothers declared lower physical activity (77.9%). Male students were physically more active compared to fathers and grandparents (91.5% vs. 76.3% vs. 80.7%, respectively) ($p < 0.05$ for the χ^2 test, Cramér's V = 0.1).

Respondents varied slightly in their preferences regarding the forms of physical activity (Figure 2). Students attended fitness classes, ran, cycled, went to the gym, played sports games, and swam (about 15% each). Students found sports games, gym exercises, and jogging more attractive than the older generations. Mothers (66.2%), fathers (53.8%), grandmothers (83.7%), and grandparents (81.1%) most often walked. This form of activity was not even mentioned by university students. Gardening was more significant for parents and grandparents, who often treated this activity as a form of recreation. Some mothers also attended fitness classes (8.1%), and some fathers played sports games (8.4%).

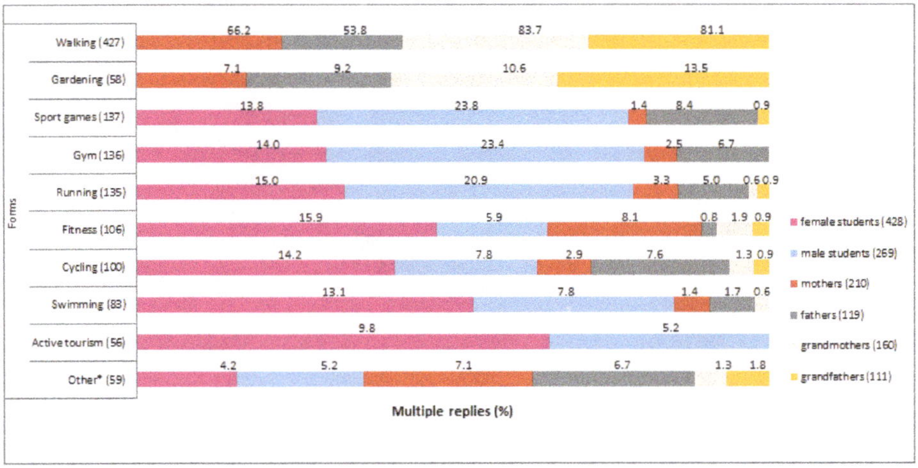

Figure 2. Physical activity patterns of students, their parents, and their grandparents (%). * Combat sports, dance, pole dance, horseback riding, tennis, table tennis, and badminton.

The evaluation of quality of life, health satisfaction, and physical, psychological, social, and environmental domains in the studied generations revealed statistically significant differences ($p < 0.001$ for the F-test). In the physical domain ($\omega^2 = 0.18$) and health satisfaction ($\omega^2 = 0.11$), the effect size was high and close to high, respectively, and in the quality of life satisfaction and in the other domains, it was moderate or close to moderate (in the environmental domain) (Table 1).

Statistically significant differences were found in satisfaction with life between female students and their grandparents, and between parents and grandparents of male students ($p < 0.001$ for the Student's t-test). Grandmothers and grandparents had the lowest scores for quality of life and in the individual domains, as well as in satisfaction with health. Grandmothers were less satisfied with their health than grandfathers ($p < 0.05$ for the Student's t-test).

Female students were more satisfied with their health than grandmothers and grandparents, and male students were more satisfied with their health than mothers ($p < 0.05$ for the Student's t-test) and fathers ($p < 0.05$ for the Student's t-test). Female students were less satisfied with their health than male students ($p < 0.05$ for the Student's t-test). In the psychological domain, we found differences between female students and fathers ($p < 0.001$ for the Student's t-test), and between mothers and fathers ($p < 0.05$ for the Student's t-test). In this domain, fathers achieved the highest scores, as well as in the social domain ($p < 0.05$ for the Student's t-test). The moderate and high effect size (Hedges' g = 0.5–1.0) confirmed the differences between students and grandparents in the assessment of satisfaction of life, satisfaction with health, and in individual domains (physical, psychological, social, and environmental). In the physical domain, the effect size was very high (g = 0.8–1.1). These differences, confirmed by the effect size, were also observed between the students' parents and grandparents (g = 0.8–0.9).

The reported quality of life differed between those who had participated in physical education, professional sport, and recreational activities in the past (Table 2).

Table 1. Total quality of life in individual domains (physical, psychological, social, environmental) and satisfaction with health among university students, their parents, and their grandparents (F-test, ω^2, Student's *t*-test, Hedges' g).

Specification	Groups *	Value of p for Student's t-Test					Hedges' g					Means
		MS	M	F	GM	GF	MS	M	F	GM	GF	
Quality of life	FS	0.578	0.949	0.751	0.000	0.000	−0.1	0.1	0.1	0.6	0.5	4.03
F = 16.87	MS		0.516	0.438	0.000	0.000		0.1	0.1	0.7	0.7	4.09
$p < 0.001$	M			0.775	0.000	0.000			0.1	0.6	0.6	4.02
$\omega^2 = 0.07$	F				0.000	0.000				0.5	0.5	4.00
	GM					0.859					−0.1	3.57
	GF											3.59
Satisfaction with health	FS	0.031	0.521	0.361	0.000	0.000	−0.3	0.1	0.1	0.8	0.5	3.83
F = 26.42	MS		0.011	0.012	0.000	0.000		0.3	0.4	1.0	0.8	4.04
$p < 0.001$	M			0.723	0.000	0.000			0.1	0.7	0.5	3.78
$\omega^2 = 0.11$	F				0.000	0.000				0.7	0.4	3.75
	GM					0.034					−0.2	3.19
	GF											3.37
Physical domain	FS	0.598	0.459	0.684	0.000	0.000	0.1	01	0.1	1.1	0.8	28.58
F = 46.21	MS		0.923	0.921	0.000	0.000		0.1	−0.1	0.9	0.9	28.33
$p < 0.001$	M			0.823	0.000	0.000			−0.1	0.9	0.8	28.28
$\omega^2 = 0.18$	F				0.000	0.000				0.9	0.8	28.39
	GM					0.294					−0.1	23.89
	GF											24.41
Psychological domain	FS	0.064	0.153	0.000	0.000	0.055	−0.2	−0.1	−0.4	0.4	0.2	21.53
F = 15.94	MS		0.466	0.282	0.000	0.001		0.1	−0.2	0.7	0.5	22.46
$p < 0.001$	M			0.032	0.000	0.001			−0.2	0.6	0.4	22.10
$\omega^2 = 0.07$	F				0.000	0.000				0.8	0.6	23.05
	GM					0.057					−0.2	19.88
	GF											20.67
Social domain	FS	0.775	0.409	0.118	0.000	0.000	0.1	0.1	−0.2	0.7	0.4	11.61
F = 20.93	MS		0.713	0.095	0.000	0.000		0.1	−0.2	0.6	0.5	11.51
$p < 0.001$	M			0.013	0.000	0.000			−0.3	0.6	0.4	11.40
$\omega^2 = 0.09$	F				0.000	0.000				0.9	0.7	12.04
	GM					0.150					−0.1	9.97
	GF											10.33
Environmental domain	FS	0.423	0.371	0.599	0.000	0.000	0.1	0.1	0.1	0.5	0.4	30.43
F = 9.34	MS		0.955	0.809	0.000	0.006		−0.1	−0.1	0.4	0.4	29.99
$p < 0.001$	M			0.813	0.000	0.000			−0.1	0.4	0.4	30.02
$\omega^2 = 0.04$	F				0.000	0.002				0.4	0.4	30.15
	GM					0.577					−0.1	28.12
	GF											28.38

* The following abbreviations were used: female students—FS, male students—MS, mothers—M, fathers—F, grandmothers—GM, grandfathers—GF.

Those willingly participating in physical education classes had a higher quality of life rating ($p < 0.01$) and scored better in the physical ($p < 0.01$) and social domain ($p < 0.01$). In a comparison of those involved in competitive sport in the past and those who were not, the former reported higher satisfaction of life ($p < 0.001$), health satisfaction ($p < 0.001$), and in physical ($p < 0.001$), psychological ($p < 0.001$), and social domains ($p < 0.01$). Those currently engaged in physical recreation were also more satisfied with their health ($p < 0.001$), performing better in the physical ($p < 0.001$), psychological ($p < 0.001$), and environmental domains ($p < 0.05$) than those who were not currently physically active. Their effect size was low (g = 0.2–0.3).

General differences in quality of life were found in relation to the marital status of the respondents ($p < 0.001$ for F-test) (Table 3).

The lowest mean scores for quality of life, satisfaction with health, and its components were characteristic of widowed and divorced persons. Unmarried persons, as well as those married and in cohabitation, rated their quality of life higher than widowed people. Divorced persons also had a higher quality of life than widowers. The effect size on the perception of quality of life among unmarried persons and those living in cohabitation (g = 0.8) compared to widows/widowers was very high, as was a higher satisfaction with health among unmarried persons in comparison with widowed persons (g = 0.8). In the physical and social domains, widowed persons had lower scores than unmarried and married persons and those living in a cohabitation (g = 0.6–1.1). In the psychological and environmental domains, widowed persons were rated the lowest in comparison with the other groups (moderate effect size).

There were statistically significant differences in quality of life among the studied students, their parents, and their grandparents, depending on the level of education (Table 4).

Table 2. Physical activity of respondents in the past and present and their quality of life (F-test, ω^2, Student's *t*-test, Hedges' g).

Specification	Physical Activity	Value of p for Student's t-Test	Hedges' g	Means
	Participation in physical education classes			
Satisfaction with life $F = 6.98, p < 0.01$ $\omega^2 = 0.01$	Willing	0.008	0.3	3.87
	Reluctant and negative			3.65
Physical domain $F = 7.72, p < 0.001$ $\omega^2 = 0.01$	Willing	0.005	0.3	26.79
	Reluctant and negative			25.34
Social domain $F = 6.84, p < 0.01$ $\omega^2 = 0.01$	Willing	0.009	0.3	11.07
	Reluctant and negative			10.37
	Practicing competitive sport			
Satisfaction with life $F = 10.12, p < 0.001$ $\omega^2 = 0.01$	Yes	0.001	0.3	4.01
	No			3.80
Satisfaction with health $F = 12.01, p < 0.001$ $\omega^2 = 0.01$	Yes	0.000	0.3	3.78
	No			3.54
Physical domain $F = 12.16, p < 0.001$ $\omega^2 = 0.01$	Yes	0.000	0.3	27.74
	No			26.35
Psychological domain $F = 13.26, p < 0.001$ $\omega^2 = 0.01$	Yes	0.000	0.3	22.33
	No			21.11
Social domain $F = 6.56, p < 0.01$ $\omega^2 = 0.01$	Yes	0.010	0.2	11.39
	No			10.87
	Current participation in physical recreation			
Satisfaction with health $F = 10.48, p < 0.001$ $\omega^2 = 0.01$	Yes	0.001	0.3	3.63
	No			3.40
Physical domain $F = 15.99, p < 0.001$ $\omega^2 = 0.01$	Yes	0.000	0.3	26.91
	No			25.32
Psychological domain $F = 12.65, p < 0.001$ $\omega^2 = 0.01$	Yes	0.000	0.3	21.58
	No			20.40
Environmental domain $F = 5.3, p < 0.05$ $\omega^2 = 0.01$	Yes	0.021	0.2	29.52
	No			28.66

Table 3. Marital status vs. quality of life reported by students, their parents, and their grandparents (F-test, ω^2, t-test, Hedges' g).

Specification	Marital Status	Value of p for Student's t-Test				Hedge's g				Means
		Married	Divorced	Widowers	Cohabitation	Married	Divorced	Widowers	Cohabitation	
Quality of life F = 16.43 $p < 0.001$ $\omega^2 = 0.06$	Unmarried	0.419	0.023	0.000	0.962	0.1	0.3	0.8	0.1	3.97
	Married		0.077	0.000	0.792		0.2	0.7	−0.1	3.92
	Divorced			0.001	0.197			0.5	−0.3	3.75
	Widowed				0.000				−0.8	3.39
	Cohabitation									3.96
Satisfaction with health F = 14.34 $p < 0.001$ $\omega^2 = 0.06$	Unmarried	0.023	0.000	0.000	0.216	0.2	0.5	0.8	0.3	3.78
	Married		0.025	0.000	0.781		0.3	0.5	0.1	3.64
	Divorced			0.057	0.280			0.3	−0.2	3.40
	Widowed				0.017				−0.5	3.17
	Cohabitation									3.59
Physical domain F = 23.36 $p < 0.001$ $\omega^2 = 0.08$	Unmarried	0.186	0.015	0.000	0.500	0.1	0.3	1.0	0.1	27.62
	Married		0.148	0.000	0.919		0.2	0.8	0.1	27.14
	Divorced			0.000	0.425			0.6	−0.2	26.25
	Widowed				0.000				−0.8	23.26
	Cohabitation									27.04
Psychological domain F = 14.78 $p < 0.001$ $\omega^2 = 0.06$	Unmarried	0.015	0.576	0.000	0.626	−0.2	0.1	0.5	0.1	21.36
	Married		0.033	0.000	0.134		0.3	0.7	0.3	22.08
	Divorced			0.002	0.923			0.4	0.1	21.05
	Widowed				0.041				−0.4	19.24
	Cohabitation									20.96
Social domain F = 32.40 $p < 0.001$ $\omega^2 = 0.01$	Unmarried	0.053	0.002	0.000	0.851	−0.1	0.4	0.8	0.1	11.21
	Married		0.000	0.000	0.311		0.6	1.1	0.2	11.56
	Divorced			0.001	0.089			0.4	−0.4	10.19
	Widowed				0.000				−0.9	9.14
	Cohabitation									11.11
Environmental domain F = 11.54 $p < 0.001$ $\omega^2 = 0.04$	Unmarried	0.528	0.070	0.000	0.795	−0.1	0.2	0.6	−0.1	29.70
	Married		0.026	0.000	0.992		0.3	0.6	−0.1	29.92
	Divorced			0.026	0.228			0.3	−0.3	28.64
	Widowed				0.002				−0.6	27.18
	Cohabitation									29.93

Table 4. Education and quality of life of students, their parents, and their grandparents (F-test, ω^2, t-test, Hedges' g).

Specification	Level of Education	Value of p for Student's t-Test		Hedges' g		Means
		Secondary	Post-Secondary	Secondary	Post-Secondary	
Quality of life F = 24.33 $p < 0.001$ $\omega^2 = 0.06$	pre-secondary	0.000	0.000	−0.4	−0.6	3.61
	secondary		0.003		−0.2	3.89
	post-secondary					4.08
Satisfaction with health F = 21.39 $p < 0.001$ $\omega^2 = 0.06$	pre-secondary	0.000	0.000	−0.4	−0.5	3.33
	secondary		0.214		−0.1	3.68
	post-secondary					3.76
Physical domain F = 28.21 $p < 0.001$ $\omega^2 = 0.06$	pre-secondary	0.000	0.000	−0.5	−0.6	24.96
	secondary		0.093		−0.1	27.16
	post-secondary					27.81
Psychological domain F = 6.83 $p < 0.001$ $\omega^2 = 0.01$	pre-secondary	0.273	0.000	−0.1	−0.3	20.93
	secondary		0.004		−0.2	21.25
	post-secondary					22.22
Social domain F = 14.71 $p < 0.001$ $\omega^2 = 0.01$	pre-secondary	0.000	0.000	−0.3	−0.4	10.38
	secondary		0.281		−0.1	11.21
	post-secondary					11.42
Environmental domain F = 12.86 $p < 0.001$ $\omega^2 = 0.01$	pre-secondary	0.007	0.000	−0.2	−0.4	28.49
	secondary		0.002		−0.3	29.38
	post-secondary					30.51

The overall variation in satisfaction with life, health, and in the physical domain (F-test) was confirmed by the moderate effect size (ω^2 = 0.06). Those with secondary education, especially post-secondary education, had higher quality of life scores, were more satisfied with their health, and achieved higher results in the remaining domains (Hedges' g within the range −0.3 to −0.6).

Employed persons (n = 440) had a higher quality of life and health satisfaction and better results in all domains in comparison to unemployed persons ($p < 0.001$ for F-test, effect size high and moderate, as shown by ω^2 = 0.15 to 0.06) (Table 5).

Table 5. Employment and quality of life for students, their parents, and their grandparents (F-test, ω^2, t-test, Hedges' g).

Specification	Employment	Value of p for Student's t-Test	Hedges' g	Means
Quality of life F = 88.27	Yes	0.000	0.6	4.10
$p < 0.001$ $\omega^2 = 0.08$	No			3.64
Satisfaction with health F = 111.57	Yes	0.000	0.7	3.89
$p < 0.001$ $\omega^2 = 0.10$	No			3.34
Physical domain F = 170.08	Yes	0.000	0.8	28.75
$p < 0.001$ $\omega^2 = 0.15$	No			24.95
Psychological domain F = 67.47	Yes	0.000	0.5	22.51
$p < 0.001$ $\omega^2 = 0.06$	No			20.43
Social domain F = 82.16	Yes	0.000	0.6	11.78
$p < 0.001$ $\omega^2 = 0.08$	No			10.38
Environmental domain F = 55.03	Yes	0.000	0.5	30.53
$p < 0.001$ $\omega^2 = 0.06$	No			28.41

In each case, statistically significant differences were found in quality of life and its components, depending on the employment status ($p < 0.001$ for the t-test). These differences were confirmed by the moderate and above-moderate effect size (g = 0.5 to 0.7), and were very high in the physical domain (g = 0.8).

6. Discussion

Scientific literature provides unequivocal evidence that physical fitness resulting from physical activity is a significant predictor of health [11–14]. Good health and fitness are important determinants of professional and social competences, as well as better quality of life. In Poland, a survey conducted at the beginning of 2019 on a representative random sample of 1858 individuals aged 15 and over revealed that 64% declared partaking in physical activity at least once a month, and 19% exercised systematically, i.e., five times a week. In comparison with the survey of 1800 Poles in the previous year, the 2019 results showed a two percentage point increase in people spending their time actively at least once a month (walking, cycling, running, gym exercises, fitness classes). However, this still means that more than one third of Poles did not engage in any physical activity, even once a month. Most physically active were young people aged 15 to 24 (80%), students (90%), those with university degrees (78%), and those with monthly incomes exceeding PLN 5000 (83%), i.e., 20% above the average wage in Poland [15].

Dependencies between physical activity and quality of life are not always unequivocal. Scientific research has repeatedly shown the positive impact of physical activity on quality of life, such as in university students. The increased physical activity resulted in an improved quality of life [17]. At the same time, some studies do not confirm any relationship between a very high level of physical activity for students and their quality of life and self-esteem, with some papers highlighting the combined influence of other factors, such as gender, age, and marital status [25,26]. A greater effect of frequent exercise on quality of life has been observed in adult pre-diabetics and women with depressive symptoms [27]. Exercise can also improve the quality of life of patients with Alzheimer's disease and following transplants of the liver, kidney, and other organs [28].

A study carried out in the years 2014–2015 among 4460 residents of Wrocław, aged 18–64, determined the relationship between physical activity and quality of life using the IPAQ (short version) and WHOQOL-BREF [29]. The authors found that the highest overall quality of life, health perception, and quality of life in the physical, psychological, social, and environmental domains were reported by respondents of both genders whose activity levels were defined as high (at least 1500 MET/min/week). In the case of low physical activity, the correlations were lower.

The results partly corroborate the hypothesis about the relationships between the levels of physical activity among students, their parents, and their grandparents, the overall quality of their lives in individual domains (physical, psychological, environmental), and their perceived health status. The quality of life of respondents was dependent upon, among other things, their current physical activity levels. This was a result of positive attitudes towards physical education lessons (in the past) and competitive sports (students at present, parents, and grandparents in the past). Those currently engaged in physical recreation were also more satisfied with their health, performing better in the physical, psychological, and environmental domains than those who were not currently physically active. The moderate effect size confirms these differences, especially concerning the physical and psychological domains. Individuals declaring some physical activity performed significantly better in these domains. There was no such dependence in the social domain.

Research carried out in many countries highlights the importance of participating in physical education classes [31–33]. Our research confirms these results. Those currently participating in physical recreation (81.7%) had positive attitudes towards physical education lessons (88.1%). They showed higher physical activity compared to all Poles (just over 70%) [17,34]. The positive attitudes of parents and grandparents likely encouraged interest in physical activity and sports activities among the surveyed students. These hypothetical relationships will be confirmed by further research.

Shaping positive attitudes towards exercise among teenagers is very important (especially among less physically active girls and women), because positive supportive attitudes towards physical activity result in sustained physical activity in later life and improved quality of life [4,35–38]. In one study, fathers and mothers had a similar effect on the level of physical activity of their children, regardless of their gender [39]. Nevertheless, girls do still exhibit lower levels of physical activity [40].

The general results of research somewhat confirm the MultiSport Index (2018), showing that the group of active people was mainly young people, educated people, people from large cities, and students. The highest inactivity rate was among pensioners (59%) and adults over 60 years of age (56%). Age is the factor that most strongly determines physical activity in Poles. The older a person is, the more likely he or she is to be less active [2,41]. The types of physical activity also differed between the sexes, regardless of the generation, which confirms the results of a previous study [37].

Physical activity among the elderly mainly consisted of walking, while sports games, jogging, and the gym were most popular among students. Cycling was declared by 7.7% of the respondents, which is probably connected with their place of residence (almost 80% of the respondents lived in cities) and the low availability of bicycle paths in Poland. During the research, it was found that, in the last six months, 29% of the respondents cycled (Multisport Index 2018). This shows that the promotion of cycling as a form of transport in Poland could increase the overall physical activity of society [42].

Studies on the relationship between physical activity and quality of life emphasize the importance of various modifying factors [43]. Kupcewicz et al. found a relationship between physical activity and quality of life of people over 50, modified by their sense of agency [44]. Boerma et al. demonstrated a relationship between recommended levels of moderate or intensive physical activity and HRQOL (175,850 Americans) and the number of sick leave days used. The percentage of adults reporting at least 14 sick leave days (regardless of age, gender, or ethnic group) was significantly lower among those who achieved the recommended level of physical activity. Achieving the recommended levels of exercise in leisure time in the French population (2333 men and 3321 women) was associated with a higher health-related quality of life. Increased intensity of exercise improved the quality of life of the respondents [10,45]. A greater effect related to the improvement of quality of life was achieved among older people exercising relatively frequently (three times a week) [46,47].

In this study, it was observed that physical activity and quality of life were also modified by age, gender, marital status, education level, and employment status. University students in study exhibited a higher level of physical activity than parents and grandparents. Women were less physically active than men, except for fathers of students, who were currently the least active among men (76.3%), and their quality of life was also lower than that of students' mothers (84.9%). Lower physical activity of students' fathers may have resulted from a high employment rate (over 72%), but, at the same time, 35.7% of fathers declared their previous involvement in competitive sports. Their low participation in physical recreation requires further research.

Many studies confirm the negative impact of age on people's quality of life, but the social, intellectual, and physical activity of older adults, while engaging in activities for the benefit of the local environment, brings a positive increase in quality of life, especially in the physical and social fields [17,29,30,48].

The greatest differences in the quality of life were found between female and male students and their grandmothers and grandfathers, and between mothers and fathers and grandmothers and grandfathers of students. Grandmothers and grandfathers had the lowest satisfaction with life and health, and in individual domains. Grandmothers were less satisfied with their health than grandfathers—otherwise, there were no differences between them in the physical, social, and environmental domains. Among men (fathers and grandfathers), the quality of life was significantly higher than among women (mothers and grandmothers). In the group of students, the impact of gender on quality of life was not so pronounced; it became more apparent in later stages of life.

In a survey conducted in 59 countries, data on the health status of people aged 18 and over (143,363 men and 115,321 women) were presented. Differences between men and women in assessing their health were found in all regions of the world. The authors suggest that combined biological and social factors (gender inequality) are the main causes of lower health assessment by women [45].

Differences between unmarried, married, and cohabiting persons were shown in comparison to widowed and divorced persons, who had the lowest quality of life, health satisfaction, and satisfaction in the physical, social, psychological, and environmental domains. Divorced persons also had a higher quality of life than widowers. These relationships were confirmed by the high and moderate strength of the effect. Villas-Boas, et al. proved that, among the predictors of quality of life for people of different ages, marital status (for middle-aged people) and income (for youth and older people) were important, but the most important was social support (important for all generations) [1].

The level of education in this study directly impacted the reported quality of life of the studied groups of students, their parents, and their grandparents. People with secondary education, especially post-secondary education, had a higher quality of life rating, were more satisfied with their health, and, in the remaining areas, achieved higher scores (moderate effect size).

Education also mitigates the differences in quality of life resulting from marital status [49]. A higher level of education is a positive determinant of quality of life.

Those employed had a higher quality of life, health satisfaction, and better scores in all areas compared to the unemployed (high and moderate effect size). In each case, statistically significant differences were found in the quality of life and its components, depending on the employment status (moderate and high effect size).

Among women, higher satisfaction with health was declared by women who were better educated, employed, or married. According to research conducted by the Ministry of Labour and Social Policy (2008), health status varies significantly depending on the employment status. The highest declared health condition was reported by employees at the age around retirement, 70% of whom described it as good or very good. According to 7% of Poles and 14% of Europeans, poor health may not only make it difficult to work, but also make it impossible to take up or continue a job. Deactivation changes their situation, deprives people of professional contacts, and thus significantly forces them to change their behavior and habits, which in turn reduces their satisfaction with life in retirement [50–52].

The results of the study showed that the grandparents of the respondents often have limitations in the field of physical activity that affect their quality of life. As shown in studies by E. Navarro-Flores et al., aging and chronic diseases, such as hyperglycemic disease, as well as musculoskeletal and cardiac processes, may cause the weakness syndrome and, consequently, become degenerative and show certain changes that may affect mental and general health. For example, aging and weakness may affect one's walking speed and increase the risk of falling due to balance changes. Moreover, the presence of weakness symptoms influences health-related quality of life (HQoL) in this population group [53].

The results of this study, which were discussed on the basis of studies carried out by other authors, clearly showed statistically significant differences in physical activity and quality of life of academic youth, their parents, and their grandparents. The essence of our study is the inclusion of the parents and grandparents of the respondents, which gives a broader picture in terms of the researched subject matter.

The continuation of this study shall include the selection of the sample consisting of students in other fields of study; it will be carried out as part of the research project at selected universities throughout Poland.

7. Conclusions

Statistically significant differences were found in physical activity and quality of life of students, their parents, and grandparents. People who were currently physically active were more satisfied with their health, and achieved better results in the physical, psychological, and environmental domains. There were no such differences in the social domain.

The quality of life of students, their parents, and their grandparents was dependent on age, gender, marital status, level of education, and employment status. Higher quality of life was characteristic of younger, better educated, married or unmarried people, remaining in cohabitation, and being employed. Widows and widowers had the lowest overall quality of life in particular domains, and were the least satisfied with their health.

To improve quality of life in a society, it is necessary to promote various forms of physical activity aimed at meeting the needs of each generation, including gender, age, marital status, level of education, and employment status. Therefore, studies in physical culture and health promotion should educate health and physical activity leaders to be followed by people of all ages.

Practical implications:

Due to the need to increase physical activity of people of different ages (with particular emphasis on academic youth), which has been shown in the study to have a significant impact on quality of life, measures should be taken to focus on health education of the society, which can effectively eliminate the problem of incorrect health behaviors and encourage children, adolescents, adults and the elderly to engage in regular physical activity.

In this study, differences (confirmed by a moderate and high effect size) were found in the quality of life between university students and their grandparents. The most important determinants of the quality of life were marital status, employment status, and level of education. Those who were currently physically active rated the quality of their lives higher than inactive persons, but the effect size was low. We used a questionnaire concerning the attitudes of respondents towards participation in physical education classes, practicing professional sport, and current recreational physical activity. The use of a standardized questionnaire in future studies would enable broader comparisons to be made between the quality of life of physically active and passive people. Moreover, extending the list of determinants of quality of life to include the frequency of family contact and financial situation would help to more precisely establish the significance of physical activity and the life choices made by respondents concerning their quality of life.

Author Contributions: Conceptualization, M.A.N. and K.S.; methodology, M.A.N.; software, K.K.; validation, L.N. and M.S.; formal analysis, P.K. and A.S.; investigation, M.S. and L.N.; resources, K.K.; data curation, K.K.; writing—original draft preparation, M.A.N. and K.S. All authors have read and agreed to the published version of the manuscript.

Funding: This research received no external funding.

Informed Consent Statement: Informed consent was obtained from all subjects involved in the study.The study was conducted according to the guidelines of the Declaration of Helsinki, and approved by the Ethics Committee of Pomeranian Medical University in Szczecin Poland (protocol code: 15/KB/V/2015 date: 10 December 2015).

Data Availability Statement: Data are not publicly available and data sharing is not applicable to this article.

Conflicts of Interest: The authors declare no conflict of interest.

Disclosure Statement: No potential conflict of interest was reported by the authors.

References

1. Villas-Boas, S.; Oliveira, A.L.; Natália Ramos, N. Predictors of Quality of Life in Different Age Groups across Adulthood. *J. Intergenerational Relatsh. (JIG)* **2019**, *17*, 42–57. [CrossRef]
2. Nowak, P.F.; Bożek, A.; Blukacz, M. Physical Activity, Sedentary Behavior, and Quality of Life among University Students. *Biomed Res. Int.* **2019**, *97*, 1281. [CrossRef]
3. Núñez-Rocha, G.M.; López-Botello, C.K.; Salinas-Martínez, A.M.; Arroyo-Acevedo, H.V.; Martínez-Villarreal, R.T.; Ávila-Ortiz, M.N. Lifestyle, Quality of Life, and Health Promotion Needs in Mexican University Students: Important Differences by Sex and Academic Discipline. *Int. J. Environ. Res. Public Health* **2020**, *17*, 8024. [CrossRef]
4. Tanabe, T.; Snyder, A.R.; Bay, R.C.; McLeod, T.C.V. Representative values of health-related quality of life among female and male adolescent athletes and the impact of gender. *Athl. Train. Sports Health Care* **2010**, *2*, 106–113. [CrossRef]
5. Snedden, T.R.; Scerpella, J.; Kliethermes, S.A.; Norman, R.S.; Blyholder, L.; Sanfilippo, J.; McGuine, T.A.; Heiderscheit, B. Sport and Physical Activity Level Impacts Health-Related Quality of Life Among Collegiate Students. *Am. J. Health Promot.* **2019**, *33*, 675–682. [CrossRef] [PubMed]
6. Solis, A.C.; Lotufo-Neto, F. Predictors of quality of life in Brazilian medical students: A systematic review and meta-analysis. *Rev. Bras. Psiquiatr.* **2019**, *41*, 556–567. [CrossRef] [PubMed]
7. Medrano-Ureña, M.D.; Ortega-Ruiz, R.; Benítez-Sillero, J.D. Physical Fitness, Exercise Self-Efficacy, and Quality of Life in Adulthood: A Systematic Review. *Int. J. Environ. Res. Public Health* **2020**, *17*, 6343. [CrossRef]
8. Miko, H.C.; Zillmann, N.; Ring-Dimitriou, S.; Dorner, T.E.; Titze, S.; Bauer, R. Effects of Physical Activity on Health. *Gesundheitswesen* **2020**, *82*, S184–S195. [CrossRef]
9. Ding, D.; Ramirez Varela, A.; Bauman, A.E.; Ekelund, U.; Lee, I.M.; Heath, G.; Katzmarzyk, P.T.; Reis, R.; Pratt, M. Towards better evidence-informed global action: Lessons learnt from the Lancet series and recent developments in physical activity and public health. *Br. J. Sports Med.* **2020**, *54*, 462–468. [CrossRef] [PubMed]
10. Sevil, J.; Práxedes, A.; Abarca-Sos, A.; Del Villar, F.; García-González, L. Levels of physical activity, motivation and barriers to participation in university students. *J. Sports Med. Phys. Fit.* **2016**, *56*, 1239–1248.
11. Brown, D.W.; Balluz, L.S.; Heath, G.W.; Moriarty, D.G.; Ford, E.S.; Giles, W.H.; Mokdad, A.H. Associations between recommended levels of physical activity and health-related quality of life. Findings from the 2001 Behavioral Risk Factor Surveillance System (BRFSS) survey. *Prev. Med.* **2003**, *37*, 520–528. [CrossRef]

12. Tittlbach, S.A.; Jekauc, D.; Schmidt, S.C.; Woll, A.; Bös, K. The relationship between physical activity, fitness, physical complaints and BMI in German adults–results of a longitudinal study. *Eur. J. Sport Sci.* **2017**, *17*, 1090–1099. [CrossRef] [PubMed]
13. Bös, K.; Tittlbach, S.; Woll, A.; Suni, J.; Oja, P. FinGer–Physicalactivity, fitness and health—An international longitudinal study in Bad Schönborn and Tampere. *Int. Sports Stud.* **2012**, *34*, 42–50.
14. Rożek-Piechura, K.; Ignasiak, Z.; Sławińska, T.; Piechura, J.; Ignasiak, T. Respiratory function, physical activity and body composition in adult rural population. *Ann. Agric. Environ. Med.* **2014**, *21*, 369–374. [CrossRef] [PubMed]
15. Physical activity of the society 2019. Physical and sporting activity of Poles. *MultiSport Index*. 2019. Available online: https://www.benefitsystems.pl/en/about-us/press-centre/press-release/multisport-index-2019-survey-of-physical-activity-among-poles/ (accessed on 11 December 2020).
16. Koistinen, S.; Olai, L.; Ståhlnacke, K.; Fält, A.; Ehrenberg, A. Oral health-related quality of life and associated factors among older people in short-term care. *Int. J. Dent. Hyg.* **2019**, *18*, 163–172. [CrossRef]
17. Çiçek, G. Quality of Life and Physical Activity among University Students. *Univers. J. Educ. Res.* **2018**, *6*, 1141–1148. [CrossRef]
18. Allender, S.; Hutchinson, L.; Foster, C. Life-change events and participation in physical activity: A systematic review. *Health Promot. Int.* **2008**, *23*, 160–172. [CrossRef]
19. Wang, L.; Qi, J. Association between Family Structure and Physical Activity of Chinese Adolescents. *BioMed Res. Int.* **2016**, *2016*, 1–7. [CrossRef]
20. Levasseur, M.; Desrosiers, J.; St-Cyr Tribble, D. Do quality of life, participation and environment of older adults differ according to level of activity? *Health Qual. Life Out.* **2018**, *6*, 30–34. [CrossRef]
21. Kruk, J. Health and Economic Costs of Physical Inactivity. *Asian Pac. J. Cancer Prev.* **2014**, *15*, 7499–7503. [CrossRef]
22. Ribeiro, O.; Teixeira, L.; Araújo, L.; Rodríguez-Blázquez, C.; Calderón-Larrañaga, A.; Forjaz, M.J. Anxiety, Depression and Quality of Life in Older Adults: Trajectories of Influence across Age. *Int. J. Environ. Res. Public Health* **2020**, *4*, 9039. [CrossRef] [PubMed]
23. Keating, X.D.; Guan, J.; Piñero, J.C.; Bridges, D.M. A meta-analysis of college students' physical activity behaviors. *J. Am. Coll. Health* **2005**, *54*, 116–125. [CrossRef] [PubMed]
24. Włodarczyk, K. Quality of life perceived by Poles in the 21st century. *Konsumpcja Rozw.* **2015**, *1*, 3–16.
25. Gruszczyńska, M.; Skorupa, P. Physical activity, quality of life and self-esteem. *Probl. Hig. Epidemiol.* **2018**, *99*, 69–73.
26. An, H.Y.; Chen, W.; Wang, C.W.; Yang, H.F.; Huang, W.T.; Fan, S.Y. The Relationships between Physical Activity and Life Satisfaction and Happiness among Young, Middle-Aged, and Older Adults. *Int. J. Environ. Res. Public Health* **2020**, *17*, 4817. [CrossRef]
27. Xu, H.; Tang, L.; Hu, Z.; Gao, F.; Yang, Y.; Qin, L.; Luo, B.-A. Association between physical activity and health-related quality of life in elderly individuals with pre-diabetes in rural Hunan Province. China: A cross-sectional stud. *BMJ Open* **2018**, *8*, 143–148. [CrossRef]
28. Kotarska, K.; Wunsch, E.; Raszeja-Wyszomirska, J.; Kempińska-Podhorodecka, A.; Wójcicki, M.; Milkiewicz, P. Leisure time physical activity and health-related behaviours after liver transplantation: A prospective, single centre study. *Gastroenterol. Rev.* **2015**, *2*, 1–5. [CrossRef]
29. Puciato, D.; Rozpara, M.; Borysiuk, Z. Physical Activity as a Determinant of Quality of Life in Working-Age People in Wrocław, Poland. *Int. J. Environ. Res. Public Health* **2018**, *15*, 623. [CrossRef]
30. Navarro-Flores, E.; Losa-Iglesias, M.E.; Becerro-de-Bengoa-Vallejo, R.; López-López, D.; Rodríguez-Sanz, D.; Palomo-López, P.; Calvo-Lobo, C. Translation and Test—Retest of the Spanish Podiatry Health Questionnaire (PHQ-S). *Int. J. Environ. Res. Public Health* **2018**, *15*, 2205. [CrossRef] [PubMed]
31. Hinkle, D.E.; Wiersma, W.; Jurs, S.G. *Applied Statistics for the Behavioral Sciences*, 5th ed.; Houghton Mifflin: Boston, MA, USA, 2003.
32. R Core Team. *R: A Language and Environment for Statistical computing*; R Foundation for Statistical Computing: Vienna, Austria, 2017. Available online: https://www.R-project.org/ (accessed on 20 December 2020).
33. Castro, L.; Svastisalee, C.; Mendes, R.; Fontaine, O.; Breda, J. School-based physical activity and good practices in Europe. *Health Probl. Civiliz.* **2019**, *13*, 9–18. [CrossRef]
34. Word Health Organization. *Global Recommendations on Physical Activity for Health*; WHO: Geneva, Switzerland, 2010.
35. Barengo, N.C.; Nissinen, A.; Tuomilehto, J.; Pekkarinen, H. Twenty-five-year trends in physical activity of 30- to 59-year-old populations in eastern Finland. *Med. Sci. Sports Exer.* **2002**, *34*, 1302–1307. [CrossRef] [PubMed]
36. Beenackers, M.A.; Kamphuis, C.B.; Giskes, K.; Brug, J.; Kunst, A.E.; Burdorf, A.; van Lenthe, F.J. Socioeconomic inequalities in occupational, leisure-time, and transport related physical activity among European adults: A systematic review. *Int. J. Behav. Nutr. Phy. Act.* **2012**, *29*, 116–120. [CrossRef]
37. Nowak, M.A.; Kotarska, K.; Nowak, L. Physical Activity, Health and Physical Fitness of Students, Their Parents and Grandparents. *Coll. Antropol.* **2019**, *43*, 1–10. Available online: https://hrcak.srce.hr/223978 (accessed on 18 December 2020).
38. Al-Naggar, R.A.; Osman, M.T.; Musa, R. Quality of Life among University Students in a Single Malaysian Institute. *Pensee J.* **2003**, *75*, 123–127.
39. Maia, J.; Gomes, T.N.; Trégouët, D.A.; Katzmarzyk, P.T. Familial Resemblance of Physical Activity Levels in The Portuguese Population. *J. Sci. Med. Sport* **2014**, *17*, 381–386. [CrossRef] [PubMed]
40. Ajman, H.; Ukić, M.; Madić, D. The relationship between family socio-economic status, family social support and adolescent physical activity. *Health Probl. Civiliz.* **2019**, *13*, 48–55. [CrossRef]

41. Central Statistical Office (CSO). *Participation in Sports and Physical Recreation in 2016–2017. Statistical Information and Elaborations*; Central Statistical Office (CSO): Warsaw, Poland. Available online: http://stat.gov.pl (accessed on 6 November 2020).
42. Hosforda, K.; Fuller, D.; Lear, A.; Teschke, K.; Gauvin, L.; Brauer, M. Evaluation of the impact of a public bicycle share program on population bicycling in Vancouver, BC. *Prev. Med. Rep.* **2018**, *12*, 176–181. [CrossRef]
43. Gill, D.L.; Hammond, C.C.; Reifsteck, E.J.; Jehu, C.M.; Williams, R.A.; Adams, M.A.; Lange, E.H.; Becofsky, K.; Rodriguez, E.; Ya-Ting, S. Physical Activity and Quality of Life. *J. Prev. Med. Public Health* **2013**, *46*, 28–34. [CrossRef]
44. Kupcewicz, E.; Grochans, E.; Kadučáková, H.; Mikla, M.; Jóźwik, M. Analysis of the Relationship between Stress Intensity and Coping Strategy and the Quality of Life of Nursing Students in Poland, Spain and Slovakia. *Int. J. Environ. Res. Public Health* **2020**, *24*, 4530. [CrossRef]
45. Boerma, T.; Hosseinpoor, A.R.; Verdes, E.; Chatterji, S. A global assessment of the gender gap in self-reported health with survey data from 59 countries. *BMC Public Health* **2016**, *16*, 675–679. [CrossRef]
46. Grasdalsmoen, M.; Eriksen, H.R.; Lønning, K.J.; Sivertsen, B. Physicalexercise, mental health problems, and suicide attempts in university students. *BMC Psychiatry.* **2020**, *20*, 175–180. [CrossRef]
47. Codina, N.; Pestana, J.V.; Valenzuela, R.; Giménez, N. Procrastination at the Core of Physical Activity (PA) and Perceived Quality of Life: A New Approach for Counteracting Lower Levels of PA Practice. *Int. J. Environ. Res. Public Health* **2020**, *17*, 3408. [CrossRef]
48. Grzanka-Tykwińska, A.; Chudzińska, M.; Kędziora-Kornatowska, K. Evaluating the quality of life of elderly people who attend classes at the University of the Third Age. *Med. Biol. Sci.* **2014**, *28*, 19–24. [CrossRef]
49. Czapiński, J.; Panek, T. Social Diagnosis 2015.The objective and subjective quality of life in Poland Report. *Social Monit. Council.* **2015**, *9*, 429–453.
50. Bouchard, C.; Blair, S.N.; Haskell, W.L. *Physicalactivity and Health*, 2nd ed.; Human Kinetics: Champaign, IL, USA, 2012; p. 456.
51. Konopack, J.F.; McAuley, E. Efficacy-media-ted effects of spirituality and physical activity on quality of life: A path analysis. *Health Qual. Life Outcomes* **2012**, *10*, 57–62. [CrossRef]
52. Rugbeer, N.; Ramklass, S.; Mckune, A.; van Heerden, J. The effect of group exercise frequency on health related quality of life in institutionalized elderly. *Pan Afr. Med. J. (PAMJ)* **2017**, *26*, 35. [CrossRef] [PubMed]
53. Navarro-Flores, E.; Romero-Morales, C.; Becerro de Bengoa-Vallejo, R.; Rodríguez-Sanz, D.; Palomo-López, P.; López-López, D.; Losa-Iglesias, M.E.; Calvo-Lobo, C. Sex Differences in Frail Older Adults with Foot Pain in a Spanish Population: An Observational Study. *Int. J. Environ. Res. Public Healthy* **2020**, *17*, 6141. [CrossRef]

International Journal of
Environmental Research and Public Health

MDPI

Article

A Pilot Study of Influence of Endurance Training on the Prooxidative and Antioxidant Status of Women after Breast Cancer

Katarzyna Domaszewska [1,*], Arkadiusz Janiak [2], Tomasz Podgórski [1], Anna Demuth [3], Jakub Kryściak [1], Paweł Perkowski [4] and Urszula Czerniak [3]

1 Department of Physiology and Biochemistry, Poznan University of Physical Education, 61-871 Poznań, Poland; podgorski@awf.poznan.pl (T.P.); j.krysciak@awf.poznan.pl (J.K.)
2 Faculty of Health Sciences, Calisia University, 62-800 Kalisz, Poland; a.janiak@akademiakaliska.edu.pl
3 Department of Anthropology and Biometry, Poznań University of Physical Education, 61-871 Poznań, Poland; demuth@awf.poznan.pl (A.D.); czerniak@awf.poznan.pl (U.C.)
4 Province Polyclinic Hospital, 62-504 Konin, Poland; paper5@tlen.pl
* Correspondence: domaszewska@awf.poznan.pl; Tel.: +48-(61)-835-51-90

Abstract: The objective of this study was to assess the effect of 8-week long endurance training on the prooxidative–antioxidative status of plasma in women treated for breast cancer. The participants of the study were 12 women after radical mastectomy aged 45 to 56 years (M = 50.6 ± 2.9 years), who had undergone full cancer treatment, on average more than 5 years after the treatment (M = 5.9 ± 0.9 years). Body mass components were measured twice using the method of bioelectric impedance analysis. In order to optimize training loads and to assess the level of exercise tolerance of the participants, the group was subject to an ergospirometric exercise test twice, before (1st) and after (2nd) the completion of the training cycle. The blood was also taken twice for biochemical analyses. Statistically significant differences were noted in the maximum exercise load, the level of which increased in the second test ($p < 0.05$). No change was observed in the level of antioxidative potential, i.e., the content of some variables, ferric reducing ability of plasma (FRAP), urea, total phenolics, thiobarbituric acid reactive substances (TBARS), and in the blood indices of the body's nutritional status during the project (total protein, albumin. Endurance training caused an increase in exercise tolerance and did not cause an aggravation of oxidative stress in women undergoing breast cancer treatment.

Keywords: FRAP; TBARS; anaerobic threshold; peakVO_2

Citation: Domaszewska, K.; Janiak, A.; Podgórski, T.; Demuth, A.; Kryściak, J.; Perkowski, P.; Czerniak, U. A Pilot Study of Influence of Endurance Training on the Prooxidative and Antioxidant Status of Women after Breast Cancer. *IJERPH* **2021**, *18*, 2822. https://doi.org/10.3390/ijerph18062822

Academic Editors: Wioletta Zukiewicz-Sobczak and Veronique Billat

Received: 6 January 2021
Accepted: 8 March 2021
Published: 10 March 2021

Publisher's Note: MDPI stays neutral with regard to jurisdictional claims in published maps and institutional affiliations.

Copyright: © 2021 by the authors. Licensee MDPI, Basel, Switzerland. This article is an open access article distributed under the terms and conditions of the Creative Commons Attribution (CC BY) license (https://creativecommons.org/licenses/by/4.0/).

1. Introduction

Breast cancer is the world's most common cancer in women, and it accounts for as many as 25% of the total number of cancer cases in this population [1]. Breast cancer treatment should be carried out by a multidisciplinary team, with the participation of an oncology surgeon, clinical oncologist, and radiation therapist in particular. Basic treatment involves surgery, which has gradually evolved towards a less aggressive form over the recent decades. The decision concerning the choice of the type of breast cancer treatment is made on the basis of the evaluation of the stage of the disease according to the Tumor–Node-Metastasis (TNM) classification of malignant tumors, the biological subtype, the patient's overall condition, and her preferences (e.g., type of surgical treatment). Adjuvant therapy is a systemic pharmacological treatment: chemotherapy, hormone therapy, and anti-HER2 therapy [2]. Early consequences of surgical treatment are related to temporary immobility and a postoperative wound, which lead to the disturbed function of the respiratory system, limited chest mobility, and weakening of the muscular strength. Circulatory insufficiency manifests itself, among others, in orthostatic intolerance [3]. Chemotherapy and radiotherapy as additional treatments result in an increased amount of free oxygen radicals and a decrease in the concentration of antioxidants in the tissue and serum. In natural conditions, such reactive oxygen species are formed during stress, in the course of diseases, during physical exercise, or as a result of harmful effects on the environment, among other things. Excessive amounts of reactive oxygen species (ROS) in damaged cells are destructive for the neighboring cells and disturb their functions [4]. A long-term inflammation leads to tissue anoxia and the production of even larger amounts of free radicals, which are not oxygen derivatives. A body constantly exposed to the action of free radicals develops a number of enzymatic and nonenzymatic mechanisms to prevent or limit the damage caused by that. The most significant enzymatic mechanisms are superoxide dismutase (SOD), catalase (CAT), glutathione peroxidase (GPX), and glutathione reductase (GR). Nonenzymatic mechanisms include, among other things, antioxidants, such as vitamin C (ascorbic acid), glutathione (GSH), tocopherol, free radical scavengers (adrenaline, bilirubin, uric acid), metal ions, i.e., iron, copper, zinc, manganese, selenium building active enzyme centers as well as ceruloplasmin, transferrin, ferritin [5,6]. Many publications indicate that a higher blood concentration of total phenolics protects the body against many civilization-related diseases (sclerosis, cancers) [7]. Moreover, it has been demonstrated that in patients receiving chemotherapy, both for hematological and solid cancers, leukocytes produce more free oxygen radicals than in healthy people. In patients treated with high-dose chemotherapy, a decrease of plasma concentration of antioxidants, such as vitamin C, alpha-tocopherol, and beta-carotene, was noted [8]. At the same time, a significant decrease in the level of GSH (significant endogenous antioxidant) was noted in patients treated with busulfan, carmustine, and cisplatin [8]. In patients, who were administered daunorubicin and cytarabine for acute myeloid leukemia, an increase in the level of MDA-malondialdehyde and a significant fall in SOD, GPX, and total antioxidant status (TAS), important components of the defense system against free radicals, was noted (second drug). At the moment, the problem of cellular antioxidant potential is not routinely considered in the planning of cancer treatment and during its course. It has, however, an indirect link with cancer patient nutrition. We know that a mixture of vitamins and microelements has strong properties supporting the antioxidant response. It is suspected that the application of this type of therapy during some methods of systemic treatment may weaken the cytotoxic effect of chemotherapy, as the escalation of oxidative stress is mentioned in the mechanism of cytotoxic action of some chemotherapeutics [9]. Studies show a close relation between free radicals, the activity of antioxidative processes and their effectiveness correlating with the effectiveness of applied chemotherapy.

The rehabilitation process following a mastectomy includes, among other things, psychological rehabilitation, physical rehabilitation, i.e., physiotherapy, and socio-professional rehabilitation. The need for rehabilitation exercises during cancer treatment is undisputed today. The objective of kinesitherapy is to restore the full range of motion in the joint on the side of the operated limb and prevent lymphedema. Over recent years, increasing attention is given to the role of physical activity undertaken by women after the completion of cancer treatment [10]. Systematic physical training results in improved fitness and physical capacity of the body increased immunity, e.g., through strengthening the antioxidative status of the body. The objective of the study was the assessment of the effect of an 8-week long endurance training on the prooxidative–antioxidative status of plasma in women treated for breast cancer.

2. Materials and Methods

The participants of the study were 12 women after radical mastectomy aged 45 to 56 years (M = 50.6 ± 2.9 years) who had undergone full cancer treatment. All participants were in the second and third stages of the disease before starting the treatment. After the surgery, they were administered chemotherapy based on anti-cyclins and taxoids. The type of treatment scheme and treatment duration depended on the stage of cancer according to the TNM staging system (stage IIB and IIIA). The next stage of treatment was radiotherapy. A dose of 50 Gy was routinely applied in 25 fractions on the whole breast and in 10–25 fractions at the site of tumor resection. Each patient had also completed hormonal treatment with tamoxifen (20 mg once a day) and additional treatment with gonadotropin-releasing hormone agonist depending on the stage, presence of risk of relapse and patient's age (for example, goserelin 3.6 mg administered subcutaneously every 28 days). On average, women taking part in the program had completed cancer treatment more than 5 years prior to the study (M = 5.9 ± 0.9 years) and did not have health contraindications for undertaking physical activity. The studied group was asked not to take vitamins or diet supplements, which could affect the measured biochemical blood indicators in the period before and during the study. The subjects were also asked not to change their dietary habits for the duration of the project and not to perform any additional physical activity, except for the implementation of the research scheme. The study was conducted according to the Declaration of Helsinki and the National Statement and Human Research Ethics Guidelines. It was also approved by the Institute for Research in Biomedicine (IRB) at the Poznan University of Medical Sciences (10 May 2012; Ethics Approval Number: 535/12). The study was performed between June and October 2012 in an accredited endurance test laboratory of the Poznan University of Physical Education. All of the patients from the sample group gave their written informed consent.

2.1. Anthropometrics Measurements

Body mass components were measured twice using the method of bioelectric impedance analysis (BIA) by means of Akern—BIA 101 (Akern Bioresearch, Florence, Italy) body composition analyzer [11]. In order to perform body composition analysis, measurements of two basic somatic properties were used (body height in cm and body weight in kg) to calculate the body mass index (BMI; kg/m^2). The following body mass components were assessed: the levels of body fat (FAT; kg), lean body mass (LBM; kg), and total body water (TBW; L). All measurements were taken in the morning from 7 am; patients were fasting; after taking the blood, they received the same light meal, and after a break, they proceeded to the exercise test. They were instructed not to consume alcohol for at least 48 h before the test, not to do vigorous physical exercise for at least 12 h before, not to use a sauna for at least 12 h before, and to void the bladder 30 min before the test. Patients with ICDs (defibrillators), pacemakers or metal implants, patients with epilepsy, hemiparesis, with wounds or skin lesions on hands or feet were excluded from the study.

2.2. International Physical Activity Questionnaire

The level of physical activity was determined in accordance with the criteria of the international physical activity questionnaire (IPAQ). Women who qualified for the project were characterized by low and moderate levels of physical activity. The level of physical activity was categorized according to the criteria of the international physical activity questionnaire. The following levels were adopted: (1) low—physical activity does not meet the criteria for moderate or intensive activity; (2) moderate—physical activity meets at least one of the following criteria: three or more days with at least 20 min of vigorous physical activity; five or more days with at least 30 min of moderate-intensity activity or walking; five or more days of any combination of physical activity with at least 600 MET-min/week, (3) high—three or more days with intensive physical activity exceeding 1500 MET-min/week or five or more days of any combination of physical exercise (walking, moderate-intensity or vigorous activities) exceeding 1500 MET-min/week [12].

2.3. Assessment of Aerobic Fitness (Graded Exercise Test Protocol)

The exercise tests were conducted between 8:00 am and noon in an air-conditioned laboratory, 2 h after consuming a light breakfast. The duration of the physical exercise was 20 min. In order to optimize training loads and to assess the level of exercise tolerance of the participants, the group was subject to the ergospirometric exercise test twice, before (1st) and after (2nd) the completion of the training cycle. The test was performed on a cycle ergometer (Kettler DX1 Pro, Ense, Germany): a three-minute warm-up with a 25 W load, after which the load was increased by 10 W every 90 s (60 revolutions per minute (RPM)).

The test was performed until refusal or impossibility to maintain the set pace. The exercise tests were conducted between 8:00 am and noon in an air-conditioned laboratory, 2 h after consuming a light breakfast (one sandwich with butter and cheese; approx. 200 kcal).

2.4. Physiological Measurements

Expired gases, minute ventilation (Ve) and heart rate (HR) during graded exercise test (GXT) were monitored continuously with an Oxycon mobile automated system (Viasys Healthcare, Höchberg, Germany). Oxygen intake (VO_2) and carbon dioxide output (VCO_2) were measured breath-by-breath and were averaged at 15 s time periods. Before each trial, the system was calibrated according to the manufacturer's instructions. Ambient conditions, i.e., temperature, humidity and barometric pressure, were recorded by the sensors. In a two-point volume calibration (0.2 and 2 L/s), the flow values were measured automatically at the set measuring points. Gas analyzer calibration was carried out with a standard gas mixture of 5% CO_2 and 16% O_2. PeakVO_2 was defined as the highest 15 s averaged VO_2 obtained during the last exercise in the test. HRmax (beats/min) was measured as the highest 15 s averaged value in the test. Time to exhaustion (TTE) and maximal work rate (WRmax) were also measured. The individual training load at the level of PPA (anaerobic threshold) was established using the "V-slope" method. To determine ventilatory threshold (VT), the V-slope method was administered using computerized regression analysis of the slopes of the CO_2 output versus O_2 uptake plot, which detects the beginning of the excess CO_2 output generated from the buffering of [H^+] [13]. The method involves the analysis of the behavior of VCO_2 as a function of VO_2 during GXT with a consequent increase in VCO_2. This resulted in a transition in the relation between the VCO_2 and VO_2. The software supplied by Viasys Healthcare was supported by visual inspection by an experienced researcher. The ventilatory equivalent method (VEQ method) was used as a secondary method, and the point when the equivalent for oxygen (VE/VO_2) raised without a concomitant rise in the equivalent for carbon dioxide (VE/VCO_2) was detected [14–19]. The VT was expressed as work rate (W), heart rate (bpm) and time when the VT was reached.

The entire study protocol and results from the cardiopulmonary exercise testing (CPET) are thoroughly described in our previous paper [20].

2.5. Blood Collection, Biochemical Measurements

The blood was taken twice for biochemical analyses -at rest, at the beginning, and after the completion of the training program. Blood samples (10 mL) were taken from a vein in the forearm using an S-Monovette blood collection system (Sarstedt, Nümbrecht, Germany). 9 mL of the blood was left to clot at room temperature, then centrifuged (1500 g, 4 °C, 5 min). The collected samples were frozen and stored at −80 °C until the analysis. Blood cell count tests were performed on fresh blood. Total protein, albumin, and urea concentrations were determined directly after taking and centrifuging the blood on the day of the test. Plasma for determining FRAP, TBARS, and total phenolics concentrations was frozen for two months. Biochemical determinations were performed on a sample defrosted once. The following parameters were determined in whole venous blood (1 mL in microtube EDTA K3; Sarstedt, Germany), using MYTHIC 18 hematological analyzer (PZ CORMAY S.A., Lublin, Poland): hemoglobin concentration, hematocrit value, the total count of erythrocytes, leukocytes, thrombocytes, and relative percentage numbers of granulocytes, lymphocytes, and monocytes. Colorimetric methods were used to determine the concentrations of the total antioxidant capacity of plasma FRAP, reference values: 600–1600 µmol/L [21], the plasma concentration of TBARS, reference values: 1–6 µmol/L) [22] and total phenolics (reference values: 2.8–4.0 g GAE/L) [23]. All reagents used for measurements of the above parameters were obtained from (Sigma-Aldrich Chemie GmbH., Steinheim, Germany). Moreover, total protein, albumin and urea concentrations were assessed using the spectrophotometric methods with commercially available reagent kits: Liquick Cor-TOTAL PROTEIN (cat. no. 2-236, sensitivity 1.0 g/L), Liquick Cor-ALBUMIN (cat. no. 2-238, sensitivity 11.4 g/L) and

Liquick Cor-UREA (cat. no. 2-261, sensitivity 0.27 mmol/L) diagnostic sets (PZ CORMAY S.A., Poland), respectively. The samples were read using a Synergy 2 SIAFRT multimode microplate reader (BioTek, Winooski, VT, USA). All pre-training and post-training specimens from each individual were analyzed in the same batch by an experienced technician, who was blinded to the origin of samples.

2.6. Training Program

The training program offered to participants was performed on a cycle ergometer for two months, three times a week. In total, 28 training sessions, each, including a 5 min warm-up (50–60% HRmax), 30–45 min of the proper part (at the level of individual PPA), a 5 min warm-down, cycling without a load. In the end, each patient performed stretching and breathing exercises for about 15 min. Exercise loads were determined individually on the basis of the ergospirometric exercise test. Each patient had undergone all designated training sessions, which always took place in the morning. The training program and exercise tests were carried out using the same ergometer. Physical training was supervised by a physiotherapist with a medical background.

2.7. Statistical Analyses

The distribution of variables was tested by Shapiro–Wilk test of normality. All the data are presented as mean (standard deviation) and median (interquartile range). The differences between paired, normally distributed variables were tested using the paired t-test and, in the case of asymmetrically distributed variables, by the Wilcoxon test. The relationship between the variables was tested while using Pearson's correlation. All results were statistically analyzed using Dell Inc. (2016) (Dell Statistica v.13, Tulsa, OK, USA).

3. Results

A group of women after a completed breast cancer treatment was examined twice, before and after an 8-week long endurance training, in order to assess its effect on the prooxidative–antioxidative status of plasma. The statistical analysis performed showed consistency with the normal distribution of most of the variables analyzed in the study, with the exception of three parameters (TBARS, neutrophils, monocytes), for which the results of parametric (paired *t*-test) and nonparametric tests (Wilcoxon matched-pairs set) were consistent. Thus, in further analysis, the results of the paired t-test were considered. The women were characterized by an average height of 164.9 cm (Me = 163.5 cm, SD = 2.9 cm, Q1 = 159 cm, Q3 = 171.5 cm). A comparison of somatic parameters (Table 1) measured in the first test and in the second one demonstrated that no significant changes in the body mass or in body composition had occurred. Endurance training resulted in an increase in the peak oxygen uptake, but no statistical significance was achieved ($p < 0.05$).

Table 1. Somatic and physical capacity parameters in sessions 1 and 2.

Parameter	1st Date ($n = 12$) Mean ± SD Me ± Q_1–Q_3	2nd Date ($n = 12$) Mean ± SD Me ± Q_1–Q_3	p-Value
Body mass (kg)	66.50 ± 8.12 65.0 ± 59.5–73.0	65.75 ± 8.18 65.5 ± 61.5–69.5	0.3568
BMI (kg/m^2)	24.45 ± 2.27 24.8 ± 23.2–25.3	24.28 ± 2.28 25.5 ± 23.6–25.2	0.4834
LBM (kg)	45.32 ± 3.78 45.5 ± 43.2–47.7	44.67 ± 3.88 45.9 ± 42.2–47.6	0.1992
TBW (L)	33.18 ± 2.76 33.4 ± 31.6–34.9	32.68 ± 2.84 33.6 ± 30.9–34.8	0.1808

Table 1. *Cont.*

Parameter	1st Date (n = 12) Mean ± SD Me ± Q_1–Q_3	2nd Date (n = 12) Mean ± SD Me ± Q_1–Q_3	p-Value
FAT (kg)	21.25 ± 5.22 19.0 ± 17.9–25.1	21.28 ± 5.50 19.5 ± 18.8–22.8	0.9722
$_{VT}$ HR (beat/min)	127.75 ± 13.07 124.0 ± 119.0–139.0	142.25 ±13.06 144.0 ± 138.0–148.0	<0.0001
$_{VT}$ Load (W)	76.67 ± 13.37 75.0 ± 75.0–85.0	94.17 ± 14.29 90.0 ± 85.0–105.0	0.011
peak HR (beat/min)	158.92 ± 15.37 162.0 ± 144.0–171.5	166.50 ± 13.56 170.0 ± 162.5–175.5	0.2134
peakVO$_2$ (mL/kg/ min)	25.74 ± 4.04 24.4 ± 22.9–28.7	27.00 ± 3.68 26.3 ± 23.4–30.6	0.1658
peak Load (W)	112.50 ± 23.01 105.0 ± 100.0–130.0	123.33 ± 22.09 120.0 ± 105.0–140.0	0.0155

Data are expressed as mean ± SD and Me ± Q1-Q3, * $p \leq 0.05$ vs. II session; abbreviations: BMI = body mass index, LBM = lean body mass, TBW = total body water, FAT = fat body mass, $_{VT}$ HR = heart rate on VT, $_{VT}$ Load = load on VT, peak HR = peak heart rate, peakVO$_2$ = peak oxygen uptake.

A comparison of values of selected hematological and biochemical blood parameters measured in the first and in the second test is shown in Table 2. Statistically significant differences were noted in the maximum exercise load, the level of which increased in the second test ($p < 0.05$). A slight increase was noted in the level of antioxidative potential, i.e., the content of some variables (FRAP, urea), with no changes or a slight decrease of other biochemical parameters of the blood (total phenolics, TBARS). No change was observed in the blood indices of the body's nutritional status during the project (total protein, albumin).

Table 2. Values of selected hematological and biochemical blood parameters in sessions 1 and 2.

Parameter	1st Date (n = 12) Mean ± SD Me ± Q_1-Q_3	2nd Date (n = 12) Mean ± SD Me ± Q_1-Q_3	p-Value
Erythrocytes (10^{12}/L)	4.50 ± 0.40 4.6 ± 4.3–4.8	4.67 ± 0.31 4.7 ± 4.5–4.9	0.0620
Hematocrit (%)	41.1±3.0 40.7 ± 40.3–42.3	42.1 ± 2.1 42.4 ± 40.4–43.5	0.1402
Hemoglobin (mmol/L)	8.89 ± 0.75 8.7 ± 8.6–9.1	9.02 ± 0.44 9.1 ± 8.7–9.4	0.4573
Leukocytes (10^9/L)	5.32 ± 1.37 5.5 ± 4.8–6.0	5.89 ± 1.53 5.6 ± 4.8–7.1	0.0765
Neutrophils (%)	58.34 ± 10.43 60.6 ± 52.7–63.2	62.27 ± 10.90 65.6 ± 56.9–67.8	0.0843
Lymphocytes (%)	36.13 ± 10.36 32.8 ± 31.7–40.3	32.70 ± 10.12 30.1 ± 27.1–38.9	0.1095
Monocytes (%)	5.53 ± 1.41 5.2 ± 4.9–5.6	5.03 ± 1.15 4.8 ± 4.3–5.4	0.2132
Total proteins (g/L)	68.30 ± 9.46 68.4 ± 60.4–75.9	72.95 ± 14.02 73.7 ± 61.8–85.4	0.2593
Albumins (g/L)	37.62 ± 3.90 36.7 ± 34.4–40.9	39.79 ± 7.29 38.9 ± 35.9–44.9	0.1857

Table 2. *Cont.*

Parameter	1st Date (n = 12) Mean ± SD Me ± Q_1-Q_3	2nd Date (n = 12) Mean ± SD Me ± Q_1-Q_3	p-Value
Total phenolics (g GAE/L)	2.44 ± 0.09 2.4 ± 2.4–2.5	2.43 ± 0.28 2.4 ± 2.3–2.7	0.9199
FRAP (μmol/L)	857.25 ± 147.17 873.8 ± 750.4–945.6	859.67 ± 148.65 846.5 ± 750.0–973.2	0.9554
TBARS (μmol/L)	5.09 ± 2.09 4.8 ± 4.1–5.3	5.02 ± 1.81 5.0 ± 3.9–5.4	0.6949
Urea (mmol/L)	3.32 ± 2.09 3.4 ± 2.6–4.0	3.71 ± 1.99 3.9 ± 3.0–4.7	0.5427

Data are expressed as mean ± SD and Me ± Q1-Q3, * $p \leq 0.05$ vs. II session; abbreviations: FRAP = ferric reducing ability of plasma, TBARS = thiobarbituric acid reactive substances.

The analysis of changes in hematological blood parameters (Table 2) did not indicate substantial significant changes after the completion of the training program by the participants. An increase in the mean concentration of most of the variables was noted, while no change was noted in lymphocytes and monocytes counts compared to the average starting level.

The analysis of relationships between the studied variables, based on the assessment of the strength and direction of relations described by Pearson's correlation, showed mutual relations. Peak VO_2 on both dates of tests showed a high negative correlation coefficient with BMI (1st test r = −0.7152, $p < 0.05$; 2nd test r = −0.6181, $p < 0.05$) and positive correlations with the value of maximum load in the 1st test (r = 0.8503, $p < 0.05$) and 2nd test (r = 0.7923, $p < 0.05$). In the first test TBARS concentration showed correlation with erythrocyte (r = −0.6256, $p < 0.05$), hematocrit (r = −0.6952, $p < 0.05$), TBARS levels (r = −0.7300, $p < 0.05$). FRAP concentration on both dates of tests correlated with the protein and total phenolics blood concentration, 1st test (r = 0.8280, $p < 0.05$) for total phenolics and (r = 0.8155, $p < 0.05$) for proteins, respectively. In 2nd test (r = 0.8293, $p < 0.05$) for total phenolics and (r = 0.8293, $p < 0.05$) for proteins, respectively. Total phenolics levels on each date of tests showed high correlation with protein blood concentration (1st test r = 0.9744, $p < 0.001$ and 2nd test r = 0.9744, $p < 0.001$). Only in the 1st test was there correlation between peakVO_2 and neutrophils (%) (r = 0.6071, $p < 0.05$) and lymphocytes (%) (r = −0.6717, $p < 0.05$). In the 1st test was correlation between Load $_{VT}$ and LBM (r = 0.8473, $p < 0.05$), HR $_{VT}$ and erythrocytes (r = −0.5867, $p < 0.05$), Load $_{VT}$ and peak Load (r = 0.8420, $p < 0.05$). In 2nd test correlation was found only between Load $_{VT}$ and peak Load (r = 0.8288, $p < 0.05$).

4. Discussion

In patients, who have undergone breast cancer treatment, impaired function of the cardiovascular and respiratory systems and weakening fatigue often occur both in the early period and a few months or years after the completion of the treatment. Studies have shown that these negative side effects resulting from the treatment process may be limited by applying moderate individualized physical activity. The oxidative stress arising in the treatment process may be reduced with appropriate pharmacotherapy, diet, but also with physical training. Changes caused by increased production of free radicals may be reversible. However, when the rate of free radical production is higher than the rate of repair mechanisms, significant contraction of mitochondria may occur with damage of mitochondrial crest, scattering of cell elements, and damage of cytoplasmatic membranes. Moreover, free radicals affect the transport of glucose, calcium-dependent ATPase activity of creatine kinase activity. In the studied group of women on each date of testing, biochemical indicators, such as FRAP, TBARS, urea, total proteins, and albumins were within the normal physiological range; only concentration of total phenolics on the

1st and second date of testing was below the reference values and amounted to 2.44 ± 0.09 and 2.43 ± 0.28 g GAE/L, respectively. Supervised individualized endurance training did not result in a significant change in any hematological and biochemical indicators, in spite of the increase in the level of exercise tolerance of the participants.

Physical activity undertaken by healthy persons, as well as persons with a low physical capacity, is of great significance for maintaining their good physical fitness and exercise tolerance. It improves the function of the circulatory and respiratory systems. Optimization of the training process and physical activity close to the anaerobic threshold (AT) are the most effective factors affecting post-training changes in the circulatory and respiratory capacity. In the case of patients after cancer treatment, physical activity contributes to an improved functional capacity, thus limiting negative consequences of the immobility or the treatment process itself [24].

The training program underwent by research participants was individualized and adjusted to the level of their circulatory and respiratory capacity. A supervised 2-month long training on a cycle ergometer resulted in no significant change in the peakVO_2 (25.74 ± 4.04 vs. 27.00 ± 3.68 mL/kg/min) with only a small decrease in the body weight. An increase in physical fitness is also indicated by obtaining a higher maximum load during the ergospirometric exercise test on the 2nd date of tests (112.50 ± 23.0 vs. 123.33 ± 22.09 W). In comparison to the results of studies on the effectiveness of applying endurance training in the rehabilitation process, the results of our study fully confirm the trends of post-training changes in physical capacity and exercise tolerance described by other authors [25–28]. A properly planned rehabilitation or sports training leads to adaptive changes in the muscle tissue, improvement of metabolic processes, and structural changes in muscles themselves. An increase in muscle strength and increased immunity of the tissue to fatigue are observed. Radical oxygen species secreted during physical exercise as well as growth and inflammatory factors produced as a result of the damage of muscle fibers take place in this process, which leads to an increase in the number of neutrophils, monocytes, and lymphocytes in the blood as well as inflammation markers in the blood [29–32]. The study of Rajneesh et al. (2008) indicated that in the blood of women with breast cancer, increased concentration of TBARS and other markers of ROS activity was observed as a result of the disease process [33]. Physical activity, in particular of to high-intensity, may cause a further increase in their concentration. In our study, on the 1st and second date, the concentration of TBARS, which is an indicator of cell membrane damage, did not change statistically significantly, which may be considered beneficial for the studied women. At the same time, we noted no statistically significant change in the total FRAP and compounds, such as total phenolics, proteins, albumins, and urea in the blood on both dates of tests, which indicates the prooxidative-antioxidative balance of the body. This may indicate a good tolerance of training loads by the studied women. No reports have been found in the literature to confirm the significance of physical activity and its effect on the concentration of total phenolics in the blood of women treated for breast cancer. However, there are reports of the anti-cancerous activity of some polyphenolic compounds (e.g., chlorogenic acid, resveratrol) by means of reducing inflammation or stimulation of apoptosis of cancerous cells [34]. Additionally, it was proved that phenolics decrease muscle damage caused by an increase in ROS concentration during physical exercise and in the period of restitution [35]. Hence, the amount of polyphenols in the blood depends on the number of consumed foodstuffs rich in phenolics, their absorption from the alimentary tract, and their use by the body [36]. In order to improve the antioxidant potential of plasma, it would seem beneficial to additionally include polyphenol supplementation in the program of patient care. Studies show that administering antioxidants results in an increase of the antioxidant potential of plasma [37]. To summarize the discussion of the significance of antioxidants and negative effects of free radicals, the theory of hormesis can be quoted, in which insignificant oxidative stress originating in cells, e.g., as a result of physical activity, may be a predictor of the development of beneficial cell changes, contributing to the tolerance of acute oxidative stress [38,39]. Drouin et al. [40] studied the effects of repeated physical exercise on hematological changes in breast cancer patients during the treatment process and after its completion. They noted that physical exercise prevents the fall in the erythrocytes counts during radiotherapy. The analysis of the number of erythrocytes, hematocrit value, and the concentration of hemoglobin in the studied group of women showed their statistically insignificant increase on the 2nd date of tests. The basis of this mechanism may be the change in the rate of erythropoiesis regulated by such factors as, among other things, oxygen deficiency, hormonal and nervous factors.

Margolis et al. [41] showed in their study that women after menopause with a larger amount of leukocytes in the blood have an increased risk of breast, large intestine, endometrial, and lung cancer, as well as higher cancer mortality. The total number of leukocytes and relative percentage of granulocytes, lymphocytes, and monocytes of the studied women were within normal ranges. The change in the number of leukocytes during physical exercise depends on its duration, intensity, and level of fitness of the subjects. The physiological mechanism of change in leukocyte count is caused by many neurohormonal factors, including cortisol, catecholamines, growth hormone, endorphins, sex hormones. Another group of factors intensifying exercise leukocytosis is secreted cytokines (TNF, IL-1, IL-6). Additionally, changes in glutamine, glucose, lipids, and heat shock protein concentrations in the blood contribute to the occurrence and intensifying of the mechanism of exercise-induced leukocytosis. In the initial stage of the exercise, the increase in the number of leukocytes is caused by the action of catecholamines, whereas the later, slower increase is due to the effect of cortisol on the bone marrow. During an intensive physical exercise, spontaneous degranulation of neutrophils occurs, with an increase in their phagocytic properties, thus stimulating neutrophil oxygen burst. Similar to other leukocytes, also NK cells are mobilized to the peripheral circulation, influenced by the physical strain. This is linked to the regulatory effect of catecholamines on their function [42]. If a body is subject to physical work with an intensity exceeding the exercise capacity of the exercising person, leukopenia occurs. The cytotoxic ability of NK lymphocytes is impaired, as is the proliferation and secretion ability of other types of lymphocytes [43]. The training program carried out by women after cancer treatment did not result in a significant change in the number of leukocytes and in their percentage.

5. Conclusions

In the group of patients after breast cancer treatment, endurance training caused an increase in exercise tolerance and did not cause aggravation of oxidative stress in women undergoing breast cancer treatment.

Limitation of the Study

The study was carried out on a specially selected group. The group was fully representative of a population of women who met the following conditions: they had undergone a radical mastectomy in the second and third stages of the disease and received full oncological treatment: chemotherapy, radiotherapy, and hormone therapy. For this reason, it is difficult to project the results for the whole population of women after cancer treatment. The limitation of the study is the small number of patients, the lack of a control group and the inability to compare results with the effects of endurance training in women at various times after the completed oncological treatment.

Author Contributions: Conceptualization, K.D and U.C.; methodology, K.D., T.P. and J.K.; software, A.J.; validation, K.D., A.D. and U.C.; resources, K.D. and A.J.; data curation, T.P., J.K. and P.P.; writing—original draft preparation, K.D; writing—review and editing, U.C.; visualization, K.D; supervision, K.D.; project administration, K.D., U.C. All authors have read and agreed to the published version of the manuscript.

Funding: This research received no external funding.

Institutional Review Board Statement: The study was conducted according to the guidelines of the Declaration of Helsinki and approved by the IRB (Institute for Research in Biomedicine) at the Poznan University of Medical Sciences (10 May 2012; Ethics Approval Number: 535/12).

Informed Consent Statement: Informed consent was obtained from all subjects involved in the study.

Data Availability Statement: The data presented in this study are available on request from the corresponding author. The data are not publicly available due to the consent provided by participants on the use of confidential data.

Conflicts of Interest: The authors declare no conflict of interest.

References

1. Domaszewska, K.; Pieńkowski, T.; Janiak, A.; Bukowska, D.; Laurentowska, M. The influence of soft tissue therapy on respiratory efficiency and chest mobility of women suffering from breast cancer. *Int. J. Environ. Res. Public Health.* **2019**, *16*, 5092. [CrossRef]
2. Mendes, D.; Alves, C.; Afonso, N.; Cardoso, F.; Passos-Coelho, J.L.; Costa, L.; Andrade, S.; Batel-Marques, F. The benefit of HER2-targeted therapies on overall survival of patients with metastatic HER2-positive breast cancer–a systematic review. *Breast Cancer Res.* **2015**, *17*, 140. [CrossRef]
3. Marianne, E.; Jensen, A.B. Late effects of breast cancer treatment and potentials for rehabilitation. *Acta Oncol.* **2011**, *50*, 187–193.
4. Halliwell, B. Free radicals, antioxidants, and human disease: Curiosity, cause, or consequence? *Lancet* **1994**, *344*, 721–724. [CrossRef]
5. Halliwell, B. Reactive oxygen species in living systems: Source, biochemistry, and role in human disease. *Am. J. Med.* **1991**, *91*, 14–22. [CrossRef]
6. Reilly, P.M.; Schiller, H.J.; Bulkley, G.B. Pharmacologic approach to tissue injury mediated by free radicals and other reactive oxygen metabolites. *Am. J. Surg.* **1991**, *161*, 488–503. [CrossRef]
7. Hertog, M.G.; Kromhout, D.; Aravanis, C.; Blackburn, H.; Bbuzina, R.; Fidanza, F.; Giampaoli, S.; Jansen, A.; Menotti, A.; Nedeljkovic, S.; et al. Flavonoid intake and long-term risk of coronary heart disease and cancer in the seven countries study. *Arch. Intern. Med.* **1995**, *155*, 381–386. [CrossRef]
8. Jonas, C.R.; Puckett, A.B.; Jones, D.P.; Griffith, D.P.; Szeszycki, E.E.; Bergman, G.F.; Furr, C.E.; Tyre, C.; Carlson, J.L.; Galloway, J.R.; et al. Plasma antioxidant status after high-dose chemotherapy: A randomized trial of parenteral nutrition in bone marrow transplantation patients. *Am. J. Clin. Nutr.* **2000**, *72*, 181–189. [CrossRef]
9. Goroshinskaya, I.A.; Kit, O.I.; Zuderman, N.; Ushakova, N.D.; Lysenko, I.B.; Nemashkalova, L.A.; Nikolaeva, N.V.; Kapuza, E.A.; Shatokhina, O.N. Oxidative processes in the blood during chemotherapy and lethality of patients with multiple myeloma. *J. Clin. Oncol.* **2018**, *36*, e20027. [CrossRef]
10. McNeely, M.L.; Campbell, K.L.; Rowe, B.H.; Klassen, T.P.; Mackey, J.R.; Courneya, K.S. Effects of exercise on breast cancer patients and survivors: A systematic review and meta-analysis. *Can. Med. Assoc. J.* **2006**, *175*, 34–41. [CrossRef]
11. Lukaski, H.C.; Johnson, P.E.; Bolonchuk, W.W.; Lykken, G.I. Assessment of fat-free mass using bioelectrical impedance measurements of the human body. *Am. J. Clin. Nutr.* **1985**, *41*, 810–817. [CrossRef]
12. Biernat, E.; Stupnicki, R.; Lebiedziński, B.; Janczewska, L. Assessment of physical activity by applying IPAQ questionnaire. *J. Phys. Educ. Sport.* **2008**, *52*, 83–89. [CrossRef]
13. Beaver, W.L.; Wasserman, K.; Whipp, B.J. A new method for detecting anaerobic threshold by gas exchange. *J. Appl. Physiol.* **1986**, *60*, 2020–2027. [CrossRef]
14. Davis, J.A.; Whipp, B.J.; Wasserman, K. The relation of ventilation to metabolic rate during moderate exercise in man. *Eur. J. Appl. Physiol. Occup. Physiol.* **1980**, *44*, 97–108. [CrossRef]
15. Gitt, A.K.; Winter, U.J.; Fritsch., J.; Pothoff, G.; Sedlak, M.; Ehmanns, S.; Ostmann, H.; Hilger, H.H. Comparison of four different methods for respiratory determination of the anaerobic threshold in normal people, and heart-and lung patients. *Z. Kardiol.* **1994**, *83* (Suppl. 3), 37–42.
16. Powers, S.K.; Dodd, S.; Garner, R. Precision of ventilatory and gas exchange alterations as a predictor of the anaerobic threshold. *Eur. J. Appl. Physiol. Occup. Physiol.* **1984**, *52*, 173–177. [CrossRef] [PubMed]
17. Reinhard, U.; Müller, P.H.; Schmülling, R.M. Determination of anaerobic threshold by the ventilation equivalent in normal individuals. *Respiration* **1979**, *38*, 36–42. [CrossRef]
18. Shimizu, M.; Myers, J.; Buchanan, N.; Walsh, D.; Kraemer, M.; McAuley, P.; Froelicher, V.F. The ventilatory threshold: Method, protocol, and evaluator agreement. *Am. Heart J.* **1991**, *122*, 509–516. [CrossRef]
19. Day, J.R.; Rossiter, H.B.; Coats, E.M.; Skasick, A.; Whipp, B.J. The maximally attainable VO_2 during exercise in humans: The peak vs. maximum issue. *J. Appl. Physiol.* **2003**, *95*, 1901–1907. [CrossRef]
20. Kryściak, J.; Domaszewska, K.; Czerniak, U. *Influence of 12-Week Endurance Training on the Physical Efficiency of Women after Mastectomy*; Civilization Health; Tygiel Scientific Publishers: Lublin, Poland, 2017; pp. 74–81.
21. Benzie, I.F.; Strain, J.J. The ferric reducing ability of plasma (FRAP) as a measure of "antioxidant power": The FRAP assay. *Anal. Biochem.* **1996**, *239*, 70–76. [CrossRef]
22. Ohkawa, H.; Ohishi, N.; Yagi, K. Assay for lipid peroxides in animal tissues by thiobarbituric acid reaction. *Anal. Biochem.* **1979**, *95*, 351–358. [CrossRef]
23. Singleton, V.L.; Rossi, J.A. Colorimetry of total phenolics with phosphomolybdic-phosphotungstic acid reagents. *Am. J. Enol. Vitic.* **1965**, *16*, 144–158.

24. Kindermann, W.; Simon, G.; Keul, J. The significance of the aerobic-anaerobic transition for the determination of work load intensities during endurance training. *Eur. J. Appl. Physiol.* **1979**, *42*, 25–34. [CrossRef] [PubMed]
25. Neilson, H.K.; Farris, M.S.; Stone, C.R.; Vaska, M.M.; Brenner, D.R.; Friedenreich, C.M. Moderate-vigorous recreational physical activity and breast cancer risk, stratified by menopause status: A systematic review and meta-analysis. *Menopause* **2017**, *24*, 322–344. [CrossRef]
26. Ekelund, U.; Steene-Johannessen, J.; Brown, W.J.; Fagerland, M.W.; Owen, N.; Powell, K.E.; Bauman, A.; Lee, I.M. Lancet Physical Activity Series 2 Executive Committee; Lancet Sedentary Behavior Working Group. Does physical activity attenuate, or even eliminate, the detrimental association of sitting time with mortality? A harmonised meta-analysis of data from more than 1 million men and women. *Lancet* **2016**, *388*, 1302–1310.
27. Egan, B.; Zierath, J.R. Exercise metabolism and the molecular regulation of skeletal muscle adaptation. *Cell Metabolism.* **2013**, *17*, 162–184. [CrossRef]
28. Kirkham, A.A.; Campbell, K.L.; McKenzie, D.C. Comparison of aerobic exercise intensity prescription methods in breast cancer. *Med. Sci. Sports Exerc.* **2013**, *45*, 1443–1450. [CrossRef]
29. Steinbacher, P.; Eckl, P. Impact of oxidative stress on exercising skeletal muscle. *Biomolecules* **2015**, *5*, 356–377. [CrossRef]
30. Yang, W.; Hu, P. Skeletal muscle regeneration is modulated by inflammation. *J. Orthop. Transl.* **2018**, *13*, 25–32. [CrossRef]
31. Sass, F.A.; Fuchs, M.; Pumberger, M.; Geissler, S.; Duda, G.N.; Perka, C.; Schmidt-Bleek, K. Immunology guides skeletal muscle regeneration. *Int. J. Mol. Sci.* **2018**, *19*, 835–854. [CrossRef]
32. Pedersen, B.K.; Hoffman-Goetz, L. Exercise and the immune system: Regulation, integration, and adaptation. *Physiol. Rev.* **2000**, *80*, 1055–1081. [CrossRef] [PubMed]
33. Rajneesh, C.P.; Manimaran, A.; Sasikala, K.R.; Adaikappan, P. Lipid peroxidation and antioxidant status in patients with breast cancer. *Singapore Med. J.* **2008**, *49*, 640–643.
34. Paszkiewicz, M.; Budzyńska, A.; Różalska, B. Immuno-modulating role of plant polyphenols. *Postepy Hig. Med. Dosw.* **2012**, *66*, 637–646. [CrossRef] [PubMed]
35. Malaguti, M.; Angeloni, C.; Hrelia, S. Polyphenols in exercise performance and prevention of exercise-induced muscle damage. *OMCL* **2013**, *2013*, 825928. [CrossRef]
36. Scalbert, A.; Williamson, G. Dietary intake and bioavailability of polyphenols. *J. Nutr.* **2000**, *130* (Suppl. 8), 2073S–2085S. [CrossRef]
37. Partyka, L.; Hartwich, J.; Dróżdż, W.; Gruca, A.; Jopek, R. Changes in oxidative stress parameters in patients with peripheral vascular disease in response to conservative and surgical treatment. *Acta Angiol.* **2001**, *7*, 29–41.
38. Radak, Z.; Chung, H.Y.; Goto, S. Exercise and hormesis: Oxidative stress-related adaptation for successful aging. *Biogerontology.* **2005**, *6*, 71–75. [CrossRef] [PubMed]
39. Radak, Z.; Naito, H.; Kaneko, T.; Tahara, S.; Nakamoto, H.; Takahashi, R.; Pelaez, F.C.; Goto, S. Exercise training decreases DNA damage and increases DNA repair and resistance against oxidative stress of proteins in aged rat skeletal muscle. *Pflugers Arch.* **2002**, *445*, 273–278. [CrossRef]
40. Drouin, J.S.; Young, T.J.; Beeler, J.; Byrne, K.; Birk, T.J.; Hryniuk, W.M.; Hryniuk, L.E. Random control clinical trial on the effects of aerobic exercise training on erythrocyte levels during radiation treatment for breast cancer. *Cancer* **2006**, *107*, 2490–2495. [CrossRef]
41. Margolis, K.L.; Rodabough, R.J.; Thomson, C.A.; Lopez, A.M.; McTiernan, A. Prospective study of leukocyte count as a predictor of incident breast, colorectal, endometrial, and lung cancer and mortality in postmenopausal women. *Arch. Intern. Med.* **2007**, *167*, 1837–1844. [CrossRef] [PubMed]
42. Timmons, B.W.; Cieslak, T. Human natural killer cell subsets and acute exercise: A brief review. *Exerc. Immunol. Rev.* **2008**, *14*, 8–23.
43. Timmons, B.W.; Tarnopolsky, M.A.; Bar-Or, O. Immune responses to strenuous exercise and carbohydrate intake in boys and men. *Pediatr. Res.* **2004**, *56*, 227–234. [CrossRef]

International Journal of
Environmental Research and Public Health

MDPI

Article

Obesity and Chosen Non-Communicable Diseases in PURE Poland Cohort Study

Katarzyna Zatońska [1], Piotr Psikus [2], Alicja Basiak-Rasała [1,*], Zuzanna Stępnicka [1], Dagmara Gaweł-Dąbrowska [1], Maria Wołyniec [1], Julia Gibka [3], Andrzej Szuba [4] and Katarzyna Połtyn-Zaradna [1]

[1] Department of Social Medicine, Wroclaw Medical University, 50-345 Wrocław, Poland; Katarzyna.zatonska@umed.wroc.pl (K.Z.); zuzanna.stepnicka@gmail.com (Z.S.); dagmara.gawel-dabrowska@umed.wroc.pl (D.G.-D.); maria.wolyniec@umed.wroc.pl (M.W.); katarzyna.poltyn-zaradna@umed.wroc.pl (K.P.-Z.)

[2] Calisia University, 62-800 Kalisz, Poland; burmistrz@um.kepno.pl

[3] Department of Health Sciences, Calisia University, 62-800 Kalisz, Poland; wydzial.medyczny@pwsz.kalisz.pl

[4] Department of Angiology, Hypertension and Diabetology, Wroclaw Medical University, 50-556 Wroclaw, Poland; andrzej.szuba@umed.wroc.pl

* Correspondence: alicja.basiak-rasala@umed.wroc.pl; Tel./Fax: +48-71-328-2145

Abstract: Introduction: Obesity has been associated with a higher risk of morbidity, disability, and death. The objective of this study was to assess the prevalence of obesity and chosen non-communicable diseases (NCDs) in the PURE Poland cohort study. Material and Methods: The study covers a group of 2035 people (1281 women and 754 men), who live in urban and rural areas of Lower Silesian voivodeship. The baseline study was conducted between 2007–2010. The data on demographic status and history of diseases were collected using questionnaires. The anthropometric parameters, blood pressure, blood lipids, and glucose level were measured. Results: Normal body weight was observed in 28.1% of participants, whereas overweight and obesity were observed in 40.1% and 31.1% of participants, respectively. Moreover, there was a significant difference in the body weight between genders. Prevalence of obesity was similar in men and women (31.0% and 31.1%, respectively). Obesity was more prevalent in rural vs. urban residents (38.5% and 26.0%, respectively). In a logistic regression analysis, the odds for obesity was two-fold higher in participants aged >64 years and rural inhabitants (OR 1.91; 95% CI 1.36–2.70; OR 1.79; 95% CI 1.48–2.16, respectively). Participants with obesity had 2.5-fold higher odds for diabetes and hypertension and two-fold higher odds for CHD in comparison with non-obese individuals (OR 2.74; 95% CI 2.01–3.73, OR 2.54; 95% CI 2.03–3.17, OR 1.88; 95% CI 1.26–2.80, respectively). Conclusions: Taken together, the prevalence of obesity was associated with particular socio-demographic factors (age, place of residence, and level of education) as well as diabetes, hypertension, and coronary heart disease.

Keywords: PURE study; noncommunicable diseases; urban; rural; obesity

Citation: Zatońska, K.; Psikus, P.; Basiak-Rasała, A.; Stępnicka, Z.; Gaweł-Dąbrowska, D.; Wołyniec, M.; Gibka, J.; Szuba, A.; Połtyn-Zaradna, K. Obesity and Chosen Non-Communicable Diseases in PURE Poland Cohort Study. *IJERPH* **2021**, *18*, 2701. https://doi.org/10.3390/ijerph18052701

Academic Editor: Yoshifumi Saisho

Received: 18 February 2021
Accepted: 5 March 2021
Published: 8 March 2021

Publisher's Note: MDPI stays neutral with regard to jurisdictional claims in published maps and institutional affiliations.

Copyright: © 2021 by the authors. Licensee MDPI, Basel, Switzerland. This article is an open access article distributed under the terms and conditions of the Creative Commons Attribution (CC BY) license (https://creativecommons.org/licenses/by/4.0/).

1. Introduction

Obesity is one of the greatest public health challenges of the 21st century in developed and developing countries. In 1997, the World Health Organization (WHO) recognized obesity as a global health problem. In 2016, more than 1.9 billion adults were overweight worldwide. Among them, more than 650 million were obese [1]. Furthermore, according to the study of NCD Risk Factor Collaboration, between 1975 and 2014, the mean body mass index (BMI) increased across both female and male population. The study included 9.2 million people from 200 countries. The mean BMI for women increased from 22.1 kg/m^2 to 24.4 kg/m^2, while the mean BMI for men increased from 21.7 to 24.2 kg/m^2. Authors concluded that if the current trend of the increase in obesity prevalence continues, the global prevalence of severe obesity will surpass the prevalence of underweight by 2025, especially in women [2].

It was estimated that in 2008, in Europe, over half of men and women were overweight, while about 23% of women and 20% of men were obese. The steady increase of excessive body weight prevalence was observed in the whole EU [3–5]. According to the latest estimates, overweight occurred in 30–70% of adults and obesity in 10–30% of adults in the European Union (EU) [6]. In Poland, according to the latest European Health Interview Survey (EHIS) conducted in 2014, within the population of 15 years old and above, over 62% of men and almost 46% of women were overweight, whereas 18% of men and 15.6% of women were obese [7,8]. A cross-sectional analysis of the first (2003–2005) and the second (2013–2014) edition of the Polish National Multicenter Health Survey (WOBASZ) revealed significant increase in obesity prevalence during the decade, especially in men [9].

The excessive body weight is a serious health problem, and it is a risk factor for many non-communicable diseases (NCD). Obesity is associated with a higher risk of morbidity, disability, and death. Its level of impact, as a risk for mortality, is similar to that of smoking [10]. Major risk factors for mortality are increased cardiovascular incidents, such as coronary heart disease, hypertension, stroke, type 2 diabetes, and some types of cancer, e.g., breast cancer in postmenopausal women, endometrial cancer, and colon and kidney cancer [11–15]. According to Fontaine et al. [16], obesity with a BMI of over 45 kg/m^2 during early adulthood (a group of 20- to 30- year-olds) can reduce life expectancy by 13 years for men, and by eight years for women.

Overweight and obesity are relevant public health problems. It was estimated that the cost of health care for people who are overweight and obese is significantly higher than those with normal weight. Moreover, in 2014, the global economic burden of obesity contributed to the global gross domestic product (GDP) by 2.8% [17]. The cost of treating obesity and its complications in Poland absorbs 21% of the budget allocated to health [18]. That amounts to nearly 3 billion zloty, based on calculations carried out in Lubelskie Voivodeship [19]. Furthermore, diseases associated with overweight and obesity are likely to be responsible for 25% of hospital admissions (which equals approximately 1,500,000 hospital admissions across the country) according to the nationwide Polish surveys of patients in hospitals [19].

The objective of this study was to assess the prevalence of obesity and chosen non-communicable diseases (NCDs) in relation to sociodemographic factors and the association between obesity and NCDs in the PURE Poland cohort study. The PURE Poland study is one of the few large, cohort studies currently conducted in Poland. Recruitment of both urban and rural population into the study cohort is a unique approach addressing health inequalities.

2. Materials and Methods

The Prospective Urban and Rural Epidemiology Study (PURE) is based on data from 21 countries with different status of economic development. Poland, at the beginning of the research, was one of the seven upper-middle income countries enrolled in the study [20]. All participants were tested in accordance with the PURE study protocol [21]. This paper presents the results of the PURE Poland study—baseline. The data were collected between 2007 and 2010. Among 2035 participants there were 1281 women and 754 men who inhabited urban and rural areas of Lower Silesia. They were divided into three age groups (<45, 45–64, and >64).

Collection of data concerning individuals was obtained at two levels [21]. First—Family/Household level included demographic information, i.e., number of family/household members, number of children, education, etc.; epidemiological information, i.e., morbidities, usage of tobacco products, etc.; and other important determinants, i.e., access to water, sanitation, household amenities, etc. All this information was collected using Family Census Questionnaire. Second—Adult participants' level, which was collected based on an Adult Questionnaire. This included data on nine INTERHEART risk factors (lipids, smoking, hypertension, diabetes, abdominal obesity, psychosocial factors (stress and symptoms of depression), consumption of fruits and vegetables, consumption of alcohol, and regular physical activity), anthropometric measures, blood pressure, spirometry, fasting blood sample, semi-quantitative food frequency questionnaires (FFQs), and physical activity (IPAQ).

The body mass of the patients was measured using Tanita Ironman Body Composition Monitor Model BC-554 with accuracy of 0.1 kg. The BMI was calculated as weight (kg) divided by height (m) squared. Subjects were classified into four BMI categories, according to the WHO guidelines: underweight (BMI < 18.5 kg/m^2), normal weight (BMI 18.5–24.9 kg/m^2), overweight (BMI 25.0–29.9 kg/m^2), and obese (BMI $\geq$ 30.0 kg/m^2). Diabetes was ascertained on the basis of (1) self-reported diabetes and/or (2) self-reported anti-diabetic medication and/or (3) fasting blood glucose measurement $\geq$126 mg/dL [22]. Hypertension was ascertained on the basis of (1) self-reported hypertension and/or (2) self-reported anti-hypertensive medication and/or (3) an average of two blood pressure measurements $\geq$140/90 mmHg, as previously described [23]. The occurrence of coronary heart disease (CHD) and stroke was self-reported by the participants. Due to lack of data regarding aforementioned variables, diabetes was assessed in a total of 1663 participants, hypertension in 2019 participants, and CHD and stroke in 2029 participants.

The aggregated information on the prevalence of obesity and NCD was presented using basic statistical parameters. The prevalence of obesity and NCD was verified with respect to epidemiological variables known to be its significant determinants and health related states correlated to obesity and/or NCD. To assess the statistical significance of the differences observed in the distribution of obesity and NCD, the chi-square test was used. The association between sociodemographic/anthropometric factors with NCDs was assessed with the use of adjusted logistic regression models after adjusting for age and/or gender. The strength of the association was measured by the odds ratio (OR) with 95% confidence intervals. For all differences, the level of the statistical significance was $p \leq 0.05$. Statistical analysis was conducted using the software Statistica 13.1 PL (StatSoft Inc., Palo Alto, CA, USA).

3. Results

A baseline characteristics of the study population in the view of body mass index categories is presented in Table 1. There were 28.1% of participants with normal body weight, while 40.1% and 31.1% of participants were overweight and obese, respectively. Moreover, there were statistically significant differences in the distribution of body weight according to sex, age, marital status, level of education, and place of residence. The occurrence of chosen NCDs (stroke, CHD, hypertension, diabetes) was also differentiated by the BMI.

The normal body weight was more prevalent in women than in men (32.3% and 20.8%, respectively). We observed similar prevalence of obesity in women and men (31.1% vs. 31.0%, respectively). Furthermore, the prevalence of overweight and obesity significantly increased in participants >44 years of age. Obesity was observed in 19.8% of participants aged 30–44 years, 32.7% in participants aged 45–64 years, and 36.7% in participants aged >64 years, respectively. Moreover, the prevalence of obesity decreased stepwise with higher level of education (45.2% of obese individuals in participants with primary education vs. 22.7% in participants with higher education, respectively). On the contrary, the prevalence of overweight increased with increasing level of education. Additionally, participants who were separated/divorced were more likely to be obese than married or never married/single participants (35.2% vs. 30.5%, 26.7%, respectively), whereas married participants were more likely to be overweight than separated or never married participants (41.2% vs. 36.0%, 38.4%, respectively). Finally, normal body weight was more prevalent in urban residents compared with those living in rural areas (30.2% and 25.0%, respectively). Obesity was much more prevalent in rural inhabitants than in urban (38.5% vs. 26.0%, respectively), whereas overweight was more prevalent in urban inhabitants (43.1% vs. 35.8%).

Table 1. A baseline characteristics of the study population in the view of body mass index categories.

Variables	N	BMI (kg/m^2) <18.5 N = 15	18.5–24.9 N = 571	25.0–29.9 N = 817	≥30.0 N = 632	p-Value *
Total	2035	0.7% (15)	28.1% (571)	40.1% (817)	31.1% (632)	
Sex:						
Men	754	0.4% (3)	20.8% (157)	47.8% (360)	31.0% (234)	<0.001
Women	1281	0.9% (12)	32.3% (414)	35.7% (457)	31.1% (398)	
Place of residence:						
Urban	1210	0.7% (9)	30.2% (365)	43.1% (522)	26.0% (314)	<0.001
Rural	825	0.7% (6)	25.0% (206)	35.8% (295)	38.5% (318)	
Age, years:						
30–44	358	1.4% (5)	43.6% (156)	35.2% (126)	19.8% (71)	<0.001
45–64	1350	0.6% (8)	25.9% (350)	40.8% (551)	32.7% (441)	
>64	327	0.6% (2)	19.9% (65)	42.8% (140)	36.7% (120)	
Level of education:						
Primary	301	0.7% (2)	18.2% (55)	35.9% (108)	45.2% (136)	<0.001
Vocational	324	1.2% (4)	25.9% (84)	37.1% (120)	35.8% (116)	
Secondary	796	0.6% (5)	28.6% (228)	40.5% (322)	30.3% (241)	
Higher	603	0.7% (4)	33.7% (203)	42.9% (259)	22.7% (137)	
Lack of information	11	0.0% (0)	9.1% (1)	72.7% (8)	18.2% (2)	
Marital status:						
Married/living together	1508	0.7% (10)	27.6% (417)	41.2% (621)	30.5% (460)	0.016
Separated/divorced	375	0.3% (1)	28.5% (107)	36.0% (135)	35.2% (132)	
Never married/single	146	2.7% (4)	32.2% (47)	38.4% (56)	26.7% (39)	
Lack of information	6	0.0% (0)	0.0% (0)	83.3% (5)	16.6% (1)	
Stroke	38	0.0% (0)	23.7% (9)	47.4% (18)	28.9% (11)	0.767
CHD	112	0.0% (0)	16.1% (18)	34.8% (39)	49.1% (55)	<0.001
Hypertension	1217	0.5% (6)	19.2% (234)	41.7% (507)	38.6% (470)	<0.001
Diabetes	200	0.0% (0)	12.5% (25)	30.0% (60)	57.5% (115)	<0.001

* Chi-square test.; BMI—Body Mass Index; CHD—Coronary Heart Disease.

In a performed logistic regression analysis, the factors that were significantly associated with increased odds for obesity included: rural place of residence, age, and education level (Table 2). In an adjusted model, the odds for obesity were over 1.5-fold higher among rural residents compared with urban residents (odds ratio [OR] = 1.79, 95% CI = 1.48–2.16) (Table 2). The was no significant difference in the occurrence of obesity between men and women. The odds for obesity increased with increasing age of participants. The highest odds for obesity were observed in the age group >64 years of age [OR 1.91, 95% CI 1.36–2.7]. It was almost two-fold higher than in the group of 30- to 44-year-olds. Furthermore, with increasing level of education, the odds of obesity significantly decreased ([OR] = 0.42, 95% CI 0.34–0.56 in participants with higher education).

Finally, the odds ratios of the factors associated with occurrence of stroke, hypertension, CHD, and diabetes among the study population were presented in Figure 1.

Table 2. The adjusted odds ratio (OR) of the occurrence of obesity in the study population.

Variable	Obesity				p-Value	$OR_{adj.}$ (95% CI)
	Yes N = 632		No N = 1403			
	n	%	n	%		
Sex:					0.974	
Women	398	63.0	883	62.9		1.00 (ref.)
Men	234	37.0	520	37.1		0.99 (0.81–1.20) [a]
Age group:						
30–44	71	11.2	287	20.5	<0.001	1.00 (ref.)
45–64	441	69.8	909	64.8		1.76 (1.42–2.17) [b]
>64	120	19.0	207	14.8		1.91 (1.36–2.70) [b]
Place of residence:						
Urban	314	49.7	896	63.9	<0.001	1.00 (ref.)
Rural	318	50.3	507	36.1		1.79 (1.48–2.16) [c]
Level of education:						
Primary	136	21.6	165	11.8		1.00 (ref.)
Vocational	116	18.4	208	14.9	<0.001	0.84 (0.75–0.93) [c]
Secondary	241	38.3	555	39.8		0.69 (0.66–0.95) [c]
Higher	137	21.7	466	33.4		0.42 (0.34–0.56) [c]
Marital status:						
Living together	460	72.9	1048	75.0	0.105	1.00 (ref.)
Separated	132	20.9	243	17.4		0.84 (0.75–1.23) [c]
Single	39	6.2	107	7.7		0.88 (0.79–1.19) [c]

[a] $OR_{adj.}$: odds ratio adjusted for age, [b] $OR_{adj.}$: odds ratio adjusted for sex, [c] $OR_{adj.}$: odds ratio adjusted for sex and age; ref.—reference value.

The odds for stroke were significantly associated with increasing age (Figure 1a). Participants aged >64 years had 5.5-fold higher odds for stroke than 30- to 44-year-olds (OR 5.49; 95% CI 1.51–19.9). The odds for stroke was not associated with obesity, gender, place of residence, education, or marital status.

Factors independently differentiating the odds for hypertension included gender, age, obesity, place of residence, and level of education (Figure 1b). The increasing age was the factor associated with the highest odds for hypertension. Participants aged >64 years had over five-fold higher odds for hypertension, and 45- to 64-year-olds had almost three-fold higher chance, than 30- to 44-year-olds (OR 5.38, 95% CI 3.71–7.80, OR 2.93, 95% CI 2.25–3.82, respectively). Men had three-fold higher odds for hypertension than women (OR 3.09; 95% CI 2.48–3.84). Obesity was associated with 2.5-fold higher odds for hypertension [OR 2.54; 95% CI 2.03–3.17]. Higher and vocational education were associated with 1.5-fold lower odds for hypertension in comparison to primary education [OR 0.66; 95% CI 0.46–0.95, OR 0.68; 95% CI 0.47–0.98, respectively]. Rural inhabitants had 1.3-fold lower odds for hypertension than urban inhabitants [OR 0.77; 95% CI 0.60–0.97].

The occurrence of coronary heart disease was independently associated with factors like gender, age, level of education, and obesity (Figure 1c). The odds for CHD were over three-fold higher in participants aged >64 years than in younger participants [OR 3.43; 95% CI 2.22–5.33]. Men had two-fold higher odds for CHD than women [OR 2.02; 95% CI 1.31–3.11]. Obese individuals had almost two-fold higher odds for CHD [OR 1.88; 95% CI 1.26–2.80]. Higher level of education was associated with 2.5-fold lower risk of CHD than primary education [OR 0.39; 95% CI 0.19–0.81]. The odds for occurrence of CHD was not differentiated by the place of residence.

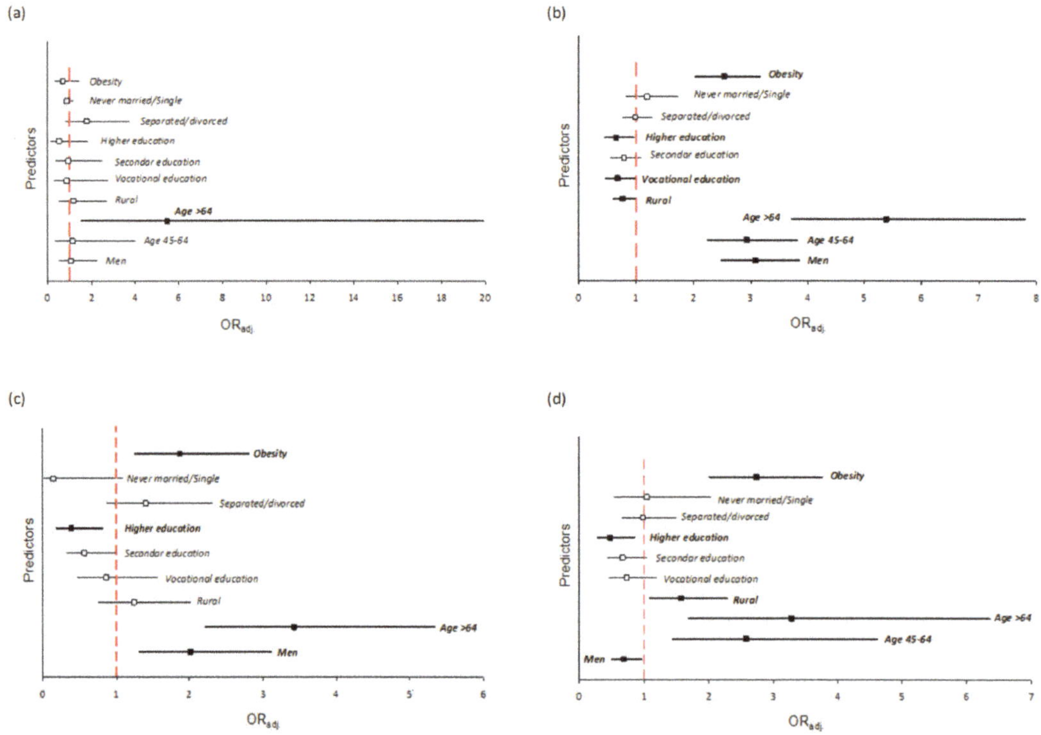

Figure 1. The odds ratios (with 95% CI) of the factors associated with occurrence of stroke (**a**), hypertension (**b**), CHD (**c**), and diabetes (**d**) among the study population (the odds ratios were adjusted to sex and/or age, as indicated previously in Table 2).

Factors independently associated with increased odds for diabetes included age, obesity, place of residence, gender, and the level of education (Figure 1d). The odds for diabetes increased with increasing age of the population. Participants aged >64 years had three-fold higher odds for diabetes than 30- to 44-year-olds [OR 3.28; 95% CI 169–6.36]. Participants with obesity had over 2.5-fold higher odds for diabetes than non-obese individuals [OR 2.74; 95% CI 2.01–3.73]. Place of residence also differentiated the risk of diabetes. Participants living in rural areas had 1.5-fold higher odds for diabetes than participants living in urban areas [OR 1.57; 95% CI 1.08–2.28].

4. Discussion

In Poland, over the last ten years, a consecutive increase in the prevalence of overweight and obesity was observed for both genders. In the representative Polish National Multicenter Health Survey (WOBASZ) conducted in 2003–2005 among 20- to 74-year-olds, overweight was diagnosed in 61% of men and in 50% of women, and obesity in 21% and 22%, respectively [24]. There was a significant increase in obesity prevalence during the decade, especially in men. In the second edition, WOBASZ II conducted in 2013–2014, obesity was observed in 24.4% of men and 25.0% of women [9]. In the Central Statistical Office report on the health state of the Polish population aged 15 and above published in 2015, 41.1% of men and 30.1% of women were overweight, whereas 18.1% of men and 15.6% of women were obese [7]. The average BMI for men in 2008 increased by 1.6 kg/m^2 compared with the year 1980 and was 26.7 kg/m^2 [25]. In women, the average BMI did not change between 1980 and 2008, and it was 25.9 kg/m^2 [25]. In a Polish-Norwegian Study (PONS) on 3854 inhabitants of Świętokrzyskie Voivodeship aged 45–64, the average BMI in men was 28.5 kg/m^2 and in women—28.2 kg/m^2. That summed up to 52% of men and 42% of women being overweight, whereas 35% of men and women being obese [26].

In our study, the prevalence of obesity was associated with particular socio-demographic factors. It was higher among people in older age groups and lower education levels. These results were consistent with previously published literature on the association of obesity with respect to age and educational level; for instance, the PONS study [26], the Vanio et al. study [27], and others [9,28,29]. Furthermore, based on the population of the NHANES study [30], obesity was much more prevalent among inhabitants of rural than urban areas, which is consistent with our findings as well. In a longitudinal cohort study conducted by Patterson et al. [31], a greater cumulative exposure to rurality during early adulthood was associated with higher risk of obesity in middle-aged adults. Similarly, in a study by Jokela et al. [32], rural residence was more associated with higher BMI than urban residence. Our results suggest that individuals with lower educational level as well as older people are statistically more likely to be obese. On the other hand, we observed higher prevalence of overweight in participants with higher level of education and participants living in urban areas.

According to the WHO, the risk of type 2 diabetes, insulin resistance, or dyslipidaemia is over three-fold higher in obese people than in those with normal body weight. The risk of the coronary heart disease and hypertension is 2–3-fold higher in obese people, while the risk of cancer is 1–2-fold higher in obese people compared with those with normal body weight [33,34]. In a study by Mongraw-Chaffin et al. [35], the increment in the BMI was associated significantly with the risk of coronary heart disease in both genders. Authors observed that in comparison to participants with normal body weight, the HR of coronary heart disease in participants with obesity was 1.6 [1.42–1.82] in women and 1.60 [1.43–1.79] in men [35]. We observed that obesity was independently associated with increased odds for diabetes, hypertension, and CHD. Similar results were obtained by Field et al. [36], who analyzed data on 10-year follow-up of Nurses' Health Study. Authors found that the incidence of diabetes, hypertension, heart disease, and stroke (only for men) elevated with increasing overweight in both men and women. Furthermore, results obtained from both NATPOL 2011 study in Poland and Polish-Norwegian Study (PONS) also suggest much higher prevalence of diabetes among obese individuals in comparison with individuals with normal body weight [26,37]. Although we cannot establish the causality, since both BMI and occurrence of NCDs were reported at the baseline study, our findings suggest a significant association between occurrence of those conditions.

The social burden of obesity is clear when comparing obesity related healthcare (44% higher) and patients with a regular body mass [38]. This is due to the obesity being a risk factor for morbidity. The obesity-related comorbidities are hypertension, reduced insulin sensitivity, diabetes mellitus, some types of cancer, and various heart diseases. Obesity is related to increased blood pressure and increased prevalence of CHD. Increase in body mass by 10 kg was associated with the increase of systolic blood pressure by 3 mmHg. Consequently, this further results in 12% increase of CHD risk [39]. The AHA (American Heart Association) recommended several important areas of further research on CVD and obesity. One of them focuses on "policy research on the impact of overweight/obesity on the future health care in people with or without CVD".

Considering the health-related and economic cost of obesity, this issue should be addressed by health policy makers. Policies aiming to decrease the social burden of obesity and reduce the number of obesity-related comorbidities can be found in series of important documents, i.e., the Polish National Healthcare Program, Polish National Civilization Diseases Prevention Program, and Global Strategy on Nutrition, Physical Activity, and Health signed at the 57th World Health Forum in May 2004, European Charter on Counteracting Obesity, and others [38]. Actions are taken in two key areas: increasing the knowledge and modification of eating habits and increasing physical activity for better energy balance. The prevalence of obesity in our study was high. Considering the high association between obesity and non-communicable diseases, our observations suggest that our study population is at the high risk of further metabolic disorders. People with overweight and obesity should be the group of special focus when planning prophylactic programs.

There are some limitations to consider. The PURE study is a longitudinal prospective cohort study, which has been continued in follow-ups every three years. In the present study, we analyzed data from baseline only. The data was collected between 2007 and 2010. The study group was selected with the use of a snowball sampling method, which can introduce a possible bias. Either in comparison to the overall population of Poland or the population of Lower Silesia, our study population is characterized by the overrepresentation of women and participants with higher education. Both the occurrence of NCDs and sociodemographic and anthropometric factors were reported at the baseline, which impede the possibility to draw conclusions about the causality of observed associations. Additionally, the number of stroke cases at the baseline was limited, and this fact potentially decreased the accuracy of statistical analysis in case of this condition. We plan to address this issue in future research of prospective data of our cohort.

5. Conclusions

The prevalence of obesity observed in the PURE Poland study baseline population was associated with particular socio-demographic factors. The increased odds for obesity were associated with older age and rural place of residence. The level of education was inversely associated with occurrence of obesity. We observed that obesity as well as increasing age were independently associated with increased odds for diabetes, hypertension, and CHD. Observed associations can be helpful to develop more adequate health promotion programs.

Author Contributions: Conceptualization: A.S. and K.P.-Z.; formal analysis: K.Z. and Z.S.; investigation: M.W.; methodology: K.Z.; visualization: Z.S.; writing—original draft: K.Z. and P.P.; writing—review and editing: A.B.-R., D.G.-D., A.S., J.G. and K.P.-Z. All authors have read and agreed to the published version of the manuscript.

Funding: The main PURE study and its components are funded by the Population Health Research Institute, the Canadian Institutes of Health Research, Heart and Stroke Foundation of Ontario, and through unrestricted grants from several pharmaceutical companies; the Poland cohort study is funded by the Polish Ministry of Science and Higher Education (grant no. 290/W-PURE/2008/0). Additionally, presented research results were conducted within the framework of the topic: "Analiza rozpowszechnienia czynników ryzyka chorób cywilizacyjnych ze szczególnym uwzględnieniem narażenia na dym tytoniowy w populacji dorosłych mieszkańców województwa dolnośląskiego oraz dzieci w wieku szkolnym" (eng. "Analysis of the prevalence of risk factors of non-communicable diseases with particular emphasis on tobacco smoke exposure in the population of adult residents of the Lower Silesia Province and school-age children"), as per records in SIMPLE system number SUB. C. 300.21.009. We declare that the funding body had no role in the design of the study; collection, analysis, or interpretation of data, and or writing the manuscript.

Institutional Review Board Statement: Not applicable.

Informed Consent Statement: Not applicable.

Data Availability Statement: Restrictions apply to the availability of these data. Cohort data was collected as a part of international Prospective Urban and Rural Epidemiological Study. Any requests regarding availability of data should be directed to the study supervisors.

Conflicts of Interest: The authors declare no conflict of interest.

References

1. World Health Organization. Obesity and Overweight Factsheets. 2020. Available online: https://www.who.int/news-room/fact-sheets/detail/obesity-and-overweight (accessed on 3 September 2020).
2. NCD Risk Factor Collaboration. Trends in adult body-mass index in 200 countries from 1975 to 2014: A pooled analysis of 1698 population-based measurement studies with 19.2 million participants. *Lancet* **2016**, *387*, 1377–1396. [CrossRef]
3. Eurostat. Overweight and Obesity—BMI Statistics in European Union. Available online: https://ec.europa.eu/eurostat/statistics-explained/pdfscache/12376.pdf (accessed on 3 September 2020).

4. Wojtyła-Buciora, P.; Bołdowski, T.; Wojtyła, C.; Żukiewicz-Sobczak, W.; Juszczak, K.; Chęcińska-Maciejewska, Z.; Wojtyła, A.; Krauss, H. National survey of body mass and dietary behaviour for middle school, high school and university students. *J. Health Inequal.* **2017**, *1*, 64–69. [CrossRef]
5. Harton, A.; Myszkowska-Ryciak, J.; Laskowski, W.; Gajewska, D. Prevalence of overweight and obesity among adolescents in Poland. *J. Health Inequal.* **2019**, *5*, 180–187. [CrossRef]
6. World Health Organization Regional Office for Europe. Overweight and Obesity in Europe: Factsheets. Available online: https://www.euro.who.int/en/health-topics/noncommunicable-diseases/obesity/data-and-statistics (accessed on 3 September 2020).
7. Główny Urząd Statystyczny. *Zdrowie i Zachowanie Zdrowotne Mieszkańców Polski w Świetle Europejskiego Ankietowego Badania Zdrowia (EHIS) 2014 r*; Statistics Poland: Warsaw, Poland, 2015.
8. Eurostat. European Health Interview Survey. Available online: https://ec.europa.eu/eurostat/web/microdata/european-health-interview-survey (accessed on 25 January 2021).
9. Stepaniak, U.; Micek, A.; Waśkiewicz, A.; Bielecki, W.; Drygas, W.; Janion, M.; Kozakiewicz, K.; Niklas, A.; Puch-Walczak, A.; Pająk, A. Prevalence of general and abdominal obesity and overweight among adults in Poland Results of the WOBASZ II study (2013–2014) and comparison with the WOBASZ study (2003–2005). *Polskie Archiwum Medycyny Wewnętrznej* **2016**, *126*, 662–671. [CrossRef]
10. Prospective Studies Collaboration. Body-mass index and cause-specific mortality in 900 000 adults: Collaborative analyses of 57 prospective studies. *Lancet* **2009**, *373*, 1083–1096. [CrossRef]
11. Chow, W.H.; Dong, L.M.; Devesa, S.S. Epidemiology and risk factors for kidney cancer. *Nat. Rev. Urol.* **2010**, *7*, 245–257. [CrossRef]
12. Knight, J.A. Diseases and Disorders Associated with Excess Body Weight. *Ann. Clin. Lab. Sci* **2011**, *41*, 107–121. Available online: http://www.annclinlabsci.org/content/41/2/107.long (accessed on 3 September 2020).
13. World Cancer Research Fund/American Institute for Cancer Research. Diet, Nutrition, Physical Activity and Cancer: A Global Perspective A Summary of the Third Expert Report. 2018. Available online: www.dietandcancerreport.org (accessed on 3 September 2020).
14. Dragomir, M.; Mullie, P.; Bota, M.; Koechlin, A.; Macacu, A.; Pizot, C.; Boyle, P. Nutrition and prostate cancer: Review of the evidence. *J. Health Inequal.* **2019**, *5*, 155–173. [CrossRef]
15. Zdrojewski, T. Completing the cardiovascular revolution: Smoking, alcohol, diet, and air pollution. *J. Health Inequal.* **2019**, *5*, 38. [CrossRef]
16. Fontaine, K.R.; Redden, D.T.; Wang, C.; Westfall, A.O.; Allison, D.B. Years of life lost due to obesity. *J. Am. Med. Assoc.* **2003**, *289*, 187–193. [CrossRef]
17. Tremmel, M.; Gerdtham, U.G.; Nilsson, P.M.; Saha, S. Economic burden of obesity: A systematic literature review. *Int. J. Environ. Res. Public Health* **2017**, *14*, 435. [CrossRef]
18. Krzyżanowska-Świniarska, B. Otyłość Prosta. In *Podręcznik Chorób Wewnętrznych*; Andrzej, S., Ed.; Medycyna Praktyczna: Krakow, Poland, 2012; pp. 1369–1373. Available online: https://medbook.com.pl/ksiazka/pokaz/id/5172/tytul/choroby-wewnetrzne-tom-1-szczeklik-medycyna-praktyczna (accessed on 3 September 2020).
19. Jarosz, M.; Respondek, W.; Rychlik, W. Economical and health-related consequences of overweight and obesity and other chronic non-communicable diseases. In *Obesity, Nutrition, Physical Activity, Health of Poles. Diagnosis of nutritional Status, Physical Activity and Nutritional Risk Factors of Obesity and Chronic Non-Communicable Diseases in Poland*; Jarosz, M., Ed.; Insytut Żywności i Żywienia: Warszawa, Poland, 2006. Available online: https://w.bibliotece.pl/2215960/Oty\T1\lość+żywienie+aktywność+fizyczna+zdrowie+Polaków (accessed on 3 September 2020).
20. Zatońska, K.; Zatoński, W.A.; Szuba, A. Prospective urban and rural epidemiology Poland—Study design. *J. Health Inequal.* **2016**, *2*, 136–141. [CrossRef]
21. Teo, K.; Chow, C.K.; Vaz, M.; Rangarajan, S.; Yusuf, S. The Prospective Urban Rural Epidemiology (PURE) study: Examining the impact of societal influences on chronic noncommunicable diseases in low-, middle-, and high-income countries. *Am. Heart J.* **2009**, *158*, 1–7.e1. [CrossRef]
22. Polskie Towarzystwo Diabetologiczne. Zalecenia Kliniczne Dotyczące Postępowania u Chorych na Cukrzycę 2019. *Diabetologia Praktyczna* **2019**, *5*, 100.
23. Szuba, A.; Martynowicz, H.; Zatońska, K.; Ilow, R.; Regulska-Ilow, B.; Różańska, D.; Wołyniec, M.; Połtyn-Zaradna, K.; Zatoński, W. Prevalence of hypertension in Polish population of PURE Poland study. *J. Health Inequal.* **2017**, *2*, 157–162. [CrossRef]
24. Biela, U.; Pajak, A.; Kaczmarczyk-Chałas, K.; Głuszek, J.; Tendera, M.; Waśkiewicz, A.; Kurjata, P.; Wyrzykowski, B. Incidence of overweight and obesity in women and men between the ages of 20–74. Results of the WOBASZ program. *Kardiologia Polska* **2005**, *63*, 632–635. Available online: https://pubmed.ncbi.nlm.nih.gov/20527435/ (accessed on 3 September 2020).
25. Matyjaszczyk, P.; Hoffmann, K.; Bryl, W. Epidemiology of chosen risk factors of cardiovascular diseases. *Przegląd Kardiodiabetologiczny* **2011**, *6*, 255–262. Available online: https://www.termedia.pl/Artykul-pogladowy-Epidemiologia-wybranych-czynnikow-ryzyka-chorob-ukladu-krazenia,47.18065.0.0.html (accessed on 3 September 2020).
26. Zatonska, K.; Janik-Koncewicz, K.; Ilow, R.; Regulska-Ilow, B.; Rozanska, D.; Szuba, A.; Einhorn, J.; Vatten, L.; Xiao-Mei, M.; Janszky, I.; et al. Prevalence of obesity—Baseline assessment in the prospective cohort 'PONS' study. *Ann. Agric. Environ. Med.* **2011**, *18*, 246–250. Available online: http://www.aaem.pl/Prevalence-of-obesity-baseline-assessment-in-the-prospective-cohort-PONS-study.71691.0.2.html (accessed on 3 September 2020).

27. Vainio, H.; Kaaks, R.; Bianchini, F. Weight control and physical activity in cancer prevention: International evaluation of the evidence. *Eur. J. Cancer Prev.* **2002**, *11*, 94–100. Available online: https://pubmed.ncbi.nlm.nih.gov/12570341/ (accessed on 3 September 2020).
28. Jarosz, M.; Rychlik, E. Overweight and obesity among adults in Poland, 1983–2005. *Adv. Med. Sci.* **2008**, *53*, 158–166. [CrossRef]
29. Ogden, C.L.; Carrol, M.D.; Kit, B.K.; Flegal, K.M. Prevalence of obesity in the United States, 2009–2010. *NCHS Data Brief.* **2012**, *82*, 1–8. Available online: https://pubmed.ncbi.nlm.nih.gov/22617494/ (accessed on 3 September 2020).
30. Befort, C.A.; Nazir, N.; Perri, M.G. Prevalence of Obesity Among Adults from Rural and Urban Areas of the United States: Findings from NHANES (2005–2008). *J. Rural Health* **2012**, *28*, 392–397. [CrossRef] [PubMed]
31. Patterson, K.A.; Gall, S.L.; Venn, A.J.; Otahal, P.; Blizzard, L.; Dwyer, T.; Cleland, V.J. Accumulated exposure to rural areas of residence over the life course is associated with overweight and obesity in adulthood: A 25-year prospective cohort study. *Ann. Epidemiol.* **2017**, *27*, 169–175.e2. [CrossRef]
32. Jokela, M.; Kivimäki, M.; Elovainio, M.; Viikari, J.; Raitakari, O.T.; Keltikangas-Järvinen, L. Urban/rural differences in body weight: Evidence for social selection and causation hypotheses in Finland. *Soc. Sci. Med.* **2009**, *68*, 867–875. [CrossRef]
33. World Health Organization. *Obesity: Preventing and Managing the Global Epidemic*; World Health Organization: Geneva, Switzerland, 2000.
34. Boyle, P.; Anderson, B.O.; Andersson, L.C.; Ariyaratne, Y.; Auleley, G.R.; Barbacid, M.; Bartelink, H.; Baselga, J.; Behbehani, K.; Belardelli, F.; et al. Need for global action for cancer control. *Ann. Oncol.* **2008**, *19*, 1519–1521. [CrossRef]
35. Mongraw-Chaffin, M.L.; Peters, S.A.; Huxley, R.R.; Woodward, M. The sex-specific association between BMI and coronary heart disease: A systematic review and meta-analysis of 95 cohorts with 1·2 million participants. *Lancet Diabetes Endocrinol.* **2015**, *3*, 437–449. [CrossRef]
36. Field, A.E.; Coakley, E.H.; Must, A.; Spadano, J.L.; Laird, N.; Dietz, W.H.; Rimm, E.; Colditz, G.A. Impact of overweight on the risk of developing common chronic diseases during a 10-year period. *Arch. Intern. Med.* **2001**, *161*, 1581–1586. [CrossRef]
37. Rutkowski, M.; Bandosz, P.; Czupryniak, L.; Gaciong, Z.; Solnica, B.; Jasiel-Wojculewicz, H.; Wyrzykowski, B.; Pencina, M.J.; Zdrojewski, T. Prevalence of diabetes and impaired fasting glucose in Poland-the NATPOL 2011 Study. *Diabet. Med.* **2014**, *31*, 1568–1571. [CrossRef]
38. Jarosz, M.; Rychlik, E. Obesity—Health and Global Challenge. 2011. Available online: http://www.pnmedycznych.pl/wp-content/uploads/2014/09/pnm_2011_712_717.pdf (accessed on 3 September 2020).
39. Artham, S.M.; Lavie, C.J.; Milani, R.V.; Ventura, H.O. Obesity and hypertension, heart failure, and coronary heart disease - Risk factor, paradox, and recommendations for weight loss. *Ochsner J.* **2009**, *9*, 124–132.

International Journal of
Environmental Research and Public Health

MDPI

Article

MALDI-TOF Protein Profiling Reflects Changes in Type 1 Diabetes Patients Depending on the Increased Amount of Adipose Tissue, Poor Control of Diabetes and the Presence of Chronic Complications

Agnieszka Zawada [1,*], Dariusz Naskręt [2], Eliza Matuszewska [3], Zenon Kokot [4], Marian Grzymisławski [1], Dorota Zozulińska-Ziółkiewicz [2], Agnieszka Dobrowolska [1] and Jan Matysiak [3]

1 Department of Gastroenterology, Dietetics and Internal Medicine, Poznan University of Medical Sciences, 60-355 Poznań, Poland; mariangrzym@ump.edu.pl (M.G.); agdob@ump.edu.pl (A.D.)
2 Department of Internal Medicine and Diabetology, Poznan University of Medical Sciences, 60-834 Poznań, Poland; dnaskret@poczta.onet.pl (D.N.); zozula@box43.pl (D.Z.-Z.)
3 Department of Inorganic and Analytical Chemistry, Poznan University of Medical Sciences, 60-780 Poznań, Poland; eliza.matuszewska@ump.edu.pl (E.M.); jmatysiak@ump.edu.pl (J.M.)
4 Faculty of Health Sciences, Calisia University Kalisz, 62-800 Kalisz, Poland; zkokot@ump.edu.pl
* Correspondence: a.zawada@ump.edu.pl; Tel.: +48-61-869-1343

Citation: Zawada, A.; Naskręt, D.; Matuszewska, E.; Kokot, Z.; Grzymisławski, M.; Zozulińska-Ziółkiewicz, D.; Dobrowolska, A.; Matysiak, J. MALDI-TOF Protein Profiling Reflects Changes in Type 1 Diabetes Patients Depending on the Increased Amount of Adipose Tissue, Poor Control of Diabetes and the Presence of Chronic Complications. *IJERPH* **2021**, *18*, 2263. https://doi.org/10.3390/ijerph18052263

Academic Editor: Wioletta Zukiewicz-Sobczak

Received: 12 January 2021
Accepted: 19 February 2021
Published: 25 February 2021

Publisher's Note: MDPI stays neutral with regard to jurisdictional claims in published maps and institutional affiliations.

Copyright: © 2021 by the authors. Licensee MDPI, Basel, Switzerland. This article is an open access article distributed under the terms and conditions of the Creative Commons Attribution (CC BY) license (https://creativecommons.org/licenses/by/4.0/).

Abstract: Introduction: Protein profiling allows the determination of the presence of proteins marking various stages of the disease, and differentiates between people at risk of various diseases. In type 1 diabetes, protein profiling had been previously used to find blood markers other than islet autoantibodies to indicate the pancreatic beta cell destruction process and to reflect the progression of type 1 diabetes mellitus (T1DM). However, T1DM is an auto-immune disease and its clinical presentation changes in time of its duration. The aim of the study: To find differences in protein profiles in patients with type 1 diabetes according to diabetes control (HbA1c > 7%) and with presence of diabetic complications or obesity. It may help to identify subgroups of patients who may need a better clinical supervision and individualized treatment. Material and methods: A group of 103 patients with auto-immunologically confirmed T1DM, and meeting the following inclusion criteria: Caucasian race, duration of diabetes >5 years, were used in the study. Criteria of exclusion: past or present cancer (treated with chemo-/radiotherapy), diseases of the liver (ALT > 3 × ULN) except for people with simple hepatic steatosis, chronic renal disease (eGFR < 30 mL/1.73 m^2/min), and acute inflammation (CRP > 5 mg/dL). The study group was divided in terms of the presence of chronic complications, obesity, or poor metabolic control (HbA1c > 7%). Protein profiling was completed by using the MALDI-TOF MS (matrix-assisted laser desorption/ionization-time of flight mass spectrometry) analyzer. Results: Differentiating proteins were identified in all of the groups. The groups burdened with complications, obesity, and poor metabolic control were characterized by increased levels of fibrinogen, complement C4 and C3. Conclusion: The groups of type 1 diabetes patients burdened with complications, obesity, and poor metabolic control were characterized by increased levels of fibrinogen, complement C4 and C3. Further detailed studies are necessary to determine more subtle changes in the proteomic profile of patients with type 1 diabetes.

Keywords: type 1 diabetes; proteomic profile; obesity; C4 complement

1. Introduction

Patients with T1DM constitute about 5–10% of the population of patients with the disease. T1DM incidence is ranging from under one to up to 60 new cases per 100,000 children per year in the under-15 age-group. Incidence varies by region. Peak incidence is usually in the 10–14-year age [1]. T1DM is also commonly recognized in adults; over half of the new cases of T1D were ≥ 20 years of age in Scotland in 2018. However, the condition often affects young people and children who will have to carefully control their glycemia for many years to prevent complications and lead a normal lifestyle. For that reason, appropriate treatment from the moment of the disease's development and preventing complications involving micro- and macroagniopathies is important. Moreover, type 1 diabetic patients are at risk of becoming overweight and obese. This risk is increased with the duration of diabetes, which is associated with high daily doses of insulin. For that reason, optimum glycaemia values should be strived for, individually adjusted minimal effective doses of insulin should be administered, and most importantly, patients should be properly educated about their condition [2]. The pathophysiology of T1DM is complex. In people with a genetic predisposition, viral infection or other environmental factors may cause activation of the immune process and clinical manifestation of the disease. In the available literature, protein profiling has been applied to type 1 diabetes to find blood markers other than islet autoantibodies to indicate the pancreatic beta cell destruction process and to reflect the progression of T1DM [3]. Other works have been performed using a mass spectrometric analysis to generate a proteomic profile of protein abundance and post-translational modifications in the aorta and kidneys of diabetic rats [4]. The majority of studies using protein profiling have determined differences in the amount of individual proteins between groups of patients with type 1 diabetes and healthy individuals. Other studies have focused on changes in the protein profile that occur during the autoimmune process during the development of type 1 diabetes and the relationship between protein profiling and the development of complications [5]. Our study is the first differentiating the protein profile in a homogeneous group of patients with autoimmune disease like T1DM. This group is well characterized, selected and treated with a unified method—the method of intensive functional insulin therapy. The patients included in the study were characterized by good compliance. The size of the group is significant, and differences in the protein profile determine the detail of the study in such a homogeneous group of patients. In this study, we focused on demonstrating the difference in protein composition in patients with T1DM, depending on the characteristics acquired during the course of the disease, such as excess body weight or the presence of complications. Demonstration of a difference in the protein profile in subgroup of T1DM patients may be clinically relevant.

In our study, we wanted to find out the differences in protein profiles in patients with T1DM related to the presence of diabetic complications, obesity, and diabetes control depending on the value of glycated hemoglobin > 7%. The demonstration of a difference in the protein profile due to presence of complications and obesity in T1DM patients may help to identify subgroups of patients who may need a better clinical supervision and individualized treatment.

2. Materials and Methods

2.1. Study Groups

The study was carried out on a group of 103 patients with T1DM treated in the Department of Gastroenterology, Dietetics, and Internal Diseases, and the Department of Internal Medicine and Diabetology in 2016–2020, who met the defined inclusion criteria. The inclusion criteria were T1DM confirmed by the presence of ICA, IA2, and GAD autoantibodies; treated with intensive functional insulin therapy or with multiple injections of constant insulin doses; age > 18 years; Caucasian race; duration of diabetes > five years; and patient informed consent to participation in the study. The exclusion criteria were other types of diabetes; past or present cancer treatment with radio-or chemotherapy; liver diseases (ALT > 3 × ULN), except for people with simple hepatic steatosis; chronic renal disease (eGFR < 30 mL/1.73 m^2/min); acute inflammation (CRP > 5 mg/dL); status post ischemic or haemorrhagic cerebral stroke (<6 months); unstable ischemic heart disease (status post STEMI with the implantation of drug-eluting stents, non-STEMI < 12 months); and other diseases (mental disorders, nutritional disorders, e.g., anorexia, bulimia), other treatment (antibiotic therapy < 3 months, and current steroid therapy), and dependence on alcohol (consumption of more than one unit of alcohol a day). The study group consisted of people with T1DM mostly treated with intensive functional insulin therapy, with mean age 34 years. The duration of diabetes was significantly longer than the inclusion criterion. All patients had a history of T1DM longer than 12 years. The clinical characteristics of the study population are given in Table 1.

Table 1. Clinical characteristic of the study group (n = 103). Data are presented as medians with interquartile ranges.

Variable	Median (IQR)
Sex [M/F], n (%)	50 (48.5)/ 53 (51.5)
Age [y]	34 (30–42)
DD [y]	17 (12–23)
IFI [y]	12 (8–18)
WHR [n]	0.9 (0.8–0.9)
BMI [kg/m^2]	26 (23–29)
TBF [kg]	19 (14–27)
VF [n]	5 (3–8)
DDI [µ/kg/d]	0.5 (0.4–0.6)
HbA1c [%]	8 (7–9)
AST [U/L]	19 (16–24)
ALT [U/L]	19 (14–26)
Creatynine, [µmol/L]	80 (71–88)
eGFR [mL/min/1.73 m^2]	88 (77–90)
hsCRP [mg/dL]	2 (1–3)
T-ch [mmol/L]	48 (42–54)
TAG [mmol/L]	1 (1–2)
HDL-ch [mmol/L]	2 (1–2)
LDL-ch [mmol/L]	3 (2–3)
non-HDL-c [mmol/L]	3 (3–4)

Abbreviations: ACR—albumin/creatinine ratio; ALT—alanine aminotransferase; AST—aspartate aminotransferase; BMI—body mass index; DD—diabetes duration; DDI—daily insulin dose, eGFR; estimated glomerular filtration rate; eGDR, HbA1c—glycated hemoglobin A1c; HDL-C—high density lipoprotein cholesterol; LDL-C—low-density lipoprotein cholesterol; TAG—triglycerides; TBF—total body fat; VF—visceral fat; WHR—waist-to-hip ratio. T-Ch—total cholesterol.

2.2. Methods

The study group was divided into three subgroups according to the presence of:

1. Obesity (BMI < 30 vs. BMI > 30 kg/m^2): BMI was calculated from the formula of the ratio of body weight to squared height.
2. Presence of diabetic complications (retinopathy, nephropathy, and neuropathy, all complications or any of them)
3. HbA1c value < 7% and > 7%: A reasonable A1C goal according to American Diabetes Association Guidelines 2020 for many nonpregnant adults [4].

2.2.1. Assessment of Diabetic Complications

Assessment of Diabetic Kidney Disease (DKD)

The elimination of albumin with urine was assessed based on a 12-h urine collection, with a determination of the albumin/creatinine index in the morning sample of urine. Albumin elimination between 30 and 300 mg a day with urine on two of three urine collection days and the albumin/creatinine index of >30 mg/g in morning urine sample were deemed positive albuminuria. Diabetic kidney disease was diagnosed in case of the presence of pathological albuminuria and at least a 10-year history of diabetes, or co-existence of diabetic retinopathy. DKD was divided into stages based on the estimated glomerular filtration rate: G1 stage (Egfr $\geq$ 90 mL/min/1.73 m^2), G2 stage (eGFR 60–89 mL/min/1.73 m^2), G3a stage (eGFR 45–59 mL/min/1.73 m^2), G3b stage (eGFR 30–45 mL/min/1.73 m^2), G4 stage (eGFR 15–29 mL/min/1.73 m^2), and G5 stage (eGFR <15 mL/min/1.73 m^2 or treatment with dialysis).

Assessment of Diabetic Retinopathy

An ophthalmoscopic examination of eye fundus was carried out following dilation of the pupil. Diabetic retinopathy was diagnosed when at least one micro-aneurysm was found in each eye. Diabetic retinopathy was classified acc. to the Polish Diabetes Association as: no evidence of diabetic retinopathy, mild non-proliferative diabetic retinopathy (NPDR), moderate non-proliferative diabetic retinopathy, severe non-proliferative diabetic retinopathy, and proliferative diabetic retinopathy (PDR) [6].

Assessment of Diabetic Neuropathy

Diabetic neuropathy was assessed with a 10 g Semmes–Weinstein monofilament test. The vibration sensation was assessed using a tuning fork (128 MHz), and the temperature sensation was assessed using a roller with a metal and plastic tip (Tiptherm), and testing the tarsal reflex. Peripheral neuropathy was assessed based on the presence of two or more components: presence of neuropathy symptoms, impaired touch and/or vibration sensation, or absence of the tarsal reflex.

2.2.2. Laboratory Analysis

Laboratory investigations were done in the "H. Święcicki" Teaching Hospital in Poznan. Laboratory measurement methods: Blood for the tests was drawn from fasting patients, using a puncture of the vein in the elbow pit. Biological materials used for the tests were full blood, plasma, and serum.

Glycated hemoglobin level (HbA1c) was assessed using the HPLC method (normal range: 4.8–7%).

Lipid profile: parameters of lipid metabolism (total cholesterol—TCh, HDL-ch fraction, LDL-ch fraction, and TAG in serum) were assayed using the standard method (laboratory standards: TCh: 130–200 mg/dL; 3.3–5.2 mmol/L, HDL-ch: M: 35–70 mg/dL; 0.9–1.8 mmol/L, HDL-ch: F: 45–80 mg/dL; 1.1–2.0 mmol/L, LDL-ch: 60–130 mg/dL; 1.5–3.4 mmol/L, TAG: 30–150 mg/dL; 0.3–1.7 mmol/L). Non-HDL-ch cholesterol level was also calculated using the formula: non-HDL-ch = total cholesterol—HDL-ch. The TAG/HDL-ch ratio was also calculated. Renal parameters: serum creatinine, normal range: F < 0.9 mg/dL, M < 1.2 mg/dL, and the estimated glomerular filtration rate (eGFR) was calculated acc. to the Modification of the Diet Renal Disease Study Equation (MDRD), normal range: 90–120 mL/min/1.73 m^2. Liver parameters: serum AST activity (reference values: F: 10–31 U/L, M: 10–35 U/L) and serum ALT values (reference values: F: 10–34 U/L, M:10–45 U/L) using the standard method.

2.2.3. MALDI-TOF MS Profiling

Protein profiling was performed in order to find proteins that were characteristic for people with diabetes complicated by obesity and inferior metabolic control (HbA1c > 7%) MALDI-TOF MS profiling and identification of discriminatory proteins and peptides.

Sample Pretreatment

Serum samples obtained from patients diagnosed with diabetes were purified and concentrated before mass spectrometry analyses. Pretreatment of the biological material was performed with ZipTip C18 (Millipore, Bedford, MA, USA) reverse phase chromatography micropipette tips. Then, 2 μL of each serum sample was mixed with 8 μL of 0.1%

trifluoroacetic acid (TFA) in water. Mixtures were then loaded onto ZipTip tips according to the manufacturer's protocol. For the tips conditioning, acetonitrile (ACN) and 0.1% TFA were used. Adsorbed peptides were first washed with 0.1% TFA in water and then eluted with 50% ACN in 0.1% TFA.

MALDI-TOF MS Analysis

For the MALDI-TOF MS (matrix-assisted laser desorption/ionization-time of flight mass spectrometry) analysis, 1 μL of each eluent obtained from the ZipTip sample pretreatment was mixed with 10 μL of matrix solution (0.3 g/L α-cyano-4-hydroxycinnamic acid (HCCA) in a 2:1 mixture of ethanol/acetone, *v/v*). The mixtures were then manually spotted onto the Anchor Chip Standard (Bruker Daltonics, Bremen, Germany) target plate in triplicates. MS analysis was performed in a linear-positive mode with Ultrafle Xtreme (Bruker Daltonics, Bremen, Germany) mass spectrometer. Ions were analyzed in the range of m/z 1000–10,000. Every MS spectrum was acquired from an average 2000 laser shots per sample. External calibration was performed with a mixture of Protein Calibration Standard I and Peptide Calibration Standard (Bruker Daltonics, Bremen, Germany) in 5:1 (*v/v*) ratio. The average mass deviation from the reference masses did not exceed 100 ppm. MALDI-TOF MS analysis was performed with the following parameters: ion source 1, 25.09 kV; ion source 2, 23.80 kV; pulsed ion extraction, 260 ns, lens 6.40 kV, matrix suppression cut off m/z 700. FlexControl 3.4 (Bruker Daltonics, Bremen, Germany) software was applied for collection and processing of the MS spectra. For the analysis of recorded MS data, Clin Pro Tools 3.0 (Bruker Daltonics, Bremen, Germany) software was applied. Statistical analyses were performed with classification algorithms (quick classifier (QC), genetic algorithm (GA), and supervised neural network (SNN)) and ROC curves [7]. QC is a chemometric algorithm that calculates peaks' average areas and uses *p*-values for peaks classification. GA, based on the process of natural selection and the idea of the evolution of the fittest individual, allows determination of the discriminatory peaks. SNN-based algorithm chooses MS data with features characteristic to studied groups and according to them, classifies spectra to the proper group. Parameters of cross-validation, external validation, and recognition capability were calculated for each algorithm. The statistical analysis resulted in the depiction of peptide candidates for the subsequent identification.

nanoLC MALDI-TOF/TOF MS Identification of Discriminatory Peaks

Identification of the peaks discriminating between the studied groups was performed with nanoLC-MALDI-TOF/TOF MS (nano-liquid chromatography-matrix-assisted laser desorption/ionization–time of flight/time of flight mass spectrometry) system. The samples were pretreated with ZipTip pipette tips and subjected to nanoLC separation. The nanoLC set consisted of EASY-nLC II (Bruker Daltonics, Bremen, Germany) nanoflow HPLC system and Proteineer-fc II (Bruker Daltonics, Bremen, Germany) collector of fractions. The nanoLC system parts were NS-MP-10 BioSphere C18 (NanoSeparations, Nieuwkoop, The Netherlands) trap column (20 mm × 100 μm I.D., particle size 5 μm, pore size 120 Å) and an Acclaim PepMap 100 (Thermo Scientific, Sunnyvale, CA, USA) column (150 mm × 75 μm I.D., particle size 3 μm, pore size 100 Å). The gradient elution method was set on 2–50% of ACN in 96 min (mobile phase A—0.05% TFA in water, mobile phase B—0.05% TFA in 90% ACN). The flow rate for separation was 300 nL/min, and the volume of the sample eluent injected into the chromatography column was 4 μL. From nanoLC separation, in total, 384 separated fractions were obtained. Each of them was mixed with a matrix solution (36 μL of HCCA saturated solution in 0.1% TFA and acetonitrile (90:10 *v/v*), 748 μL of acetonitrile and 0.1% TFA (95:5 *v/v*) mixture, 8 μL of 10% TFA, and 8 μL of 100 mM ammonium phosphate) and spotted automatically onto the AnchorChip Standard (Bruker Daltonics, Bremen, Germany) target plate by the collector of fractions. HyStar 3.2 (Bruker Daltonics, Bremen, Germany) software was used for the nanoLC system operating. For the MS analysis, UltrafleXtreme (Bruker Daltonics, Bremen, Germany) mass spectrometer working in a reflector mode in the range of m/z 700–3500 was used. External calibration was performed with a mixture of Peptide Calibration Standard (Bruker Daltonics, Bremen, Germany). A list of the precursor ions for the identification was established with WARP-LC (Bruker Daltonics, Bremen, Germany) software. Applied settings for MS and MS/MS mode were ion source 1, 7.50 kV; ion source 2, 6.75 kV; reflectron 1, 29.50 kV; reflectron 2, 14.00 kV; lens, 3.50 kV; lift 1, 19.00 kV; lift 2, 3.00 kV; and pulsed ion extraction time, 80 ns. For the spectra acquisition, processing and evaluation FlexControl 3.4, FlexAnalysis 3.4, and BioTools 3.2 (Bruker Daltonics, Bremen, Germany) software were used. For the identification of discriminative proteins and peptides, a SwissProt database and Mascot 2.4.1 search engine with taxonomic restriction to *Homo sapiens* were applied. The protein search parameters were as follows: fragment ion mass tolerance m/z $\pm$ 0.7, precursor ion mass tolerance $\pm$ 50 ppm, peptide charge + 1, and monoisotopic mass.

2.2.4. Statistical Analysis

The statistical analysis was performed using Statistica PL version 13.3. The conformity of the interval data distribution with the normal distribution was assessed using the Kolmogorov–Smirnov test. In most of the data, no normal distribution was observed. In the analysis, a statistical method for non-parametric variables, Mann–Whitney's U-test, was used. The results were presented as numbers and percentages, as well as medians and interquartile range (IQR). The value of $p < 0.05$ was assumed to be statistically significant.

3. Results

Incidence of complications are presented in Table 2. The presence of one complication qualified the patient to a group with complications.

Table 2. Incidence of complication in the study group.

Complications	Incidence Frequency in Numbers (n = 103)	Incidence Frequency as a Ratio (103 = 100%)
Diabetic retinopathy	39	37.9%
Diabetic renal disease	9	8.7%
Autonomic neuropathy	21	20.4%
Peripheral neuropathy	32	31.1%
The group with any complications	54	52.4

Characteristics of the study group according to presence of excess body fat, complication, and value of HbA1c are presented in Table 3.

Table 3. Clinical characteristics of the study groups (excess body fat vs. normal body fat), (diabetes control, HbA1c < 7) and (presence of complication vs. absence of complication). Data are presented as medians with interquartile ranges.

Value	EBF N = 46	NBF N = 57	EBF/NBF P < 0.05	HbA1c > 7% N = 80	HbA1c < 7% N = 23	HbA1cp < 0.05	PofC N = 55	AofC N = 48	PofC/ AofC P < 0.05
Age [y]	35.0 (31.0–46.0)	33.0 (29.0–40.0)	0.09	34.0 (25.0–40.0)	35.0 (31.0–42.0)	0.64	38.0 (32.0–47.0)	32.0 (24.0–37.0)	0.00
Sex [M/F]	23/23	27/30	0.82	35/45	15/8	0.07	25/30	25/23	0.51
DD [y]	18.0 (12.0–23.5)	16.0 (11.0–23.0)	0.90	15.0 (11.0–20.0)	19.0 (7.0–21.0)	0.38	20.0 (15.0–29.0)	13.0 (8.0–17.0)	0.00
TBF [kg]	27.6 (23.1–33.2)	14.1 (10.9–18.2)	0.00	21.7 (14.1–29.7)	17.3 (11.8–20.0)	0.07	20.0 (14.1–27.1)	18.2 (13.5–25.0)	0.32
BMI [kg/m^2]	29.3 (27.8–30.6)	23.7 (21.4–25.7)	0.00	27.1 (24.0–29.8)	24.2 (22.1–28.3)	0.06	26.4 (23.4–30.3)	27.1 (23.0–28.8)	0.66
VF [n]	8.0 (6.5–10.0)	4.0 (2.0–5.0)	0.00	5.0 (3.0–8.0)	6.0 (3.0–7.0)	0.93	6.0 (4.0–8.0)	5.0 (2.0–7.0)	0.12
WHR [n]	0.9 (0.8–0.9)	0.8 (0.8–0.9)	0.00	0.9 (0.8–0.9)	0.8 (0.9–1.0)	0.04	6.0 (4.0–8.0)	5.0 (2.0–7.0)	0.05

Table 3. *Cont.*

Value	EBF N = 46	NBF N = 57	EBF/NBF P < 0.05	HbA1c > 7% N = 80	HbA1c < 7% N = 23	HbA1cp < 0.05	PofC N = 55	AofC N = 48	PofC/ AofC P < 0.05
HbA1c [%]	8.4 (7.3–8.9)	7.8 (6.8–8.9)	0.18	8.9 (8.4–9.8)	6.5 (6.2–6.8)	0.00	7.9 (6.9–9.0)	8.4 (7.2–8.9)	0.75
AST [IU/L]	19.0 (15.5–27.0)	19.0 (16.0–22.0)	0.54	18.0 (16.0–27.0)	19.0 (15.0–22.0)	0.66	19.0 (15.0–24.0)	19.0 (16.0–27.0)	0.38
ALT [IU/L]	21.0 (14.0–28.5)	17.0 (14.0–25.0)	0.13	18.0 (13.0–27.0)	21.0 (15.0–25.0)	0.61	18.0 (14.0–24.0)	21.0 (14.0–29.0)	0.25
Creatinine [μmol/L]	70.7 (61.9–88.4)	77.8 (70.7–88.4)	0.19	70.7 (61.9–79.6)	79.6 (70.7–88.4)	0.14	79.6 (70.7–88.4)	79.6 (70.7–88.4)	0.18
GFR [mL/min/1.72 m^2]	88.2 (82.2–90.0)	86.0 (75.7–90.0)	0.31	90.0 (75.5–90.0)	84.1 (76.5–90.0)	0.38	83.1 (73.4–90.0)	90.0 (84.5–90.0)	0.01
CRP [mg/dL]	2.2 (1.0–4.5)	1.2 (0.6–2.1)	0.00	2.1 (1.0–4.1)	1.0 (0.4–2.0)	0.03	1.4 (0.7–3.1)	1.7 (0.8–3.1)	0.85
TCh [mmol/L]	4.9 (4.5–5.7)	4.6 (3.9–5.1)	0.00	4.8 (4.2–5.3)	4.7 (4.1–5.5)	0.96	4.9 (4.3–5.4)	4.6 (4.0–5.2)	0.31
TAG [mmol/L]	1.2 (1.0–1.8)	0.9 (0.7–1.3)	0.00	1.1 (0.9–1.7)	0.9 (0.7–1.2)	0.00	1.3 (0.9–1.5)	1.3 (0.8–1.4)	0.45
HDL-ch [mmol/L]	1.5 (1.3–1.9)	1.7 (1.4–2.0)	0.11	1.6 (1.3–1.9)	1.7 (1.4–2.3)	0.13	1.6 (1.3–2.0)	1.6 (1.3–2.0)	0.96
LDL-ch [mmol/L]	2.7 (2.5–3.3)	2.4 (1.9–2.9)	0.01	2.6 (2.1–3.1)	2.5 (1.9–3.4)	0.84	2.7 (2.6–3.2)	2.5 (1.9–3.1)	0.18
non-HDL-ch [mmol/L]	3.2 (2.9–4.0)	2.7 (2.2–3.3)	0.00	3.1 (2.6–3.6)	2.7 (2.3–3.9)	0.41	3.2 (2.7–3.6)	2.8 (2.4–3.8)	0.33
ACR [mg/d]	3.5 (2.5–5.2)	3.8 (2.7–5.4)	0.82	3.8 (2.5–5.4)	3.7 (2.9–5.3)	0.88	4.3 (2.9–8.1)	3.3 (2.3–4.3)	0.00

Abbreviations: ACR—albumin/creatinine ratio; AofC—absence of complication; ALT—alanine aminotransferase; AST—aspartate aminotransferase; BMI—body mass index; DD—diabetes duration; DDI—daily insulin dose, EBF—excess body fat; eGFR—estimated glomerular filtration rate; eGDR, HbA1c—glycated hemoglobin A1c; HDL-C—high-density lipoprotein cholesterol; LDL-C—low-density lipoprotein cholesterol; NBF—normal body fat; PofC—presence of complication; TAG—triglycerides; TBF—total body fat; VF—visceral fat; WHR—waist-to-hip ratio; T-Ch—total cholesterol.

The results of MALDI-TOF MS analysis are presented in Tables 4–6 according to division due to excess fat, diabetes control (HbA1c > 7%), and presence of diabetic complication. Spectra recorded for fibrinogen alpha chain, complement C3, and complement C4A are presented in Figures 1–3.

Table 4. Division due to excess body fat.

Division Due to Excess Fat				
Model	**Cross Validation [%]**	**Recognition Capability [%]**	**External Validation—Correct Classified Part of Valid Spectra [%]—TEST**	**External Validation—Correct Classified Part of Valid Spectra [%]—CONTROL**
GA	49.5	93.8	52.9	82.1
SNN	59.2	67.1	60.8	38.5
QC	58.0	63.4	43.1	74.4
Identified peaks (m/z) classified as discriminatory based on **GA**				
	1537.88		fibrinogen alpha chain	
	1519.99		complement C3 (oxidation)	
	1449.61		complement C4A	
Identified peaks (m/z) classified as discriminatory based on **SNN**				
	1519.99		complement C3 (oxidation)	
	1537.88		fibrinogen alpha chain	
Identified peaks (m/z) classified as discriminatory based on **QC**				
	1435.73		complement C4A	
	1449.61		complement C4A	
	1519.99		complement C3 (oxidation)	
	1537.88		fibrinogen alpha chain	

Table 5. Division due to diabetic control (HbA1c > 7%).

Model	Cross Validation [%]	Recognition Capability [%]	External Validation—Correct Classified Part of Valid Spectra [%]—TEST	External Validation—Correct Classified Part of Valid Spectra [%]—CONTROL
Division Due to Diabetes Control (HbA1c > 7%)				
GA	62.2	85.8	63.6	30.6
SNN	64.6	53.7	0	88.9
QC	66.9	66.9	56.8	55.6
Identified peaks (m/z) classified as discriminatory based on **GA**				
	1537.88		fibrinogen alpha chain	
	1449.61		complement C4A	
	1520.00		complement C3 (oxidation)	
Identified peaks (m/z) classified as discriminatory based on **SNN**				
	1519.99		complement C3 (oxidation)	
	1537.88		fibrinogen alpha chain	
Identified peaks (m/z) classified as discriminatory based on **QC**				
	1537.88		fibrinogen alpha chain	

Table 6. Division due to diabetes complications.

Model	Cross Validation [%]	Recognition Capability [%]	External Validation—Correct Classified Part of Valid Spectra [%]—TEST	External Validation—Correct Classified Part of Valid Spectra [%]—CONTROL
Division Due to Diabetes Complications				
GA	48.2	84.6	46.2	67.6
SNN	48.1	65.6	53.8	79.4
QC	38.0	63.1	46.2	58.8
Identified peaks (m/z) classified as discriminatory based on **GA**				
	1537.88		fibrinogen alpha chain	
	1617.79		fibrinogen alpha chain (peak 1537 phosphorylation)	
	1435.73		complement C4A	
Identified peaks (m/z) classified as discriminatory based on **SNN**				
	1537.88		fibrinogen alpha chain	
	1435.73		complement C4A	
	1520.00		complement C3 (oxidation)	
	1617.79		fibrinogen alpha chain (peak 1537 phosphorylation)	
Identified peaks (m/z) classified as discriminatory based on **QC**				
	1537.88		fibrinogen alpha chain	

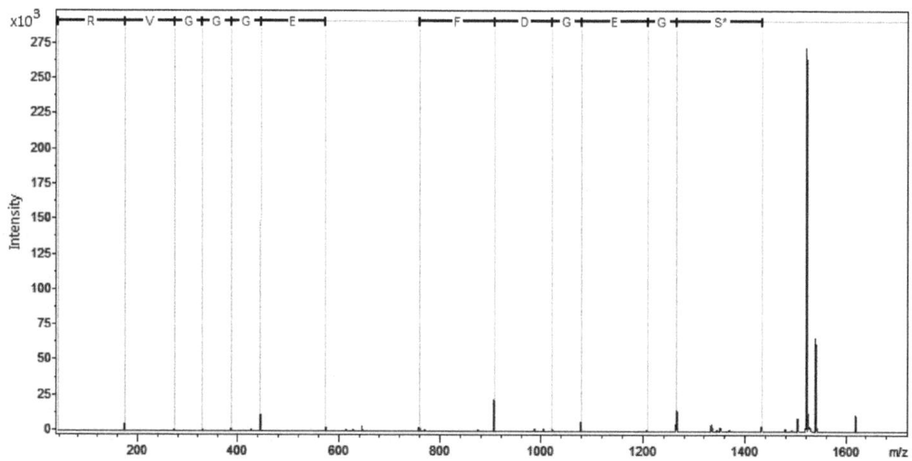

Figure 1. Fragmentation spectrum of peak 1617.79, identified as fibrinogen alpha chain.

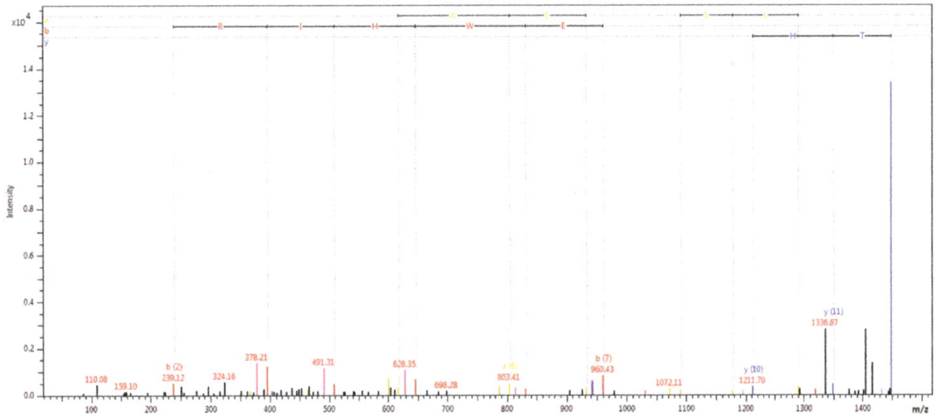

Figure 2. Fragmentation spectrum of peak 1449.61, identified as complement C4A.

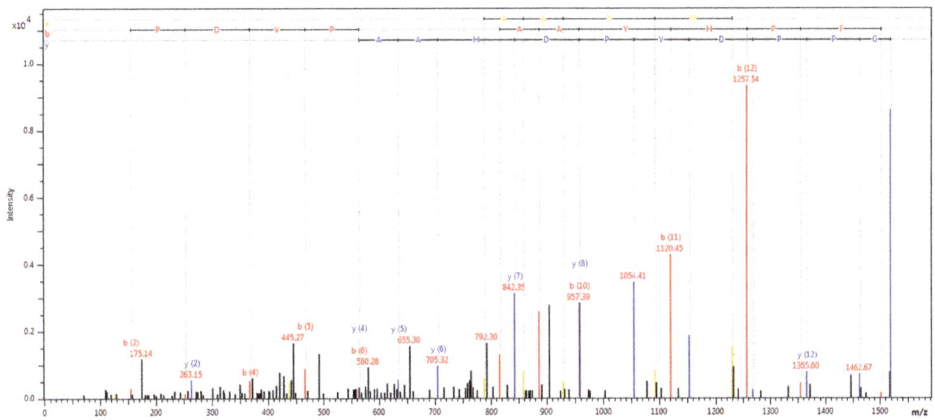

Figure 3. Fragmentation spectrum of peak 1519.99, identified as complement C3.

4. Discussion

There are not many studies that would analyze the protein profile in patients with T1DM. In our study, we wanted to find differences in metabolic profile in a homogeneous group of people with T1DM in accordance with the presence of diabetic complications, obesity, and diabetes control.

It is well known that the metabolic control of diabetes (mean glucose levels, glycated haemoglobin value, and the time when glucose levels does not remain within their normal ranges) is directly associated with the risk of the development of chronic complications [8,9]. However, there are diabetic patients who do not rapidly develop any chronic complications despite having an inferior diabetes control [10]. It is essential to find proteomic features that could allow for the differentiation of patients with the good and poor control of diabetes, which could increase the risk and accelerate the development of chronic complications. Increasing the level of C3, C4, and fibrinogen in the blood may lead the diabetologist to a more frequent diabetic follow-up visit, patient's compliance, and better glycemic control. This may inhibit the progression of chronic complications.

Based on the available literature, we were the first to assess the protein/peptide profiles in a group of patients with T1DM, depending on the presence of complications: obesity and metabolic imbalance. Various proteins were identified in all groups of patients, but changed levels of fibrinogen complement C4 and C3 were a common element found in groups burdened with obesity and poor metabolic control. Ostergaard et al. observed that circulating a complement activation product C3a level was increased in streptozotocin-induced diabetic mice as compared to control mice [11]. However, the authors did not perform protein profiling in relation to HbA1c value or the presence of complications of diabetes. It is also stressed that the complement system was emerging as a new potential target in diabetic kidney disease. In our study, we also found a changed level of C3a complement in all groups of patients with excess fat (EBF excess body fat), complication (presence of complication), and a higher value of HbA1c (HbA1c >7%). The presence of these proteomic markers could be connected with inappropriate diabetes treatment.

Despite the fact that occurrence of T1DM is associated with a genetic predisposition, its development is influenced by various environmental factors; the exact etiology of the disease is not fully known. HLA antigens are proteins found on the body's cells' surface and constitute a specific system for every human being. In white people, the presence of the HLA-DR3 and HLA-DR4 genotypes predisposes to the development of diabetes. However, even in people with this genetic predisposition, it is uncertain whether they will develop T1DM. This is dependent on the presence of factors initiating the trigger of an autoimmune process (e.g., severe infection, which initiates the process of destruction of the pancreatic beta cells). This is confirmed by studies carried out in identical twins. With the presence of T1DM in one of the twins, the probability of developing the other is 50 percent. The presence of genetic determinants is not a standard area of study in patients with T1DM, but it can affect changes in the protein profile between individual patients.

The presence of antibodies is associated with the development of type 1 diabetes mellitus, but research programs, including DASP—Diabetes Antibody Standardization Program and Environmental Determinants of Diabetes in Young Consortium, confirm that not all islet auto-antibody-positive subjects progress to T1DM.

Burch et al. attempted to identify new protein markers present in the pancreas of T1DM and type 2 diabetes (T2DM) patients, and in diabetes-free individuals who are carriers of antibodies characteristic for the disease [12]. The study identified proteins allowing differentiation between type 1 and type 2 diabetes in pancreatic tissue. Additionally, proteins' characteristics for the immunological process, independent on hyperglycemia, were identified by their absence in patients with T2DM. Among proteins differentiating between patients with T1DM and non-diabetic patients, there are complement C5, C7, C8, and C9. On the other hand, complement C3 and C9 proteins ensure differentiation between non-diabetic individuals with positive antibodies and non-diabetic individuals without antibodies [12]. In our study, complement C3, C4, and fibrinogen were proteins that differentiated the group of T1DM patients into subgroups with good (HbA1c < 7%) and poor (HbA1c > 7%) metabolic control and the occurrence of complications.

In a similar study, Nyalwidhe et al. indicated five proteins, including the C9 factor, playing an important role in the development of immunization, inflammation, and metabolic control processes in the course of the development of T1DM [13].

Comparing the metabolomic profile of patients with T1DM to healthy individuals and T2DM patients, Zhang et al. found differences in 33 peptides [14]. Among those proteins, complement C3 showed a significant down-regulation. Notably, peptides from the C1 inhibitor demonstrated a significant up-regulation in T1DM but were down-regulated in T2DM compared to healthy controls. C1 inhibitor is well known to regulate the activation of the C1 complex and may play a crucial role in blood coagulation, fibrinolysis, and suppression of inflammation. In de Oliveira's study performed on patients with T1DM compared to healthy controls, eight serum proteins were identified as being differentially expressed. C4 complement was down-regulated in diabetic patients [15]. Our study only looked at a group of diabetic patients. We did not compare them with the healthy ones and hence the possibility of different conclusions.

However, Rowe et al. found that complement activation occurred in the pancreas of patients with T1DM and that C4d might be a biomarker for T1DM. In this study, pancreatic C4d antigen expression was more prevalent in patients with T1DM than in diabetes-free subjects (diabetes-free with T1DM-associated islet auto-antibodies, auto-antibodies-negative control subjects, or those with T2DM). C4d density did not differ between diabetes-free subjects [16]. However, these studies mark the proteomic differences between healthy and diabetic subjects or type 1 and type 2 diabetes. Our study focused exclusively on type 1 diabetes.

Christine von Toerne in the study of 45 islet auto-antibody-positive and -negative children from the BABYDIAB/BABYDIET birth cohorts performed a proteomic analysis [17]. Two peptides (from apolipoprotein M and apolipoprotein C-IV) were sufficient to discriminate between auto-antibody-positive and auto-antibody-negative children. Hepatocyte growth factor activator, complement factor H, ceruloplasmin, and age predicted progression time to type 1 diabetes significantly better than age alone.

M. García-Ramírez et al. compared the protein profile of vitreous fluid from diabetic patients with proliferative diabetic retinopathy (PDR) to that from non-diabetic patients with idiopathic macular holes. Eight proteins were highly produced in PDR patients in comparison to non-diabetic subjects: zincα2-glycoprotein (ZAG), apolipoprotein (apo) A1, apoH, fibrinogen A, and the complement factors C3, C4b, C9, and factor B [18].

In our previous study on amino-acid profiling in patients subjected to controlled body weight reduction, it was found that levels of 10 AAs (α-amino-n-butyric acid, alanine, citrulline, glutamine, glycine, hydroxyproline, isoleucine, proline, sarcosine, and threonine) were significantly increased after weight loss compared to their values before the program, while aspartic acid level was decreased [19]. This confirms that changes in protein and amino-acid profile constitute an important element in the pathway of metabolic disorders occurring in the course of the development of chronic complications of diabetes. Changes in the protein profile presented in the study may also be a result of improper glycemic control.

Zhi found differences in the protein profiles of 31 proteins between T1DM patients and healthy individuals [20]. At least 21 of 31 proteins could be functionally relevant to T1DM, as they might be involved in innate immunity, inflammation, and immune response (lymphocyte activation and proliferation and glucose regulation). Complement C3b, complement C4, CRP, and adiponectin are four of those proteins. After a final validation, the differentiating value was attributed to adiponectin and myeloperoxidase. Both proteins are commonly known factors associated with the development of metabolic disorders.

5. Conclusions

The molecular mechanism underlying diabetes and its progression is not fully known. Proteomic tools are helping to advance our understanding of the origin, onset, development, treatment, and prevention of diabetes. Moreover, proteomic technologies are becoming more specific and sensitive, and their employment is a significant opportunity to expand our knowledge of T1DM. Protein profiling may also have clinical benefits for a patient with type 1 diabetes. Increased levels of C3, C4, and fibrinogen may draw the attention of a diabetologist for more frequent follow-up at the Diabetes Outpatient Clinic. The patient benefit is that individuals with proteomic markers may increase their interest in glycemic control, which may inhibit the progression of chronic complications. To sum up, advances in proteomic approaches and complete sequences of the human proteome will allow us to unravel changes in the proteomic profile of clinical samples from diabetic patients.

Author Contributions: Conceptualization: A.Z., D.N., E.M., and J.M.; methodology: A.Z., E.M., and D.N.; software: A.Z. and E.M.; validation, formal analysis: E.M. and A.Z.; investigation: A.Z., D.N., and M.G.; resources: A.D. and M.G.; data curation: A.Z., D.N., E.M., and D.Z.-Z.; writing—original draft preparation: A.Z.,E.M., D.N., M.G., A.D., Z.K., and D.Z.-Z; visualization: A.Z.; supervision: A.D., Z.K., M.G., D.Z.-Z., and J.M.; project administration: A.Z., D.N., and D.Z.-Z. All authors have read and agreed to the published version of the manuscript.

Funding: The study was supported by a Young Scientists Grant of the Poznan University of Medical Science (No. 502-14-0222-3800-10838).

Institutional Review Board Statement: The study protocol was approved by the Poznan University Institutional Review Board (5 May 2016), No. 504/16. The study protocol conforms to the ethical guidelines of the 1975 Declaration of Helsinki as reflected in a prior approval by the institution's human research committee.

Informed Consent Statement: Informed consent was obtained from all subjects involved in the study.

Conflicts of Interest: The authors declare no conflict of interest.

References

1. Diaz-Valencia, P.A.; Bougnères, P.; Valleron, A.J. Global epidemiology of type 1 diabetes in young adults and adults. A systematic review. *BMC Public Health* **2015**, *15*, 255. [CrossRef]
2. Araszkiewicz, A.; Zozulinska-Ziolkiewicz, D.; Trepinska, M.; Wierusz-Wysocka, B. Knowledge after five-day teaching program in intensive insulin therapy performed at the onset of type 1 diabetes influence the development of late diabetic complications. *Diabetes Res. Clin. Pract.* **2008**, *81*, 61–67. [CrossRef]
3. Liu, C.W.; Bramer, L.; Webb-Robertson, B.J.; Waugh, K.; Rewers, M.J.; Zhang, Q. Temporal profiles of plasma proteome during childhood development. *J. Proteom.* **2017**, *30*, 321–328. [CrossRef] [PubMed]
4. Hariri, M.A.; Elmedawar, M.; Zhu, R.; Jaffa, M.A.; Zhao, J.; Mirzaei, P.; Ahmed, A.; Kobeissy, F.; Ziyadeh, F.N.; Mechref, Y.; et al. Proteome profiling in the aorta and kidney of type 1 diabetic rats. *PLoS ONE* **2017**, *12*, e0187752. [CrossRef]
5. Merchant, M.L.; Klein, J.B. Proteomics and diabetic nephropathy. *Semin. Nephrol.* **2007**, *27*, 627–636. [CrossRef] [PubMed]
6. Polish Diabetes AssociationStandards of Medical Care in Diabetesd. *Diabetes Care* **2020**, *6*, 1–106.
7. Matuszewska, E.; Matysiak, J.; Bręborowicz, A.; Olejniczak, K.; Kycler, Z.; Kokot, Z.J.; Matysiak, J. Proteomic features characterization of Hymenoptera venom allergy. *Allergy Asthma Clin. Immunol.* **2019**, *15*, 77. [CrossRef]
8. Nordwall, M.; Arnqvist, H.J.; Bojestig, M.; Ludvigsson, J. Good glycemic control remains crucial in prevention of late diabetic complications – the Linköping Diabetes Complications Study. *Pediatric Diabetes* **2009**, *10*, 168–176. [CrossRef] [PubMed]
9. Zhang, L.; Krzentowski, G.; Albert, A.; Lefebvre, P.J. Risk of Developing Retinopathy in Diabetes Control and Complications Trial Type 1 Diabetic Patients With Good or Poor Metabolic Control. *Diabetes Care* **2001**, *24*, 1275–1279. [CrossRef] [PubMed]
10. Hoferab, S.E.; Bennettsc, B.K.; Chana, A.; Hollowaya, B.; Karschimkusd, C.J.; Jenkinsd, A.; Silinkae, M.; CDonaghueae, K. Association between PON 1 polymorphisms, PON activity and diabetes complications. *J. Diabetes Its Complicat.* **2006**, *20*, 322–328. [CrossRef] [PubMed]
11. Løstergaard, J.A.; Ruseva, M.M.; Malik, T.H.; Hoffmann-Petersen, I.T.; Pickering, M.C.; Thiel, S.; Hansen, T.K. Increased Autoreactivity of the Complement-Activating Molecule Mannan-Binding Lectin in a Type 1 Diabetes Model. *J. Diabetes Res.* **2016**, *2016*, 1–7. [CrossRef] [PubMed]
12. Burch, T.C.; Morris, M.A.; Campbell-Thompson, M.; Pugliese, A.; Nadler, J.L.; Nyalwidhe, J.O. ProteomicAnalysisofDiseaseStratified HumanPancreasTissueIndicatesUnique SignatureofType1Diabetes. *PLoS ONE* **2015**, *10*, e0135663. [CrossRef]
13. Nyalwidhe, J.O.; Grzesik, W.J.; Burch, T.C.; Semeraro, M.L.; Waseem, T.; Gerling, I.C.; Mirmira, R.G.; Morris, M.A.; Nadler, J.L. Comparative quantitative proteomic analysis of disease stratified laser captured microdissected human islets identifies proteins and pathways potentially related to type 1 diabetes. *PLoS ONE* **2017**, *12*, e0183908. [CrossRef] [PubMed]
14. Zhang, Q.; Fillmore, T.L.; Schepmoes, A.A.; Clauss, T.R.W.; Gritsenko, M.A.; Mueller, P.W.; Rewers, M.; Atkinson, M.A.; Smith, R.D.; Metz, T.O. Serum proteomics reveals systemic dysregulation of innate immunity in type 1 diabetes. *J. Exp. Med.* **2013**, *210*, 191–203. [CrossRef] [PubMed]
15. De Oliveira, V.N.; Limaneto, A.B.M.; Vantilburg, M.F.; Deoliveira, A.C.; Moreira, M.; Duartepintolobo, M.; Rondina, D.; Fernandes, V.O.; Diasrangelmontenegro, A.P.; Montenegrojúnior, R.M.; et al. Proteomic analysis to identify candidate biomarkers associated with type 1 diabetes. *Diabetes Metab. Syndr. Obes.* **2018**, *11*, 289–301. [CrossRef] [PubMed]
16. Rowe, P.; Wasserfall, C.; Croker, B.; Thompson, M.C.-; Pugliese, A.; Atkinson, M.; Schatz, D. IncreasedComplementActivationin HumanType1DiabetesPancreata. *Diabetes Care* **2013**, *36*, 3815–3817. [CrossRef] [PubMed]
17. Vontoerne, C.; Laimighofer, M.; Achenbach, P.; Beyerlein, A.; Lasherasgala, T.; Krumsiek, J.; Theis, F.J.; Ziegler, A.G.; Hauck, S.M. Peptide serum markers in islet autoantibody-positive children. *Diabetologia* **2017**, *60*, 287–295. [CrossRef] [PubMed]
18. García-Ramírez, M.; Canals, F.; Hernández, C.; Colomé, N.; Ferrer, C.; Carrasco, E.; García-Arumí, J.; Simó, R. Proteomic analysis of human vitreous fluid by fluorescencebased difference gel electrophoresis (DIGE): A new strategy for identifying potential candidates in the pathogenesis of proliferative diabetic retinopathy. *Diabetologia* **2007**, *50*, 1294–1303.
19. Moszak, M.; Klupczynska, A.; Kanikowska, A.; Kokot, Z.J.; Zawada, A.; Grzymisławska, M.; Grzymisławski, M. The influence of 3-week body mass reduction program on the metabolic parameters and free amino acid profiles in adult Polish people with obesity. *Adv. Clin. Exp. Med.* **2018**, *27*, 749–757. [CrossRef]
20. Zhi, W.; Sharma, A.; Purohit, S.; Miller, E.; Bode, B.; Anderson, S.W.; Reed, J.C.; Steed, R.D.; Steed, L.; Hopkins, D.; et al. Discovery and Validation of Serum Protein Changes in Type 1 Diabetes Patients Using High Throughput Two Dimensional Liquid Chromatography-Mass Spectrometry and Immunoassays. *Mol. Cell. Proteom.* **2011**, *10*, M111.02203–M111.02210. [CrossRef] [PubMed]

International Journal of
Environmental Research and Public Health

MDPI

Article

A Spatio-Temporal Analysis of the Health Situation in Poland Based on Functional Discriminant Coordinates

Mirosław Krzyśko [1], Waldemar Wołyńki [2,*], Marcin Szymkowiak [3,4] and Andrzej Wojtyła [5]

1 Interfaculty Institute of Mathematics and Statistics, Calisia University-Kalisz, 62-800 Kalisz, Poland; mkrzysko@amu.edu.pl
2 Faculty of Mathematics and Computer Science, Adam Mickiewicz University, 61-614 Poznań, Poland
3 Institute of Informatics and Quantitative Economics, Poznań University of Economics and Business, 61-875 Poznań, Poland; m.szymkowiak@ue.poznan.pl
4 Statistical Office in Poznań, 60-624 Poznań, Poland
5 Health Sciences Faculty, Calisia University-Kalisz, 62-800 Kalisz, Poland; a.wojtyla55@gmail.com
* Correspondence: wolynski@amu.edu.pl

Abstract: The aim of this study was to investigate if the provinces of Poland are homogeneous in terms of the observed spatio-temporal data characterizing the health situation of their inhabitants. The health situation is understood as a set of selected factors influencing inhabitants' health and the healthcare system in their area of residence. So far, studies concerning the health situation of selected territorial units have been based on data relating to a specific year rather than longer periods. The task of assessing province homogeneity was carried out in two stages. In stage one, the original spatio-temporal data space (space of multivariate time series) was transformed into a functional discriminant coordinates space. The resulting functional discriminant coordinates are synthetic measures of the health situation of inhabitants of particular provinces. These measures contain complete information regarding 8 diagnostic variables examined over a period of 6 years. In the second stage, the Ward method, commonly used in cluster analysis, was applied in order to identify groups of homogeneous provinces in the space of functional discriminant coordinates. Sixteen provinces were divided into four clusters. The homogeneity of the clusters was confirmed by the multivariate functional coefficient of variation.

Keywords: health policy; health inequalities; healthcare access; spatial distribution of the health situation; cluster analysis; functional discriminant coordinates; multivariate functional coefficient of variation; spatio-temporal data

Citation: Krzyśko, M.; Wołyński, W.; Szymkowiak, M.; Wojtyła, A. A Spatio-Temporal Analysis of the Health Situation in Poland Based on Functional Discriminant Coordinates. *IJERPH* **2021**, *18*, 1109. https://doi.org/10.3390/ijerph18031109

Academic Editor: Paul B. Tchounwou
Received: 7 December 2020
Accepted: 22 January 2021
Published: 27 January 2021

Publisher's Note: MDPI stays neutral with regard to jurisdictional claims in published maps and institutional affiliations.

Copyright: © 2021 by the authors. Licensee MDPI, Basel, Switzerland. This article is an open access article distributed under the terms and conditions of the Creative Commons Attribution (CC BY) license (https://creativecommons.org/licenses/by/4.0/).

1. Introduction

Health is universally regarded as one of the most highly appreciated values. Good health is the main factor contributing to people's well-being, which enhances their opportunities to participate in social life and to benefit from economic and employment growth. Better health is also consistently associated with greater life satisfaction (Ngamaba et al. [1]). A good health situation is instrumental in achieving good labor market outcomes. By reducing the individual's capacity to work long hours, a deteriorated health status decreases their chance of getting employed and being productive at work and has a strong impact on the labor market situation (James et al. [2], OECD [3]).

For the purpose of this study, the term "health situation" is understood as inhabitants' health described by the set of selected indicators and healthcare access in their area of residence. According to Penchansky and Thomas [4], healthcare access can be defined as a multi-faceted concept expressing the "degree of fit" between clients (patients) and the healthcare system according to five important dimensions: availability, accessibility, accommodation, affordability, and acceptability. In our article, we focus mainly on the first dimension, i.e., availability, which represents the "spatial" component of healthcare access. This choice is motivated by the fact that our understanding of availability is the same as that presented by Penchansky and Thomas [4], i.e., as the relationship between the number and type of existing services (and resources) and the number of patients and types of their needs. In other words, it represents the adequacy of the supply of physicians, dentists, and other providers; of facilities, such as clinics and hospitals; and of specialized programs and services, such as mental health and emergency care. The second dimension highlighted by Penchansky and Thomas [4], i.e., accessibility, also represents the "spatial" component of healthcare access and is understood as the relationship between the location of supply and the location of clients, while accounting for clients' transportation resources and travel time, distance and cost. In our approach, however, we do not take this spatial component into account, mainly due to limited data availability. However, it should be emphasized that the literature on accessibility in the context of healthcare access is wide (see, for instance, Wang [5] and Neutens [6]). There is also a rich literature devoted to applications of the access concept to various health care services, e.g., regarding health care accessibility analyzed in connection with availability (Barbarisi et al. [7], Bruno et al. [8], Lu et al. [9], Okuyama et al. [10], Pu et al. [11]).

The importance attached to health by different organizations is reflected in the way health protection is implemented in public policies, particularly in health policy, which aims to have a positive influence on population health (De Leeuw et al. [12]). This policy can be understood as government decisions and plans of action to make progress towards achieving the goals of the health system: improved health status of the population, better financial risk protection, and better client satisfaction; or intermediate outcomes for health systems, which include: quality, access, and efficiency (Campos and Reich [13]). In many countries, the main objective of public health policy is to create the conditions for good and equitable health for the entire population and within specific groups and to eliminate avoidable health inequalities. It should also be underlined that decisions made in sectors outside of public health and health care, such as education, transportation, and criminal justice, strongly affect health and well-being (Pollack et al. [14]).

As mentioned above, the main objective of activities in the area of health policy is to improve the health of the population, and this improvement is nowadays understood in two ways (Łyszczarz [15]). First, in terms of improving the average health status, e.g., measured in terms of life expectancy or premature mortality. Second, increasing the importance attached to the issue of inequalities in health. The term 'health inequality' refers to differences in the health of individuals or specific subgroups; any measurable aspect of health that varies across individuals or according to socially relevant groupings can be called a health inequality (Boyle [16], Kawachi et al. [17], Arcaya et al. [18]). The aim of actions in the field of health policy is to reduce such inequalities. Large inequalities in health status exist across population groups, countries and specific regions within countries (Wojtyła-Buciora et al. [19], Wojtyła et al. [20]). These health inequalities are linked to many factors, including differential exposure to health risk factors and access to health care (Samet [21]). Inequalities in health are mainly manifested as differences in the health status between socioeconomic groups, but they can also described in terms of employment status, sex or geographic location (Crombie et al. [22]). The prospect of reducing inequalities seems to be increasingly important in contemporary research trends on the implementation of health policy in the world (see, e.g., Spinakis et al. [23]). It should also be emphasized that health inequalities have to be considered as a global problem, which not only affects populations of the poorest countries and regions but also those of the richest ones; persistent health inequalities are among the most serious and challenging health problems worldwide (Barreto [24]). How policies can reduce the main factors of health inequality and promote health equality will be a key challenge for public health in the future.

Monitoring population health and eliminating health inequalities are essential activities aimed at maintaining and improving public health. The main goals of health monitoring involve measuring the extent of health problems, their trends, and the degree of variation between different population groups, including spatial distribution, as well as identifying priority areas for public health (Pizot et al. [25]). Another objective of monitoring is to track the current health situation at the national and local level. This is especially important today, in the era of the Covid-19 pandemic, when up-to-date information is required at lower levels of spatial aggregation.

The unequal spatial distribution of resources, such as clinics, hospitals, nurses, pharmacies, or doctors, could make entire communities more vulnerable and less resilient to adverse health effects. That is why the health situation needs to be investigated by accounting for spatial differences to gain a deeper understanding of why and how some geographical areas experience different health than others (Ozdenerol [26]). Understanding the role played by location in shaping the geographic distribution of the health situation within countries is critical for informing appropriate public health policy regarding prevention and treatment (Casper et al. [27]).

There are numerous articles about the spatial variation in the health situation, health inequalities or health conditions at the local or national level (see, for instance, Gilliland et al. [28], Wang and Nie [29], Chen et al. [30]). In the case of Poland, various analyses have been conducted to investigate regional inequalities in the health status of the population (Wierzbicka [31], Bem et al. [32]).

Interestingly, all studies mentioned above were based on data for a specific year or, in some cases where comparative analysis was involved, for two years (e.g., Shi et al. [33], Hübelová et al. [34]). If the authors of these articles chose, say, p variables describing the health situation of a given territorial unit, then the obtained data were p-dimensional vectors or points in a p-dimensional Euclidean space.

This article presents a more general approach to investigating the health situation across territorial units, which is based on spatio-temporal data. This kind of data is more general than static vector data as it takes into account changes that happen over time. The statistical methodology involving the use of functional discriminant coordinates and cluster analysis is applied to available data for Poland. However, this approach can be used to investigate the health situation or other phenomena at lower levels of spatial aggregation in other countries. For this reason, its results may be useful for policy-makers in the field of public health. The data to measure the health situation in Poland come from the Local Data Bank (LDB). Several important variables related to the health situation observed at the level of districts (LAU—also called poviats) located within provinces (approximately equivalent to NUTS2 (regions) level and also called voivodships) (see Section 2) in the period 2013–2018 were taken into account in the analysis. More specifically, each district is described by 8 variables representing the situation over 6 years. The data for 380 districts were arranged in the form of a matrix with 6 rows and 8 columns, containing a total of 18,240 numerical values.

The main aim of this article is to determine whether Polish provinces are homogeneous in terms of spatio-temporal data characterizing their health situation. In order to answer this question, three multivariate statistical methods were used: multivariate functional discriminant coordinates analysis (MFDCA), functional cluster analysis (FCA), and the multivariate functional coefficient of variation (MFCV).

In the first step, spatio-temporal data were transformed into functional data by applying a continuous function of time t (see, e.g., Górecki and Krzyśko [35]). Functional data can be regarded as realizations of the random process $X(t)$. Then, functional discriminant coordinates were constructed in the functional data space, and further calculations were performed in the functional discriminant coordinate space.

At this point, an important question arises: do the functional data recorded as continuous functions really exist and can these multivariate functions actually be derived? This question is critical because, in practice, values of an observed random process are always recorded in discrete moments in time, sparsely or densely distributed in the interval of variability over time. Thus, in this case, we encounter a time series or, in other words, a highly-dimensional vector of observations. However, there are numerous reasons why it is useful to model a time series as a continuous function (elements of a certain functional space); one of them is that functional data have many advantages in comparison to other representations of time series. In particular, the MFDCA derived in the present study has the following statistical advantages:

- Firstly, functional data are normally used to cope with the problem of missing observations, which is inevitable in many areas of applied research. Unfortunately, most methods concerning data analysis require complete time series. The removal of a time series with missing observations from a data set is one of popular solutions, but this can lead, and in most cases does lead, to serious data loss. Another possibility is to use one of the many methods of missing data prediction, but, in that case, the results will depend on the interpolation method. Contrary to these approaches, in the case of functional data, the problem of missing observations is resolved by expressing a given time series in the form of a continuous function set.
- Secondly, in the statistical development of MFDCA, the structure of observations is naturally retained when using functional data, i.e., the temporal link is maintained and the information regarding any measurement is taken into account. Consequently, results are assumed to be robust.
- Thirdly, moments of observation do not have to be equally spaced in a particular time series, which can be a major advantage in online applications.
- Fourthly, when using functional data, one avoids the problem of dimensionality. When the total number of time points in which observations are made exceeds the number of time series under analysis, most statistical methods do not provide satisfactory results because of misleading false estimates. In the case of functional data, this problem can be avoided because the time series are replaced by a set of continuous representative functions, which are independent of the time points in which observations are made.

The construction of functional discriminant coordinates is described in Górecki et al. [36], and their application to fruit data can be found in Hanusz et al. [37]. Two other proposals are: kernel discriminant coordinates (Krzyśko et al. [38]) and discriminant coordinates with the additional condition imposed on the covariance matrix (Krzyśko et al. [39]).

In the second step, cluster analysis was used to distinguish between groups of homogeneous provinces. Ward's hierarchical clustering method was chosen as a commonly used technique in cluster analysis. Moreover, to determine whether obtained clusters are homogeneous, a functional multivariate coefficient of variation was applied.

The main value of this article, according to its authors, is the proposed statistical methodology. Despite the use of country-specific data for the purpose of spatial analysis of the health situation, the presented methods are universal and can be successfully applied to any territorial unit and spatio-temporal dynamic data connected to other phenomena (e.g., poverty or the labor market situation at lower levels of spatial aggregation).

This article is organized as follows. Section 2 contains a short description of data used to analyze differences in the health situation across Polish provinces. The section also provides a description of the administrative division of Poland and details of the procedure of data standardization, as well as their transformation into functional data. Section 3 presents the statistical methodology involving the use of functional discriminant coordinates, cluster analysis, and the functional multivariate coefficient of variation. How this approach was applied to real data describing the health situation in Poland is described in Section 4. Finally, concluding remarks and further steps to be taken in the future are provided in Section 5.

2. The Data

The original data set contains values of $p = 8$ variables characterizing the health situation of the population (see Table 1). All variables come from the LDB, which is Poland's largest database of information relating to the economy, society and the environment. Data and statistical indicators in the LDB describe entire country, as well as units representing three NUTS levels: macroregions (NUTS1), regions (NUTS2), and subregions (NUTS3).

Table 1. List of variables used in analysis.

Variable	Description	Type of Variable
1	Nurses and midwives per 10,000 population	S
2	Doctors per 10,000 population	S
3	Population per generally available pharmacy	D
4	Deaths of people due to cardiovascular disease per 100,000 population	D
5	Total deaths due to cancer per 100,000 population	D
6	Health out-patient departments per 10,000 population	S
7	Number of doctors consultations per 10,000 population	S
8	Infant deaths per 1000 live births	D

Table 1 also contains information about variable type, with S denoting the so-called stimulant, where a higher value means a better situation (in terms of health), and D denoting the so-called destimulant, where lower values represent a better situation (Walesiak and Dudek [40]).

The variables were selected with a view to obtaining a relatively comprehensive description of the health situation of the population and taking into account their availability and completeness. The data cover the period 2013–2018, i.e., $T = 6$ years and describe $n = 380$ districts located within 16 provinces (see Table 2).

Table 2. The composition of Polish provinces.

Number	Province Name	Number of Districts
1	dolnośląskie	30
2	kujawsko-pomorskie	23
3	lubelskie	24
4	lubuskie	14
5	łódzkie	24
6	małopolskie	22
7	mazowieckie	42
8	opolskie	12
9	podkarpackie	25
10	podlaskie	17
11	pomorskie	20
12	śląskie	36
13	świętokrzyskie	14
14	warmińsko-mazurskie	21
15	wielkopolskie	35
16	zachodniopomorskie	21
	Total	**380**

Provinces are essentially equivalent to NUTS2 units, while districts are the upper level of local administrative units, which are currently not part of the NUTS system. The NUTS classification (Nomenclature of territorial units for statistics) is a geographical standard used for a statistical division of the EU Member States economic territories into three regional levels of specified classes of the population. It was established in order to enable the collection, compilation, and dissemination of harmonized regional statistics in the European Union. More information about the administrative division of Poland can be found at https://stat.gov.pl/en/regional-statistics/classification-of-territorial-units/administrative-division-of-poland/. Figure 1 shows the administrative division of Poland into provinces and districts (the left panel) and the division of OPOLSKIE (as an example) into districts (the right panel).

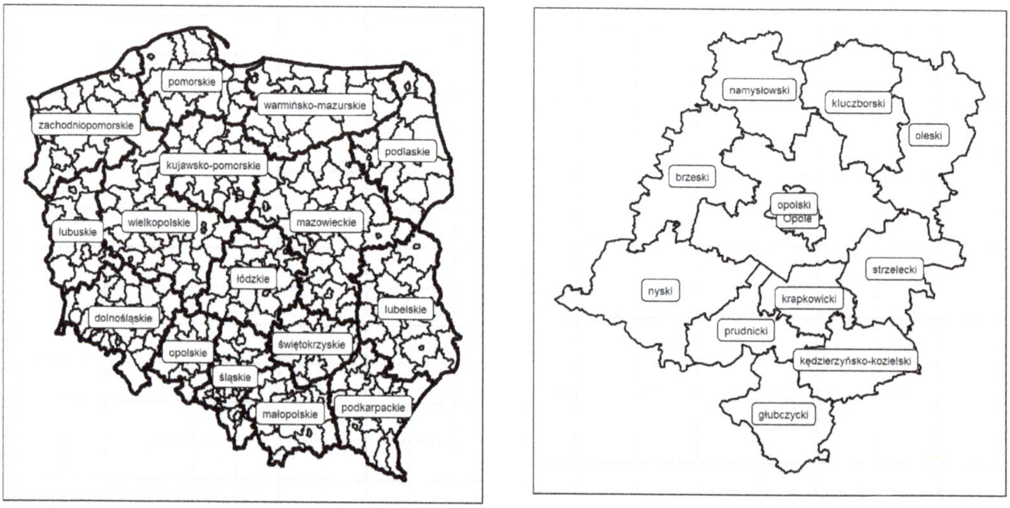

Figure 1. Administrative division of Poland—provinces and districts.

The values of the selected variables, expressed in different measurement units and having different ranges of variation, were standardized using the method of zero unitization (see, for example, Jajuga and Walesiak [41]).

Subsequently, the unitized data were transformed into functional data using the least squares method (see, e.g., Górecki and Krzyśko [35]).

Now, let us assume that the d-th component of the $\mathbf{Z}$ process can be represented by a finite number of orthonormal basis functions $\{\varphi_b\}$:

$$Z_d(t) = \sum_{b=0}^{B_d} \alpha_{db}\varphi_b(t),\ t \in I,\ d = 1, 2, \ldots, p,$$

where α_{db} are random variables such that $\text{Var}(\alpha_{db}) < \infty$ for $d = 1, 2, \ldots, p$ and $b = 0, 1, \ldots, B_d$.

Let

$$\boldsymbol{\alpha} = (\alpha_{10}, \ldots, \alpha_{1B_1}, \ldots, \alpha_{p0}, \ldots, \alpha_{pB_p})'$$

and

$$\boldsymbol{\Phi}(t) = \begin{bmatrix} \boldsymbol{\varphi}'_{B_1}(t) & \mathbf{0}' & \ldots & \mathbf{0}' \\ \mathbf{0}' & \boldsymbol{\varphi}'_{B_2}(t) & \ldots & \mathbf{0}' \\ \ldots & \ldots & \ldots & \ldots \\ \mathbf{0}' & \mathbf{0}' & \ldots & \boldsymbol{\varphi}'_{B_p}(t) \end{bmatrix},$$

where $\boldsymbol{\varphi}_{B_d} = (\varphi_0, \ldots, \varphi_{B_d})'$, $d = 1, \ldots, p$, $\boldsymbol{\alpha} \in \mathbb{R}^{K+p}, \boldsymbol{\Phi} \in \mathbb{R}^{p\times(K+p)}, K = B_1 + \ldots + B_p$. Then,

$$\mathbf{Z}(t) = \boldsymbol{\Phi}(t)\boldsymbol{\alpha},\ t \in I. \tag{1}$$

Individual years (time points) were assigned the following values: $t_1 = 0.5$ (2013), $t_2 = 1.5$ (2014), $\ldots, t_6 = 5.5$ (2018). The ϕ functions are considered on the interval $I = [0, T] = [0, 6]$. The Fourier base of the form

$$\phi_0(t) = 1/\sqrt{T},\ \phi_{2k-1}(t) = \sqrt{2/T}\sin(2\pi kt/T),\ \phi_{2k}(t) = \sqrt{2/T}\cos(2\pi kt/T),$$

where $t \in [0, T]$, $k = 1, 2, \ldots$, was adopted as the orthonormal basis. Górecki and Krzyśko [35] showed that the Fourier base leads to a minimal number of terms in the expansion of a given function into a series, which is a desirable feature because expansion coefficients play the role of new variables in the functional approach. Given the small number of time points, for each of the 8 variables, the number of expansion terms was the same and equal to 5. Hence, $B_1 = \ldots = B_8 = 4$, $K = B_1 + \cdots + B_8 = 32$, $K + p = 40$. Thus, $\alpha \in \mathbf{R}^{40}$ and $\mathbf{\Phi} \in \mathbf{R}^{8 \times 40}$.

Figure 2 shows the functional data (average values) for 8 variables and 16 provinces. One can see how the values of individual variables vary over time and between provinces.

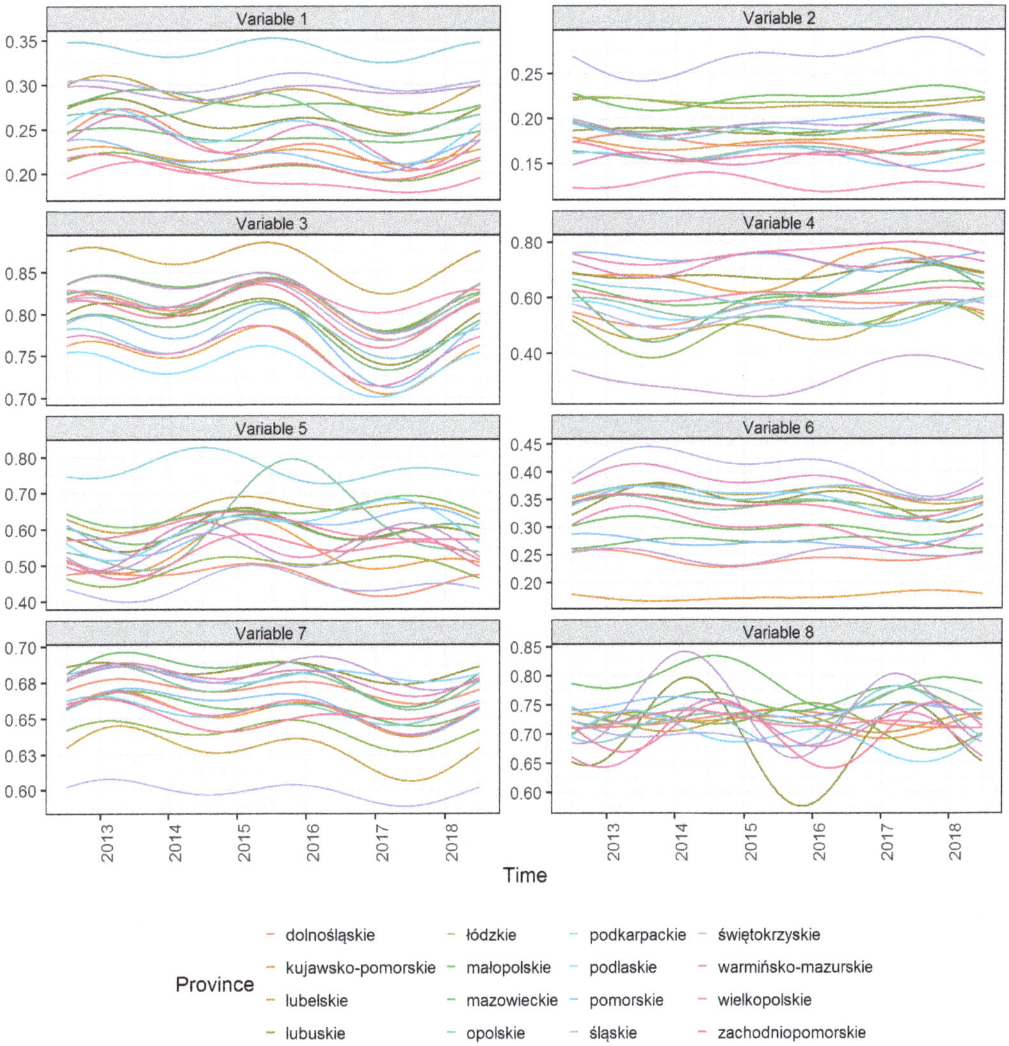

Figure 2. Average values of 8 variables calculated from functional data for districts included in each of the 16 provinces. **Note:** The ordinate axis shows the unitized values of a given variable.

3. Statistical Methodology

3.1. Functional Discriminant Coordinates

Our purpose is to construct a discriminant coordinate based on multivariate functional data, i.e., to construct

$$U = <\boldsymbol{u}, \mathbf{Z}> = \int_I \boldsymbol{u}'(t)\mathbf{Z}(t)dt$$

such that their between-class variance is maximal compared with the within-class variance, where

$$\boldsymbol{u}(t) = \boldsymbol{\Phi}(t)\boldsymbol{\gamma}.$$

The construction of functional discriminant coordinates is described in Górecki et al. [36] and Hanusz et al. [37].

The construction of discriminant coordinates for the random process $\mathbf{Z}$ essentially consists in constructing classical discriminant coordinates for a random vector $\boldsymbol{\alpha}$ because the discriminant component U_k has the form $U_k = \boldsymbol{\gamma}_k'\boldsymbol{\alpha}$, where $\boldsymbol{\alpha}$ is the random vector in the representation $\mathbf{Z}(t) = \boldsymbol{\Phi}(t)\boldsymbol{\alpha}$ of the random process $\mathbf{Z}$, and $\boldsymbol{\gamma}_k$ is an eigenvector in the generalized eigenproblem $(\mathbf{B} - \lambda_k\mathbf{W})\boldsymbol{\gamma}_k = \mathbf{0}$, where $\mathbf{B}$ and $\mathbf{W}$ are the between-class and within-class matrices, respectively.

Remark 1. *The examination of the elements of the vector weight function for the original processes in each discriminant coordinate (elements of the vectors $\boldsymbol{u}_k$) helps to interpret the principal axes of between-class variation.*

At a given time point t, the greater the absolute value of a component of the vector weight function, the greater the contribution in the structure of the given functional discriminant coordinate, from the process $\mathbf{Z}$ corresponding to that component. The total contribution of a particular original process Z_i in the structure of a particular functional discriminant coordinate is equal to the area under the module weighting function corresponding to this process.

In practice, vector $\boldsymbol{\alpha}$ is unknown and must be estimated based on the sample. Let $\mathbf{z}_{i1}, \mathbf{z}_{i2}, \ldots, \mathbf{z}_{in_i}$ be a sample belonging to the i-th class, where $i = 1, 2, \ldots, L$. The function $\mathbf{z}_{ij}$ has the form

$$\mathbf{z}_{ij}(t) = \boldsymbol{\Phi}(t)\boldsymbol{a}_{ij},$$

where $\boldsymbol{a}_{ij} = (a_{10}^{(ij)}, \ldots, a_{1K_1}^{(ij)}, \ldots, a_{p0}^{(ij)}, \ldots, a_{pK_p}^{(ij)})'$, $i = 1, 2, \ldots, L$, $j = 1, 2, \ldots, n_i$.

Let

$$\bar{\boldsymbol{a}}_i = \frac{1}{n_i}\sum_{j=1}^{n_i}\boldsymbol{a}_{ij},\ \bar{\boldsymbol{a}} = \frac{1}{n}\sum_{i=1}^{L}n_i\bar{\boldsymbol{a}}_i,\ i = 1, \ldots, L,\ n = n_1 + \cdots + n_L.$$

Then,

$$\hat{\mathbf{B}} = \frac{1}{L-1}\sum_{i=1}^{L}n_i(\bar{\boldsymbol{a}}_i - \bar{\boldsymbol{a}})(\bar{\boldsymbol{a}}_i - \bar{\boldsymbol{a}})',$$

$$\hat{\mathbf{W}} = \frac{1}{n-L}\sum_{i=1}^{L}\sum_{j=1}^{n_i}(\boldsymbol{a}_{ij} - \bar{\boldsymbol{a}}_i)(\boldsymbol{a}_{ij} - \bar{\boldsymbol{a}}_i)'.$$

Next, we find the non-zero eigenvalues $\hat{\lambda}_1^\star \geq \hat{\lambda}_2^\star \geq \ldots \geq \hat{\lambda}_s^\star$ and the corresponding eigenvectors $\hat{\boldsymbol{\gamma}}_1, \hat{\boldsymbol{\gamma}}_2, \ldots, \hat{\boldsymbol{\gamma}}_s$ of the matrix $\hat{\mathbf{W}}^{-1}\hat{\mathbf{B}}$, where $s = min(K + p, L - 1)$. Hence,

$$\hat{\boldsymbol{u}}_k(t) = \boldsymbol{\Phi}(t)\hat{\boldsymbol{\gamma}}_k,$$

and the coefficients of the projection of the j-th realization $\mathbf{z}_{ij}$ of the process $\mathbf{Z}$ belonging to the i-th class on the k-th functional discriminant coordinate are equal to:

$$\hat{U}_{ijk} = <\hat{\boldsymbol{u}}_k, \mathbf{z}_{ij}> = \hat{\boldsymbol{\gamma}}_k'\boldsymbol{a}_{ij},$$

for $i = 1,2,\ldots,L, j = 1,2,\ldots,n_i, k = 1,2,\ldots,s$.

The plots of the pairs $(\hat{U}_{ij1}, \hat{U}_{ij2})$ provide a visual representation of the relative position of groups in the two-dimensional space. Since the configuration obtained is deemed to be optimal in terms of the ability to discriminate between the groups, wide overlaps are to be considered as a sign of no or small differences between the groups involved.

3.2. Cluster Analysis

Provinces that are homogeneous in terms of the considered variables were identified using cluster analysis. More precisely, we applied Ward's hierarchical clustering method (see, for example, Seber [42], Chapter 7; Mirkin [43]; Krzyśko et al. [44], Chapter 12). The clustering procedure is based on the Mahalanobis distance between the provinces.

Let $\hat{\mathbf{U}}_{ij} = (\hat{U}_{ij1},\ldots,\hat{U}_{ijs})'$. This distance is defined by the following formula:

$$d_{ij}^2 = (\bar{\mathbf{U}}_i - \bar{\mathbf{U}}_j)'\mathbf{S}^{-1}(\bar{\mathbf{U}}_i - \bar{\mathbf{U}}_j),$$

where

$$\bar{\mathbf{U}}_i = \frac{1}{n_i}\sum_{j=1}^{n_i}\hat{\mathbf{U}}_{ij},\ i = 1,2,\ldots,L,$$

and

$$\mathbf{S} = \frac{1}{n-L}\sum_{i=1}^{L}\sum_{j=1}^{n_i}(\hat{\mathbf{U}}_{ij} - \bar{\mathbf{U}}_i)(\hat{\mathbf{U}}_{ij} - \bar{\mathbf{U}}_i)',\ n = n_1 + \ldots + n_L.$$

The Mahalanobis distance takes into account not only the difference between the mean vectors of two provinces; the difference is also weighted by the variances and covariances of the examined variables estimated for all provinces (the differentiation of districts around the mean provinces was taken into account).

3.3. Functional Multivariate Coefficient of Variation

Let $\mathbf{Z} = (Z_1,\ldots,Z_p)'$, be a p-dimensional random process with mean function $\boldsymbol{\mu} = (\mu_1,\ldots,\mu_p)' \neq \mathbf{0}$. We assume that the process $\mathbf{Z}$ belongs to the Hilbert space $L_2^p(I)$ of p-dimensional vectors of square integrable functions on I.

The functional multivariate coefficient of variation (MFCV) for the random process $\mathbf{Z}$ is defined as follows (Krzyśko and Smaga [45])

$$\text{MFCV} = \frac{\sqrt{\text{Var}(<\boldsymbol{\mu}_*, \mathbf{Z}>}}{||\boldsymbol{\mu}||},$$

where $\boldsymbol{\mu}_*(t) = \boldsymbol{\mu}(t)/||\boldsymbol{\mu}||, t \in I$.

If process $\mathbf{Z}$ has the form (1), then

$$\text{MFCV} = \sqrt{\frac{\boldsymbol{a}'\boldsymbol{J}_\Phi\boldsymbol{\Sigma}_\alpha\boldsymbol{J}_\Phi\boldsymbol{a}}{(\boldsymbol{a}'\boldsymbol{J}_\Phi\boldsymbol{a})^2}},$$

where $\boldsymbol{J}_\Phi = \text{diag}(\boldsymbol{J}_{\phi_1},\ldots,\boldsymbol{J}_{\phi_p})$, $\boldsymbol{J}_{\phi_k} = \int_I \boldsymbol{\phi}_k(t)\boldsymbol{\phi}_k'(t)dt$, and $\boldsymbol{\Sigma}_\alpha = \text{Cov}(\boldsymbol{\alpha})$ is the $B_k \times B_k$ cross product matrix corresponding to the basis $\{\phi_{kl}\}_{l=1}^{\infty}, k = 1,\ldots,p$. For the orthonormal basis, for instance the Fourier basis, the cross product matrix is equal to the identity matrix. Then (Albert and Zhang [46]),

$$\text{MFCV} = \sqrt{\frac{\boldsymbol{a}'\boldsymbol{\Sigma}_\alpha\boldsymbol{a}}{(\boldsymbol{a}'\boldsymbol{a})^2}}.$$

4. Results

To construct functional discriminant coordinates, we calculated the estimates $\boldsymbol{a}_i$ of the vectors $\boldsymbol{\alpha}_i, i = 1,\ldots,L$.

The vectors $\boldsymbol{a}_i$ were then used to construct the estimator $\hat{\boldsymbol{B}}$ of the matrix of between-class variability and the estimator $\hat{\boldsymbol{W}}$ of the matrix of within-class variability. Next, the non-zero eigenvalues of $\hat{\lambda}_k^\star$ of the matrix $\hat{\boldsymbol{W}}^{-1}\hat{\boldsymbol{B}}$ and the corresponding eigenvectors $\hat{\boldsymbol{\gamma}}_k$, $k = 1, \ldots, 15$, were calculated.

Multivariate functional discriminant coordinates have the form:

$$\hat{U}_k = < \hat{\boldsymbol{u}}_k, \mathbf{Z} >,$$

where

$$\hat{\boldsymbol{u}}_k(t) = \boldsymbol{\Phi}(t)\hat{\boldsymbol{\gamma}}_k,\ k = 1, \ldots, s,$$

$$s = min(K + p, L - 1) = min(40, 15) = 15,$$

are vectors of weight functions.

We treat the resulting multivariate functional discriminant coordinates as indicators (synthetic measures) of the health situation of inhabitants of Polish provinces. These indicators contain full information on the values of 8 diagnostic variables measured over 6 years. They are, therefore, composite indicators of the health situation.

These 15 composite indicators have a different power of differentiating between the provinces (these indicators have different variances (eigenvalues); see Table 3).

Table 3. Eigenvalues and related statistics.

Number	Eigenvalue	% Total Variance	% Cumulative Variance
1	48.2682	27.6745	27.6745
2	28.6024	16.3991	44.0735
3	26.3461	15.1055	59.1790
4	15.1611	8.6926	67.8716
5	11.6214	6.6631	74.5347
6	9.4012	5.3901	79.9248
7	8.5436	4.8984	84.8232
8	5.6155	3.2197	88.0429
9	4.9859	2.8587	90.9016
10	4.8982	2.8083	93.7099
11	3.4169	1.9591	95.6690
12	2.7111	1.5544	97.2234
13	2.0385	1.1687	98.3921
14	1.4576	0.8357	99.2279
15	1.3467	0.7721	100.0000

The first indicator is the strongest and the fifteenth is the least powerful. It is not possible to see the mutual position of provinces in the 15-dimensional space of these indicators, but it is possible in the space of the first two composite indicators that differentiate between the provinces most clearly.

It can be noticed that 44.1% of total variability is attributed to the first two multivariate functional coordinates.

The mean values of the 16 provinces in the system of the first two functional discriminant coordinates are presented in Table 4.

Table 4. The mean values of the 16 provinces (first two).

Number	Variable 1	Variable 2
1	−1.0418	−0.1272
2	−0.9538	−1.8010
3	2.1856	0.8277
4	−0.2247	−0.0520
5	−0.3005	0.4795
6	1.0408	0.2084
7	0.4584	−1.0042
8	0.9223	1.5262
9	2.5934	−1.5963
10	0.5307	2.2876
11	−1.0424	0.8810
12	−0.7883	0.7896
13	3.0622	0.1767
14	−1.6026	0.6644
15	−1.5847	−1.2004
16	−0.9763	0.6872

The location of the 16 provinces in the system of the first two functional discriminant coordinates is shown in Figure 3.

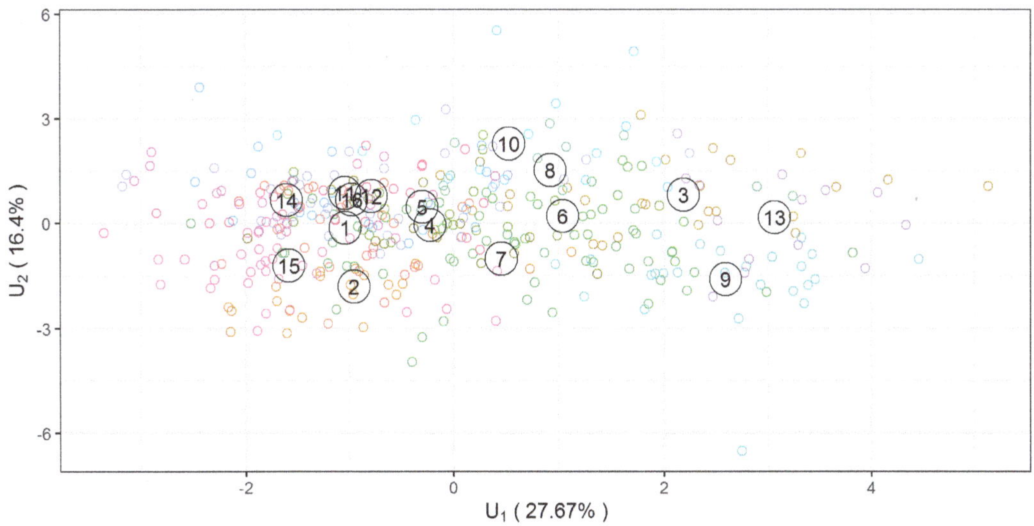

Figure 3. Plotted values of the first two functional discriminant coordinates.

The total contribution of the individual variables to the structure of the particular functional discriminant coordinates can be estimated using the area under the absolute value of the weight functions corresponding to a given variable. The graphs of the eight components of the vector weight function for the first and second functional discriminant coordinates are shown in Figure 4.

These contributions, for the first and second functional discriminant coordinates for 8 variables are also given in Table 5. Table 5 shows that the largest share in the construction of the first functional discriminant coordinate is played by variable No. 2 (Doctors per 10,000 population)—32.0%—and variable No. 7 (Number of doctor consultations per 10,000 population)—14.7%. On the other hand, variable No. 4 (Deaths of people due to cardiovascular disease per 100,000 population)—21.0%—and variable No. 2 (Doctors per 10,000 population)—20.2%—have the greatest share in the construction of the second functional discriminant coordinate. Values of coefficients of the vector weight functions are also presented in Table 5.

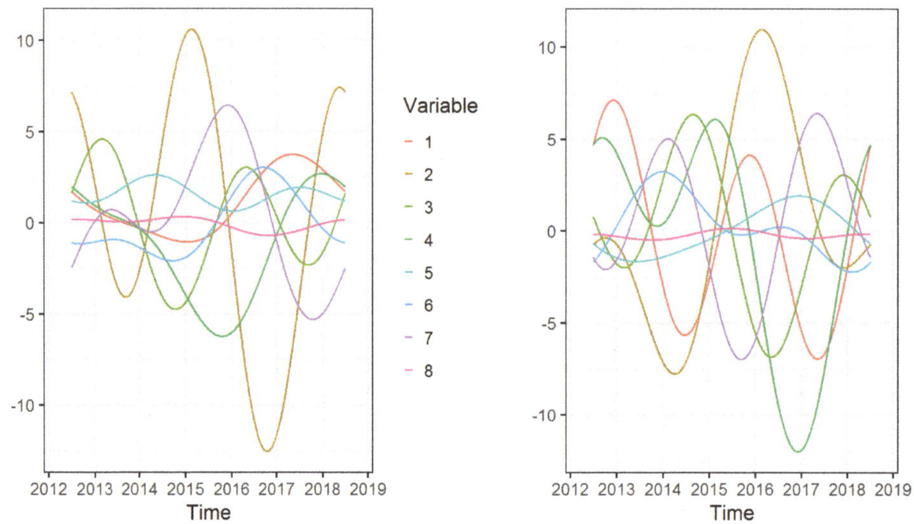

Figure 4. Weight functions for the first (left) and the second (right) functional discriminant coordinate.

Table 5. Values of coefficients of the vector weight functions.

First functional discriminant coordinate							
Variable	$\hat{\gamma}_{10}$	$\hat{\gamma}_{11}$	$\hat{\gamma}_{12}$	$\hat{\gamma}_{13}$	$\hat{\gamma}_{14}$	Area	Area (%)
1	2.5407	−3.2389	2.0978	-0.3165	−0.9303	9.0039	8.2381
2	0.5792	7.8019	−0.7947	−7.3340	12.8083	34.9609	31.9876
3	0.2994	−0.5582	2.4586	6.3065	0.1062	14.4837	13.2519
4	−3.0143	0.3007	6.7926	−1.9114	−1.2470	15.1264	13.8399
5	3.8320	0.7027	0.1091	−0.9202	−0.7656	9.3864	8.5882
6	−0.1795	−3.5460	−0.4691	1.3195	−1.3298	8.7501	8.0059
7	0.9525	0.6116	−6.6553	4.7278	1.7190	16.0264	14.6634
8	−0.1802	0.6553	0.0063	−0.1974	0.3950	1.5574	1.4249
Second functional discriminant coordinate							
Variable	$\hat{\gamma}_{20}$	$\hat{\gamma}_{21}$	$\hat{\gamma}_{22}$	$\hat{\gamma}_{23}$	$\hat{\gamma}_{24}$	Area	Area (%)
1	−0.8113	2.1021	1.9010	7.7922	6.7172	23.1732	17.3879
2	1.3009	−9.6903	−5.7988	6.8425	3.5944	26.8581	20.1529
3	0.2159	5.0384	0.8509	−7.6787	0.3243	18.8570	14.1493
4	−1.4582	10.6698	0.1764	−1.8355	8.9551	28.0178	21.0231
5	−0.0163	−2.8675	−0.7596	−0.2515	−0.4281	6.5687	4.9288
6	0.7540	2.9261	−1.3371	0.7459	−2.1031	7.4906	5.6206
7	1.2655	−0.1701	4.4019	−4.0271	−7.7531	20.7457	15.5665
8	−0.5597	−0.0570	−0.2735	−0.0484	0.3602	1.5604	1.1709

In the next step, cluster analysis was used to select groups of homogeneous provinces in a fifteen-dimensional space of functional discriminant coordinates. The Ward method was selected as a commonly used technique. The Mahalanobis distance was chosen as a measure of the distance between the mean vectors of individual provinces. The obtained dendrogram is presented in Figure 5.

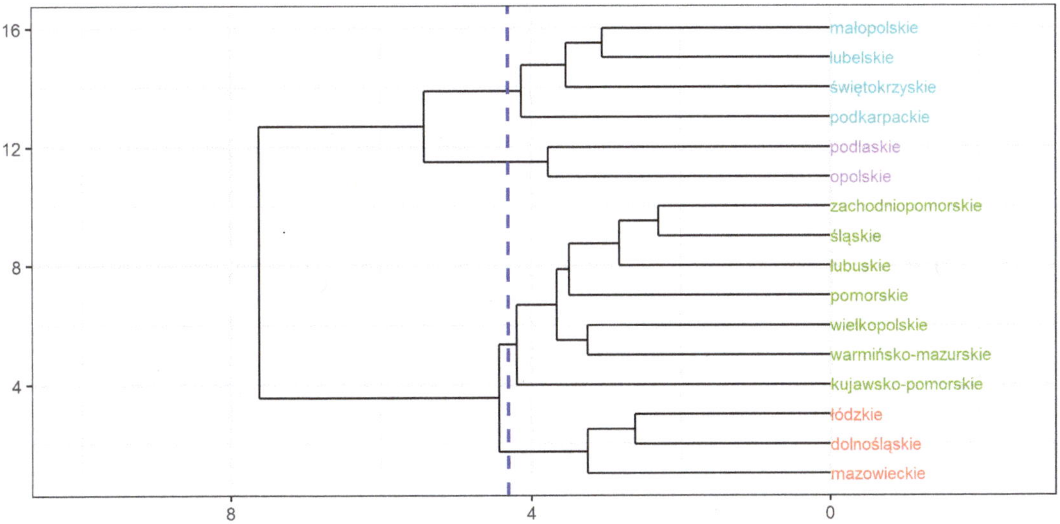

Figure 5. Dendrogram for 16 Polish provinces (the Ward method).

We obtained four homogeneous clusters. Which cluster individual provinces belong to is shown in Table 6 and in Figure 6 (spatial distribution).

Table 6. Membership of provinces in the four clusters.

Number	Province	Cluster
1	dolnośląskie	I
2	kujawsko-pomorskie	II
3	lubelskie	III
4	lubuskie	II
5	łódzkie	I
6	małopolskie	III
7	mazowieckie	I
8	opolskie	IV
9	podkarpackie	III
10	podlaskie	IV
11	pomorskie	II
12	śląskie	II
13	świętokrzyskie	III
14	warmińsko-mazurskie	II
15	wielkopolskie	II
16	zachodniopomorskie	II

Taking into account the spatial variation in the health situation of the provinces, it is possible to distinguish four spatial clusters. The first one, denoted as II, consists of six provinces located in the north-western part of Poland. At the opposite end of the country, there are four provinces that make up cluster III. Finally, in the approximate middle belt, one can see cluster I, consisting of three provinces. Cluster IV consists of two non-contiguous provinces (podlaskie and opolskie), located in different parts of Poland.

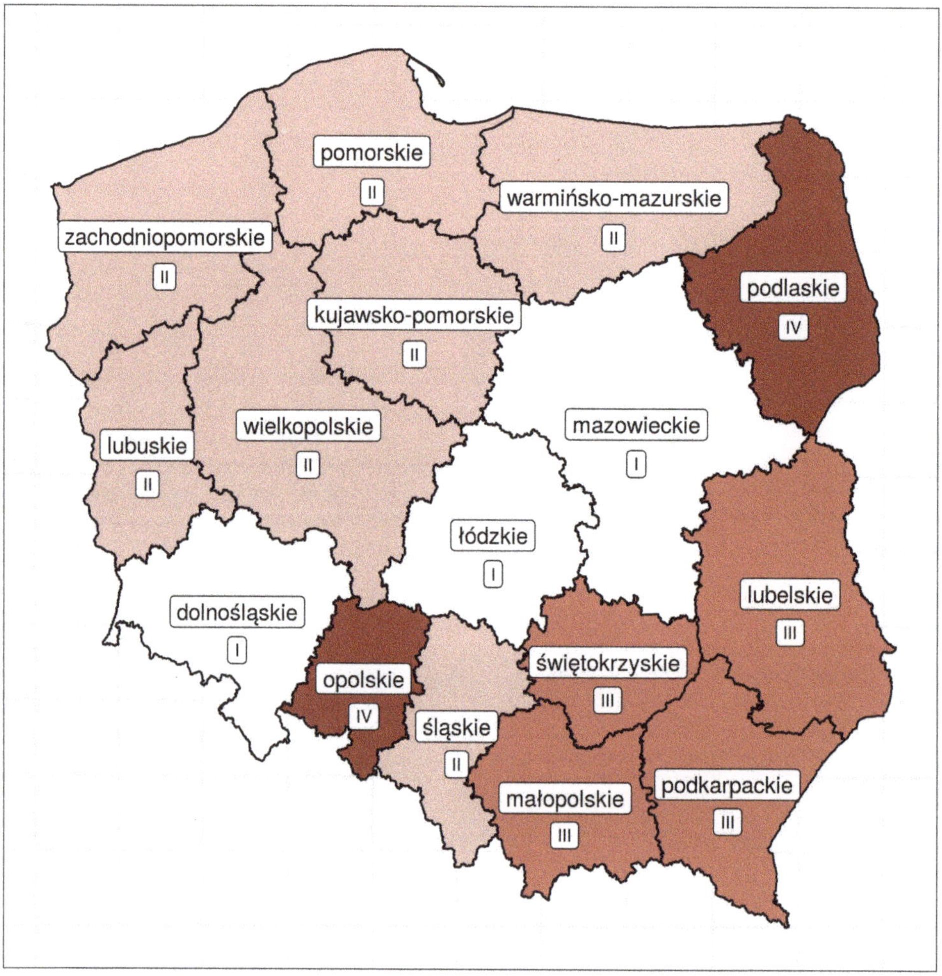

Figure 6. Spatial variation in the health situation across the provinces of Poland.

Figure 6 shows that Poland can be divided into two part: Western Poland (provinces belonging to clusters I and II) and Eastern Poland (provinces belonging to clusters III and IV). The health of inhabitants living in the provinces belonging to cluster I is the best, while that of people living in the provinces belonging to cluster IV is the worst. All the previous studies conducted in Poland show that Western Poland is better developed in socio-economic terms than Eastern Poland (see, for instance, Szymkowiak et al. [47], Marchetti et al. [48], Roszka [49]). Current research shows that this division is also valid as regards the health situation.

Decision-makers at national and local government levels should be advised to redirect more funds to improve the health situation of inhabitants of Easter Poland.

To verify that the obtained four clusters are really homogeneous, a multivariate functional coefficient of variation (MFCV) was calculated for all provinces together and for each cluster separately (see Table 7).

Table 7. Values of the multivariate functional coefficient of variation (MFCV).

Provinces	MFCV
All	0.3705
Cluster I	0.2975
Cluster II	0.2890
Cluster III	0.1728
Cluster IV	0.2047

As can be seen, the coefficients of variation for individual clusters are lower than that for all indeed homogeneous.

5. Conclusions

The above statistical analysis provides evidence for the conclusion that the provinces are not homogeneous in terms of the selected variables characterizing the health situation of their inhabitants. The analysis consisted of multiple steps. Values of the selected variables, which are expressed in different measurement units and have different ranges of variation, were standardized using the method of zero unitization. Then, the unitized data were transformed into functional data in order to enable the construction of discriminant coordinates in the functional data space. The multivariate functional discriminant coordinates were treated as composite indicators (synthetic measures) of the health situation of inhabitants of Polish provinces. These indicators contain full information on the values of 8 diagnostic variables measured over 6 years.

In the next step, cluster analysis was applied to select groups of homogeneous provinces in the space of functional discriminant coordinates using the Ward method. The Mahalanobis distance was chosen as a measure of distance between the mean vectors of individual provinces. The homogeneity of the resulting four clusters was analyzed using a multivariate functional coefficient of variation, which was calculated for all provinces together and for each cluster separately. It turned out that the coefficients of variation for individual clusters are smaller than the corresponding value for combined provinces, which confirms that the clusters are indeed homogeneous.

The obtained clusters illustrate changes in the situation of the provinces (over a period of six analyzed years). In previous studies, data for each year are analyzed separately using classical statistical methods. However, one must not forget that one deals with spatio-temporal data that change over time.

The authors realize that the choice of diagnostic variables may be a weakness of this study. These particular diagnostic variables were selected with a view to obtaining relatively comprehensive description of the health situation of the population, given their availability and completeness. Therefore, the selection should be treated mainly as an illustration of the proposed statistical methodology for processing spatio-temporal data.

The statistical methods used for multivariate functional data were suggested earlier by the authors of this paper.

Author Contributions: Conceptualization, M.K.; Methodology, M.K. and W.W.; Software, W.W. and M.S.; Validation, A.W.; Formal analysis, M.K., W.W., and M.S.; Investigation, M.K., W.W., M.S., and A.W.; Data curation, M.S.; Writing—original draft preparation, M.K., W.W., M.S., and A.W.; Writing—review and editing, M.K., W.W., M.S., and A.W.; Visualization, W.W. and M.S.; Funding acquisition, A.W. All authors have read and agreed to the published version of the manuscript.

Funding: This research received no external funding.

Institutional Review Board Statement: Not applicable.

Informed Consent Statement: Not applicable.

Data Availability Statement: The data presented in this study are openly available in the Local Data Bank, which is Poland's largest database of information relating to the economy, society and the environment.

Conflicts of Interest: The authors declare no conflict of interest.

References

1. Ngamaba, K.H.; Panagioti, M.; Armitage, C.J. How strongly related are health status and subjective well-being? Systematic review and meta-analysis. *Eur. J. Public Health* **2010**, *27*, 879–885. [CrossRef] [PubMed]
2. James, C.; Devaux, M.; Sassi, F. *Inclusive Growth and Health*; OECD Health Working Papers, No. 103; OECD Publishing: Paris, France, 2017.
3. OECD. *Health for Everyone? Social Inequalities in Health and Health Systems*; OECD Health Policy Studies; OECD Publishing: Paris, France, 2019. [CrossRef]
4. Penchansky, R.; Thomas, J.W. The concept of access: Definition and relationship to consumer satisfaction. *Med. Care* **1981**, *19*, 127–140. [CrossRef] [PubMed]
5. Wang, F. Measurement. optimization. and impact of health care accessibility: A methodological review. *Ann. Assoc. Am. Geogr.* **2012**, *102*, 1104–1112. [CrossRef] [PubMed]
6. Neutens, T. Accessibility, equity and health care: Review and research directions for transport geographers. *J. Transp. Geogr.* **2015**,*43*, 14–27. [CrossRef]
7. Barbarisi, I.; Bruno, G.; Diglio, A.; Elizalde, J.; Piccolo, C. A spatial analysis to evaluate the impact of deregulation policies in the pharmacy sector: Evidence from the case of Navarre. *Health Policy* **2019**, *123*, 1108–1115. [CrossRef]
8. Bruno, G.; Cavola, M.; Diglio, A.; Piccolo, C. Improving spatial accessibility to regional health systems through facility capacity management. *Socio-Econ. Plan. Sci.* **2020**, *71*, 100881. [CrossRef]
9. Lu, C.; Zhang, Z.; Lan, X. Impact of China's referral reform on the equity and spatial accessibility of healthcare resources: A case study of Beijing. *Soc. Sci. Med.* **2019**, *235*, 112386. [CrossRef]
10. Okuyama, K.; Akai, K.; Kijima, T.; Abe, T.; Isomura, M.; Nabika, T. Effect of geographic accessibility to primary care on treatment status of hypertension. *PLoS ONE* **2019**, *14*, e0213098. [CrossRef]
11. Pu, Q.; Yoo, E.H.; Rothstein, D.H.; Cairo, S.; Malemo, L. Improving the spatial accessibility of healthcare in North Kivu, Democratic Republic of Congo. *Appl. Geogr.* **2020**, *121*, 102262. [CrossRef]
12. De Leeuw, E.; Clavier, C.; Breton, E. Health policy–why research it and how: Health political science. *Health Res. Policy Syst.* **2014**, *12*, 55. [CrossRef]
13. Campos, P.A.; Reich, M.R. Political analysis for health policy implementation. *Health Syst. Reform* **2019**, *5*, 224–235. [CrossRef] [PubMed]
14. Pollack Porter, K.M.; Rutkow, L.; McGinty, E.E. The importance of policy change for addressing public health problems. *Public Health Rep.* **2018**, *133*, 9S–14S. [CrossRef] [PubMed]
15. Łyszczarz, B. Dynamika regionalnych nierówności w zdrowiu w Polsce (Dynamics of Regional Health Inequalities in Poland). *Nierówności Społeczne Wzrost Gospod.* **2014**, *38*, 191–200.
16. Boyle, P. Global public health—Challenges and leadership. *J. Health Inequalities* **2018**, *4*, 55–61. [CrossRef]
17. Kawachi, I.; Subramanian, S.V.; Almeida-Filho, N. A glossary for health inequalities. *J. Epidemiol. Community Health* **2002**, *56*, 647–652. [CrossRef] [PubMed]
18. Arcaya, M.C.; Arcaya, A.L.; Subramanian, S.V. Inequalities in health: Definitions, concepts, and theories. *Glob. Health Action* **2015**, *8*, 1–12. [CrossRef] [PubMed]
19. Wojtyła-Buciora, P.; Stawińska-Witoszyńska, B.; Wojtyła, K.; Klimberg, A.; Wojtyła, C.; Wojtyła, A.; Samolczyk-Wanyura, D.; Marcinkowski, J.T. Assessing physical activity and sedentary lifestyle behaviours for children and adolescents living in a district of Poland. What are the key determinants for improving health? *Ann. Agric. Environ. Med.* **2014**, *21*, 606–612. [CrossRef] [PubMed]
20. Wojtyła, C.; Biliński, P.; Paprzycki, P.; Warzocha, K. Haematological parameters in postpartum women and their babies in Poland—Comparison of urban and rural areas. *Ann. Agric. Environ. Med.* **2011**, *18*, 380–385.
21. Samet, J.M. The environment and health inequalities: Problems and solutions. *J. Health Inequalities* **2019**, *5*, 21–27. [CrossRef]
22. Crombie, I.K.; Irvine, L.; Elliott, L.; Wallace, H. *Closing the Health Inequalities Gap: An International Perspective*; No. EUR/05/5048925; WHO Regional Office for Europe: Copenhagen, Denmark, 2005.
23. Spinakis, A.; Anastasiou, G.; Panousis, V.; Spiliopoulos, K.; Palaiologou, S.; Yfantopoulos, J. *Expert Review and Proposals for Measurement of Health Inequalities in the European Union—Full Report*; European Commission Directorate General for Health and Consumers: Luxembourg, 2011.
24. Barreto, M.L. Health inequalities: A global perspective. *Ciência Saúde Coletiva* **2017**, *22*, 2097–2108. [CrossRef]
25. Pizot, C.; Dragomir, M.; Macacu, A.; Koechlin, A.; Bota, M.; Boyle, P. Global burden of pancreas cancer: Regional disparities in incidence, mortality and survival. *J. Health Inequalities* **2019**, *5*, 96–112. [CrossRef]

26. Ozdenerol, E. *Spatial Health Inequalities: Adapting GIS Tools and Data Analysis*; CRC Press: Boca Raton, FL, USA, 2016.
27. Casper, M.; Kramer, M.R.; Peacock, J.M.; Vaughan, A.S. Population Health, Place, and Space: Spatial Perspectives in Chronic Disease Research and Practice. *Prev. Chronic Dis.* **2019**, *16*, E123. [CrossRef] [PubMed]
28. Gilliland, J.A.; Shah, T.I.; Clark, A.; Sibbald, S.; Seabrook, J.A. A geospatial approach to understanding inequalities in accessibility to primary care among vulnerable populations. *PLoS ONE* **2019**, *14*, e0210113. [CrossRef] [PubMed]
29. Wang, Z.; Nie, K. Measuring Spatial Patterns of Health Care Facilities and Their Relationships with Hypertension Inpatients in a Network-Constrained Urban System. *Int. J. Environ. Res. Public Health* **2019**, *16*, 3204. [CrossRef]
30. Chen, J.; Bai, Y.; Zhang, P.; Qiu, J.; Hu, Y.; Wang, T.; Xu, C.; Gong, P. A Spatial Distribution Equilibrium Evaluation of Health Service Resources at Community Grid Scale in Yichang, China. *Sustainability* **2020**, *12*, 52. [CrossRef]
31. Wierzbicka, A. Taxonomic analysis of the Polish public health in comparison with selected European countries. *Stat. Transition. New Ser.* **2012**, *13*, 343–364.
32. Bem, A.; Ucieklak-Jeż, P.; Siedlecki, R. The spatial differentiation of the availability of health care in polish regions. *Procedia-Soc. Behav. Sci.* **2016**, *220*, 12–20. [CrossRef]
33. Shi, L.; Starfield, B.; Kennedy, B.; Kawachi, I. Income inequality, primary care, and health indicators. *J. Fam. Pract.* **1999**, *48*, 275–284.
34. Hübelová, D.; Kozumpliková, A.; Jadaczková, V.; Rousová, G. Spatial differentiation of selected healyh factors of the South Moravian Region population. *Geogr. Cassoviensis* **2018**, *XII*, 33–52.
35. Górecki, T.; Krzyśko, M. Functional Principal Components Analysis. In *Data Analysis Methods and Its Applications*; Pociecha, J., Decker, R., Eds.; C.H. Beck: Munich, Germany, 2012; pp. 71–87.
36. Górecki, T.; Krzyśko, M.; Waszak, Ł.; Wołyński, W. Selected statistical methods of data for multivariate functional data. *Stat. Pap.* **2018**, *59*, 153–182.
37. Hanusz, Z.; Krzyśko, M.; Nadulski, R.; Waszak, Ł. Discriminant coordinates analysis for multivariate functional data. *Commun. Stat. Theory Methods* **2020**, *49*, 4506–4519. [CrossRef]
38. Krzyśko, M.; Wołyński, W.; Ratajczak, W.; Kierczyńska, A.; Wenerska, B. Sustainable development of Polish macroregions-study by means of the kernel discriminant coordinates method. *Int. J. Environ. Res. Public Health* **2020**, *17*, 7021. [CrossRef] [PubMed]
39. Krzyśko, M.; Łukaszonek, W.; Wołyński, W. Discriminant coordinates analysis in the case of multivariate repeated measures data. *Stat. Transit. New Ser.* **2018**, *19*, 495–506. [CrossRef]
40. Walesiak, M.; Dudek, A. The Choice of Variable Normalization Method in Cluster Analysis. In *Education Excellence and Innovation Management: A 2025 Vision to Sustain Economic Development During Global Challenges*; Soliman, K.S., Ed.; International Business Information Management Association: Seville, Spain, 2020; pp. 325–340.
41. Jajuga, K.; Walesiak, M. Standardisation of data set under different measurement scales. In *Classification and Information Processing at the Turn of the Millennium*; Springer: Berlin/Heidelberg, Germany, 2000; pp. 105–112.
42. Seber, G.A.F. *Multiple Observations*; John Wiley and Sons, Inc.: New York, NY, USA, 2004.
43. Mirkin, B. *Clustering: A Data Recovery Approach*, 2nd ed.; Taylor and Francis Group, LLC.: London, UK, 2013.
44. Krzyśko, M.; Wołyński, W.; Górecki, T.; Skorzybut, M. *Learning Systems: Pattern Recognition, Cluster Analysis and Dimensional Reduction*; Scientific and Technical Publishers: Warsaw, Poland, 2008.
45. Krzyśko, M.; Smaga, Ł. A multivariate coefficient of variation for functional data. *Stat. Its Interface* **2019**, *12*, 647–658.
46. Albert, A.; Zhang, L. A novel definition of the multivariate coefficient of variation. *Biom. J.* **2010**, *52*, 667–675. [CrossRef]
47. Szymkowiak, M.; Młodak, A.; Wawrowski, Ł. Mapping poverty at the level of subregions in Poland using indirect estimation. *Stat. Transit. New Ser.* **2017**, *18*, 609–635. [CrossRef]
48. Marchetti, S.; Beręsewicz, M.; Salvati, N.; Szymkowiak, M.; Wawrowski, Ł. The use of a three-level M-quantile model to map poverty at local administrative unit 1 in Poland. *J. R. Stat. Soc. Ser. A* **2018**, *181*, 1077–1104. [CrossRef]
49. Roszka, W. Spatial microsimulation of personal income in Poland at the level of subregions. *Stat. Transition. New Ser.* **2019**, *20*, 133–153. [CrossRef]

International Journal of
Environmental Research and Public Health

MDPI

Article

Perinatal Outcomes in a Population of Diabetic and Obese Pregnant Women—The Results of the Polish National Survey

Cezary Wojtyla [1,2,3,*], Pawel Stanirowski [2,4,†], Pawel Gutaj [2,5,†], Michal Ciebiera [2,6,†] and Andrzej Wojtyla [7]

1 International Prevention Research Institute—Collaborating Centre, Calisia University, 62-800 Kalisz, Poland
2 Club 35, Polish Society of Gynecologists and Obstetricians, 02-677 Warsaw, Poland; stanirowski@gmail.com (P.S.); pgutaj@o2.pl (P.G.); michal.ciebiera@gmail.com (M.C.)
3 Department of Oncological Gynecology and Obstetrics, Centre of Postgraduate Medical Education, 00-416 Warsaw, Poland
4 First Department of Obstetrics and Gynecology, Medical University of Warsaw, 02-015 Warsaw, Poland
5 Department of Reproduction, Poznan University of Medical Sciences, St, 60-535 Poznan, Poland
6 Second Department of Obstetrics and Gynecology, Centre of Postgraduate Medical Education, 01-809 Warsaw, Poland
7 World Institute for Family Health, Calisia University, 62-800 Kalisz, Poland; a.wojtyla55@gmail.com
* Correspondence: czwo@op.pl
† Equal contributors.

Abstract: Obesity and diabetes increase the risk of complications during gestation and at delivery. The aim of this study was to compare the perinatal outcomes in the populations of diabetic and obese Polish women, based on the results of a national survey performed in years 2012 and 2017, as well as to determine the risk factors of the gestational diabetes mellitus (GDM). Questionnaires from 6276 women were collected. Obese women constituted 5.5% and 7.5% of study population in years 2012 and 2017, respectively. Among women whose pregnancies were complicated by diabetes mellitus, GDM constituted the most common type of glucose intolerance during both time periods (2012: 89% vs. 2017: 85.6%). In the group of obese women an insignificant increase in the rate of induced deliveries was noted (2012: 9.9% vs. 2017: 11.7%), whereas the fetal birth-weight decreased significantly (2012: 3565 g vs. 2017: 3405 g, $p < 0.05$). In the group of diabetic pregnant women the percentage of cesarean sections, labour inductions and fetal birth defects was characterized by an insignificant upward trend. Risk of GDM was significantly increased in women aged over 35 years—(2012: OR 1.9 (95% CI: 1.1–2.9) and 2017: OR = 2.1 (95% CI: 1.5–2.9), $p < 0.05$—, as well as in overweight women—2012: OR 1.8 (95% CI: 1.2–2.7) and 2017: OR 2.6 (95% CI: 1.9–3.4), $p < 0.05$—during both analysed time periods. Based on the study results, it is necessary to develop population-based programmes to prevent obesity and to introduce and enforce the rules of appropriate screening for glucose tolerance disorders during pregnancy.

Keywords: gestational diabetes mellitus; diabetes; obesity; pregnancy; cesarean section

Citation: Wojtyla, C.; Stanirowski, P.; Gutaj, P.; Ciebiera, M.; Wojtyla, A. Perinatal Outcomes in a Population of Diabetic and Obese Pregnant Women—The Results of the Polish National Survey. *IJERPH* **2021**, *18*, 560. https://doi.org/10.3390/ijerph18020560

Received: 27 October 2020
Accepted: 8 January 2021
Published: 11 January 2021

Publisher's Note: MDPI stays neutral with regard to jurisdictional claims in published maps and institutional affiliations.

Copyright: © 2021 by the authors. Licensee MDPI, Basel, Switzerland. This article is an open access article distributed under the terms and conditions of the Creative Commons Attribution (CC BY) license (https://creativecommons.org/licenses/by/4.0/).

1. Introduction

1.1. Diabetes Mellitus—Overview

Diabetes mellitus (DM), commonly known as diabetes, is a group of metabolic disorders, whose common feature is a high blood glucose level over a prolonged period, being the result of absolute or relative insulin deficiency [1,2]. Detrimental effect of the chronic hyperglycemia on metabolic pathways of proteins, lipids and electrolytes is well-established in the literature [2,3].

According to the Global Report on Diabetes published by the World Health Organization (WHO), in 2014 nearly 422 million of adults suffered from DM worldwide [4]. In addition, recent research performed by the American Diabetes Association (ADA) revealed that 1.5 million Americans are diagnosed with DM every year [5].

1.2. Diabetes Mellitus in Pregnancy—Classification

DM in pregnancy may be divided into two major sub-types. One of them is pregestational DM (PGDM) which preexists in women who get pregnant [2,3,6]. The second one, hyperglycemia which was first detected during pregnancy should be categorized using the WHO criteria as DM in pregnancy (DIP) or gestational DM (GDM) [4,7]. According to the current standards, DIP is a condition that may be diagnosed, if standard DM criteria are met during screening, whereas GDM is diagnosed when women meet at least one of the criteria during 75g oral glucose tolerance test (OGTT) [3,7,8]. GDM occurs in pregnant women who develop hyperglycemia, but do not have a previous history of diabetes [2,3,9].

According to DeSisto et al. the prevalence of GDM in 2010 was reported at 4.6% when analysed using birth certificates and reached 8.7% when reported by standardized questionnaires [10]. Another study from the United States conducted by Fong et al. revealed the prevalence of GDM of 5.34%, while the PGDM equaled 0.82% [11].

It is of importance that women with DM and GDM are at an increased risk of multiple complications during pregnancy and at delivery [12–14].

1.3. Obesity and Obesity in Pregnancy—Overview

Obesity is a pathological condition characterized by an excessive accumulation of fat [15]. Body mass index (BMI) is the most popular tool used in obesity classification [16]. According to current standards, obesity is diagnosed when patient's BMI exceeds 30 kg/m^2 [16,17]. Most commonly obesity is caused by an excessive energy intake, lack of physical activity in conjunction with epigenetic and genetic predispositions [18–20]. According to Global Burden of Diseases 2015 Obesity Collaborators study including data from 195 countries, obesity occurs in about 600 million of adults and, what is even more worrying, in almost 100 million of children and adolescents [18]. During pregnancy obesity has a major impact on both maternal metabolism and fetal development [21]. The metabolic pathways that are most commonly altered in obese patients include glucose metabolism, increased insulin resistance and disrupted fat oxidation [21]. According to a recent study by Chen et al. in 2014 the estimated number of overweight and obese pregnant women equaled 38.9 million, with only obese women accounting for 14.6 million [22]. In the course of pregnancy obese women are at an increased risk of multiple perinatal complications, including preeclampsia, GDM, DIP and others [23–26].

1.4. Diabetes and Obesity during Pregnancy—Polish Guidelines and Standards

Education, optimal self-management and medical personnel support are crucial to prevent serious complications and reduce the risk of long-term complications [7,27]. Different countries developed various guidelines adapted to their systems of healthcare. Accordingly, Poland has its own guidelines concerning hyperglycemia in pregnancy. Interestingly, nowadays in Poland three documents tackle the issue of standardization and guidelines for hyperglycemia in pregnancy. They were developed by the Polish Society of Gynecologists and Obstetricians (PSGO) [2], Diabetes Poland [3] and by the Ministry of Health of Poland [28]. According to PSGO and Diabetes Poland the categorization and the diagnostic criteria of hyperglycemia during pregnancy are in accordance with WHO guidelines. Both documents advise screening in the first as well as in the late second or early third trimester of gestation [2,3]. The new standards of healthcare during pregnancy issued by the Polish Ministry of Health in 2018 include the recommendation to perform standard screening in the first trimester or at the time of the first obstetric visit. However, the second round of screening differs, as it should be performed between the 24th and 26th week of gestation, instead of the 24–28th week period as recommended by the WHO [28].

Pregnancy management practice in Poland changed significantly between the years 2012 and 2017. According to a 2012 standard by the Polish Ministry of Health on pregnancy management a single fasting glucose measurement should be performed in the first trimester, whereas in the late second/early third trimester it is advised to perform a two-step 75 g OGTT for the diagnosis of GDM [29]. Due to subsequent clinical trial results, some changes were introduced and many clinicians started to use new WHO guidelines (2013) [4,7]. The Hyperglycemia and Adverse Pregnancy Outcome (HAPO) study included a recommendation of a three-step 75 g OGTT as a new equalization criterion [30]. PSGO and Diabetes Poland guidelines were both up-to-date and evolved annually in case of Diabetes Poland [31], and in case of PSGO the guidelines appeared in 2011 [32], were updated in 2014 [33] and then re-edited and updated in 2017 [2]. As a result, in 2017, the current strategy for women in Poland included fasting glucose assessment in the first trimester and a three-step 75 g OGTT performed between 24th and 28th pregnancy weeks [2].

PSGO also published their own guidelines concerning perinatal care over obese pregnant women in year 2012 [34]. Importantly, as regards women with BMI $\geq$ 30 kg/m^2, 75 g OGTT was recommended as early as the first trimester of pregnancy, instead of a single fasting glucose measurement performed in the general obstetric population [34].

Numerous recommendations published over the years show how difficult it is to manage pregnant women with DM or obesity. The aim of the present study was to compare the perinatal outcomes in the populations of diabetic and obese Polish women, and those with both conditions, based on the results of a national survey performed in years 2012 and 2017, as well as to determine the risk factors of GDM. We would like to evaluate how those outcomes and correlated measurements changed with the implementation of the new guidelines as standards.

2. Materials and Methods

The analyses of pregnant women were carried out in years 2012 and 2017 within the Polish Pregnancy-related Assessment Monitoring System (Pol-PrAMS). This population-based study was conducted in all of the hospitals in Poland. Groups of Polish women and their newborns were surveyed during postpartum hospitalization. Thus, all of the women hospitalized postpartum on the designated days of the study were deemed eligible for the study. Informed consent was verbally obtained from all women, which was approved by the Ethics Committee. Participation was anonymous and voluntary, and the surveys were completed by the women after consent. Thus, each completed questionnaire was a documentation of consent to the study.

The survey was carried out once in each hospital, which had at least one of the following units in its structure: maternity ward, department of gynecology and obstetrics, department of obstetrics with rooming-in, labor ward or neonatal department. The survey was conducted simultaneously throughout the country, using the structures of the Poviat Sanitary and Epidemiological Stations, as units subordinate to the Chief Sanitary Inspectorate. These types of Stations are located in every poviat in Poland (poviat is the second level of the administrative division of Poland), which allowed for the efficient conduction of research throughout the country within a few weeks of the year. In 2012 the study was conducted on one day in each hospital, during the third week of March. In 2017 the study was conducted between the 2 February and 22 March.

The evaluation was preceded by obtaining the consent from principals of each hospital. In 2012, 3555 mothers were hospitalized in 395 units on designated days for the study. The consent to conduct the study was obtained from the directors of 377 institutions where 2905 mothers were hospitalized. A total of 2825 questionnaires were qualified for the statistical analysis. In 2017, births took place in 397 hospitals and consent was obtained from 380 directors. A total of 3627 women were hospitalized and 3451 questionnaires were qualified for the statistical analysis.

The questionnaire was divided into two parts. The first part contained 77 questions concerning: maternal age, place of residence, education, marital, social and economic status, maternity profile (i.e., earlier births, miscarriages, possible difficulties in conceiving), as well as risky health behaviours prior to and during pregnancy (e.g., smoking, alcohol, drugs and other psychoactive substance abuse). This part also contained data on the course of pregnancy (i.e., reasons for hospitalization, pregnancy complications, diagnostic tests performed during pregnancy). There was no defined framework for information on socioeconomic data. They were the subjective opinion of women themselves. However, social conditions can be described as not only features of individuals and households, such as income, wealth, educational attainment, family structure, housing, and transportation resources, but also features of communities, such as the prevalence and depth of poverty, rates of crime, accessibility of safe places to play and exercise, availability of transportation to jobs that provide a living wage, and availability of good schools and sources of nutritious food in a neighborhood [35]. Mothers who stayed in hospitals after birth completed the first part of the survey. The second part had nine questions that were filled-in by the medical personnel providing healthcare to the mother and newborn, with the use of medical records (pregnancy cards and patient's medical history). The questions in this part concerned the mode of delivery, newborn's health status after birth and birth defects. It also included questions about the results of laboratory tests performed on mothers and newborns after birth. The design of Pol-PrAMS study is presented in detail in another paper [36].

The survey aimed to compare the obstetric outcomes in a group of women with the diagnosis of DM and obesity, and to determine the risk factors of GDM. Women with BMI ≥ 25 kg/m^2 or ≥ 30 kg/m^2 were defined as overweight or obese, respectively. BMI was calculated on the basis of data, such as height and weight before pregnancy, provided by the women in the survey. Currently in Poland, the diagnosis of GDM is based on the three-step 75 g OGTT (fasting glucose $\geq$ 92—5.1 mmol/L; in 60 min $\geq$ 180 mg/dL—10 mmol/L and/or in 120 min $\geq$ 153 mg/dL—8.5 mmol/L]) [2]. These were also the criteria for the diagnosis of GDM in year 2017. In 2012, GDM was diagnosed based on the recommendations of the Polish Ministry of Health [29]. According to the standard on pregnancy management, GDM diagnosis was based on the two-step 75 g OGTT (fasting glucose $\geq$ 100 mg/dL—5.5 mmol/L—and/or in 120 min $\geq$ 140 mg/dL—7.8 mmol/L) performed between 24–28 gestational weeks.

Statistical Analysis

Overall, 2825 women in year 2012 and 3451 women in year 2017 were included in the study. Continuous variables were compared using a Student's *t*-test, while a chi-square test was applied for categorical variables. The results were expressed as the mean and standard deviation, or as a frequency (%). Logistic regression models were created to estimate the odds ratios (OR) and 95% confidence intervals (95% CI) for associations between selected variables and the risk of GDM development. All statistical analyses were performed using IBM SPSS software version 25 (IBM, Armonk, NY, USA). A p-value of < 0.05 was considered statistically significant.

The aim of the study was to compare the obstetric outcomes in groups of women with the diagnosis of DM and obesity, and to determine the risk factors of GDM. Therefore, the groups of women were standardized in terms of different ages and places of residence in both populations. As a result, we minimized the chance of bias due to differences in the structure of two analysed populations of women. Supplementary Table S1 presents the raw characteristics of the groups of women prior to the standardization. Further analyses were performed with the use of weighted data, in which the structure of the population studied in 2012 was matched to the population studied in 2017 in terms of the age structure and place of residence. The matching was carried out with the rim weighting method (SPSSINC RAKE procedure). The analysed proportions, before and after weighting, did not reveal statistically significant differences as regards the place of residence. The implemented method involves the calculation of the specific weight for each record, so a value larger than "one" may be attributed to a single record. It also provides explanation for the differences in the numbers of subjects in each group during both study periods. Table 1 presents the characteristics of women in both groups after the standardization.

Table 1. Characteristics of the study population.

Characteristics	2012			2017		p
	N	%		N	%	
Age (years)						ns
≤25	543	19.2		656	19.2	
26–30	974	34.5		1177	34.5	
Over 30	1307	46.3		1580	46.3	
Education						ns
Primary	167	6.1		187	5.6	
Secondary	1240	45.1		1431	43.1	
Tertiary	1341	48.8		1704	51.3	
Place of residence (inhabitants)						ns
City (≥100,000)	725	25.8		804	24.1	
Town/city (<100,000)	977	34.8		1154	34.7	
Rural area	1103	39.3		1373	41.2	
Social conditions						<0.05
Very good	735	26.0		1295	37.8	
Good	1610	57.0		1821	53.2	
Average/poor	478	16.9		308	9.0	
Economic status						<0.05
Very good	284	10.1		652	18.9	
Good	1675	59.4		2188	63.4	
Average/poor	863	30.6		611	17.7	
BMI before pregnancy (kg/m^2)						<0.05
<18.5	228	8.3		267	8.0	
18.5–24.99	1907	69.5		2211	66.5	
25.0–29.99	458	16.7		599	18.0	
≥30	151	5.5		247	7.5	
Parameter	**N**	**Mean**	**S.D.**	**Minimum**	**Maximum**	**p**
Height (cm)						ns
2012	2752	166.0	6.1	139.0	188.0	
2017	3344	165.9	5.9	146.0	198.0	
Gestational weight gain (kg)						<0.05
2012	2753	15.5	5.7	0.0	102.0	
2017	3307	14.7	6.2	0.0	89.0	

BMI—body mass index; ns—non-significant, $p > 0.05$.

3. Results

Following the standardization, the age structure was similar in both groups of participants with the highest percentage of women being aged over 30 years. Women who completed tertiary education constituted the highest percentage of participants in both analysed groups—48.8% in 2012 and 51.3% in 2017. The rates of women with the primary education were the lowest: 6.1% and 5.6% in years 2012 and 2017, respectively. The largest percentage of women in both groups lived in rural areas: 39.3% in 2012 and 41.2% in 2017. Slightly over 34% of participants in both groups lived in towns/cities with up to 100,000 inhabitants. The largest percentage of women in both groups described their social and economic conditions as good. Normal body weight expressed as BMI 18.5–24.99 kg/m^2 was noted in 69.5% of women in 2012, while in 2017 the respective percentage reached 66.5%. Overweight women constituted 16.7% and 18.0%, while obese 5.5% and 7.5% of studied population in years 2012 and 2017, respectively ($p < 0.05$). The mean height of women in both groups was approx. 166 cm. A slightly increased gestational weight gain was observed in the group of women who gave birth in 2012. It reached 15.5 kg, while in 2017 it was 14.7 kg, ($p < 0.05$). Both groups of women differed significantly with respect to pregestational BMI, gestational weight gain, social conditions and the economic status ($p < 0.05$). Table 1 presents the characteristics of women in both analysed years.

The percentage of women whose pregnancy was complicated by DM increased significantly over the 5-year period ($p < 0.05$) (Table 2). The respective percentages for the years 2012 and 2017 amounted to 4.5% and 7.2% of studied population. GDM prevailed in both analysed years. It accounted for 89.0% of DM cases in 2012 and 85.6% in 2017. The percentage of women in whom GDM was diagnosed in the 1st trimester of pregnancy increased significantly from 3.5% in 2012 to 13.6% in 2017 ($p < 0.05$).

Table 2. The distribution of diabetes mellitus and obesity in Polish pregnant women in years 2012 and 2017 (the numbers in brackets refer to the percentage of GDM/PGDM subtypes in all patients diagnosed with diabetes mellitus).

Variable	2012		2017		*p*
	N	%	N	%	
Diabetes (total number of patients)	2799	100	3451	100	
No	2672	95.5	3201	92.8	<0.05
Yes	127	4.5 (100.0)	250	7.2 (100.0)	
PGDM	14	0.5 (11.0)	36	1.0 (14.4)	
GDM	113	4.0 (89.0)	214	6.2 (85.6)	
Period when GDM was diagnosed					<0.05
1st trimester	4	3.5	29	13.6	
2nd or 3rd trimester	109	96.5	185	86.4	
Obese (BMI $\geq$ 30 kg/m^2)					<0.05
No	2592	94.5	3077	92.6	
Yes	151	5.5	247	7.5	
Obese with GDM					<0.05
No	2795	99.7	3409	99.1	
Yes	9	0.3	32	0.9	

PGDM—pregestational diabetes mellitus; GDM—gestational diabetes mellitus; BMI—body mass index.

Both analysed populations differed in terms of the percentage of women who had been obese before the pregnancy. In 2012 women with BMI $\geq$ 30 kg/m^2 constituted 5.5%, while in 2017—7.5% of the total study group ($p < 0.05$). The differences were also observed in rates of obese women in whom the diagnosis of GDM was made during the pregnancy. The respective percentages were 0.3% in 2012 and 0.9% in 2017 ($p < 0.05$). The distribution of diabetes and obesity in the study populations is presented in Table 2.

Table 3 presents selected perinatal outcomes in individual groups of women in the years 2012 and 2017. No statistically significant differences were noted in terms of diagnosed fetal defects. The comparison of variables in 2012 and 2017 revealed an insignificant increase in the percentage of congenital defects in the offspring of women suffering from DM (0.0% vs 1.0%). On the contrary, the percentage of birth defects decreased in obese pregnant women (3.3% vs. 2.1%). Between the years 2012 and 2017 a significant increase in the percentage of induced deliveries and instrumental deliveries/cesarean section (CS) in the whole study group was noted ($p < 0.05$). At the same time, no changes referring to the fetal birth-weight were observed in the total study population.

In the group of obese women the percentage of induced deliveries increased insignificantly over the analysed 5-year period (9.9% vs. 11.7%), whereas the fetal birth-weight decreased significantly (3565 g vs. 3405 g, $p < 0.05$). In 2012 almost 45% of obese women with concurrent GDM delivered via CS or with the use of a vacuum extractor or obstetric forceps. The respective percentage in 2017 was 64.5%. The percentage of induced deliveries in this subgroup of patients was comparable in both study periods (approx. 22%). The mean fetal birth-weight reached 3603 g and 3419 g, in years 2012 and 2017, respectively. However, the difference was not statistically significant.

Table 3. Selected perinatal outcomes in the individual groups of women analysed in years 2012 and 2017.

Variable	All Women			Obese Women *			Obese * with GDM			Diabetes **		
	2012 N (%)	2017 N (%)	*p*	2012 N (%)	2017 N (%)	*p*	2012 N (%)	2017 N (%)	*p*	2012 N (%)	2017 N (%)	*p*
Type of delivery			<0.05			ns			ns			ns
Vaginal	1706 (61.5)	1879 (58.0)		71 (48.0)	109 (48.0)		5 (55.6)	11 (35.5)		72 (58.5)	134 (55.4)	
C-section/assisted	1070 (38.5)	1360 (42.0)		77 (52.0)	118 (52.0)		4 (44.4)	20 (64.5)		51 (41.5)	108 (44.6)	
Labor induction			<0.05			ns			ns			ns
No	2563 (90.8)	3069 (88.9)		136 (90.1)	218 (88.3)		7 (77.8)	25 (78.1)		112 (88.2)	211 (84.4)	
Yes	261 (9.2)	382 (11.1)		15 (9.9)	29 (11.7)		2 (22.2)	7 (21.9)		15 (11.8)	39 (15.6)	
Fetal birth defects			ns			ns			-			ns
No	2785 (98.6)	2735 (98.5)		146 (96.7)	185 (98.9)		9 (100.0)	24 (100.0)		127 (100.0)	202 (99.0)	
Yes	39 (1.4)	42 (1.5)		5 (3.3)	2 (1.1)		0 (0.0)	0 (0.0)		0 (0.0)	2 (1.0)	
Fetal birth-weight (g)	3373	3368	ns	3565	3405	<0.05	3603	3419	ns	3285	3342	ns
Standard deviation	572	567		679	575		516	608		640	615	
Minimum	380	300		1200	695		2900	1720		870	338	
Maximum	5220	5470		5220	4870		4550	4850		4550	5000	

ns—non-significant, $p > 0.05$; * body mass index before pregnancy ≥ 30 kg/m^2; ** diagnosis of pregestational or gestational diabetes mellitus in the current pregnancy.

In the group of women with a history of DM the percentage of all the perinatal outcomes was characterized by an upward trend over the years. However, statistically significant differences were not observed (Table 3). The exact levels of statistical significance were as follows: type of delivery, $p = 0.56$; labor induction, $p = 0.32$; fetal birth defects, $p = 0.26$ and fetal birth-weight, $p = 0.59$.

Figure 1 presents the risk of developing GDM in the years 2012 and 2017 depending on selected variables. In both study periods the risk of GDM occurrence was not increased in case of at least 2 or 4 deliveries in the past, as well as in the group of women who delivered fetuses with the birth-weight equal to or higher than 4500 g. Additionally, in 2012 the risk was not increased in a group of women who delivered fetuses weighing 4000 g or more, women aged over 40 years and with pregestational BMI exceeding 30 kg/m^2. A statistically significant increase in the risk of GDM was observed in women aged over 35 years both in 2012 (OR = 1.9, 95% CI: 1.1–2.9) and in 2017 (OR = 2.1, 95% CI: 1.5–2.9). Moreover, an increased risk of GDM was noted in women with the pregestational BMI exceeding 25 kg/m^2 in both study periods. In 2012 the OR was 1.8 (95% CI: 1.2–2.7), whereas in 2017 the OR was 2.6 (95% CI: 1.9–3.4). As regards the population studied in 2017 a significant increase in the risk of GDM was observed in those women who had delivered fetuses with the birth-weight of 4000 g or more (OR = 1.7, 95% CI: 1.1–2.7), aged over 40 years (OR = 2.4, 95% CI: 1.2–4.8) and with pregestational BMI $\geq$ 30 kg/m^2 (OR = 2.4, 95% CI: 1.6–3.6).

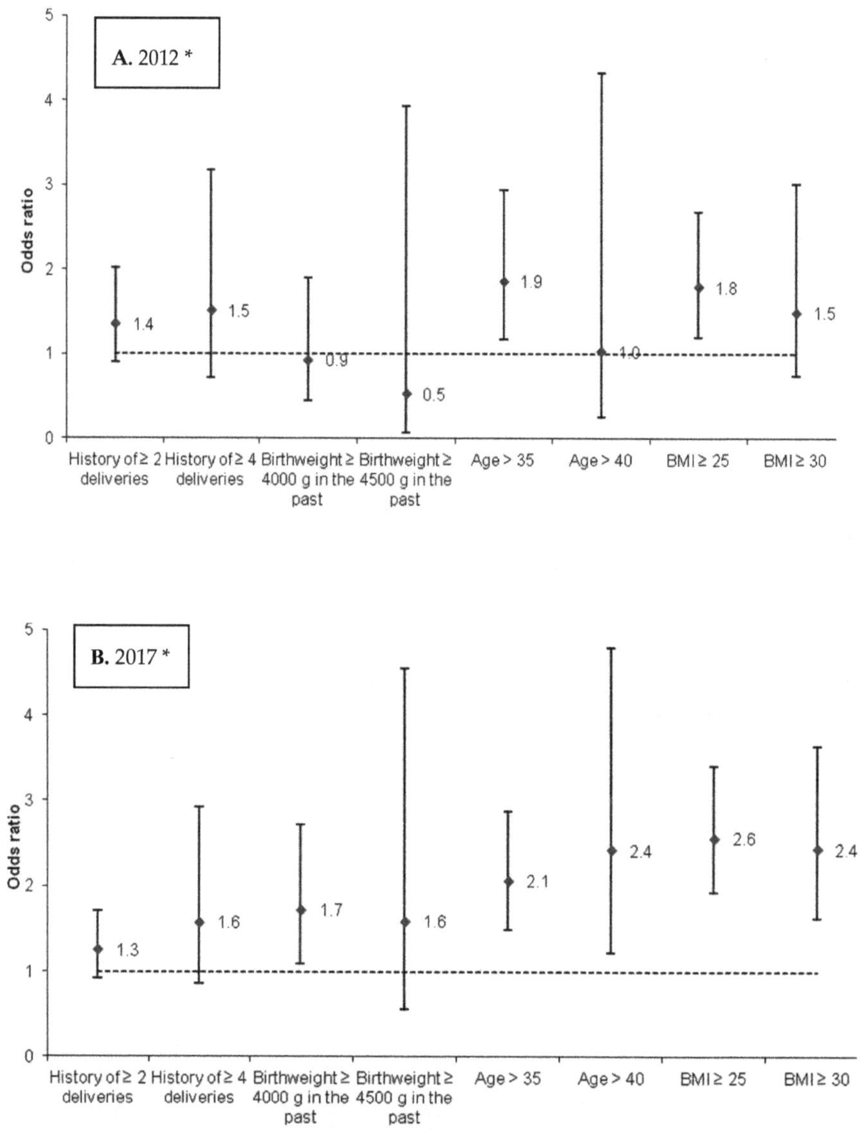

Figure 1. Risk factors of gestational diabetes mellitus in years 2012 (**A**) and 2017 (**B**). (* a group of women belonging to the complementary category of an individual variable was a reference category for each of the variables).

4. Discussion

In the present study we investigated perinatal outcomes in a population of Polish women based on the results of a national survey conducted in the years 2012 and 2017. The analysis of data obtained over this 5-year period indicated a significant increase in the percentages of pregnancies complicated by obesity and DM. At the same time, national recommendations published in years 2011/2012 together with their later updates, concerning the management of obese and/or diabetic pregnant women contributed to the significant reduction in the fetal birthweight in the first group of patients and an increased detectability of GDM in the first trimester of gestation [2,32–34]. Both the patient's age >35 at the moment of conception and pre-gestational BMI >25 kg/m^2 constituted significant risk factors for developing GDM in both analysed time periods.

The observed upward trend referring to the percentage of CS, patients with pre-pregnancy BMI exceeding 30 kg/m^2, and pregnancies complicated by GDM/PGDM in the population of Polish women reflects worldwide tendencies [22,37–40]. The common risk factor for the above-mentioned complications is the growing maternal age at conception associated with a 2- to 3-fold higher perinatal morbidity [41]. Apart from the higher rate of CS and GDM, a positive correlation between the advanced maternal age and the increased risk of fetal chromosomal aberrations, preterm delivery, stillbirth, multiple pregnancy, pre-eclampsia, thrombosis, postpartum hemorrhage, hysterectomy and stroke was noted [41,42]. In the Polish population in year 2017, over 46% of women were over 30 years of age, while 5 years before the respective percentage was slightly over 32% (see: Supplementary Table S1). The observed changes in the age structure of Polish pregnant women indicate the necessity of updating domestic recommendations concerning perinatal care with particular attention paid to suitable prophylaxis, diagnostics and treatment in patients older than 35 years. To date in Poland only two documents have been published aiming to improve perinatal care in case of advanced maternal age: PSGO recommendations concerning the indications for induced delivery in patients over 40 years issued in 2017 [43] and the programme of prenatal tests covered by the National Health Fund for patients older than 35 years which has been in operation since 2005 [44].

Based on the results of the national survey in 2017 in Poland the percentage of CS amounted to 42% and exceeded the "ideal rate" suggested by the WHO by almost 30% [38,45]. Apart from the ageing of the obstetric population and its related morbidity, the main reasons for the increasing number of CS in Poland include wide list of indications for surgical deliveries, as well as the preferences of women themselves. Direct and indirect interrelations between those factors create an urgent necessity for developing new or updating the already-existing national strategies concerning the indications for CS. PSGO recommendations issued in 2018 constitute the response to the above-mentioned demands and include the majority of obstetric and extra-obstetric indications for a surgical intervention during pregnancy [46]. Nonetheless, the assessment of their effectiveness requires a population-based survey to be conducted in subsequent years.

Over the 5-year period the percentage of overweight and obese women in Poland increased from 16.7% to 18% and from 5.5% to 7.5%, respectively. According to the literature, an increased caloric supply, urbanization and gross national income are the main factors responsible for the excessive weight gain and obesity in pregnant women in high-income countries [22]. Polish data grossly confirmed the worldwide observations. The comparison of national survey performed in 2017 with data obtained in 2012 showed a significant increase in the percentage of women with tertiary education and those who described their social conditions and economic status as very good (see: Supplementary Table S1). At the same time no differences were observed with regard to the place of residence with the percentage of women living in cities with over 100,000 inhabitants at around 25% in both analysed time periods.

Observed tendencies revealed through the national survey are unfavorable, as they indicate a strong negative trend towards the increased rate of obese women over the past few years despite a significant improvement concerning the level of education and the socioeconomic status (see: Supplementary Table S1). Although large population-based studies concerning the socioeconomic risk factors of obesity have not been conducted in the Polish population so far, research performed by other authors demonstrated their strong inter-relations. In the study by Cutler et al. authors observed the positive correlation between years of schooling and a reduced risk of being overweight and obese. Simultaneously, persons with higher levels of education were more physically active. The relationship between the level of education and obesity seems to be non-linear with the increasing effects of additional years of schooling [47]. In addition, a negative association between the level of education and the probability of being overweight was demonstrated in a cross-sectional study of twins [48]. According to some authors' suggestions, the patient's sex significantly modifies the association between socioeconomic factors and obesity. Among women, a higher level of education presented the strongest correlation with low BMI and waist-to-hip ratio (WHR), whereas, in men both parameters were more affected by the income [49].

Despite the unfavorable upward tendency referring to the percentage of pregnancies with concomitant obesity, the recommendations published by PSGO in 2012 contributed to the improvement in perinatal care in the affected group of patients, i.e., obese women [34]. The effectiveness of the measures taken is reflected by the significant decrease in the neonatal birthweight and the simultaneous increase in the number of induced deliveries. The observed changes regarding perinatal outcomes are directly associated with the recommendations of a Polish group of experts in which they emphasized the necessity and determined the aims of the preconception care of obese patients, set the upper limit of weight gain during pregnancy as well as suggested performing an additional pre-delivery ultrasound examination to estimate the fetal birth-weight. Furthermore, despite the fact that isolated obesity does not constitute direct indication for labor induction, such management was suggested after 38 gestational weeks in patients with BMI over 40 kg/m^2. As regards the main assumptions of the perinatal care in obese pregnant women Polish recommendations are similar to those published by other national obstetric societies [50]. Nonetheless, issues such as folic acid and vitamin D supplementation, the prophylaxis of venous thromboembolism or pre-eclampsia require updating.

One of the assumptions of the recommendations published by PSGO was the proper control of body weight in pregnancy [2,32,33]. Polish recommendations remained in line with international guidelines, which indicate that a woman with a normal body weight expressed in BMI should gain between 11.5 kg and 15.9 kg during pregnancy [51]. As a consequence, it was possible to reduce the average weight gain in pregnancy from 15.5 kg in 2012 to 14.7 kg in 2017. Noteworthy, in year 2012 the average weight gain of pregnant women in Poland was close to the specified upper limit.

Certain concerns are raised as regards the group of obese pregnant women in whom GDM was diagnosed. This population of patients noted significant increase from 0.3% to 0.9% over the 5-year period. Moreover, despite more effective perinatal care leading to reduced neonatal birthweight, the percentage of induced deliveries did not change, whereas the percentage of CS increased by almost 20% in the group of women with BMI > 30 kg/m^2 and concomitant GDM. The association between obesity/GDM and an increased risk of an operative delivery is well-documented and, most probably, the observed tendencies in the Polish population are related to the synergistic effect of both diseases [52–54]. Nevertheless, due to the increasing percentage of CS performed in Poland, publication of separate recommendations for labor induction in obese diabetic women seems to be justified.

Apart from the advanced maternal age, obesity is one of the most important risk factors of GDM, which was confirmed by the results of the Polish national survey. Data obtained in the years 2012 and 2017 reflect the worldwide tendency as regards the increased incidence of GDM [55,56]. From the clinical perspective the 4-fold increase in GDM detectability in the first trimester constitutes one of the most important observations. The presumable reason for that is the increasing awareness of healthcare providers regarding diagnostic standards of DM in pregnancy. According to Polish recommendations, patients presenting the risk factors of GDM (including obesity) should be offered a three-step 75 g OGTT during the first antenatal visit. Thus, the increasing frequency of OGTTs performed in the 1st trimester of gestation may be associated with an increased detectability of hyperglycemia in early pregnancy observed in year 2017 as compared to 2012. An increased GDM detectability in the 1st trimester of pregnancy seems to be a favorable phenomenon, as it provides early care to patients with glucose tolerance disorders, which were undetected prior to gestation. A large multicenter DALI trial revealed that patients with an early GDM diagnosis are characterized by a significantly poorer metabolic profile as compared to those diagnosed in the second/third trimester. An early intervention may therefore contribute to the better glycemia control, reduced gestational weight gain and ultimately reduced neonatal adiposity [57].

The main limitation of our study is the fact that diagnostic criteria for gestational diabetes mellitus have changed between 2012 and 2017. Nevertheless, it allowed us to assess the impact of the new diagnostic strategy on the perinatal outcomes. Another limitation is the occurrence of differences in the characteristics of analysed cohorts of women in both study periods. We tried to eliminate this bias using rim weighting method (SPSSINC RAKE procedure). Limitation is also the retrospective character of our study in conjunction with the voluntary participation. In addition, some of the variables evaluated in the study relied on the subjective assessment of study participants, such as the assessment of socio-economic conditions.

Despite that, the present study is the most extensive analysis of the perinatal outcomes in obese and diabetic pregnant women in Poland conducted so far. It's undeniable advantage is associated with the size of the study group and the fact that it was performed in the majority of Polish obstetric centers. As a consequence, it minimized the risk of selection bias, which might be due to the fact that the participation in the study was voluntary. Furthermore, the clinical value of the study is increased by the fact that the data collection was performed over analogous time periods with a 5-year interval. The latter facilitated the analysis of the trends in the epidemiology of obesity and diabetes in the population of Polish pregnant women. Study results reflected tendencies observed in many developed countries as regards the increasing frequency of obesity and GDM, as well as the fact that women of reproductive age decide to conceive later. It seems, however, that apart from the increased rate of CS, the above-mentioned trends did not have significant impact on the other adverse perinatal outcomes. Nonetheless, it is necessary to develop population-based programmes to prevent obesity and to introduce and enforce the rules of appropriate screening for glucose tolerance disorders during pregnancy.

5. Conclusions

Over the 5 years period a significant increase in the percentage of obese women of reproductive age was observed in Poland. This fact, along with the change in the diagnostic criteria for GDM, contributed to the increase in the percentage of diabetes diagnoses, in particular in the first trimester of pregnancy. Noteworthy, in the population of obese women we observed a significant decrease in the fetal birthweight. In the analysed period of time, several documents appeared in Poland that defined the standards of care for pregnancies complicated by diabetes and obesity. These recommendations provide the most probable explanation of the observed national trends.

Supplementary Materials: The following are available online at https://www.mdpi.com/1660-4601/18/2/560/s1, Table S1: Characteristics of the study population—raw data.

Author Contributions: Conceptualization, C.W., P.S., P.G. and M.C.; data curation, C.W.; formal analysis, A.W.; funding acquisition, A.W.; investigation, C.W., P.S., P.G. and M.C.; methodology, A.W.; project administration, A.W.; resources, C.W., P.S., P.G. and M.C.; supervision, A.W.; validation, P.S., P.G. and M.C.; visualization, C.W.; writing—original draft, C.W., P.S., P.G. and M.C.; writing—review and editing, A.W. All authors have read and agreed to the published version of the manuscript.

Funding: This research received no external funding.

Institutional Review Board Statement: The study was conducted according to the guidelines of the Declaration of Helsinki and approved by the Ethics Committee of Institute of Rural Health in Lublin (reference number 03/2011).

Informed Consent Statement: Informed consent was obtained from all subjects involved in the study.

Data Availability Statement: The data presented in this study are available on request from the corresponding author. The data are not publicly available due to privacy.

Conflicts of Interest: The authors declare no conflict of interest.

References

1. American Diabetes Association. Introduction: Standards of Medical Care in Diabetes—2020. *Diabetes Care* **2019**, *43*, S1–S2. [CrossRef]
2. Wender-Ozegowska, E.; Bomba-Opoń, D.; Brązert, J.; Celewicz, Z.; Czajkowski, K.; Gutaj, P.; Malinowska-Polubiec, A.; Zawiejska, A.; Wielgos, M. Standards of Polish Society of Gynecologists and Obstetricians in management of women with diabetes. *Ginekol. Pol.* **2018**, *89*, 341–350. [CrossRef]
3. Araszkiewicz, A.; Bandurska-Stankiewicz, E.; Budzyński, A.; Cypryk, K.; Czech, A.; Czupryniak, L.; Drzewoski, J.; Dzida, G.; Dziedzic, T.; Franek, E.; et al. 2019 Guidelines on the management of diabetic patients. A position of Diabetes Poland. *Clin. Diabetol.* **2019**, *8*, 1–95. [CrossRef]
4. World Health Organization (WHO). Global Report on Diabetes. Available online: https://www.who.int/diabetes/global-report/en/ (accessed on 20 October 2020).
5. American Diabetes Association. Economic Costs of Diabetes in the U.S. in 2017. *Diabetes Care* **2018**, *41*, 917–928. [CrossRef] [PubMed]
6. Negrato, C.A.; Mattar, R.; Gomes, M.B. Adverse pregnancy outcomes in women with diabetes. *Diabetol. Metab. Syndr.* **2012**, *4*, 41. [CrossRef] [PubMed]
7. World Health Organization (WHO). Diabetes Fact Sheet WHO. Available online: https://www.who.int/news-room/fact-sheets/detail/diabetes (accessed on 20 March 2020).
8. International Association of Diabetes and Pregnancy Study Groups Consensus Panel. International Association of Diabetes and Pregnancy Study Groups Recommendations on the Diagnosis and Classification of Hyperglycemia in Pregnancy. *Diabetes Care* **2010**, *33*, 676–682. [CrossRef]
9. American Diabetes Association. Diagnosis and Classification of Diabetes Mellitus. *Diabetes Care* **2008**, *32*, S62–S67. [CrossRef]
10. DeSisto, C.L.; Kim, S.Y.; Sharma, A.J. Prevalence Estimates of Gestational Diabetes Mellitus in the United States, Pregnancy Risk Assessment Monitoring System (PRAMS), 2007–2010. *Prev. Chronic Dis.* **2014**, *11*, E104. [CrossRef]
11. Fong, A.; Serra, A.; Herrero, T.; Pan, D.; Ogunyemi, D. Pre-gestational versus gestational diabetes: A population based study on clinical and demographic differences. *J. Diabetes Complicat.* **2014**, *28*, 29–34. [CrossRef]
12. Buchanan, T.A.; Xiang, A.H.; Page, K.A. Gestational diabetes mellitus: Risks and management during and after pregnancy. *Nat. Rev. Endocrinol.* **2012**, *8*, 639–649. [CrossRef]
13. National Institute of Clinical Excellence (NICE). Diabetes in Pregnancy: Management of Diabetes and Its Complications from Preconception to the Postnatal Period. Available online: https://www.nice.org.uk/guidance/ng3 (accessed on 20 October 2020).
14. Kim, C.; Newton, K.M.; Knopp, R.H. Gestational Diabetes and the Incidence of Type 2 Diabetes: A systematic review. *Diabetes Care* **2002**, *25*, 1862–1868. [CrossRef] [PubMed]
15. Hruby, A.; Hu, F.B. The Epidemiology of Obesity: A Big Picture. *PharmacoEconomics* **2015**, *33*, 673–689. [CrossRef] [PubMed]
16. Nuttall, F.Q. Body Mass Index. *Nutr. Today* **2015**, *50*, 117–128. [CrossRef] [PubMed]
17. Pi-Sunyer, X. The Medical Risks of Obesity. *Postgrad. Med.* **2009**, *121*, 21–33. [CrossRef]
18. The GBD 2015 Obesity Collaborators. Health effects of overweight and obesity in 195 countries over 25 years. *N. Engl. J. Med.* **2017**, *377*, 13–27. [CrossRef]
19. Haslam, D.; James, W.P.T. Obesity. *Lancet* **2005**, *366*, 1197–1209. [CrossRef]
20. Bleich, S.N.; Cutler, D.; Murray, C.; Adams, A. Why Is the Developed World Obese? *Annu. Rev. Public Health* **2008**, *29*, 273–295. [CrossRef]
21. Nelson, S.M.; Matthews, P.; Poston, L. Maternal metabolism and obesity: Modifiable determinants of pregnancy outcome. *Hum. Reprod. Update* **2009**, *16*, 255–275. [CrossRef]
22. Chen, C.; Xu, X.; Yan, Y. Estimated global overweight and obesity burden in pregnant women based on panel data model. *PLoS ONE* **2018**, *13*, e0202183. [CrossRef]
23. Smith, G.C.; Shah, I.; Pell, J.P.; Crossley, J.A.; Dobbie, R. Maternal Obesity in Early Pregnancy and Risk of Spontaneous and Elective Preterm Deliveries: A Retrospective Cohort Study. *Am. J. Public Health* **2007**, *97*, 157–162. [CrossRef]
24. Athukorala, C.; Rumbold, A.R.; Willson, K.; Crowther, C.A. The risk of adverse pregnancy outcomes in women who are overweight or obese. *BMC Pregnancy Childbirth* **2010**, *10*, 56. [CrossRef] [PubMed]
25. Stothard, K.J.; Tennant, P.W.G.; Bell, R.; Rankin, J. Maternal Overweight and Obesity and the Risk of Congenital Anomalies. *JAMA* **2009**, *301*, 636–650. [CrossRef] [PubMed]
26. Correa, A.; Marcinkevage, J. Prepregnancy obesity and the risk of birth defects: An update. *Nutr. Rev.* **2013**, *71*, S68–S77. [CrossRef] [PubMed]
27. Kitzmiller, J.L.; Block, J.M.; Brown, F.M.; Catalano, P.M.; Conway, D.L.; Coustan, D.R.; Gunderson, E.P.; Herman, W.H.; Hoffman, L.D.; Inturrisi, M.; et al. Managing Preexisting Diabetes for Pregnancy: Summary of evidence and consensus recommendations for care. *Diabetes Care* **2008**, *31*, 1060–1079. [CrossRef] [PubMed]
28. Polish Ministry of Health. Law Act. Rozporządzenie Ministra Zdrowia z Dnia 16 Sierpnia 2018 r. W Sprawie Standardu Organizacyjnego Opieki Okołoporodowej. Available online: http://prawo.sejm.gov.pl/isap.nsf/DocDetails.xsp?id=WDU20180001756 (accessed on 21 March 2020).

29. Polish Ministry of Health. Law Act. Rozporządzenie Ministra Zdrowia z Dnia 20 Września 2012 r. W Sprawie Standardów Postępowania Medycznego Przy Udzielaniu Świadczeń Zdrowotnych z Zakresu Opieki Okołoporodowej Sprawowanej Nad Kobietą w Okresie Fizjologicznej Ciąży, Fizjologicznego Porodu, Połogu Oraz Opieki Nad Noworodkiem. Available online: http://prawo.sejm.gov.pl/isap.nsf/DocDetails.xsp?id=WDU20120001100 (accessed on 20 March 2020).
30. Coustan, D.R.; Lowe, L.P.; Metzger, B.E.; Dyer, A.R. The Hyperglycemia and Adverse Pregnancy Outcome (HAPO) study: Paving the way for new diagnostic criteria for gestational diabetes mellitus. *Am. J. Obstet. Gynecol.* **2010**, *202*, 654.e1–654.e6. [CrossRef]
31. Diabetes Poland. Polskie Towarzystwo Diabetologiczne—Zalecenia Kliniczne. Available online: https://cukrzyca.info.pl/zalecenia_kliniczne (accessed on 23 October 2020).
32. Wender-Ozegowska, E.; Bomba-Opoń, D.; Brazert, J.; Celewicz, Z.; Czajkowski, K.; Karowicz-Bilińska, A.; Malinowska-Polubiec, A.; Meczekalski, B.; Zawiejska, A. Polish Gynecological Society standards of medical care in management of women with diabetes. *Ginekol. Pol.* **2011**, *82*, 474–479.
33. Bomba-Opoń, B.B.; Celewicz, Z.; Czajkowski, K.; Karowicz-Bilińska, A.; Malinowska-Polubiec, A.; Męczekalski, B.; Zawiejska, A.; Wender-Ożegowska, W. Polish Gynecological Society standards of medical care in management of women with diabetes—Update. *Ginekol. Pol.* **2014**, *85*, 476–478.
34. Wender-Ozegowska, E.; Bomba-Opoń, D.; Brazert, J.; Celewicz, Z.; Czajkowski, K.; Karowicz-Bilińska, A.; Malinowska-Polubiec, A.; Meczekalski, B.; Zawiejska, A. Recommendations of Polish Gynecological Society concerning perinatal care in obese pregnant women. *Ginekol. Pol.* **2012**, *83*, 795–799.
35. Braveman, P.A. Social conditions, health equity, and human rights. *Health Hum. Rights* **2010**, *12*, 31–48.
36. Wojtyła, C.; Wojtyła-Buciora, P. Polish Pregnancy-related Assessment Monitoring System (Pol-PrAMS): Research on lifestyle health behaviours of Polish women during gestation—Study design. *J. Health Inequalities* **2016**, *2*, 185–191. [CrossRef]
37. Declercq, E.; Young, R.; Cabral, H.; Ecker, J. Is a Rising Cesarean Delivery Rate Inevitable? Trends in Industrialized Countries, 1987 to 2007. *Birth* **2011**, *38*, 99–104. [CrossRef] [PubMed]
38. Ye, J.; Zhang, J.; Mikolajczyk, R.; Torloni, M.R.; Gülmezoglu, A.M.; Betran, A.P. Association between rates of caesarean section and maternal and neonatal mortality in the 21st century: A worldwide population-based ecological study with longitudinal data. *BJOG Int. J. Obstet. Gynaecol.* **2016**, *123*, 745–753. [CrossRef] [PubMed]
39. Ferrara, A. Increasing Prevalence of Gestational Diabetes Mellitus: A public health perspective. *Diabetes Care* **2007**, *30*, S141–S146. [CrossRef] [PubMed]
40. Bell, R.; Bailey, K.; Cresswell, T.; Hawthorne, G.; Critchley, J.A.; Lewis-Barned, N. Trends in prevalence and outcomes of pregnancy in women with pre-existing type I and type II diabetes. *BJOG Int. J. Obstet. Gynaecol.* **2008**, *115*, 445–452. [CrossRef]
41. Sheen, J.-J.; Wright, J.D.; Goffman, D.; Kern-Goldberger, A.; Booker, W.; Siddiq, Z.; D'Alton, M.E.; Friedman, A.M. Maternal age and risk for adverse outcomes. *Am. J. Obstet. Gynecol.* **2018**, *219*, 390.e1–390.e15. [CrossRef]
42. Schmidt, L.; Sobotka, T.; Bentzen, J.G.; Andersen, A.N. Demographic and medical consequences of the postponement of parenthood. *Hum. Reprod. Update* **2011**, *18*, 29–43. [CrossRef]
43. Bomba-Opoń, D.; Drews, K.; Huras, H.; Laudański, P.; Paszkowski, T.; Wielgos, M. Polish Gynecological Society Recommendations for Labor Induction. *Ginekol. Pol.* **2017**, *88*, 224–234. [CrossRef]
44. Prenatal Screening Program of the Polish National Health Fund. Available online: https://www.pacjent.gov.pl/program-profilaktyczny/program-badan-prenatalnych (accessed on 13 May 2020).
45. World Health Organization Human Reproduction Programme. WHO Statement on caesarean section rates. *Reprod. Health Matters* **2015**, *23*, 149–150. [CrossRef]
46. Wielgos, M.; Bomba-Opoń, D.; Breborowicz, G.H.; Czajkowski, K.; Debski, R.; Leszczynska-Gorzelak, B.; Oszukowski, P.; Radowicki, S.; Zimmer, M. Recommendations of the Polish Society of Gynecologists and Obstetricians regarding caesarean sections. *Ginekol. Pol.* **2018**, *89*, 644–657. [CrossRef]
47. Cutler, D.M.; Lleras-Muney, A. *Education and Health: Evaluating Theories and Evidence*; NBER Working Paper No. 12352; NBER: Cambridge, MA, USA, 2006.
48. Amin, V.; Dunn, P.; Spector, T. Does education attenuate the genetic risk of obesity? Evidence from U.K. Twins. *Econ. Hum. Biol.* **2018**, *31*, 200–208. [CrossRef]
49. Park, H.S.; Oh, S.W.; Park, H.S. Socioeconomic Status in Relation to Obesity and Abdominal Obesity in Korean Adults: A Focus on Sex Differences. *Obesity* **2006**, *14*, 909–919. [CrossRef]
50. Vitner, D.; Harris, K.; Maxwell, C.; Farine, D. Obesity in pregnancy: A comparison of four national guidelines. *J. Matern. Neonatal Med.* **2018**, *32*, 2580–2590. [CrossRef] [PubMed]
51. Institute of Medicine (US); National Research Council (US); Committee to Reexamine IOM Pregnancy Weight Guidelines. *Weight Gain during Pregnancy: Reexamining the Guidelines*; National Academies Press: Washington, DC, USA, 2009.
52. Nasrat, H.; Fageeh, W.; Abalkhail, B.; Yamani, T.; Ardawi, M. Determinants of pregnancy outcome in patients with gestational diabetes. *Int. J. Gynecol. Obstet.* **1996**, *53*, 117–123. [CrossRef]
53. Bo, S.; Menato, G.; Signorile, A.; Bardelli, C.; Lezo, A.; Gallo, M.; Gambino, R.; Cassader, M.; Massobrio, M.; Pagano, G. Obesity or diabetes: What is worse for the mother and for the baby? *Diabetes Metab.* **2003**, *29*, 175–178. [CrossRef]
54. Pettersen-Dahl, A.; Murzakanova, G.; Sandvik, L.; Laine, K. Maternal body mass index as a predictor for delivery method. *Acta Obstet. Gynecol. Scand.* **2018**, *97*, 212–218. [CrossRef] [PubMed]

55. Yen, I.-W.; Lee, C.-N.; Lin, M.-W.; Fan, K.-C.; Wei, J.-N.; Chen, K.-Y.; Chen, S.-C.; Tai, Y.-Y.; Kuo, C.-H.; Lin, C.-H.; et al. Overweight and obesity are associated with clustering of metabolic risk factors in early pregnancy and the risk of GDM. *PLoS ONE* **2019**, *14*, e0225978. [CrossRef]
56. Egan, A.M.; Vellinga, A.; Harreiter, J.; Simmons, D.; Desoye, G.; Corcoy, R.; Adelantado, J.M.; Devlieger, R.; Van Assche, A.; Galjaard, S.; et al. Epidemiology of gestational diabetes mellitus according to IADPSG/WHO 2013 criteria among obese pregnant women in Europe. *Diabetologia* **2017**, *60*, 1913–1921. [CrossRef]
57. Harreiter, J.; Desoye, G.; Van Poppel, M.; Kautzky-Willer, A.; Dunne, F.; Corcoy, R.; Devlieger, R.; Fracp, D.S.; Adelantado, J.M.; Damm, P.; et al. The Effects of Lifestyle and/or Vitamin D Supplementation Interventions on Pregnancy Outcomes: What Have We Learned from the DALI Studies? *Curr. Diabetes Rep.* **2019**, *19*, 162. [CrossRef]

International Journal of
Environmental Research and Public Health

MDPI

Article

MALDI-MSI—A Step Forward in Overcoming the Diagnostic Challenges in Ovarian Tumors

Dagmara Pietkiewicz [1], Agnieszka Horała [2], Szymon Plewa [1], Piotr Jasiński [3], Ewa Nowak-Markwitz [2], Zenon J. Kokot [4] and Jan Matysiak [1,*]

[1] Department of Inorganic and Analytical Chemistry, Poznan University of Medical Sciences, 6 Grunwaldzka Street, 60-780 Poznan, Poland; dagmarapietkiewicz3@gmail.com (D.P.); splewa@ump.edu.pl (S.P.)
[2] Gynecologic Oncology Department, Poznan University of Medical Sciences, 33 Polna Street, 60-535 Poznan, Poland; ahorala@ump.edu.pl (A.H.); ewamarkwitz@ump.edu.pl (E.N.-M.)
[3] Department of Pathology Gynecological and Obstetric Clinical Hospital, Poznan University of Medical Sciences, 33 Polna Street, 60-535 Poznan, Poland; pjasinski@gpsk.ump.edu.pl
[4] Faculty of Health Sciences, Calisia University, 13 Kaszubska Street, 62-800 Kalisz, Poland; z.kokot@akademiakaliska.edu.pl
* Correspondence: jmatysiak@ump.edu.pl

Received: 15 September 2020; Accepted: 16 October 2020; Published: 18 October 2020

Abstract: This study presents the use of matrix-assisted laser desorption and ionization mass spectrometry imaging (MALDI-MSI) directly on the tissue of two ovarian tumors that often present a diagnostic challenge, a low-grade serous borderline ovarian tumor and ovarian fibrothecoma. Different spatial distribution of m/z values within the tissue samples was observed, and regiospecific peaks were identified. Among the 106 peaks in the borderline ovarian tumor five, regiospecific peaks (m/z: 2861.35; 2775.79; 3368.34; 3438.43; 4936.37) were selected using FlexImaging software. Subsequently, the distribution of those selected peaks was visualized on the fibrothecoma tissue section, which demonstrated the differences in the tissue homo-/heterogeneous structure of both tumors. The comparison with the histopathological staining of the ovarian borderline tumor tissue section, obtained during serial sectioning, showed a close correlation of the molecular map with the morphological and histopathological features of the tissue and allowed the identification of different tissue types within the sample. This study highlights the potential significance of MSI in enabling morphological characterization of ovarian tumors as well as correct diagnosis and further prognosis than thus far seen in the literature. Osteopontin, tropomyosin and orosomucoid are only a couple of the molecules investigated using MALDI-MSI in ovarian cancer research. This study, in line with the available literature, proves the potential of MALDI-MSI to overcome the current limitations of classic histopathological examination giving a more in-depth insight into the tissue structure and thus lead to the more accurate differential diagnosis of ovarian tumors, especially in the most challenging cases.

Keywords: proteomics; mass spectrometry; MALDI-MSI; tissue imaging; ovarian tumors

1. Introduction

Ovarian tumors are a common gynecological health problem. Diagnostic methods currently used in clinical practice in order to detect ovarian tumors encompass a basic gynecological bimanual examination and a transvaginal ultrasound scan. Most of the detected ovarian tumors are benign, and some do not require any treatment. There are several diagnostic tools used to assess the probability of ovarian malignancy, such as two serum biomarkers, cancer antigen 125 (CA125) and human epididymis protein 4 (HE4), ultrasound features, clinical information and combinations of all listed above. However, the sensitivity and specificity of the available diagnostic tests are limited [1]. Therefore, it is often difficult to assess preoperatively if an ovarian tumor is benign or malignant. Most of the patients are qualified for surgical treatment as histopathological examination of the resected ovarian tissue, as this is a golden standard in confirming the diagnosis and so far the only reliable method of differentiating between benign and malignant tumors of the ovary. An intraoperative histopathological examination is also widely used to assess the type of tumor, allowing the surgeon to choose the proper treatment during the actual surgery. This information is crucial for deciding if fertility-sparing treatment can be applied or if radical oncological surgery is needed. Thus, a precise intraoperative tissue examination is often indispensable and has significant clinical consequences.

Ovarian tumors are a heterogeneous group of neoplasms which develop from different kinds of tissue: the epithelium (65–70%), the germ cells (15–20%) and the ovarian stroma (5–10%); or which can be metastatic from other organs (5%). The vast majority of the malignant ovarian tumors (about 90%) are of epithelial origin and can be further divided into the following subtypes: serous, endometrioid, clear cell, mucinous [2]. Various studies have shown that even though these subtypes all originate from the epithelial tissue, they have unique sequences of tumorigenesis. In consequence, ovarian cancer is a highly heterogenic disease composed of a diverse group of tumors that have distinctive morphological and molecular genetic features [3].

Borderline ovarian tumors (BOTs), also called atypical proliferative tumors, are a special group of ovarian tumors characterized by atypical cellular proliferation but without stromal invasion [4]. As BOTs can be associated with microinvasion, intraepithelial carcinoma, lymph node involvement or non-invasive peritoneal implants [5], the histopathological distinction of BOTs from malignant tumors can be challenging. Moreover, the diagnostic criteria for the less common histologic subtypes are not clearly defined [6].

Histopathological examination is often complemented by immunohistochemistry (IHC) to identify tumor-specific proteins—a time–consuming method but often essential for establishing the correct diagnosis. Another significant problem in histopathology is tissue heterogeneity. Areas with low tumor differentiation determine prognosis, even if most of the tumor area is well differentiated. Methods used in traditional biomarkers discovery studies (e.g., peptide/protein profiling) that examine blood, serum or even tissue homogenate results in the loss of crucial histological information and significant differences that might be observed only as a localized histological target concentration [7].

Due to the above-mentioned considerations, the histopathological examination of ovarian tumor tissue can be very challenging. To overcome the limitations of classic histological imaging methods, in the era of 'multi-omics', molecular imaging was proposed as a tool for acquiring spatial and quantitative information about thousands of molecules without labelling potential targets [8]. While tumor tissue reflects all proteomic, metabolomic, and genetic changes, direct investigation of the tissue sample provides comprehensive information about tumor origin and heterogeneity [9] and may also provide prognostic information. Therefore, the matrix-assisted laser desorption and ionization mass spectrometry imaging (MALDI-MSI) technique, which has added to proteomics this anatomical dimension, is rapidly developing and has received significant attention in recent years in the field of biomarker search. This two-dimensional mass spectrometry technique is intended for the visualization of the spatial distribution of the biomolecules within the tissue. It is suited for the study of various types of molecular compounds, with proteins, peptides and lipids being the most commonly investigated. One of the diseases investigated so far was ovarian cancer, where serous ovarian cancer was the most common subject of research It was indicated that some compounds like osteopontin [10], tropomyosin [11], orosomucoid [12] and N-glycans [13,14] have an altered expression and might be correlated with the presence of ovarian cancer or ovarian cancer progression.

MALDI-MSI allows tissue integrity to be maintained while obtaining information about the spatial distribution of multiple peptide ions objectively [15]. Compared to IHC, the amount of tissue needed for MSI analysis is small, which is a great advantage considering the amount of additional molecular testing performed to ensure the best patient care [16]. The spectral data obtained from the MALDI-MSI experiments are then calculated and provide mapping information about the distribution of the selected molecules in the tissue sections, making it possible to distinguish between different tissue types within the same section. The obtained data is subsequently correlated with the results of the histopathological studies. Because of the possibility of combining the mass spectrometry (MS) results with morphological features, MALDI-MSI could be a helpful tool for fast and objective tumor classification and grading.

For this pilot study, we selected two ovarian tumors that often present a diagnostic challenge both for the clinicians and for the histopathologists. The presented research aimed to investigate the usefulness of MALDI-MSI as a complementary tool for histopathological examination in diagnosing ovarian tumors.

2. Materials and Methods

2.1. Sample Information

The ovarian tumor tissue samples were obtained during surgeries at the Department of Gynecologic Oncology (Poznan University of Medical Sciences, Poznan, Poland), in accordance with the Declaration of Helsinki, after the approval of the study protocol and of the written information for patients by the Local Bioethical Committee of Poznan University of Medical Sciences, Poland (decision no. 139/20). The tumors were diagnosed during standard histopathological examination and additional immunohistochemistry staining by a pathologist, an expert in ovarian tumors as a low-grade serous borderline tumor (also referred to as atypical proliferative serous tumor) and ovarian fibrothecoma.

2.2. Sample Preparation

For the MALDI Imaging experiment, a fresh frozen tissue sample was sectioned at a thickness of 10 µm using a cryostat (Leica Biosystems, Wetzlar, Germany) chilled to −20 °C. The sections were transferred onto pre-cooled indium tin oxide (ITO)-coated glass slides (Bruker Daltonics, Bremen, Germany). Subsequently, each section was thaw mounted on the cooled slide until all visible moisture on it disappeared. The slides carrying the tissue sections were transferred into a vacuum desiccator to dry for 30 min. The tissue sections were treated with wash steps of 70% ethanol (EtOH) (twice, 1 min each) and 96% ethanol (1 min). Then, the tissue sections were dehydrated using a vacuum desiccator for 15 min. The peptide calibration standard II (Bruker Daltonics, Bremen, Germany) was manually spotted (1µl) next to the tissue for calibration of the acquisition method. ITO glass slides were marked and scanned at 3200 dpi on a Reflecta MF 5000 scanner for instrument teaching purposes. The α-Cyano-4-hydroxycinnamic acid (HCCA) matrix stock solution (7g/L) in 50% acetonitrile (ACN), 0.2% (v/v) trifluoroacetic acid (TFA) was further diluted with ACN in a volume ratio of 3:1 and then applied onto the prepared glass slide with a tissue section using a method of the ImagePrep station (Bruker Daltonics, Bremen, Germany). Matrix solution application method was optimized to achieve a homogeneous matrix layer and then obtain good quality MS analysis results (Table 1).

Table 1. ImagePrep parameters of the optimized matrix solution application method.

ImagePrep Parameter	Value
Matrix thickness	0.5 V
Incubation time	30 s
Wetness	40%
Spray power	10%
Spray modulation	30%

2.3. MALDI-MSI

MS data were acquired using an UltrafleXtreme MALDI-TOF/TOF instrument (Bruker Daltonics, Bremen, Germany) controlled by flexControl (v3.4, Bruker Daltonics, Bremen, Germany) and flexImaging (v4.1, Bruker Daltonics, Bremen, Germany) in linear positive ion mode. Settings for MALDI-TOF MS analysis were as follows: 2000–20,000 m/z range. To suppress unwanted ions, the matrix suppression parameter was set at 1500Da. Laser parameters were as follows: pulsed ion extraction 260 ns and lens 6.40 kV. 2000 shots were acquired at each position using the Smartbeam 2_small laser diameter within a 200 µm raster width.

2.4. Imaging Data Analysis

The selection of peaks characterized by the highest regiospecificity in the analyzed ovarian tumor tissue was carried out manually in flexImaging software dedicated to tissue imaging analyses. We defined regions of interest (ROI) for borderline ovarian tumor tissue corresponding to the different tissue subtypes visible on the hematoxylin and eosin (H&E) stained tissue section. Then in the flexImaging we displayed the density plot, and we manually chose the m/z values that seemed to be the most differentiating between ROIs in the borderline ovarian tumor tissue section. The averaged spectra (Figure 1) obtained during analysis of the serous borderline tumor, and the ovarian fibrothecoma, were analyzed, and then, based on the manually selected peaks, the differentiating masses were visualized with the flexImaging software.

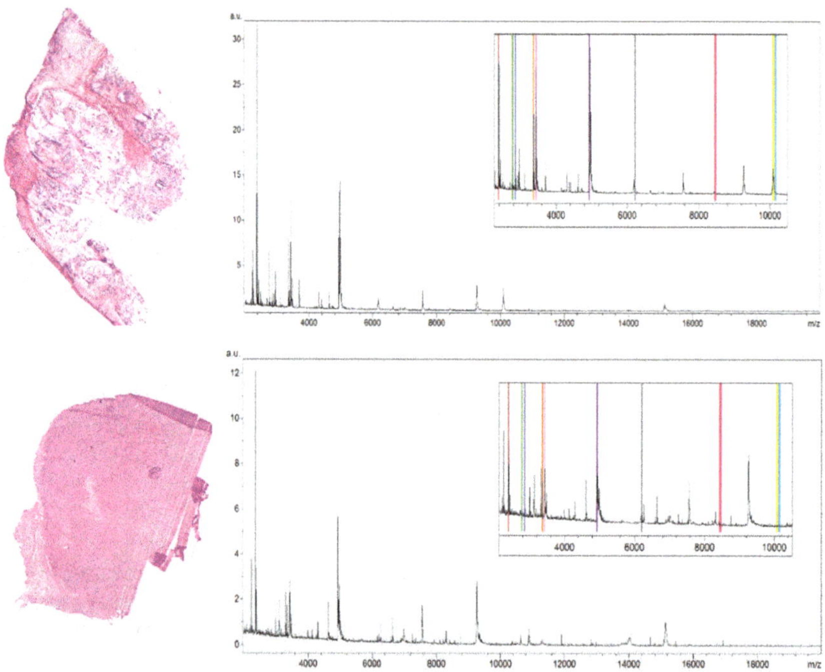

Figure 1. Hematoxylin and eosin stained tissue sections of a low-grade serous borderline ovarian tumor (**top picture**) and ovarian fibrothecoma (**bottom picture**) next to the averaged spectra of the analyzed tissue sections. For each spectrum, the zoomed fragment is presented with color-labeled regiospecific peaks within the borderline tumor tissue section (m/z: 2775.79 (green), 2861.35 (navy blue), 3368.34 (orange), 3438.43 (pink), 4936.37 (purple)) and peaks differentiating these two tissue sections (m/z: 2386.53 (red), 6204.735 (grey), 8453.171 (magenta), 10,146.87 (blue), 10,092.39 (yellow)).

2.5. Hematoxylin and Eosine Staining

For the hematoxylin and eosin staining, one of the sections obtained during serial sectioning was taken to perform the correlation of MALDI-MSI data with histology. The staining procedure was multi-stage. First, the slide was dip-washed in deionized water. Next, the tissue section was stained in hematoxylin solution for 5 min and dip-washed in deionized water once again to remove an excess of the staining solution. The slide was rinsed in running tap water for 5 min and then washed in deionized water for 1 min. In the next step, the slide was put into eosine solution until the section was sufficiently stained (approximately 1 min) and was washed in deionized water. Then, the tissue sections were washed for two min each, in 70% EtOH, 80% EtOH, 90% EtOH, 96% EtOH, 96% EtOH again and xylene. One droplet of mounting medium was spread using a glass stir bar on the slide, and a coverslip was placed onto the tissue.

3. Results

The methodology proposed in this study allowed the measurement of 106 m/z values in the tissue section of the serous borderline tumor and 133 m/z values in the ovarian fibrothecoma tissue section in the 2000–20,000 m/z range (Figure 1). Among the 106 ions from the serous borderline tumor tissue, five regiospecific peaks (m/z: 2861.35; 2775.79; 3368.34; 3438.43; 4936.37) were selected using FlexImaging software and then compared with the distribution of the same masses in the fibrothecoma tissue section. Different distribution of the selected peaks was shown in molecular images (Figure 2). Due to the homogeneous structure of the fibrothecoma tissue, no regiospecific masses were found. Visualization of the regiospecific peaks of the serous borderline tumor on the fibrothecoma tissue section also shows the lack of regiospecific distribution of these masses in the fibrothecoma tissue (Figure 2). The visualization of the differentiated distribution of the selected regiospecific peaks in serous borderline tumor tissue and then the comparison of the results with the histopathological staining of the borderline tumor tissue section, obtained during serial sectioning, allowed the identification of mesenchymal and epithelial tissue within the borderline tumor tissue. The peaks with m/z: 4936.37 and 2775.79 are proposed as characteristic of mesenchymal tissue (Figure 3) and the peaks with m/z: 2861.35, 3368.34 and 3438.43 are proposed as characteristic of epithelial tissue (Figure 3). Epithelial and mesenchymal tissue were assigned based on the microscopic examination of histopathological features in H&E stained sections that were characterized by an expertise pathologist. In addition, the MALDI MSI images were analyzed and discussed with the pathologist, who helped to recognize particular parts of the tissue on the tissue section. In the next step, the serous borderline tumor was compared with the ovarian fibrothecoma. The peaks differentiating both tissues were selected. The peaks at m/z: 2386.53, 6204.735, 8453.171, 10,146.87, 10,092.39 showed the greatest differentiation between the tissues. These m/z values were abundant in the regions of a well-differentiated ovarian tissue whereas in the regions of the borderline tumor with atypical cell proliferation, they were scarce. Differentiating compounds were then visualized and shown on the molecular maps of ovarian fibrothecoma and borderline ovarian tumor (Figure 4).

Immunohistochemistry (IHC)

In the borderline ovarian tumor sample, three immunohistochemistry (IHC) markers were used: Wilms' tumor protein (WT-1), tumor protein p53 (p53) and paired box gene 8 (PAX-8) (Figure 5). There was a positive reaction for WT-1 and PAX-8 and a weak positive reaction for p53.

The WT-1 gene was originally identified as a tumor suppressor gene and is overexpressed in a variety of neoplasms. WT1 is expressed in the ovarian surface epithelium and stromal and granulosa cells of a healthy ovary [17]. In the tumoral ovary, WT1 is characteristic for the serous subtype of ovarian tumors [17]. PAX-8 is characteristic for the epithelial phenotypes (serous, clear cell and endometrioid) of ovarian tumors and is useful in differentiation between primary ovarian and metastatic tumors of the ovary [17]. The mutated form of p53, a tumor suppressor gene, is indicative of high-grade serous ovarian cancer.

The presence of WT-1 and PAX-8 proteins confirmed the origin of the tumor from the serous epithelium and a weak positive reaction in the few cells for p53 confirms a borderline.

In the case of the ovarian fibrothecoma, a positive reaction with inhibin alpha was indicative of the steroid activity of the tumor cells and confirmed the diagnosis.

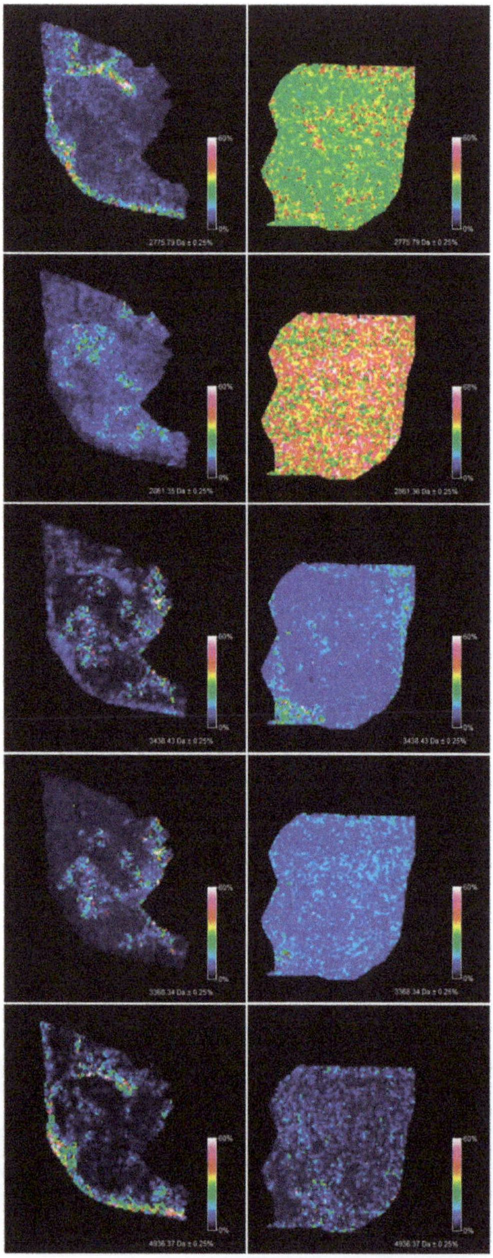

Figure 2. Pictures visualize the homogenous distribution of the regiospecific compounds from the borderline ovarian tumor tissue section (**left**) in the fibrothecoma tissue section (**right**).

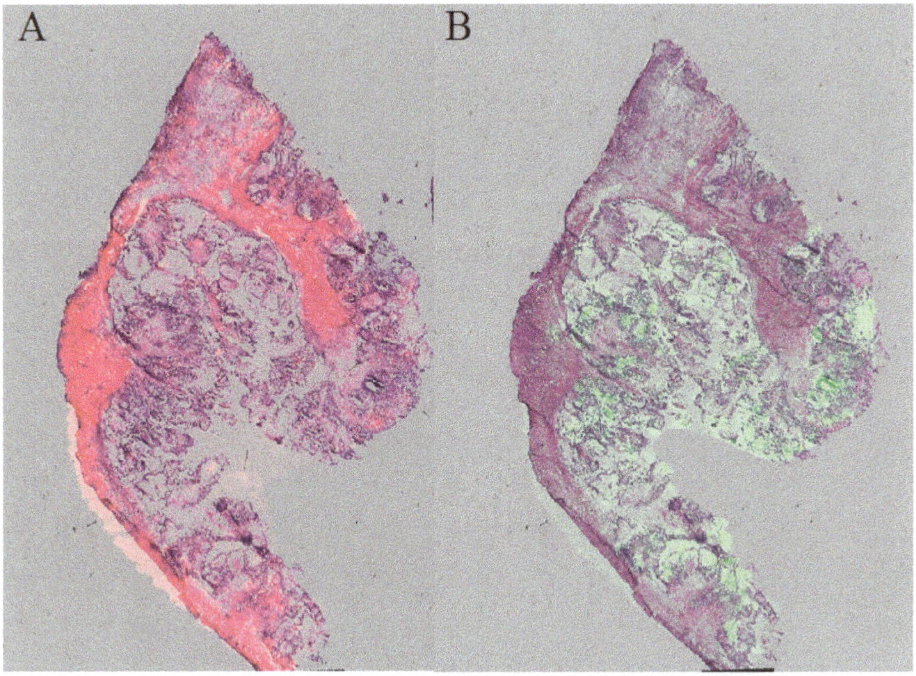

Figure 3. Picture **A** visualizing m/z: 4936.37 and 2775.79 characteristic of mesenchymal tissue and picture **B** visualizing m/z: 2861.35, 3368.34 and 3438.43 characteristic of epithelial tissue.

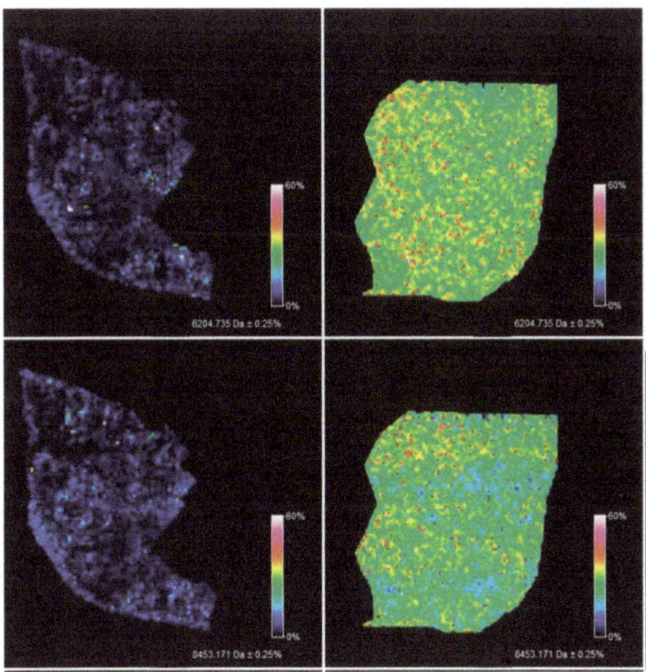

Figure 4. *Cont.*

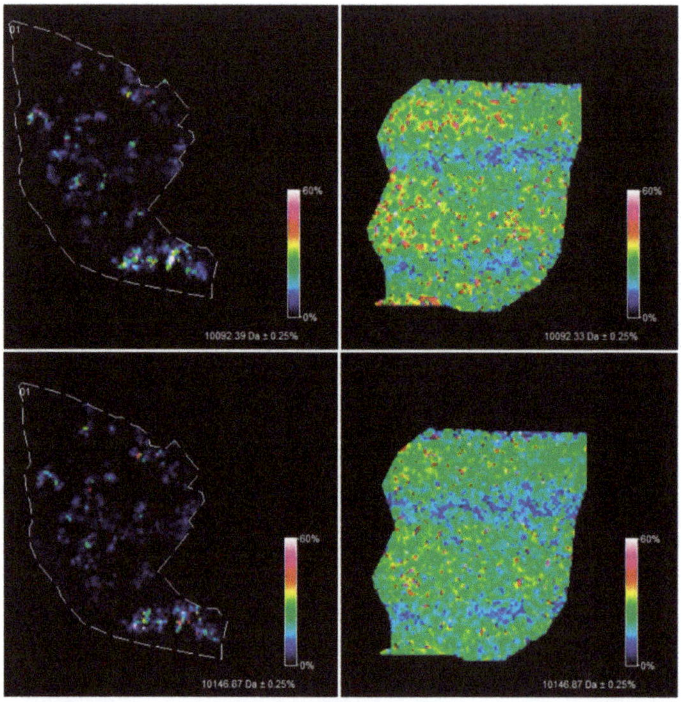

Figure 4. Visualization of the peaks (m/z: 6204.735, 8453.171, 10,146.87, 10,092.39) differentiating the borderline ovarian tissue (**left**) from the ovarian fibrothecoma tissue (**right**) based on the average spectra.

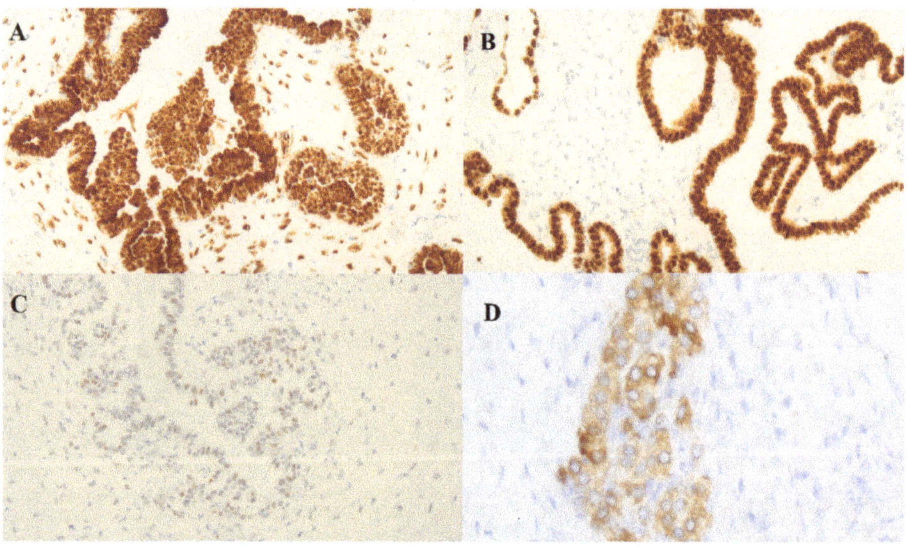

Figure 5. Immunohistochemistry (IHC) detection of (**A**) WT-1, (**B**) PAX-8, (**C**) p53 in a borderline ovarian tumor sample and (**D**) inhibin alpha in an ovarian fibrothecoma sample.

4. Discussion

MALDI-MSI has been widely used in research on various diseases (e.g., Parkinson's disease [18], diabetes [19], Alzheimer's disease [20]) and the number of its possible applications is constantly growing. Among its potential applications, the analysis of cancer tissue is highly ranked. Cancerous tissue can be analyzed in terms of diagnostics, i.e., differentiation of cancerous tissue from non-cancerous tissue and prognosis, i.e., the ability to predict the disease outcomes. The current use of tissue imaging techniques focuses on the possibility of transferring mass spectrometry to clinical practice, especially when tissue information can be obtained during surgery. Multiple studies have shown that the MSI technique can be used for rapid diagnosis or prognostic testing using patient tissues [21–23]. Most of these studies were carried out ex vivo performing the MSI analysis of tissue sections. However, some in vivo studies using alternative technologies, such as the iKnife, are also available [24]. The results obtained in this work, as well as those presented in the available literature, indicate that the future of clinical diagnostics will be strongly influenced by "omic" techniques as a basis for the detection of potential tissue biomarkers. Unlike antibody-based methods, MSI provides a unique opportunity to detect differentiating compounds or other biomolecules found in tissue without knowing the structure of these molecules in advance.

For the purpose of this study, the tissue samples of a low-grade serous borderline ovarian tumor and ovarian fibrothecoma were selected as examples of ovarian lesions whose correct diagnosis is challenging and which might be misdiagnosed as ovarian cancer.

BOTs are a heterogeneous group of ovarian neoplasms characterized by atypical cell proliferation but no destructive invasion of the stroma. Their correct diagnosis can be challenging as the transformation from a benign cystadenoma or cystadenofibroma through BOT into the low-grade serous ovarian cancer is gradual, and it can be hard to draw the line between one and another. According to the World Health Organization (WHO) 2014 classification, tumors with more than 10% of borderline histology within a cystadenoma or cystadenofibroma qualifies as BOT. In contrast, serous cystadenomas with borderline foci that occupy less than 10% of the epithelial volume are referred to as "cystadenoma/fibroma with focal epithelial proliferation". In addition, these tumors are difficult to differentiate from truly invasive ovarian cancer due to their microscopic features like microinvasion and non-destructive omental implants [25,26].

Although occasionally BOTs may give implants in the omentum and peritoneum, they are considered a distinct clinical entity than ovarian cancer. On the other hand, BOTs share the staging system with invasive ovarian cancer and carry a significant risk of recurrence after conservative surgery. The risk of progression to invasive ovarian cancer seems significant and reached up to 4% in some reports [27]. BOTs generally have an excellent prognosis with a 10-year survival rate of 97% for all stages combined [6], although, as mentioned before, recurrences and malignant transformation can occur. The standard treatment includes complete surgical staging (i.e., hysterectomy with bilateral salpingo-oophorectomy, omentectomy, peritoneal biopsies, cytology of peritoneal washings and appendectomy in case of mucinous BOT), similarly to early-stage ovarian cancer, but adjuvant chemotherapy is not necessary. A large multicenter study revealed that overdiagnosis of a borderline tumor was made in 11.5% (92/803) of patients [28]. As explained above, the diagnosis of BOT has severe clinical implications, and thus its accurate identification is of crucial importance.

Fibrothecomas are rare ovarian neoplasms originating from the sex cord-stroma. They are often found accidentally during a transvaginal ultrasound, magnetic resonance or computer tomography. Large tumors are often associated with pleural effusion and ascites symptoms typically suggestive of advanced-stage ovarian cancer. Diagnosing ovarian fibrothecomas preoperatively can be difficult also due to their solid nature and elevated serum CA-125 levels [29].

In our study, both patients were qualified for surgery due to being suspected of ovarian cancer. The correct diagnosis was only possible after the histopathological examination of the excised lesions complemented by immunohistochemical tests. However, as explained before, these techniques have their limitations. This emphasizes the need to search for a different diagnostic method that could potentially characterize various ovarian neoplasms more objectively, with higher specificity and sensitivity. Our results suggest that MALDI-MSI is a promising tool in identifying biomarkers or biomarker tissue patterns indicative of specific ovarian tumors as well as a specific tissue's regions. In the literature, there are several articles aimed at either correct ovarian tumor identification on MALDI-MSI or searching for biomarkers. Most of them, however, are focused on ovarian cancer rather than benign ovarian tumors. In the study by Ayed et al. [12], the researchers noticed, similarly to our findings, significant differences between the analyzed tissues and were able to identify putative ovarian cancer biomarkers by the complementary use of liquid chromatography coupled with mass spectrometry of the digested cancerous tissue. Nevertheless, this study compared benign ovarian tumors with advanced-stage ovarian cancer and did not take into account early-stage ovarian cancer, whose diagnosis is the most challenging. Another interesting study in the field concentrated on the correct classification of ovarian cancer histotypes (serous/endometrioid/mucinous/clear cell) and additionally included BOTs arguing that the differential diagnosis between BOT and truly invasive carcinoma can be very challenging (which was also the rationale to choose this particular tissue for our research) [30]. In this study, proteomic information from MALDI-MSI was combined with machine learning approaches to create a model discriminating between different epithelial ovarian cancer subtypes. Although the methods used in both studies were similar, it is impossible to compare the results with our research due to differences in the study design and data processing. Other available studies have rather been focused on the investigation of the distribution of specific markers in the ovarian cancer tissue [13,31]. In the literature search, we did not find any previous studies on the MALDI-MSI used in diagnosing benign ovarian tumors.

In this pilot study, we were able to demonstrate some differentiation not only between ovarian borderline tumor and fibrothecoma but also within the ovarian borderline tumor tissue itself, identifying the regions with increased cell proliferation and the regions of different tissue origin. Moreover, the differences in the tissue homo-/heterogeneous structure of both tumors were shown. Further research could define characteristic m/z peak patterns for each of the ovarian tumors that would allow quick and reliable diagnosis of the ovarian lesion. If, with the advancements in the technique, the time of the MALDI-MSI examination is reduced, it could also be possible to apply it intraoperatively to assist the surgeons. MALDI-MSI could be used to correctly identify if the ovarian lesion is benign, borderline or malignant, to identify its histological subtype (important as, e.g., in mucinous borderline/malignant tumors of the ovary, appendectomy is recommended) or to define the optimal surgical margins.

The results obtained in this pilot study suggest that the MALDI-MSI method is not only complementary to histopathology but might be a technological step forward in the precise diagnosis of ovarian tumors. In addition, MALDI-MSI provides comprehensive information combining current histopathological knowledge with molecular information obtained with the cutting-edge imaging methods, and by identification of the compounds representing specific m/z peaks it could contribute to tissue biomarker discoveries. MSI enables simultaneous, label-free imaging of 100's–1000's of compounds directly from the tissue section. Imaging diverse classes of biomolecules (e.g., metabolites, lipids, glycans) gives the opportunity to develop diagnostic/prognostic assays that complement established histopathological and molecular methods. The MALDI-MSI method provides comprehensive visualization of the spatial distribution of potential differentiating compounds between ovarian tumors, which significantly improves the possibilities of biomarker analysis. Performing histological staining after MSI data acquisition enables seamless integration of the molecular information provided by MSI with established histopathological methods [32]. This approach is likely to be used in clinical studies from bench to bedside and may eventually lead to a new quality of diagnosis of ovarian tumors.

While most MALDI-MSI studies investigated ovarian cancer tissue, the distinctive feature of our research is the application of MALDI-MSI technique for the characterization of benign ovarian tumors. We examined two rare ovarian neoplasms, that have not been characterized in the previous studies, but which might often be misdiagnosed as ovarian cancer. As a pilot study, it was restricted to only two samples and naturally, the findings require verification in larger patient cohorts to estimate the robustness of our observations. In the next step, the identification of the selected m/z values should be performed to understand the findings better and to analyze them along with the tumor biology. This study shows the complementarity of the MALDI-MSI technique to the methods used so far in clinical practice, e.g., H&E staining and the results of the performed analyses are promising in identifying biomarkers of specific ovarian tumors or specific tissue regions. Addressing the above-mentioned limitations can be a starting point for future research.

5. Conclusions

The MALDI-MSI technique enabled the creation of a molecular map of the spatial distribution of the m/z peaks representing peptides/proteins in the borderline ovarian tumor and ovarian fibrothecoma tissues, both of which may pose a diagnostic challenge for clinicians and for histopathologists. The analysis of the ovarian tumor tissue using MALDI-MSI showed a close correlation of the molecular map with its morphological and histopathological features and allowed us to identify different tissue types within the borderline ovarian tumor tissue section. Moreover, creating molecular maps has shown the differences in the distribution of regiospecific m/z values within the examined samples, which could be useful for a quick and accurate intraoperative diagnosis of defining resection margins. The presented technique may open the door for overcoming the current limitations of classic histological imaging methods, thus leading to the more accurate differential diagnosis of the most challenging cases. This pilot study gives grounds for further research on the application of MALDI-MSI in the differential diagnosis of ovarian tumors.

Author Contributions: Conceptualization, D.P., J.M., S.P., A.H., P.J., E.N.-M., Z.J.K.; methodology, D.P., S.P., P.J.; software, D.P., S.P., J.M.; validation, D.P.; formal analysis, D.P., A.H., E.N.-M., Z.J.K., J.M.; investigation, D.P., A.H.; resources, A.H., E.N.-M., P.J.; data curation D.P., S.P., A.H., P.J.; writing—original draft preparation, D.P., A.H., S.P., P.J., E.N.-M., Z.J.K., J.M.; visualization, D.P., P.J.; supervision, J.M., E.N.-M., Z.J.K.; project administration, J.M.; funding acquisition Z.J.K., J.M. All authors have read and agreed to the published version of the manuscript.

Funding: This research was financially supported by the Polish National Science Centre (grant number: 2014/15/B/NZ7/00964). The funders did not participate in the study design, data collection and analysis, decision to publish and manuscript preparation.

Conflicts of Interest: The authors declare no conflict of interest.

References

1. Ebell, M.H.; Culp, M.; Lastinger, K.; Dasigi, T. A Systematic Review of the Bimanual Examination as a Test for Ovarian Cancer. *Am. J. Prev. Med.* **2015**, *48*, 350–356. [CrossRef] [PubMed]
2. Kroeger, P.T., Jr.; Drapkin, R. Pathogenesis and heterogeneity of ovarian cancer. *Curr. Opin. Obstet. Gynecol.* **2017**, *29*, 26–34. [CrossRef] [PubMed]
3. Manuscript, A.; Origin, T. The Origin and Pathogenesis of Epithelial Ovarian Cancer—A Proposed Unifying Theory. *Am. J. Surg. Pathol.* **2011**, *34*, 433–443. [CrossRef]
4. Silverberg, S.G.; Bell, D.A.; Kurman, R.J.; Seidman, J.D.; Prat, J.; Ath, F.; Ronnett, B.M.; Copeland, L.; Silva, E.; Gorstein, F.; et al. Borderline Ovarian Tumors: Key Points and Workshop Summary. *Hum. Pathol.* **2004**, *35*, 910–917. [CrossRef] [PubMed]
5. Seidman, J.D.; Kurman, R.J. Ovarian Serous Borderline Tumors:A Critical Review of the Literature With Emphasis on Prognostic Indicators. *Hum. Pathol.* **2000**, *31*, 539–557. [CrossRef]
6. Hauptmann, S.; Friedrich, K.; Redline, R.; Avril, S. Ovarian borderline tumors in the 2014 WHO classification: Evolving concepts and diagnostic criteria. *Virchows Archiv.* **2017**, *470*, 125–142. [CrossRef] [PubMed]
7. Scott, A.J.; Jones, J.W.; Orschell, C.M.; Macvittie, T.J.; Kane, M.A.; Ernst, R.K. Mass spectrometry imaging enriches biomarker discovery approaches with Candidate mapping. *Health Phys.* **2014**, *106*, 120–128. [CrossRef]
8. Imaging, P.; Classic, F.; Mass, M.; Histology, S.M. *Proteome Imaging: From Classic to Modern Mass Spectrometry-Based Molecular Histology*; Springer: Berlin/Heidelberg, Germany, 2019; ISBN 9783030159504.

9. Schöne, C.; Höfler, H.; Walch, A. MALDI imaging mass spectrometry in cancer research: Combining proteomic profiling and histological evaluation. *Clin. Biochem.* **2013**, *46*, 539–545. [CrossRef]
10. Longuespée, R.; Gagnon, H. Proteomic analyses of serous and endometrioid epithelial ovarian cancers: Cases studies: Molecular insights of a possible histological etiology of serous ovarian cancer. *Proteom.-Clin. Appl.* **2013**, *7*, 337–354. [CrossRef]
11. Schwamborn, K. Discerning the Primary Carcinoma in Malignant Peritoneal and Pleural Effusions Using Imaging Mass Spectrometry—A Feasibility Study. *Proteom.-Clin. Appl.* **2018**, *13*, 1–17. [CrossRef]
12. El Ayed, M. MALDI imaging mass spectrometry in ovarian cancer for tracking, identifying, and validating biomarkers. *Med. Sci. Monit.* **2010**, *9*, 291–300.
13. Briggs, M.T.; Condina, M.R.; Ho, Y.Y.; Everest-Dass, A.V.; Mittal, P.; Kaur, G.; Oehler, M.K.; Packer, N.H.; Hoffmann, P. MALDI Mass Spectrometry Imaging of Early- and Late-Stage Serous Ovarian Cancer Tissue Reveals Stage-Specific N-Glycans. *Proteomics* **2019**, *19*, e1800482. [CrossRef] [PubMed]
14. Everest-Dass, A.V.; Briggs, M.T.; Kaur, G.; Oehler, M.K. N-glycan MALDI Imaging Mass Spectrometry on Formalin-Fixed Paraffin-Embedded Tissue Enables the Delineation of Ovarian Cancer Tissues. *Mol. Cell. Proteom.* **2016**, *15*, 3003–3016. [CrossRef] [PubMed]
15. Yang, F.Y.; Saqib, H.S.A.; Ruan, Q.Q.; You, M.S. Mass spectrometry imaging: An emerging technology for the analysis of metabolites in insects. *Arch. Insect Biochem. Physiol.* **2019**, *103*, e21643. [CrossRef] [PubMed]
16. Aichler, M.; Walch, A. MALDI Imaging mass spectrometry: Current frontiers and perspectives in pathology research and practice. *Lab. Investig.* **2015**, *95*, 422–431. [CrossRef] [PubMed]
17. Iliac, L.U.L.; Arcangiu, M.A.L.U.C.; Anevari, S.I.C.; Anciu, M.D.; Nofriescu, M.O.; Linei, C.O.A.M.Ă. The value of PAX8 and WT1 molecules in ovarian cancer diagnosis. *Rom. J. Morphol. Embryol.* **2013**, *54*, 17–27.
18. Hulme, H.; Fridjonsdottir, E.; Gunnarsdottir, H.; Vallianatou, T.; Zhang, X.; Wadensten, H.; Shariatgorji, R.; Nilsson, A.; Bezard, E.; Svenningsson, P.; et al. Simultaneous mass spectrometry imaging of multiple neuropeptides in the brain and alterations induced by experimental parkinsonism and L-DOPA therapy. *Neurobiol. Dis.* **2020**, *137*, 104738. [CrossRef]
19. Bunney, P.; Zink, A.; Holm, A.; Billington, C.; Kotz, C. Imaging mass spectrometry reveals direct albumin fragmentation within the diabetic kidney. *Physiol. Behav.* **2017**, *176*, 139–148. [CrossRef] [PubMed]
20. Kaya, I.; Zetterberg, H.; Blennow, K.; Hanrieder, J. Shedding Light on the Molecular Pathology of Amyloid Plaques in Transgenic Alzheimer's Disease Mice Using Multimodal MALDI Imaging Mass Spectrometry. *ACS Chem. Neurosci.* **2018**, *9*, 1802–1817. [CrossRef]
21. Addie, R.D.; Ballu, B.; Bove, J.V.M.G.; Morreau, H. Current State and Future Challenges of Mass Spectrometry Imaging for Clinical Research. *Anal. Chem.* **2015**, *87*, 6426–6433. [CrossRef]
22. Basu, S.S.; Regan, M.S.; Randall, E.C.; Abdelmoula, W.M.; Clark, A.R.; Gimenez-Cassina Lopez, B.; Cornett, D.S.; Haase, A.; Santagata, S.; Agar, N.Y.R. Rapid MALDI mass spectrometry imaging for surgical pathology. *NPJ Precis. Oncol.* **2019**, *3*, 17. [CrossRef] [PubMed]
23. Brégeon, F.; Brioude, G.; De Dominicis, F.; Atieh, T.; D'Journo, X.B.; Flaudrops, C.; Rolain, J.M.; Raoult, D.; Thomas, P.A. MALDI-ToF mass spectrometry for the rapid diagnosis of cancerous lung nodules. *PLoS ONE* **2014**, *9*, e97511. [CrossRef]
24. Balog, J.; Sasi-szabó, L.; Kinross, J.; Lewis, M.R.; Muirhead, L.J.; Veselkov, K.; Mirnezami, R.; Dezs, B.; Damjanovich, L.; Darzi, A.; et al. Intraoperative Tissue Identification Using Rapid Evaporative Ionisation Mass Spectrometry. *Sci. Transl. Med.* **2013**, *5*, 194ra93. [CrossRef] [PubMed]
25. Seong, S.J.; Kim, D.H.; Kim, M.K.; Song, T. Controversies in borderline ovarian tumors. *J. Gynecol. Oncol.* **2015**, *26*, 343–349. [CrossRef]
26. Sherman, M.E.; Berman, J.; Birrer, M.J.; Cho, K.R.; Ellenson, L.H.; Gorstein, F.; Seidman, J.D. Current challenges and opportunities for research on borderline ovarian tumors. *Hum. Pathol.* **2004**, *35*, 961–970. [CrossRef] [PubMed]
27. Vang, R.; G Hannibal, C.; Junge, J.; Frederiksen, K.; K Kjaer, S.; Robert, J.K. Long-term Behavior of Serous Borderline Tumors Subdivided Into Atypical Proliferative Tumors and Noninvasive Low-grade Carcinomas: A Population-based Clinicopathologic Study of 942 Cases. *Am. J. Surg. Pathol.* **2018**, *41*, 502–532. [CrossRef] [PubMed]

28. Ewald-riegler, N.; De Gregorio, N.; Reuss, A.; Mahner, S.; Fotopoulou, C.; Kommoss, F.; Schmalfeldt, B.; Hilpert, F.; Fehm, T.; Burges, A.; et al. Borderline tumours of the ovary: A cohort study of the Arbeitsgmeinschaft Gynäkologische Onkologie (AGO) Study Group. *Eur. J. Cancer* **2013**, *49*, 1905–1914. [CrossRef]
29. Shen, Y.; Liang, Y.; Cheng, X.; Lu, W.; Xie, X.; Wan, X. Ovarian fibroma/fibrothecoma with elevated serum CA125 level: A cohort of 66 cases. *Medicine* **2018**, *97*, 1–4. [CrossRef]
30. Klein, O. MALDI-Imaging for classification of epithelial ovarian cancer histotypes from a tissue microarray using machine learning methods. *Proteom.-Clin. Appl.* **2018**, *13*, e1700181. [CrossRef]
31. Hernandez, Y.C.; Boskamp, T.; Casadonte, R.; Hauberg-Lotte, L.; Oetjen, J.; Lachmund, D.; Peter, A.; Trede, D.; Kriegsmann, K.; Kriegsmann, M.; et al. Targeted feature extraction in MALDI mass spectrometry imaging to discriminate proteomic profiles of breast and ovarian cancer. *Proteom.- Clin. Appl.* **2012**, *13*, 1–49. [CrossRef]
32. Walch, A.; Rauser, S.; Deininger, S.-O.; Höfler, H. MALDI imaging mass spectrometry for direct tissue analysis: A new frontier for molecular histology. *Histochem. Cell Biol.* **2008**, *130*, 421–434. [CrossRef] [PubMed]

Publisher's Note: MDPI stays neutral with regard to jurisdictional claims in published maps and institutional affiliations.

© 2020 by the authors. Licensee MDPI, Basel, Switzerland. This article is an open access article distributed under the terms and conditions of the Creative Commons Attribution (CC BY) license (http://creativecommons.org/licenses/by/4.0/).

International Journal of
Environmental Research and Public Health

MDPI

Article

Sustainable Development of Polish Macroregions—Study by Means of the Kernel Discriminant Coordinates Method

Mirosław Krzyśko [1], Waldemar Wołyński [2,*], Waldemar Ratajczak [3], Anna Kierczyńska [4] and Beata Wenerska [3]

1 Interfaculty Institute of Mathematics and Statistics, Calisia University-Kalisz, 62-800 Kalisz, Poland; mkrzysko@amu.edu.pl
2 Faculty of Mathematics and Computer Science, Adam Mickiewicz University, 61-614 Poznań, Poland
3 Faculty of Social Sciences, Calisia University-Kalisz, 62-800 Kalisz, Poland; walrat@amu.edu.pl (W.R.); b.wenerska@pwsz.kalisz.pl (B.W.)
4 Faculty of Economic, The Great Poland Socio-Economic University, 63-000 Środa Wlkp., Poland; a.kierczynska@onet.pl
* Correspondence: wolynski@amu.edu.pl

Received: 30 July 2020; Accepted: 18 September 2020; Published: 25 September 2020

Abstract: The aim of this study was to investigate if the macroregions of Poland are homogeneous in terms of the observed spatio-temporal data characterizing their sustainable development. So far, works related to the sustainable development of selected territorial units have been based on data relating to a specific year rather than many years. The solution to the problem of macroregion homogeneity goes through two stages. In step one, the original spatio-temporal data space (matrix space) was transformed into a kernel discriminant coordinates space. The obtained kernel discriminant coordinates function as synthetic measures of the level of sustainable development of Polish macroregions. These measures contain complete information on the values of 27 diagnostic features examined over 15 years. In the second step, cluster analysis was used in order to identify groups of homogeneous macroregions in the space of kernel discriminant coordinates. The agglomeration method and the Ward method were chosen as commonly used methods. By means of both methods, three super macroregions composed of homogeneous macroregions were identified. Within the kernel discriminant coordinates, the differentiating power of a selected set of 27 features characterizing the sustainable development of macroregions was also assessed. To this end, five different and most commonly used methods of discriminant analysis were used to test the correctness of the classification. Depending on the method, the classification errors amounted to zero or were close to zero, which proves a well-chosen set of diagnostic features. Although the data relate only to a specific country (Poland), the presented statistical methodology is universal and can be applied to any territorial unit and spatial-temporal dynamic data.

Keywords: sustainable development; spatio-temporal data; NUTS-1; kernel discriminant coordinates method; super macroregions

1. Introduction

The concept of sustainable development played an important role in the second half of the 20th century. The concept was first defined in a report of the World Commission on Environment and Development, "Our Common Future", under the supervision of G. Brundtland (Our Common Future, UN New York, 1987). Sustainable development is "development that meets the needs of today and does not limit the ability of future generations to meet their own needs." According to the report, this means that "all future generations have the right to live and enjoy all the values of the environment known to them, just like you, or even better." In other words, it is a socio-economic development in which the forms and dynamics of economic activity, institutions, lifestyles and the population size are such that the existing and future generations will be provided with an adequate standard of living, and all aspects of this development are secured by the availability of natural resources, ecosystems and life support systems. This means that the economic development of the current generation should not happen at the cost of the depletion of non-renewable stocks (Strategy for the sustainable development of Poland until 2025, Ministry of the Environment, Warsaw, 1999).

Sustainable development has been defined in a variety of ways (Awan et al. [1], Awan et al. [2], Borys [3]). In our opinion, it is based on the following four pillars: 1. Demography and social capital, 2. Production, services and trade, 3. Public finance, and 4. The environment and protection thereof. "Sustainability" has become a global buzzword as a potential solution to many international, regional, and local problems of contemporary society: over-population, diseases, political conflicts, infrastructure deterioration, pollution, and unlimited urban expansion accompanied by limited availability of resources. The United Nations World Commission on Environment and Development (WECD, 1987) coined a definition of sustainable development that is probably the most well-known in all of the literature on sustainability: "development that meets the needs of the present without compromising the ability of future generations to meet their own needs".

Along with the concept of sustainable development, the related term sustainable regional development appeared as a traditional goal of the state's regional policy. This applies to activities of public administration bodies that should aim at counteracting excessive or unjustified interregional differences.

Regional development is usually defined as integral community development (social, economic, environmental and healthcare, technological, cultural and recreational) in a particular territory. Regional development must be based on its optimal expansion constituents (social, natural and economic development aspects) aimed at maintaining a certain standard of living and improvement of its quality through the mentioned constituents. Regional development encompasses not only traditional policy in a specific territory, but also a socioeconomic process taking place in a specific political and cultural context (Atkinson [4], Bourdeau [5], Spangenberg [6], OECD [7], Wheeler and Beatley [8], Wong [9]). Regional development in today's context is at a critical juncture, with multiple crises (financial, food and energy) forcing us to re-assess the economic paradigm of our time and to evaluate how to better address the unfulfilled promises that we are currently leaving to future generations in the areas of employment, social progress, quality of life and respect for nature. While there is no doubt about the importance of integrating the pillars of sustainable development on the regional level, implementation of this concept has proved challenging in practice. In fact, integration of the environmental, economic, and social dimensions of sustainable development on a regional level implies the implementation of complimentary and coordinated actions in different areas. This results in economic growth that is also supposed to achieve social objectives, without posing a threat to the rare resources of the planet. Effective integration of these four dimensions (pillars) requires implementation of a set of focused and specific actions that are complementary and fit into an overarching sustainable development framework.

Previous papers related to the sustainable development of regions were based on data for a specific year or, for comparison purposes, for two specific years. If the authors of these papers chose, say, p features describing sustainable development of regions, then the obtained data were p-dimensional vectors or points in p-dimensional Euclidean space. Then, from these vectors numerical indicators (linear combinations) were constructed, considered to be the characteristics of the sustainable development of regions (Kontostanou-Karalivanou et al. [10], United Nations [11], Streimikiene [12], Dočekalová et al. [13], Jovovic et al. [14], Roszkowska and Wyszkowska [15]). An extensive overview of sustainability performance evaluation, including a literature review and future directions, was presented by Büyüközkan and Karabulut [16], Krasławski et al. [17].

This paper deals with a more general issue and presents a proposal for the regionalization of spatio-temporal data, which are more general than vector data considered at a given moment. We also propose a new way of constructing numerical indicators describing the sustainable development of specific territorial units.

The paper is based on data concerning Poland. The country is administratively divided into seven macroregions composed of voivodships (16 voivodships in total), and each voivodeship is further divided into poviats. The term poviat, defining a given territorial unit, is used by Statistics Poland, which is associated with EUROSTAT. In total, there are 379 poviats in Poland. Each poviat has been described by means of 27 features, divided into four pillars that characterize the sustainable development of this territorial unit. These features were observed in the years 2002–2016, i.e., for 15 years. Our data can be written in the form of a matrix with 15 rows and 27 columns. This type of data is called spatio-temporal data. They are dynamic and more general than static vector data. The choice of Poland as the research area stemmed from the availability of data. The data concerns 379 poviats characterized by 27 features observed over 15 years. The data therefore consist of 153,495 numbers, a very large data set that is not easy to collect. The data were provided by the Local Data Bank of Statistics Poland. On the other hand, when selecting Poland as the research area, the authors of the study had in mind the usefulness of the obtained results for Polish governmental decision-makers.

The aim of this paper is to investigate if Polish macroregions are homogenous in terms of spatio-temporal data characterizing their sustainable development.

To solve this problem by means of statistical methodology, two multidimensional statistical methods were used: discriminant analysis and cluster analysis. The following solutions were adopted:

In the first step, the original spatio-temporal data space (matrix space) was transformed into a vector space of discriminant coordinates. This transformation takes place in two ways known in the discriminant analysis literature: classical discriminant coordinates (Fisher [18], Rao [19]) and kernel discriminant coordinates, as first described by Baudat and Anouar [20]. Both transformations were developed for fixed vector data. Since our data are a matrix, it was necessary to extend these transformations to the matrix case. This extension proved troublesome in the case of Fisher's classical discriminant coordinates. Hence the choice of kernel discriminant coordinates, since in their case the said extension was possible. The space of kernel discriminant coordinates is a space of dimension $\min(p, c-1)$, where p is the number of observed features and c is the number of macroregions. Our case consists of a six-dimensional space, so the dimension of the space in which further inference will take place is relatively low.

We treated the obtained kernel discriminant coordinates further as synthetic measures of the level of sustainable development of Polish macroregions. These measures contain complete information on the values of 27 diagnostic features measured over 15 years. Each of the six kernel discriminant coordinates was given a different power to differentiate macroregions. In the space of the first two kernel discriminant coordinates, it is possible to graphically present the mutual position of the studied macroregions.

In the space of kernel discriminant coordinates, it is also easy to assess the differentiating power of a selected set of features characterizing individual poviats. To this end, various methods of discriminant analysis can be used, verifying the correctness of the classification of individual poviats into the seven distinguished macroregions. In this paper, five different, most commonly used methods of discriminant analysis were used. The percentage of misclassifications of individual poviats to seven macroregions was calculated using these methods. Zero or close to zero classification errors indicate a well-chosen set of diagnostic features.

In the second step, cluster analysis was used in order to select groups of homogeneous macroregions in the six-dimensional space of kernel discriminant coordinates. The agglomeration method and the Ward method were chosen as commonly used methods. As a result, three super macroregions composed of homogeneous macroregions were identified.

According to the authors, the proposed statistical methodology is the main value of this paper. Although the data relate only to a specific country, the presented statistical methodology is universal and can be applied to any territorial unit and spatio-temporal dynamic data.

The paper is organized as follows: Section 2 contains a presentation of spatial units and data. Section 3 describes the statistical methodology. Section 4 contains the research results. Concluding remarks are provided in Section 5.

2. Spatial Units and Data

Achieving the research goal adopted in this work required collecting data characterizing various aspects of sustainable development for $N = 379$ poviats belonging to one of the $c = 7$ macroregions. The macroregions are NUTS-1 units according to the EUROSTAT nomenclature (2018).

The structure of NUTS-1 in Poland is presented in Figure 1 and Table 1.

Figure 1. Macroregions in Poland (NUTS-1. 2018). Source: Statistics Poland.

Table 1. Structure of Polish macroregions including poviats.

Number	Name of the Macroregion	Number of Poviats in a Macroregion
1	Southern	58
2	Northwestern	70
3	Southwestern	41
4	Northern	64
5	Central	38
6	Eastern	66
7	Masovian voivodship	42

A set of $p = 27$ variables (local data bank information of Statistics Poland was used) was collected for each poviat, for the period of 2002–2016, i.e., $T = 15$ years. They were divided into four pillars relating to various areas of sustainable development, i.e., 1. Social capital, 2. Production, services and trade, 3. Finance, 4. The environment and protection thereof (see Table 2). The choice of the variables was made on the basis of their merits and in accordance with the authors' previous experience. The accuracy of the selection of variables was confirmed using discriminant analysis methods (see Section 4).

Table 2. List of the variables used in the research divided into pillars.

	Pillar		Variable
1.	Demography and Social Capital	1.	Population at working age
		2.	Femininity ratio
		3.	Registered unemployment
		4.	Library books borrowers
2.	Production, Services and Trade	5.	General secondary schools
		6.	Primary schools
		7.	Pre-schools
		8.	Out-patient clinics
		9.	Pharmaceutical outlets
		10.	Libraries and branches
		11.	Hotels, motels and boarding houses
		12.	Tourist accommodation establishments
		13.	The length of the water supply network
		14.	The length of the sewerage network
		15.	Boiler houses
		16.	Poviat hard surface roads
3.	Public Finance	17.	Poviats' own revenue
		18.	General subsidies
		19.	Average monthly gross wages and salaries
4.	Environment and its Protection	20.	Municipal biological wastewater treatment plants
		21.	Industrial biological wastewater treatment plants
		22.	Industrial chemical wastewater treatment plants
		23.	Industrial wastewater untreated, discharged
		24.	Forests
		25.	Renewals and afforestation—communal and private forests
		26.	Legal protected area
		27.	Monuments of nature

Therefore each poviat was characterized by an matrix of 15×27 size.

3. Statistical Methodology

3.1. Kernel Discriminant Coordinates for Spatio-Temporal Data

The main idea of the kernel-based methods is to map the input data to a feature space through nonlinear mapping where the inner products in the feature space can be computed by a kernel function while the nonlinear mapping is not known explicitly. Kernel Fisher discriminant analysis (KFD) provided by Baudat and Anouar [20] and the generalized discriminant analysis (GDA) provided by Mika et al. [21] are two independently developed approaches for kernel-based nonlinear extensions of discriminant coordinates. They are essentially equivalent. The method is described in a book by Shawe-Taylor and Cristianini [22]. To avoid confusion, we will refer to this approach as kernel discriminant coordinates analysis (KDCA). These new variables are also sometimes called kernel canonical variates yet the name is misleading because kernel canonical variables with completely different properties occur in the kernel canonical analysis. Another name is "kernel discriminant functions", which is inappropriate because discriminant functions are surfaces that separate the c classes from one another.

In this paper, the authors present a proposal of extending vector kernel analysis to matrix kernel analysis in their original way. Obviously, in the present case the data are spatio-temporal. The kernel discriminant coordinates analysis is a method developed for fixed vector data. This paper includes an extension of this method to spatio-temporal data. In this case, each object is characterized by a $(T \times p)$-size matrix $\mathbf{X}$ containing the p values observed at T moments.

Let $\{(\boldsymbol{x}_1, \boldsymbol{y}_1), \ldots, (\boldsymbol{x}_N, \boldsymbol{y}_N)\}$, $\boldsymbol{x}_i \in \mathbb{R}^{T \times p}$, $\boldsymbol{y}_i \in \mathbb{R}^c$, $i = 1, \ldots, N$, be a data set.

The input space $\mathbb{R}^{T\times p}$ is now mapped nonlinearly into a feature space $\mathcal{H}_k$:

$$\phi\colon \mathbb{R}^{T\times p} \to \mathcal{H}_k$$

where ϕ is the mapping function that maps the input space to the reproducing kernel Hilbert space (RKHS) $\mathcal{H}_k$. Please note that $\mathcal{H}_k$ could have arbitrarily large, possibly infinite dimensionality. The vector $\phi(\boldsymbol{x}_i) = \tilde{\boldsymbol{x}}_i$ is called the feature vector corresponding to the observation $\boldsymbol{x}_i \in \mathbb{R}^{T\times p}$, $i = 1, \ldots, N$.

The nonlinear transformation ϕ is in general unknown; however, we select a known form of the nonnegative definite kernel function:

$$k(\boldsymbol{x}_i, \boldsymbol{x}_j) = \langle \phi(\boldsymbol{x}_i), \phi(\boldsymbol{x}_j)\rangle_{\mathcal{H}_k}, \; i, j = 1, 2, \ldots, N.$$

A kernel function can be interpreted as a kind of similarity measure between the matrices $\boldsymbol{x}_i$ and $\boldsymbol{x}_j$.

Throughout this paper we use the Gaussian kernel:

$$k(\boldsymbol{x}_i, \boldsymbol{x}_j) = \exp(-\gamma \|\boldsymbol{x}_i - \boldsymbol{x}_j\|_F^2), \; \gamma > 0,$$

where

$$\|\boldsymbol{A}\|_F = \sqrt{\operatorname{tr}(\boldsymbol{A}^\top \boldsymbol{A})}$$

is the Frobenius norm. The constant $\gamma > 0$ is appropriately selected from the data. We take into account the lower-triangular matrix, which has its (i, j)-th element given by $\|\boldsymbol{x}_i - \boldsymbol{x}_j\|_F^2$, $i, j = 1, \ldots, N$. The value of γ was used as the reciprocal of the arithmetic mean of the elements of this matrix.

Similarly, as in the classical case of DCA, in kernel discriminant analysis we find, for a c-class problem, the $c-1$ vectors onto which the projections of the data of one class are maximally separated from the remaining classes in the feature space.

Let $\tilde{\boldsymbol{m}} = \frac{1}{n}\sum_{i=1}^{n} \tilde{\boldsymbol{x}}_i$ be the sample mean, and let $\tilde{\boldsymbol{m}}_j = \frac{1}{n_j}\sum_{i\in V_j} \tilde{\boldsymbol{x}}_i$ be the mean of the jth class, for $j = 1, \ldots, c$, in the feature space.

By $\tilde{\boldsymbol{S}}_b$ and $\tilde{\boldsymbol{S}}_t$ we denote the between-class and total scatter matrices in the feature space, respectively.

We have:

$$\tilde{\boldsymbol{S}}_b = \sum_{j=1}^{c} n_j \langle \boldsymbol{\mu}_j, \boldsymbol{\mu}_j\rangle_{\mathcal{H}_k}, \; \tilde{\boldsymbol{S}}_t = \sum_{j=1}^{n} \langle v_j, v_j\rangle_{\mathcal{H}_k},$$

where

$$\boldsymbol{\mu}_j = \tilde{\boldsymbol{m}}_j - \tilde{\boldsymbol{m}}, \; v_j = \tilde{\boldsymbol{x}}_j - \tilde{\boldsymbol{m}}_j, \; j = 1, \ldots, n.$$

KDCA seeks vectors $\boldsymbol{b}_i$ $(i = 1, \ldots, c-1)$ that maximize the ratio of between-class scatter and total scatter for maximum class separation.

Namely, we want to maximize the objective function:

$$J(\boldsymbol{b}_i) = \frac{\boldsymbol{b}_i^\top \tilde{\boldsymbol{S}}_b \boldsymbol{b}_i}{\boldsymbol{b}_i^\top \tilde{\boldsymbol{S}}_t \boldsymbol{b}_i},$$

subject to an additional restriction:

$$\boldsymbol{b}_i^\top \tilde{\boldsymbol{S}}_t \boldsymbol{b}_j = \delta_{ij}, \; i, j = 1, \ldots, c-1.$$

Finding the directional vectors $\boldsymbol{b}_i$ reduces to solving the following generalized eigenvalue problem:

$$\tilde{\boldsymbol{S}}_b \boldsymbol{b}_i = \lambda_i \tilde{\boldsymbol{S}}_t \boldsymbol{b}_i, \; \boldsymbol{b}_i^\top \tilde{\boldsymbol{S}}_t \boldsymbol{b}_j = \delta_{ij}. \quad (1)$$

Now, we show how to solve problem (1) without knowing the explicit representation of the mapping ϕ and the feature space $\mathcal{H}_k$ and without forming $\tilde{\boldsymbol{S}}_b$ and $\tilde{\boldsymbol{S}}_t$ explicitly.

Please note that any vector $\boldsymbol{b} \in \mathbb{R}^N$ can be represented as:

$$\boldsymbol{b} = \boldsymbol{b}_1 + \boldsymbol{b}_2$$

where $\boldsymbol{b}_1 \in \text{span}\{\tilde{\boldsymbol{x}}_1, \ldots, \tilde{\boldsymbol{x}}_N\}$ and $\boldsymbol{b}_2 \in \text{span}\{\tilde{\boldsymbol{x}}_1, \ldots, \tilde{\boldsymbol{x}}_N\}^{\perp}$, and $\tilde{\boldsymbol{S}}_b\boldsymbol{b}_2 = 0$ and $\tilde{\boldsymbol{S}}_t\boldsymbol{b}_2 = 0$ for any $\boldsymbol{b}_2 \in \text{span}\{\tilde{\boldsymbol{x}}_1, \ldots, \tilde{\boldsymbol{x}}_N\}^{\perp}$.

Therefore, for any vector $\boldsymbol{b}$ satisfying (1),

$$\tilde{\boldsymbol{S}}_b\boldsymbol{b}_1 = \tilde{\boldsymbol{S}}_b(\boldsymbol{b}_1 + \boldsymbol{b}_2) = \lambda\tilde{\boldsymbol{S}}_t(\boldsymbol{b}_1 + \boldsymbol{b}_2) = \lambda\tilde{\boldsymbol{S}}_t\boldsymbol{b}_1.$$

Hence we can restrict the solution space for (1) to $\text{span}\{\tilde{\boldsymbol{x}}_1, \ldots, \tilde{\boldsymbol{x}}_N\}$.

Let $\boldsymbol{b}$ be represented as a linear combination of $\tilde{\boldsymbol{x}}_i$, $i = 1, \ldots, N$, $\boldsymbol{b} = \sum_{i=1}^{N} \alpha_i\tilde{\boldsymbol{x}}_i$ and $\boldsymbol{\alpha} = (\alpha_1, \ldots, \alpha_N)'$.

Hence, the generalized eigenvalue problem

$$\tilde{\boldsymbol{S}}_b\boldsymbol{b} = \lambda_i\tilde{\boldsymbol{S}}_t\boldsymbol{b}$$

is equivalent to

$$\tilde{\boldsymbol{K}}\boldsymbol{D}\tilde{\boldsymbol{K}}\boldsymbol{\alpha} = \lambda\tilde{\boldsymbol{K}}\tilde{\boldsymbol{K}}\boldsymbol{\alpha},$$

where

$$\tilde{\boldsymbol{K}} = \boldsymbol{H}\boldsymbol{K}\boldsymbol{H},$$

$\boldsymbol{K} = (k(\boldsymbol{x}_i, \boldsymbol{x}_j))$, $i, j = 1, \ldots, N$, $\boldsymbol{H} = \boldsymbol{I}_N - \frac{1}{N}\boldsymbol{1}_N\boldsymbol{1}_N^{\top}$, $\boldsymbol{1}_N \in \mathbb{R}^N$ and the matrix $\boldsymbol{D}$ is defined by

$$D_{ij} = \begin{cases} \frac{1}{n_k}, & \text{if } \boldsymbol{x}_k \text{ and } \boldsymbol{x}_j \text{ belong to the } k\text{th class,} \\ 0, & \text{otherwise.} \end{cases}$$

Suppose the rank of $\boldsymbol{K}$ is r $(r \leq N)$. The projection of observation $\boldsymbol{x}$ onto $\boldsymbol{b}$ in $\mathcal{H}_k$ is given by:

$$\langle \boldsymbol{b}, \phi(\boldsymbol{x}) \rangle = \sum_{i=1}^{N} \alpha_i \langle \phi(\boldsymbol{x}_i), \phi(\boldsymbol{x}) \rangle = \sum_{i=1}^{N} \alpha_i k(\boldsymbol{x}_i, \boldsymbol{x}).$$

Solving the generalized eigenvalue problem presents certain difficulties because both matrices $\tilde{\boldsymbol{K}}\boldsymbol{D}\tilde{\boldsymbol{K}}$ and $\tilde{\boldsymbol{K}}\tilde{\boldsymbol{K}}$ are nonnegative definite. Krzyśko et al. [23] compared six algorithms that overcome these difficulties in different ways. The used performance quality criterion was the percentage of misclassification using the linear discriminant function in the spaces of kernel discriminant coordinates. The running time of the procedures was also recorded.

The mentioned authors recommended the algorithms proposed by Baudat and Anouar [20], Cai et al. [24] and two algorithms proposed by Zhang et al. [25]. They present similar classification errors and have comparable running times.

3.2. Cluster Analysis

Cluster analysis methods were applied to identify areas with similar degrees of sustainable development. Two methods were used, namely the Ward method and the agglomeration method (see, for example, Seber [26], Chapter 7, Mirkin [27], and Krzyśko et al. [28], Chapter 12).

The cluster procedure is based on the Mahalanobis distance between the macroregions to which individual poviats were satisfactorily qualified. This distance is defined by the following formula:

$$d_{ij}^2 = (\bar{\boldsymbol{x}}_i - \bar{\boldsymbol{x}}_j)^\top \boldsymbol{S}^{-1}(\bar{\boldsymbol{x}}_i - \bar{\boldsymbol{x}}_j),$$

where

$$\bar{\boldsymbol{x}}_i = \frac{1}{N_i}\sum_{j=1}^{N_i} \boldsymbol{x}_{ij},\ i = 1, 2, \ldots, c,$$

and

$$\boldsymbol{S} = \frac{1}{N-c}\sum_{i=1}^{c}\sum_{j=1}^{N_i}(\boldsymbol{x}_{ij} - \bar{\boldsymbol{x}}_i)(\boldsymbol{x}_{ij} - \bar{\boldsymbol{x}}_i)^\top,\ N = N_1 + \ldots + N_c.$$

4. Research Results

In the first step, the original data written in the form of $N = 379$ matrices (corresponding to the individual poviats) of the size of $T \times p$, where $p = 27$ is the number of features characterizing various aspects of sustainable development of poviats and $T = 15$ is the number of subsequent years in which these features were observed (years 2002–2016), were transformed into $c - 1 = 6$ kernel discriminant coordinates, where $c = 7$ is the number of macroregions in question. The structure of kernel discriminant coordinates for spatio-temporal data is described in Section 3.1. We went from matrix space to ordinary Euclidean space. The space of kernel discriminant coordinates is a space of dimension $\min(p, c - 1)$, where p is the number of observed features, and c is the number of macroregions. In our case, it is a six-dimensional space, so the dimension of the space in which further inference will take place is relatively small, as was previously mentioned.

We treat the obtained kernel discriminant coordinates further as indicators (synthetic measures) of the level of sustainable development of Polish macroregions. These indicators contain full information on the values the 27 diagnostic features measured over 15 years, which characterize the sustainable development of poviats and consequently the macroregions that include individual poviats. They are therefore composed indicators of sustainable development.

These six synthetic indicators have a different power of differentiating the studied macroregions (these indicators have different variances). The first indicator is the strongest and the sixth indicator is the least powerful. It is not possible to see the mutual position of the most important macroregion in the six-dimensional space of these indicators, but it is possible in the space of the first two composed indicators that differentiate the macroregions most potently. Figure 2 shows the mutual position of the seven macroregions in the system of the first two composed indicators (kernel discriminant coordinates).

In the full six-dimensional space of kernel discriminant coordinates, it is also easy to assess the differentiating power of a selected set of 27 features characterizing individual poviats. To this end, various methods of discriminant analysis can be used, examining the correctness of the classification of individual poviats into the seven identified macroregions.

In this paper, five different, most commonly used, methods of discriminant analysis were used (see, for example, Wasserman [29], Chapter 22, Krzyśko et al. [28], Chapters 1 and 4–6, Hastie et al. [30]): linear discriminant function (LDF), naive Bayes normal classifier (NB (Normal)), K-nearest neighbors method (KNN), classification trees (CART) (Tree (CART)) and support vector machines (SVM). Table 3 presents the percentage of misclassifications of individual poviats into seven macroregions, using these methods, and the estimated ten-fold cross validation method (10-cv). Clearly, zero or close to zero classification errors indicate a well-chosen set of 27 diagnostic features. This result was corroborated by our substantive selection of features characterizing sustainable development of macroregions with statistical arguments.

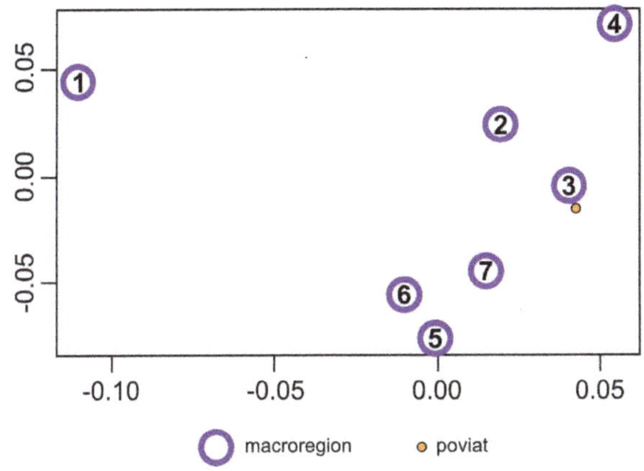

Figure 2. Macroregions in the first and second KDC systems and poviats recognized as belonging to specific macroregions. Source: own compilation.

Table 3. Percentage of misclassification of poviats in the 6-dimensional space of KDC.

Number	Classifiers	10-cv (%)
1	LDF	0.00
2	NB (Normal)	0.79
3	KNN ($K = 2, \ldots, 8$)	0.00
4	Tree (CART)	0.00
5	SVM	0.00

In the second step, cluster analysis was used in order to select groups of homogeneous macroregions in the six-dimensional space of kernel discriminant coordinates. The agglomeration method and the Ward method were selected as commonly used methods. The Mahalanobis distance was chosen as a measure of the distance between the mean vectors of individual macroregions. It takes into account not only the difference between the mean vectors of two macroregions; the difference is also weighted by the variances and covariances of the examined features estimated for a total of seven macroregions (the differentiation of poviats around the mean macroregions was taken into account). The Mahalanobis distances between macroregions are shown in Table 4.

Table 4. Mahalanobis distances between macroregions.

Macroregion	1	2	3	4	5	6	7
1	0.0000	538.5115	631.1308	**717.4584**	588.5517	513.6408	**574.5919**
2	538.5115	0.0000	**128.1944**	**245.4427**	396.7969	339.3620	261.7809
3	631.1308	128.1944	0.0000	**295.5870**	340.7639	306.7737	199.0511
4	717.4584	245.4427	295.5870	0.0000	628.0889	579.4410	486.7190
5	588.5517	396.7969	340.7639	628.0889	0.0000	**81.6527**	**142.1345**
6	513.6408	339.3620	306.7737	579.4410	81.6527	0.0000	**117.2979**
7	574.5919	261.7809	199.0511	486.7190	142.1345	117.2979	0.0000

On the basis of the Mahalanobis distances, dendrograms were constructed using the agglomeration method and the Ward method to identify areas with similar degrees of sustainable development. Both provided results identical to the ones illustrated in Figure 3.

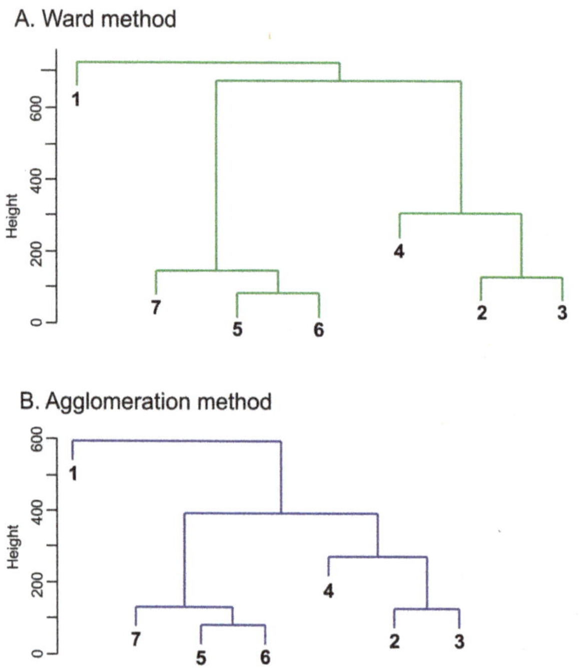

Figure 3. Grouping of macroregions with cluster methods (1—southern, 2—northwestern, 3—southwestern, 4—northern, 5—central, 6—eastern, 7—Masovian voivodship). Source: own compilation.

Three areas were found with a high degree of internal similarity considered for sustainability. Two of them can be referred to as super macroregions.

Super macroregion (area) I consist of the following macroregions:

2—northwestern,

3—southwestern,

4—northern.

Super macroregion (area) II:

5—central,

6—eastern,

7—Masovian voivodship.

The third area represents a separate southern macroregion (I).

The delimitated super macroregions vary with respect to area, territorial capital and socio-economic development specified by the values of the Human Development Index (HDI).

Super macroregion I consists of three macroregions, eight regions (voivodships) and 175 sub-regions (poviats). It represents 50.1% of Poland's total area.

Its three regions, i.e., Great Poland, Lower Silesia and Pomerania, belong to a set of five regions of Poland that enjoy the highest GDP in the long run. By contrast, the Poznań poviat is the most developed poviat in Poland in socio-economic terms (Kierczyńska [31]).

Notably, super macroregion I includes regions with a lower degree of economic development but with enormous natural environment resources. These regions include Warmian-Masurian, Pomeranian, West-Pomeranian and Lubusz. This diversity within the super macroregion translates into the average HDI value of super macroregion I, determined for 2015, amounting to 0.864. It is lower than the average HDI value of super macroregion II and the southern macroregion (Hozer-Koćmiel [32]).

Super macroregion II also consists of three macroregions, including six voivodships and 146 poviats. It covers 41.1% of Central and Eastern Poland. Its seventh macroregion (Masovian voivodship) in terms of GDP generation, economic and social development is the strongest macroregion in Poland. In addition, due to the determinants of sustainable development, super macroregion II is metaphorically called the "Eastern Wall". It refers to regions with unique natural resources—not only on a Polish, but also a European, and even a global scale.

The average HDI value of super macroregion II is relatively high and amounts to 0.884. This is due to the extremely high position of macroregion 7 on the national scale and valuable environmental resources of the other macroregions.

In addition to two super macroregions, cluster analysis also identified the third area, i.e., the southern macroregion (8.8% of Poland's area). It consists of two regions (voivodships): Lesser Poland and Silesia. These are voivodships (two out of five) with the highest GDP in Poland in the long term.

Due to the high level of economic and social development of both regions, the southern macroregion has a higher average HDI value than both super macroregions i.e., HDI = 0.889.

Equally interesting is the fact that the HDI growth dynamics in the period 1995–2015 in the southern macroregion were the highest among the three newly delimitated spatial units, and amounted to 17.4%.

For the first super macroregion it was 12.5% and the second macroregion it was 12.6%.

Notably, the southern macroregion consists of only two regions with high parameters of socio-economic development.

The research objective adopted in the study was thus achieved. Three new spatial units consisting of macroregions (NUTS-1) were distinguished. They are characterized by a high degree of similarity in terms of sustainable development (the attributes of sustainable development were characterized by 27 variables divided into four pillars).

The accuracy of the obtained division results from the application of the kernel discriminant coordinates method. As a multidimensional method, it allows to achieve an extremely precise division of the examined objects.

Nevertheless, the super macroregions presented in Figure 4 should be treated as a scientific proposal subject to modification—taking into account other aspects of sustainable development and a different time interval.

Figure 4. Super macroregions and macroregion of Poland. Source: own compilation.

5. Conclusions

This paper presented a proposal of structuring synthetic indicators for characterizing the sustainable development of macroregions described by spatio-temporal data. The values of kernel discriminant coordinates were adopted as these indicators. From a statistical point of view, this is an objective solution. The number of synthetic indicators is $\min(p, c-1)$, where p is the number of selected features characterizing the sustainable development of macroregions and c is the number of the studied macroregions. In the space of the first two indicators, one can see the mutual position of macroregions, while in the full space of the indicators, one can examine the strength differentiating macroregions by a selected set of features characterizing their sustainable development. The complete space of the indicators also allows to distinguish homogeneous subgroups of macroregions by means of the methods of cluster analysis. The operation of the proposed statistical methodology has been illustrated with spatio-temporal data concerning seven Polish macroregions.

The selection of 27 features characterizing the sustainable development of these regions turned out to be accurate, because in the six-dimensional space of synthetic indicators, errors in classifying 379 poviats to seven macroregions turned out to be zero or close to zero, depending on the selected statistical classifier. Consequently, the characteristics that optimally differentiate the studied macroregions were selected. The action of the proposed statistical methodology was illustrated with spatio-temporal data on seven Polish macroregions.

In the full six-dimensional space of synthetic indicators, three homogeneous subgroups of macroregions, called super regions, were separated using cluster analysis methods.

Although the data concern only one country, Poland, the presented statistical methodology is universal and can be applied to any territorial units in the world described by spatio-temporal data.

To the best of our knowledge, the proposed statistical methodology is new in the scientific literature. The justification may be the fact that we do not know any other solution for spatio-temporal data. In each study, the obtained results will always be a compromise between attempted assessment of the differentiation of the level of territorial units due to their sustainable development on the one hand, and the weakness of some diagnostic variables or the choice of the research method on the other hand.

Currently, research is being conducted on the construction of synthetic indicators describing the sustainable development of macroregions by replacing kernel discriminant coordinates with functional discriminant coordinates. Matrix data can be transformed into continuous functions (elements of a certain functional space). The transformed data are called functional data (Ramsay and Silverman [33], Górecki et al. [34]).

Author Contributions: Conceptualization, M.K.; Data curation, A.K.; Formal analysis, W.W. and M.K.; Investigation, B.W.; Methodology, W.R.; Software, W.W.; Validation, B.W.; Writing—original draft, A.K.; Writing—review & editing, W.W. and W.R. All authors have read and agreed to the published version of the manuscript.

Funding: This research received no external funding.

Conflicts of Interest: The authors declare no conflict of interest.

References

1. Awan, U.; Abbasi, A.S.; Humayon, A.A. The Concept of Civic Sustainability is Need of Hour. *Res. J. Environ. Earth Sci.* **2014**, *6*, 347–352. [CrossRef]
2. Awan, U.; Krasławski, A.; Huiskonen, J. Understanding influential factors on implementing social sustainability practices in Manufacturing Firms: An interpretive structural modelling (ISM) analysis. *Procedia Manuf.* **2018**, *17*, 1039–1048. [CrossRef]
3. Borys, T. (Ed.) *Indicators of Sustainable Development*; Economics and Environment: Białystok, Poland, 1999. (In Polish)

4. Atkinson, A. Developing indicators of sustainable community: Lessons from sustainable Seattle. *Environ. Impact Assess. Rev.* **1996**, *16*, 337–350. [CrossRef]
5. Bourdeau, L. Sustainable development and future of construction: A comparison of visions from various countries. *Build. Res. Inf.* **1999**, *27*, 354–366. [CrossRef]
6. Spangenberg, J.H. Environmental space and prism of sustainability: Frameworks for indicators measuring sustainable development. *Ecol. Indic.* **2002**, *2*, 295–309. [CrossRef]
7. *OECD Territorial Reviews. Poland*; OECD: Paris, France, 2008.
8. Wheeler, S.M.; Beatley, T. (Eds.) *The Sustainable Urban Development*; Routledge: London, UK; New York, NY, USA, 2009.
9. Wong, C. *Indicators for Urban and Regional Planning*; Routledge: London, UK; New York, NY, USA, 2006.
10. Kontostanou-Karalivanou, O.; Maxon, P.A.; Sauerborn, K.; Scoullos, M.J. Criteria and indicators for regional sustainable development. In *Sustainable Development of European Cities and Regions*; Vonkeman, G.H., Ed.; Springer: Berlin/Heidelberg, Germany, 2000; pp. 63–83.
11. United Nations. *Indicators of Sustainable Development: Guidelines and Methodologies*, 3rd ed.; United Nations: New York, NY, USA, 2007.
12. Streimikiene, D. Comparative assessment indicators of quality of life in Romania and Lithuania. *Econ. Sociol.* **2014**, *7*, 11–21. [CrossRef] [PubMed]
13. Dočekalova, M.P.; Kocmanová, A.; Koleňák, J. Determination of economic indicators in the context of corporate sustainability performance. *Bus. Theory Pract.* **2015**, *16*, 15–24. [CrossRef]
14. Jovovic, R.; Draskovic, M.; Delibasic, M.; Jovovic, M. The concept of sustainable regional development-institutional aspects, policies and prospects. *J. Int. Stud.* **2017**, *10*, 255–266. [CrossRef] [PubMed]
15. Roszkowska, E.; Wyszkowska, D. Spatial differentiation of the elderly population in Poland's voivodeships in the years 2000–2016. *Optim. Econ. Stud.* **2020**, *1*, 89–112. [CrossRef]
16. Büyüközkan, G.; Karabulut, Y. Sustainability performance evaluation: Literature review and future directions. *J. Environ. Manag.* **2018**, *217*, 253–267. [CrossRef] [PubMed]
17. Krasławski, A.; Huiskonen, J.; Awan, U. Progress from blue to the green world: Multilevel governance for pollution prevention planning and sustainability. In *Handbook of Environmental Materials Management*; Hussain, C.M., Ed.; Springer: Berlin/Heidelberg, Germany, 2019; Chapter A3.
18. Fisher, R.A. The use of multiple measurements in taxonomic problems. *Ann. Eugen.* **1936**, *5*, 179–188. [CrossRef]
19. Rao, C.R. The utilization of multiple measurements in problems of biological classification. *J. R. Stat. Soc. Ser. B* **1948**, *10*, 158–203. [CrossRef]
20. Baudat, G.; Anouar, F. Generalized discriminant analysis using a kernel approach. *Neural Comput.* **2000**, *12*, 2385–2404. [CrossRef] [PubMed]
21. Mika, S.; Ratsch, G.; Weston, J.; Schölkopf, B.; Müller, K.R. Fisher discriminant analysis with kernels. In *Neural Networks for Signal Processing IX*; Hu, Y.H., Larsen, J., Wilson, E., Douglas, S., Eds.; IEEE Signal Processing Society: Piscataway, NJ, USA, 1999; pp. 41–48.
22. Shawe-Taylor, J.; Cristianini, N. *Kernel Methods for Pattern Analysis*; Cambridge University Press: Cambridge, UK, 2004.
23. Krzyśko, M.; Waszak, Ł.; Wołyński, W. Comparison of kernel discriminant coordinates algorithms. *Commun. Stat. Simul. Comput.* **2014**, *43*, 2138–2148.
24. Cai, D.; He, X.; Han, J. *Efficient Kernel Discriminant Analysis via Spectral Regression*; Technical Report; Computer Science Department, University of Illinois, Urbana and Champaign, IL, USA, 2007.
25. Zhang, Z.; Dai, G.; Xu, C.; Jordan, M.I. Regularized discriminant analysis, ridge regression and beyond. *J. Mach. Learn. Res.* **2010**, *11*, 2199–2228.
26. Seber, G.A.F. *Multiple Observations*; John Wiley and Sons, Inc.: New York, NY, USA, 2004.
27. Mirkin, B. *Clustering. A Data Recovery Approach*, 2nd ed.; Taylor and Francis Group, LLC.: London, UK, 2013.
28. Krzyśko, M.; Wołyński, W.; Górecki, T.; Skorzybut, M. *Learning Systems: Pattern Recognition, Cluster Analysis and Dimensional Reduction*; Scientific and Technical Publishers: Warsaw, Poland, 2008. (In Polish)
29. Wasserman, L. *All of Statistics. A Concise Course in Statistical Inference*; Springer Science+Business Media, Inc.: New York, NY, USA, 2004.

30. Hastie T.; Tibshirani, R.; Friedman, J.R. *The Elements of Statistical Learning: Data Mining, Inference, and Prediction*, 2nd ed.; Springer: New York, NY, USA, 2009.
31. Kierczyńska, A. *Typology of Polish Poviats in View of the Variability of Economic and Social Development in 2002–2016*; The Great Poland Socio-Economic University: Środa Wlkp., Poland; Forthcoming. (In Polish)
32. Hozer-Koćmiel, M. Ocena rozwoju społeczno-ekonomicznego województw za pomocą HDI (Assessement of the socio-economic development of voivodeships using HDI). *WiadomoŚci Stat.* **2018**, *3*, 40–49.
33. Ramsay, J.O.; Silverman, B.W. *Functional Data Analysis*, 2nd ed.; Springer: New York, NY, USA, 2005.
34. Górecki, T.; Krzyśko, M.; Waszak, Ł. Functional discriminant coordinates. *Commun. Stat. Theory Methods* **2014**, *43*, 1013–1025. [CrossRef]

© 2020 by the authors. Licensee MDPI, Basel, Switzerland. This article is an open access article distributed under the terms and conditions of the Creative Commons Attribution (CC BY) license (http://creativecommons.org/licenses/by/4.0/).

International Journal of
Environmental Research and Public Health

MDPI

Article

Prenatal Fine Particulate Matter ($PM_{2.5}$) Exposure and Pregnancy Outcomes—Analysis of Term Pregnancies in Poland

Cezary Wojtyla [1,*], Karolina Zielinska [1], Paulina Wojtyla-Buciora [2] and Grzegorz Panek [1]

1 Department of Oncologic Gynecology and Obstetrics, Centre of Postgraduate Medical Education, 231 Czerniakowska St., 00-416 Warsaw, Poland; karolina.anna.zielinska@gmail.com (K.Z.); gmpanek@wp.pl (G.P.)
2 Faculty of Health Sciences, State University of Applied Sciences, 16 Kaszubska St., 62-800 Kalisz, Poland; paulinawojtyla@gmail.com
* Correspondence: czwo@op.pl

Received: 20 June 2020; Accepted: 7 August 2020; Published: 11 August 2020

Abstract: Air pollution is currently one of the greatest threats to global health. Polish cities are among the most heavily polluted in Europe. Due to air pollution 43,100 people die prematurely in Poland every year. However, these data do not take into account the health consequences of air pollution for unborn children. Thus, the aim of this study was to evaluate the effects of the fine particulate matter air pollution (less than 2.5 µm in diameter) on pregnancy outcomes. An analysis of pregnant women and their children was made using a questionnaire survey from a nationwide study conducted in 2017. Questionnaires from 1095 pregnant women and data from their medical records were collected. An analysis of air pollution in Poland was conducted using the air quality database maintained by the Chief Inspectorate for Environmental Protection in Poland. A higher concentration of $PM_{2.5}$ was associated with a decrease in birth weight and a higher risk of low birthweight (i.e., <2500 g). We also observed lower APGAR scores. Thus, all possible efforts to reduce air pollution are critically needed.

Keywords: air pollution; pregnancy; $PM_{2.5}$; pregnancy outcomes

1. Introduction

In 2019, the World Health Organization (WHO) placed both air pollution and climate change among the top ten global health hazards. Simultaneously, air pollution was pointed by the WHO as the single greatest environmental risk to health [1]. Each year, 7,000,000 people die prematurely from diseases that may be linked to indoor and outdoor air pollution [1].

Polish cities are among the most heavily polluted in Europe. According to the WHO Ambient Air Pollution Database, 33 out of the 50 most polluted cities in Europe are located in Poland [2]. Consequently, 43,100 people are estimated to die prematurely due to air pollution in Poland every year [3]. However, the number may be much higher when ramifications of air pollution for future generations' life and health are considered. The developing fetus is highly susceptible to the hazardous chemical compounds contained in the air inhaled by a pregnant woman. Toxic compounds contained in the particulate matter of less than 2.5 µm in diameter ($PM_{2.5}$) are transferred through the pulmonary alveoli to the blood flow of the exposed woman and, together with blood, reach placenta and fetal circulation, emitting a toxic influence to the fetal cells [4]. The developing fetus is highly susceptible to toxins, particularly during specific windows of gestation. The fetus is more susceptible than adults to the negative effects of the toxic and carcinogenic compounds due to its high absorption and retention of the aforementioned compounds and the difficulties in their metabolism and detoxification caused by the immature liver and immunological system. Furthermore, inefficient are also the DNA repair mechanisms of the fetus with simultaneous intensive cell proliferation at this stage of life. Moreover, these compounds may interfere with placenta and disrupt its transport mechanisms [5], which may cause the intrauterine growth restriction, preterm birth, birth defects, as well as DNA changes, leading to mutations and epigenetic changes [6]. This type of influence on the genetic material is multiplied during subsequent cell divisions, increasing the risk of chronic diseases and tumors in the extrauterine life [5].

In the global perspective, 25% of the $PM_{2.5}$ air pollution is caused by fuel combustion. Another 15% is contributed to by industry—especially coal-based energy production. About 20% of the aforementioned fine particulate matter in the global air pollution is generated by households and the burning of coal, wood and various house waste. The same percentage comes from unspecified sources of human origins and another 18% from natural dust and salt [7]. However, considering Poland, it is believed that even 50% of the $PM_{2.5}$ air pollution is attributed to the household sector [8]. Consequently, the cold months of the year are a period of significantly increased concentration of the fine particulate matter in the air—especially on windless days. One such situation was observed in Poland at the turn of the years 2016 and 2017, when the concentration of $PM_{2.5}$ reached levels comparable to Beijing and New Delhi [2].

The objective of this study is to evaluate the effects of the particulate matter air pollution of less than 2.5 μm in diameter on the pregnancy outcomes.

2. Materials and Methods

An analysis of pregnant women and their children was conducted using survey questionnaires from the Polish pregnancy-related assessment monitoring system (Pol-PrAMS) program. This population-based study was conducted between 2 February and 22 March 2017 in all hospitals in Poland. Groups of Polish women and their newborns were surveyed in all of the hospitals in Poland during postpartum hospitalization. Thus, all of the women hospitalized postpartum on the designated days of the study were deemed eligible for the study. The ethics committee of the Institute of Rural Health in Lublin, Poland, approved the study (reference number 03/2011). Participation was anonymous and voluntary. The survey was divided into two parts. The first contained over 70 questions, namely, the age, place of residence, education, social and economic status and health behaviors before and during pregnancy. Mothers who stayed in hospitals after birth completed the first part of the survey. The second part had nine questions that were completed by the medical personnel providing healthcare to the mother and newborn, with the use of medical records (pregnancy cards and patient medical history). The questions in this part concerned the childbirth method, newborn status after birth and birth defects. It also included questions about the results of laboratory tests performed on mothers and newborns after birth. The methodology of this study is described in detail in a separate articles [9,10].

Overall, 3451 women who gave birth and their newborns were hospitalized in obstetric wards of the surveyed hospitals in Poland. Our sample consisted of women who gave birth in cities where the fine particulate matter monitoring stations were located. Additionally, only term pregnancies were included, i.e., those who ended between the full 37th and the incomplete 42nd week of gestation. Thus, our sample consisted of 1095 women.

An analysis of air pollution in Poland in the period from 15 April 2016 to 22 March 2017 was made using the air quality database maintained by the Chief Inspectorate for Environmental Protection in Poland. The database contains records of the average daily concentration of $PM_{2.5}$ throughout the entire year in the cities where monitoring stations are located. There are 64 stations of this type in Poland, but we included in the analysis 50 cities in which women gave birth during the study period. For each city, the average concentration of $PM_{2.5}$, for the period between the 15 April 2016 and the 22 March 2017, was calculated. The reason for selecting this period was the fact that all of the women in our sample were pregnant at that time. It was assumed that women who were admitted to the hospital at the time of the survey on the 2 February 2017 and who gave birth before the 42nd week of pregnancy, had their last period on the 15 April 2016, as per Naegele's rule. The last woman applicable for the survey would most likely give birth on the 22 March 2017, on the last day of the Pol-PrAMS study. Women taking part in the survey were divided into two groups. The first consisted of those who gave birth in cities where the average concentration of $PM_{2.5}$ was no higher than 25 μg/m^3, which is the acceptable average level per annum, as indicated in the directive of the European Union [11]. The other group consisted of women who gave birth in cities polluted with the average fine particulate matter concentration higher than 25 μg/m^3.

Continuous variables were compared using Mann–Whitney–Wilcoxon Test, while a chi-squared test was applied for categorical variables. The results were expressed as the mean and standard deviation and 95% confidence intervals (95% CI) or as a frequency (%). Logistic regression models were created to estimate the odds ratios and 95% CI. For each analyzed outcome, the crude odds ratio was computed for the single predictor: index of exposure to fine particulate matter ($PM_{2.5}$) concentration higher than 25 μg/m^3. In multivariable models the effect of exposure was adjusted to the effects of education, economic status, social conditions, cigarette smoking, alcohol consumption and presented as adjusted odds ratio (aOR). All statistical analyses were performed using IBM SPSS software version 25 (IBM Corp., Armonk, NY, USA). A p-value of <0.05 was considered statistically significant.

Each time we discuss gestational diabetes mellitus (GDM) in the results of our work, it refers to diabetes diagnosed for the first time during pregnancy in patients who did not suffer from this disease before pregnancy [12]. At the same time, we only consider pregnancy-induced hypertension (PIH), i.e., hypertension that was diagnosed for the first time after the 20th week of the current pregnancy [13].

3. Results

Supplementary Figure S1 presents the concentration of $PM_{2.5}$ in the period between the 15 April 2016 and the 22 March 2017 in the analyzed cities in Poland.

Approximately 1 in 3 women was between 31 and 35 years old (34.4%). Almost the same percentage referred to women aged between 26 and 30 years old (34.1%) The highest percentage corresponded to women with a higher education degree (62%), over 33% were women who had completed secondary education, and almost 5% had finished primary school education. More than half of the women considered their financial situation to be good. Over 40% of women described their social conditions to be very good or good (respectively 40.7% and 49.9%). Almost 69% of the women declared to have never smoked, the second largest group were women who had stopped smoking for the duration of pregnancy (14.5%). Simultaneously, 5.1% reported that they have been smoking during pregnancy. Slightly more than 40% of women admitted to consuming a certain amount of alcohol during pregnancy. The pregnancy of 7.1% had complications linked to GDM and 5.7% had high blood pressure-related complications. Both analyzed groups differed in regard to education, financial conditions and social conditions. Patient characteristics are shown in Table 1.

The average weight and birth length of newborns that were born in the areas of the average fine particulate matter pollution ≤25 μg/m^3 were 3478 g and 55.0 cm, respectively. In newborns born in areas where the concentration of the fine particulate matter was >25 μg/m^3, the birthweight was lower by 115 g ($p < 0.05$) and the length of the infant was 0.4 cm shorter ($p > 0.05$). The aforementioned details are shown in Table 2.

Table 1. Patient characteristics.

	Average Annual $PM_{2.5}$ Concentration ($\mu g/m^3$)						
	≤25 $\mu g/m^3$		>25 $\mu g/m^3$		All		p
	N	%	N	%	N	%	
Age							ns
<25	104	16.1%	71	16.2%	175	16.2%	
26–30	222	34.4%	147	33.6%	369	34.1%	
31–35	228	35.4%	144	33.0%	372	34.4%	
>35	91	14.1%	75	17.2%	166	15.3%	
Education							<0.05
Basic	32	5.1%	19	4.4%	51	4.8%	
Secondary	188	29.8%	164	38.2%	352	33.2%	
Higher	410	65.1%	246	57.4%	656	62.0%	
Economic Status							<0.05
Very good	156	24.1%	77	17.4%	233	21.4%	
Good	372	57.5%	287	64.8%	659	60.4%	
Average/bad	119	18.4%	79	17.8%	198	18.2%	
Social Conditions							<0.05
Very good	289	44.7%	154	34.8%	443	40.7%	
Good	298	46.2%	245	55.5%	543	49.9%	
Average/bad	59	9.1%	43	9.7%	102	9.4%	
Cigarette Smoking							ns
Yes	39	6.1%	16	3.7%	55	5.1%	
Quit during pregnancy	99	15.5%	56	13.0%	155	14.5%	
No, from several years	71	11.1%	52	12.0%	123	11.5%	
No, never	429	67.3%	308	71.3%	737	68.9%	
Alcohol Consumption							ns
Never	374	57.5%	265	59.6%	639	58.3%	
At most once per month	169	26.0%	99	22.2%	268	24.5%	
Twice per month and more often	107	16.5%	81	18.2%	188	17.2%	
Gestational Diabetes Mellitus							ns
No	598	92.0%	419	94.2%	1017	82.9%	
Yes	52	8.0%	26	5.8%	78	7.1%	
Pregnancy-induced Hypertension							ns
No	571	94.2%	396	94.5%	967	94.3%	
Yes	35	5.8%	23	5.5%	58	5.7%	

p-value refers to two groups of women with analyzed $PM_{2.5}$ concentration; ns—statistically insignificant, $p < 0.05$.

Table 2. Birth weight and newborn length depending on $PM_{2.5}$ concentration.

Variable	N	Mean	Mean Difference	S.D.	95% CI Lower Limit	95% CI Upper Limit	p-Value
Weight (g)							<0.05
$PM_{2.5} \leq 25\ \mu g/m^3$	634	3478	0	446	3444	3513	
$PM_{2.5} > 25\ \mu g/m^3$	432	3363	−115	490	3317	3409	
Length (cm)							ns
$PM_{2.5} \leq 25\ \mu g/m^3$	608	55.0	0	2.7	54.7	55.1	
$PM_{2.5} > 25\ \mu g/m^3$	418	55.0	−0.4	3.1	54.3	54.9	

Ns—statistically insignificant, $p < 0.05$.

Heightened $PM_{2.5}$ concentration was associated with an increase in the number of births in which the newborn weight was lower than 2500 g ($p < 0.05$). Infants born in cities where fine particulate matter pollution was >25 $\mu g/m^3$ had Apgar scores of less than 8 more frequently, tested between 1 and 10 min after birth ($p < 0.05$). However, in this group there was an increase in the percentage of women whose pregnancy or labor was complicated by one the following pathologies: birthweight lower than 2500 g, premature rupture of membranes (PROM), the Apgar scores of less than 8 (tested between 1 and 10 min after the birth) or diagnosed birth defects (<0.05). Nevertheless, no statistically significant effect of increased $PM_{2.5}$ concentration was observed in relation to the following aspects: premature rupture of membranes, type of labor, birth defects, sex of the child, hospitalization during pregnancy ($p > 0.05$). Table 3 presents the unadjusted data on influence of air pollution on the obstetrical outcomes.

Table 3. Pregnancy outcomes depending on average annual $PM_{2.5}$ concentration.

	Average Annual $PM_{2.5}$ Concentration (μg/m³)				p
	≤25 μg/m³		**>25 μg/m³**		
	N	%	N	%	
Pregnancy Outcome					
Birthweight (g)					<0.05
>2500	629	99.2	419	97.0	
<2500	5	0.8	13	3.0	
PROM					ns
No	454	71.6	282	66.4	
Yes	180	28.4	143	33.6	
Type of Labor					ns
Vaginal	387	59.8	250	57.1	
Cesarean section	260	40.2	188	42.9	
APGAR					<0.05
≥8	590	97.0	415	94.3	
<8	18	3.0	25	5.7	
Birth Defects					ns
No	520	98.7	363	99.5	
Yes	7	1.3	2	0.5	
Sex of the Child					ns
Female	306	47.9	202	47.8	
Male	333	52.1	221	52.2	
Hospitalization					ns
No	372	62.0	257	62.8	
Yes	228	38.0	152	37.2	
Pathology					<0.05
No	451	69.5	274	61.7	
Yes	198	30.5	170	38.3	

Legend: PROM—premature rupture of membranes; Pathology—low birthweight or PROM or APGAR score < 8 or birth defects; ns—statistically insignificant ($p < 0.05$).

Results from models adjusted for maternal characteristics are reported in Table 4. These present the risk factors of the occurrence of different pregnancy outcomes. Children born in the areas of heightened air pollution were 4 times more likely to have a birthweight lower than 2500 g (OR = 4.3, 95%; CI: 1.5–12.3; $p < 0.05$) and almost twice more frequently scored less than 8 points in the Apgar score (OR = 2.4, 95% CI: 1.2–4.6 $p < 0.05$). In this group the following pathologies appeared more often: low birthweight or PROM or APGAR < 8 or birth defects (OR = 1.4, 95% CI: 1.1–1.9; $p < 0.05$). Results of tests of difference in PROM and cesarean section between areas with lower and higher levels of air pollution were not statistically significant. Surprisingly, a lower risk of the GDM development was observed (OR = 0.5, 95% CI: 0.3–0.9; $p < 0.05$).

Table 4. Adjusted odds ratios and 95% confidence intervals for different pregnancy outcomes among women exposed to fine particulate matter ($PM_{2.5}$) concentration greater than 25 μg/m^3.

	Adjusted Odds Ratio		
	aOR	**95% CI**	***p***
Birthweight (g)			<0.05
>2500	Reference		
<2500	4.3	1.5–12.3	
PROM			ns
No	Reference		
Yes	1.2	0.9–1.6	
Type of Labor			ns
Vaginal	Reference		
Cesarean section	1.2	0.9–1.5	
APGAR			
≥8	Reference		<0.05
<8	2.4	1.2–4.6	
Birth Defects			
No	Reference		ns
Yes	0.4	0.1–1.9	
GDM			<0.05
No	Reference		
Yes	0.5	0.3–0.9	
PIH			ns
No	Reference		
Yes	1.0	0.6–1.7	
Hospitalization			ns
No	Reference		
Yes	0.9	0.7–1.2	
Pathology			<0.05
No	Reference		
Yes	1.4	1.1–1.9	

Legend: PROM—premature rupture of membranes; GDM—gestational diabetes mellitus; PIH—pregnancy induced hypertension; PPH—pregnancy hypertension; pathology—low birthweight or PROM or APGAR < 8 or birth defects; ns—statistically insignificant ($p < 0.05$).

4. Discussion

Air pollution with fine particulate matter of less than 2.5 μm in diameter has become a global health challenge. Over recent years, the number of studies analyzing the link between prenatal risks and further development of the fetus and child has increased.

Increased concentration of $PM_{2.5}$ during the entirety of the pregnancy has a negative impact on birthweight [14]. This finding is consistent with results reported elsewhere [15,16]. Attention is also paid to the impact of air pollution during specific periods of pregnancy on lower birthweight. In his analysis of 400,000 children, Kurman observed a stronger negative impact of higher air pollution on the birth weight of children in the first trimester of pregnancy than in the remaining two trimesters [17]. Higher risk of a low birth weight (i.e., <2500 g) was also observed [18]. Details of this process are not yet fully understood. It may be due to the increased alveolar ventilation, which is a physiological phenomenon during pregnancy, but may cause increased absorption of air pollutants [19].

Research carried out by Maciel-Rutz et al. does not indicate any significant impact of $PM_{2.5}$ on the Apgar scores in newborns [20]. In our study, the Apgar score of less than 8 was observed in newborns born in the cities more heavily polluted than the average 25 μg/m^3. Nevertheless, it needs to be emphasized that the evaluation of the direct correlation between air pollution and the Apgar score is difficult to unambiguously conclude. The conditions of newborns are affected by numerous factors, like the course of birth or the accompanying illnesses of the mother or child.

Different conclusions were drawn from the analysis of the influence of air pollution on birth defects. There is a plentitude of research carried out with the goal of finding the link between these two factors. However, the results are inconclusive and require a further in-depth analysis in order to understand the mechanism of this phenomenon. It seems that the influence of air pollution is greatest in the critical periods of formation of particular internal organs during embryogenesis that is in the first trimester. Thus, our use of average ambient levels of $PM_{2.5}$ over the previous year may not have captured meaningful variations in pollutant levels during periods of critical fetal development relevant for these outcomes

Vrijheid, M. et al. proposed a hypothesis that inhalation of pollution and its particles leads to their absorption into the blood flow, which causes the oxidative stress that damages the cells' DNA. Furthermore, the inhaled compounds may contribute to a local inflammatory response in placenta, cause changes in the coagulation system, as well as influence the processes of migration and differentiation of neural crest cells [21].

There are many research papers describing the effects of air pollution on the structure and performance of the hearth. Tanwar et al. and Hall et al. pointed out that the exposure to $PM_{2.5}$ during the prenatal period causes significant disruptions of the heart structure and activity in adulthood. The risk of these complications is especially high when the exposure takes place between the 2nd and the 8th week of pregnancy, during the period of development of the cardiovascular system [22,23]. Other research indicates an increased risk of pulmonary stenosis or ventricular septal defect [24] as well as coarctation of the aorta [25]. Other observations include a heightened risk of defects in the digestive system [25], formation cleft palate [26] or an increased risk of genital defects [27]. Nevertheless, our study does not conclude the existence of the link between air pollution and heightened risk of birth defects in a child.

Many studies indicate a link between air pollution and increased risk of developing gestational diabetes. The risk is connected to the increased concentration of NO during the first and the second trimester of the pregnancy [28], SO_2 during the first trimester (particularly between the 4th and the 10th pregnancy week) [29] and $PM_{2.5}$ in the second trimester [30]. Our results are inconsistent with previously reported findings, as this research revealed a lowered risk of gestational diabetes mellitus in women living in cities with a high $PM_{2.5}$ concentration. The reason may be the methodological differences of abovementioned studies. Two of the three previous studies cited, used different pollutants (NO and SO_2) and all used different time periods for indicators of exposure (trimester versus year-long period). It seems likely that these differences could contribute to differences in patterns of results.

The limitations of this study include the estimation of $PM_{2.5}$ exposure in pregnant women. According to the methodology, the risk stems from the average concentration of $PM_{2.5}$ in the place of delivery. This does not necessarily mean that the expectant woman stayed within this place for the entire pregnancy. The women were not asked for how long they had stayed in a city where a monitoring station was placed. Thus, our results are only an estimate of women's exposure to air pollution. Moreover, the air pollution we took into account is the mean air pollution from the entire period under study, i.e., between the 15 April 2016 and the 22 March 2017. Additionally, it needs to be remembered that both of the samples were not identical. We tried to eliminate the differences in the structures of both groups by taking into account such variables as education, economic status, social conditions, smoking and alcohol consumption in the logistic regression models. Despite these limitations, this is the first and only research of this type in Poland that analyses the effects of air pollution on pregnancy outcomes in such a large sample.

5. Conclusions

Our study adds to a larger body of evidence showing adverse effects of air pollution for birth outcomes. Air pollution may impact pregnancy outcomes in terms of lower birth weight, higher risk of giving birth to a child weighing <2500 g and also to a child in a worse general condition defined as APGAR scores. Yet, this study limitations do not allow us to draw indisputable conclusions. Further study is warranted to address some inconsistencies in findings across studies and gaps in the literature, but that based on the body of evidence a failure to address excess exposure to $PM_{2.5}$ contributes to poor birth outcomes among children exposed in utero. These poor birth outcomes have been associated with heightened medical risk throughout the life course [30,31]. Thus, efforts to reduce air pollution are likely to not only promote better birth outcomes, but also contribute to improved health over the life course.

Supplementary Materials: The following are available online at http://www.mdpi.com/1660-4601/17/16/5820/s1, Figure S1: Concentration of $PM_{2.5}$ in analyzed cities in Poland, between 15 April 2016 and 22 March 2017.

Author Contributions: Conceptualization, C.W., K.Z., P.W.-B. and G.P.; Methodology, C.W. and P.W.-B.; Software, C.W. and K.Z.; Validation, P.W.-B. and G.P.; Formal Analysis, C.W., K.Z., P.W.-B. and G.P.; Resources, C.W.; Data Curation, C.W.; Writing—Original Draft Preparation, C.W., K.Z. and P.W.-B.; Writing—Review & Editing, C.W. and G.P.; Visualization, C.W., K.Z. and P.W.-B.; Supervision, G.P. Project administration, C.W. and G.P. All authors have read and agreed to the published version of the manuscript.

Funding: This research received no external funding.

Conflicts of Interest: The authors declare no conflict of interest.

References

1. Ten threats to global health in 2019. Available online: https://www.who.int/emergencies/ten-threats-to-global-health-in-2019 (accessed on 29 April 2020).
2. Parascandola, M. Ambient air pollution and lung cancer in Poland: Research findings and gaps. *J. Health. Inequal.* **2018**, *4*, 3–8. [CrossRef]
3. European Environment Agency. *Air Quality in Europe—2019 Report*; Publications Office of the European Union: Luxembourg, 2019.
4. Jedrychowski, W.A.; Majewska, R.; Spengler, J.D.; Camann, D.; Roen, E.L.; Perera, F.P. Prenatal exposure to fine particles and polycyclic aromatic hydrocarbons and birth outcomes: A two-pollutant approach. *Int. Arch. Occup. Environ. Health.* **2017**, *90*, 255–264. [CrossRef] [PubMed]
5. Perera, F. Molecular epidemiology, prenatal exposure and prevention of cancer. *Environmen. Health.* **2011**, *10*, 55. [CrossRef] [PubMed]
6. Jirtle, R.L.; Skinner, M.K. Environmental epigenomics and disease susceptibility. *Nat. Rev. Genet.* **2007**, *8*, 253–262. [CrossRef]
7. Karaguliana, F.; Belisd, C.A.; Francisco, C.; Dorab, C.; Prüss-Ustünb, A.M.; Bonjourb, S.; Adair-Rohanib, H.; Amannc, M. Contributions to cities' ambient particulate matter (PM): A systematic review of local source contributions at global level. *Atmos. Environ.* **2015**, *120*, 475–483. [CrossRef]
8. European Environment Agency. Poland—Air pollution country fact sheet 2019. Available online: https://www.eea.europa.eu/themes/air/country-fact-sheets/2019-country-fact-sheets/poland (accessed on 29 April 2020).
9. Wojtyla, C.; Wojtyla-Buciora, P. Polish Pregnancy-related Assessment Monitoring System (Pol-PrAMS): Research on lifestyle health behaviours of Polish women during gestation—Study design. *J. Health. Inequal.* **2016**, *2*, 185–191. [CrossRef]
10. Wojtyła, C.; Biliński, P.; Paprzycki, P.; Warzocha, K. Haematological parameters in postpartum women and their babies in Poland - comparison of urban and rural areas. *Ann. Agric. Environ. Med.* **2011**, *18*, 380–385.
11. European Commission. *Directive 2008/50/EC of the European Parliament and of the Council of 21 May 2008 on Ambient Air Quality and Cleaner Air for Europe OJ L 152, 11.6.2008*; European Commission: Brussels, Belgium, 2008; pp. 1–44.
12. American College of Obstetricians and Gynecologists. Committee on Practice Bulletins —Obstetrics. ACOG Practice Bulletin Number 190: Gestational Diabetes Mellitus. *Obstet. Gynecol.* **2018**, *131*, e49–e64. [CrossRef]

13. American College of Obstetricians and Gynecologists. Committee on Practice Bulletins—Obstetrics. ACOG Practice Bulletin Number 222: Gestational Hypertension and Preeclampsia. *Obstet. Gynecol.* **2020**, *135*, e237–e260. [CrossRef]
14. Yuan, L.; Zhang, Y.; Gao, Y.; Tian, Y. Maternal fine particulate matter (PM2.5) exposure and adverse birth outcomes: An updated systematic review based on cohort studies. *Environ. Sci. Pollut. Res. Int.* **2019**, *26*, 13963–13983. [CrossRef]
15. Rosa, M.J.; Pajak, A.; Just, A.C.; Sheffield, P.E.; Kloog, I.; Schwartz, J.; Coull, B.; Enlow, M.B.; Baccarelli, A.A.; Huddleston, K.; et al. Prenatal exposure to PM 2.5 and birth weight: A pooled analysis from three North American longitudinal pregnancy cohort studies. *Environ. Int.* **2017**, *107*, 173–180. [CrossRef] [PubMed]
16. Pedersen, M.; Giorgis-Allemand, L.; Bernard, C.; Aguilera, I.; Andersen, A.M.N.; Ballester, F.; Beelen, R.M.J.; Chatzi, L.; Cirach, M.; Danileviciute, A.; et al. Ambient air pollution and low birthweight: A European cohort study (ESCAPE). *Lancet Respir. Med.* **2013**, *1*, 695–704. [CrossRef]
17. Kumar, N. The Exposure Uncertainty Analysis: The Association between Birth Weight and Trimester Specific Exposure to Particulate Matter (PM2.5 vs. PM10). *Int. J. Environ. Res. Public Health.* **2016**, *13*, 906. [CrossRef] [PubMed]
18. Liu, Y.; Xu, J.; Chen, D.; Sun, P.; Ma, X. The association between air pollution and preterm birth and low birth weight in Guangdong, China. *BMC Public Health* **2019**, *19*, 3. [CrossRef] [PubMed]
19. Smarr, M.M.; Vadillo-Ortega, F.; Castillo-Castrejon, M.; O'Neill, M.S. The use of ultrasound measurements in environmental epidemiological studies of air pollution and fetal growth. *Curr. Opin. Pediatr.* **2013**, *25*, 240–246. [CrossRef] [PubMed]
20. Maciel-Ruiz, J.A.; López-Rivera, C.; Robles-Morales, R.; Veloz-Martínez, M.G.; López-Arellano, R.; Rodríguez-Patiño, G.; Petrosyan, P.; Govezensky, T.; Salazar, A.M.; Ostrosky-Wegman, P.; et al. Prenatal exposure to particulate matter and ozone: Bulky DNA adducts, plasma isoprostanes, allele risk variants, and neonate susceptibility in the Mexico City Metropolitan Area. *Environ. Mol. Mutagen.* **2019**, *60*, 428–442. [CrossRef]
21. Vrijheid, M.; Martinez, D.; Manzanares, S.; Dadvand, P.; Schembari, A.; Rankin, J.; Nieuwenhuijsen, M. Ambient Air Pollution and Risk of Congenital Anomalies: A Systematic Review and Meta-analysis. *Environ. Health Perspect.* **2010**, *119*, 598–606. [CrossRef]
22. Tanwar, V.; Gorr, M.W.; Velten, M.; Eichenseer, C.M.; Long, V.P.; Bonilla, I.M.; Shettigar, V.; Ziolo, M.T.; Davis, J.P.; Baine, S.H.; et al. In Utero Particulate Matter Exposure Produces Heart Failure, Electrical Remodeling, and Epigenetic Changes at Adulthood. *J. Am. Heart Assoc.* **2017**, *6*, e005796. [CrossRef]
23. Hall, K.C.; Robinson, J.C. The association between maternal exposure to pollutant particulate matter 2.5 and neonatal congenital heart defects. *JBI Database of Syst. Rev. Implement. Rep.* **2017**, *15*, 2257–2264. [CrossRef]
24. Padula, A.M.; Tager, I.B.; Carmichael, S.L.; Hammond, S.K.; Yang, W.; Lurmann, F.; Shaw, G.M. Ambient Air Pollution and Traffic Exposures and Congenital Heart Defects in the San Joaquin Valley of California. *Paediatr. Perinat. Epidemiol.* **2013**, *27*, 329–339. [CrossRef]
25. Schembari, A.; Nieuwenhuijsen, M.J.; Salvador, J.; de Nazelle, A.; Cirach, M.; Dadvand, P.; Beelen, R.; Hoek, G.; Basagaña, X.; Vrijheid, M. Traffic-Related Air Pollution and Congenital Anomalies in Barcelona. *Environ. Health. Perspect.* **2014**, *122*, 317–323. [CrossRef] [PubMed]
26. Tanner, J.P.; Salemi, J.L.; Stuart, A.L.; Yu, H.; Jordan, M.M.; DuClos, C.; Cavicchia, P.; Correia, J.A.; Watkins, S.M.; Kirby, R.S. Associations between exposure to ambient benzene and PM2.5 during pregnancy and the risk of selected birth defects in offspring. *Environ. Res.* **2015**, *142*, 345–353. [CrossRef] [PubMed]
27. Salavati, N.; Strak, M.; Burgerhof, J.G.M.; de Walle, H.E.K.; Erwich, J.J.H.M.; Bakker, M.K. The association of air pollution with congenital anomalies: An exploratory study in the northern Netherlands. *Int. J. Hyg. Environ. Health.* **2018**, *221*, 1061–1067. [CrossRef] [PubMed]
28. Pan, S.C.; Huang, C.C.; Lin, S.J.; Chen, B.Y.; Chang, C.C.; Leon Guo, Y.L. Gestational diabetes mellitus was related to ambient air pollutant nitric oxide during early gestation. *Environ. Res.* **2017**, *158*, 318–323. [CrossRef] [PubMed]
29. Zhang, H.; Dong, H.; Ren, M.; Liang, Q.; Shen, X.; Wang, Q.; Huang, C. Ambient air pollution exposure and gestational diabetes mellitus in Guangzhou, China: A prospective cohort study. *Sci. Total Environ.* **2020**, *699*, 134390. [CrossRef]

30. Fleisch, A.F.; Kloog, I.; Luttmann-Gibson, H.; Gold, D.R.; Oken, E.; Schwartz, J.D. Air pollution exposure and gestational diabetes mellitus among pregnant women in Massachusetts: A cohort study. *Environ. Health* **2016**, *15*, 40. [CrossRef]
31. Samet, J.; Buran, M. The burden of avoidable disease from air pollution: Implications for prevention. *J. Health Inequal.* **2020**, *6*, 2–6. [CrossRef]

© 2020 by the authors. Licensee MDPI, Basel, Switzerland. This article is an open access article distributed under the terms and conditions of the Creative Commons Attribution (CC BY) license (http://creativecommons.org/licenses/by/4.0/).

International Journal of
Environmental Research and Public Health

MDPI

Article

Changes in Knowledge about Umbilical Cord Blood Banking and Genetic Tests among Pregnant Women from Polish Urban and Rural Areas between 2010–2012 and 2017

Maria Szubert [1,2,*], Malwina Ilowiecka [1,2], Jacek Wilczynski [1,2], Monika Szpotanska-Sikorska [3] and Cezary Wojtyla [4,5]

1. Clinic of Surgical and Oncologic Gynecology, 1st Department of Gynecology and Obstetrics, Medical University of Lodz, 94-029 Lodz, Poland; malilo1@o2.pl (M.I.); jacek.wilczynski@umed.lodz.pl (J.W.)
2. M. Pirogow Teaching Hospital, Wilenska 37 St., 94-029 Lodz, Poland
3. 1st Department of Obstetrics and Gynecology, Medical University of Warsaw; 1/3 Starynkiewicza Sq, 02-015 Warsaw, Poland; mszpotanska@wp.pl
4. International Prevention Research Institute–Collaborating Centre, State University of Applied Sciences, Kaszubska 16 St., 62-800, Kalisz, Poland; czwo@op.pl
5. Department of Oncologic Gynecology and Obstetrics, Centre of Postgraduate Medical Education, Czerniakowska 231 St., 00-416 Warsaw, Poland

* Correspondence: maria.szubert@umed.lodz.pl; Tel.: +48-42-680-47-22 or +48-60-440-27-86

Received: 29 June 2020; Accepted: 5 August 2020; Published: 8 August 2020

Abstract: The aim of this study was to evaluate knowledge of umbilical cord blood (UBC) banking and prenatal genetic diagnosis among pregnant women from rural and urban areas, and how this knowledge changed within a five-year period. A survey by questionnaire was conducted between 2010 and 2012, and in 2017 in public hospitals; the study population comprised 6128 women, with 2797 patients from the years 2010–2012 and 3331 from the year 2017. 41% of the studied population declared that they were living in rural areas. In the 2010–2012 period, fewer women from rural areas knew about UBC banking. In 2017 that same relative difference in knowledge persisted, but the percentage of women who now knew about this procedure rose significantly in both studied groups. Prenatal diagnosis was more familiar for urban inhabitants both in 2010 and 2017 but as with the UBC data, a trend of growing awareness was also seen in pregnant women from rural areas. Knowledge of new techniques such as UBC banking and genetic tests has grown among pregnant women during the time frame of our study, but there is still a need to emphasize the benefits of these two possibilities to pregnant women, especially among rural inhabitants.

Keywords: genetic testing; umbilical cord blood banking; pregnancy; women health

1. Introduction

Dynamic technological changes and development of new, innovative diagnostic methods are discernable in every medical branch, especially in genetics and transplantology [1,2]. Contemporary society has also gained broad access to a variety of medical information through the internet. This unfiltered stream of knowledge can cause difficulties in understanding the relevance and reliability of the acquired information, especially in the medical field. Two recent milestones relating to pregnant women that have added to our knowledge are prenatal genetic diagnosis methods using genetic screening, and umbilical cord blood (UBC) banking of stem cells [3,4]. UBC collecting can be undertaken only once—just after delivery. Stem cell transplants are now used to treat numerous types of immune- and blood-related disorders and genetic diseases [5]. There are, however, significant gaps in parents' knowledge and awareness of cord blood banking which have been identified in different studies [6]. Women also have limited knowledge about prenatal screening according to Seven et al. [7]. Demographic factors may also affect women's knowledge about genetic syndromes and prenatal testing [8].

In this study we assessed the knowledge of pregnant women from rural and urban areas regarding UBC banking and genetic testing for medical and scientific use. Our special concern was to identify any changes in knowledge levels during the 5-year timeframe of our study, from the period 2010–2012 to 2017, within rural and urban populations.

2. Materials and Methods

Our study was conducted by the Chief Sanitary Inspectorate and the Institute of Rural Health in Lublin in the years 2010–2012 and in 2017 across all Polish public hospitals, and involved a randomized group of women and their children. The study analyzed responses to a questionnaire developed within the Pol-PrAMS program (Polish Pregnancy-related Assessment Monitoring System) using the PRAMS (Pregnancy Risk Assessment Monitoring System) model developed in the United States. The Bioethics Committee of the Institute of Rural Health in Lublin approved the study (permission no. 03/2011).

The participants in the whole study comprised 12,066 pregnant women, including 8625 subjects from the years 2010–2012 and 3441 from 2017. The questionnaire was divided into two parts. The first part (in paper) was completed by the pregnant women and included their personal information and data on pro-health behaviors and the course of pregnancy. The second part was filled in by healthcare professionals based on the medical records of the patients and their children and was gathered after delivery. The results of other parts of the study have been published elsewhere [9].

Calculations for the aspect of the study reported here were made after elimination of entries which did not include information on a place of residence, and after correlation of the population characteristics for the years 2010–2012 with that of the 2017 population in terms of age, place and region of residence. The demographic population structure in 2010–2012 was significantly different to that studied in 2017 in terms of age and region of residence. A compromise method of correction was adopted by randomly withdrawing those cases that were overrepresented. The population from the 2010–2012 study was referenced to the population structure from 2017 in terms of age, place of residence (city/village), and region of residence (16 provinces). This data constituted a trivariate table, covering 3331 women subjects from 2017 divided into $3 \times 3 \times 16 = 144$ subgroups. The number of subjects in each subgroup was expressed as a percentage of the total. An analogous table was prepared for the population studied in 2010–2012. Based on the coefficients from the 2017 table, the expected counts in each subgroup for 2010–2012 were calculated, from which the overcounts and undercounts could be derived. The population size was reduced so that there were no undercounts present in the table, only overcounts. This was achieved by reducing the total size of the interviewed population from the 2010–2012 study to 2797 cases. Current and newly expected counts were provided for each subgroup. The required number of cases was then randomly selected from the SPSS database program, containing the data for each subgroup of the 2010–2012 study. This procedure randomly selects an m-element sample of cases from an n-element population. There were 144 such procedures separately performed for the subgroups. The adjusted study population of 2797 cases did not differ significantly from the 2017 study population in terms of the unified variable structure. According to this description we analyzed data from 6128 women, comprising 2797 patients from the years 2010–2012 and 3331 from the year 2017. A detailed description of the groups has already been published elsewhere [10]. Flow chart 1 illustrates the whole process of establishing sample size (flow chart near here).

Flow chart 1: Process of inclusion of participants in the study. Number of births in Poland is given for the studied timeframe (for the exact day in each year when procedures of the Pol-PrAMS study were conducted).

Obtained data were compared between years of the survey in women residing in urban and rural areas. The statistical analysis of the collected data was made based on a chi-squared test and Mann–Whitney test using IBM SPSS software version 25 (IBM Corp., Armonk, NY, USA).

3. Results

3.1. Sociodemographic Characteristics

The largest proportion of the study group surveyed during both periods (2010–2012 and 2017) comprised women aged between 26 and 30 years, and the smallest proportion consisted of women aged over 35 years. Among the respondents, city residents made up 59% of the examined group and women living in rural areas accounted for 41% of the total group (Table 1).

Table 1. Characteristics of the study population.

	2010–2012		2017		p
	n	%	n	%	
Age (years)					ns
≤25	532	19.0	627	19.0	
26–30	965	34.6	1139	34.6	
31–35	925	33.1	1073	32.6	
>35	371	13.3	457	13.8	
Place of residence (inhabitants)					ns
City ≥ 100,000	676	24.2	804	24.2	
City < 100,000	967	34.6	1154	34.6	
Rural area	1154	41.2	1373	41.2	
Education					ns
Primary/lower secondary	52	4.8	86	4.5	
Upper secondary/ post-secondary	403	37.2	665	34.6	
Higher	596	54.9	1139	59.2	
Other	34	3.1	34	1.7	

ns—non-significant, $p > 0.05$.

3.2. Provision of Specialist Care to the Pregnant Population in Urban and Rural Areas

Comparing the provision of specialist care to pregnant women in rural and urban areas in the studied time periods of 2010–2012 and 2017, we found that women living in rural areas had gynecological examinations in the 1st, 2nd and 3rd trimester of pregnancy relatively less often than women from urban areas ($p < 0.05$). Meanwhile, the data from 2017 showed that the general number of medical visits during the 1st, 2nd and 3rd trimester of pregnancy increased for women living in rural and those resident in urban areas when compared with data from the years 2010–2012 (Table 2).

Table 2. Frequency of gynecological checkups during pregnancy in urban and rural populations in the years 2010–2012 and 2017.

			YEARS 2010–2012				YEAR 2017			
Visit to the Gynecologist			Valid n	Mean	SD	Median	Valid n	Mean	SD	Median
Visits in the 1st trimester of pregnancy	Place of residence	Rural	722	2.89	1.59	3.28	1187	3.28	1.29	3.00
		Urban	1022	3.20	1.57	3.49	1737	3.49	1.46	3.00
		Total	1744	3.07	1.53	3.40 *	2924	3.40	1.40	3.00 *
Visits in the 2nd trimester of pregnancy	Place of residence	Rural	715	3.43	1.48	3.71	1184	3.71	1.34	3.00
		Urban	1019	3.60	1.88	3.87	1728	3.87	1.49	3.00
		Total	1734	3.53	1.63	3.80 *	2912	3.80	1.43	3.00 *
Visits in the 3rd trimester of pregnancy	Place of residence	Rural	715	4.16	2.09	4.54	1176	4.54	1.96	4.00
		Urban	1018	4.40	2.22	4.79	1708	4.79	2.05	4.00
		Total	1733	4.30	2.12	4.69 *	2884	4.69	2.02	4.00 *

*—$p < 0.05$ (Mann–Whitney test).

3.3. Women's Knowledge of Umbilical Cord Blood Banking

Across the whole study period, levels of knowledge about UBC banking among pregnant women resident in urban areas were higher than among those living in rural areas. In 2017, compared with those surveyed in the earlier period, fewer women said that they had not heard of UBC banking at all (1.2% of urban residents and 6.5% ($p < 0.05$) of rural residents respectively). In the answers of urban residents from 2010–2012, 57.2% of the respondents thought that this procedure was useful (vs. 49.4%, $p < 0.05$ of the women from rural areas), whereas 3.1% considered it to be useless (vs. 1.6%, $p < 0.05$ of the women from rural areas) (Figures 1 and 2). In 2017, despite increased awareness about UBC banking, fewer women, from both populations, considered it useful.

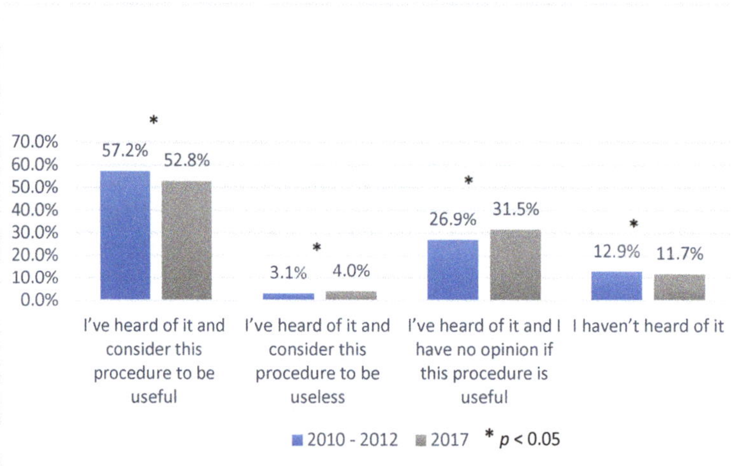

Figure 1. Knowledge and opinions of women from urban areas about umbilical cord blood banking in the years 2010–2012 and 2017; *—$p < 0.05$.

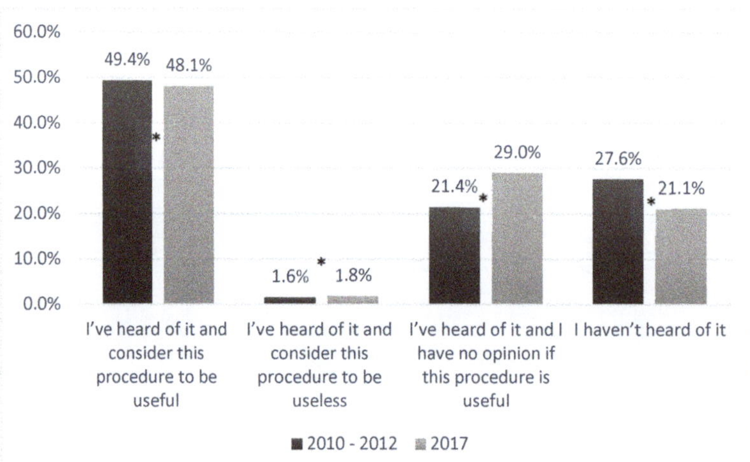

Figure 2. Knowledge and opinions of women from rural areas about umbilical cord blood banking in the years 2010–2012 and 2017; *—$p < 0.05$.

In 2017, there were 4.4% more urban women and 7.6% more rural women who answered that they had no opinion about this procedure ($p < 0.05$).

3.4. Women's Knowledge about Genetic Testing

In 2017, in both urban and rural populations, the number of women who had never heard of using genetic material for medical or scientific purposes decreased when compared with the answers given in the years 2010–2012 (Figures 3–6). In 2017, among women living in urban areas, there was an 8.3% decrease in patients unconscious of the possibility of genetic testing, whereas among women living in rural areas there was an 8% decrease ($p < 0.05$). On the other hand, the number of women who regarded these tests as necessary increased in both populations in 2017. Among women from urban areas this increase was 4.3%, whereas among women from rural areas it was 6.2% ($p < 0.05$). Moreover, the number of women who answered "I have no opinion on this issue" increased in that same year by 4.3% and 2% in the urban and rural areas respectively ($p < 0.05$).

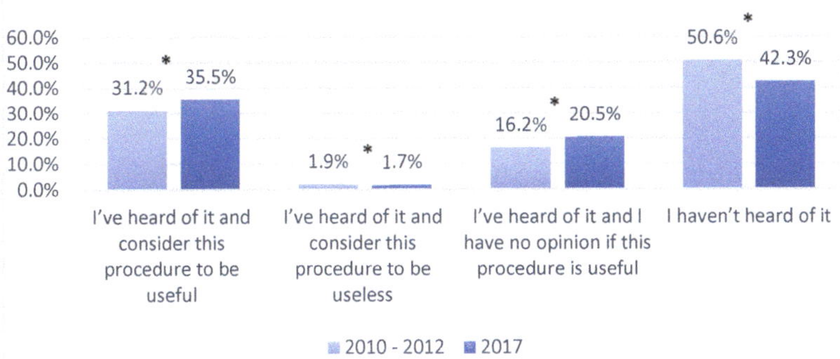

Figure 3. Knowledge and opinions of women from urban populations on using genetic material for medical or scientific purposes in the years 2010–2012 and 2017; *—$p < 0.05$.

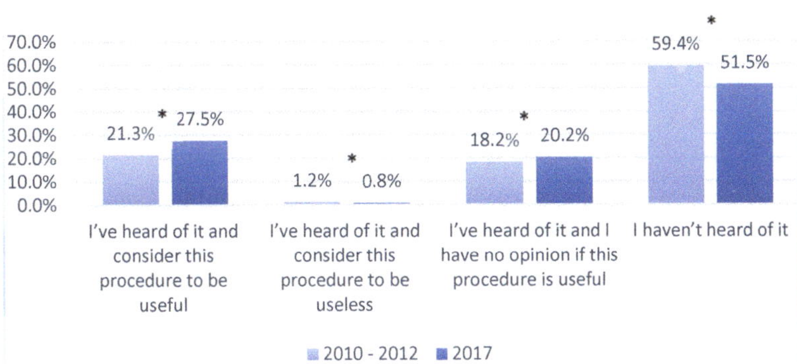

Figure 4. Knowledge and opinions of women from rural populations on using genetic material for medical or scientific purposes in the years 2010–2012 and 2017; *—$p < 0.05$.

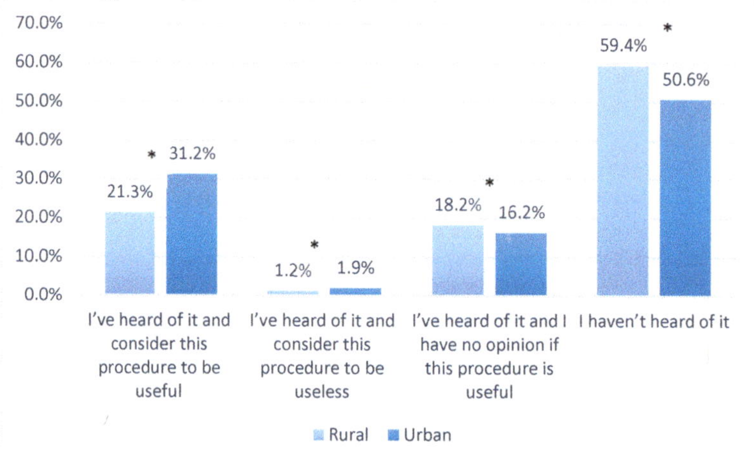

Figure 5. Knowledge and opinions of women from rural and urban populations in the years 2010–2012 on using genetic material for medical or scientific purposes; *—$p < 0.05$.

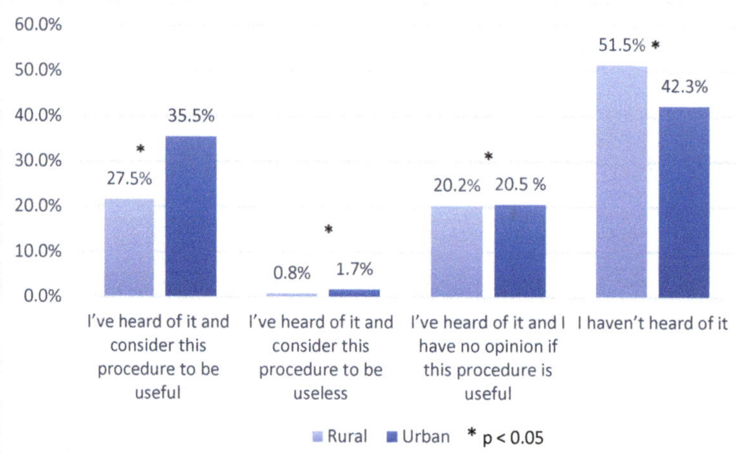

Figure 6. Knowledge and opinions of women from rural and urban populations in 2017 on using genetic material for medical or scientific purposes; *—$p < 0.05$.

A similar trend to that observed in the survey on umbilical cord blood was recorded in the survey on genetic tests. In both survey periods, 2010–2012 and 2017, more women from urban areas had heard of genetic testing compared with those from rural areas. Nevertheless, the data included in Table 3 show that the percentage share of the patients who had never heard of genetic testing is more than 50% of the answers analyzed.

Table 3. Knowledge and opinions of all examined women on using genetic material for medical or scientific purposes.

Knowledge of Using Genetic Material from Blood		Year of Study			
		(1) 2010–2012		(2) 2017	
		Number	% of Valid n in Column	Number	% of Valid n in Column
Have you heard of using genetic material for medical or scientific purposes?	1. I have heard of it and consider this procedure to be useful	500	27.1%	1041	32.2%
	2. I have heard of it and consider this procedure to be useless	30	1.6%	44	1.4%
	3. I have heard of it and I have no opinion whether this procedure is useful	315	17%	659	20.4%
	4. I have not heard of it	1003	54.3%	1488	46%
	In total	1848	100%	3232	100%

4. Discussion

Our assessment of pregnant women's knowledge and attitudes to cord blood storing is the first in which place of residence and changes in opinions over a 5-year period have been analyzed. It is also the first study conducted in the Polish population concerning both innovative techniques (UBC and genetic testing) that are important for pregnant women and wider society. Genetic testing during pregnancy with the use of cell-free fetal DNA is recommended in clinical obstetric guidelines worldwide [11–14]. Cell-free fetal DNA represents extracellular DNA which originates from trophoblastic cells that contain the entire fetal genotype. Assessment of cell-free fetal DNA from mother's blood is useful in the calculation of the risk of congenital genetic syndromes [15]. Women surveyed in our study were asked to give answers about their knowledge of the use of genetic testing in general, and not only about cell-free DNA. Surprisingly more than half of the patients surveyed had not heard about genetic testing at all in 2010–2012. By 2017 that situation had changed, but the knowledge of rural inhabitants was still poorer than that of urban residents. As genetic testing is nowadays the most important part of perinatal counselling, knowledge about this possibility is of significant importance for public health. Women's rights could be affected if they are not properly tested and counseled for genetic disorders in early pregnancy.

The first recipient of umbilical cord blood stem cells was a 5-year old boy with severe Fanconi anemia in 1988 [16]. In Poland, a successful frozen and banked umbilical stem cell transplantation in a boy with acute myeloblastic leukemia was performed eight years later [17]. Initially, UBC transplantation was limited to children, and both related and unrelated cord blood transplants have been performed with high rates of success for a variety of hematologic disorders and metabolic storage diseases [16].

Nowadays, UBC is used for adults too, for instance for myelodysplastic syndrome or secondary acute myeloblastic leukemia [18,19]. The use of stem cells from umbilical cord blood has also recently been highlighted in relation to the COVID-19 pandemic. Several case reports about the effective treatment of severe COVID-19 cases have been published, where mesenchymal stem cells (MSCs) have been shown to halt and reverse the cytokine storm [20,21].

Studies exploring pregnant women's and expectant parents' knowledge and awareness of cord blood donation and banking have already been conducted in 15 countries according to a systematic review by Peberdy et al. [6]. Our study is the first in Poland and was conducted as a part of a larger study on the lifestyle health behaviors of pregnant women.

According to our data, between 2010–2012 and 2017 the total number of pregnancy monitoring visits during pregnancy significantly increased in both studied populations. This should suggest that the knowledge of pregnant women about new possibilities such as UBC banking and genetic testing ought to have increased too. Polish guidelines for pregnancy management state that gynecologists should inform patients about genetic testing as well as about the possibility of UBC banking or donation [22,23]. While more patients said they had "heard" about UBC banking by 2017 than in the 2010–2012 survey period, regardless of the place of residence, a smaller percentage of patients were convinced about necessity of this procedure. The explanation for this fact could be found in the study by Hatzistilli et al. of health professionals' knowledge about umbilical cord blood donation. They stated that only 15.6% of the health care providers they surveyed (doctors, nurses, and midwives) declared they were quite well- or well-informed about the collection methods and the uses of stored blood [24]. One can conclude that the less knowledge the health professional has, the poorer the consultation will be, and the rate of UBC banking will remain low. These conclusions are also suggested by others [25,26]. In most countries there are two possibilities for cord blood banking: by donation in public banks or by user-pays storage for private use [27,28]. Lack of exact information about costs, and misunderstanding about who can benefit from UBC banking can, in our opinion, influence the percentage of patients convinced about the value of this procedure; but this aspect was not analyzed in our study.

5. Conclusions

We conclude therefore that it is very important that obstetricians keep up to date with UBC collection and its storage guidelines and financing through educational programs, so that they can share their knowledge with pregnant women, especially in rural areas. The importance of UBC collection and genetic testing should be highlighted during check-up visits from the beginning of pregnancy. Proper genetic testing may have an impact on the rate of severe congenital malformations at birth. Education of nurses, midwives, and doctors could help improve the percentages of pregnant women who understand the value of these two modern techniques.

Author Contributions: Conceptualization, M.S., M.I. and J.W.; methodology, C.W. and M.S.; formal analysis, J.W. and M.S.-S.; investigation, M.S., M.I., J.W. and C.W.; resources, C.W.; data curation, C.W.; writing—original draft preparation, M.S. and M.I.; writing—review and editing, C.W., M.S.-S. and J.W.; visualization, M.S., M.I. and M.S.-S.; supervision, J.W.; project administration, M.S. and C.W. All authors have read and agreed to the published version of the manuscript.

Funding: This research received no external funding.

Conflicts of Interest: The authors declare no conflict of interest.

References

1. Sterkowicz, S. HISTORIA MEDYCYNY. Historia pierwszych transplantacji serca w polsce. *Kardiochirurgia Torakochirurgia Pol. Pol. J. Thorac. Cardiovasc. Surg.* **2009**, *6*, 313–316.
2. Szpotanska-Sikorska, M.; Mazanowska, N.; Madej, A.; Kociszewska-Najman, B.; Wielgos, M.; Pietrzak, B. Reproductive life planning in women after kidney or liver transplantation. *Clin. Transplant.* **2018**, *32*. [CrossRef] [PubMed]
3. Navarrete, C.; Contreras, M. Cord blood banking: A historical perspective. *Br. J. Haematol.* **2009**, *147*, 236–245. [CrossRef] [PubMed]
4. Gluckman, E. Milestones in umbilical cord blood transplantation. *Blood Rev.* **2011**, *25*, 255–259. [CrossRef]
5. Matsumoto, M.M.; Dajani, R.; Matthews, K.R. Cord blood banking in the arab world: Current status and future developments. *Biol. Blood Marrow Transplant.* **2015**, *21*, 1188–1194. [CrossRef]
6. Peberdy, L.; Young, J.; Massey, D.L.; Kearney, L. Parents' knowledge, awareness and attitudes of cord blood donation and banking options: An integrative review. *BMC Pregnancy Childbirth* **2018**, *18*. [CrossRef]
7. Seven, M.; Akyüz, A.; Eroglu, K.; Daack-Hirsch, S.; Skirton, H. Womens knowledge and use of prenatal screening tests. *J. Clin. Nurs.* **2017**, *26*, 1869–1877. [CrossRef]

8. Suriadi, C.; Jovanovska, M.; Quinlivan, J.A. Factors affecting mothers knowledge of genetic screening. *Aust. New Zealand J. Obstet. Gynaecol.* **2004**, *44*, 30–34. [CrossRef]
9. Wojtyła, C.; Wojtyła-Buciora, P. Polish pregnancy-related assessment monitoring system (Pol-PrAMS): Research on lifestyle health behaviours of polish women during gestation—Study design. *J. Health Inequalities* **2016**, *2*, 185–191. [CrossRef]
10. Szubert, M.; Ilowiecka, M.; Wilczynski, J.; Bilinski, P.; Wojtyla, C. Health-related behaviors of pregnant women residing in urban and rural areas in poland. *Int. J. Envion. Res. Public Health* **2020**, *17*, 4395. [CrossRef]
11. Goldwaser, T.; Klugman, S. Cell-free DNA for the detection of fetal aneuploidy. *Fertil. Steril.* **2018**, *109*, 195–200. [CrossRef] [PubMed]
12. Veyver, I.V.D. Faculty opinions recommendation of committee opinion No. 640: Cell-free DNA screening for fetal aneuploidy. *Fac. Opin. Post Publ. Peer Rev. Biomed. Lit.* **2016**. [CrossRef]
13. Sieroszewski, P.; Wielgos, M.; Radowicki, S.; Sasiadek, M.; Borowiec, M.; Borowski, D.; Jakubowski, L.; Kaczmarek, P.; Latos-Bielenska, A.; Laudanski, P.; et al. Cell-free fetal DNA testing in prenatal diagnosis: Recommendations of the polish gynecological society and the polish human genetics society. *Eur. J. Obstet. Gynecol. Reprod. Biol.* **2017**, *214*, 190–191. [CrossRef] [PubMed]
14. ACOG. Committee Opinion No. 545: Noninvasive prenatal testing for fetal aneuploidy. *Obstet. Gynecol.* **2012**, *120*, 1532–1534. [CrossRef] [PubMed]
15. Moczulska, H.; Borowiec, M.; Sieroszewski, P. Zastosowanie testu oceny wolnego DNA płodu w diagnostyce prenatalnej na podstawie rekomendacji Polskiego Towarzystwa Ginekologicznego i Polskiego Towarzystwa Genetyki Człowieka—Uwagi praktyczne. *Ginekol. I Perinatol. Prakt.* **2016**, *1*, 10–12.
16. Ballen, K.K.; Gluckman, E.; Broxmeyer, H.E. Umbilical cord blood transplantation: The first 25 years and beyond. *Blood* **2013**, *122*, 491–498. [CrossRef]
17. Borys-Wójcik, S.; Józkowiak, M.; Stefańska, K.; Knap, S.; Pieńkowski, W.; Gutaj, P.; Bruska, M.; Kempisty, B. Historical background of umbilical stem cell culture. *Med J. Cell Biol.* **2019**, *7*, 11–14. [CrossRef]
18. Brunstein, C.G.; Baker, K.S.; Wagner, J.E. Umbilical cord blood transplantation for myeloid malignancies. *Curr. Opin. Hematol.* **2007**, *14*, 162–169. [CrossRef]
19. Ballen, K.K.; Barker, J.N. Has umbilical cord blood transplantation for AML become mainstream? *Curr. Opin. Hematol.* **2013**, *20*, 144–149. [CrossRef]
20. Leng, Z.; Zhu, R.; Hou, W.; Feng, Y.; Yang, Y.; Han, Q.; Shan, G.; Meng, F.; Du, D.; Wang, S.; et al. Transplantation of ACE2-mesenchymal stem cells improves the outcome of patients with COVID-19 pneumonia. *Aging Dis.* **2020**, *11*, 216. [CrossRef]
21. Atluri, S.; Manchikanti, L.; Hirsch, J.A. Expanded umbilical cord mesenchymal stem cells (UC-MSCs) as a therapeutic strategy in managing critically Ill COVID-19 patients: The case for compassionate use. *Pain Physician* **2020**, *23*, 71–83.
22. Kotarski, J.; Wielgoś, M.; Brązert, J.; Ryszard, C.; Artur, C.; Krzysztof, D.; Tomasz, N.; Ryszard, P.; Krzysztof, P.; Stanisław, R.; et al. Rekomendacje polskiego towarzystwa ginekologicznego dotyczące postępowania w zakresie diagnostyki prenatalnej. *Ginekol. Pol.* **2009**, *80*, 390–393.
23. Rozporządzenie Ministra Zdrowia z dnia 4 października 2012 w sprawie standardów postępowania medycznego przy udzielaniu świadczeń zdrowotnych z zakresu opieki okołoporodowej sprawowanej nad kobietą w okresie fizjologicznej ciąży, fizjologicznego porodu, połogu oraz opieki nad noworodkiem (Dz. U. Nr 112, poz. 654, z późn. zm.). Available online: https://isap.sejm.gov.pl/isap.nsf/download.xsp/WDU20180001756/O/D20181756.pdf (accessed on 25 April 2020).
24. Hatzistilli, H.; Zissimopoulou, O.; Galanis, P.; Siskou, O.; Prezerakos, P.; Zissimopoulos, A.; Kaitelidou, D. Health professionals' knowledge and attitude towards the umbilical cord blood donation in Greece. *Hippokratia* **2014**, *18*, 110–115. [PubMed]
25. Gupta, V.; Agarwal, L.; Ballal, P.; Pandey, D. Cord blood banking: Antenatal care provider's roles and responsibilities. *Stem Cells Int.* **2019**, *2019*, 1–7. [CrossRef] [PubMed]
26. Roh, E.Y.; Shin, S.; Kim, B.J.; Kim, M.; Hong, J.J.; Yoon, J.H. Roles of obstetricians in quality management of cord blood collection and informing potential donors: A survey on obstetricians' understanding of cord blood in Korea. *Transfusion* **2014**, *54*, 3164–3172. [CrossRef] [PubMed]

27. Screnci, M.; Murgi, E.; Pirrè, G.; Valente, E.; Gesuiti, P.; Corona, F.; Girelli, G. Donating umbilical cord blood to a public bank or storing it in a private bank: Knowledge and preference of blood donors and of pregnant women. *Blood Transfus.* **2012**, *10*, 331–337. [CrossRef]
28. Manegold, G.; Meyer-Monard, S.; Tichelli, A.; Granado, C.; Hösli, I.; Troeger, C. Controversies in hybrid banking: Attitudes of swiss public umbilical cord blood donors toward private and public banking. *Arch. Gynecol. Obstet.* **2010**, *284*, 99–104. [CrossRef]

© 2020 by the authors. Licensee MDPI, Basel, Switzerland. This article is an open access article distributed under the terms and conditions of the Creative Commons Attribution (CC BY) license (http://creativecommons.org/licenses/by/4.0/).

International Journal of
Environmental Research and Public Health

Article

How to Screen Suitable Service Improve Community Health Care Services by University Students in Taiwan

Guey-Shin Shyu [1], Shinn-Jou Lin [2], Wei-Ta Fang [3,*] and Bai-You Cheng [2,*]

[1] Department of Tourism, Tungnan University, Shenkeng District, New Taipei City 22202, Taiwan; gsshyu@mail.tnu.edu.tw

[2] Graduate Institute of Environmental Resources Management, TransWorld University, Douliu City, Yunlin County 64063, Taiwan; shinnjou@gmail.com

[3] Graduate Institute of Environmental Education, National Taiwan Normal University, Taipei City 11677, Taiwan

* Correspondence: wtfang@ntnu.edu.tw (W.-T.F.); biyocheng@twu.edu.tw (B.-Y.C.); Tel.: +886-2-7734-6558 (W.-T.F.); +886-5-5370988 (ext. 8233) (B.-Y.C.)

Received: 24 June 2020; Accepted: 23 July 2020; Published: 27 July 2020

Abstract: Engaging in social contributions to enhance social participation and attending community experiential service learning or internship courses have become an essential learning experience for university students. On the basis of postmodern education theories, this study adopted images and oral accounts involving personal experiences to construct a postmodern education research scheme by using the method of collaborative ethnography. This study selected and performed the following services: filming a community documentary, administering community health dance classes, and archiving community cultural artifacts in databases. Interviews were also administered to facilitate implementation of the actual services. Community health services commonly seen in Taiwan and abroad were compiled, and the resources required for each service were examined. Subsequently, factor analysis was performed to explore the characteristic of these services in order to recommend feasible services for university students to undertake. The results indicated that the eight resources required for the 59 common community health services were (1) a designated space or venue, (2) materials, (3) monetary resources, (4) human resources, (5) expertise, (6) professional equipment, (7) patience, and (8) empathy. The results revealed three principal components, namely labor services, high-resource services, and professional services, for a total explanatory power of 67.99%; the individual explanatory power of these components accounted for 25.04%, 21.81%, and 21.15%, respectively. Next, community health care services suitable for university students to perform were selected and implemented, and these services were well received. The study results indicated that community and environmental justice can be realized by identifying with the value of community health services and promoting postmodern education theories and social norms. The research results are suitable for implementation after the COVID-19 pandemic.

Keywords: community health services; postmodern education theories; factor analysis

1. Introduction

The Taiwanese government proposed the concept of comprehensive community development in 1993, aiming to establish community culture, cohere community consensus, and construct the concept of community life to formulate new thinking and policies for cultural administration [1]. This policy-based term originated from the goal of integrating the five major aspects of community development, namely people, culture, land, landscape, and production. In fact, this concept pertains to the movement of transforming community culture [2–5]. In other words, community residents voluntarily participate in related activities, exercise their creativity, and manage community affairs comprehensively to build a cultural scene of their own community. Many sociologists have emphasized that community affairs should be managed by local residents, who then combine community resources and reach consensus through public discussion to determine community development collaboratively. In short, handling community affairs involves bottom-up process, public participation, and local autonomy [6–10].

Regarding the implications of community, we observed value transformation in community cases that arises from changes in a modernized lifestyle [11,12]. Community development involves sociology, psychology, and cultural anthropology; hence, community-related topics are complex issues that cannot be solved simply through scientific management [13]. In other words, adequate community management does not accentuate verifying quantitative hypotheses; instead, it regards research participants as a being of mutual subjectivity, replacing mechanics-based Cartesian perspectives of instrumental rationality with an organic community worldview. In the past, modernism was represented by Cartesian instrumental rationality and Newtonian mechanics and involved a reductionistic process to understand management tools [14–16]. Contemporary society has entered a period of pluralistic postmodernism when modernism is overloaded with data such that rational thinking cannot guarantee effective management strategies [4]. In the postmodern era, people generally value the opinions of the disadvantaged in communities, respect different cultures and ethnic groups, and recognize the value and status of dissenting opinions [17]. Postmodernism mostly adopts a pluralistic argument that enables humans to learn to respect the differences between each other [15]. Postmodernists value interpersonal differences, oppose the objectivity and universality of knowledge emphasized in modernism, adopt small narratives for thinking, and view things from a pluralistic perspective [16]. Humans do not follow the Western scientific trajectory of development in building their societies; instead, they start to identify with their own ethnic groups and cultural features, as well as fight for their own rights. Peters and Waterman [18] argued that postmodern management should adopt a Back to Basics approach to conducting reformation [18]. The art of management should permeate every detail of society, using symbols such as consciousness, emblem, drama, vision, and love to thoroughly subvert the strict system, control, and structure of modernist management [19–22].

Community development is essentially a spontaneous activity of community residents. However, most public topics must be handled by public authorities or financially supported by the public sector. To support the particular requirements of community development, governments organize specific incentives and grants to enhance training programs and foster public and community awareness. Because community development differs from other policies in implementation methods, the Council for Cultural Affairs (now the Ministry of Culture) under the Executive Yuan proposed the concept of comprehensive community development and launched multiple grant programs. Accordingly, Taiwan's community cooperation policies gradually extended from the Council for Cultural Affairs to the other central ministries [23].

In 2005, the Taiwanese government proposed the Six-Star Plan for a Healthy Community to increase the range of community development and government participation, implement the bottom-up approach, and improve public participation through comprehensively promoting the plan, training community development talents, and building community databases [1]. The six major aspects of the plan and their implications are as follows: (a) humanistic education: develop community awareness, strengthen the operation of community organizations, and implement community lifelong learning; (b) industrial development: promote industrial transformation and upgrading and facilitate local job opportunities; (c) social welfare and health care: develop community health care services and community health; (d) community policing: build a community security system, implement a community disaster prevention system, and develop a domestic violence prevention system; (e) environmental landscape: develop community styles and facilities as well as repurpose community space; and (f) environmental and ecological protection: promote community cleaning and improve ecological conservation. These are the principles for implementing this policy [1].

After 20 years, ideals regarding the participation and autonomous management of communities are no longer limited to the comprehensive community development policies [23]. Programs implemented by departments related to environmental protection, internal affairs, agriculture, culture, policing, economy, commerce, health, and social affairs also focus on community residents to facilitate bottom-up operation, promote spontaneity, as well as transform communities and create local atmospheres according to residents' actual needs [24–28].

The aforementioned literature review indicates that postmodern education management is characterized by anti-rationalism, anti-collectivism, cultural pluralism, and the use of digital images and Internet information for management [29–32]. In discourses related to the use of digital images and internet information, community management involves records compiled through photo-voice and photo elicitation. Regardless of their level of visual and auditory aesthetics and whether the artistry of the recorded experience is accentuated, these records provide information of static material culture and dynamic spoken arts and performing arts in community life that can be examined through content analysis. Therefore, we applied the visual and auditory communication contexts of phenomenology and incorporated other phenomenological elements including on-site reading, observation, listening, conversation, reflection, and introspection [32]. How to return to the ontology of education and body, extract knowledge from the direct intuition and a priori nature of phenomenology, as well as develop a sense of love, respect, and appreciation in and incorporate self-talk into community activity observation are crucial topics for postmodern education and community studies [29–32].

In recent years, the reduced costs of video and image recording have enabled it to become a prevalent method to facilitate community development and retain community records. Local government has provided filming-related professional training courses to implement a bottom-up autonomous management scheme in communities, thereby highlighting the self-worth of the communities. Therefore, training courses for creating community documentaries have been provided. This enables community groups to develop autonomous media and express their civil power [33,34].

This study proposed an analysis process that involved recording dialogues and images and performing statistics commonly used in social science to examine the target plans and provide information for decision making. The purposes of this study had three topics: (1) summarizing the main types of community health services; (2) recommending that university students can operate these services; and (3) adding actual implementation and recording the final results from university students' social works.

2. Materials and Methods

2.1. Research Target: Lunfeng Community at Douliu City in Central Taiwan

The target community is located in southern Douliu city in center Taiwan. Several traditional Chinese three-sided courtyard houses built in the 1900s and a sugar factory area developed during Japanese colonization still remain in the northern part of the community. Because the community boasts ample sunshine and plenty of water resources, Japanese colonizers built a sugar factory here, and the sugar refining industry once thrived in the area. After the industry declined, the area became famous for producing fruits such as pomelo and orange. The target community has approximately 2912 residents and 1107 households (Table 1). Its main agricultural products include honey tangerine, orange, pomelo, and Taiwan Giant Bamboo (*Dendrocalamus latiflorus*) shoot.

Figure 1 presents the organization overview of the target community, revealing that most of the counseling organizations established under government support have ceased to operate. Currently, the Community Health Care Base and the Community Environmental Volunteer Team are the major and most active organizations that remain operational. In addition, the Longevity Club regularly organizes trips and meal gatherings for older residents in the community. Although these activities are held occasionally, the organization has a steady operation procedure.

Since there had occurred a community-based health care operated by government departments, there could be detected social development associations to run this business. For some reasons, community activity centers often have only detected resident gatherings. Our research team, therefore, usually has been hired four university students who have been involved in the community for a long stay. The internship period of the students has accounted for about one year, and often every now and then another student group could replace their duties after one year. When they were running some activities on holidays, other interns could be involved in supporting up to more than fifteen volunteer students.

Table 1. Basic information of population in the study area.

	Male	Age	Female
Age Distribution	99 (7.0%)	0–9	113 (7.6%)
	130 (9.1%)	10–19	134 (9.0%)
	195 (13.7%)	20–29	196 (13.2%)
	213 (15.0%)	30–39	210 (14.1%)
	219 (15.4%)	40–49	212 (14.2%)
	222 (15.6%)	50–59	211 (14.1%)
	189 (13.3%)	60–69	203 (13.6%)
	154 (10.8%)	>70	214 (14.3%)
Male	1419	-	-
Female	1493	-	-
Totals	2912	-	-
Number of Households	1107	-	-

(Information updated: June 2020).

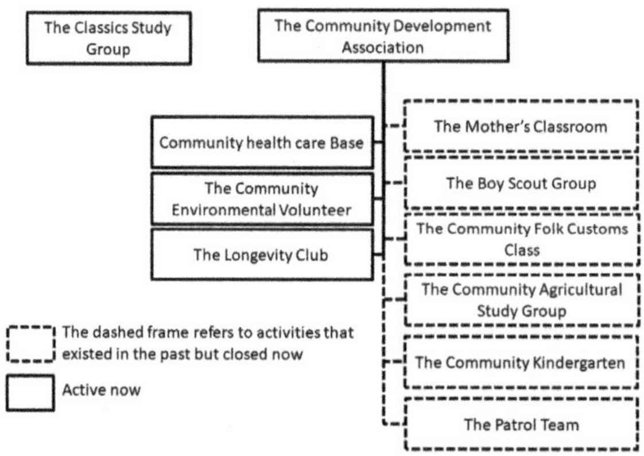

Figure 1. Organization overview of the target community.

The Community Health Care Base was formally established in 2015 and holds activities including health-promoting and care visits, phone greetings, regular Monday gatherings (providing blood pressure measurement services), karaoke, exercises (for promoting blood circulation), and health-promoting courses for improving older-adult services. These activities facilitate the eliciting of a sense of cohesion between the residents. The organizations of the community are listed as follows:

(1) The Community Environmental Volunteer Team cooperates with the Irrigation Association to maintain community environment and reconstruct landscapes. This team is currently one of the most active organizations in the community.

(2) The Longevity Club is an organization for older community residents to interact with each other. It organizes one trip and the Double Ninth Festival annually.
(3) The Classics Study Group was established privately with a focus on Yiguandao scriptures.
(4) The Mother's Classroom used to offer art courses but is currently closed due to teacher shortage.
(5) The Boy Scout Group was established according to the education act promulgated by the provincial government of Taiwan and is disbanded now.
(6) The Community Folk Customs Class shut down due to teacher shortage.
(7) The Community Agricultural Study Group was established jointly with the Council of Agriculture and the 4-H Club but has terminated now.
(8) The Community Kindergarten has terminated.
(9) The Patrol Team was established in 2006 and disbanded after 2008.

2.2. Multivariate Factor Analysis

The purpose of factor analysis is to define latent constructs. Because latent factors involve concepts such as natural backgrounds, social justice, and values that cannot be measured directly, factor analysis can be performed to explore the structural components of such concepts, define the various constructs related to such concepts, and determine the variables associated with each construct [35,36].

Factor analysis does not differentiate dependent variables from independent variables; instead, it analyzes the relationships between all variables and utilizes their correlations to maximize the variance of all the variables. Therefore, after the structural components of the data are determined, factor analysis is usually used to summarize and reduce data. Relational concepts can be derived after the variables are examined through factor analysis. These concepts summarize all the variables without losing excessive information and become the so-called constructs after being named appropriately [34–36].

Factor analysis enables data reduction to select representative variables that retain most of the explanatory power of the original variables, in addition to maintaining the original data structure [35]. Moreover, because the extracted principal factors do not exist in the original data but serve as new variable synthesized through examining the data structure, they can be used to represent latent factors that cannot be categorized directly but have significant influences. This enables complementing analysis items that are left out or unclassifiable. This procedure is known as exploratory factor analysis [36].

The mathematical model of factor analysis is the Equation (1). There n denotes the number of variable ($X_1, X_2, \ldots X_n$), and m denotes the number of underlying factors ($F_1, F_2, \ldots F_m$).

Xi is the variable represented in latent factors.

This model assumes that there are m underlying factors whereby each observed variable is a linear function of these factors together with a residual variate.

This model intends to reproduce the maximum correlations.

$$X_i = a_{i1}F_1 + a_{i2}F_2 + \cdots a_{im}F_m e_i \quad (1)$$

where $i = 1, 2, \ldots, n$.

The factor loadings are a_{i1}, a_{i2}, a_{im}, which denotes that a_{i1} is the factor loading of ith variable on the 1st factor. The specific or unique factor is denoted by e_i.

The factor loadings were how much the variable has contributed to the factor. The basic statistic used in factor analysis is the correlation coefficient, which determines the relationship between two variables and uses matrix algebra calculated by computing. For all pairs X_i, we want to find factors such that when they are extracted, there is an absence of partial correlation between the tests, that is, the partial correlations are zero [37,38].

In matrix notation, factor analysis can be described by the equation $R = PCP' + U^2$, where R is the matrix of correlation coefficients among observed variables, P is the primary factor pattern or loading matrix (P' is the transpose), C is the matrix of correlations among common factors, and U^2

is the diagonal matrix or unique variances [37–39]. All statistical work was done using IBM® SPSS® Statistics for Windows, Version 23.0. Armonk, NY: IBM Corp [40].

3. Results

3.1. Community Health Care Services and Invested Resources

This study compiled a total of 59 care services prevalent in Taiwan and foreign countries. Resources invested in these care services were categorized as follows: (1) a designated space or venue for regular gatherings (hereafter referred to as designated space); (2) materials that cannot be exchanged for other resources after they are provided; (3) monetary resources that can be used to purchase materials; (4) human resources that contain at least three people; (5) expertise required to organize related activities; (6) professional equipment required in related activities; (7) patience needed to participate in activities that last for more than 1 h; and (8) empathy that enables participants to cope with undesirable situations (e.g., odors and mental disorders of service recipients) [7,8]. Therefore, each care service corresponds to at least one resource. For example, sheltered farms (Service 1) that provide jobs for homeless people require eight resources, whereas employment assistance (Service 12) only requires one resource (i.e., expertise). Care services and their required resources are presented in Table 2. However, these 59 kinds of care services were not applicable in every community. In the future, university students can choose service items through the screening mechanism detected from this research. However, the venue, culture, and basic equipment of the implementation targets ought to be considered in future studies, such as network, power supply, etc. Thereafter, comprehensive evaluation should be involved to assess which ones are the most appropriate service activities. This is, therefore, also a limitation of this study due to locality.

3.2. Using Factor Analysis to Categorize Characteristics

Initially, matrix calculation was performed to determine the correlations between the resources invested in the 59 care services. The correlation analysis involved 59 samples, each having eight variables (Table 3). As shown in Table 2, except for human resources and patience, which attained a significant correlation coefficient of 0.73, other variables were slightly correlated when analyzed in pairs. This indicated that using each variable to independently examine the care services was infeasible to determine the main resources required by the services. Subsequently, the results of the Kaiser–Meyer–Olkin (KMO) test revealed a KMO index of 0.516 (>0.5), implying the existence of latent principal factors [37–40]. Therefore, factor analysis was performed to reduce the number of variables and compile the principal factor components.

This study calculated the total amount of explained variance and extracted factors from the data (Table 4). Examining the eigenvalues (>1) of the unrotated component loadings showed that the first three components explained 67.99% of the data. In particular, the first component explained the greatest amount of variance (27.238%), and the second and third components explained 22.325% and 18.429% of the variance, respectively. Analyzing the eigenvalues (>1) of the rotated component loadings revealed that the first three components also explained 67.99% of the data. Particularly, the first component accounted for the greatest amount of variance (25.036%), and the second and third components explained 21.809% and 21.147% of the variance, respectively. In the latter case, although differences between the variances explained were reduced, the total amount of variance explained did not increase. Subsequently, a scree plot was adopted to determine the number of extracted factors. This plot involves a scatter plot that presents factor number on the x-axis and unrotated eigenvalues on the y-axis [37–40]. By observing this plot, researchers can subjectively determine a threshold value for factor loading. Eigenvalues vary greatly above the threshold value but slightly below the threshold value. The factor number corresponding to the threshold value denotes the number of factors to be extracted. Adopting a threshold of eigenvalue >1 in the scree plot is reasonable; therefore, three factors were extracted in this study [38].

Table 2. Common community health care services in Taiwan and abroad.

1. Sheltered Farms that Provide Jobs for Homeless People (1)(2)(3)(4)(5)(6)(7)(8)	21. Building Friendly Environments (2)(4)(5)	41. Providing New and Reliable Health Knowledge (5)
2. Reviving regular community activities (2)(3)(5)	22. Physical fitness services (2)(4)(5)(6)(7)	42. Promoting social participation (2)(3)(4)(7)
3. Repairing summer camp facilities for disabled children (2)(3)(4)(5)(6)(7)(8)	23. Reminiscence therapy (4)(5)(7)	43. Painting the shelters of homeless people or abused women (2)(4)(7)
4. Building new community facilities for the needed (1)(2)(3)(4)(5)(6)(7)	24. Art therapy (4)(5)(7)	44. Painting/maintaining public parks or playgrounds (2)(4)(7)
5. Organizing special festival activities for the community (2)(3)(4)(5)(6)(7)	25. Community seminars (4)(5)	45. Charity shops collecting and selling secondhand goods (1)(2)
6. Improving schools (1)(2)(3)(4)(5)(6)(7)	26. Medical counseling (4)(5)(6)(8)	46. Cleaning the environment (4)(7)(8)
7. Enhancing community health care network (3)(5)	27. Spiritual care (religious belief) (5)	47. Care visits (2)(4)(7)(8)
8. Providing sufficient and sound computer equipment (1)(2)(5)	28. Promoting dementia prevention policies (4)(5)(7)(8)	48. Support services for disabled people who lead an independent life (4)(7)(8)
9. Organizing arts nights (1)(2)(4)(5)(6)	29. Home economics education (1)(4)(5)	49. Study groups (1)(2)(4)(7)
10. Service centers for foreign spouses (1)(2)(5)(7)	30. Health education (5)(8)	50. Filming community documentaries (4)(5)(7)
11. Recycling (2)(4)(5)(7)(8)	31. Teaching foreign languages (5)	51. Providing sports and entertainment facilities (1)(2)
12. Employment assistance (5)	32. Providing instructions on how to use the internet (5)	52. Establishing community libraries (1)(2)(4)(7)
13. Dance performances (4)(5)(7)	33. Encouraging lifelong learning (4)(5)(7)	53. Food provision (2)(4)(7)
14. Pottery DIY courses (1)(5)(6)	34. Health seminars (5)(8)	54. Shower provision (2)(4)
15. Painting DIY courses (1)(5)(6)	35. Spiritual restoration for homeless people (4)(5)(7)	55. Offering community health dance classes (4)(5)(7)
16. Free clinics (4)(5)(6)(7)	36. Promoting breast cancer prevention in the community (5)	56. Lending assistive devices (e.g., wheelchairs, hospital beds, crutches, and walkers) (1)(2)
17. Free haircuts (4)(5)(6)(7)	37. Archiving community cultural artifacts in databases (4)(5)(7)	57. Providing open spaces (1)
18. Providing psychological counseling services (5)(7)(8)	38. Promoting preventive health care (5)	58. Providing storage rooms (1)
19. Health-promoting services (4)(5)(6)(7)(8)	39. Advocating healthy diets (5)	59. Phone greetings (4)(7)
20. Massage services (4)(5)(7)(8)	40. Publicizing appropriate concepts of medication and seeking medical advice (5)	-

Note: Resources required by each service are numbered as follows: (1) designated space; (2) materials; (3) monetary resources; (4) human resources; (5) expertise; (6) professional equipment; (7) patience; and (8) empathy. Whether it is executable still needs to be considered on site.

Table 3. Correlation matrix of service resources required.

Item	Item (1)	Item (2)	Item (3)	Item (4)	Item (5)	Item (6)	Item (7)	Item (8)
Item (1)	1.00	0.37 *	0.09	−0.22 *	−0.23 *	0.23 *	−0.20	−0.23 *
Item (2)	-	1.00	0.39 **	0.21	−0.37 *	0.16	0.18	−0.09
Item (3)	-	-	1.00	0.11	0.13	0.39 **	0.17	0.03
Item (4)	-	-	-	1.00	−0.10	0.26 *	0.73 ***	0.17
Item (5)	-	-	-	-	1.00	0.32 **	−0.10	0.05
Item (6)	-	-	-	-	-	1.00	0.16	0.11
Item (7)	-	-	-	-	-	-	1.00	0.24 *
Item (8)	-	-	-	-	-	-	-	1.00

Note: Resources required by each service are numbered as follows: (1) designated space; (2) materials; (3) monetary resources; (4) human resources; (5) expertise; (6) professional equipment; (7) patience; and (8) empathy. Single tailed test. p Value < 0.05 *, p Value < 0.01 **, p Value < 0.001 ***; Kaiser–Meyer–Olkin Measure of Sampling Adequacy (KMO) is 0.516; Bartlett's Test of Sphericity is very significant $p < 0.001$.

Table 4. Total variance analysis and factors loading.

	Initial Eigenvalues			Extraction Sums of Squared Loadings			Rotation Sums of Squared Loadings		
Factors	Total	% of Variance	Cumulative (%)	Total	% of Variance	Cumulative (%)	Total	% of Variance	Cumulative (%)
1	2.18	27.24	27.24	2.18	27.24	27.24	2.00	25.04	25.04
2	1.79	22.33	49.56	1.79	22.33	49.56	1.75	21.81	46.85
3	1.47	18.43	67.99	1.47	18.43	67.99	1.69	21.15	67.99
4	0.84	10.54	78.53	-	-	-	-	-	-
5	0.76	9.46	87.99	-	-	-	-	-	-
6	0.38	4.79	92.78	-	-	-	-	-	-
7	0.36	4.43	97.21	-	-	-	-	-	-
8	0.22	2.79	100.00	-	-	-	-	-	-

Note: extraction method: principal component analysis.

3.3. *Factor Naming*

Table 5 presents the component loadings of the factors. In the table, human resources and patience were extracted as the variables of the first factor. Designated space and materials were extracted as the variables of the second factor. Expertise and professional devices were extracted as the variables of the third factor [37–39].

The first factor was named labor services. This type of care service refers to labor intensive activities that provide care services. The second factor was named high-resource services, which require substantial money and appeal and provide a large amount of materials to service recipients. The third factor was named professional services. This type of service requires expertise, provides expertise required for related activities, and is delivered in various ways.

Table 5. Component loading.

Resources	Component		
	1	2	3
Human Resources	* **0.803**	−0.272	−0.254
Patience	* **0.793**	−0.301	−0.283
Monetary Resources	0.526	0.331	0.445
A Designated Space or Venue	−0.032	* **0.811**	0.077
Materials	0.499	* **0.681**	−0.204
Empathy	0.309	−0.459	0.027
Expertise	−0.069	−0.393	* **0.808**
Professional Equipment	0.528	0.160	* **0.656**

Note: 3 components extracted; extraction method: principal component analysis. The bold word with * is main construction variable.

According to the aforementioned analysis results, community health care services suitable for university students to provide include labor services and professional services. After discussing with the director-general of the community development association, we selected Services 50 (filming community documentaries), 55 (offering community health dance classes), and 37 (archiving community cultural artifacts in databases) as the actual services provided to the Lunfeng community.

4. Discussion

4.1. Filming a Community Documentary

During the service, the participating university students outlined the filming plan, conducted fieldwork, cultivated their filming skills, learned to utilize storyboards for filming, wrote a script, and practiced film editing techniques. In addition to theories taught in classes, the students acquired valuable experiences when they explored touching true stories, witnessed the ordinary happiness reflected by these stories, and filmed a documentary integrating the local cultural landscapes and histories [32,33].

First, the students sought community stories for the film. They adopted the sugar factory in the community as the main theme of the stories because the sugar factory had long been crucial to the community's industrial development before it was shut down. In addition, the factory was a critical source of income for multiple local families. Moreover, the childhood memories commonly shared by older adults in the community are closely related to the sugar factory. Accordingly, the stories of the sugar story were used as the filming materials. Initially, the story setting was designed to focus on the history of the sugar industry associated with the sugar factory when it was still in operation. However, this setting was slightly relevant to the community. Therefore, the students revised the filming plan, sought community residents who had worked (either part time or full time) at and retired from the sugar factory, and collected their stories through fieldwork. Subsequently, the story outline and subsections were determined. During data collection, the students found that the prosperity of the sugar factory had impeded the interactions between ordinary community children and factory employees' children because the factory dormitory was walled off. In addition, the factory paid particular attention to providing benefits for its employees. For example, the provision of a public bathhouse, an old television in the Zhongshan Hall, and train-fare concessions had excited envy from local residents who were unrelated to the factory employees. Furthermore, interviews were administered to the children of the employees in the Special Police Corps. Overall, 14 local residents were interviewed to complete the entire film.

4.2. Offering Community Health Dance Classes

This study designed a dancing exercise program suitable for older adults. The dance was composed of simple and repetitive movements designed at a level of difficulty for kindergarten children; hence, older adults should have no problem performing it [41]. The dancing music spanned 8 minutes. The participants of this activity comprised nine older adults and six community residents, who practiced 2 h daily. During the practice, taping marks were placed on the ground for instructing the positions of the participants. Subsequently, the taping marks were removed to examine how the participants performed with the marks. After a week of intensive practice, the participants performed in a large event held by Taiwanese government. The performance was well received, and the participants gained confidence through this activity. Therefore, this care service exerted a positive effect on the community development.

4.3. Archiving Community Cultural Artifacts in Databases

In 2014, the Lunfeng community established a cultural living museum that achieved and organized the community's meaningful cultural artifacts. The archiving process revealed the distribution of cultural artifacts collected from the Lunfeng community. Specifically, the artifacts constituted 11 production tools (41%), 4 cooking utensils (15%), 6 household tools (22%), 2 transportation-related objects (7%), and 4 musical instruments (15%). The statistics revealed that production tools accounted for the highest number of the collected and intact artifacts, indicating that the community mainly focuses on agricultural production. Household artifacts accounted for the second largest category, and musical instruments and cooking utensils each accounted for 15% of the collection, indicating that the residents also pursue various hobbies to entertain themselves in leisure time [42].

5. Conclusions

This study pointed out that postmodern community management pertains to an autonomous managerial mechanism. It requires broad-minded thinking, dialogue, problem-solving skills, and cooperation to shape the context of community relationships. During this period (2015–2020), university students were recruited to deliver community health care services selected in this study (i.e., filming community documentaries, offering community health dance classes, and archiving community cultural artifacts in databases). Moreover, interviews were administered to facilitate the implementation of the actual services. The results showed that three principal components occurred, namely labor services, high-resource services, and professional services, for a total explanatory power of 67.99%. The individual explanatory power of these components accounted for 25.04%, 21.81%, and 21.15%, respectively. After factor extractions, three factor strata were determined and individually named as labor services, high-resource services, and professional services. This study adopted a qualitative method involving audio/video recording and adopted photo-voice and photo elicitation. Community health care services suitable for university students were selected using a multivariate method. After the filming the community documentary and teaching the health dance program, the participating students established a friendly relationship with the local residents, indicating the favorable effect of the community health care services performed in this study. In addition, the establishment of community cultural artifact databases rendered the overall care services more feasible for continuation. These care services enabled the university students to experience social concerns that were otherwise unobservable in academic theories. Overall, these activities have both theoretical and practical implications.

Author Contributions: Data curation, G.-S.S. and S.-J.L.; Investigation, G.-S.S.; Methodology, W.-T.F. and B.-Y.C.; Resources, S.-J.L.; Writing—original draft, B.-Y.C.; Writing—review and editing, W.-T.F. All authors have read and agreed to the published version of the manuscript.

Funding: This research received no external funding.

Conflicts of Interest: The authors declare no conflict of interest.

References

1. Peng, L.-H.; Wu, Y.-P. Research of Youth Community Education and Development of Local Cultural Industries. *Procedia Soc. Behav. Sci.* **2016**, *217*, 414–420. [CrossRef]
2. Payne, J.; Razi, S.; Emery, K.; Quattrone, W.; Tardif-Douglin, M. Integrating Community Health Workers (CHWs) into Health Care Organizations. *J. Community Health* **2017**, *42*, 983–990. [CrossRef]
3. Assefa, Y.; Gelaw, Y.A.; Hill, P.S.; Taye, B.W.; Van Damme, W. Community health extension program of Ethiopia, 2003–2018: Successes and challenges toward universal coverage for primary healthcare services. *Glob. Health* **2019**, *15*, 24. [CrossRef] [PubMed]
4. Austin-Evelyn, K.; Rabkin, M.; Macheka, T.; Mutiti, A.; Mwansa-Kambafwile, J.; Dlamini, T.; El-Sadr, W.M. Community health worker perspectives on a new primary health care initiative in the Eastern Cape of South Africa. *PLoS ONE* **2017**, *12*, e0173863. [CrossRef] [PubMed]
5. Komaramy, M.; Ceballos, V.; Zurawski, A.; Bodenheimer, T.; Thom, D.H.; Arora, S. Extension for Community Healthcare Outcomes (ECHO): A new model for community health worker training and support. *J. Public Health Policy* **2017**, *39*, 203–216. [CrossRef] [PubMed]
6. Berkowitz, S.A.; Parashuram, S.; Rowan, K.; Andon, L.; Bass, E.B.; Bellantoni, M.; Brotman, D.J.; Deutschendorf, A.; Dunbar, L.; Durso, S.C.; et al. Association of a Care Coordination Model With Health Care Costs and Utilization: The Johns Hopkins Community Health Partnership (J-CHiP). *JAMA Netw. Open* **2018**, *1*, e184273. [CrossRef]

7. Prinja, S.; Gupta, A.; Verma, R.; Bahuguna, P.; Kumar, D.; Kaur, M.; Kumar, R. Cost of Delivering Health Care Services in Public Sector Primary and Community Health Centres in North India. *PLoS ONE* **2016**, *11*, e0160986. [CrossRef]
8. Panday, S.; Bissell, P.; Van Teijlingen, E.; Simkhada, P. The contribution of female community health volunteers (FCHVs) to maternity care in Nepal: A qualitative study. *BMC Health Serv. Res.* **2017**, *17*, 623. [CrossRef]
9. Grant, M.; Wilford, A.; Haskins, L.; Phakathi, S.; Mntambo, N.; Horwood, C. Trust of community health workers influences the acceptance of community-based maternal and child health services. *Afr. J. Prim. Health Care Fam. Med.* **2017**, *9*, e1–e8. [CrossRef]
10. Choi, B.Y.; Blumberg, C.; Williams, K.; Information, P.E.K.F.C. Mobile Integrated Health Care and Community Paramedicine: An Emerging Emergency Medical Services Concept. *Ann. Emerg. Med.* **2016**, *67*, 361–366. [CrossRef]
11. Litzelman, D.K.; Inui, T.S.; Griffin, W.J.; Perkins, A.; Cottingham, A.H.; Schmitt-Wendholt, K.M.; Ivy, S.S. Impact of Community Health Workers on Elderly Patients' Advance Care Planning and Health Care Utilization. *Med. Care* **2017**, *55*, 319–326. [CrossRef] [PubMed]
12. Olaniran, A.; Smith, H.; Unkels, R.; Bar-Zeev, S.; Broek, N.R.V.D. Who is a community health worker?—A systematic review of definitions. *Glob. Health Action* **2017**, *10*, 1272223. [CrossRef] [PubMed]
13. Nxumalo, N.; Goudge, J.; Manderson, L. Community health workers, recipients' experiences and constraints to care in South Africa—A pathway to trust. *AIDS Care* **2016**, *28*, 61–71. [CrossRef]
14. Farris, G.; Sircar, M.; Bortinger, J.; Moore, A.; Krupp, J.E.; Marshall, J.; Abrams, A.; Lipsitz, L.; Mattison, M. Extension for Community Healthcare Outcomes-Care Transitions: Enhancing Geriatric Care Transitions Through a Multidisciplinary Videoconference. *J. Am. Geriatr. Soc.* **2016**, *65*, 598–602. [CrossRef] [PubMed]
15. Shi, L.; Lee, D.; Chung, M.; Liang, H.; Lock, D.; Sripipatana, A. Patient-Centered Medical Home Recognition and Clinical Performance in U.S. Community Health Centers. *Health Serv. Res.* **2016**, *52*, 984–1004. [CrossRef]
16. Song, Y.; Zhang, T.; Chen, L.; Yi, B.; Hao, X.; Zhou, S.; Zhang, R.; Greene, C. Increasing seasonal influenza vaccination among high risk groups in China: Do community healthcare workers have a role to play? *Vaccine* **2017**, *35*, 4060–4063. [CrossRef]
17. Surjaningrum, E.R.; Minas, H.; Jorm, A.; Kakuma, R. The feasibility of a role for community health workers in integrated mental health care for perinatal depression: A qualitative study from Surabaya, Indonesia. *Int. J. Ment. Health Syst.* **2018**, *12*, 27. [CrossRef]
18. Peters Thomas, J.; Waterman Robert, H. *In Search of Excellence: Lessons from America's Best-Run Companies*; Harper & Row, Publishers Inc.: New York, NY, USA, 1982.
19. Henderson, S.; Horne, M.; Hills, R.; Kendall, E. Cultural competence in healthcare in the community: A concept analysis. *Health Soc. Care Community* **2018**, *26*, 590–603. [CrossRef]
20. Bailey, J.E.; Surbhi, S.; Bell, P.C.; Jones, A.M.; Rashed, S.; Ugwueke, M.O. SafeMed: Using pharmacy technicians in a novel role as community health workers to improve transitions of care. *J. Am. Pharm. Assoc.* **2016**, *56*, 73–81. [CrossRef]
21. Rogers, E.A.; Manser, S.T.; Cleary, J.; Joseph, A.M.; Harwood, E.; Call, K.T. Integrating Community Health Workers Into Medical Homes. *Ann. Fam. Med.* **2018**, *16*, 14–20. [CrossRef]
22. Kuule, Y.; Dobson, A.E.; Woldeyohannes, D.; Zolfo, M.; Najjemba, R.; Edwin, B.M.R.; Haven, N.; Verdonck, K.; Owiti, P.; Wilkinson, E. Community Health Volunteers in Primary Healthcare in Rural Uganda: Factors Influencing Performance. *Front. Public Health* **2017**, *5*, 62. [CrossRef] [PubMed]
23. Davies, B.; Challis, D. *Matching Resources to Needs in Community Care: An Evaluated Demonstration of a Long-Term Care Model*; Routledge Revivals: New York, NY, USA, 2018.
24. Kangovi, S.; Mitra, N.; Grande, D.; Huo, H.; Smith, R.A.; Long, J.A. Community Health Worker Support for Disadvantaged Patients with Multiple Chronic Diseases: A Randomized Clinical Trial. *Am. J. Public Health* **2017**, *107*, 1660–1667. [CrossRef] [PubMed]
25. Pandey, J.; Singh, M. Donning the mask: Effects of emotional labour strategies on burnout and job satisfaction in community healthcare. *Health Policy Plan.* **2015**, *31*, 551–562. [CrossRef]
26. Ballard, M.; Schwarz, R. Employing practitioner expertise in optimizing community healthcare systems. *Healthcare* **2019**, *7*, 100334. [CrossRef]
27. Gregg, A.; Tutek, J.; Leatherwood, M.D.; Crawford, W.; Friend, R.; Crowther, M.; McKinney, R. Systematic Review of Community Paramedicine and EMS Mobile Integrated Health Care Interventions in the United States. *Popul. Health Manag.* **2019**, *22*, 213–222. [CrossRef]

28. Sommanustweechai, A.; Putthasri, W.; Nwe, M.L.; Aung, S.T.; Theint, M.M.; Tangcharoensathien, V.; Wynn, S.S. Community health worker in hard-to-reach rural areas of Myanmar: Filling primary health care service gaps. *Hum. Resour. Health* **2016**, *14*, 64. [CrossRef] [PubMed]
29. Wadge, H.; Bhatti, Y.; Carter, A.; Harris, M.; Parston, G.; Darzi, A. *Brazil's Family Health Strategy: Using Community Health Workers to Provide Primary Care*; Commonw Fund: New York, NY, USA, 2016.
30. Snell-Rood, C.; Feltner, F.; Schoenberg, N. What Role Can Community Health Workers Play in Connecting Rural Women with Depression to the "De Facto" Mental Health Care System? *Community Ment. Health J.* **2018**, *55*, 63–73. [CrossRef] [PubMed]
31. Davis, M.M.; Lindberg, P.; Cross, S.; Lowe, S.; Gunn, R.; Dillon, K. Aligning systems science and community-based participatory research: A case example of the Community Health Advocacy and Research Alliance (CHARA). *J. Clin. Transl. Sci.* **2018**, *2*, 280–288. [CrossRef]
32. Prinja, S.; Nimesh, R.; Gupta, A.; Bahuguna, P.; Gupta, M.; Thakur, J. Impact of m-health application used by community health volunteers on improving utilisation of maternal, new-born and child health care services in a rural area of Uttar Pradesh, India. *Trop. Med. Int. Health* **2017**, *22*, 895–907. [CrossRef]
33. Pla, V.G. Filming Responsibly: Ethnicity, Community, and the Nation in Ana Lucía Cuevas's El eco del dolor de mucha gente. In *Latin American Documentary Film in the New Millennium*; Palgrave Macmillan: New York, NY, USA, 2016; pp. 259–274.
34. Krzychała, S. Professional praxis community in a dialogical perspective: Towards the application of Bakhtinian categories in the documentary method. *Forum Qual. Soz. Forum Qual. Social Res.* **2019**, *20*, 17.
35. Keith, T.Z.; Reynolds, M.R. Using confirmatory factor analysis to aid in understanding the constructs measured by intelligence tests. In *Contemporary Intellectual Assessment: Theories, Tests, and Issues*; Flanagan, D.P., McDonough, E.M., Eds.; The Guilford Press: Washington, DC, USA, 2018; pp. 853–900.
36. Crede, M.; Harms, P. Questionable research practices when using confirmatory factor analysis. *J. Manag. Psychol.* **2019**, *34*, 18–30. [CrossRef]
37. Lewis, T.F. Evidence Regarding the Internal Structure: Confirmatory Factor Analysis. *Meas. Eval. Couns. Dev.* **2017**, *50*, 239–247. [CrossRef]
38. Orcan, F. Exploratory and confirmatory factor analysis: Which one to use first. *J. Meas. Eval. Educ. Psychol.* **2018**, *9*, 414–421. [CrossRef]
39. Wu, H.; Estabrook, R. Identification of confirmatory factor analysis models of different levels of invariance for ordered categorical outcomes. *Psychometrika* **2016**, *81*, 1014–1045. [CrossRef]
40. Wagner, W.E., III. *Using IBM® SPSS® Statistics for Research Methods and Social Science Statistics*; Sage Publications: Armonk, NY, USA, 2019.
41. Iuliano, J.E.; Lutrick, K.; Maez, P.; Nacim, E.; Reinschmidt, K. Dance for Your Health: Exploring Social Latin Dancing for Community Health Promotion. *Am. J. Health Educ.* **2017**, *48*, 142–145. [CrossRef]
42. Copeland, A.; Lipinski, T.; Jones, K. *Collecting, Organizing, and Preserving Diverse Publication Sources for the Good of One Community Archive: Legal Challenges and Recommendations*; Data Archiving and Networked Services (DANS): Indianapolis, IN, USA, 2017.

© 2020 by the authors. Licensee MDPI, Basel, Switzerland. This article is an open access article distributed under the terms and conditions of the Creative Commons Attribution (CC BY) license (http://creativecommons.org/licenses/by/4.0/).

International Journal of
Environmental Research and Public Health

MDPI

Article

Alterations in Serum-Free Amino Acid Profiles in Childhood Asthma

Joanna Matysiak [1,*], Agnieszka Klupczynska [2], Kacper Packi [2], Anna Mackowiak-Jakubowska [2], Anna Bręborowicz [3], Olga Pawlicka [2], Katarzyna Olejniczak [3], Zenon J. Kokot [1] and Jan Matysiak [2]

1 Faculty of Health Sciences, The President Stanisław Wojciechowski State University of Applied Sciences in Kalisz, 62-800 Kalisz, Poland; zkokot@ump.edu.pl
2 Department of Inorganic and Analytical Chemistry, Poznan University of Medical Sciences, 60 -780 Poznan, Poland; aklupczynska@ump.edu.pl (A.K.); kacperpacki1@wp.pl (K.P.); amackowiak@gmail.com (A.M.-J.); olga_szklanny@interia.pl (O.P.); jmatysiak@ump.edu.pl (J.M.)
3 Department of Pulmonology, Pediatric Allergy and Clinical Immunology, Poznan University of Medical Sciences, 60-572 Poznan, Poland; abreborowicz@wp.pl (A.B.); kasiatomeko@gmail.com (K.O.)
* Correspondence: jkamatysiak@gmail.com

Received: 17 June 2020; Accepted: 28 June 2020; Published: 2 July 2020

Abstract: Asthma often begins in childhood, although making an early diagnosis is difficult. Clinical manifestations, the exclusion of other causes of bronchial obstruction, and responsiveness to anti-inflammatory therapy are the main tool of diagnosis. However, novel, precise, and functional biochemical markers are needed in the differentiation of asthma phenotypes, endotypes, and creating personalized therapy. The aim of the study was to search for metabolomic-based asthma biomarkers among free amino acids (AAs). A wide panel of serum-free AAs in asthmatic children, covering both proteinogenic and non-proteinogenic AAs, were analyzed. The examination included two groups of individuals between 3 and 18 years old: asthmatic children and the control group consisted of children with neither asthma nor allergies. High-performance liquid chromatography combined with tandem mass spectrometry (LC-MS/MS technique) was used for AA measurements. The data were analyzed by applying uni- and multivariate statistical tests. The obtained results indicate the decreased serum concentration of taurine, L-valine, DL-β-aminoisobutyric acid, and increased levels of γ-amino-n-butyric acid and L-arginine in asthmatic children when compared to controls. The altered concentration of these AAs can testify to their role in the pathogenesis of childhood asthma. The authors' results should contribute to the future introduction of new diagnostic markers into clinical practice.

Keywords: metabolomics; metabolites; amino acids; biomarkers; asthma; diagnosis; children

1. Introduction

Asthma is a serious problem in contemporary medicine. It is a chronic illness, caused by both genetic and environmental factors and characterized by the inflammation of the lower airways. Inflammation leads to oedema and the infiltration of the bronchial mucosa, bronchospasm, and an excessive mucus secretion, resulting in reversible airway obstruction. However, in a significant percentage of patients, chronic inflammation leads to the remodeling of the bronchial walls and fixed airflow limitation. The highest incidence rates of asthma are recorded in developed countries [1]. The recent global epidemiological report has shown that nowadays this ailment is affecting more than 300 million people all over the world. In many patients, asthma begins in childhood [2]. Recurrent wheezing episodes triggered by viral infection are the most common manifestation of asthma in infancy and preschool children. However, the nonspecific symptoms, such as wheezing, cough, and chest tightness, make it difficult to distinguish asthma from other diseases [3,4]. An atopic background predisposes to asthma development but is not an essential condition. Objective tests supporting preschool asthma diagnoses are not commonly available and clinical criteria remain a major diagnostic tool in clinical practice. There are significant differences between childhood and adulthood asthma, concerning immunology, histopathology, and clinical manifestations.

Many researchers have tried to identify specific biological markers of childhood asthma that would aid in the diagnosis of this disorder. Finding such compounds is extremely important because accurate diagnosis and optimal treatment play a crucial role in the proper functioning of asthmatic children and can greatly improve their quality of life.

One of the disciplines being used to discover new biomarkers of various diseases and to identify biochemical pathways involved in their pathogenesis is metabolomics. This latest of the so-called omic techniques is focused on low-molecular-weight intermediates and the products of metabolism, such as amino acids, fatty acids, carbohydrates, or nucleotides. The composition of the metabolome is determined by the current condition of the organism, which is affected by external and internal factors, including pathology, applied therapy, or diet [5,6]. It was found that changes in the human metabolome precede clinical symptoms' occurrence [7]. Therefore, the metabolomic approach offers a powerful tool in the development of new diagnostic tests, including respiratory tract diseases.

In metabolomic research, two strategies are used: targeted and non-targeted analysis. A targeted approach is used for the precise quantitative analysis of a limited number of metabolites with known biochemical properties. By using highly sensitive and selective analytical techniques, a certain group of metabolites can be detected and quantified, even if they are present in very low concentrations. A non-targeted analysis, also called metabolite fingerprinting, enables the classification of the tested biological material on the basis of the complete metabolic profile, rather than individual substances. The non-targeted strategy aims to structurally detect diverse compounds present in the metabolome, however, without quantitative data [5,6,8,9].

Metabolomics has provided potential biomarkers of asthma belonging to different groups of compounds that take part in various biological pathways. Metabolomics was applied to urine [10,11], plasma/serum [12–17], and exhaled breath condensate (EBC) samples [18]. By applying the metabolomic profiling of urine, the significant differences in the concentrations of phenylalanine, glycolic acid, 2-oxoglutarate, threonine, and phenylacetic acid were observed between asthmatic children and healthy ones [10]. In 2013, another research group explicated that the serum of asthmatics is characterized by increased levels of glutamine, histidine, and methionine and by decreased levels of acetate, arginine, choline, formate, glucose, methanol, and O-phosphocholine [15]. Abnormalities were also found in the serum levels of lysophosphatidylcholines and phosphatidylcholines [16]. In 2013, the metabolome of breath condensate was also tested. That study indicated that compounds related to adenosine, retinoic acid, and vitamin D may be used in the profiling of different asthma phenotypes [18].

Free amino acids (AAs) represent a particularly interesting group of metabolites worth examining among asthmatic children [19,20]. Amino acids can be divided into proteinogenic and non-proteinogenic. Proteinogenic AAs are incorporated into proteins during translation and encompass twenty-two AAs. Non-proteinogenic AAs are not naturally encoded or found in the genetic code of any organism, but most of them also play important functions in the organism (i.e., carnitine, creatine, ornithine, taurine, hydroxyproline, hydroxylysine). Non-proteinogenic amino acids often occur as intermediates in the metabolic pathways for standard AAs [21,22]. Abnormalities in free AA levels in body fluids have been reported in various disorders, i.a., liver diseases [23], chronic renal failure [24,25], neoplastic diseases [26–30], diabetes mellitus [31,32], and obesity [31,33].

A variety of methods allow the determination of AAs in biological samples. The most commonly used analytical techniques include high-performance liquid chromatography (HPLC), gas chromatography combined with mass spectrometry (GC-MS), and high-performance liquid chromatography coupled with mass spectrometry (LC-MS/MS). The ideal technique for determining the amino acid profile should be characterized by: high resolution, selectivity, the ability to determine the largest amount of amino acids in one analytical cycle, short duration, high sensitivity, repeatability, and low cost of analysis. Poschke et al. [34] used the HPLC technique to determine free AAs in blood serum. The method used by the researchers allowed the simultaneous determination of 15 AAs. The duration of a single analytical cycle was a total of 31 min (28 min analysis and 3 min column regeneration). In addition, before each analytical cycle, a derivatization reaction was necessary. Shi et al. also used the LC technique [35]. This method enabled the determination of 19 AA concentrations. The duration of a single cycle was 30 min. The analysis was preceded by the column equilibration, lasting 30 min. Deng et al. [36] used gas chromatography coupled with mass spectrometry to determine free AAs in the blood of newborns. The technique enabled the simultaneous determination of five AAs and required two-step derivatization (esterification and acylation). The duration of a single analytical run was 15 min. All described methods were characterized by high resolution, repeatability, accuracy, and selectivity [35,36]. We applied high-performance liquid chromatography coupled with mass spectrometry (LC-MS/MS) that, in addition to mentioned advantages, allows the determination of the widest spectrum of free amino acids (42 analytes. LC-MS/MS is also characterized by a short duration of a single analysis-18 min-which is particularly important when determining a large number of samples, as well as the ability to analyze body fluids of various origins.

The examination of free serum AAs would contribute not only to estimating their diagnostic utility in asthmatic children but will also contribute to broadening the knowledge about the mechanisms of this disorder. Scientists attempted several times to find the relation between selected AA levels and asthma occurrence [14,37]. However, the research projects which have been conducted so far failed to clearly answer the question of the impact of asthma on the profile of free AAs in human blood. Therefore, it is necessary to analyze the broad profile of free AAs in asthmatics and compare the levels of these metabolites with healthy subjects. Children are a particularly interesting research group, due to the different course of the disease in comparison with adults. To date, only a few metabolomic studies have been devoted to the pediatric population [10,11,17,18], and none of them showed a wide range of free AAs in serum. The current research aimed at applying the targeted metabolomic approach in order to find the differences with potential diagnostic significance in asthma. Previously conducted serum research has focused mainly on the altered levels of arginine and arginase activity [12]. This is the first study that presents the analysis of a wide panel of free AAs in asthmatic children, covering both proteinogenic and non-proteinogenic AAs.

2. Methods

2.1. Chemicals and Reagents

The analysis of free AAs was performed using an aTRAQ kit (Sciex, Framingham, MA, USA) that allows for quantifying of the 42 free amino acids (proteinogenic and non-proteinogenic) in a range of biological fluids. Deionized water was obtained from a Simplicity UV (Merck Millipore, Darmstadt, Germany) purifying system. HPLC gradient grade methanol was supplied by J.T. Baker (Center Valley, PA, USA).

2.2. Patients

The study was conducted involving two groups of subjects between 3 and 18 years of age: asthmatic children (n = 13) and the control group, which consisted of children with neither asthma nor allergies (n = 17). Children were recruited from the Department of Pediatric Pneumonology, Allergology and Clinical Immunology, K. Jonscher Clinical Hospital, Poznan University of Medical Sciences, after the written consent of their parents. Thirty serum samples were the test material, and they were stored at −80 °C until analysis. The study was approved by the Local Ethical Committee of Poznan University of Medical Sciences, Poland (Decision No. 530/12), and was consistent with the requirements of the Helsinki declaration.

2.3. Sample Collection and Preparation

Blood samples from both groups (asthmatic children and control group) were collected, processed, and stored in the same way. Blood was collected during medical examination into tubes with a clotting activator (S-Monovette system, Sarstedt, Nümbrecht, Germany). Then, sera were obtained by centrifugation at 300 × 10 RPM for 20 min. The sera were aliquoted and stored at −80 °C until analysis. All samples used in the study were analyzed on the first freeze-thaw cycle.

For AA determination in the collected sera, an aTRAQ reagent kit was used. The reliability of the method was confirmed in the literature [38–40]. The advantages of the aTRAQ methodology include a low sample volume required for analysis, a broad range of analytes (Table S1), and the use of labeled internal standards for each amino acid, which provide accurate quantitative results. The first step of the preparation procedure was the thawing of samples and transferring 40 µL of serum into Eppendorf tubes. Then 10 µL of 10% sulfosalicylic acid were added to precipitate proteins. The samples were mixed and centrifuged (2 min; 10,000× *g*). Then 10 µL of the supernatant were transferred to a clean tube and mixed with 40 µL of borate buffer (pH = 8.5). After mixing and centrifuging, 10 µL of the solution were transferred to a clean tube and 5 µL of labeling reagent solution (aTRAQ Reagent Δ8) were added. The samples were incubated for 30 min at room temperature. After that, 5 µL of hydroxylamine were added to stop the labeling reaction and the samples were incubated at room temperature for 15 min. Then samples were mixed with 32 µL of the internal standard solution. The tube contents were concentrated to a volume of approximately 20 µL using a vacuum concentrator (miVac Duo Concentrator, Genevac, Stone Ridge, NY, USA). Then samples were mixed with 20 µL of water and transferred to an autosampler vial.

2.4. LC-MS/MS Instrumentation

The measurements of AA concentrations were conducted using liquid chromatography-tandem mass spectrometry (LC-MS/MS) and a fully validated, highly selective method [38,39]. The applied method offers a valid alternative to the most conventional method of AA quantification that employs ion exchange chromatography (IEC) followed by post column ninhydrin derivatization and UV detection [41]. The LC-MS/MS method offers several advantages compared to the IEC method: decreased run time, a high amount of analytes quantified in one analytical run, low limits of quantification, and superior specificity via the use of the scheduled multiple reaction monitoring (sMRM) mode [38–40]. The analyses were performed on the 1260 Infinity HPLC system (Agilent Technologies, Santa Clara, CA, USA) coupled with a 4000 QTRAP triple quadrupole mass spectrometer (Sciex, Framingham, MA, USA) with an electrospray ionization (ESI) source. The chromatographic separation was carried out using a C18 column (4.6 mm × 150 mm, 5 µm) (Sciex, Framingham, MA, USA) with a flow rate of 0.8 mL/min. The eluent A was composed of water with 0.1% formic acid and 0.01% heptafluorobutyric acid and the eluent B contained methanol with 0.1% formic acid and 0.01% heptafluorobutyric acid. The gradient elution program was as follows: 0 min, 2% B; 0–6 min linear from 2 to 40% B; 6–10 min, 40% B; 10–11 min linear from 40 to 90% B; 11–12 min, 90% B; 12–13 min linear from 90 to 2% B; 13–18 min, 2% B. The separation of the temperature and injection volume was set at 50 °C and 2 µL, respectively. The analyses were performed in a positive ionization mode. The MS parameters were as follows: ion spray voltage, 4500 V; declustering potential, 30 V; entrance potential, 10 V; collision energy, 30 V (except from an argininosuccinic acid (Asa), cystathionine (Cth), L-cystine (Cys), L-homocystine (Hcy), δ-hydroxylysine (Hyl), L-lysine (Lys), and L-ornithine (Orn), 50 V) and collision cell exit potential, 50 V. Nitrogen was used as both a curtain gas and a collision gas. The temperature of an ion source, gas 1, and gas 2 was set at 600 °C, 60 psig, and 50 psig, respectively. The device operated in scheduled multiple reaction monitoring (sMRM) mode. The list of MRM transitions for 42 analytes and their corresponding internal standards is included in Table S1 in the Supplementary Materials.

2.5. Statistical Analysis

To perform statistical analyses, Statistica 10.0 (StatSoft Inc., Tulsa, OK, USA) and MetaboAnalyst 3.0 web platform (www.metaboanalyst.ca) were used [42]. A value of $p < 0.05$ was considered statistically significant. The data were analyzed by applying uni-and multivariate statistical tests. Firstly, the Shapiro-Wilk test was used to check the normality. The Mann-Whitney U test was used to compare variables without a normal distribution and the Levene's test was applied to examine the equality of variances for variables with a normal distribution. When the Levene's test result was not statistically significant ($p > 0.05$), which indicated the homogeneity of variance between groups, the Student's t-test was performed. When the Levene's test result was statistically significant ($p < 0.05$) Welch's t-test was applied.

The multivariate statistical analyses consisted of a partial least squares discriminant analysis (PLS-DA) and a ROC curve analysis. Before using multivariate statistical analyses, the data were subjected to a process of normalization, transformation, and scaling. PLS-DA is a supervised multivariate analysis, in which after determining the number of factors needed to create a model, it is possible to assign samples to one of the two groups. PLS-DA also gives the opportunity to choose the variables most relevant to the classification of samples. To rank the AAs according to their importance in discrimination between groups, a variable importance in projection (VIP) score is frequently used. The higher the value of the VIP, the more important the variable is in the classification of patients. The ROC curves are used to assess the sensitivity and specificity of the discriminator. The greater the area under the curve (AUC), the better is the classification of the samples to one of the groups by the model. The ROC curves were created using the random forest algorithm and Monte Carlo cross-validation.

3. Results

3.1. Patients' Characteristics

Thirty patients were enrolled: 13 in the asthmatic group and 17 in the control group. Control subjects were matched to the studied group in terms of age, sex, and ethnic origin. The characteristics of asthma patients and the control group are presented in Table 1.

Table 1. Characteristics of the subjects.

Characteristics	Asthma Patients	Control Subjects
No. of Subjects	13	17
Sex		
Male	7 (53.8%)	11 (64.7%)
Female	6 (46.2%)	6 (35.3%)
Age		
Median	12	10
Range	4–16	3–18
Asthma Severity		
Mild	7 (53.8%)	
Moderate	4 (30.8%)	
Severe	2 (15.4%)	
The Daily dose of Corticosteroids (Budesonide or Equivalent)		
100–200 µg	6 (46.2%)	
250–350 µg	2 (15.4%)	
400–500 µg	3 (23.0%)	
>500 µg	1 (7.7%)	
Unknown	1 (7.7%)	

Table 1. *Cont.*

	Characteristics		Asthma Patients	Control Subjects
	Comorbidities			
	Hypoacusia		1	
	Atopic Dermatitis		1	
	Allergic Rhinitis		3	
	Coeliac Disease		1	
	Cholecystitis		1	
	Lung Function			
FEV1/VC		Mean	85.25%	
		Range	67–94%	
FEV1		Mean	82.5%	
		Range	40–97%	
FVC EX		Mean	87.33%	
		Range	49–95%	
	Total IgE			
	Mean		278.61 kU/I	96.29 kU/I
	Range		15.5–1035 kU/I	12.2–797 kU/I

3.2. Alterations in Serum-Free AA Profiles in Childhood Asthma

The applied methodology allowed a broad panel of free AA concentrations, including both proteinogenic and non-proteinogenic amino acids, to be measured in the studied serum samples. Thirty-five of 42 amino acids were detected and quantified in analyzed samples. The remaining seven amino acids (O-phospho-L-serine, O-phosphoethanolamine, L-homocitrulline, argininosuccinic acid, L-anserine, L-carnosine, L-homocysteine) occurred below the lower level of quantification in all analyzed samples. Some of the amino acids (δ-hydroxylysine, cystathionine, L-cystine) were detected only in part of the samples and were therefore excluded from further data analysis. Finally, the concentrations of the 32 amino acids were subjected to statistical analysis. The concentrations of AAs both in the asthmatic and control groups are shown in Table 2.

Table 2. The determined concentrations of 32 amino acids in serum samples collected from children with asthma and the control group. Concentration values are given in μM.

Amino Acid	Abbreviation	Children with Asthma (n = 13) Median	Mean	SD	Control Group (n = 17) Median	Mean	SD
1-Methyl-L-Histidine	1MHis	3.09	4.42	4.4	1.74	1.93	1.73
3-Methyl-L-Histidine	3MHis	2.38	2.52	0.84	2.04	2.32	1.06
L-A-Aminoadipic Acid	Aad	0.64	0.66	0.2	0.66	0.78	0.29
L-A-Amino-N-Butyric Acid	Abu	15.56	16.72	5.24	18.98	19.53	6.16
L-Alanine	Ala	443.71	423.32	89.17	368.45	396.24	106.97
L-Arginine	Arg	112.59	109.97	12.66	87.67	93.15	22.77
L-Asparagine	Asn	46.87	49.32	5.23	51.71	52.24	11.23
L-Aspartic Acid	Asp	11.71	11.94	3.65	9.89	11.65	7.89
D,L-B-Aminoisobutyric Acid	bAib	1.13	1.11	0.31	1.26	1.47	0.61
B-Alanine	bAla	16.51	13.84	6.61	8.95	9.64	3.9
L-Citrulline	Cit	24.37	23.77	5.46	24.71	24.31	3.79
Ethanolamine	EtN	8.94	8.96	2.17	8.35	8.96	1.65
Y-Amino-N-Butyric Acid	GABA	1.25	1.18	0.49	0.67	0.75	0.31
L-Glutamine	Gln	424.51	429.37	56.74	459.64	458.41	71.53
L-Glutamic Acid	Glu	52.97	58	19.77	64.58	62.66	18.27
Glycine	Gly	278.17	283.25	32.45	255.44	268.74	46.85
L-Histidine	His	69.11	68.38	7.93	61.62	66.83	13.1
Hydroxy-L-Proline	Hyp	22.02	22.44	14.17	13.32	15.94	6.71
L-Isoleucine	Ile	63.9	61.95	12.54	63.32	68.1	16.01
L-Leucine	Leu	92.42	97.89	18.21	104.96	111.25	25.25

Table 2. *Cont.*

Amino Acid	Abbreviation	Children with Asthma (n = 13) Median	Mean	SD	Control Group (n = 17) Median	Mean	SD
L-Lysine	Lys	142.42	147.4	25.22	150.06	157.79	40.29
L-Methionine	Met	21.32	22.06	5.39	22.45	23.44	7.17
L-Ornithine	Orn	59.45	64.63	15.76	61.35	65.67	18.49
L-Phenylalanine	Phe	55.46	52.55	8.97	54.67	54.85	9.84
L-Proline	Pro	197.03	181.4	67.24	171.13	177.63	51.23
Sarcosine	Sar	1.28	1.24	0.74	1.26	1.26	0.51
L-Serine	Ser	133.87	140.51	14.25	140.57	139.22	22.81
Taurine	Tau	66.95	70.75	17.75	86.33	85.95	16.55
L-Threonine	Thr	94.75	101.29	27.81	97.24	96.91	20.15
L-Tryptophan	Trp	51.47	51.61	7.49	57.34	59.19	13.14
L-Tyrosine	Tyr	52.31	52.57	11.82	51.51	55.81	18.07
L-Valine	Val	165.8	173.07	29.58	195.68	204.36	38.34

In order to evaluate differences in the serum metabolic profiles between asthmatic children and healthy subjects, univariate statistical analysis was used (Table 3). The applied tests demonstrated that the statistically significant differences between the studied groups occurred between the levels of the following five amino acids: taurine (Tau), ɣ-amino-n-butyric acid (GABA), DL-β-aminoisobutyric acid (bAib), L-arginine (Arg), and L-valine (Val). In asthmatics, the levels of taurine, L-valine, and DL-β-aminoisobutyric acid were reduced compared to the control group, while the levels of ɣ-amino-n-butyric acid and L-arginine were increased (Table 2).

Table 3. Results of univariate statistical analysis of serum-free amino acids in children with asthma and the control group. Bold type for p values indicates statistical significance.

Amino Acid	Shapiro–Wilk Test: Children with Asthma (n = 13)	Shapiro–Wilk Test: Control Group (n = 17)	Levene's Test	Mann–Whitney U Test	Student's t-Test	Welch's t-Test
	p Value					
1MHis	**0.052421**	**0.003404**		0.276532		
3MHis	0.875277	0.234623	0.327524		0.591304	
Aad	0.496739	0.112430	0.07644		0.216463	
Abu	0.621549	0.972163	0.644789		0.198044	
Ala	0.178613	0.371844	0.669371		0.467315	
Arg	0.288521	0.071223	**0.043268**			**0.016265**
Asn	0.469455	0.584648	**0.009499**			0.353783
Asp	0.753162	**0.000012**		0.241259		
bAib	0.694310	0.326845	**0.020441**			**0.045451**
bAla	0.385620	0.326401	**0.034083**			0.056388
Cit	0794372	0.951293	0.206713		0.753887	
EtN	0.128147	0.172044	0.617651		0.997393	
GABA	0.227097	0.609400	0.099223		**0.006383**	
Gln	0.389567	0.267469	0.442514		0.239702	
Glu	0.572162	0.911833	0.656593		0.509076	
Gly	0.437805	0.194367	0.324685		0.348727	
His	0.813939	0.357040	**0.015545**			0.690912
Hyp	0.093057	**0.017077**		0.167247		
Ile	0.738028	0.211595	0.309006		0.263792	
Leu	0.212462	0.279189	0.302909		0.118475	
Lys	0.752616	0.270535	0.367141		0.422490	
Met	0.792258	0.308677	0.289639		0.566910	
Orn	**0.026378**	0.288794		0.769551		
Phe	0.185401	0.541752	0.844958		0.515613	
Pro	0.173926	0.379691	0.219423		0.862817	
Sar	0.547304	0.953359	0.354722		0.953930	

Table 3. *Cont.*

Amino Acid	p Value					
	Shapiro–Wilk Test Children with Asthma (n = 13)	Control Group (n = 17)	Levene's Test	Mann–Whitney U Test	Student's t-Test	Welch's t-Test
Ser	0.132290	0.220079	0.300592		0.859334	
Tau	**0.011918**	0.966747		**0.012036**		
Thr	0.615151	0.330271	0.439175		0.619944	
Trp	0.301341	**0.046520**		0.131898		
Tyr	0.562556	0.220185	0.094828		0.580349	
Val	0.773498	0.170362	0.652018		**0.021465**	

3.3. Discrimination between Asthmatic Children And Healthy Subjects by AA Profiles

To determine the differentiation ability of AA profiles, the multivariate statistical analysis was performed. A model for discrimination between two analyzed groups was created using PLS-DA analysis. Figure 1a is a diagram of distance vectors for the three components obtained by performing PLS-DA analysis (a score plot). The points corresponding to the samples belonging to the study group (asthmatics) have been marked in red, and the samples of people in the control group (healthy peers) have been marked by green. The figure shows a clear separation between the asthmatic children and controls. The PLS-DA analysis also revealed which variables have the greatest importance in the sample grouping. Figure 1b shows the variables listed according to their contribution in sample classification. The most differentiating AAs were as follows: taurine (Tau), L-valine (Val), L-arginine (Arg), γ-amino-n-butyric acid (GABA), L-leucine (Leu), and L-tryptophan (Trp).

In the last step of statistical analysis, ROC curves were plotted for models, consisting of a different number of variables (32, 20, 10, 5, 3, 2) (Figure 2). It is noteworthy that increasing the number of AAs included in the model resulted in increasing the area under the ROC curves. The largest increase was observed when the number of variables was changed from two to five, while the addition of further variables into the model caused a slight increase in the area under the curve.

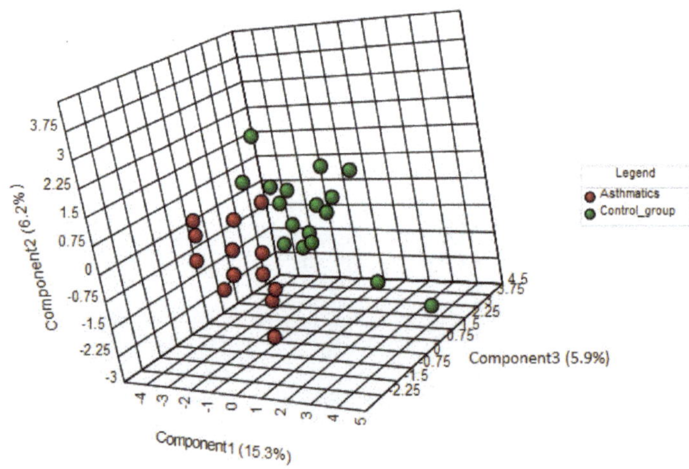

(a)

Figure 1. *Cont.*

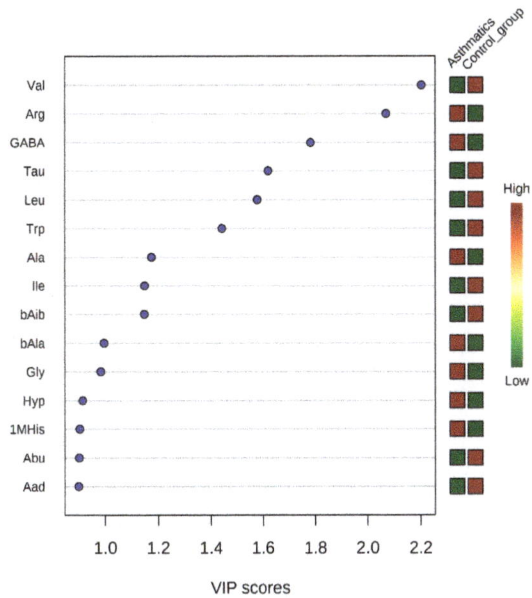

(b)

Figure 1. Results of a partial least squares discriminant analysis (PLS-DA) of free amino acid profiles determined in children with asthma and the control group. (**a**): Three-dimensional score plot (with explained variances shown in brackets). (**b**): The variables having the greatest importance in sample classification (amino acids with the highest variable importance in projection (VIP) score). Green and red boxes on the right hand side of the figure show if the levels of the respective amino acids are decreased or increased in the group.

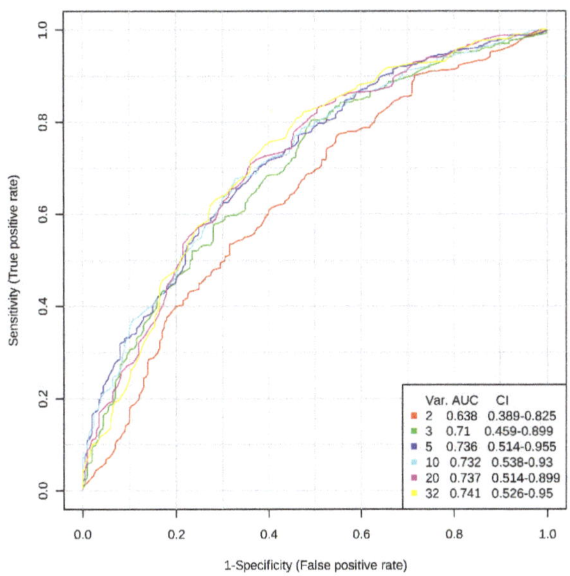

Figure 2. Receiver operating characteristics (ROC) curve analysis obtained for models based on different numbers of variables (serum levels of free amino acids), with area under the curve (AUC) values.

4. Discussion

In the face of the high prevalence of asthma in society and the need for the rapid implementation of treatment, it is important to have quantifiable indicators that correlate with the development of the disease. The aim of the study was to search for serum metabolic biomarkers among free AAs which may provide a distinction between asthmatic children and healthy ones. This paper applies the modern LC-MS/MS-based methodology that ensures the high specificity and sensitivity of the performed measurements and enables the simultaneous quantification of a broad AA profile (Table 2) in a small sample volume (40 μL).

Amino acids are considered to be mediators of immune activity in asthma and can act as antioxidants. In particular, amino acids such as glutamine, glutamate, glycine, and taurine have potentially protective effects, while others, like phenylalanine, may have adverse effects. Among the analyzed AAs, arginine deserves particular attention. Alterations in arginine homeostasis may lead to typical asthma symptoms, such as inflammation, airway hyperresponsiveness, and remodeling. Changes in arginine bioavailability may contribute to decreasing levels of bronchodilation NO and the excessive formation of peroxynitrite, which has procontractile properties [43,44].

In the available literature, there are conflicting reports on concentrations of this AA in both asthmatic children and adults, as well as information about the increased activity of arginase in asthmatics. Morris et al. [14] showed significantly lower levels of arginine in the serum of patients with asthma (45 ± 22 μM) compared to the control group (92 ± 29 μM). In turn, Lara et al. [12] and Fogarty et al. [13] did not confirm this association. In this study, the concentrations of arginine in the asthmatic children were significantly higher (110 ± 13 μM) than in healthy participants (93 ± 23 μM). It should be noted that the concentrations of this amino acid in both control groups analyzed by Morris et al. [14], and subjected to this study's experiment, achieved a similar level. The different serum concentrations of arginine in asthmatic patients between this paper and Morris et al.'s study may result from a different age range of included individuals. Morris et al. [14] included patients aged from 2 to 52 years, while children from 3 to 18 years old participated in the present study. The increased concentration of arginine in asthmatic children may result from an immature immune system. Asthma is accompanied by the chronic inflammation of the airways. It is caused by immune cells, which include, inter alia, T helper cells. These cells are also directly involved in the metabolism of arginine, inducing the expression of the enzyme responsible for its distribution. Activated type 2 T helper cells (Th2) secrete cytokines, stimulating arginase activity [14,45]. Therefore, it may be assumed that the immature immune system in children determines the decreased activity of arginase, which is manifested by elevated levels of arginine in the blood (Figure 3). A recent study conducted by Kelly et al. [46] in a group of 411 three-year-old children provides strong evidence for dysregulated arginine metabolism in asthmatic children. Arginine levels in human serum depend on other metabolites within the network. It has been shown that arginine can interact with other metabolites and has joint influences on asthma. Arginine is produced from citrulline and acts as the substrate for nitric oxide (NO) synthesis, while its uptake can be inhibited by L-ornithine, L-lysine, and other metabolites involved in the pathways that regulate the levels of these metabolites (Figure 3). Kraj et al. [37] also revealed the altered metabolism of arginine in children with controlled asthma. The profile of AAs crucial for arginine metabolism significantly differed from that observed for acute asthma in adults and healthy individuals, although arginase activity remained unchanged when compared to the healthy group. Thus, effective disease control and the introduction of optimal treatment could have a significant impact on the maintenance of arginase enzymatic activity. Certainly, the dysregulated metabolism of arginine, its changing serum level, and its interaction with exogenous metabolites deserve further investigation as potential causative agents in asthma [45].

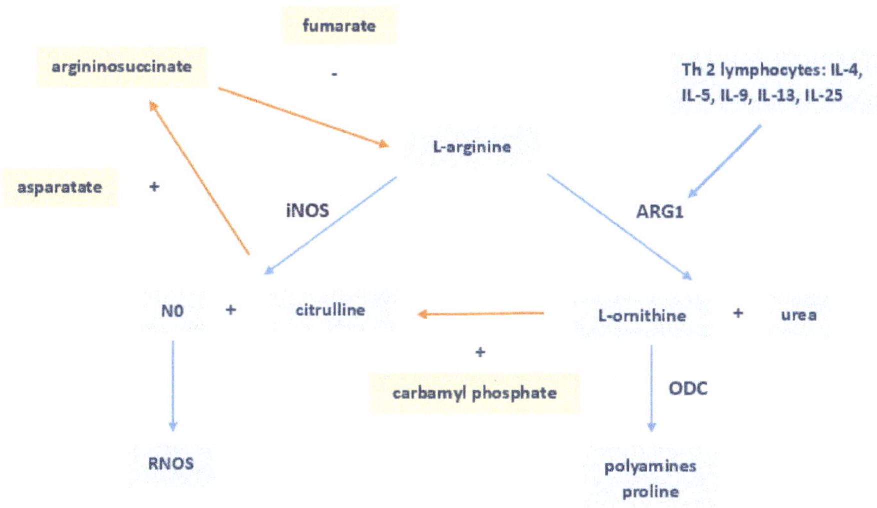

Figure 3. Metabolism of arginine in asthmatic children. The immature immune system in children manifested by decreased levels of T helper cells (Th 2) and inflammatory cytokines (IL-4, IL-5, IL-9, IL-13, IL-25) determines the decreased activity of arginase and the elevated level of arginine in the blood. Arginine cooperates with other metabolites within the network that affect each other's concentrations. ARG1-arginase; iNOS-inducible nitric oxide synthase; RNOS-reactive oxygen species; ODC-ornithine decarboxylase.

Another amino acid whose level was decreased in asthmatic children's serum is taurine. It is a sulfur amino acid and it is not incorporated into proteins [47]. Taurine has a wide range of biological functions, serving, for example, as a cryoprotectant, antioxidant, and membrane stabilizer. Meanwhile, its deficiency leads to developmental abnormalities in the brain, heart, and skeletal muscles [47,48]. Taurine also plays a crucial role in the lungs [49]. It has been suggested to be an agonist of $\alpha4$ subunit-containing GABAA receptors and may cause the relaxation of airway smooth muscle [50]. Comhair et al. [51] found a significantly higher level of taurine in asthmatic adults in the untargeted metabolomic study of plasma samples. The researchers suggested that the altered level of taurine is related to alterations in NO-associated transport [52]. The results obtained by Comhair et al. are contrary to the ones within this paper, which may have arisen from the age disparity of the studied populations (mean age 35 vs. 10 years). Zinellu et al. considered taurine as a potential biomarker of oxidative stress in asthma [53]. The mean concentration of plasma taurine was lower in asthmatics adults (mean age 58.8 ± 13.6 years, range 38–79) than in healthy ones, but this difference was not statistically significant. It is noteworthy that Zinellu et al.'s study focused exclusively on patients with mild stable asthma.

In this study, valine occurred at a decreased level in asthmatic children. Valine, one of the branched-chain amino acids (BCAA), is classified as an essential amino acid. BCAAs have many diverse functions in the human organism and they are crucial for energy, stress, and muscle metabolism (www.hmdb.ca). Valine was suggested as a biomarker of breast cancer [28,54], oral cancer [54], pancreatic cancer [54,55], liver disorders [56,57], colorectal adenoma [58], and type 2 diabetes [59], but there are not many publications about its role in asthma pathophysiology. Motta et al. [60] investigated the impact of different condensation temperatures on the exhaled breath condensate (EBC) metabolome in asthmatic adults. They demonstrated the decreased level of valine in the exhaled breath condensate of asthmatic patients (mean age 35 ± 1.2 years) compared to controls regardless of temperature. Researchers also suggested that the EBC metabolome depends on asthma severity. However, they did not profile metabolites in the serum/plasma of both adults and children suffering from asthma. Univariate and multivariate analyses conducted by Ghosh et al. [61] showed that serum levels of valine and 11 other metabolites (histidine, phenylalanine, lysine, asparagine, L-leucine, glutamate, lipid, isoleucine, N-acetylglycoproteins, citric acid, and glucose) were dysregulated in asthma–chronic obstructive pulmonary disease overlap (ACO) patients. This disease is similar to asthma but it is characterized by a faster decline in lung function, more frequent exacerbations, and a poorer quality of life when compared to asthma or chronic obstructive pulmonary disease alone [61]. Due to the lack of clear diagnostic and therapeutic guidelines, this may be an important finding.

Any discrepancies in the results between available studies are explainable. Metabolite profiles can vary depending upon multiple factors. A large number of results are not replicated due to the variety of biospecimens used in studies, like exhaled breath condensate, urine, plasma, and serum. Furthermore, there are not enough studies profiling metabolites in more than one biospecimen type, so it is not possible to determine the relationship between metabolites in different biological samples from the same person. Recently, Chiu et al. [62] simultaneously analyzed the metabolic profile of blood and urine related to IgE reactions for childhood asthma. Their research showed a significantly higher level of histidine in plasma and a lower concentration of 1-methylnicotinamide and trimethylamine N-oxide (TMAO) in the urine of asthmatic children compared to healthy individuals. Furthermore, Chiu et al. [62] correlated the plasma level of 3-hydroxybutyric acid, leucine, and valine to the urine concentration of hydroxy acids. However, this was a single comparative study, and more advanced research using different biospecimens in parallel, like exhaled breath condensate, urine, plasma, and serum, is needed. This issue could be one of the future directions in asthma metabolomic studies. Processing procedures and collection conditions can also affect the metabolome. Another important element that may account for the differences in study findings is the technique of metabolomic profiling. Nuclear magnetic resonance is a spectroscopic technique which uses the magnetic properties of atomic nuclei to generate structure information, and thus can identify metabolites in a studied biological fluid based on their unique chemical shift pattern and the intensity of the peak [63]. Meanwhile, tandem mass spectrometry combines chromatography, a technique used to separate metabolites, with MS, to measure their abundances. Research showed that measurements of NMR and MS may not always be comparable [10,15,63]. The results may also vary due to the heterogeneity of the asthma diagnostic criteria, with the variable use of medical diagnosis and/or spirometry criteria. In addition, it is known that the metabolome is very sensitive to external influences, including smoking, diet, and treatment regimen, and it changes with factors like BMI [19]. Certainly, more advanced studies are needed to identify all metabolic shifts caused by various environmental factors and physiological characteristics. In the nearest future, there will be a great need for standardization in the field of metabolomics and for the development of strict criteria for conducting and reporting metabolomic studies.

Among AAs with the highest discriminatory ability in childhood asthma, two derivatives of butyric acid, ɣ-amino-n-butyric acid (GABA) and DL-β-aminoisobutyric acid (bAib, BAIBA) were found. This is the first study that indicates alterations in the serum levels of these two metabolites in childhood asthma and asthma in general. In this study, the level of ɣ-amino-n-butyric acid was increased in asthmatic children compared to the control group, and the level of DL-β-aminoisobutyric acid was reduced. In the recent literature, there are findings that the GABAergic system exists in the airway epithelium and has a role in mucus overproduction in asthma [64]. Meanwhile, DL-β-aminoisobutyric acid is involved in the regulation of carbohydrate and lipid metabolism and also decreases inflammatory reactions [65]. Thus, the occurrence of an inflammatory process in asthma may be associated with a reduced serum level of DL-β-aminoisobutyric acid. However, in the future, advanced and detailed studies are needed to clarify if the dysregulation of DL-β-aminoisobutyric acid production and/or its action is engaged in the pathogenesis of childhood asthma.

Due to the small size of the study groups in the performed AA profiling, various statistical analyses were used. It should be emphasized that both the uni- and multivariate statistics indicated the following AAs as the most differentiated: taurine, ɣ-amino-n-butyric acid, L-arginine, DL-β-aminoisobutyric acid, and L-valine. The altered concentration of these compounds can testify to their role in the pathogenesis of childhood asthma. Kelly et al. summarized the 21 metabolomic studies of adulthood and childhood asthma, all of which reported significant findings and concluded that individual metabolites and metabolic profiles measured in EBC, urine, plasma, and serum could identify people with asthma and even distinguish asthma phenotypes with a high discriminatory ability. Therefore, the introduction of the monitoring of the level of serum AAs in clinical practice is worth considering. This can be used both as a method of diagnostics and prognostics for controlling the course of therapy. Based on the plotted ROC curves, it can be assumed that a classification model is better if more variables are included. However, the inclusion of several of the most discriminative compounds is sufficient to carry out the classification. This is an important conclusion for the possible use of metabolomic analysis in the diagnosis of childhood asthma, considering that it will significantly reduce costs while maintaining adequate sensitivity and specificity.

5. Conclusions

The analysis of the human metabolic profile is a very promising tool for clinical applications since it is a sensitive indicator of both endogenous and exogenous factors affecting the patient's body [66]. Changes in the human metabolome occur before changes in laboratory parameters and long before the onset of clinical symptoms. Due to the ability to measure these differences, metabolomics give hope for the detection of many diseases in the early stages of their development, which in turn can significantly increase the level of treatment effectiveness. Despite placing great hope in nitric oxide as a diagnostic marker for asthma in the past, today it can only be used to monitor treatment. Therefore, our results regarding the altered concentration of taurine, L-valine, DL-β-aminoisobutyric acid, γ-amino-n-butyric acid, and L-arginine in the serum of asthmatic children may contribute to the future introduction of new diagnostic markers. The involvement of AAs in the metabolic pathways proposed in this paper justifies the continuation of AA determination in various biospecimens collected from asthmatic patients. In the future, data on metabolomics in asthma should be integrated with the proteomic approach, as this may be the most informative integrative strategy.

Supplementary Materials: The following are available online at http://www.mdpi.com/1660-4601/17/13/4758/s1, Table S1: MRM transitions for the analyzed amino acids (Q1 > Q3).

Author Contributions: For research articles with several authors, a short paragraph specifying their individual contributions must be provided. The following statements should be used "Conceptualization, J.M. (Joanna Matysiak), A.K., J.M. (Jan Matysiak), A.B. and Z.J.K.; methodology, J.M. (Joanna Matysiak), A.K., A.M.-J.; software, A.K. and J.M. (Jan Matysiak); validation, A.K.; formal analysis, J.M. (Joanna Matysiak) and A.K.; investigation, A.K., A.M.-J. and J.M. (Jan Matysiak); resources, J.M. (Joanna Matysiak), K.O. and A.B.; data curation, J.M. (Joanna Matysiak), A.K., K.P., A.M.-J., O.P. and J.M. (Jan Matysiak); writing—original draft preparation, J.M. (Joanna Matysiak), A.K., K.P., A.M.-J., A.B., O.P., Z.J.K. and J.M. (Jan Matysiak); writing—review and editing, J.M. (Joanna Matysiak), A.K., K.P. and J.M. (Jan Matysiak); visualization, A.K. and K.P.; supervision, J.M. (Joanna Matysiak), J.M. (Jan Matysiak) and Z.J.K.; project administration, J.M. (Jan Matysiak); funding acquisition, J.M. (Joanna Matysiak), J.M. (Jan Matysiak) and Z.J.K. All authors have read and agreed to the published version of the manuscript".

Funding: This research was funded by Polish Ministry of Science and Higher Education, grant number DWD/3/10/2019. Agnieszka Klupczynska was supported by the Foundation for Polish Science (FNP).

Conflicts of Interest: The authors declare no conflict of interest.

References

1. Russell, R.J.; Brightling, C. Pathogenesis of asthma: Implications for precision medicine. *Clin. Sci. (Lond. Engl. 1979)* **2017**, *131*, 1723–1735. [CrossRef]
2. Epidemiology of Asthma: Current Opinion in Otolaryngology & Head and Neck Surgery. Available online: https://journals.lww.com/co-otolaryngology/Abstract/2016/06000/Epidemiology_of_asthma.14.aspx (accessed on 2 January 2020).
3. Adamko, D.J.; Sykes, B.D.; Rowe, B.H. The metabolomics of asthma: Novel diagnostic potential. *Chest* **2012**, *141*, 1295–1302. [CrossRef]

4. Adamko, D.J.; Nair, P.; Mayers, I.; Tsuyuki, R.T.; Regush, S.; Rowe, B.H. Metabolomic profiling of asthma and chronic obstructive pulmonary disease: A pilot study differentiating diseases. *J. Allergy Clin. Immunol.* **2015**, *136*, 571–580. [CrossRef]
5. Drexler, D.M.; Reily, M.D.; Shipkova, P.A. Advances in mass spectrometry applied to pharmaceutical metabolomics. *Anal. Bioanal. Chem.* **2011**, *399*, 2645–2653. [CrossRef]
6. Kim, Y.M.; Heyman, H.M. Mass Spectrometry-Based Metabolomics. In *Fungal Genomics*; de Vries, R.P., Tsang, A., Grigoriev, I.V., Eds.; Part of the Methods in Molecular Biology Methods Book Series; Humana Press: New York, NY, USA, 2018; Volume 1775, pp. 107–118.
7. Dettmer, K.; Hammock, B. Metabolomics—A new exciting field within the "omics" sciences. *Environ. Health Perspect.* **2004**, *112*, A396–A397. [CrossRef]
8. Schuhmacher, R.; Krska, R.; Weckwerth, W.; Goodacre, R. Metabolomics and metabolite profiling. *Anal. Bioanal. Chem.* **2013**, *405*, 5003–5004. [CrossRef] [PubMed]
9. Sun, J.; Charron, C.S.; Novotny, J.A.; Peng, B.; Yu, L.; Chen, P. Profiling glucosinolate metabolites in human urine and plasma after broccoli consumption using non-targeted and targeted metabolomic analyses. *Food Chem.* **2020**, *309*, 125660. [CrossRef] [PubMed]
10. Saude, E.J.; Skappak, C.D.; Regush, S.; Cook, K.; Ben-Zvi, A.; Becker, A.; Moqbel, R.; Sykes, B.D.; Rowe, B.H.; Adamko, D.J. Metabolomic profiling of asthma: Diagnostic utility of urine nuclear magnetic resonance spectroscopy. *J. Allergy Clin. Immunol.* **2011**, *127*, 757–764. [CrossRef] [PubMed]
11. Mattarucchi, E.; Baraldi, E.; Guillou, C. Metabolomics applied to urine samples in childhood asthma; differentiation between asthma phenotypes and identification of relevant metabolites. *Biomed. Chromatogr. BMC* **2012**, *26*, 89–94. [CrossRef]
12. Lara, A.; Khatri, S.B.; Wang, Z.; Comhair, S.A.A.; Xu, W.; Dweik, R.A.; Bodine, M.; Levison, B.S.; Hammel, J.; Bleecker, E.; et al. Alterations of the arginine metabolome in asthma. *Am. J. Respir. Crit. Care Med.* **2008**, *178*, 673–681. [CrossRef]
13. Fogarty, A.; Broadfield, E.; Lewis, S.; Lawson, N.; Britton, J. Amino acids and asthma: A case-control study. *Eur. Respir. J.* **2004**, *23*, 565–568. [CrossRef] [PubMed]
14. Morris, C.R.; Poljakovic, M.; Lavrisha, L.; Machado, L.; Kuypers, F.A.; Morris, S.M. Decreased arginine bioavailability and increased serum arginase activity in asthma. *Am. J. Respir. Crit. Care Med.* **2004**, *170*, 148–153. [CrossRef] [PubMed]
15. Jung, J.; Kim, S.-H.; Lee, H.-S.; Choi, G.S.; Jung, Y.-S.; Ryu, D.H.; Park, H.-S.; Hwang, G.-S. Serum metabolomics reveals pathways and biomarkers associated with asthma pathogenesis. *Clin. Exp. Allergy J. Br. Soc. Allergy Clin. Immunol.* **2013**, *43*, 425–433. [CrossRef] [PubMed]
16. Ried, J.S.; Baurecht, H.; Stückler, F.; Krumsiek, J.; Gieger, C.; Heinrich, J.; Kabesch, M.; Prehn, C.; Peters, A.; Rodriguez, E.; et al. Integrative genetic and metabolite profiling analysis suggests altered phosphatidylcholine metabolism in asthma. *Allergy* **2013**, *68*, 629–636. [CrossRef]
17. Checkley, W.; Deza, M.P.; Klawitter, J.; Romero, K.M.; Klawitter, J.; Pollard, S.L.; Wise, R.A.; Christians, U.; Hansel, N.N. Identifying biomarkers for asthma diagnosis using targeted metabolomics approaches. *Respir. Med.* **2016**, *121*, 59–66. [CrossRef]
18. Carraro, S.; Giordano, G.; Reniero, F.; Carpi, D.; Stocchero, M.; Sterk, P.J.; Baraldi, E. Asthma severity in childhood and metabolomic profiling of breath condensate. *Allergy* **2013**, *68*, 110–117. [CrossRef]
19. Kelly, R.S.; Dahlin, A.; McGeachie, M.J.; Qiu, W.; Sordillo, J.; Wan, E.S.; Wu, A.C.; Lasky-Su, J. Asthma Metabolomics and the Potential for Integrative Omics in Research and the Clinic. *Chest* **2017**, *151*, 262–277. [CrossRef]
20. Turi, K.N.; Romick-Rosendale, L.; Ryckman, K.K.; Hartert, T.V. A review of metabolomics approaches and their application in identifying causal pathways of childhood asthma. *J. Allergy Clin. Immunol.* **2018**, *141*, 1191–1201. [CrossRef]
21. Rodnina, M.V.; Beringer, M.; Wintermeyer, W. How Ribosomes Make Peptide Bonds. *Trends Biochem. Sci.* **2007**, *32*, 20–26. [CrossRef]
22. Curis, E.; Nicolis, I.; Moinard, C.; Osowska, S.; Zerrouk, N.; Bénazeth, S.; Cynober, L. Almost All about Citrulline in Mammals. *Amino Acids* **2005**, *29*, 177–205. [CrossRef]
23. Jin, R.; Banton, S.; Tran, V.T.; Konomi, J.V.; Li, S.; Jones, D.P.; Vos, M.B. Amino Acid Metabolism Is Altered in Adolescents with Nonalcoholic Fatty Liver Disease—An Untargeted, High Resolution Metabolomics Study. *J. Pediatr.* **2016**, *172*, 14–19.e5. [CrossRef] [PubMed]

24. Fadel, F.I.; Elshamaa, M.F.; Essam, R.G.; Elghoroury, E.A.; El-Saeed, G.S.M.; El-Toukhy, S.E.; Ibrahim, M.H. Some Amino Acids Levels: Glutamine, Glutamate, and Homocysteine, in Plasma of Children with Chronic Kidney Disease. *Int. J. Biomed. Sci.* **2014**, *10*, 36–42. [PubMed]
25. Sawy, M.E.; Zaki, M.M.; EL-Hakim, I.Z.; Mowafy, M.E.; Al-Abd, H.S. Serum amino acid abnormalities in pediatric patients with chronic renal failure with and without history of thromboembolic manifestations. *Egypt. J. Med. Hum. Genet.* **2012**, *13*, 73–80. [CrossRef]
26. Klupczyńska, A.; Dereziński, P.; Dyszkiewicz, W.; Pawlak, K.; Kasprzyk, M.; Kokot, Z.J. Evaluation of serum amino acid profiles' utility in non-small cell lung cancer detection in Polish population. *Lung Cancer* **2016**, *100*, 71–76. [CrossRef] [PubMed]
27. Matysiak, J.; Dereziński, P.; Klupczyńska, A.; Hajduk, J.P.; Swiatly, A.; Plewa, S.; Horała, A.; Jaźwiński, P.; Banach, P.; Nowak-Markwitz, E.; et al. Proteomic and metabolomic strategy of searching for biomarkers of genital cancer diseases using mass spectrometry methods. *J. Med. Sci.* **2016**, *85*, 330. [CrossRef]
28. Cheng, F.; Wang, Z.; Huang, Y.; Duan, Y.; Wang, X. Investigation of salivary free amino acid profile for early diagnosis of breast cancer with ultra performance liquid chromatography-mass spectrometry. *Clin. Chim. Acta Int. J. Clin. Chem.* **2015**, *447*, 23–31. [CrossRef]
29. Hasim, A.; Aili, A.; Maimaiti, A.; Mamtimin, B.; Abudula, A.; Upur, H. Plasma-free amino acid profiling of cervical cancer and cervical intraepithelial neoplasia patients and its application for early detection. *Mol. Biol. Rep.* **2013**, *40*, 5853–5859. [CrossRef]
30. Liu, X.; Zhang, M.; Liu, X.; Sun, H.; Guo, Z.; Tang, X.; Wang, Z.; Li, J.; Li, H.; Sun, W.; et al. Urine Metabolomics for Renal Cell Carcinoma (RCC) Prediction: Tryptophan Metabolism as an Important Pathway in RCC. *Front. Oncol.* **2019**, *9*, 663. [CrossRef]
31. Takashina, C.; Tsujino, I.; Watanabe, T.; Sakaue, S.; Ikeda, D.; Yamada, A.; Sato, T.; Ohira, H.; Otsuka, Y.; Oyama-Manabe, N.; et al. Associations among the plasma amino acid profile, obesity, and glucose metabolism in Japanese adults with normal glucose tolerance. *Nutr. Metab.* **2016**, *13*, 5. [CrossRef]
32. Yamakado, M.; Nagao, K.; Imaizumi, A.; Tani, M.; Toda, A. Plasma Free Amino Acid Profiles Predict Four-Year Risk of Developing Diabetes, Metabolic Syndrome, Dyslipidemia, and Hypertension in Japanese Population. *Sci. Rep.* **2015**, *5*, 11919. [CrossRef]
33. Gao, X.; Zhang, W.; Wang, Y.; Pedram, P.; Cahill, F.; Zhai, G.; Randell, E.; Gulliver, W.; Sun, G. Serum metabolic biomarkers distinguish metabolically healthy peripherally obese from unhealthy centrally obese individuals. *Nutr. Metab.* **2016**, *13*, 33. [CrossRef] [PubMed]
34. Tumor-Dependent Increase of Serum Amino Acid Levels in Breast Cancer Patients Has Diagnostic Potential and Correlates with Molecular Tumor Subtypes. Available online: https://www.ncbi.nlm.nih.gov/pmc/articles/PMC3835137/ (accessed on 25 June 2020).
35. Shi, T.; Tang, T.; Qian, K.; Wang, F.; Li, J.; Cao, Y. High-performance liquid chromatographic method for determination of amino acids by precolumn derivatization with 4-chloro-3,5-dinitrobenzotrifluoride. *Anal. Chim. Acta* **2009**, *654*, 154–161. [CrossRef]
36. Deng, C.; Li, N.; Zhang, X. Rapid determination of amino acids in neonatal blood samples based on derivatization with isobutyl chloroformate followed by solid-phase microextraction and gas chromatography/mass spectrometry. *Rapid Commun. Mass Spectrom.* **2004**, *18*, 2558–2564. [CrossRef]
37. Kraj, L.; Krawiec, M.; Koter, M.; Graboń, W.; Kraj, G.; Chołojczyk, M.; Kulus, M.; Barańczyk-Kuźma, A. Altered L-arginine metabolism in children with controlled asthma. *Allergy Asthma Proc.* **2014**, *35*, 80–83. [CrossRef] [PubMed]
38. Matysiak, J.; Dereziński, P.; Klupczyńska, A.; Matysiak, J.; Kaczmarek, E.; Kokot, Z.J. Effects of a Honeybee Sting on the Serum Free Amino Acid Profile in Humans. *PLoS One* **2014**, *9*, e103533. [CrossRef]
39. Held, P.K.; White, L.; Pasquali, M. Quantitative urine amino acid analysis using liquid chromatography tandem mass spectrometry and aTRAQ reagents. *J. Chromatogr. B Anal. Technol. Biomed. Life Sci.* **2011**, *879*, 2695–2703. [CrossRef]
40. Filee, R.; Schoos, R.; Boemer, F. Evaluation of Physiological Amino Acids Profiling by Tandem Mass Spectrometry. *JIMD Rep.* **2014**, *13*, 119–128.
41. Hogenboom, J.A.; D'Incecco, P.; Fuselli, F.; Pellegrino, L. Ion-Exchange Chromatographic Method for the Determination of the Free Amino Acid Composition of Cheese and Other Dairy Products: An Inter-Laboratory Validation Study. *Food Anal. Methods* **2017**, *10*, 3137–3148. [CrossRef]

42. Xia, J.; Sinelnikov, I.V.; Han, B.; Wishart, D.S. MetaboAnalyst 3.0—Making metabolomics more meaningful. *Nucleic Acids Res.* **2015**, *43*, W251–W257. [CrossRef]
43. Maarsingh, H.; Zaagsma, J.; Meurs, H. Arginine homeostasis in allergic asthma. *Eur. J. Pharmacol.* **2008**, *585*, 375–384. [CrossRef]
44. Zimmermann, N.; Rothenberg, M.E. The arginine-arginase balance in asthma and lung inflammation. *Eur. J. Pharmacol.* **2006**, *533*, 253–262. [CrossRef] [PubMed]
45. Morris, C.R. Arginine and asthma. *Nestle Nutr. Inst. Workshop Ser.* **2013**, *77*, 1–15. [PubMed]
46. Kelly, R.S.; McGeachie, M.J.; Lee-Sarwar, K.A.; Kachroo, P.; Chu, S.H.; Virkud, Y.V.; Huang, M.; Litonjua, A.A.; Weiss, S.T.; Lasky-Su, J. Partial Least Squares Discriminant Analysis and Bayesian Networks for Metabolomic Prediction of Childhood Asthma. *Metabolites* **2018**, *8*, 68. [CrossRef] [PubMed]
47. Lambert, I.H.; Kristensen, D.M.; Holm, J.B.; Mortensen, O.H. Physiological role of taurine—From organism to organelle. *Acta Physiol. (Oxf. Engl.)* **2015**, *213*, 191–212. [CrossRef]
48. Ripps, H.; Shen, W. Review: Taurine: A "very essential" amino acid. *Mol. Vis.* **2012**, *18*, 2673–2686.
49. Mongin, A.A. Potential link between cysteinyl-leukotriene receptors and release of bioactive amino acids in regulation of lung function. Focus on "Volume-sensitive release of organic osmolytes in the human lung epithelial cell line A549: Role of the 5-lipoxygenase". *Am. J. Physiol. Cell Physiol.* **2013**, *305*, C24–C25. [CrossRef]
50. Gallos, G.; Yim, P.; Chang, S.; Zhang, Y.; Xu, D.; Cook, J.M.; Gerthoffer, W.T.; Emala, C.W. Targeting the restricted α-subunit repertoire of airway smooth muscle $GABA_A$ receptors augments airway smooth muscle relaxation. *Am. J. Physiol. Lung Cell. Mol. Physiol.* **2012**, *302*, L248–L256. [CrossRef]
51. Comhair, S.A.A.; McDunn, J.; Bennett, C.; Fettig, J.; Erzurum, S.C.; Kalhan, S.C. Metabolomic Endotype of Asthma. *J. Immunol. (Baltim. Md. 1950)* **2015**, *195*, 643–650. [CrossRef]
52. Bridges, C.C.; Ola, M.S.; Prasad, P.D.; El-Sherbeny, A.; Ganapathy, V.; Smith, S.B. Regulation of taurine transporter expression by NO in cultured human retinal pigment epithelial cells. *Am. J. Physiol. Cell Physiol.* **2001**, *281*, C1825–C1836. [CrossRef]
53. Zinellu, A.; Fois, A.G.; Sotgia, S.; Zinellu, E.; Bifulco, F.; Pintus, G.; Mangoni, A.A.; Carru, C.; Pirina, P. Plasma protein thiols: An early marker of oxidative stress in asthma and chronic obstructive pulmonary disease. *Eur. J. Clin. Investig.* **2016**, *46*, 181–188. [CrossRef]
54. Sugimoto, M.; Wong, D.T.; Hirayama, A.; Soga, T.; Tomita, M. Capillary electrophoresis mass spectrometry-based saliva metabolomics identified oral, breast and pancreatic cancer-specific profiles. *Metab. Off. J. Metab. Soc.* **2010**, *6*, 78–95. [CrossRef] [PubMed]
55. Fang, F.; He, X.; Deng, H.; Chen, Q.; Lu, J.; Spraul, M.; Yu, Y. Discrimination of metabolic profiles of pancreatic cancer from chronic pancreatitis by high-resolution magic angle spinning 1H nuclear magnetic resonance and principal components analysis. *Cancer Sci.* **2007**, *98*, 1678–1682. [CrossRef] [PubMed]
56. Ascha, M.; Wang, Z.; Ascha, M.S.; Dweik, R.; Zein, N.N.; Grove, D.; Brown, J.M.; Marshall, S.; Lopez, R.; Hanouneh, I.A. Metabolomics studies identify novel diagnostic and prognostic indicators in patients with alcoholic hepatitis. *World J. Hepatol.* **2016**, *8*, 499–508. [CrossRef] [PubMed]
57. Kinny-Köster, B.; Bartels, M.; Becker, S.; Scholz, M.; Thiery, J.; Ceglarek, U.; Kaiser, T. Plasma Amino Acid Concentrations Predict Mortality in Patients with End-Stage Liver Disease. *PLoS ONE* **2016**, *11*, e0159205. [CrossRef]
58. Budhathoki, S.; Iwasaki, M.; Yamaji, T.; Yamamoto, H.; Kato, Y.; Tsugane, S. Association of plasma concentrations of branched-chain amino acids with risk of colorectal adenoma in a large Japanese population. *Ann. Oncol. Off. J. Eur. Soc. Med. Oncol.* **2017**, *28*, 818–823. [CrossRef]
59. Yu, D.; Moore, S.C.; Matthews, C.E.; Xiang, Y.-B.; Zhang, X.; Gao, Y.-T.; Zheng, W.; Shu, X.-O. Plasma metabolomic profiles in association with type 2 diabetes risk and prevalence in Chinese adults. *Metab. Off. J. Metab. Soc.* **2016**, *12*, 3. [CrossRef]
60. Motta, A.; Paris, D.; D'Amato, M.; Melck, D.; Calabrese, C.; Vitale, C.; Stanziola, A.A.; Corso, G.; Sofia, M.; Maniscalco, M. NMR metabolomic analysis of exhaled breath condensate of asthmatic patients at two different temperatures. *J. Proteome Res.* **2014**, *13*, 6107–6120. [CrossRef]
61. Ghosh, N.; Choudhury, P.; Subramani, E.; Saha, D.; Sengupta, S.; Joshi, M.; Banerjee, R.; Roychowdhury, S.; Bhattacharyya, P.; Chaudhury, K. Metabolomic signatures of asthma-COPD overlap (ACO) are different from asthma and COPD. *Metab. Off. J. Metab. Soc.* **2019**, *15*, 87. [CrossRef]

62. Chiu, C.-Y.; Cheng, M.-L.; Chiang, M.-H.; Wang, C.-J.; Tsai, M.-H.; Lin, G. Metabolomic Analysis Reveals Distinct Profiles in the Plasma and Urine Associated with IgE Reactions in Childhood Asthma. *J. Clin. Med.* **2020**, *9*, 887. [CrossRef]
63. Smith, I.C.P.; Blandford, D.E. Nuclear magnetic resonance spectroscopy. *Anal. Chem.* **1995**, *67*, 509–518. [CrossRef]
64. Xiang, Y.-Y.; Wang, S.; Liu, M.; Hirota, J.A.; Li, J.; Ju, W.; Fan, Y.; Kelly, M.M.; Ye, B.; Orser, B.; et al. A GABAergic system in airway epithelium is essential for mucus overproduction in asthma. *Nat. Med.* **2007**, *13*, 862–867. [CrossRef] [PubMed]
65. Tanianskii, D.A.; Jarzebska, N.; Birkenfeld, A.L.; O'Sullivan, J.F.; Rodionov, R.N. Beta-Aminoisobutyric Acid As a Novel Regulator of Carbohydrate and Lipid Metabolism. *Nutrients* **2019**, *11*, 524. [CrossRef] [PubMed]
66. Trifonova, O.P.; Lokhov, P.G.; Archakov, A.I. Metabolic profiling of human blood. *Biomeditsinskaia Khimiia* **2014**, *60*, 281–294. [CrossRef] [PubMed]

© 2020 by the authors. Licensee MDPI, Basel, Switzerland. This article is an open access article distributed under the terms and conditions of the Creative Commons Attribution (CC BY) license (http://creativecommons.org/licenses/by/4.0/).

International Journal of
Environmental Research and Public Health

MDPI

Article

Bone Metabolism in Patients Treated for Depression

Elżbieta Skowrońska-Jóźwiak [1,2,*], Piotr Gałecki [3], Ewa Głowacka [4], Cezary Wojtyła [5,6], Przemysław Biliński [7,8] and Andrzej Lewiński [1,2]

1 Department of Endocrinology and Metabolic Diseases, Medical University of Lodz, 93-338 Lodz, Poland; andrzej.lewinski@umed.lodz.pl
2 Department of Endocrinology and Metabolic Diseases, Polish Mother's Memorial Hospital—Research Institute, 93-338 Lodz, Poland
3 Department of Adult Psychiatry, Medical University of Lodz, 91-229 Lodz, Poland; piotr.galecki@umed.lodz.pl
4 Department of Laboratory Diagnostics, Polish Mother's Memorial Hospital—Research Institute, 93-338 Lodz, Poland; ewa.biol@interia.pl
5 Department of Oncological Gynecology and Obstetrics, Center of Postgraduate Medical Education, 00-416 Warsaw, Poland; czwo@op.pl
6 International Prevention Research Institute—Collaborating Centre, State University of Applied Sciences, 62-800 Kalisz, Poland
7 Faculty of Heath Sciences, State University of Applied Sciences, 62-800 Kalisz, Poland; bildom@gmail.com
8 Copernicus Memorial Comprehensive Cancer Center and Traumatology, 93-513 Lodz, Poland
* Correspondence: elzbieta.skowronska-jozwiak@umed.lodz.pl; Tel.: +48-422-7111-42

Received: 25 May 2020; Accepted: 30 June 2020; Published: 2 July 2020

Abstract: Background: Depression and osteoporosis are severe public health problems. There are conflicting findings regarding the influence of depression on bone metabolism. The aim of the presented study was to compare bone turnover markers and vitamin D levels between patients treated for depression and healthy controls. Patients and Methods: We determined a concentration of osteocalcin, carboxy-terminal telopeptide of type I collagen (β-CTX), 25-hydroxyvitamin D (25OHD) and $1,25(OH)_2D_3$ in 99 patients, aged 46.9 ± 11 years, treated for depression, as well as in 45 healthy subjects. Depressive status was determined with the Hamilton Depression Scale (HDRS). Results: In patients treated for depression, we demonstrated significantly lower osteocalcin concentrations ($p < 0.03$) and higher concentration of β-CTX (result on the border of significance; $p = 0.08$). Those relationship were stronger in women. The level of 25OHD and $1,25(OH)_2D_3$ did not differ significantly between the examined groups. We observed a negative correlation between the 25OHD and HDRS score after treatment in all patients treated for depression and in subgroups of women and subjects with recurrent depression. Conclusions: Our results indicate that depression is related to disturbances in bone metabolism, especially in women and patients with recurrent depression, suggesting its role in context of osteoporosis development.

Keywords: depression; bone markers; osteocalcin; β-CTX; vitamin D

1. Introduction

Osteoporosis is a severe public health problem. The essence of the disease is reduced bone strength, predisposing to an increased risk for fractures. As many as 9 million low-energy fractures are recorded worldwide every year [1–3]. According to the International Osteoporosis Foundation (IOF) data, over 200 million women in the world suffer from osteoporosis, the majority being above 60 years of age. In 2017, new fragility fractures in Europe were estimated at 2.7 million with an associated annual cost of €37.5 billion and a loss of 1.0 million quality-adjusted life years, with a projected increase to 3.3 million in 2030 [3]. Osteoporotic fractures are a major cause of morbidity and mortality in the population [2,3]. Data obtained in 2005 from the Polish Health Care Fund gives information about 17,625 hip fracture (mean incidence 224/100,000) [4]. A total of 289,230 hip fractures were observed in Poland in the years 2008–2015, with the incidence increasing from 192/100,000 in 2008 to 444/100,000 in 2015 [5]. The one-year mortality ranged between 30.45% and 32.8% for men and 26.2% and 28% for women [5]. In Europe, osteoporosis accounted for more disability and life years lost than rheumatoid arthritis, but less than osteoarthritis. With regard to neoplastic diseases, the burden of osteoporosis was greater than for all sites of cancer, with the exception of lung cancer [2]. A large number of risk factors for fractures have been identified: age, gender, postmenopausal status, body mass index, and smoking [6,7]. Identification of risk factors is important for early diagnosis and treatment, resulting in a significant public health benefit [7]. Recently, depression and its treatment were suspected as risk factors for osteoporosis development [8–10]. The problem is a significant public health issue since depression affects 98.7 million people globally, being a major cause of disability worldwide [11]. Depression has been associated with a low bone mineral density [8,9]. Moreover, the increased risk of fracture in patients with depression was demonstrated; 26% when the hazard risk was assessed (n = 309,862) and 39% when the risk ratio was determined (n = 64,975) [9]. Data on the influence of depression therapy on bone metabolism, especially with SSRIs (selective serotonin reuptake inhibitors), are conflicting [10,12]. There are many hypotheses associating depression with osteoporosis [13–20]. The first mechanism is similar to steroid-induced osteoporosis and affects the hypothalamic–pituitary–adrenal axis; in subjects suffering from depression elevated ACTH and cortisol levels have been demonstrated [13,14]. Moreover, in depressed individuals, glucocorticoid tissue hypersensitivity due to a greater prevalence of Bcl1 polymorphism has been observed [15]. The second mechanism is related to overproduction of inflammatory cytokines, such as interleukins IL-1β, IL-2, IL-6, and tumor necrosis factor-alpha, in patients with depression [16]. Other, probably coexisting factors to consider are the propensity to fall [17], low mobility [18], adoption of poor health behaviors (e.g., smoking and physical inactivity), and drugs, especially SSRIs [10,12,18]. Vitamin D deficiency seems to be an important factor, connecting osteoporosis with depression [8–10]. Vitamin D has a dual role in the human body, acting as a hormone and fat-soluble vitamin, regulating the expression of over 900 genes by binding to the vitamin D receptor (VDR) [19]. The presence of VDR receptors has been confirmed in more than 30 cell types in the human body, including nerve cells as well as in the glia of the cerebral cortex and hippocampus [20], locations relevant to depression [21]. The enzyme necessary for the hydroxylation of 25OHD to its active form, 1,25$(OH)_2D_3$, is present in the hypothalamus, cerebellum, and black substance (substantia nigra) [22]. Moreover, deficiency of vitamin D is a global problem [23], what also has been confirmed both in the general Polish population [24] and the population of Polish patients suffering from depression [25].

Bone metabolism has never been assessed in a population of Polish patients suffering from depression. The aim of the present study was to compare bone metabolism and vitamin D supply between patients treated for depression and the healthy controls.

2. Patients and Methods

The procedures used in the study were approved by Bioethics Committee of the Polish Mother's Memorial Hospital—Research Institute (consent number: No.75/2015).

Patients with depression participating in the study were recruited from patients undergoing hospital treatment in the Department of Adult Psychiatry of the Medical University of Lodz.

The inclusion criteria for participation in the depression group were based on the diagnostic criteria of depressive episodes and recurrent depressive disorder according to International Classification of Diseases (ICD-10) (F32.0-7.32.2, F33.0-F33.8) [ICD-10]. Exclusion criteria for participation in the depression group were as follows: central nervous system injuries; history of inflammatory, autoimmune, and cancer diseases; abuse or addiction to psychoactive substances; lack of consent to participate in the study diseases; and/or treatments that affect bone turnover (except vitamin D supplementation—during hospitalization patients were taking 1000 IU of vitamin D daily.) The list of factors affecting bone metabolism consisted of a history of osteoporosis; thyroid diseases; rheumatoid arthritis; malabsorption; hypogonadism; undergoing treatment with the usage of substances, such as steroids, bisphosphonates, and denosumab; and abnormal result of any of the following laboratory tests: complete blood count, creatinine, calcium, phosphates, CRP (C-reactive protein), TSH (thyroid stimulating hormone), and testosterone. Patients with depression were divided into two subgroups—those with a first episode of depression and those with recurrent depression.

Healthy, non-depressed controls were recruited from the Outpatient Clinic of Osteoporosis Polish Mother's Memorial Hospital—Research Institute, Lodz, after exclusion of osteoporosis.

Exclusion criteria to participate in the control group were a negative history of depression, osteoporosis, and diseases and/or treatments that affect bone turnover (corresponding to group of patients with depression). Nine subjects were excluded from the study due to a diagnosis of primary hyperthyroidism ($n = 2$), hypothyroidism ($n = 2$), renal failure ($n = 1$), hypocalcemia ($n = 1$), steroid treatment ($n = 1$), failed vitamin D assessment ($n = 1$), and lack of consent ($n = 1$). Qualification to participate in the study was random, based on non-returnable draw rules. Participation in the study did not affect the principles of treatment. Before proceeding with the study, each of the subjects signed an informed consent to participate in the study in accordance with the protocol.

The Hamilton Depression Rating Scale (HDRS) was used to assess the severity and dynamics of the symptoms of depression [26] on the admission to hospital (HDRS I) and after obtaining clinical improvement, on average after 8 weeks of treatment (HDRS II). We used the Polish version of the HDRS, validated by Wiglusz and et al. [27], containing 17 items, conducted by qualified personnel. HDRS-17 demonstrated the best psychometric properties for a cutoff score of 11 with a sensitivity of 100%, specificity of 89.3%, positive predictive value of 72.4%, and negative predictive value of 100% [27].

Regarding the biochemical measurements, osteocalcin (OC), a non-collagenous protein of the bone matrix, synthesized by osteoblasts, was chosen as a marker of bone formation. As a marker of bone resorption, we used the carboxy-terminal telopeptide of type I collagen (β-CTX) serum assay. OC, β-CTX, and vitamin D (25OHD) were determined by commercially available electrochemiluminescence immunoassays (ECLIA Cobas e601, Roche, Rotkreuz, Switzerland). The $1{,}25(OH)_2D_3$ measurement was performed by automated direct immunoassay on the Liaison XL (DiaSorin S.p.A, Saluggia, Italy). The intra-assay and inter-assay CVs were less than 4.0%.

Statistics: Shapiro–Wilk's test was used to assess the distribution of the variables. A one-way ANOVA was applied for statistical analysis for continuous variables with the subsequent use of a post-hoc test in order to assess differences between particular groups. In case of a non-normal distribution of data, Mann–Whitney's test was used. Associations between HDRS and 25OHD were investigated with Spearman's coefficient of correlation. Statistically significant differences were accepted when the p value was below 0.05. The analyses were carried out using STATISTICA 13 software (StatSoft, Kraków, Poland).

3. Results

We recruited 144 subjects to the present study; 99 patients treated for depression and 45 healthy subjects constituting the control group. Characteristics of the studied population are presented in Table 1. The compared groups do not differ significantly according to age, gender, BMI, and smoking ($p > 0.1$).

Table 1. Characteristics of the studied group presented as the mean ± SD. HDRS I (Hamilton Depression Rating Scale score on admission); HDRS II (Hamilton Depression Rating Scale score after obtaining clinical improvement).

	Controls $n = 45$	All Patients with Depression $n = 99$	The First Episode of Depression $n = 31$	Recurrent Depression $n = 68$
Age (years)	45.8 ± 10.8	46.9 ± 11	43.3 ± 12.5	49.5 ± 9.9
Women/men	24/21	55/42	17/14	38/27
HDRS I	-	22.9 ± 7.2	21.9 ± 7.1	23.4 ± 7.1
HDRS II	-	6.6 ± 4.1	5.4 ± 4.0	6.2 ± 3.9
Number of depressive episodes	-	5.9 ± 3.9	-	6.8 ± 3.5
Duration of disease (years)	-	6.9 ± 5.5	1.9 ± 1.8	9.1± 8.4

We demonstrated significantly lower osteocalcin concentrations in patients treated for depression when compared to the healthy controls—Table 2. We also found a higher concentration of βCTX in depression, however it not did reach the level of significance. The level of 25OHD and 1,25(OH)$_2$D$_3$ did not differ significantly between the studied groups—Table 2.

Table 2. Vitamin D and bone markers in patients with depression and the healthy controls. Data presented as the mean ± SD; 25OHD—25-hydroxyvitamin D; 1,25(OH)$_2$D$_3$—1,25-dihydroxyvitamin D$_3$; β-CTX—carboxy-terminal telopeptide of type I collagen; p-level of significance.

	Reference Range	Depression	Controls	p
25OHD (ng/mL)	Optimal > 30 Insufficiency 20–30 Deficiency < 20	18.84 ± 7.09	17.56 ± 6.6	0.32
1,25(OH)$_2$D$_3$ (ng/L)	29–83.6	51.33 ± 15.37	55.95 ± 7.07	0.37
β-CTX (ng/dL)	130–710	507 ± 320	389 ± 158.8	0.08
Osteocalcin (ng/mL)	15–46	19.17 ± 7.86	22.88 ± 9.44	0.03

When women and men were analyzed separately, we observed a significantly higher β-CTX concentration in women with depression than in healthy women—Table 3. Similarly, we noticed a difference between the concentration of 25OHD and osteocalcin, but the results were on the level of significance. In contrast, in the group of men we did not find such relations (p = NS). Women and men with depression did not differ according to age ($p = 0.42$), duration of disease ($p = 0.85$), and HDRS I or HDRS II results ($p = 0.4$ and $p = 0.99$, respectively).

Table 3. Vitamin D and bone markers in women with depression and healthy controls. Data presented as the mean ± SD; 25OHD—25-hydroxyvitamin D; 1,25(OH)$_2$D$_3$—1,25-dihydroxyvitamin D$_3$; β-CTX—carboxy-terminal telopeptide of type I collagen; p-value level of significance.

	Women with Depression	Healthy Women	p
25OHD (ng/mL)	19.34 ± 6.31	16.57 ± 6.74	0.055
1,25(OH)$_2$D$_3$ (ng/L)	51.67 ± 15.32	50.33 ± 18.31	0.9
β-CTX (ng/dL)	521.08 ± 290.34	314.59 ± 115.78	0.01
Osteocalcin (ng/mL)	19.37 ± 7.34	23.17 ± 10.22	0.059

Mean concentration of 25OHD did not differ in the examined groups. The optimal concentration of 25OHD (>30 ng/mL) was found only in 4.3%, despite cholecalciferol supplementation. Deficiency of vitamin D was frequently observed in the studied population, both in the depression group and the control group. We found significant correlation ($p < 0.05$) between the 25OHD levels and HRDS II score in all studied patients with depression (rs = −0.33,), in all women with depression (rs = −0.56), in subjects with recurrent depression (rs = −0.39), and in women with recurrent depression (rs = −0.45). Moreover, in all mentioned groups, 25OHD was significantly higher in the groups of patients with an HDRS II score below 7 points compared to patients with an HDRS II score higher or equal to 7 points ($p < 0.03$).

4. Discussion

We demonstrated that markers of bone turnover in patients treated for depression took values characteristic of accelerated bone loss (lower osteocalcin and elevated β-CTX), which might lead to osteoporosis development in that group of patients. That relationship was stronger in women with depression compared to the healthy controls. These results are similar to those observed by other researchers [14,18,28–31], who also studied young women with major depression. Generally, women are most likely to develop osteoporosis [1–3]. However, this statement applies rather to postmenopausal women. Interestingly, in our study young women in the premenopausal period were at risk. The data we collected did not confirm increased bone turnover in men, while data from the literature concerning males are divergent, both according to age and drug use [29,31].

A lot of data on vitamin D deficiency on the course of depression has been published [25,32,33]. According to [32] (n = 5607 subjects) and [33] (n = 1602 subjects), vitamin D deficiency can be a significant cause of late-life depression in older people, especially among men over 65 years of age. In addition, in a 10-year longitudinal study, the relationship between the 25OHD level and the effectiveness of cognitive processes dependent on the prefrontal cortex was confirmed [34]. This is further supported by findings showing effects of 1,25-OH_2D_3 on the synthesis of nerve growth factor (NGF). 1,25-Dihydroxyvitamin D3 is a potent inducer of nerve growth factor synthesis [35]. The synthesis of tryptophan is also transcriptionally activated by vitamin D; thus, low 25OHD levels may accordingly cause low levels of serotonin [36].

Surprisingly, in our study, unlike previous data [25,37], neither 25OHD nor 1,25$(OH)_2D_3$ levels in patients with depression were different from the healthy controls. Even paradoxically, in women with depression, the 25OHD concentration was higher than in the controls. It was related to supplementation of vitamin D by patients with depression. This supplementation, while conducted irregularly without dose titration, led to increasing vitamin D concentration and achieving the level similar to the control group not taking vitamin D. However, in all studied subjects, the 25OHD levels were lower than the optimal according to recommendations of the Endocrine Society [23]. Since we observed the negative correlation between 25OHD and the Hamilton score after treatment in all depressed patients, particularly in women and in patients with recurrent depression, we advocate for more careful supplementation of vitamin D in that group of patients. We are aware of conflicting results of vitamin D supplementation in depression [37–40], so we treat this supplementation rather as a necessary element of osteoporosis prophylactics. Nevertheless, it is noteworthy that the beneficial effects of vitamin D supplementation among those with a replete vitamin D status were suggested [37,40].

We are aware of the limitations of our study, which include a lack data according diet habits—in particular calcium and protein intake, alcohol consumption, and physical activity. There is no doubt that a relatively strong association exists between osteoporosis and the mentioned factors, but we would like to emphasize that the survey methods of their assessment still do not fulfill the criteria of being valid, reliable, and practical, which was shown in relation to physical activity [41] and especially to alcohol consumption [42]. Other limitations are a lack of densitometry results and a small number of studied patients; however, the number is comparable with other studies [18,28]. Strengths of the study included an investigation carried out on a large group of well-defined patients with depression, seeming to be representative enough for the Polish population.

In conclusion, our study denotes a relationship between depression and bone metabolism, particularly in women and in patients with recurrent depression, likely predisposing osteoporosis development. However, future trials are needed, targeting depressed patients for developing an optimal model of osteoporosis prophylactics in these patients.

Author Contributions: Conceptualization E.S.-J. and P.G.; investigation E.G., C.W., P.G., P.B.; methodology E.S.-J. and P.G.; supervision, E.S.-J. and A.L.; validation, E.S.-J. and A.L.; writing—original draft, E.S.-J.; writing—review and editing, P.G. and A.L. All authors have read and agreed to the published version of the manuscript.

Funding: This work was supported by statutory funds from the Medical University of Lodz and the Polish Mother's Memorial Hospital—Research Institute, Lodz, Poland.

Conflicts of Interest: The authors declare that they have no competing interests.

References

1. Hernlund, E.; Svedbom, A.; Ivergård, M.; Compston, J.; Cooper, C.; Stenmark, J.; McCloskey, E.V.; Jönsson, B.; Kanis, J.A. Osteoporosis in the European Union: Medical management, epidemiology and economic burden. A report prepared in collaboration with the International Osteoporosis Foundation (IOF) and the European Federation of Pharmaceutical Industry Associations (EFPIA). *Arch. Osteoporos.* **2013**, *8*, 136. [CrossRef] [PubMed]
2. Johnell, O.; Kanis, J.A. An estimate of the worldwide prevalence and disability associated with osteoporotic fractures. *Osteoporos. Int.* **2006**, *17*, 1726–1733. [CrossRef]
3. Borgström, F.; Karlsson, L.; Ortsäter, G.; Norton, N.; Halbout, P.; Cooper, C.; Lorentzon, M.; McCloskey, E.V.; Harvey, N.C.; Javaid, M.K.; et al. International Osteoporosis Foundation. Fragility fractures in Europe: Burden, management and opportunities. *Arch. Osteoporos.* **2020**, *15*, 59. [CrossRef]
4. Czerwinski, E.; Kanis, J.A.; Trybulec, B.; Johansson, H.; Borowy, P.; Osieleniec, J. The incidence and risk of hip fracture in Poland. *Osteoporos. Int.* **2009**, *20*, 1363–1367. [CrossRef] [PubMed]
5. Glinkowski, W.; Narloch, J.; Krasuski, K.; Śliwczyński, A. The increase of osteoporotic hip fractures and associated one-year mortality in Poland: 2008–2015. *J. Clin. Med.* **2019**, *8*, 1487. [CrossRef] [PubMed]
6. Brandt, A.; Zatoński, M. The tobacco endgame: Where are we, and what are the challenges ahead? *J. Health Inequalities* **2019**, *5*, 28–30. [CrossRef]
7. Zatoński, W.A.; Wojtyła, A. From the editors. *J. Health Inequalities* **2019**, *5*, 121. [CrossRef]
8. Cizza, G.; Primma, S.; Coyle, M.; Gourgiotis, L.; Csako, G. Depression and osteoporosis: A research synthesis with meta-analysis. *Horm. Metab. Res.* **2010**, *42*, 467–482. [CrossRef]
9. Wu, Q.; Liu, B.; Tonmoy, S. Depression and risk of fracture and bone loss: An updated meta-analysis of prospective studies. *Osteoporos. Int.* **2018**, *29*, 1303–1312. [CrossRef]
10. Williams, L.J.; Berk, M.; Hodge, J.M.; Kotowicz, M.A.; Stuart, A.L.; Chandrasekaran, V.; Cleminson, J.; Pasco, J.A. Selective serotonin reuptake inhibitors (SSRIs) and markers of bone turnover in men. *Calcif. Tissue Int.* **2018**, *103*, 125–130. [CrossRef]
11. World Health Organization. *The global Burden of Disease: 2004 Update*; World Health Organization: Geneva, Switzerland, 2008; ISBN 92 4 156257 9.
12. Schweiger, J.U.; Schweiger, U.; Hüppe, M.; Kahl, K.G.; Greggersen, W.; Jauch-Chara, K.; Fassbinder, E. The use of antidepressive agents and bone mineral density in women: A meta-analysis. *Int. J. Environ. Res. Public Health.* **2018**, *15*, 1373. [CrossRef] [PubMed]
13. Nandam, L.S.; Brazel, M.; Zhou, M.; Jhaveri, D.J. Cortisol and major depressive disorder—Translating findings from humans to animal models and back. *Front. Psychiatry* **2019**, *10*, 974. [CrossRef] [PubMed]
14. Altindag, O.; Altindag, A.; Asoglu, M.; Gunes, M.; Soran, N.; Deveci, Z. Relation of cortisol levels and bone mineral density among premenopausal women with major depression. *Int. J. Clin. Pract.* **2007**, *61*, 416–420. [CrossRef] [PubMed]
15. van Rossum, E.F.; Binder, E.B.; Majer, M.; Koper, J.W.; Ising, M.; Modell, S.; Salyakina, D.; Lamberts, S.W.; Holsboer, F. Polymorphisms of the glucocorticoid receptor gene and major depression. *Biol. Psychiatry* **2006**, *59*, 681–688. [CrossRef] [PubMed]

16. Simon, N.M.; McNamara, K.; Chow, C.W.; Maser, R.S.; Papakostas, G.I.; Pollack, M.H.; Nierenberg, A.A.; Fava, M.; Wong, K.K. A detailed examination of cytokine abnormalities in major depressive disorder. *Eur. Neuropsychopharmacol.* **2008**, *18*, 230–233. [CrossRef]
17. Stuart, A.L.; Pasco, J.A.; Jacka, F.N.; Berk, M.; Williams, L.J. Falls and depression in men: A population-based study. *Am. J. Men's Health* **2018**, *12*, 14–18. [CrossRef]
18. Petronijević, M.; Petronijević, N.; Ivković, M.; Stefanović, D.; Radonjić, N.; Glisić, B.; Ristić, G.; Damjanović, A.; Paunović, V. Low bone mineral density and high bone metabolism turnover in premenopausal women with unipolar depression. *Bone* **2008**, *42*, 582–590. [CrossRef]
19. Lu, M.; Taylor, B.V.; Körner, H. Genomic effects of the vitamin d receptor: Potentially the link between vitamin D, immune cells, and multiple sclerosis. *Front. Immunol.* **2018**, *9*, 477. [CrossRef]
20. Fernandes de Abreu, D.; Eyles, D.; Feron, F. Vitamin D, a neuro-immunomodulator: Implications for neurodegenerative and autoimmune diseases. *Psychoneuroendocrinology* **2009**, *34*, S265–S277. [CrossRef]
21. Gałecki, P.; Talarowska, M.; Anderson, G.; Berk, M.; Maes, M. Mechanisms underlying neurocognitive dysfunctions in recurrent major depression. *Med. Sci. Monit.* **2015**, *21*, 1535–1547. [CrossRef] [PubMed]
22. Obradovic, D.; Gronemeyer, H.; Lutz, B.; Rein, T. Cross-talk of Vitamin D and glucocorticoids in hippocampal cells. *J. Neurochem.* **2006**, *96*, 500–509. [CrossRef]
23. Holick, M.F.; Binkley, N.C.; Bischoff-Ferrari, H.A.; Gordon, C.M.; Hanley, D.A.; Heaney, R.P.; Murad, M.H.; Weaver, C.M. Endocrine Society. Evaluation, treatment, and prevention of vitamin D deficiency: An Endocrine Society clinical practice guideline. *J. Clin. Endocrinol. Metab.* **2011**, *96*, 1911–1930. [CrossRef] [PubMed]
24. Płudowski, P.; Ducki, C.; Konstantynowicz, J.; Jaworski, M. Vitamin D status in Poland. *Pol. Arch. Med.* **2016**, *126*, 530–539. [CrossRef]
25. Józefowicz, O.; Rabe-Jabłońska, J.; Woźniacka, A.; Strzelecki, D. Analysis of vitamin D status in major depression. *J. Psychiatr. Pract.* **2014**, *20*, 329–337. [CrossRef] [PubMed]
26. Hamilton, M. A rating scale for depression. *J. Neurol. Neurosurg. Psychiatry* **1960**, *23*, 56–62. [CrossRef] [PubMed]
27. Wiglusz, M.S.; Landowski, J.; Michalak, L.; Cubała, W.J. Validation of the Polish Version of the Hamilton Rating Scale for Depression in patients with epilepsy. *Epilepsy Behav.* **2016**, *62*, 81–84. [CrossRef]
28. Aydin, H.; Mutlu, N.; Akbas, N.B. Treatment of a major depression episode suppresses markers of bone turnover in premenopausal women. *J. Psychiatr. Res.* **2011**, *4*, 1316–1320. [CrossRef] [PubMed]
29. Rawson, K.S.; Dixon, D.; Civitelli, R.; Peterson, T.R.; Mulsant, B.H.; Reynolds, C.F., 3rd; Lenze, E.J. Bone turnover with venlafaxine treatment in older adults with depression. *J. Am. Geriatr. Soc.* **2017**, *65*, 2057–2063. [CrossRef] [PubMed]
30. Lee, C.W.; Liao, C.H.; Lin, C.L.; Liang, J.A.; Sung, F.C.; Kao, C.H. Increased risk of osteoporosis in patients with depression: A population-based retrospective cohort study. *Mayo Clin. Proc.* **2015**, *90*, 63–70. [CrossRef]
31. Rauma, P.H.; Pasco, J.A.; Berk, M.; Stuart, A.L.; Koivumaa-Honkanen, H.; Honkanen, R.J.; Hodge, J.M.; Williams, L.J. The association between major depressive disorder, use of antidepressants and bone mineral density (BMD) in men. *J. Musculoskelet. Neuronal Interact.* **2015**, *5*, 177–185.
32. de Oliveira, C.; Hirani, V.; Biddulph, J.P. Associations between vitamin D levels and depressive symptoms in later life: Evidence from the English Longitudinal Study of Ageing (ELSA). *J. Gerontol. A Biol. Sci. Med. Sci.* **2018**, *73*, 1377–1382. [CrossRef] [PubMed]
33. Vidgren, M.; Virtanen, J.K.; Tolmunen, T.; Nurmi, T.; Tuomainen, T.P.; Voutilainen, S.; Ruusunen, A. Serum concentrations of 25-hydroxyvitamin D and depression in a general middle-aged to elderly population in Finland. *J. Nutr. Health Aging* **2018**, *22*, 159–164. [CrossRef] [PubMed]
34. Goodwill, A.M.; Campbell, S.; Simpson, S., Jr.; Bisignano, M.; Chiang, C.; Dennerstein, L.; Szoeke, C. Vitamin D status is associated with executive function a decade later: Data from the Women's Healthy Ageing Project. *Maturitas* **2018**, *107*, 56–62. [CrossRef] [PubMed]
35. Wion, D.; MacGrogan, D.; Neveu, I.; Jehan, F.; Houlgatte, R.; Brachet, P. 1,25-Dihydroxyvitamin D3 is a potent inducer of nerve growth factor synthesis. *J. Neurosci. Res.* **1991**, *28*, 110–114. [CrossRef]
36. Shaffer, J.A.; Edmondson, D.; Wasson, L.T.; Falzon, L.; Homma, K.; Ezeokoli, N.; Li, P.; Davidson, K.W. Vitamin D supplementation for depressive symptoms: A systematic review and meta-analysis of randomized controlled trials. *Psychosom. Med.* **2014**, *76*, 190–196. [CrossRef]
37. Oude Voshaar, R.C.; Derks, W.J.; Comijs, H.C.; Schoevers, R.A.; de Borst, M.H.; Marijnissen, R.M. Antidepressants related to vitamin D3. *Transl. Psychiatry* **2014**, *4*, e383. [CrossRef]

38. Patrick, R.P.; Ames, B.N. Vitamin D and the omega-3 fatty acids control serotonin synthesis and action, part 2: Relevance for ADHD, bipolar disorder, schizophrenia, and impulsive behavior. *FASEB J.* **2015**, *29*, 2207–2222. [CrossRef]
39. Scott, D.; Ebeling, P.R. Vitamin D and Public Health. *Int. J. Environ. Res. Public Health.* **2019**, *16*, E848. [CrossRef]
40. Li, G.; Mbuagbaw, L.; Samaan, Z.; Falavigna, M.; Zhang, S.; Adachi, J.D.; Cheng, J.; Papaioannou, A.; Thabane, L. Efficacy of vitamin D supplementation in depression in adults: A systematic review. *J. Clin. Endocrinol. Metab.* **2014**, *99*, 757–767. [CrossRef]
41. Sallis, J.F.; Haskel, W.L.; Wod, P.D.; Fortmann, S.P.; Rogers, T.; Blair, S.N.; Paffenbarger, R.S., Jr. Physical activity assessment methodology in the five-city project. *Am. J. Epidemiol.* **1985**, *121*, 91–106. [CrossRef]
42. Northcote, J.; Livingston, M. Accuracy of self-reported drinking: Observational verification of 'last occasion' drink estimates of young adults. *Alcohol Alcohol.* **2011**, *46*, 709–713. [CrossRef] [PubMed]

© 2020 by the authors. Licensee MDPI, Basel, Switzerland. This article is an open access article distributed under the terms and conditions of the Creative Commons Attribution (CC BY) license (http://creativecommons.org/licenses/by/4.0/).

International Journal of
Environmental Research and Public Health

MDPI

Article

Cervical Cancer Mortality in East-Central European Countries

Cezary Wojtyla [1,*], Michal Ciebiera [2], Dariusz Kowalczyk [3] and Grzegorz Panek [1]

1 Department of Oncologic Gynecology and Obstetrics, The Center of Postgraduate Medical Education, 00-416 Warsaw, Poland; gmpanek@wp.pl
2 Second Department of Obstetrics and Gynecology, The Center of Postgraduate Medical Education, 01-809 Warsaw, Poland; michal.ciebiera@gmail.com
3 Faculty of Health Sciences, State University of Applied Sciences, 62-800 Kalisz, Poland; dariusz.w.kowalczyk@wco.pl
* Correspondence: czwo@op.pl

Received: 29 May 2020; Accepted: 25 June 2020; Published: 28 June 2020

Abstract: Changes that took place in Europe in the early 1990s had an impact on health-associated issues. They were an impulse for the changes in healthcare systems and, consequently, also for the changes in cancer control programmes. Those changes also had an effect on mortality rates due to cervical cancer (CC). Therefore, the aim of this study is to analyse CC mortality trends in east-central Europe after 1990. Data on deaths due to CC were retrieved from the WHO Mortality Database. Trends in east-central European countries between 1990 and 2017 were assessed using Joinpoint Regression Program software. CC mortality decreased in the majority of analysed countries. However, an increase was observed in Latvia and Bulgaria. Despite decreasing mortality in the majority of the analysed countries, significant differences were observed. In order to improve the epidemiological situation, effective early detection programmes for cervical cancer ought to be rearranged and based not only on pap smears but also on molecular methods, as well as on introducing widespread programmes of vaccination against HPV.

Keywords: cervix uteri; epidemiology; screening; mortality; time trends; east-central Europe

---

1. Introduction

Cervical cancer (CC) is the ninth most frequently diagnosed neoplastic disease in European women and the second most common tumour occurring in women aged 15–44 years [1]. As the cause of death from cancer, it occupies the eleventh place in Europe and holds second place in younger women [1]. However, significant differences occur between European countries in terms of the incidence and mortality of cervical cancer. According to the Catalan Institute of Oncology and the International Agency for Research on Cancer, incidence rates in 2018 were 16.0/100,000 for women in Eastern Europe, and 6.8/100,000 for women in Western Europe, and mortality rates were 6.1/100,000 and 2.1/100,000, respectively [1].

The east-central European region has 80 million female inhabitants aged over 15 years, who are the future risk group for cervical cancer [1]. The region may be defined by the following countries: Albania, Bosnia and Herzegovina, Belarus, Bulgaria, Croatia, the Czech Republic, Estonia, Hungary, Latvia, Lithuania, Macedonia, Montenegro, Moldova, Poland, Romania, Serbia, Slovakia, Slovenia, and Ukraine [1]. This regional division is not only geographically important, but it also encompasses the countries belonging to the former Eastern Bloc, a system that had governed health policy for several decades very differently from Western Europe. It has undoubtedly contributed to the inequalities regarding the epidemiology of cervical cancer in Europe. Therefore, the aim of this study is to analyse CC mortality trends in east-central Europe after 1990.

2. Materials and Methods

CC mortality trends were investigated in this study for all the countries of central and eastern Europe (CEE), as aforementioned, except some parts of Albania, Bosnia and Herzegovina, and Montenegro due to incomplete data. A commonly adopted term used in comparative analyses is EU10, referring to the ten east-central countries which joined the European Union (EU) in 2004 and 2007 (i.e., Bulgaria, the Czech Republic, Estonia, Hungary, Latvia, Lithuania, Poland, Romania, Slovakia, and Slovenia), whereas the term EU15 refers to the EU Member States prior to 2004 (i.e., Austria, Belgium, Denmark, Finland, France, Greece, Spain, the Netherlands, Ireland, Luxembourg, Germany, Portugal, Sweden, Great Britain, and Italy).

Data on CC mortality for every analysed country were obtained from the World Health Organization Mortality Database between 1990 (or in the first year after 1990 for which such data were available) and 2017 (or in the last available year before 2017). This database provides information concerning the number of deaths by causes defined in the International Classification of Diseases (ICD) according to gender, age and calendar year. Our study analysed the standardised mortality rates due to CC in European countries according to ICD-9 diagnosis code 180 and ICD-10 code C53. Cases encoded as cervical cancer, cancer of the corpus uteri and unspecified uterine cancer were analysed together for the group of women aged 15–44. It stems from the fact of a discrepancy, which was observed when coding the causes of deaths due to uterine cancers and the fact that the majority of deaths due to uterine cancers were caused by CC among women in this age group [2]. The standard world population was used to calculate standardised mortality rates. No extrapolation was made for any missing data.

The analysis of the total cervical cancer mortality rates in CEE countries between 1990 and 2015 was performed using the Joinpoint Regression Program ver. 4.5.0.1. This software was used to test whether time trend factors or their changes within the studied time interval were statistically significant ($p < 0.05$). The maximum number of joinpoints was set to 3. In addition, we calculated the average annual percentage change (APC) between 1990 and 2017, with the number of joinpoints set to 0.

3. Results

The mortality rates of CC are shown in Table 1 for women of all ages (0+) in CEE from the beginning of the observation period (either in 1990 or in the first year after 1990 for which the data were available) till the end of the observation in 2017 (or in the last available year before 2017). The rates of change between 1990 and 2017 are also expressed as APCs. In 1990, the highest rates were noted in women from Romania, Poland, Hungary and Serbia. They ranged from 7.0/100,000 women in Serbia and Hungary to 10.1/100,000 women in Romania. The mortality rate for EU10 was 9.6/100,000 women. At that time, EU15 presented the mortality rate at the level of 3.5/100,000 women. Macedonia and Croatia were the only countries with the rates lower than the EU15 mean. In 2017, the rates were lower for the majority of CEE countries compared to 1990. We also observed a decrease in APC for most of them. However, it increased in Latvia and Bulgaria. The largest decreases in the mortality rate expressed as APC occurred in Poland, the Czech Republic, and Slovenia and were twice as high as the APC for EU10. In 2017, the highest rates were noted in Romania, Serbia, Moldova, Lithuania, and Latvia, whilst the lowest ones in Slovenia, Croatia, and Macedonia. CC mortality is still lower in Western Europe, and its rate of decrease is higher compared to the CEE countries.

The joinpoint regression analysis results on CC mortality trends for CEE countries are shown in Table 2 and Figure 1. The countries studied may be divided into several categories according to cervical cancer mortality trends. The first one (Figure 1A) is observed in those countries where mortality in women of all ages decreased (i.e., the Czech Republic, Estonia, Slovenia, Lithuania, Moldova, Macedonia, Serbia, Poland and Romania). In the Czech Republic, Estonia, Poland and Slovenia, such decreases had been noted since the beginning of the observation. In Lithuania, Moldova, Macedonia, Serbia, and Romania, the trends changed during the observation. The second group comprised Latvia (Figure 1B), where cervical cancer mortality increased in all age groups. However, mortality in the oldest women slightly decreased, but the change was insignificant. The remaining countries demonstrated disproportionate mortality trends for various age groups of women. Nevertheless, they were defined according to the category of trend change observed in all women (0+). In Hungary, Belarus, Slovakia, Ukraine and Croatia, we observed a decrease in mortality in this age group (Figure 1C). In Hungary, the oldest group of women (65+) was the only one with an increasing trend. However, it was insignificant. In Belarus, we observed an increasing trend in the youngest (15–44 years old) and middle-aged (45–54 years old) groups of women. The trend in the oldest group was decreasing. In Slovakia, an insignificant increasing trend was observed only in the middle-aged group. The situation was the same in Ukraine, but the increasing trend in middle-aged women was statistically significant. We also included Croatia into this group, but an insignificant change of trend in the 0+ group was observed. Moreover, an increasing trend of mortality had been present in the middle-aged group since 1999. In the remaining groups, the trend was decreasing. Finally, Bulgaria had been characterised by increasing mortality in the 0+ group of women until 1999, when it reached a plateau (Figure 1D). The only group with a decreasing trend was the youngest one.

Table 1. Cervical cancer data availability, mortality rates and annual percentage change (APC) in all age groups of women (0+).

Country	Years	Cervical Cancer Mortality Rates (per 100,000 Population)		
		1990 [a]	2017 [b]	APC 1990 [a] and 2017 [b]
Latvia	1990–2015	4.4	6.1	1.7
Bulgaria	1990–2015	3.9	5.7	1.1
Ukraine	1990–2017	5.7	5.2	−0.2
Serbia	1998–2016	7.0	6.7	−0.4
Lithuania	1900–2017	6.9	6.6	−0.4
Belarus	1990–2014	4.4	4.1	−0.4
Croatia	1990–2016	3.4	2.6	−0.5
Slovakia	1992–2014	5.0	4.8	−0.5
Romania	1990–2017	10.1	8.5	−0.6
Moldova	1990–2017	6.7	6.1	−0.7
Macedonia	1991–2014	1.5	2.4	−1.2
Estonia	1990–2016	5.3	4.5	−1.6
Hungary	1990–2017	7.0	4.0	−1.8
the Czech Republic	1990–2017	5.0	3.1	−2.3
Poland	1990–2016	7.7	4.1	−2.3
Slovenia	1990–2015	3.6	2.2	−2.7
UE10	1990–2014	9.6	7.3	−1.2
UE15	1990–2015	3.5	1.9	−2.5

[a] Or the first available year after 1990; [b] or the last available year before 2017.

Table 2. Joinpoint analysis for cervical cancer mortality of women at all ages (0+), 15–44, 45–64 and over 65 (65+) years old, by country.

	0+	15–44	45–54	65+
Belarus				
APC 1 (Period 1)	−0.4 * (1990–2014)	1.3 * (1990–2014)	0.3 (1990–2014)	−3.9 * (1990–2014)
Bulgaria				
APC 1 (Period 1)	3.8 * (1990–1999)	1.0 (1990–2004)	5.0 * (1990–1999)	0.2 (1990–2015)
APC 2 (Period 2)	0.0 (1999–2015)	−2.8 * (2004–2015)	0.7 * (1999–2015)	
Croatia				
APC 1 (Period 1)	−2.4 * (1990–2001)	−2.5 (1990–2016)	−3.3 (1990–1999)	−1.1 * (1990–2016)
APC 2 (Period 2)	0.8 (2001–2016)		2.1 * (1999–2016)	
the Czech Republic				
APC 1 (Period 1)	−2.3 * (1990–2017)	−3.3 * (1990–2017)	−2.9 * (1990–2017)	−1.8 (1990–2017)
Estonia				
APC 1 (Period 1)	−1.6 * (1990–2016)	−4.1 * (1990–2016)	−0.4 (1990–2016)	−2.8 * (1990–2016)
Hungary				
APC 1 (Period 1)	−1.8 * (1990–2017)	2.2 (1990–1996)	−0.5 * (1990–2017)	−3.8 * (1990–2008)
APC 2 (Period 2)		−4.5 * (1996–2017)		0.4 (2008–2017)
Latvia				
APC 1 (Period 1)	1.7 * (1990–2015)	2.4 * (1990–2015)	2.0 * (1990–2015)	−0.5 (1990–2015)
Lithuania				
APC 1 (Period 1)	2.9 * (1990–2000)	5.1 * (1990–2000)	2.4 * (1990–2002)	−1.1 (1990–2017)
APC 2 (Period 2)	−1.9 * (2000–2017)	−3.2 * (2000–2017)	−1.7 * (2002–2017)	
Macedonia				
APC 1 (Period 1)	6.2 * (1991–2000)	95.0 (1990–1993)	11.8 * (1991–1998)	−1.4 (1991–2014)
APC 2 (Period 2)	−4.7 * (2000–2014)	−4.0 (1993–2014)	−2.6 * (1998–2014)	
Moldova				
APC 1 (Period 1)	−0.7 * (1990–2017)	3.1 * (1990–2005)	−0.5 (1990–2017)	−1.9 * (1990–2017)
APC 2 (Period 2)		−5.0 * (2005–2017)		
Poland				
APC 1 (Period 1)	−2.1 * (1990–2009)	−2.6 * (1990–1998)	−1.0 * (1990–2009)	−2.7 * (1990–2005)
APC 2 (Period 2)	−3.5 * (2009–2016)	−6.1 * (1998–2016)	−4.4 * (2009–2016)	−1.0 * (2005–2016)
Romania				
APC 1 (Period 1)	1.1 * (1990–2002)	0.6 (1990–2000)	1.7 * (1990–2007)	−0.4 * (1990–2017)
APC 2 (Period 2)	−1.9 * (2002–2017)	−5.5 * (2000–2017)	−4.4 * (2007–2011)	
APC 3 (Period 3)			2.7 (2011–2014)	
APC 4 (Period 4)			−5.4 * (2014–2017)	
Serbia				
APC 1 (Period 1)	1.0 (1998–2007)	−2.2 * (1998–2016)	2.6 * (1998–2007)	−1.1 * (1998–2016)
APC 2 (Period 2)	−1.8 (2007–2016)		−1.8 * (2007–2016)	
Slovakia				
APC 1 (Period 1)	−0.5 * (1992–2014)	−3.2* (1992–2014)	0.2 (1992–2014)	0.7 (1992–2001)
APC 1 (Period 2)				−17.3 (2001–2004)
APC 3 (Period 3)				18.1 (2004–2007)
APC 4 (Period 4)				−0.2 (2007–2014)
Slovenia				
APC 1 (Period 1)	−2.7 * (1990–2015)	−4.5 * (1990–2015)	−2.0 * (1990–2015)	−2.5 * (1990–2015)
Ukraine				
APC 1 (Period 1)	0.0 (1990–2014)	7.1 * (1990–1995)	0.5* (1990–2017)	−4.3 * (1990–2007)
APC 2 (Period 2)	−3.3 * (2014–2017)	1.5 * (1995–2009)		−1.6 * (2007–2017)
APC 3 (Period 3)		−2.7 * (2009–2017)		

* statistically significant, $p < 0.05$.

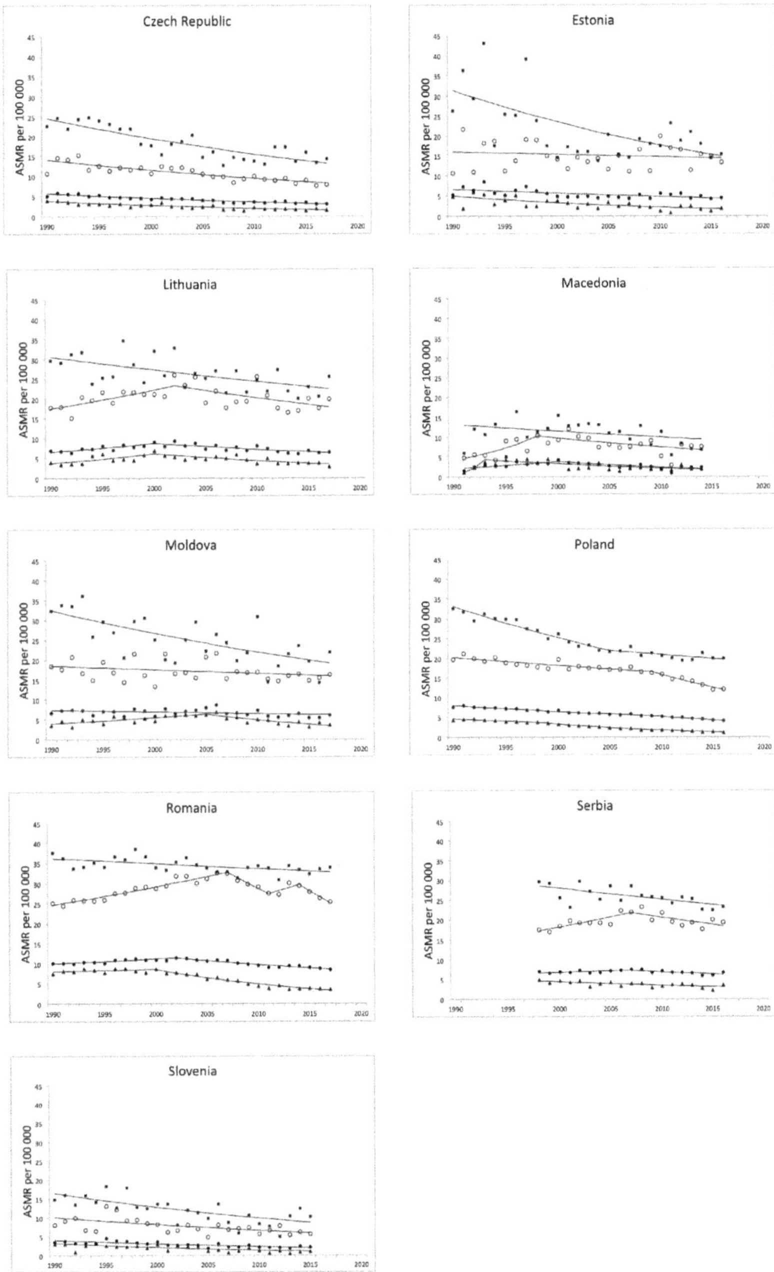

(A)

Figure 1. *Cont.*

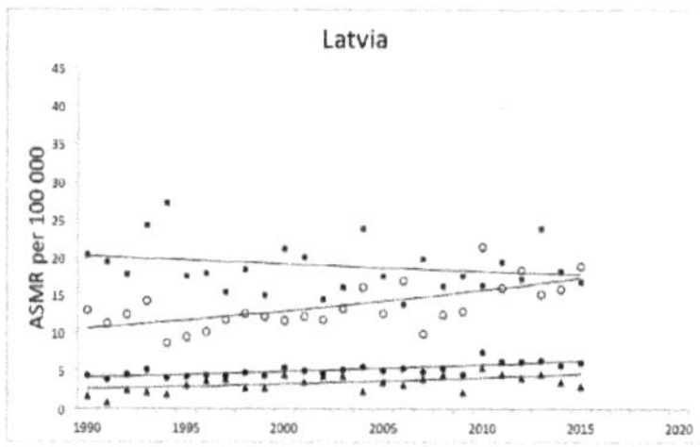

(B)

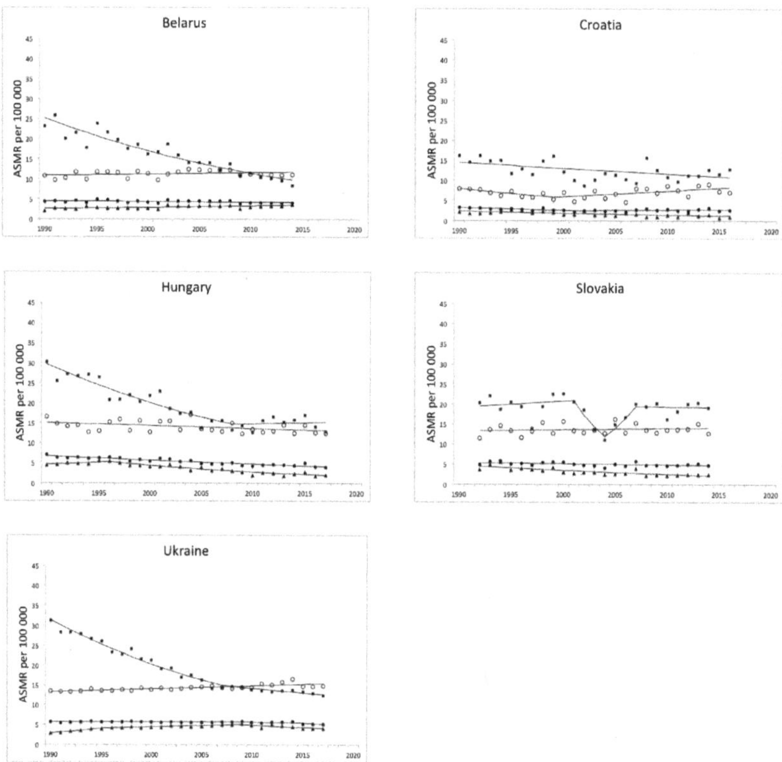

(C)

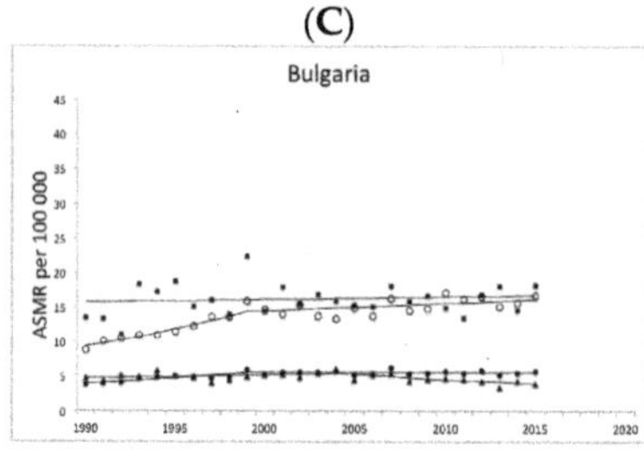

(D)

Figure 1. Joinpoint analysis of trends in age-standardised mortality rates due to cervical cancer in women of all ages (a solid circle), 15–44 (a triangle), 45–64 (an empty circle) and over 65 (a square) years old during the period 1990–2017. ASMR—age-standardised mortality rates. (**A**)—first pattern; (**B**)—second pattern; (**C**)—third pattern; (**D**)—fourth pattern.

4. Discussion

Our study began in 1990, when the Iron Curtain in Europe collapsed, which marked the beginning of numerous political, economic, social, and public health-related changes. However, those changes were not rapid. For many years, east-central European countries have been struggling with many problems left over from the previous political system. Some changes in healthcare made in the 1980s [3] in the Eastern Bloc had persisted. Furthermore, the importance of clinical medicine had begun to be emphasised regarding public health. It survived the collapse of the Iron Curtain in many central and eastern European countries [4–6]. However, at that time, important population-based and organised early-detection programmes for cervical cancer were not undertaken, as was the case in numerous western European countries [7]. Therefore, cervical cancer mortality rates remained high in CEE countries at the beginning of the 1990s, contributing to a significant disproportion between western and CEE countries [8]. In 1993, the first edition of the European Guidelines for cervical cancer screening was issued, which set out the basic principles for detecting cervical cancer [9]. The health ministers of EU countries signed the European Council Recommendation on Cancer Screening in 2003, which included the organisation of CC screening [10]. It was then backed up with the second edition issued in 2008, with recommendations for decision-makers on how a well-functioning screening programme should be organised [11]. In 2004, 2007 and 2013, numerous CEE countries joined the EU and were obliged to resolve the problem of high CC mortality. As a result, many countries launched programmes for the early detection of this disease [12]. However, their effectiveness depended largely on the coverage of the target population [13]. Moreover, organised screening, as opposed to opportunistic screening, is more cost-effective, because it only examines a defined target population at appropriate time intervals. It also provides quality assurance at all levels and evaluates programme effectiveness over time [11,14].

Not all CEE countries have managed or decided to introduce such screening programmes for reducing CC mortality, despite their proven benefits. Those countries which did found differences in how they were organised, their target populations, screening methods (e.g., conventional cytology or liquid-based cytology) and whether HPV was investigated in the programme [15]. Studies by Elfström et al. [15] and by Poljaket et al. [12] investigated the functioning of screening programmes in selected European countries and found that opportunistic programmes were still adopted in Bosnia and Herzegovina, Macedonia, Montenegro, Serbia, Bulgaria, and Slovakia. The programme introduced in the Czech Republic was non-population-based. The screening coverage of the east-central target population of Europe was also lower than in western Europe and exceeded 60% only in Slovenia [15]. Only in the Czech Republic, Latvia, and Slovenia, all quality indicators of the screening register were tracked (i.e., comparability, completeness, validity and timelines). An organised screening programme is one where monitoring and evaluation are audited. Unfortunately, it has not been undertaken in the majority of CEE countries [15].

The epidemiological situation in CEE countries has changed considerably over the last decades. A study by Arbyn et al. [16] on the Baltic countries indicated that CC mortality had increased in Latvia and Lithuania between the early 1980s and 2004 in almost all age groups, but it was stable in Estonia. A similar upward trend (APCs) had been observed in Bulgaria and Romania since 1980. Between 1989 and 2004, an upward trend still continued in Bulgaria, but it became insignificant. A slight downward trend had been observed in the mortality of Bulgarian and Romanian women aged over 55 years from 1985 onwards, whilst those aged 20–59 overall showed a concomitant increase, with the exception of the 20–29 year age subdivision.

As regards some other countries of the tested region, CC mortality for all ages decreased in Bosnia and Herzegovina, Croatia, the Czech Republic, Hungary, Poland, Slovakia, and Slovenia [17–19]. However, the trend increased in Belarus, Moldova (until the first part of the 2000s), and Ukraine. In Albania, it remained relatively stable [17,18].

CC mortality trends were investigated in another study by Arbyn et al. [17]. They analysed selected age groups of women, dating back to between the first part of the 1950s and 2005. The groups were found to vary significantly over the decades, according to the country and age group. A long-standing downward trend in all age groups was observed in the Czech Republic, Poland, and Hungary. The present study demonstrates that this trend continued on in the Czech Republic and Poland. Furthermore, other countries have joined this decreasing trend (Estonia, Slovenia, Lithuania, Moldova, Macedonia, Serbia and Romania). Hungary was excluded because trends in women aged above 65 were insignificantly changed. The aforementioned study also revealed decreasing mortality in young Croatian women, below 44 years, and in those aged 60–64 years until the 1980s, but an upward trend in all the other age groups [17]. Our observations confirm that the downward trend in young adult women continued but was also apparent in those aged over 65 years. In our study, women were divided into slightly different age groups. Nevertheless, we also found upward trends in other age groups. The study by Arbyn et al. also showed increased mortality rates in Macedonia and Slovakia in almost all age groups of women between 1990 and 2000, except for women aged 30–34 and 70–74 years, where the mortality was decreased [17]. Our study reveals that the changed trends in Macedonia qualifies this country for the group with decreasing mortality trends in all age groups. In Slovakia, the trend was upward in women aged over 44 years, whilst it was downward for younger women. The study by Arbyn et al. also investigated mortality trends in Bosnia and Herzegovina (1985–1990), as well as Serbia, Montenegro (1995–2000), and Albania. Women's mortality increased in the group under 44 years in Bosnia and Herzegovina, Serbia and Montenegro, but decreased in all other age groups. In Albania, the mortality decreased in the groups under 34 and 80–84 years between 1985 and 2000, whereas it increased at ages 40–44 years, and was unstably irregular for the other age groups [17]. In Serbia, we observed decreasing trends for all groups of women. The study by Arbyn et al. showed increased mortality in Belarus (between 1980–2000) in women under 54 years, but a downward trend in all the other age groups for the last few years of the study period [17]. We also found declining mortality trends over our entire observation period in women over 65 years of age, as well as the all-women group (0+). Increases were seen for women aged up to 65 years. In Slovenia and Moldova, decreases were observed by Arbyn et al. in women under 54 and in those aged 80–84 years, whereas an upward trend was found for all the other age groups [17]. In our study, both those countries qualify as presenting falling mortality trends for all age groups of women. Finally, the study by Arbyn et al. demonstrated increased mortality rates in Ukraine in almost all age groups of women, except for those aged 70–84 years [17]. The situation in Ukraine has changed in recent years. Our observations indicate a recent drop in the mortality trend for the youngest group of women aged 15–44 years and also a downward trend in the oldest group of women in our study (65+ years) and for the 0+ group. Nonetheless, the upward trend is still observed in women aged 45–64 years.

When investigating trends in CC mortality dating back several decades, several problems are encountered, such as inconsistencies in classifying reproductive organ tumors and the fact that the International Classification of Diseases has been modified several times over recent years. Death certificates do not precisely distinguish between cervical and corpus uteri cancers since they were mostly classified as being unspecified uterine cancers. Upon reviewing the data of specific countries over recent decades, it is noticeable that up to 93% of all uterine cancers were diagnosed in such a way, for example, in Spain in 1960 [8]. In contrast, the UK classified unspecified uterine cancers only in 6% of cases at that time. However, in 1995, such diagnoses in Spain were made in 32% of cases, whereas in the UK, the rate was 21% of cases [8]. Mortality due to cervical cancer is, therefore, affected by misclassification bias. Some studies, for example, by Levi at al. and Wojtyla et al., have tried to minimise the problem by investigating the mortality only in young women, in whom cervical tumors constitute the overwhelming majority of all uterine tumors [2,8]. To our knowledge, this paper is the first one to illustrate the mortality of women due to CC in all age groups of women living in the region of central and eastern Europe using a separate method of classifying deaths among young adult women. In this way, it allows the minimisation of the classification bias and more reliable comparison of the results between this group and other groups of women.

Our study has several limitations. A change of the International Classifications of Diseases was made during the period covered by our analysis. Individual countries introduced the new classification at different times. In addition, uterine cancers may be classified as cancer of the corpus uteri, cervix or unspecified parts of the uterus. Both of the above factors make the epidemiological analysis of cervical cancer prone to classification bias. It is irrelevant in case of changes in the last two International Classifications of Diseases because all three abovementioned codes have their equivalents in individual classifications. However, it affects the loss of some cervical cancers classified as unspecified uterine cancer. The bias can only be minimised for younger women by assuming that uterine cancers in this age group (under 45 years old) are cervical cancers. As mentioned before, the vast majority of uterine cancers in this age group are cervical cancers [2,8]. However, this methodology cannot be applied to other groups of women. Another limitation stems from the comparison of countries with different populations. Obviously, mortality in our analysis is presented in the form of standardised mortality rates using the world standard population. Nevertheless, the number of deaths in some countries is low and small changes in the number of deaths may exert a significant impact on observed trends.

Despite a great deal of effort devoted to reducing and achieving low rates of CC mortality, as observed in most western and northern European countries, it is still high in most CEE countries. This study confirms previous observations on the inequalities in female mortality from cervical cancer in Europe. Nevertheless, the observations are based on the latest data. However, epidemiological studies of this type are used to shape the current health policy of individual countries and individual health areas. The observations described in our paper should encourage decision-makers to increase their efforts to improve the situation, especially where there are available tools to achieve it. Poorly functioning screening programmes are the main reason responsible. However, the case of Slovenia is exceptional. It is the only country in the region with a well-functioning and effective screening programme. It included 82.1% of women (data from 2004–2008) and led to a 40% fall in the cervical cancer incidence rate during 2003–2009 [20]. Great hopes are currently pinned upon protective HPV vaccines. The vaccines, coupled with fully functioning and organised early-detection programmes for CC, may result in this disease being no longer regarded as a public health issue in the world. Hall MT et al. [21]. performed a simulation of what the expected CC incidence and mortality rates in Australia could be if high-coverage protective HPV vaccination programmes were introduced. They found that in a short time, CC could be eliminated from the population. The time for eradication is much shorter than that achieved by screening alone, where the obtained levels of incidence and mortality are significantly lower. Therefore, it appears that only the simultaneous introduction of appropriate screening programmes coupled with high-coverage protective vaccinations may lead to the situation when the level of CC mortality in CEE countries approximates the one observed in Western Europe at the moment.

Nevertheless, there is one more issue that may have an enormous impact on perpetuating the positive mortality trend generated by the above factors. It is associated with the organisation of appropriate educational programmes concerning cervical cancer prevention for young women. Those women are mostly included in the current cervical cancer prevention programme in a given country because it shapes their awareness of cervical cancer prevention for their entire future life. Sudden changes in the organisation of screening will have a lower impact on women from older age groups because their attitude towards cervical cancer prevention has already been shaped. Young women may not be fully educated and aware of the threat yet, because they have not developed the habit of undergoing routine screening. New diagnostic methods and changes in screening programmes will be ineffective if women are unaware of the threat and the possibilities of its prevention.

5. Conclusions

Although CC mortality has decreased in the majority of CEE countries over recent decades, there are still disproportions between particular countries, as well as between age groups of women within individual countries from this region of Europe. These countries present significantly higher mortality rates than western Europe. In order to improve the epidemiological situation for the coming years, early detection programmes ought to be reorganised and based not only on pap smears but also on molecular methods, together with widespread programmes of vaccination against HPV.

Author Contributions: Conceptualization, C.W., M.C., D.K. and G.P.; Methodology, C.W. and G.P.; Software, C.W. and D.K.; Validation, M.C. and G.P.; Formal Analysis, C.W., M.C., D.K. and G.P.; Resources, C.W.; Data Curation, C.W.; Writing—Original Draft Preparation, C.W., M.C. and D.K.; Writing—Review & Editing, C.W. and G.P.; Visualization, C.W., M.C. and D.K.; Supervision, G.P. Project administration, C.W. and G.P. All authors have read and agreed to the published version of the manuscript.

Funding: The APC was funded by the State University of Applied Sciences, 16 Kaszubska St., 62-800 Kalisz, Poland.

Conflicts of Interest: The authors declare no conflict of interest.

References

1. Bruni, L.; Albero, G.; Serrano, B.; Mena, M.; Gómez, D.; Muñoz, J.; Bosch, F.X.; de Sanjosé, S. Human Papillomavirus and Related Diseases in the World. Summary Report 17 June 2019. ICO/IARC Information Centre on HPV and Cancer. Available online: https://www.hpvcentre.net/statistics/reports/XWX.pdf (accessed on 28 April 2020).
2. Wojtyla, C.; Janik-Koncewicz, K.; La Vecchia, C. Cervical cancer mortality in young adult European women. *Eur. J. Cancer* **2020**, *126*, 56–64. [CrossRef] [PubMed]
3. Zatonski, W.; Pisarska-Krawczyk, M.; Wojtyla, C.; Janik-Koncewicz, K. Patterns of cervical cancer mortality in young adult women in three countries of the European Union: Finland, Poland, and Latvia. *J. Health Inequal.* **2016**, *2*, 95–100. [CrossRef]
4. Zatonski, W.; Zatonski, M. Health in the Polish People's Republic. *J. Health Inequal.* **2016**, *2*, 7–16. [CrossRef]
5. Zatonski, W.; HEM Project Team. *Closing the Health Gap in European Union*; The Maria Skłodowska-Curie Memorial Cancer Center and Institute of Oncology: Warsaw, Poland, 2008.
6. Zatonski, W.; HEM Project Team. Epidemiological analysis of health situation development in Europe and its causes until 1990. *Ann. Agric. Environ. Med.* **2011**, *18*, 194–202. [PubMed]
7. Anttila, A.; Ronco, G.; Clifford, G.; Bray, F.; Hakama, M.; Arbyn, M. Weiderpass, Cervical cancer screening programmes and policies in 18 European countries. *Br. J. Cancer* **2004**, *91*, 935–941. [CrossRef] [PubMed]
8. Levi, F.; Lucchini, F.; Negri, E.; Francesch, S.; La Vecchia, C. Cervical cancer mortality in young women in Europe: Patterns and trends. *Eur. J. Cancer* **2000**, *36*, 2266–2271. [CrossRef]
9. Coleman, D.; Day, N.; Douglas, G.; Farmery, E.; Lynge, E.; Philip, J.; Segnan, N. European guidelines for quality assurance in cervical cancer screening. Europe against cancer programme. *Eur. J. Cancer* **1993**, *29A*, S1–S38. [PubMed]
10. Council of the European Union. Council recommendation of 2 December 2003 on cancer screening. *Off. J. Eur. Union* **2003**, *878*, 34–38.
11. Arbyn, M.; Anttila, A.; Jordan, J.; Ronco, G.; Schenck, U.; Segnan, N.; Wiener, H.G.; Herbert, A.; Daniel, J.; von Karsa, L. *European Guidelines for Quality Assurance in Cervical Cancer Screening*, 2nd ed.; Office for Official Publications of the European Communities: Luxembourg, Luxembourg, 2008.
12. Poljak, M.; Seme, K.; Maver, P.J.; Kocjan, B.J.; Cuschieri, K.S.; Rogovskaya, S.I.; Arbyn, M.; Syrjänen, S. Human papillomavirus prevalence and type-distribution, cervical cancer screening practices and current status of vaccination implementation in Central and Eastern Europe. *Vaccine* **2013**, *31*, H59–H70. [CrossRef] [PubMed]
13. Laara, E.; Day, N.E.; Hakama, M. Trends in mortality from cervical cancer in the Nordic countries: Association with organised screening programmes. *Lancet* **1987**, *1*, 1247–1249. [CrossRef]
14. International Agency for Research on Cancer. *Cervix Cancer Screening*; IARC: Lyon, France, 2005.
15. Elfström, K.M.; Arnheim-Dahlström, L.; von Karsa, L.; Dillner, J. Cervical cancer screening in Europe: Quality assurance and organisation of programmes. *Eur. J. Cancer* **2015**, *51*, 950–968. [CrossRef] [PubMed]
16. Arbyn, M.; Antoine, J.; Mägi, M.; Smailyte, G.; Stengrevics, A.; Suteu, O.; Valerianova, Z.; Bray, F.; Weiderpass, E. Trends in cervical cancer incidence and mortality in the Baltic countries, Bulgaria and Romania. *Int. J. Cancer* **2011**, *128*, 1899–1907. [CrossRef] [PubMed]
17. Arbyn, M.; Raifu, A.O.; Antoine, J. *Trend of Cervical Cancer Mortality in Europe*; Scientific Institute of Public Health: Brussels, Belgium, 2008.
18. Bray, F.; Lortet-Tieulent, J.; Znaor, A.; Brotons, M.; Poljak, M.; Arbyn, M. Patterns and trends in human papillomavirus-related diseases in Central and Eastern Europe and Central Asia. *Vaccine* **2013**, *31*, H32–H45. [CrossRef] [PubMed]

19. Arbyn, M.; Raifu, A.O.; Weiderpass, E.; Bray, F.; Anttila, A. Trends of cervical cancer mortality in the member states of the European Union. *Eur. J. Cancer* **2009**, *45*, 2640–2648. [CrossRef] [PubMed]
20. Kesic, V. Prevention of cervical cancer in Central and Eastern Europe and Central Asia: A challenge for the future. *Vaccine* **2013**, *31*, vii–ix. [CrossRef] [PubMed]
21. Hall, M.T.; Simms, K.T.; Lew, J.B.; Smith, M.A.; Brotherton, J.M.; Saville, M.; Frazer, I.H.; Canfell, K. The projected timeframe until cervical cancer elimination in Australia: A modelling study. *Lancet Public Health* **2019**, *1*, e19–e27. [CrossRef]

© 2020 by the authors. Licensee MDPI, Basel, Switzerland. This article is an open access article distributed under the terms and conditions of the Creative Commons Attribution (CC BY) license (http://creativecommons.org/licenses/by/4.0/).

International Journal of
Environmental Research and Public Health

MDPI

Article

Health-Related Behaviors of Pregnant Women Residing in Urban and Rural Areas in Poland

Maria Szubert [1,2], Malwina Ilowiecka [1], Jacek Wilczynski [1], Przemyslaw Bilinski [3] and Cezary Wojtyla [2,4,5,*]

1 Clinic of Surgical and Oncologic Gynecology, 1st Department of Gynecology and Obstetrics, Medical University of Lodz, M. Pirogow Teaching Hospital, Wilenska 37 St., 94-029 Lodz, Poland; maja.szubert@gmail.com (M.S.); malilo1@o2.pl (M.I.); jrwil@post.pl (J.W.)
2 Club 35, Polish Society of Gynecologists and Obstetricians, Cybernetyki 7F/87 St., 02-677 Warsaw, Poland
3 Faculty of Health Sciences, State University of Applied Sciences, Kaszubska 16 St., 62-800 Kalisz, Poland; bildom@gmail.com
4 International Prevention Research Institute—Collaborating Centre, State University of Applied Sciences, Kaszubska 16 St., 62-800 Kalisz, Poland
5 Department of Oncologic Gynecology and Obstetrics, Centre of Postgraduate Medical Education, Czerniakowska 231 St., 00-416 Warsaw, Poland
* Correspondence: czwo@op.pl

Received: 24 April 2020; Accepted: 18 June 2020; Published: 18 June 2020

Abstract: The aim of this study was to evaluate the knowledge regarding a healthy lifestyle and prophylaxis during pregnancy among women from rural and urban areas and how this changed within a 5-year period. Analyses of the population of pregnant women in Poland were made using a questionnaire survey. The survey was conducted in the years 2010–2012 and 2017. Questionnaires from 6128 pregnant women were collected. The statistical analyses were conducted using IBM SPSS. The examined population was comprised of 41% women from rural areas and 59% women from urban areas. Alcohol consumption was lower among women from rural areas than among urban inhabitants in 2010–2012; in 2017 a trend of even lower consumption was observed. Folic acid supplementation was more broadly developed in the urban population; however, in 2017, higher percentage rates of both populations admitted taking folates before pregnancy. More women in urban than in rural areas performed physical activity during pregnancy, but the differences decreased in 2017. Knowledge of a healthy lifestyle and prophylaxis during pregnancy increased regardless of place of residence; however, the most evident change could be observed among women from rural areas.

Keywords: health behaviors; pregnancy; rural area; urban area

1. Introduction

Contemporary medicine offers access to tools and options that may not only reduce the prevalence of certain diseases or modify disease-related complications but also extend lifespan and improve its quality. According to the report published by M. Lalonde in 1974, our lifestyle determines our health by as much as 50%, whereas the influence of healthcare itself is limited only to 10% [1]. Almost 45 years have passed since this theory was presented, and although many new statements have been made since that time, Lalonde's health field concept clearly reflects the significance of health prevention. An advanced development in medicine may not be an argument for diminishing the impact that lifestyle exerts on overall health. Reasonable behaviors aimed at the protection and improvement of health are particularly important in pregnant women since each mother's lifestyle has a great impact on the course of pregnancy and fetal development [2]. Therefore, both worldwide and in Poland, research studies have been conducted regularly with the aim of evaluating knowledge on pro-health factors and behavior among future mothers [3,4]. The identification of risk factors and assessment of health awareness, particularly in populations of pregnant women, creates the option to implement specific actions and projects aimed at ensuring proper care or at filling in knowledge gaps related to the aforementioned issues. Thus, we undertook this study in two specific subgroups: rural and urban populations of pregnant women. It is worthwhile to mention that, from 2009, recommendations regarding nearly all aspects of pregnancy were released in Poland which were intended to be implemented irrespective of the place of residence of the future mother. Such studies are also important because they show whether such actions bring certain measurable benefits, such as a reduced number of diseases and complications, which in turn may be translated into lower financial outlays for secondary prevention. Based on the published analyses on health awareness among pregnant women, it may be concluded that the results are influenced by similar factors to those among non-pregnant women, such as age, place of residence, level of education and marital and socioeconomic status. These factors indicate that there is a positive correlation between a more advanced age of a pregnant woman, a higher education level, a higher socioeconomic status and a greater care for health [5,6]. The noticeable differences in the abovementioned observations resulted in the focusing of educational activities on populations in which the awareness of health prevention importance is still too low. One of the population groups in East-Central Europe whose economic situation has greatly improved in the recent years is the inhabitants of rural regions. The Regulation of the Minister of Health of 16 August 2018 on the organizational standard of perinatal care obliged medical professionals in Poland to take the same pro-health actions among pregnant patients regardless of their place of residence or financial status [7].

The aim of this study was to analyze health awareness, expressed both as knowledge and pro-health behaviors, in the populations of pregnant women living in urban and rural areas. Based on these data, we attempted to assess changes in health awareness in the two populations within a five-year observation period; i.e., between the years 2010–2012 and 2017. The working hypothesis assumed that, despite place of residence, women should have a comparable health awareness, which should increase in the studied timeframe.

2. Materials and Methods

The study was conducted by the Chief Sanitary Inspectorate and the Institute of Rural Health in the years 2010–2012 and in 2017 in all Polish public hospitals and involved a group of women and their children. The analysis was made based on a questionnaire developed within Pol-PrAMS program (Polish Pregnancy-Related Assessment Monitoring System) using the model of the PRAMS (Pregnancy Risk Assessment Monitoring System) applied in the United States. All women gave their informed consent for inclusion before they participated in the study. The Bioethics Committee of the Institute of Rural Health in Lublin approved the study (permission no. 03/2011).

The participants in the Pol-PrAMS study were 12.066 pregnant women, including 8625 subjects in the years 2010–2012 and 3441 in 2017. Women hospitalized in public hospitals in Poland for delivery were asked to fill in the paper version of the questionnaire, which was divided into two parts. The first one was filled in by the patients and referred to their personal data, pro-health behaviors and awareness of a healthy lifestyle during pregnancy and the course of pregnancy, whereas the second part was filled in by healthcare professionals based on medical records of the patients and their children. The Pregnancy Physical Activity Questionnaire (PPAQ) was used to estimate the physical activity of pregnant women in both groups. The use of PPAQ allowed the presentation of the physical activity of pregnant women as their energy expenditure, measured in Metabolic Equivalents of Task (METs). Physical activity is presented in MET-h/week units. The Mann–Whitney–Wilcoxon test was used to compare groups in relation to physical activity. The detailed methodology of the Pol-PrAMS study and the use of PPAQ has been presented in separate studies [8,9].

The demographic population structure in 2010–2012 was significantly different to that studied in 2017 in terms of age and region of residence. A compromise method of correction was adopted by randomly withdrawing those cases that were overrepresented [10]. The population from the 2010–2012 study was referenced to the population structure from 2017 in terms of age, place of residence (city/village) and region of residence (16 provinces). This data constituted a trivariate table, covering 3331 women subjects from 2017 divided into $3 \times 3 \times 16 = 144$ subgroups. The number of subjects in each subgroup was expressed as a percentage of the total. An analogous table was prepared for the population studied in 2010–2012. Based on the coefficients from the 2017 table, the expected counts in each subgroup for 2010–2012 were calculated, from which the overcounts and undercounts could be derived. The population size was reduced so that there were no undercounts present in the table; i.e., only overcounts were present. This was achieved by reducing the total size of the interviewed population from the 2010–2012 study to 2797 cases. Current and newly expected counts were provided for each subgroup. The required number of cases was then randomly selected using the "SAMPLE n from m" procedure from the SPSS database program, containing the data for each subgroup of the 2010–2012 study. This procedure randomly selects an m-element sample of cases from an n-element population [11]. In total, 144 of these procedures were separately performed for the subgroups. The adjusted study population of 2797 cases did not differ significantly from the 2017 study population in terms of the unified variable structure. Obtained data were compared between the years of the survey in women residing urban and rural areas. The statistical analysis of the collected data was made based on a chi-squared test using IBM SPSS software version 25 (IBM Corp., Armonk, NY, USA).

Due to the fact that some women did not answer some questions, the sample size in the description of the results is smaller than the actual number of women included for the study; thus, the percentage value is the so-called valid percentage, calculated for the number of answers and not the number of completed questionnaires.

3. Results

3.1. Characteristics of the Study Population

The largest part of the study group, both in the years 2010–2012 and in 2017, included women aged between 26 and 30 years; the smallest one comprised women aged over 35 years. Among the surveyed participants, women living in urban areas accounted for around 59%, whereas those from rural regions accounted for around 41% of the study group. The highest percentage in both groups was represented by women with a normal body weight before pregnancy—i.e., their Body Mass Index (BMI) was between 18.5–24.9 kg/m^2—and they constituted about 67%. The average weight gain of pregnant women in the years 2010–2012 was 14.4 kg, while in 2017 it was 14.6 kg. Characteristics of the study population are presented in Table 1.

Table 1. Characteristics of the study population.

Variable	2010–2012		2017		*p*
	N	%	N	%	
Age (years)					0.91
Under 25	532	19.0	627	19.0	
26–30	965	34.6	1139	34.6	
31–35	925	33.1	1073	32.6	
Over 35	371	13.3	457	13.8	
Place of residence (inhabitants)					0.99
City ≥ 100,000	676	24.2	804	24.1	
City < 100,000	967	34.6	1154	34.7	
Rural area	1154	41.2	1373	41.2	
BMI before pregnancy (kg/m^2)					0.2

Table 1. *Cont.*

Variable	2010–2012		2017		*p*
	N	%	N	%	
<18.5	230	8.4	256	7.9	
18.5–24.9	1838	67.6	2147	66.6	
25.0–29.9	471	17.3	581	18.0	
30.0–34.9	143	5.3	176	5.5	
35.0–39.9	25	0.9	53	1.6	
≥40	13	0.5	13	0.4	
Parameter	**Mean**	**S.D.**	**Median**		*p*
Weight before pregnancy (kg)					0.29
2010–2012	63.5	12.2	61.0		
2017	63.9	12.5	62.0		
Weight gain in pregnancy (kg)					0.4
2010–2012	14.4	5.6	13.0		
2017	14.6	6.2	14.0		

3.2. *Alcohol and Stimulant Use During Pregnancy*

Cigarette smoking by pregnant women living in both urban and rural areas decreased over the years (Table 2). In the group of women living in urban areas, about 16% of women quit smoking because of pregnancy, both in 2010–2012 and in 2017. This percentage increased from 14.0% to 15.9% in the group of women living in rural areas.

Table 2. Alcohol consumption, stimulant use, folic acid supplementation and consumption of various types of drinks in connection with pregnancy in urban and rural areas in Poland, in the period of 2010–2012 and 2017.

	Urban Area			Rural Area		
	2010–2012 % (N)	2017 % (N)	*p*	2010–2012 % (N)	2017 % (N)	*p*
Smoking			0.1			0.65
Yes	7.1 (77)	5.4 (104)		6.4 (49)	6.0 (80)	
No	76.6 (829)	78.6 (1508)		79.6 (611)	78.1 (1042)	
Quit in pregnancy	16.3 (176)	16.0 (306)		14.0 (107)	15.9 (212)	
Alcohol						
Before pregnancy	45.1 (675)	40.8 (798)	0.01	36.8 (370)	33.1 (454)	0.06
During pregnancy	18.7 (276)	6.0 (118)	0.00	15.4 (152)	3.5 (48)	0.00
Legal highs			0.43			0.62
Yes	0.4 (4)	0.2 (4)		0.1 (1)	0.2 (3)	
No	99.6 (1075)	99.8 (1867)		99.9 (764)	99.8 (1304)	
Folic acid supplementation						
Before and during pregnancy	29.5 (445)	32.3 (598)	0.08	21.0 (211)	27.1 (332)	0.00
During pregnancy	62.0 (937)	62.8 (1164)	0.63	67.4 (675)	66.9 (818)	0.81
Without supplementation	8.5 (129)	4.9 (91)	0.00	11.6 (116)	6.0 (73)	0.00
Water			0.2			0.38
Never	2.4 (38)	1.6 (31)		3.2 (34)	2.3 (30)	
Sometimes	13.6 (214)	12.8 (241)		18.5 (197)	17.8 (227)	
Every day	84.0 (1319)	85.6 (1612)		78.2 (831)	79.9 (1020)	
Sweetened beverages			0.00			0.00
Never	30.1 (440)	38.7 (673)		28.0 (268)	32.5 (378)	
Sometimes	51.3 (750)	51.8 (900)		49.0 (469)	57.4 (667)	

Table 2. *Cont.*

	Urban Area			Rural Area		
	2010–2012 % (N)	2017 % (N)	p	2010–2012 % (N)	2017 % (N)	p
Everyday	18.6 (272)	9.4 (164)		23.0 (220)	10.1 (117)	
Tea			0.00			0.00
Never	4.8 (75)	4.4 (81)		3.8 (40)	3.1 (40)	
Sometimes	27.5 (425)	37.4 (695)		22.1 (232)	32.8 (418)	
Everyday	67.7 (1048)	58.3 (1083)		74.1 (777)	64.1 (817)	
Cofee			0.00			0.00
Never	40.6 (600)	41.1 (730)		45.0 (434)	42.7 (510)	
Sometimes	38.0 (561)	43.6 (775)		34.0 (328)	42.2 (503)	
Everyday	21.4 (317)	15.4 (273)		21.0 (203)	15.1 (180)	

Women living in rural areas consumed alcohol less frequently than those living in cities, both before and during pregnancy (Table 2). In 2010–2012 and 2017, 36.8% and 33.1% of women living in rural areas consumed alcohol before pregnancy, respectively, while the percentages among women living in the city were 45.1% and 40.8%, respectively. Alcohol consumption during pregnancy in rural areas was 15.4% in 2010–2012 and 3.5% in 2017, and for city inhabitants, these values were 18.7% and 6.0%, respectively.

In the study population, 0.1–0.4% of women used legal highs during pregnancy; the percentage rate remained at a very low level both in urban and rural areas (Table 2).

3.3. Folic Acid Supplementation

A majority of the study participants from both rural and city regions began folic acid supplementation during pregnancy—in the years 2010–2012, 67.4% of the women living in rural areas and 62% of those living in urban areas supplemented their diet with folic acid, and in 2017, the values were 66.9% and 62.8%, respectively (Table 2). However, a significantly higher percentage of women living in urban areas started to supplement folic acid before pregnancy and continued supplementation during pregnancy compared to women from rural regions. The dose of folic acid taken was not studied.

3.4. Types of Beverages

Considering the types of beverages consumed, the women most frequently chose water. As many as 84% of urban population studied in the years 2010–2012 and 85.6% of individuals analyzed in 2017 declared drinking water every day, whereas the women living in rural areas declared drinking higher amounts of sweetened beverages and tea. In the years 2010–2012, the percentage rate was 23% for sweetened beverages and 74.1% for tea, and in 2017, the values were 10.1% for sweetened beverages and 64.1% for tea. Coffee consumption decreased among women living in rural and urban areas during the analyzed years (Table 2).

3.5. Physical Activity

The median for the total value of physical activity in the years 2010–2012 was 200.3 MET-h/week in women living in urban areas and 188.1 MET-h/week in those living in rural regions. In 2017, the respective figures were 195.6 MET-h/week in urban areas and 178.4 in rural areas. Over the analyzed years, the only differences observed in the rural population were related to sedentary physical activity. The energy expenditure related to household activity among women living in urban areas increased significantly in the analyzed period of time; however, it decreased significantly in the case of occupational physical activity. Women living in urban areas more often engaged in occupational physical activity in pregnancy as well as sport activity, expressed in MET-h/week, compared to rural inhabitants. Physical activity of analyzed groups of women is presented in Table 3.

Table 3. Physical activity of women during pregnancy living in rural and urban areas in Poland in the years 2010–2012 and 2017.

	N	Mean	S.D	Median	First Quartile	Third Quartile	*p*
Urban area							
Total physical activity							0.97
2010–2012	1059	215.3	116.3	200.3	140.2	265.6	
2017	1897	217.3	117.5	195.6	139.8	270.8	
Household activity							0.00
2010–2012	1048	102.8	77.3	80.5	49.6	136.9	
2017	1880	112.1	81.7	92.1	52.3	150.3	
Occupational activity							0.00
2010–2012	595	38.4	62.0	14	0.0	67.2	
2017	1094	33.2	59.0	0.0	0.0	56.0	
Sport/execrice activity							0.88
2010–2012	1029	7.6	8.8	4.8	2.4	9.6	
2017	1858	8.0	9.8	4.8	2.4	9.6	
Sedentary activity							0.00
2010–2012	1048	69.9	41.5	65.9	39.6	97.8	
2017	1872	60.4	38.0	54.1	30.6	87.0	
Moderate-intensity activity							0.04
2010–2012	1056	53.7	64.1	34.4	9.6	71.0	
2017	1890	58.6	70.4	36.0	13.1	78.6	
Vigorous-intensity activity							0.00
2010–2012	928	0.9	3.1	0.0	0.0	0.0	
2017	1747	1.25	3.9	0.0	0.0	0.0	
Rural area							
Total physical activity							0.39
2010–2012	762	206.2	118.2	188.1	126.5	268.3	
2017	1313	204.2	119.9	178.4	125.6	251.5	
Household activity							0.2
2010–2012	753	110.9	85.5	92.1	50.9	145.8	
2017	1302	113.2	81.6	93.6	54.4	150.3	
Occupational activity							0.16
2010–2012	429	31.0	49.5	0.0	0.0	56.0	
2017	710	30.4	60.5	0.0	0.0	52.9	
Sport/execrice activity							0.22
2010–2012	732	5.9	7.1	3.6	1.7	8.1	
2017	1268	6.2	9.2	3.4	0.8	8.0	
Sedentary activity							0.02
2010–2012	742	58.1	40.3	48.2	28.7	84.7	
2017	1292	52.6	35.3	47.3	25.6	75.7	
Moderate-intensity activity							0.69
2010–2012	754	56.6	66.0	35.3	10.1	77.8	
2017	1306	55.5	69.4	32.6	10.5	74.6	
Vigorous-intensity activity							0.09
2010–2012	647	0.64	2.73	0.0	0.0	0.0	
2017	1175	0.94	3.5	0.0	0.0	0.0	

3.6. Body Mass Index

The distribution of BMI among women living in urban and rural areas was comparable in 2010–2012, whereas in 2017, in the rural population, both before pregnancy and in the prenatal period, the percentage rate of overweight women as well as those with class I (BMI 30–34.9 kg/m^2), class II (BMI 35–39.9 kg/m^2) and class III obesity (BMI $\geq$ 40 kg/m^2) was higher than among those living in urban areas. Comparing the groups of women from both analyzed periods of time, there were no statistically significant differences observed in relation to BMI before delivery in the rural and urban population (Table 4).

Table 4. Body Mass Index (BMI) of women living in rural and urban areas in Poland, before pregnancy and before labor, in the years 2010–2012 and 2017.

	Urban Area			Rural Area		
	2010–2012 % (N)	2017 % (N)	*p*	2010–2012 % (N)	2017 % (N)	*p*
BMI before pregnancy (kg/m^2)			0.8			0.12
<18.5	8.7 (139)	8.5 (159)		8.2 (91)	7.3 (97)	
18.5–24.9	68.2 (1094)	68.8 (1306)		66.7 (744)	63.3 (841)	
25.0–29.9	16.8 (270)	16.4 (311)		18.0 (201)	20.3 (270)	
30.0–34.9	5.0 (80)	4.8 (92)		5.7 (63)	6.3 (84)	
35.0–39.9	0.8 (13)	1.2 (23)		1.1 (12)	2.3 (30)	
≥40.0	0.6 (9)	0.4 (7)		0.4 (4)	0.5 (6)	
BMI before labor (kg/m^2)			0.39			0.05
<18.5	0.1 (1)	0.1 (2)		0.1 (1)	0.1 (1)	
18.5–24.9	25.1 (402)	24.3 (458)		21.7 (241)	19.3 (254)	
25.0–29.9	46.8 (751)	45.7 (860)		49.2 (547)	45.9 (605)	
30.0–34.9	21.6 (347)	21.5 (404)		21.6 (240)	24.9 (328)	
35.0–39.9	5.4 (86)	6.9 (130)		6.0 (67)	7.4 (98)	
≥40.0	1.1 (17)	1.5 (28)		1.4 (16)	2.5 (33)	

4. Discussion

To the best of our knowledge, this is the first study on changes in health awareness in the population of pregnant women residing in urban and rural areas in Poland. Understanding these changes is important for healthcare providers to improve counselling during pregnancy and improve maternal and newborn health outcomes. Rural areas in Poland occupy more than 93% of the country's area; this high share automatically becomes a source of causes and effects for differences both at the regional and local level [12]. Despite this fact, health awareness in the rural population, both expressed as better knowledge and healthier behaviors declared by surveyed patients, has been increasing dynamically. These changes in the rural population of pregnant women are probably related to improvements in road infrastructure and access to the Internet, but these aspects were not studied in our survey [13]. The increasing health awareness might be also influenced by an increase in gross domestic product (about 230% in 2016 in comparison to 2002) in the rural population [14].

The analysis of the rural population clearly demonstrates a decrease in the number of births and professional activity and an increase in expenditure on health and education (in the years 1999–2004). An additional aspect is the constantly improving access to out-patient health care in rural areas [15]. Additionally, the nature of migration from rural to urban areas has been changing; increasingly often, people living in rural regions give up permanent migration and decide to choose temporary migration [14].

Currently, the healthcare offered to pregnant women focuses on the promotion of the so-called healthy lifestyle already in the preconception period. In pregnancy, the first medical appointment should be arranged by the 10th week of gestation, and it should include counselling with regard to diet, folic acid intake, non-use of stimulants and smoking cessation [7,16,17]. It is particularly important that a healthcare professional and a pregnant woman should cooperate in order to prepare a schedule of prenatal care that should be modified as required by the patient's health condition and needs. Thus, it is possible to achieve the best possible results and avoid mistakes that could lead to complications and failures.

A permanent problem faced by the Polish society, despite numerous social campaigns, is increasing alcohol consumption [18]. The problem of consuming both hard and weak liquors is also present for pregnant women. This is confirmed by the fact that the National Health Plan for the Years 2016–2020 includes a point on the dissemination of knowledge regarding the harmful effects resulting from alcohol consumption by pregnant women [19]. Moreover, although this aspect remains ignored, it is crucial for women to realize the importance of abstaining from drinking alcohol even in the preconception period; in particular, it should be stressed that the organogenesis stage lasts from the second until the end of the eighth week, which is particularly important given that a large number of women find it difficult to define the exact day of conception [20].

A survey performed in 2012 showed that urban inhabitants consume alcohol more frequently than the rural population [21]; this is probably the reason why this trend is also observed in the group of pregnant women included in our study. A positive phenomenon, however, is the fact that the consumption of alcohol among women living in rural as well as in urban areas has decreased over the analyzed years.

Studies conducted in Poland show that smoking only depends on place of residence to a small extent. In 2016, it was proved that women living in small towns with up to 20,000 inhabitants were addicted to smoking less frequently. The highest percentage rate of female smokers was reported in cities with 100,000–500,000 inhabitants [22].

Our study did not find any statistically significant differences regarding changes in smoking habits in pregnancy in the analyzed period of time, either in rural and urban areas. In addition, the percentage of smoking women living in rural and urban areas remained at a comparable level. In literature sources, it can increasingly often be seen that a negative factor in smoking cessation is not only a low socioeconomic status but also the fact that women ignore the harmful effects of the addiction if their previous child was born healthy. Factors that limit smoking cessation also include a partner who is a smoker, as well as a belief that the addiction may be a stress-reducing activity [23,24]. Due to the large scale and topicality of the problem, not only does an attending doctor play a great role in increasing the awareness of women, but also midwives, childbirth education institutions and individuals are important advocates in a pregnant woman's immediate environment. Importantly, a woman's attention should also be focused on reducing passive smoking as a negative factor for the health of the offspring [25].

Some studies show that women are more often successful at resisting such addictions as cigarettes, alcohol or caffeinated drinks than, for example, at increasing amounts of vegetables or fruit in their diet [26]. Our study partly confirmed these data. About 14–16% of women quit smoking during pregnancy in the analyzed period, while a significant part of them stopped drinking alcohol in connection with pregnancy, especially among women participating in the study in 2017. Apart from personal attitude and motivation, healthy lifestyle is also influenced by individual (such as financial and time-related) and infrastructural options. Although a size of a place of residence determines physical activity to a small extent only, the greatest percentage rate of active individuals was observed among people living in cities with more than 500,000 inhabitants, which may result from the better sport facilities available [27]. In our study, the mean total energy expenditure of women living in cities increased over the analyzed years, in contrast to women living in rural areas. In addition, this expenditure among urban residents was higher than among women living in rural areas.

A similar dependency exists regarding the consumption of sweetened beverages in the rural and urban populations. This may be partly explained by the unfavorable economic situation of rural areas; however, the phenomena that may influence this situation have not yet been fully investigated [28]. Nevertheless, it needs to be emphasized that the opinions given by the women participating in our survey show that the consumption of sweetened beverages among the rural population has decreased significantly in the last few years, and the figures are at present only slightly higher than those for the women representing the urban population.

An alarming occurrence is an increasing number of rural inhabitants who are overweight or suffer from class I, II and III obesity, which reflects a general worldwide trend [29,30]. Studies show that obese individuals are not able to classify their body mass or do not realize that it is abnormal. Moreover, it has been found that partners of overweight or obese persons deal with the same problem, which may be an additional factor causing difficulties in maintaining normal bodyweight before and during pregnancy [31]. Consultation with an attending doctor, proper preparation and the discussion of common objectives, including the maximum weight gain during pregnancy, should be one of the main issues within a preconception care scheme offered to women who plan to become pregnant. Obesity also generates disturbances in levels of microelements crucial for fetal development; e.g., folic acid [32].

The influence of folic acid on the development of the fetal nervous system was observed as early as in the 1960s by the British physician Bryan Hibbard, while, in 1991, the Medical Research Council proved in its study that daily intake of 4 mg of folic acid may prevent defects of the fetal neural tube in women who had previously given birth to a child with such an anomaly [33]. More recent research studies have demonstrated that women of reproductive age planning pregnancy should commence folic acid supplementation already at 12 weeks before the planned pregnancy [34,35]. According to our data, a majority of women begin to take folic acid after they have their pregnancy confirmed; this is particularly evident in women representing the rural population. Only every third women surveyed was aware of the importance of folic acid supplementation before conception and the recommended dose; this may suggest that only 30% of pregnancies were planned, but this aspect was not assessed in this study. In contrast, among female transplant recipients, only every third pregnancy is unplanned [36]. Family planning is therefore of high importance when discussing health awareness among pregnant women. The targeted promotion of folic acid supplement use should be conducted periodically by gynecologists and primary care physicians during annual medical screenings [37].

Our study suggests that, despite the universal access to information for both rural and urban inhabitants and development of prenatal care, there is still a need for education about proper nutrition and other health-related issues regardless of place of residence. These suggestions are also supported by Ługowska et al. and Dereń et al. [38,39] and in other populations [40]. Progress in antenatal care has been made worldwide, which is also supported by our results [41]. Some studies indicate that not only should recommendations be implemented but also "active clinical mentoring" (including such activities as organizing and training a national team of mentors including senior midwives, obstetricians and pediatricians) should be used to improve health awareness even before pregnancy [42]. The meta-analysis by Miteniece E et al. indicated other issues that should be improved, with the geographical distance to healthcare institutions being one of the most important [43].

5. Conclusions

In the last decade, there has been a considerable change in pro-health behaviors in the population of women living in rural areas and a slight change among those living in cities. Although the results of studies seem to be satisfactory, there are still aspects that should be changed in both of the groups. At the same time, as a result of globalization, other problems may arise in Polish rural and urban populations, and they could occur paradoxically in both of the groups, posing new challenges to doctors and other medical professionals providing care to pregnant women.

Author Contributions: Conceptualization, M.S., M.I. and J.W.; Methodology, C.W. and P.B.; Formal Analysis, J.W.; Investigation, M.S., M.I., J.W. and C.W.; Resources, C.W.; Data Curation, C.W.; Funding Acquisition, P.B.; Writing—Original Draft Preparation, M.S. and M.I.; Writing—Review & Editing, C.W., P.B. and J.W.; Visualization, M.S., M.I. and C.W.; Supervision, J.W.; Project Administration, P.B. All authors have read and agreed to the published version of the manuscript.

Funding: This research received no external funding.

Conflicts of Interest: The authors declare no conflict of interest.

References

1. MacDougall, H. Reinventing public health: A new perspective on the health of Canadians and its international impact. *J. Epidemiol. Community Health* **2007**, *61*, 955–959. [CrossRef] [PubMed]
2. Banaszak-Żak, B.; Dobrzyn, D. Czynniki warunkujące zdrowie kobiety ciężarnej. *Ann. UMCS* **2004**, *59* (Suppl. 14), 56–60.

3. Parveen, R.; Latifa, S.A.S. Health Education needs for pregnancy: A study among woman attending primary health centers. *J. Fam. Community Med.* **2003**, *10*, 31–38.
4. Inskip, H.M.; Crozier, S.R.; Godfrey, K.M.; Borland, S.E.; Cooper, C.; Robinson, S.M. Southampton Women's Survey Study Group. Women's compliance with nutrition and lifestyle recommendations before pregnancy: General population cohort study. *BMJ* **2009**, *338*, b481. [CrossRef] [PubMed]
5. Baron, R.; Manniën, J.; Te Velde, S.J.; Klomp, T.; Hutton, E.K.; Brug, J. Socio-demographic inequalities across a range of health status indicators and health behaviours among pregnant women in prenatal primary care: A cross-sectional study. *BMC Pregnancy Childbirth* **2015**, *15*, 261. [CrossRef] [PubMed]
6. Gacek, M. Niektóre zachowania zdrowotne oraz wybrane wskaźniki stanu zdrowia grupy kobiet ciężarnych. *Probl. Hig. Epidemiol.* **2010**, *91*, 48–53.
7. The Regulation of the Minister of Health of 16th August 2018 on the Organizational Standard of Perinatal Care (Dz.U. 2018 poz. 1756). Available online: http://prawo.sejm.gov.pl/isap.nsf/download.xsp/WDU20180001756/O/D20181756.pdf (accessed on 21 April 2020).
8. Wojtyla, C.; Wojtyla-Buciora, P. Polish Pregnancy-related Assessment Monitoring System (Pol-PrAMS): Research on lifestyle health behaviours of Polish women during gestation—Study design. *J. Health Inequal.* **2016**, *2*, 185–191. [CrossRef]
9. Wojtyla, C.; Ciebiera, M.; Wojtyla-Buciora, P.; Janaszczyk, A.; Brzecka, P.; Wojtyla, A. Physical activity patterns in third trimester of pregnancy—Use of pregnancy physical activity questionnaire in Poland. *Ann. Agric. Environ. Med.* **2019**. [CrossRef]
10. Lavrakas, P.J. *Encyclopedia of Survey Research Methods*; Sage Publications: Thousand Oaks, CA, USA, 2008.
11. IBM. SPSS Sample Procedure. Available online: https://www.ibm.com/support/knowledgecenter/en/SSLVMB_25.0.0/statistics_reference_project_ddita/spss/base/syn_sample_overview.html (accessed on 24 May 2020).
12. Bartkowiak, N. The diversity of socioeconomic development of rural areas in Poland in the Western borderland and the problem of post-state farm localities. *Oeconomia Copernic.* **2017**, *3*, 417–430. [CrossRef]
13. Guminski, M.; Information Society in Poland. Results of Statistical Surveys in the Years 2014–2018. Available online: http://stat.gov.pl/obszary-tematyczne/nauka-i-technika-spoleczenstwo-informacyjne/spoleczenstwo-informacyjne/ (accessed on 11 November 2019).
14. The Ministry of Agriculture and Rural Development. *Diagnoza Sytuacji Społeczno-Gospodarczej Rolnictwa, Obszarów Wiejskich i Rybactwa w Polsce. Strategia Zrównoważonego Rozwoju Wsi, Rolnictwa i Rybactwa 2030*; Ministerswto Rolnictwa i Rozwoju Wsi.: Warsaw, Poland, 2019. Available online: https://www.gov.pl/web/rolnictwo/strategia-zrownowazonego-rozwoju-wsi-rolnictwa-i-rybactwa-2030 (accessed on 11 November 2019).
15. Rural Areas in Poland. 2011. Available online: https://stat.gov.pl/en/topics/agriculture-forestry/agriculture/rural-areas-in-poland,3,1.html (accessed on 5 November 2019).
16. ACOG Committee Opinion No. 762: Prepregnancy Counseling. *Obstet. Gynecol.* **2019**, *133*, e78–e89.
17. Farahi, N.; Zolotor, A. Recommendations for preconception counseling and care. *Am. Fam. Physician* **2013**, *88*, 499–506. [PubMed]
18. The State Agency for the Prevention of Alcohol-Related Problems. Consumption of Alcoholic Beverages in Liters Per Capita in Poland over the Years. Available online: https://www.parpa.pl/index.php/badania-i-informacje-statystyczne/statystyki (accessed on 21 April 2020).
19. Regulation of the Council of Ministers of 4 August 2016 Regarding the National Health Program for 2016-2020 (Dz U 2016 poz. 1492). Available online: http://prawo.sejm.gov.pl/isap.nsf/download.xsp/WDU20160001492/O/D20161492.pdf (accessed on 21 April 2020).
20. Bręborowicz, G.H. *Obstetrics and Gynecology*, 2nd ed.; PZWL: Warsaw, Poland, 2015; pp. 1–14.
21. Kantar TNS. Alcohol Consumption in Poland in 2012. Research Report. Available online: http://www.tnsglobal.pl/jakpijapolacy/pdf/raport.pdf (accessed on 21 April 2020).
22. Chief Sanitary Inspectorate. Report from a Nationwide Survey on Attitudes towards Smoking. Available online: https://gis.gov.pl/wp-content/uploads/2018/04/Postawy-Polak%C3%B3w-do-palenia-tytoniu-Raport-2017.pdf (accessed on 21 April 2020).
23. Schneider, S.; Huy, C.; Schütz, J.; Diehl, K. Smoking cessation during pregnancy: A systematic literature review. *Drug Alcohol. Rev.* **2010**, *29*, 81–90. [CrossRef] [PubMed]

24. Chaaya, M.; Awwad, J.; Campbell, O.M.; Sibai, A.; Kaddour, A. Demographic and psychosocial profile of smoking among pregnant women in Lebanon: Public health implications. *Matern. Child Health J.* **2003**, *7*, 179–186. [CrossRef] [PubMed]
25. Wojtyla, C.; Wojtyla-Buciora, P.; Ciebiera, M.; Orzechowski, S.; Wojtyla, A. The effect of active and passive maternal smoking before and during pregnancy on neonatal weight at birth. *Arch. Med. Sci.* **2018**, *14*. [CrossRef]
26. Crozier, S.R.; Robinson, S.M.; Borland, S.E.; Godfrey, K.M.; Cooper, C.; Inskip, H.M.; SWS Study Group. Do women change their health behaviours in pregnancy? Findings from the southampton women's survey. *Paediatr. Perinat. Epidemiol.* **2009**, *23*, 446–453. [CrossRef]
27. Ministry of Sport and Tourism. Sport and Recreation: Nationwide Study of Poles. Available online: https://www.msit.gov.pl/pl/sport/badania-i-analizy/aktywnosc-fizyczna-spol/575,Aktywnosc-fizyczna-spoleczenstwa.html (accessed on 21 April 2020).
28. Sharkey, J.R.; Johnson, C.M.; Dean, W.R. Less-healthy eating behaviors have a greater association with a high level of sugar-sweetened beverage consumption among rural adults than among urban adults. *Food Nutr. Res.* **2011**, *55*, 5819. [CrossRef]
29. McCormack, L.; Martin, S.; McGlade, C.; Meendering, J.; Foster, J.; Miller, A. Differences in Overweight/Obesity Among Youth in a Midwest State by Rural-Urban Continuum Codes. *SD Med.* **2019**, *72*, 419–423.
30. Trivedi, T.; Liu, J.; Probst, J.; Merchant, A.; Jhones, S.; Martin, A.B. Obesity and obesity-related behaviors among rural and urban adults in the USA. *Rural Remote Health* **2015**, *15*, 3267.
31. Thompson, M.W. Pregnant women's knowledge of obesity and ideal weight gain in pregnancy, and health behaviours of pregnant women and their partners. *Aust. N. Z. J. Obstet. Gynaecol.* **2011**, *51*, 460–463. [CrossRef]
32. Chitayat, D.; Matsuil, D.; Amitai, Y.; Kennedy, D.; Vohra, S.; Rieder, M.; Koren, G. Folic acid supplementation for pregnant women and those planning pregnancy: 2015 update. *J. Clin. Pharmacol.* **2016**, *56*, 170–175. [CrossRef]
33. Hoffbrand, A.V.; Weir, D.G. The History of folic acid. *Br. J. Haematol.* **2001**, *113*, 579–589. [CrossRef] [PubMed]
34. Polish Society of Gynecologists and Obstetricians. Folate supplementation during the preconception period, pregnancy and puerperium. *Ginekol. Pol.* **2017**, *88*, 633–636. [CrossRef] [PubMed]
35. Wilson, R.D.; Audibert, F.; Brock, J.A.; Carroll, J.; Cartier, L.; Gagnon, A.; Johnson, J.A.; Langlois, S.; Murphy-Kaulbeck, L.; Okun, N.; et al. Pre-conception Folic Acid and Multivitamin Supplementation for the Primary and Secondary Prevention of Neural Tube Defects and Other Folic Acid-Sensitive Congenital Anomalies. *J. Obstet. Gynaecol. Can.* **2015**, *37*, 534–552. [CrossRef]
36. Szpotanska-Sikorska, M.; Mazanowska, N.; Kociszewska-Najman, B.; Wielgos, M.; Pietrzak, B. Unintended pregnancies and family planning among women after kidney or liver transplantation. *J. Matern. Fetal Neonatal Med.* **2020**, *33*, 1346–1352. [CrossRef] [PubMed]
37. Wegner, C.; Kancherla, V.; Lux, A.; Köhn, A.; Bretschneider, D.; Freese, K.; Heiduk, M.; Redlich, A.; Schleef, D.; Jorch, G.; et al. Periconceptional folic acid supplement use among women of reproductive age and its determinants in central rural Germany: Results from a cross sectional study. *Birth Defects Res.* **2020**. [CrossRef]
38. Ługowska, K.; Kolanowski, W. The nutritional behaviour of pregnant women in Poland. *Int. J. Environ. Res. Public Health* **2019**, *16*, 4357. [CrossRef]
39. Dereń, K.; Gaweł, M.; Łuszczki, E.; Jarmakiewicz, S.; Sokal, A.; Polak, E.; Wyszyńska, J. Nutritional behawior of pregnant women from the Podkarpacie province. *Eur. J. Clin. Exp. Med.* **2017**, *15*, 322–329. [CrossRef]
40. Miteniece, E.; Pavlova, M.; Rechel, B.; Rezeberga, D.; Murauskienė, L.; Groot, W. Barriers to accessing adequate maternal care in Latvia: A mixed-method study among women, providers and decision-makers. *Health Policy (Amst. Neth.)* **2019**, *123*, 87–95. [CrossRef]
41. Moller, A.B.; Petzold, M.; Chou, D.; Say, L. Early antenatal care visit: A systematic analysis of regional and global levels and trends of coverage from 1990 to 2013. The Lancet. *Glob. Health* **2020**, *5*, e977–e983. [CrossRef]

42. Xiong, X.; Carter, R.; Lusamba-Dikassa, P.S.; Kuburhanwa, E.C.; Kimanuka, F.; Salumu, F.; Clarysse, G.; Tutu, B.K.; Yuma, S.; Iyeti, A.M.; et al. Improving the quality of maternal and newborn health outcomes through a clinical mentorship program in the Democratic Republic of the Congo: Study protocol. *Reprod. Health* **2019**, *16*, 147. [CrossRef]
43. Miteniece, E.; Pavlova, M.; Rechel, B.; Groot, W. Barriers to accessing adequate maternal care in Central and Eastern European countries: A systematic literature review. *Soc. Sci. Med. (1982)* **2017**, *177*, 1–8. [CrossRef] [PubMed]

© 2020 by the authors. Licensee MDPI, Basel, Switzerland. This article is an open access article distributed under the terms and conditions of the Creative Commons Attribution (CC BY) license (http://creativecommons.org/licenses/by/4.0/).

International Journal of
Environmental Research and Public Health

MDPI

Article

The Quality of Life of Seniors Hospitalized Due to Cardiovascular Diseases in Poland

Katarzyna Sygit [1,*], Katarzyna Siedlecka-Pasierbiak [2], Marian Sygit [1] and Elżbieta Cipora [3]

1 President Stanislaw Wojciechowski State Vocational Academy in Kalisz, Faculty of Health Sciences, 62-800 Kalisz, Poland; msygit@onet.pl

2 Poddębickie Centrum Zdrowia Hospital, 99-200 Łódź, Poland; katarzyna.pasierbiak@gmail.com

3 State Vocational Academy in Sanok, Sanok Medical Institute, 38-500 Sanok, Poland; elacipora@interia.pl

* Correspondence: k.sygit@pwsz-kalisz.edu.pl

Received: 19 April 2020; Accepted: 20 May 2020; Published: 25 May 2020

Abstract: Introduction: In the light of the increased ageing of the world population, social policy needs to be focused on actions aimed at improving the quality of life of older people. Objective: The main objective of this study was to assess the quality of life in a population of seniors hospitalized due to cardiovascular disease. Materials and methods: The study included 408 elderly patients hospitalized for cardiovascular diseases in the Poddębickie Centrum Zdrowia Hospital in Poddębice, Łódzkie voivodship, Poland. The study used two survey questionnaires: the author's survey questionnaire and the standardized SF36v2 Questionnaire. Statistical analysis of the obtained test results was carried out in the R program, version 3.5.1. Results: Having analyzed the health status of the study group, it was found that the largest group of subjects (84.07%) were treated due to hypertension. Among the ailments that hindered daily functioning, the respondents indicated primarily poor eyesight (53.68%). Patients assessed their own health as 'mediocre' (average) (58.58%). The analysis of the study results from the SF36v2 Questionnaire showed that the highest quality of life was in the limited activity due to emotional problems (RE) dimension, social functioning (SF), and physical functioning (PF); the weakest scores were observed in vitality (VT), general health perception (GH), and health transition (HT) dimensions. Conclusions: The significant demographic, social and socio-medical factors that determined respondents' quality of life were: age, gender, marital status, education and health situation. The analysis of quality of life according to the SF36v2 Questionnaire showed that the study group functioned better in the mental dimension (MCS—mental component summary, overall mental health) than in the physical one (PCS—physical component summary, total physical health).

Keywords: quality of life; the elderly; seniors; health; cardiovascular diseases; physical and psychological well-being

1. Introduction

The forecasts of the United Nations (UN) predict that by 2030, the percentage of Europe's population aged 65+ will reach 23.8%. Currently, Italy, Germany and Greece are leaders among countries with the highest senior citizen proportion, i.e., people aged 65+ in the general population [1–3]. Although in Poland the elderly (65+) constitute over 15% of general population (while the EU average is almost 19%), it should be emphasized that the situation of Poland is only seemingly favorable compared to Europe, because the situation will change radically in the coming decades. By 2050, Poland will become one of the European countries with the highest old age ratio—it will double and amount to over 30%. The oldest age group (80+), which today amounts to about 4% of the general population, also needs to be considered. According to Eurostat, it will reach almost 10% of the general population in 2050 [1]. Research shows that 80+ individuals suffer from an increasing number of diseases, becoming more and more dependent and in need of help [1,4–6].

Cardiovascular disease (CVD) is the most frequently mentioned issue among health problems of the elderly; it encompasses ischemic heart disease, stroke, and atherosclerosis, which are the main causes of morbidity, disability and premature mortality in developed countries. The results of the Global Burden of Disease Study indicate that cardiovascular disease has caused about 15.6 million deaths worldwide (29.6%). It is estimated that in Europe, cardiological diseases are the cause of 45% of all deaths, i.e., over 4 million per year [2,3,7,8].

Since 1991, mortality caused by cardiovascular diseases in Poland has systematically decreased; however, it is still considered 'high'. In 2015, cardiological diseases were the cause of approximately 46.8% of deaths, while they constituted as much as approx. 52% in the early 1990s [1,3,7].

The most pressing problems of the elderly include not only health problems, but also loneliness, disability, poverty, and a sense of uselessness. All these problems point to the existing marginalization of seniors as a group, which may be exemplified by their gradual elimination from active professional and social life [9–11].

From the psychosocial perspective, old age is a set of various interrelated social, economic, family and cultural conditions whose impact varies both environmentally and individually; it is a consequence of losses to which the old person is exposed: the loss of social prestige, impoverishment, and excess of free time which is difficult to fill up satisfactorily [3,11].

Old age is a difficult period in human life, and one needs to be properly prepared for it. The development of geriatrics and gerontology results in progress in improving the quality of life in old age. Therefore, extensive actions should be taken to help the elderly with such numerous health, social or economic problems that greatly affect their quality of life [7,10,11].

The quality of seniors' life has become the subject of numerous studies (especially in recent years) and a topic often raised in scientific articles [12–17]. The concept of quality of life is difficult to define, as its perception depends on the worldview, experience and education.

According to the World Health Organization (WHO) definition, quality of life is 'an individual's perception of their position in life in the context of the culture and value systems in which they live and in relation to their goals, expectations, standards and concerns' [9]. The indicators of quality of life include the ability to continue to play one's life roles, the ability to adapt, psychological well-being and functioning within social groups [9,10].

Undoubtedly, the quality of life changes over time and is significantly susceptible to internal and external factors such as: family life, friendship, work, neighbors, place of residence, home, education, health, standard of living, and nationality [9–11]. All these internal and external conditions mean that the quality of life of the elderly is sometimes expressed by:

- ✓ clinical assessment of physical and mental fitness, susceptibility to diseases, disability;
- ✓ environmental conditions—housing situation, standard of living, group membership, family status, support systems;
- ✓ ecological environment—relaxation and recreation facilities in the place of residence, transport conditions;
- ✓ lifestyle—diet, smoking, drinking alcohol, exercising;
- ✓ economic and living conditions—income from pension, other forms of financial support [11].

Undoubtedly, the quality of life changes over time and is significantly susceptible to internal and external factors. Currently in the developed world, a new lifestyle has emerged: one with a special focus on social life, expanding one's interests and doing sports. The demand for education and training is on the increase, which is why various forms of activities for seniors (e.g., Third Age Universities) are currently so popular [15,16,18–20]. A number of standardized research tools on quality of life have been developed. The most commonly used tool is the SF36v2 Questionnaire, intended for subjective assessment of one's health, and it was used in this study. The quality of life of the respondents was assessed in 11 dimensions (Figure 1) [21–23].

Figure 1. Dimensions used to assess the quality of life in SF36v2 questionnaire.Source: authors' own work.

The main objective of this study was to assess the quality of life in a population of seniors hospitalized due to cardiovascular disease.

The main hypothesisis as follows: demographic and social factors, health, as well as social and environmental problems in everyday life influence the quality of life of seniors hospitalized due to cardiovascular disease.

2. Materials and Methods

The study was conducted in 2018 among 408 elderly patients hospitalized for cardiovascular diseases in the Poddębickie Centrum Zdrowia Hospital in Poddębice, Łódzkie voivodship, Poland.

The criteria for the study group were as follows: age 60+; gender women and men diagnosed with cardiovascular disease and hospitalized in the Poddębickie Centrum Zdrowia Hospital Health Centre in Poddębice, admitted to hospital on schedule and based on a referral from a doctor, agreeing to be included in the study. The exclusion criteria were: age below 60, no cardiovascular disease, or no consent for participation in the study.

The method used in the study was a diagnostic survey. The research tools used in the study includedthe authors' 2-part survey questionnaire. The first part focused on social information, while the second part aimed at investigating health of the study group. The second tool was the SF36v2 Questionnaire, intended for subjective assessment of health. It consisted of 11 questions, containing 36 statements, which helped specify the following elements: physical fitness, limitations due to health condition, pain, general health perception, vitality, social functioning, sense of mental health, limitations caused by emotional problems, change of health condition, physical functioning total physical health, mental functioning total mental health.

Participation in the study was voluntary and involved completing an anonymous survey questionnaire. The research toolkit was accompanied by the respondent's statement regarding consent to participate in the study; respondents were given information about the study. The study was conducted by the authors of the paper. Respondents could refrain from completing the questionnaire at any stage. Fromthe moment of agreeing to participate in the study, the collected data about the respondents were stored in a secure place with no access by unauthorized persons.

Based on the SF36v2 questionnaire (Quality Of Life Questionnaire), the quality of life of the study group in 11 dimensions was assessed: physical functioning (PF); role limitations due to physical problems (RP); bodily pain (BP); general health perception (GH); vitality (VT); social functioning (SF); mental health (MH); role limitation due to emotional problems (RE); health transition (HT); physical component summary (PCS); and mental component summary (MCS).

The quality of life in each dimension was expressed by a number 0–100. Higher numbers indicate a better quality of respondents' life. There are no standards for SF-36, so it was not possible to assess whether the results achieved by the respondents meant high or low quality of life. However, a comparison of dimensions was made to identify areas with the highest and poorest quality of life.

Questionnaire SF36v2 was made available by the Office of Grants and Scholarly Research and obtained license No. QM046214.

The study was approved by the Bioethics Committee of the Medical University of Lodz, under number RNN/155/18/KE.

The analysis of the obtained test results was carried out in the R program, version 3.5.1.

The analysis of quantitative variables (i.e., expressed in numbers) was performed by calculating the mean, standard deviation, median, quartiles, minimum and maximum values. The analysis of qualitative variables (i.e., not expressed in numbers) was conducted by calculating the number and percentage of occurrences of each value. Comparison of qualitative variables in groups was made using the chi-square test (with Yates correction for 2 × 2 tables) or Fisher's exact test (only when expected numbers in tables were low). The comparison of the values of quantitative variables in two groups was made using Student's t test (when the variable had normal distribution in these groups) or the Mann–Whitney test (non-normal distribution). The comparison of the values of quantitative variables in three or more groups was made using ANOVA variance analysis (when the variable had normal distribution in these groups) or the Kruskal–Wallis test (non-normal distribution). After detecting statistically significant differences, posthoc analysis was carried out with Fisher's Least Significant Difference LSD test (normal distribution) or Dunn's test (non-normal distribution) to identify statistically significant differences between groups.

Correlations between quantitative variables were analyzed using the Pearson correlation coefficient (when both variables had normal distribution) or Spearman correlation coefficient (otherwise). The strength of correlation was interpreted according to the following scheme:

$|r| \geq 0.9$—very strong correlation
$0.7 \leq |r| < 0.9$—strong correlation
$0.5 \leq |r| < 0.7$—medium correlation
$0.3 \leq |r| < 0.5$—weak correlation
$|r| < 0,3$—very week correlation (negligible) [24].

The normality of variable distribution was tested using the Shapiro–Wilk test.

The analysis adopted the significance level of 0.05. Thus, all p values below 0.05 were interpreted as significant correlations.

3. Results

3.1. The Overview of the Subjects

The group1 was dominated by women: 227 (55.64%) versus 181 men (44.36%). The age structure of the study group was as follows: the average age was 70.92 (SD = 6.51) and ranged from 64 to 95; the median was 68, therefore half of the group was younger than 68, and the other half was older. In terms of place of residence, the most numerous groups were inhabitants of cities below 100,000—177 people (43.38%), and over 100,000—126 people (30.88%). One hundred and two individuals (25.00%) lived in a rural environment. When analyzing education of the respondents, secondary education (including post-secondary, non-tertiary education) was the most common—reported by 148 people (36.27%), followed by 105 people (25.74%) with vocational education, 102 people (25.00%) with higher education and 43 people (2.45%) with primary education. The structure of marital status of the studied population was as follows: the largest group comprised married people—222 people (54.41%), followed by widows and widowers—92 people (22.55%), the divorced—50 people (12.25%), single people—26 people (6.37%), while cohabitation partners accounted for 10 people (2.45%). The main source of income for the majority of respondents was disability pension or old-age pension (non-work sources)—263 individuals (64.46%). Other sources of income included work in private institutions and companies—91 people (22.30%), work in budgetary institutions—56 people (13.73%), agriculture—20 people (4.90%), while 7 people (1.72%) were family-dependent. Among other sources of income indicated by the respondents (47 people, 11.52%), the family allowance and rehabilitation allowance were the most common (Table 1).

Table 1. The overview of the study group.

No.	The Overview of the Study Group		N	%
1.	Gender	Woman	227	55.64
		Man	181	44.36
2.	Age (in years)	Average—70.92 SD—6.52 Median—68 Min—64 Max—95 Q1—66 Q3—74	-	-
3.	Place of residence	City < 100,000 inhabitants	177	43.38
		City > 100,000 inhabitants	126	30.88
		Rural area	102	25.00
		No answer	3	0.74
4.	Education of the subjects	Primary (incomplete)	10	2.45
		Primary	43	10.54
		Vocational	105	25.74
		Secondary, including post-secondary	148	36.27
		Higher. including bachelor's degree	102	25.00
5.	Marital status	Single	26	6.37
		Married	222	54.41
		In separation	2	0.49
		Divorced	50	12.25
		Widow/widower	92	22.55
		In a partnership	6	1.47
		Cohabitation	10	2.45
6.	Source of income	Work in budgetary institutions	56	13.73
		Work in private institutions and companies	91	22.30
		Agriculture work	20	4.90
		Family-dependent	7	1.72
		Casual work	12	2.94
		Non-profit	263	64.46
		Other	47	11.52

N—number of respondents; %—percentage of respondents; SD—average age of respondents.

3.2. Health Condition of the Respondents

Analyzing the incidence of treated cardiovascular diseases in the study group, it was found that 343 people (84.07%) had hypertension, followed by varicose veins and venous thrombosis—163 people (39.43%), arrhythmia and cardiac conduction disorders—133 people (32.60%), atherosclerosis—124 people (30.39), coronary heart disease—71 people (17.40%), and heart defects—31 people (7.60%). Respondents also reported other diseases such as stroke and myocardial infarction. Among the ailments that hindered their daily functioning, respondents indicated primarily poor eyesight—219 people (53.68%), hearing impairment—114 people (27.94%), as well as imbalance—90 people (22.06%). Respondents also reported other issues that hindered their everyday life, such as urinary incontinence, headaches, abdominal pains, and chest pains. The results show that the majority of the study group had persistent pain that hampered their everyday life (the most common complaints were back and knee pain)—261 people (63.97%). The majority of the study group—89.46%—took medication prescribed by a doctor. Worryingly, 42 people (10.29%) stated that they did not follow their doctor's prescriptions. The majority of patients assessed their health as 'mediocre' (average)—239 people (58.58%). A group of 91 people (22.30%) assessed their health as 'good', 69 people (16.91%) as 'poor', and only 2 people (0.49%) as 'very good' (Table 2).

Table 2. Health condition of the respondents.

No.	Health Condition		N	%
1.	Treatment connected to cardiovascular disease (CVD) *	Hypertension	343	84.07
		Atherosclerosis	124	30.39
		Ischemic heart disease	71	17.40
		Rhythm and cardiac conduction disorders	133	32.60
		Heart defects (congenital and acquired)	31	7.60
		Varicose veins. venous thrombosis	163	39.95
		Other	116	28.43
2.	Ailments hindering everyday life *	Eyesight impairment	219	53.68
		Hearing impairment	114	27.94
		Speech disorders	6	1.47
		Speech disorders	90	22.06
		Other	121	29.66
3.	Pain hindering everyday life	Yes	261	63.97
		No	146	35.78
		No answer	1	0.25
4.	Taking medications prescribed by a doctor	Yes	365	89.46
		No	42	10.29
		No answer	1	0.25
5.	Subjective health assessment	Very good	2	0.49
		Good	91	22.30
		Mediocre (average)	239	58.58
		Poor	69	16.91
		Very poor	6	1.47
		No answer	1	0.25

* Percentage does not add up to 100%, as this was a multiple-choice question. N—number of respondents; %—percentage of respondents;

3.3. Assessment of the Respondents' Quality of Life with the SF36v2 Questionnaire

The analysis results show that the study group had the best quality of life in the RE, SF and PF dimensions, while it was the lowest in the VT, GH and HT dimensions. The respondents functioned somewhat better in the mental dimension (MCS) than in the physical one (PCS) (Table 3).

Table 3. Quality of respondents' life in individual dimensions.

Dimension	N *	Mean	SD	Median	Min	Max	Q1	Q3
PF	405	56.14	29.82	60	0	100	30	85
RP	408	49.42	25.42	50	0	100	31.25	68.75
BP	407	50.49	20.1	50	10	100	30	70
GH	408	44.47	16.49	45	0	85	35	55
VT	408	46.58	14.75	50	10	90	35	60
SF	408	56.31	23.9	62.5	0	100	37.5	75
RE	408	60.6	27.99	66.67	0	100	50	75
MH	405	52.59	16.13	52	8	88	44	64
HT	408	38.66	20.48	50	0	100	25	50
PCS	404	50.14	19.32	53.03	1.52	90.91	36.36	65.15
MCS	405	52.71	16.05	53.85	6.15	90.77	41.54	66.15

* Some respondents omitted certain questions, thus preventing calculation of some dimensions; PF—physical functioning; RP—role limitations due to physical problems; BP—bodily pain; GH—general health perception; VT—vitality; SF—social functioning; MH—mental health; RE—role limitation due to emotional problems; HT—health transition; PCS—physical component summary; MCS—mental component summary (MCS).

There were statistically significant correlations between the quality of life in the PF and PCS dimensions and the gender of respondents ($p < 0.05$). They were significantly higher for men than for women (Table 4).

Table 4. Quality of life in individual dimensions versus gender of respondents.

Dimension		Women	Men	p *
PF	mean ± SD	52.69 ± 29.69	60.44 ± 29.49	0.006
	Median	55	65	NP
	Quartile	30–80	35–90	
RP	mean ± SD	47.47 ± 24.75	51.86 ± 26.1	0.067
	Median	50	50	NP
	Quartile	25–62.5	31.25–68.75	
BP	mean ± SD	49.29 ± 19.83	51.99 ± 20.4	0.198
	Median	50	50	NP
	Quartile	30–60	30–70	
GH	mean ± SD	44.82 ± 16.87	44.03 ± 16.04	0.996
	Median	45	45	NP
	Quartile	35–55	35–55	
VT	mean ± SD	45.51 ± 14.57	47.93 ± 14.92	0.112
	Median	50	50	NP
	Quartile	35–55	40–60	
SF	mean ± SD	54.79 ± 24.44	58.22 ± 23.14	0.159
	Median	50	62.5	NP
	Quartile	37.5–75	50–75	
RE	mean ± SD	58.52 ± 27.54	63.21 ± 28.41	0.117
	Median	58.33	66.67	NP
	Quartile	50–75	50–83.33	
MH	mean ± SD	52.67 ± 15.84	52.49 ± 16.53	0.88
	Median	56	52	NP
	Quartile	44–64	4467	
HT	mean ± SD	38.55 ± 21.71	38.81 ± 18.89	0.826
	Median	50	50	NP
	Quartile	25–50	25–50	
PCS	mean ± SD	48.58 ± 19	52.07 ± 19.6	0.042
	Median	51.52	55.3	NP
	Quartile	34.85–63.64	37.88–67.05	
MCS	mean ± SD	51.81 ± 15.44	53.86 ± 16.77	0.164
	Median	52.31	55.38	NP
	Quartile	40–64.62	43.08–66.15	

* p = Normal distribution in groups, Student's t-test; NP = No normal distribution in groups, Mann-Whitney test.

Analysis of the results found that X correlates significantly and negatively with the quality of life in the PF, RP, BP, VT, SF, RE, PCS and MCS dimensions (since $p < 0.05$); therefore, the older the age of the respondents, the lower the quality of life in these dimensions (Table 5).

Table 5. Quality of life in individual dimensions versus age of respondents.

Dimension	Correlation with Age			
	Correlation Coefficient	p *	Correlation Direction	Correlation Strength
PF	−0.356	$p < 0.001$ NP	negative	weak
RP	−0.237	$p < 0.001$ NP	negative	very weak
BP	−0.173	$p < 0.001$ NP	negative	very weak
GH	−0.052	$p = 0.294$ NP	—	—
VT	−0.165	$p = 0.001$ NP	negative	very weak
SF	−0.232	$p < 0.001$ NP	negative	very weak
RE	−0.229	$p < 0.001$ NP	negative	very weak
MH	−0.023	$p = 0.65$ NP	—	—
HT	−0.059	$p = 0.238$ NP	—	—
PCS	−0.289	$p < 0.001$ NP	negative	very weak
MCS	−0.174	$p < 0.001$ NP	negative	very weak

* p = normal distribution of both correlated variables, Pearson correlation coefficient; NP = no normal distribution of at least one of the correlated variables, Spearman's correlation coefficient.

The study found that quality of life in all dimensions depended on the education of respondents ($p < 0.05$).

A posthoc analysis was performed, which helped discover that:

- in the BP dimension, respondents with higher education had a higher quality of life than other respondents; and those with secondary education had a higher quality of life than individuals with vocational education;
- in the RP, VT and SF dimensions, respondents with higher education had a higher quality of life than other respondents, and those with secondary education had a higher quality of life than individuals with primary (incomplete) education;
- in the GH and MH dimensions, respondents with higher and secondary education had a higher quality of life than individuals with vocational and primary (incomplete) education;
- in the HT dimension, respondents with higher and secondary education had a higher quality of life than respondents with vocational and primary (incomplete) education, while respondents with vocational education had a higher quality of life than individuals with primary (incomplete) education;
- in the PF, RE, PCS and MCS dimensions, respondents with higher education had a higher quality of life than other respondents; and those with secondary education had a higher quality of life than individuals with vocational and primary (incomplete) education (Table 6).

Table 6. Quality of respondents' life in individual dimensions versus their education.

Dimension		(Incomplete) Primary	Vocational	Secondary	Higher	p *
PF	Mean ± SD	38.92 ± 30.84	49.33 ± 29.01	58.71 ± 28.59	68.04 ± 26.21	< 0.001
	Median	40	50	65	75	NP
	Quartile	10–65	30–75	35–85	51.25–90	H > S,V,P; S > V,P
RP	Mean ± SD	39.27 ± 23.17	45.42 ± 24.43	48.73 ± 26.79	59.8 ± 22.1	< 0.001
	Median	37.5	50	50	62.5	NP
	Quartile	25–56.25	25–62.5	31.25–68.75	50–75	H > S,V,P; S > P
BP	Mean ± SD	46.98 ± 18.04	43.62 ± 17.33	50.41 ± 20.77	59.51 ± 19.67	< 0.001
	Median	50	50	50	60	NP
	Quartile	30–60	30–50	30–70	50–70	H > S,P,V; S > V
GH	Mean ± SD	38.11 ± 19.91	41.52 ± 16.29	45.71 ± 14.96	49.02 ± 15.46	< 0.001
	Median	40	40	45	50	NP
	Quartile	30–50	30–50	35–55	40–58.75	H,S > V,P

Table 6. *Cont.*

Dimension		(Incomplete) Primary	Vocational	Secondary	Higher	p *
VT	Mean ± SD	40.28 ± 16.85	43.81 ± 14.34	47.03 ± 14.56	52.06 ± 12.32	< 0.001
	Median	40	45	50	55	NP
	Quartile	30–50	30–55	35–56.25	45–60	H > S,V,P; S > P
SF	Mean ± SD	48.35 ± 23	51.67 ± 23.13	56.08 ± 23.21	65.56 ± 23.5	< 0.001
	Median	50	50	62.5	75	NP
	Quartile	25–62.5	37.5–62.5	37.5-75	50–87.5	H > S,V,P; S > P
RE	Mean ± SD	48.27 ± 29.48	55.71 ± 28.62	62.27 ± 27.41	69.61 ± 24.14	< 0.001
	Median	50	50	66.67	75	NP
	Quartile	25–75	41.67–75	50–75	50–83.33	H > S.,V,P; S > V,P
MH	Mean ± SD	43.62 ± 18.7	48.23 ± 15.75	54.66 ± 15.37	58.75 ± 12.72	< 0.001
	Median	44	50	56	60	NP
	Quartile	28–56	36–57	48–64	48–68	H,S > V,P
HT	Mean ± SD	27.36 ± 23.65	35.24 ± 21.29	40.71 ± 19.4	45.1 ± 16.14	< 0.001
	Median	25	25	50	50	NP
	Quartile	0–50	25–50	25–50	50–50	H,S > V,P; V > P
PCS	Mean ± SD	40.05 ± 19.57	45.15 ± 17.8	51.06 ± 19.05	58.99 ± 17.2	< 0.001
	Median	40.91	43.94	54.55	65.15	NP
	Quartile	24.24–56.06	34.85–60.61	36.36–65.15	47.73–71.21	H > S,V,P; S > V,P
MCS	Mean ± SD	44.03 ± 18.26	48.71 ± 14.89	53.94 ± 15.22	59.53 ± 13.98	< 0.001
	Median	44.62	49.23	55.38	61.54	NP
	Quartile	33.85–55.38	36.92–60	44.62–64.62	50.77–69.23	H > S,V,P; S > V,P

* p = normal distribution in groups, ANOVA + results of posthoc analysis (Fisher's LSD test); NP = no normal distribution in groups, Kruskal–Wallis test + posthoc analysis results (Dunn's test); education: H-higher, S-secondary, V-vocational, P-primary.

The results analysis showed that the quality of life in the PF and RE dimensions depended on the marital status of the respondents ($p < 0.05$); it was relatively higher among respondents in relationships (Table 7).

Table 7. Quality of life in individual dimensions versus marital status of respondents.

Dimension		In a Relationship	Single	p *
PF	Mean ± SD	58.86 ± 28.13	52.29 ± 31.74	0.05
	Median	65	55	NP
	Quartile	35–85	25–80	
RP	Mean ± SD	50.29 ± 25.91	48.2 ± 24.73	0.425
	Median	50	50	NP
	Quartile	31.25–68.75	31.25–62.5	
BP	Mean ± SD	50.04 ± 20.53	51.12 ± 19.53	0.43
	Median	50	50	NP
	Quartile	30–70	40–60	
GH	Mean ± SD	44.6 ± 15.36	44.29 ± 17.99	0.659
	Median	45	45	NP
	Quartile	35–55	35–55	
VT	Mean ± SD	47.29 ± 14.95	45.59 ± 14.46	0.292
	Median	50	50	NP
	Quartile	35–60	35–55	
SF	Mean ± SD	56.88 ± 23.71	55.51 ± 24.21	0.785
	Median	62.5	62.5	NP
	Quartile	37.5–75	37.5–75	
RE	Mean ± SD	62.99 ± 29.16	57.25 ± 25.98	0.032
	Median	66.67	58.33	NP
	Quartile	50–89.58	50–75	

Table 7. *Cont.*

Dimension		In a Relationship	Single	p *
MH	Mean ± SD	52.93 ± 16.49	52.12 ± 15.64	0.617
	Median	54	52	NP
	Quartile	44–65	44–64	
HT	Mean ± SD	38.13 ± 20.11	39.41 ± 21.02	0.553
	Median	50	50	NP
	Quartile	25–50	25–50	
PCS	Mean ± SD	51.09 ± 18.51	48.78 ± 20.4	0.407
	Median	53.03	51.52	NP
	Quartile	37.88–65.15	34.85–66.67	
MCS	Mean ± SD	53.6 ± 16.31	51.46 ± 15.65	0.217
	Median	54.62	52.31	NP
	Quartile	42.69–66.15	41.54–64.62	

* p = normal distribution in groups, Student's t-test; NP = no normal distribution in groups, Mann–Whitney test.

4. Discussion

The term 'quality of life' is understood as physical and mental well-being, satisfaction with life, happiness, and fulfilment of desires. Its perception is multidimensional and based on the subjective assessment of an individual [15,16,25,26].

The objective of the study was to assess the quality of life in a population of seniors hospitalized due to cardiovascular disease. The research results help formulate a number of valuable conclusions which are presented in the final part of the paper.

Along with the socio-economic development, the increase in the standard of living, the development of production, the appearance of new goods and services, as well as civilizational and cultural changes, the patterns of behavior of the individuals began to change and their life needs increased. Therefore, there is a justified need to conduct research on the quality of life, both from the point of view of the individual and the entire society, which determines the ways, styles and standards of living [12–14,27–30].

The health condition of the respondents was an important element of their quality of life [31–33]. Research showed that the majority of respondents (84.07%) were treated for hypertension, while 39.95% were diagnosed with varicose veins and venous thrombosis.

Nowadays, hypertension is recognized as one of the most pressing health problems in Poland and worldwide [2,7,9–11,31,33–36]. According to a report by the World Health Organization, almost 1 billion people aged 25+ suffer from hypertension. It is a disease with a complex etiology and may be caused by many factors. Hypertension complications are the cause of 12.8% of all deaths worldwide [2,8]. Many studies have observed that in high-income countries, the percentage of people with hypertension is reduced due to the promotion of a healthy lifestyle and access to appropriate therapies [4,5,9,37,38]. In other countries, it is necessary to implement measures to improve the detection and treatment of hypertension [29,37–40].

Other cardiovascular diseases that are relatively common among seniors include atherosclerosis, coronary heart disease, arrhythmia and heart conduction disorders, and heart defects. In the study group, these were relatively rare, but other studies have shown that atherosclerosis is relatively common in adults; over time, it leads to clinically observable cardiovascular diseases, such as coronary artery disease, stroke, and peripheral atherosclerosis. These diseases are the main cause of death in Poland and around the world [2,3,7,8]. In Europe, cardiovascular diseases are responsible for around 40% of deaths before the age of 75 [2,5,41]. Due to effective prevention and treatment, a significant reduction in mortality due to atherosclerotic diseases has been observed in many European countries [3,5,9,11,36].

The needs and possibilities to access them, as well as associated problems, are an important quality of life factor for people aged 60+ [11,35,36,40].

A significant problem for the respondents was pain, which hindered everyday functioning. This was reported by more than half of the respondents—63.97%. Pain is often found in patients with pre-existing cardiovascular disease [2,5,7,8,34,35]. Pain significantly worsens patients' quality of life. Numerous studies have yielded interesting results regarding the occurrence of chest pain, lower limbs (oedema of the lower legs), and joint pain [2,3,7,8,35,36,40]. This is a huge problem for patients and their immediate environment.

To improve patients' health, and thus their quality of life, it is important to take medication prescribed by a doctor. Among the patients, 89.46% indicated that they followed doctor's instructions regarding medication. However, numerous studies have shown that patients do not take drugs prescribed by their physician due to, e.g., material difficulties, forgetting about it or ignoring medical recommendations [4,6,39,40,42].

When self-assessing their health, 58.58% of respondents declared their health to be mediocre (average), while 22.30% described it as 'good'. Subjective health self-assessment was important in assessing patients' quality of life. Studies by other authors have also shown that most patients hospitalized due to cardiovascular disease assess their health worse than 'mediocre'.

Based on the SF36v2 Questionnaire, the quality of life of the subjects was assessed in 11 dimensions. It was found that the highest quality of life was in the limited activity due to emotional problems(RE) dimension, social functioning (SF), and physical functioning (PF); the weakest scores were observed in vitality (VT), general health perception (GH), and health transition (HT) dimensions. The respondents functioned somewhat better in the mental dimension (MCS) than in the physical one (PCS). In studies on the quality of life with the SF36v2 Questionnaire, other authors came to the conclusion that in the population of people aged 50+, the best quality of life was in the physical fitness (PF) and general health perceptions (GH) dimensions [21–23,43,44].

The results of this study show that there is a relationship between the quality of life and the gender of the subjects. This relationship was discovered in two dimensions: physical functioning (PF) and PCS (functioning in the physical dimension, total physical health). It was significantly higher in the group of men. In a study conducted by Hopman et al., a similar relationship may be noticed, in particular regarding the highest quality of life in the physical functioning (PF) dimension [23].

The research also confirmed that the quality of life depended on the education of the respondents in each of the dimensions. Similar conclusions were reached by Lam et al. and Wang et al. Education (higher education in particular) and high awareness of health or well-being resulted in a higher quality of life of the respondents [21,22].

The quality of life, as observed in this study, depended on the marital status of respondents. The quality of life was relatively higher among the respondents in relationships. This is confirmed by other studies, which proves that a life in companionship, sharing one's problems, and support of a loved one in times of illness are priceless and affect both physical and mental health [22,25].

It should be noted that the perception of quality of life by older people is slightly different compared to the rest of the population before the age of 65. When seniors assess their quality of life, they include their physical and mental well-being, satisfaction with life, happiness and fulfilment of desires [15,21,22,45].

Research on the quality of life of the elderly should be a source of information on how to assess patients' life situation—preferably by means of self-assessment. Such research may provide valuable information on how to proceed with patients [2,3,7,8,32,33,35,46,47].

When assessing seniors' quality of life, various aspects related to health and physical and mental well-being should be taken into account, as well as social functioning aspects. These reflect a sense of happiness, a vision of oneself, social comparison, coping with changes and limitations, a positive attitude, and determination, which in turn reflect the scope of necessary support [25–29].

5. Conclusions

Significant demographic, social and socio-medical factors which determined the quality of subjects' life included age, gender, marital status, education and health situation, in particular cardiovascular diseases, visual impairment, hearing impairment, balance disorders and backache.The study also showed that most of the respondents assessed their health as 'average' and their quality of life was closely correlated with their health.The analysis of quality of life of seniors with pre-existing cardiovascular disease (assessed with the SF36v2 Questionnaire) indicatedthe highest quality of life in terms of physical fitness and social functioning, as well as limited activity caused by emotional problems, andthe lowest quality of life in terms of vitality, perception of one's own health and health transition. The study group functioned better in the mental dimension (MCS, overall mental health) than in the physical one (PCS, total physical health).

Practical implications: It would be advisable to continue research on the quality of life of seniors in order to discover the social and environmental problems which significantly determine quality of life of the elderly population. Due to the diverse quality of seniors' lives, depending on gender, age, health, education and marital status, measures should be taken to improve their quality of life, in particular to facilitate the functioning of the elderly in everyday life; additionally, an effective health education implemented in schools from an early age is recommended to reduce socio-medical and financial problems before reaching old age, as these problems may have a significant impact on the quality of life of seniors in the future.

Author Contributions: Data curation, K.S.; Formal analysis, K.S.-P.; Funding acquisition, E.C.; Investigation, M.S.; Methodology, K.S.-P.; Project administration, K.S.; Resources, M.S.; Software, K.S.-P.; Supervision, K.S.; Validation, K.S.; Visualization, K.S.; Writing—original draft, M.S.; Writing—review & editing, E.C. All authors have read and agreed to the published version of the manuscript.

Funding: This research received no external funding.

Conflicts of Interest: The authors declare no conflict of interest.

References

1. Ageing and Health. Available online: http://www.who.int/mediacentre/factsheets/fs404/en/ (accessed on 8 July 2019).
2. Writing, G.M.; Mozaffarian, D.; Benjamin, E.J.; Go, A.S.; Arnett, D.K.; Blaha, M.J.; Cushman, M.; Das, S.R.; de Ferranti, S.; Després, J.P.; et al. Executive Summary: Heart Disease and Stroke Statistics—2016 Update: A Report From the American Heart Association. *Circulation* **2016**, *133*, 447–454. [CrossRef]
3. Ponikowski, P.; Voors, A.A.; Anker, S.D.; Bueno, H.; Cleland, J.G.F.; Coats, A.J.S.; Falk, V.; González-Juanatey, J.R.; Harjola, V.P.; Jankowska, E.A.; et al. 2016 ESC guidelines for the diagnosis and treatment of acute and chronic heart failure: The task force for the diagnosis and treatment of acute and chronic heart failure of the European Society of Cardiology (ESC). Developed with the special contribution of the Heart Failure Association (HFA) of the ESC. *Eur. Heart J.* **2016**, *37*, 2129–2200. [CrossRef]
4. Zachariah, D.; Stevens, D.; Sidorowicz, G.; Spooner, C.; Rowell, N.; Taylor, J.; Kay, R.; Salek, M.S.; Kalra, P.R. LIVE:LIFE study investigators. Quality of life improvement in older patients with heart failure initiated on ivabradine: Results from the UK multi-centre LIVE:LIFE prospective cohort study. *Int. J. Cardiol.* **2017**, *249*, 313–318. [CrossRef] [PubMed]
5. Crişan, S.; Petrescu, L.; Lazăr, M.A.; Văcărescu, C.; Nicola, A.R.; Cozma, D.; Mornoş, C.; Luca, C.T. Reduced ejection fraction heart failure—New data from multicenter studies and national registries regarding general and elderly populations: Hopes and disappointments. *Clin. Interv. Aging* **2018**, *13*, 651–656. [CrossRef] [PubMed]
6. Murad, K.; Goff, D.C.; Morgan, T.M.; Burke, G.L.; Bartz, T.M.; Kizer, J.R.; Chaudhry, S.I.; Gottdiener, J.S.; Kitzman, D.W. Burden of comorbidities and functional and cognitive impairments in elderly patients at the initial diagnosis of heart failure and their impact on total mortality: The cardiovascular health study. *JACC Heart Fail.* **2015**, *3*, 542–550. [CrossRef] [PubMed]

7. Díez-Villanueva, P.; Alfonso, F. Heart failure in the elderly. *J. Geriatr. Cardiol.* **2016**, *13*, 115–117. [CrossRef] [PubMed]
8. Ragnar, D.; Gudmundur, T.; Haukur, E.; Örn, Ó.; Thor, A.; Tamara, B.H.; Lenore, L.; Vilmundur, G. Prevalence of heart failure in the elderly and future projections: The AGES-Reykjavík study. *Scand. Cardiovasc. J.* **2017**, *51*, 183–189. [CrossRef]
9. WHOQOL Group. The World Health Organization quality of life assessment (WHOQOL): Position paper from the world health organization. *Soc. Sci. Med.* **1995**, *41*, 1403–1409. [CrossRef]
10. Cell, D.F. Quality of life: Concepts and definition. *J. Pain Symptom Manag.* **1994**, *9*, 186–192. [CrossRef]
11. Wu, J.R.; Lennie, T.A.; Frazier, S.K.; Moser, D.K. Health-related quality of life, functional status, and cardiac event-free survival in patients with heart failure. *J. Cardiovasc. Nurs.* **2016**, *31*, 236–244. [CrossRef]
12. Anderson, K.L.; Burckhardt, C.S. Conceptualization and measurement of quality of life as an outcome variable for health care intervention and research. *J. Adv. Nurs.* **1999**, *29*, 298–306. [CrossRef] [PubMed]
13. Kato, N.; Kinugawa, K.; Seki, S.; Shiga, T.; Hatano, M.; Yao, A.; Hirata, Y.; Kazuma, K.; Nagai, R. Quality of life as an independent predictor for cardiac events and death in patients with heart failure. *Circ. J.* **2011**, *75*, 1661–1669. [CrossRef]
14. Schubert, C.R.; Cruickshanks, K.J.; Dalton, D.S.; Klein, B.E.K.; Klein, R.; Nondahl, D.M. Prevalence of sleep problems and quality of life in an older population. *Sleep* **2002**, *25*, 889–893. [PubMed]
15. Silverstein, M.; Parker, M.G. Leisure Activities and Quality of Life among the Oldest Old in Sweden. *Res. Aging* **2002**, *24*, 528–547. [CrossRef]
16. Power, M.; Quinn, K.; Schmidt, S. Development of the WHOQOL-Old module. *Qual. Life Res.* **2005**, *14*, 2197–2214. [CrossRef] [PubMed]
17. Haywood, K.L.; Garratt, A.M.; Fitzpatrick, R. Quality of life in older people: A structured review of self-assessed health instruments. *Expert Rev. Pharm. Outcomes Res.* **2006**, *6*, 181–194. [CrossRef]
18. Balboa-Castillo, T.; León-Muñoz, L.M.; Graciani, A.; Rodríguez-Artalejo, F.; Guallar-Castillón, P. Longitudinal association of physical activity and sedentary behavior during leisure time with health-related quality of life in community-dwelling older adults. *Heal. Qual. Life Outcomes* **2011**, *9*, 47. [CrossRef]
19. Brajša-Žganec, A.; Merkaš, M.; Šverko, I. Quality of Life and Leisure Activities: How do Leisure Activities Contribute to Subjective Well-Being? *Soc. Indic. Res.* **2011**, *102*, 81–91. [CrossRef]
20. Denfeld, Q.E.; Winters-Stone, K.; Mudd, J.O.; Gelow, J.M.; Kurdi, S.; Lee, C.S. The prevalence of frailty in heart failure: A systematic review and meta-analysis. *Int. J. Cardiol.* **2017**, *236*, 283–289. [CrossRef]
21. Lam, C.L.K.; Tse, E.Y.Y.; Gandek, B.; Fong, D.Y.T. The SF-36 summary scales were valid, reliable, and equivalent in a Chinese population. *J. Clin. Epidemiol.* **2005**, *58*, 815–822. [CrossRef]
22. Wang, R.; Wu, C.; Zhao, Y.; Yan, X.; Ma, X.; Wu, M.; Liu, W.; Gu, Z.; Zhao, J.; He, J. Health related quality of life measured by SF-36: A population-based study in Shanghai, China. *BMC Public Health* **2008**, *8*, 292. [CrossRef] [PubMed]
23. Hopman, W.M.; Towheed, T.; Anastassiades, T.; Tenenhouse, A.; Poliquin, S.; Berger, C.; Joseph, L.; Brown, J.P.; Murray, T.M.; Adachi, J.D.; et al. Canadian normative data for the SF-36 health survey. *CMAJ* **2000**, *163*, 265–271. [PubMed]
24. Hinkle, D.E.; Wiersma, W.; Jurs, S.G. *Applied Statistics for the Behavioral Sciences*, 5th ed.; Houghton Mifflin: Boston, MA, USA, 2003.
25. Zhou, B.; Chen, K.; Wang, J.; Wang, H.; Zhang, S.; Zheng, W. Quality of Life and Related Factors in the Older Rural and Urban Chinese Populations in Zhejiang Province. *J. Appl. Gerontol.* **2010**, *30*, 199–225. [CrossRef]
26. Peel, N.M.; McClure, R.J.; Bartlett, H.P. Behavioral determinants of healthy aging. *Am. J. Prev. Med.* **2005**, *28*, 298–304. [CrossRef] [PubMed]
27. Sun, W.; Aodeng, S.; Tanimoto, Y.; Watanabe, M.; Han, J.; Wang, B.; Yu, L.; Kono, K. Quality of life (QOL) of the community-dwelling elderly and associated factors: A population-based study in urban areas of China. *Arch. Gerontol. Geriatr.* **2015**, *60*, 311–316. [CrossRef]
28. Paskulin, L.; Vianna, L.; Molzahn, A.E. Factors associated with quality of life of Brazilian older adults. *Int. Nurs. Rev.* **2009**, *56*, 109–115. [CrossRef]
29. Yahaya, N.; Abdullah, S.S.; Momtaz, Y.A.; Hamid, T.A. Quality of Life of Older Malaysians Living Alone. *Educ. Gerontol.* **2010**, *36*, 893–906. [CrossRef]
30. Nilsson, J.; Rana, A.K.; Kabir, Z.N. Social capital and quality of life in old age: Results from a cross-sectional study in rural Bangladesh. *J. Aging Health* **2006**, *18*, 419–434. [CrossRef]

31. Rector, T.S. A conceptual model of quality of life in relation to heart failure. *J. Card Fail.* **2005**, *11*, 173–176. [CrossRef]
32. Zuluaga, M.C.; Guallar-Castillón, P.; López-García, E.; Banegas, J.R.; Conde-Herrera, M.; Olcoz-Chiva, M.; Rodríguez-Pascual, C.; Rodriguez-Artalejo, F. Generic and disease-specific quality of life as a predictor of long-term mortality in heart failure. *Eur. J. Heart Fail.* **2010**, *12*, 1372–1378. [CrossRef]
33. Garin, O.; Ferrer, M.; Pont, A. Disease-specific health-related quality of life questionnaires for heart failure: A systematic review with meta-analyses. *Qual. Life Res.* **2009**, *18*, 71–85. [CrossRef] [PubMed]
34. Heo, S.; Moser, D.K.; Lennie, T.A.; Zambroski, C.H.; Chung, M.L. A comparison of health-related quality of life between older adults with heart failure and healthy older adults. *Heart Lung.* **2007**, *36*, 16–24. [CrossRef] [PubMed]
35. Merryn, G.; Sarah, B.; Chris, P.; Sheila, P.; David, S.; Salah, G. SmallPredictors of the quality of life of older people with heart failure recruited from primary care. *Age Ageing* **2006**, *35*, 172–177. [CrossRef]
36. Erceg, P.; Despotovic, N.; Milosevic, D.P.; Soldatovic, I.; Zdravkovic, S.; Tomic, S.; Markovic, I.; Mihajlovic, G.; Brajovic, M.D.; Bojovic, O.; et al. Health-related quality of life in elderly patients hospitalized with chronic heart failure. *Clin. Interv. Aging* **2013**, *8*, 1539–1546. [CrossRef] [PubMed]
37. Iqbal, J.; Francis, L.; Reid, J.; Murray, S.; Denvir, M. Quality of life in patients with chronic heart failure and their carers: A 3-year follow-up study assessing hospitalization and mortality. *Eur. J. Heart Fail.* **2010**, *12*, 1002–1008. [CrossRef] [PubMed]
38. Rodriguez-Artalejo, F.; Guallar-Castillon, P.; Pascual, C.R.; Otero, C.M.; Montes, A.O.; Garcia, A.N.; Conthe, P.; Chiva, M.O.; Banegas, J.R.; Herrera, M.C. Health-related quality of life as a predictor of hospital readmission and death among patients with heart failure. *Arch. Intern. Med.* **2005**, *165*, 1274–1279. [CrossRef] [PubMed]
39. Lazzarini, V.; Mentz, R.J.; Fiuzat, M.; Metra, M.; O'Connor, C.M. Heart failure in elderly patients: Distinctive features and unresolved issues. *Eur. J. Heart Fail.* **2013**, *15*, 717–723. [CrossRef]
40. Garin, O.; Herdman, M.; Vilagut, G.; Ferrer, M.; Ribera, A.; Rajmil, L.; Valderas, J.M.; Guillemin, F.; Revicki, D.; Alonso, J. Assessing health-related quality of life in patients with heart failure: A systematic, standardized comparison of available measures. *Heart Fail. Rev.* **2014**, *19*, 359–367. [CrossRef]
41. Lupón, J.; Gastelurrutia, P.; de Antonio, M.; González, B.; Cano, L.; Cabanes, R.; Urrutia, A.; Díez, C.; Coll, R.; Altimir, S.; et al. Quality of life monitoring in ambulatory heart failure patients: Temporal changes and prognostic value. *Eur. J. Heart Fail.* **2013**, *15*, 103–109. [CrossRef]
42. Schipper, H.; Clinch, J.J.; Olweny, C.L.M. Quality of life studies: Definitions and conceptual issues. In *Quality of Life and Pharmacoeconomics in Clinical Trial*, 2nd ed.; Spilker, B., Ed.; Lippincott-Raven: New York, NY, USA, 1996; pp. 11–23.
43. Carvalho, M.V.; Siqueira, L.B.; Sousa, A.L.; Jardim, P.C. The influence of hypertension on quality of life. *Arq Bras Cardiol* **2013**, *100*, 164–174. [CrossRef]
44. Liu, R.; Wu, S.; Hao, Y.; Gu, J.; Fang, J.; Cai, N.; Zhang, J. The Chinese version of the world health organization quality of life instrument-older adults module (WHOQOL-OLD): Psychometric evaluation. *Health Qual. Life Outcomes* **2013**, *11*, 156. [CrossRef] [PubMed]
45. Magilvy, J.K.; Congdon, J.G. The Crisis Nature of Health Care Transitions for Rural Older Adults. *Public Health Nurs.* **2000**, *17*, 336–345. [CrossRef] [PubMed]
46. Deaton, A. Income, Health, and Well-Being around the World: Evidence from the Gallup World Poll. *J. Econ. Perspect.* **2008**, *22*, 53–72. [CrossRef] [PubMed]
47. Gu, D.; Dupre, M.E.; Liu, G. Characteristics of the institutionalized and community-residing oldest-old in China. *Soc. Sci. Med.* **2007**, *64*, 871–883. [CrossRef]

© 2020 by the authors. Licensee MDPI, Basel, Switzerland. This article is an open access article distributed under the terms and conditions of the Creative Commons Attribution (CC BY) license (http://creativecommons.org/licenses/by/4.0/).

International Journal of
Environmental Research and Public Health

MDPI

Article

Exposure to Environmental Tobacco Smoke (ETS) among Employees of Hospitality Venues in the Light of Changes in Anti-Tobacco Legislation in Poland

Emilia Krakowiak [1], Katarzyna Sygit [2,*], Marian Sygit [2], Elżbieta Cipora [3] and Jan Krakowiak [1]

[1] Social Medicine Unit, Department of Social and Preventive Medicine, Medical University of Łódź, 90-647 Łódź, Poland; emilia.krakowiak@wp.pl (E.K.); jan.krakowiak@umed.lodz.pl (J.K.)

[2] Faculty of Health Sciences, President Stanislaw Wojciechowski State Vocational Academy in Kalisz, 62-800 Kalisz, Poland; msygit@onet.pl

[3] Sanok Medical Institute, State Vocational Academy in Sanok, 38-500 Sanok, Poland; elacipora@interia.pl

* Correspondence: k.sygit@pwsz-kalisz.edu.pl

Received: 19 April 2020; Accepted: 20 May 2020; Published: 23 May 2020

Abstract: *Introduction:* Numerous studies conducted in Europe and worldwide have indicated that employees of hospitality venues are the most exposed professional group to environmental tobacco smoke (ETS) in the workplace. The purpose of this study was to assess the exposure of employees of hospitality venues to ETS in the light of changes in anti-tobacco legislation in Poland. *Materials and methods:* The study consisted of two stages. The first stage was conducted in 2010, while the second in 2015. The study was conducted among employees of 300 randomly selected hospitality venues in the city of Łódź (Poland). In total, 2607 questionnaires were analysed. The study used two survey questionnaires created and recommended by the Institute for Global Tobacco Control to study exposure to ETS. Statistical analysis was made with Statistica 13.1 PL (StatSoft, Poland). *Results:* In the group of all nonsmoking employees, individuals exposed to ETS at work in 2010 accounted for 72.6%; while in 2015 it was 51.8%. Factors affecting exposure to ETS in the workplace included, among others: age, marital status, education, position held, presence of a smoking room on the premises, and noncompliance with the provisions of the anti-tobacco laws. *Conclusions:* The prevalence of tobacco smoking among employees of hospitality venues decreased in 2010–2015, however, it remained high. More than half of nonsmoking employees were exposed to ETS at work.

Keywords: smoking; tobacco smoke; exposure to tobacco smoke; ETS; anti-tobacco law; hospitality venues

1. Introduction

Environmental tobacco smoke (ETS) is the sum of second-hand smoke (SHS) and third-hand smoke (THS). SHS is the combination of the side-stream cigarette smoke in the intervals between puffs as a result of cigarette smouldering and the smoke exhaled by the smoker, while THS is the smoke persisting in the environments long after the active smoking is ceased. Thus, second-hand smoke exposure consists of an unintentional inhalation of smoke that occurs close to people smoking and/or in indoor environments where tobacco was recently used, while THS exposure occurs in enclosed environments where tobacco was used hours or days before [1,2].

Exposure to ETS has serious health consequences. The increased risk of developing malignant neoplasms, noncancer respiratory diseases, cardiovascular diseases, or pregnancy failure needs to be highlighted. Due to numerous health threats, promotional and educational activities, as well as new legal solutions, to protect nonsmokers against passive exposure to ETS have been introduced for many years [3–6].

In 1998, the state of California was the first in the United States to introduce comprehensive legal solutions prohibiting smoking in all workplaces, including bars and restaurants, and in 2004, Ireland was the first European country to introduce a total ban on smoking in public places [7–12].

Legal measures introduced in other European countries differ in terms of the degree of limitation of exposure to ETS—ranging from a total smoking ban in all workplaces (including hospitality venues) to partial protection by means of creating designating areas in bars or restaurants for smokers (separate rooms where smoking is allowed) [13–17].

Numerous studies conducted in USA, Europe, and elsewhere indicate that it is the employees of bars, cafes, restaurants, nightclubs, and music venues that are the most exposed professional group to ETS in the workplace; therefore, this group desperately needs legal protection of their health interests [18–23].

In recent years, a very large impact on shaping tobacco policy in Poland has been made by the European Parliament and the European Council. The Polish Anti-Tobacco Act (Act of July 22, 2016, amending the Act on Protection of Health against the Consequences of Use of Tobacco and Tobacco Products, Journal of Laws 2016, item 1331) in its current form contains many provisions aimed at protecting life and health against the consequences of exposure to ETS [17,24,25].

In Poland, the Anti-Tobacco Act (Act of July 22, 2016, amending the Act on Protection of Health against the Consequences of Use of Tobacco and Tobacco Products, Journal of Laws 2016, item 1331), despite its restrictions on exposure to ETS, does not fully protect from such exposure [14,17].

Unfortunately, the Polish legislator, despite many amendments made to the Act on Health Protection against the Consequences of Use of Tobacco and Tobacco Products, did not decide to introduce an absolute ban on smoking in hospitality venues. It means that these establishments are not completely smoke-free and safe for employees and customers alike [17,21,26].

Another aspect of tobacco law is their proper implementation, compliance, and enforcement. Unfortunately, in many hospitality venues, smoking ban is frequently violated, often with the indifferent attitude of employers, venue owners, staff, and even people exposed to ETS [10,27].

Importantly, according to WHO's recommendations, the most effective protective measures against exposure to ETS are only those that lead to complete elimination of smoking and tobacco smoke, while all other technical solutions, such as ventilation, air filtration, separate smoking rooms, do not provide complete protection for nonsmokers [3,15,19]. Researches conducted in countries such as Ireland, Norway, Scotland, and France, whose authorities have decided to introduce a total ban on smoking in bars, restaurants, and clubs, confirm the decrease in employees' exposure to ETS, improvement of their health, and thus, the effectiveness of such strict tobacco laws. In these countries, legal changes were accompanied by anti-tobacco social campaigns to make people aware of the seriousness of the problem of exposure to ETS and of health benefits brought by a total ban on smoking in public places [1,2,28,29].

The new legislative solutions are therefore aimed at protecting individuals from exposure to ETS in public places and protect employees of hospitality venues against such exposure. It is important, however, that only a full ban on smoking in workplaces (hospitality venues) can effectively protect employees from exposure to ETS [3,5,7].

The purpose of this study was to assess the exposure of employees of hospitality venues to ETS in the light of changes in anti-tobacco legislation in Poland.

2. Materials and Methods

The study consisted of two stages. The first stage was conducted in 2010 among 1360 employees of 300 randomly selected hospitality venues in the city of Łódź (Poland). The second stage was conducted in 2015 in the same establishments. It should be emphasized that in 2015, out of 300 premises included in the first study, 57 had been closed. Thus, the second stage of the study was conducted among 1247 employees from 243 hospitality venues in Łódź.

To collect empirical data, the study used two survey questionnaires created and recommended by the Institute for Global Tobacco Control to study exposure to ETS.

The survey questionnaire used in the first stage of the study contained 41 questions. It covered basic demographic and work-related questions, as well as questions about smoking and exposure to ETS at work, at home, and in other public places. The respondents were also asked about their smoking behaviours, exposure to ETS, and their health. The 2015 survey questionnaire contained 86 questions and was extended for the purposes of the second stage with additional questions to enable inter alia assessment of the functioning of the amended anti-tobacco law, opinion on the harmfulness of the use of regular and smokeless tobacco, and opinion on the introduction of the smoking ban in various situations and various places. Both survey questionnaires were anonymous and voluntary, intended for self-completion by respondents.

Our epidemiological study, which assessed the exposure to ETS among employees of hospitality venues in the light of changes in anti-tobacco legislation in Poland, is one of the first such large population studies to be carried out in Poland using two questionnaires created and recommended by the Institute for Global Tobacco Control for ETS exposure testing.

The use of the questionnaire-survey recommended for assessing exposure to ETS and treated as diagnostic tests in epidemiological studies was verified in pilot studies which confirmed the credibility, reliability, and validity of the results obtained. They were assessed by experts as appropriate for use in this epidemiological study. Assessment of ETS exposure is also performed in clinical-laboratory studies by testing u-cotinine in urine.

However, in epidemiological and population studies (such as our study), a good research tool is the diagnostic test used by us and recommended by the Institute for Global Tobacco Control, which helps achieve the intended research goal.

In total, 2607 questionnaires were analysed. The project of the study received a positive opinion from the Bioethics Committee at the Medical University of Łódź (no. RNN/117/15/KE of April 21, 2015)

Statistical analysis was made with Statistica 13.1 PL (StatSoft, Poland).

The structure of the study groups analysed according to variables was described with structural indicators (%). To test the relationship between the analysed variables, a χ^2 test was used.

Due to the size of groups in individual analyses, the test was modified with *Yates's correction for continuity*.

Single-factor and multifactor regression models were used in the statistical analysis. Single-factor logistic regression allowed to assess the relationship between exposure to ETS (dependent variable) and selected demographic and social characteristics, occupation, and working conditions declared by the respondents (independent variables). Independent variables were also included in the statistical analysis using a multifactor regression model to assess their simultaneous impact on the dependent variable.

Additionally, the odds ratios (OR) were calculated with their corresponding 95% confidence intervals (95% CI). The odds ratio is the ratio of a chance of occurrence of a given phenomenon in one group to the chance of its occurrence in another group. OR ≈ 1 means that the odds are similar, OR < 1 means that the chance of occurrence in the study group is smaller than in the reference (control) group, while OR > 1 means that the chance is greater.

3. Results

3.1. The Overview of the Subjects

In the 2010 study group of 1360 individuals, 17.2% of respondents (234 people) were aged 20 and less, 63.7% (867 people) were aged 21–30, while 13.1% of respondents (178 persons) were aged 31–40. Among 1247 respondents of the 2015 survey, the percentage of people aged 20 and less was 10.9% (136 people), 67.2% (838 people) were aged 21–30, and 14% of respondents (175 people) were aged 31–40. There was no statistically significant difference in the structure of respondents by age in 2010 and 2015 ($p > 0.05$).

In the group of 1360 subjects in 2010, 34.9% of subjects (475 people) were men, while 65.1% (885 people) were women.

Among 1247 people participating in the 2015 survey, the proportion of men was 35% (437 people), while 65% of respondents (810 people) were women. There was no statistically significant difference in the structure of respondents by gender in 2010 and 2015 ($p > 0.05$).

In the 2010 study group of 1360 individuals, 1.2% of respondents (17 people) completed primary education, 10.2% (139 people) completed vocational education, while 59.6% (810 people) completed secondary education. Higher education, bachelor's or master's degree, was declared by 29% of respondents (394 people).

Among 1247 respondents of the 2015 survey, the percentage of people with primary education was 1.2% (15 people), 10% (125 people) reported vocational education, while 58.8% of respondents (733 people) declared secondary education. In particular, 30% of respondents (374 people) declared higher education, Bachelor's or Master's degree. There was no statistically significant difference in the structure of respondents by education in 2010 and 2015 ($p > 0.05$).

In the 2010 study group of 1360 respondents, 73.5% of them (999 people) declared being single, 24.6% (334 people) were married, 1.2% of respondents (16 people) were divorced, while 0.8% of respondents (11 people) declared to be a widow or a widower.

Among 1247 participants of the 2015 survey, 45.4% (566 people) declared to be single, 39.9% (498 people) were married, 13.3% of respondents (166 people) were divorced, while 1.4% of respondents (17 people) reported being a widow or a widower. The observed differences were statistically significant ($p < 0.001$; $chi^2 = 272.655$) (Table 1).

Table 1. The overview of the study group.

The Overview of the Study Group		2010		2015		Statistical Differences
		N	%	N	%	
Respondents' age	20 years and less	234	17.2	136	10.9	$p > 0.05$
	21–30	867	63.7	838	67.2	
	31–40	178	13.1	175	14	
	41+	81	6.0	98	7.9	
Gender	F	885	65.1	810	65	$p > 0.05$
	M	475	34.9	437	35	
Education	Primary	17	1.2	15	1.2	$p > 0.05$
	Vocational	139	10.2	125	10	
	Secondary	810	59.6	733	58.8	
	Higher	394	29	374	30	
Marital status	Single	999	73.5	566	45.4	$p > 0.001$; $chi^2 = 272.655$
	Married	334	24.6	498	39.9	
	Divorced	16	1.2	166	13.3	
	Widow (er)	11	0.8	17	1.4	

3.2. *Analysis of Respondents' Answers on Tobacco Smoking*

In 2010, in the study group of 1360 respondents, 24.9% of them (339 people) reported smoking on daily basis, while 10.4% (141 people) declared occasional smoking. Nonsmokers accounted for 64.7% (880 people). In 2015, in the study group of 1247 respondents, 23% of them (287 people) reported smoking on daily basis, while 6.7% (83 people) declared occasional smoking. Nonsmokers accounted for 70.3% (877 people). The observed differences were statistically significant ($p < 0.001$; $chi^2 = 14.472$).

In 2010, in a group of 480 smokers, 25.4% of respondents (122 people) said that they smoked 10 or fewer cigarettes a day, 51.9% of respondents (249 people) stated that they smoked from 11 to 20 cigarettes a day, 1% of respondents (5 people) smoked between 21 and 30 cigarettes per day, 0.2% of respondents (1 person) smoked 31 or more cigarettes a day, while 21.5% of respondents (103 people) chose 'I don't know' as their response.

In 2015, in a group of 370 smokers, 46% of respondents (170 people) said that they smoked 10 or fewer cigarettes a day, 31% of respondents (115 people) stated that they smoked from 11 to 20 cigarettes a day, 22.2% of respondents (82 people) smoked between 21 and 30 cigarettes per day, while 0.8% of respondents (3 people) smoked 31 or more cigarettes a day. The observed differences were statistically significant ($p < 0.001$; chi^2 = 212.883).

In 2010, in a group of 1360 respondents, 31.1% (423 people) declared working in a smoking room, 37.6% (511 people) in a nonsmoking room, and 31.3% (426 people) respondents reported working in both rooms. In 2015, in the group of 1247 respondents, 4.7% (59 people) declared working in a smoking room, 73.5% (917 people) in a nonsmoking room, and 21.7% (271 people) respondents reported working in both rooms. The observed differences were statistically significant ($p < 0.001$; chi^2 = 420.681).

In 2010, in the study group of 1360 people, 96.5% (1312 people) declared that the smoking ban in the nonsmoking room was observed, while 3.5% (48 people) said that the smoking ban was not observed. In 2015, in a group of 1247 respondents, 88.5% (1104 people) declared that the smoking ban in the nonsmoking room was observed, while 11.5% (143 people) said it was not observed. The observed differences were statistically significant ($p < 0.001$; chi^2 = 60.374).

In the group of 1360 respondents in 2010, 89.9% of respondents (1222 people) declared that there was a room for smokers in their establishment. Among 1247 participants of the 2015 survey, affirmative answer to this question was given by 56% (698 people). The observed differences were statistically significant ($p < 0.001$; chi^2 = 384.715).

In the group of 880 nonsmokers surveyed in 2010, 49% (431 people) reported that they stayed in a room for smokers in their workplace every day, 8.5% (75 people) declared such stay once a week, 15.1% (133 persons) once a month, while 27.4% of respondents (241 persons) replied that they never stayed in such rooms.

In the group of 877 nonsmokers surveyed in 2015, 25% (219 people) reported that they stayed in a room for smokers in their workplace every day, 10.3% (90 people) declared such stay once a week, 16.5% (145 people) once a month, while 48.2% of respondents (423 people) replied that they never stayed in such rooms. The observed differences were statistically significant ($p < 0.001$; chi^2 = 120.907) (Table 2).

Table 2. Analysis of smoking by hospitality venues staff and their exposure to environmental tobacco smoke (ETS).

Analysis of Smoking and Exposure to ETS		**2010**		**2015**		**Statistical Differences**
		N	%	N	%	
Smoking among the staff	Every day	339	24.9	287	23	$p < 0.01$; chi^2 = 14.472
	Sometimes	141	10.4	83	6.7	
	I do not smoke	880	64.7	877	70	
The number of cigarettes smoked per day by an employee	Up to 10	122	25.4	170	46	$p < 0.01$; chi^2 = 212.883
	11–20	249	51.9	115	31	
	21–30	5	1	82	22.2	
	31+	1	0.2	3	0.8	
	I don't know	103	21.5	–	–	
Workplace—type of room	Smoking room	423	31.1	59	4.7	$p < 0.01$; chi^2 = 420.681
	Nonsmoking room	511	37.6	917	73.5	
	Both rooms	426	31.3	271	21.7	

Table 2. *Cont.*

Analysis of Smoking and Exposure to ETS		2010		2015		Statistical Differences
		N	%	N	%	
Compliance with the smoking ban in nonsmoking rooms	The regulations are followed	1312	96.5	1194	88.5	$p < 0.01$; chi^2 = 60.374
	The regulations are not followed	35	48	143	11.5	
Presence of a smoking room in the workplace	Yes	1222	89.9	698	56	$p < 0.01$; chi^2 = 384.715
	No	138	10.1	549	44	
Staying in smoking rooms by nonsmoking employees	Every day	431	49	219	25	$p < 0.001$; chi^2 = 120.907
	Once a week	8.5	75	90	10.3	
	Once a month	133	15.1	145	16.5	
	Never	241	27.4	423	48.2	

3.3. *Impact of Selected Factors which Expose Hospitality Venues Staff to ETS in the Workplace in 2010 and 2015—Single-Factor and Multifactor Analysis*

In the group of all nonsmoking employees, individuals exposed to ETS at work in 2010 accounted for 72.6% (639 people). The one-factor logistic regression analysis shows that factors affecting exposure to ETS at work in 2010 included: age, marital status, education, position held, presence of the smoking room, noncompliance with the anti-tobacco laws, and the type of room in which a respondent worked.

The largest increase in the odds ratio of exposure to ETS at work was observed among bartenders/waiters (OR 72.29; 95% CI 17.30–30.20; $p < 0.001$) and cooks (OR 55.45; 95% CI 13.06–235.46; $p < 0.001$).

The highest exposure to ETS at work was reported in the group of people under 20 years old (OR 7.87; 95% CI 3.48–17.82; $p < 0.001$) and in the group of people with primary and vocational education (OR 17.76; 95% CI 5.42–58.17; $p < 0.001$).

A significant increase in the odds ratio of exposure to ETS was observed among respondents working in the smoking rooms (OR 6.80; 95% CI 4.38–10.54; $p < 0.001$).

The analysis also showed a significant increase in the odds ratio of exposure to ETS at work in a group of people whose workplace did not respect provisions of the anti-tobacco law (OR 5.89; 95% CI 1.39–24.88; $p < 0.001$) (Table 3).

In the group of all nonsmoking employees, individuals exposed to ETS at work in 2015 accounted for 51.8% (454 people). The one-factor logistic regression analysis showed that factors affecting exposure to ETS in the workplace in 2015 included: whether anti-tobacco laws were followed, whether there was a smoking ban outside the room intended for customers and whether these provisions were observed, type of venue, working hours, the rules regarding smoking in the workplace, whether there was a smoking room in the establishment, whether the venue had its own regulations prohibiting the use of tobacco products by employees indoors, whether these regulations had been communicated to employees, whether the venue had regulations prohibiting the use of tobacco products by clients indoors, awareness that exposure to ETS might cause heart disease in nonsmokers, and awareness that exposure to ETS might cause cancer in nonsmokers.

The largest increase in the odds ratio of exposure to ETS at work was observed among employees of venues where there was a room for smokers and nonsmokers (OR 42.74; 95% CI 19.24–94.97; $p < 0.001$) and among employees of establishments where smoking was allowed in all indoor rooms (OR 9.68; 95% CI 4.54–20.65; $p < 0.001$) and at a designated place (OR 3.48; 95% CI 2.33–5.19; $p < 0.001$).

A significant increase in the odds ratio of exposure to ETS was also observed among people working in establishments where smoking was not prohibited outside the smoking room intended for customers (OR 18.64; 95% CI 8.06–43,08; $p < 0.001$) and in venues whose regulations on smoking ban in rooms other than the room intended for customers were not observed (OR 6.27; 95% CI 3.72–10.58; $p < 0.001$).

Table 3. Impact of selected factors which expose hospitality venues staff to ETS in the workplace in 2010—single-factor analysis.

Variable	Exposed $N = 639$		Not Exposed $N = 241$		One-Factor Logistic Regression 2010		
Age	N	%	N	%	**OR**	**95% CI**	p
20	174	89.2	21	10.8	7.87	3.48–17.82	**$p < 0.001$**
21–25	257	76.3	80	23.7	3.05	1.48–6.30	**$p < 0.01$**
26–30	128	59.8	86	40.2	1.41	0.68–2.94	$p > 0.05$
31–35	42	61.8	26	38.2	1.53	0.66–3.55	$p > 0.05$
36–40	18	66.7	9	33.3	1.90	0.66–5.44	$p > 0.05$
41+	20	51.3	19	48.7	1.00	Ref.	
Gender	N	%	N	%	**OR**	**95% CI**	p
Men	152	70.7	63	29.3	0.88	0.63–1.24	$p > 0.05$
Women	487	73.2	178	26.8	1.00	Ref.	
Marital status	N	%	N	%	**OR**	**95% CI**	p
Single	539	80.7	129	19.3	4.88	3.47–6.88	**$p < 0.001$**
Married	89	46.1	104	53.9	1.00	Ref.	
Divorced	4	36.4	7	63.6	0.67	0.19–2.54	$p > 0.05$
Widow (er)	7	87.5	1	12.5	8.18	0.98–67.97	$p > 0.05$
Education	N	%	N	%	**OR**	**95% CI**	p
Primary and vocational	65	95.6	3	0.4	17.76	5.42–58.17	**$p < 0.001$**
Secondary	441	77.4	129	22.6	2.80	2.03–3.86	**$p < 0.001$**
Higher	133	55.0	109	45.0	1.00	Ref.	
Position	N	%	N	%	**OR**	**95% CI**	p
Owner/manager	2	0.2	123	98.0	1.00	Ref.	
Bartender/waiter	576	92.2	49	7.8	72.29	17.30–30.20	**$p < 0.001$**
Cook	55	47.4	61	52.6	55.45	13.06–235.46	**$p < 0.001$**
Other	6	42.9	8	57.1	46.13	7.97–266.85	**$p < 0.001$**
Is there a smoking room in the establishment	N	%	N	%	**OR**	**95% CI**	p
Yes	591	73.6	212	26.4	1.68	1.03–2.74	**$p < 0.05$**
No	48	62.3	29	37.7	1.00	Ref.	
Is the anti-tobacco law followed	N	%	N	%	**OR**	**95% CI**	p
Yes	609	71.8	239	28.2	1.00	Ref.	
No	30	93.8	2	6.2	5.89	1.39–24.88	**$p < 0.05$**
Room where the respondent works	N	%	N	%	**OR**	**95% CI**	p
For smokers	210	87.5	30	12.5	6.80	4.38–10.54	**$p < 0.001$**
For nonsmokers	173	50.7	168	49.3	1.00	Ref.	
Both rooms	256	85.6	43	14.4	5.78	3.92–8.52	**$p < 0.001$**
Is there a ban on smoking outside the room?	N	%	N	%	**OR**	**95% CI**	p
Yes	375	72.5	142	27.5	1.00	Ref.	
No	264	72.7	99	27.3	1.01	0.73–1.39	$p > 0.05$
Is the ban on smoking outside the smoking room observed?	N	%	N	%	**OR**	**95% CI**	p
Yes	162	70.7	67	29.3	1.00	Ref.	
No	477	73.3	174	26.7	1.13	0.81–1.58	$p > 0.05$

The results of the analysis showed a significant increase in the odds ratio of exposure to ETS among people working in night clubs and music clubs (OR 3.63; 95% CI 1.96–6.75; $p < 0.001$) and working after 6:00 p.m. (OR 2.41; 95% CI 1.66–3.49; $p < 0.001$).

The analysis also showed a significant increase in the odds ratio of exposure to ETS at work in establishments where there was a smoking room (OR 3.84; 95% CI 2.90–5.08; $p < 0.001$).

A significant increase in the odds ratio of exposure to ETS was also observed among people working in establishments where the provisions of the anti-tobacco law were not followed (OR 6.90; 95% CI 3.49–13.63; $p < 0.001$) and in establishments which did not have regulations prohibiting the use of tobacco products by employees (OR 3.27; 95% CI 1.57–6.79; $p < 0.001$) and customers (OR 5.29; 95% CI 2.31–12.11; $p < 0.001$) indoors.

Higher exposure to ETS occurred among workers who disagreed with the statement that exposure to ETS caused heart disease (OR 3.63; 95% CI 2.11–6.24; $p < 0.001$) and cancer in nonsmokers (OR 3.78; 95% CI 2.33–6.13; $p < 0.001$) (Table 4).

Table 4. Impact of selected factors which expose hospitality venues staff to ETS in the workplace in 2015—single-factor analysis.

Variable	Exposed N = 454		Not Exposed N = 423		One-Factor Logistic Regression 2015		
Age	*N*	%	*N*	%	**OR**	**95% CI**	*p*
<20	72	52.6	65	47.4	1.10	0.55–2.25	$p > 0.05$
21–25	150	51.2	143	48.8	1.05	0.54–2.04	$p > 0.05$
26–30	154	52.0	142	48.0	1.08	0.56–2.11	$p > 0.05$
31–35	38	60.3	25	39.7	1.52	0.68–3.39	$p > 0.05$
36–40	19	42.3	27	58.7	0.70	0.30–1.66	$p > 0.05$
41+	21	50.0	21	50.0	1.00	Ref.	
Gender	*N*	%	*N*	%	**OR**	**95% CI**	*p*
Men	150	53.2	132	46.8	1.09	0.82–1.45	$p > 0.05$
Women	304	51.1	291	48.9	1.00	Ref.	
Marital status	*N*	%	*N*	%	**OR**	**95% CI**	*p*
Single	214	57.5	158	42.5	1.43	1.08–1.91	$p < 0.05$
Married	187	48.6	198	51.4	1.00	Ref.	
Divorced	48	44.9	59	55.1	0.86	0.56–1.33	$p > 0.05$
Widow (er)	5	38.5	8	61.5	0.66	0.21–2.06	$p > 0.05$
Education	*N*	%	*N*	%	**OR**	**95% CI**	*P*
Primary and vocational	50	58.1	36	41.9	0.90	0.54–1.51	$p > 0.05$
Secondary	260	47.0	293	53.0	0.58	0.43–0.79	$p > 0.05$
Higher	144	60.5	94	39.5	1.00	Ref.	
Position	*N*	%	*N*	%	**OR**	**95% CI**	*p*
Owner/manager	44	61.1	28	38.9	1.00	Ref.	
Bartender/waiter	343	50.5	336	49.5	1.54	0.94–2.53	$p > 0.05$
Cook	59	51.3	56	48.7	1.49	0.82–2.72	$p > 0.05$
Other	8	72.7	3	27.3	0.59	0.14–2.42	$p > 0.05$
Is there a ban on smoking outside the room for customers	*N*	%	*N*	%	**OR**	**95% CI**	*p*
Yes	358	46.2	417	53.8	1.00	Ref.	
No	96	94.1	6	5.9	18.64	8.06–43.08	$\mathbf{p < 0.001}$

Table 4. *Cont.*

Variable	Exposed N = 454		Not Exposed N = 423		One-Factor Logistic Regression 2015		
Is the ban on smoking outside the room for customers observed?	*N*	%	*N*	%	**OR**	**95% CI**	*p*
Yes	355	46.7	405	53.3	1.00	Ref.	
No	99	84.6	18	15.4	6.27	3.72–10.58	$p < 0.001$
Type of establishment	*N*	%	*N*	%	**OR**	**95% CI**	*p*
Restaurant/café	268	48.6	284	51.4	1.00	Ref.	
Bar	138	52.5	125	47.5	1.17	0.87–1.57	$p > 0.05$
Night club or music club	48	77.4	14	22.6	3.63	1.96–6.75	$p < 0.001$
Working hours	*N*	%	*N*	%	**OR**	**95% CI**	*p*
Until 6:00 p.m.	279	52.0	258	48.0	1.00	Ref.	
After 6:00 p.m.	175	51.5	165	48.5	2.41	1.66–3.49	$p < 0.001$
Is the respondent concerned about the effects of ETS on health	*N*	%	*N*	%	**OR**	**95% CI**	*p*
Yes	395	54.3	333	45.7	1.00	Ref.	
No or not much	59	39.6	90	60.4	0.56	0.39–0.79	$p < 0.01$
Workplace smoking rules	*N*	%	*N*	%	**OR**	**95% CI**	*p*
Smoking is allowed	46	83.6	9	16.4	9.68	4.54–20.65	$p < 0.001$
Smoking is allowed in designated areas	112	64.7	61	35.3	3.48	2.33–5.19	$p < 0.001$
Smoking is prohibited everywhere	94	34.6	178	65.4	1.00	Ref.	
Smoking is prohibited in rooms for customers	44	20.8	168	79.2	0.50	0.33–0.75	$p < 0.001$
There is a separate smoking room and a nonsmoking room	158	95.8	7	4.2	42.74	19.24–94.97	$p < 0.001$
Is there a designated smoking room in the establishment	*N*	%	*N*	%	**OR**	**95% CI**	*p*
Yes	310	67.1	152	32.9	3.84	2.90–5.08	$p < 0.001$
No	144	34.7	271	65.3	1.00	Ref.	
Is there a policy for employees	*N*	%	*N*	%	**OR**	**95% CI**	*p*
Yes	312	48.7	329	51.3	1.00	Ref.	
No	31	75.6	10	24.4	3.27	1.57–6.79	$p < 0.01$
I don't know	111	56.9	84	43.1	1.39	1.01–1.93	$p < 0.05$
Was the policy communicated to employees	*N*	%	*N*	%	**OR**	**95% CI**	*p*
Yes	225	44.3	283	55.7	1.00	Ref.	
No/I don't know	229	62.1	140	37.9	2.06	1.56–2.71	$p < 0.001$
Is there a policy for customers	*N*	%	*N*	%	**OR**	**95% CI**	*p*
Yes	345	47.9	376	52.1	1.00	Ref.	
No	34	82.9	7	17.1	5.29	2.31–12.11	$p < 0.001$
I don't know	75	65.2	40	34.8	2.04	1.35–3.08	$p < 0.001$
No smoking signs in the establishment	*N*	%	*N*	%	**OR**	**95% CI**	*p*
Yes	410	50.4	403	49.6	1.00	Ref.	
No	13	68.4	6	31.6	2.13	0.80–5.67	$p > 0.05$
I don't know	31	68.9	14	31.1	2.18	1.14–4.16	$p < 0.05$

Table 4. *Cont.*

Variable	Exposed N = 454		Not Exposed N = 423		One-Factor Logistic Regression 2015		
Exposure to ETS causes heart disease in nonsmokers	*N*	%	*N*	%	**OR**	**95% CI**	*p*
I don't agree	63	77.8	18	22.2	3.63	2.11–6.24	$p < 0.001$
I agree	391	49.1	405	50.9	1.00	Ref.	
Exposure to ETS causes cancer in nonsmokers	*N*	%	*N*	%	**OR**	**95% CI**	*p*
I don't agree	81	77.9	23	22.1	3.78	2.33–6.13	$p < 0.001$
I agree	373	48.3	400	51.7	1.00	Ref.	

A multifactor logistic regression analysis conducted in 2010 shows that the odds ratio of exposure to ETS among employees of hospitality venues was increased the most by noncompliance with the provisions of the anti-tobacco laws (OR 24.73; 95% CI 3.54–172.88; $p < 0.01$).

A significant increase in the odds ratio of exposure to ETS was observed among employees with primary, vocational, and secondary education (OR 4.78; 95% CI 3.10–7.37; $p < 0.001$) and among those working in a smoking room (OR 11.10; 95% CI 6.50–18.94; $p < 0.001$) (Table 5).

Table 5. Impact of selected factors which expose hospitality venues staff to ETS in the workplace in 2010—multifactor analysis.

Variable	Exposed N = 639		Not Exposed N = 241		Multifactor Logistic Regression 2010		
Age	*N*	%	*N*	%	**OR**	**95% CI**	*p*
<20	174	89.2	21	10.8	1.08	0.43–2.76	$p > 0.05$
21–25	257	76.3	80	23.7	0.76	0.19–1.03	$p > 0.05$
26–30	128	59.8	86	40.2	0.43	0.19–0.99	0.05
31–35	42	61.8	26	38.2	1.01	0.39–2.59	$p > 0.05$
36–40	18	66.7	9	33.3	1.28	0.38–4.29	$p > 0.05$
41+	20	51.3	19	48.7	1.00	Ref.	
Marital status	*N*	%	*N*	%	**OR**	**95% CI**	*p*
Single	539	80.7	129	19.3	3.86	2.47–6.03	$p < 0.001$
Married	89	46.1	104	53.9	1.00	Ref.	
Divorced	4	36.4	7	63.6	0.03	0.01–0.20	$p > 0.05$
Widow(er)	7	87.5	1	12.5	8.47	0.93–77.59	$p > 0.05$
Education	*N*	%	*N*	%	**OR**	**95% CI**	*p*
Primary + vocational + secondary	506	79.3	132	20.7	4.78	3.10–7.37	$p < 0.001$
Higher	133	55.0	109	45.0	1.00	Ref.	
Is there a smoking room in the establishment	*N*	%	*N*	%	**OR**	**95% CI**	*p*
Yes	591	73.6	212	26.4	0.96	0.51–1.81	$p > 0.05$
No	48	62.3	29	37.7	1.00	Ref.	
Is the anti-tobacco law followed	*N*	%	*N*	%	**OR**	**95% CI**	*p*
Yes	609	71.8	239	28.2	1.00	Ref.	
No	30	93.8	2	6.2	24.73	3.54–172.88	$p < 0.001$
Room where the respondent works	*N*	%	*N*	%	**OR**	**95% CI**	*p*
For smokers	210	87.5	30	12.5	11.10	6.50–18.94	$p < 0.001$
For nonsmokers	173	50.7	168	49.3	1.00	Ref.	
Both rooms	256	85.6	43	14.4	7.05	4.33–11.48	$p < 0.001$

A multifactor logistic regression analysis carried out in 2015 showed that the odds ratio of exposure to ETS among employees of hospitality venues was increased the most when the establishment had both smoking and nonsmoking rooms (OR 10.88; 95% CI 3.94–30.03; $p < 0.001$).

A significant increase in the odds ratio of exposure to ETS was observed among employees of establishments that did not have regulations prohibiting the use of tobacco products by employees indoors and among employees who were unaware of whether such regulations were in force (OR 5.11; 95% CI 1.99–13.15; $p < 0.001$).

A factor that significantly increased the odds ratio of exposure to ETS was the lack of knowledge that exposure to ETS caused cancer in nonsmokers (OR 7.95; 95% CI 3.64–17.34; $p < 0.001$) (Table 6).

Table 6. Impact of selected factors which expose hospitality venues staff to ETS in the workplace in 2015—multifactor analysis.

Variable	Exposed N = 454		Not Exposed N = 423		Multifactor Logistic Regression 2015		
Age	*N*	%	*N*	%	**OR**	**95% CI**	*p*
<20	72	52.6	65	47.4	1.01	0.31–3.25	$p > 0.05$
21–25	150	51.2	143	48.8	1.27	0.44–3.69	$p > 0.05$
26–30	154	52.0	142	48.0	1.67	0.58–4.80	$p > 0.05$
31–35	38	60.3	25	39.7	2.63	0.80–8.71	$p > 0.05$
36–40	19	42.3	27	58.7	2.81	0.81–9.75	$p > 0.05$
41+	21	50.0	21	50.0	1.00	Ref.	
Marital status	*N*	%	*N*	%	**OR**	**95% CI**	*p*
Single	214	57.5	158	42.5	1.41	0.91–2.18	$p > 0.05$
Married	187	48.6	198	51.4	1.00	Ref.	
Divorced	48	44.9	59	55.1	1.20	0.66–2.19	$p > 0.05$
Widow(er)	5	38.5	8	61.5	1.14	0.27–4.81	$p > 0.05$
Is the anti-tobacco law followed	*N*	%	*N*	%	**OR**	**95% CI**	*p*
Yes	389	48.5	413	51.5	1.00	Ref.	
No	65	86.7	10	13.3	0.50	0.13–1.83	$p > 0.05$
Is there a ban on smoking outside the restaurant room	*N*	%	*N*	%	**OR**	**95% CI**	*p*
Yes	358	46.2	417	53.8	1.00	Ref.	
No	96	94.1	6	5.9	3.53	1.55–8.04	$\mathbf{p < 0.01}$
Is the ban on smoking outside the room observed?	*N*	%	*N*	%	**OR**	**95% CI**	*p*
Yes	355	46.7	405	53.3	1.00	Ref.	
No	99	84.6	18	15.4	6.63	2.05–21.54	$\mathbf{p < 0.01}$
Working hours	*N*	%	*N*	%	**OR**	**95% CI**	*p*
Until 6:00 p.m.	279	52.0	258	48.0	1.00	Ref.	
After 6:00 p.m.	175	51.5	165	48.5	0.62	0.34–1.15	$p > 0.05$
Is the respondent concerned about the effects of ETS on health	*N*	%	*N*	%	**OR**	**95% CI**	*p*
Yes	395	54.3	333	45.7	1.00	Ref.	
No or not much	59	39.6	90	60.4	0.30	0.16–0.56	$\mathbf{p < 0.001}$

Table 6. *Cont.*

Variable	Exposed N = 454		Not Exposed N = 423		Multifactor Logistic Regression 2015		
Workplace smoking rules	*N*	%	*N*	%	**OR**	**95% CI**	*p*
Smoking is allowed	46	83.6	9	16.4	0.81	0.20–3.36	$p > 0.05$
Smoking is allowed in designated areas	112	64.7	61	35.3	1.87	0.97–3.63	$p > 0.05$
Smoking is prohibited everywhere	94	34.6	178	65.4	1.00	Ref.	
Smoking is prohibited in rooms for customers	44	20.8	168	79.2	0.58	0.34–0.99	$\mathbf{p < 0.05}$
There is a separate smoking room and a nonsmoking room	158	95.8	7	4.2	10.88	3.94–30.03	$\mathbf{p < 0.001}$
Is there a designated smoking room in the establishment	*N*	%	*N*	%	**OR**	**95% CI**	*p*
Yes	310	67.1	152	32.9	1.38	0.79–2.41	$p > 0.05$
No	144	34.7	271	65.3	1.00	Ref.	
Is there a policy for employees	*N*	%	*N*	%	**OR**	**95% CI**	*p*
Yes	312	48.7	329	51.3	1.00	Ref.	
No/I don't know	142	60.2	94	39.8	5.11	1.99–13.15	$\mathbf{p < 0.001}$
Was the policy communicated to employees	*N*	%	*N*	%	**OR**	**95% CI**	*p*
Yes	225	44.3	283	55.7	1.00	Ref.	
No/I don't know	229	62.1	140	37.9	1.96	1.25–3.06	$\mathbf{p < 0.01}$
Is there a policy for customers	*N*	%	*N*	%	**OR**	**95% CI**	*p*
Yes	345	47.9	376	52.1	1.00	Ref.	
No	109	69.9	47	30.1	1.45	0.49–4.25	$p > 0.05$
No smoking signs in the establishment	*N*	%	*N*	%	**OR**	**95% CI**	*p*
Yes	410	50.4	403	49.6	1.00	Ref.	
No	13	68.4	6	31.6	1.21	0.24–6.02	$p > 0.05$
I don't know	31	68.9	14	31.1	1.84	0.60–5.64	$p > 0.05$
Exposure to ETS causes heart disease in nonsmokers	*N*	%	*N*	%	**OR**	**95% CI**	*p*
I don't agree	63	77.8	18	22.2	1.72	0.72–4.08	$p > 0.05$
I agree	391	49.1	405	50.9	1.00	Ref.	
Exposure to ETS causes cancer in nonsmokers	*N*	%	*N*	%	**OR**	**95% CI**	*p*
I don't agree	81	77.9	23	22.1	7.95	3.64–17.34	$\mathbf{p < 0.001}$
I agree	373	48.3	400	51.7	1.00	Ref.	

4. Discussion

The issue of exposure of hospitality venues staff to ETS at work is a very important social and health problem, directly related to the lack of full legal protection for nonsmokers [5,18,28–30].

In 2010, the percentage of nonsmokers exposed to ETS in the workplace was over 72%, while in 2015 it was close to 52%. The analysis of the results obtained from the studies carried out in 2010 and 2015 confirmed a significant decrease in the exposure to ETS and existence of a significant problem in terms of the exposure of hospitality venues employees to ETS.

In 2010, nonsmoking women declared more frequent exposure to ETS at work, while in 2015 it was nonsmoking men who were more likely to be exposed. Therefore, it should be stated that hospitality venues are a unique type of space where exposure to ETS is very high and this threat applies to both women and men [8–10].

Despite the introduction of laws prohibiting smoking in hospitality venues, in 2015, more than half of nonsmoking staff stayed in rooms where cigarettes were smoked, and therefore, were exposed to ETS at work. Comparing the results of this study with the results of the GATS study (Global Adult Tobacco Survey), employees of hospitality venues were more often exposed to inhalation of ETS (environmental tobacco smoke) than those participating in the study in the years 2009–2010 [14,31,32]. Over 33% of them declared exposure to inhalation of ETS in the last month indoors at work. Additionally, respondents of the GATS survey declared exposure to ETS: over 98% in bars, pubs, music clubs and discos, and almost 54% in restaurants, cafes, and bistros [24,32]. When analysing results of studies presented in 2011, 2013, 2015, and 2017 in reports from a nationwide survey on smoking attitudes, it may be noticed that the percentage of people exposed to ETS in hospitality venues regularly decreased in 2011–2017 [21,33]. In 2015, 16% of respondents stated that they were exposed to ETS in bars and pubs, while in 2017, only 12% of them were exposed. In cafes, these values were 9% and 6%, respectively, while in restaurants, the percentage did not decrease and continued to be 7% [24,27].

In discos and music clubs, the proportion of respondents exposed to ETS also dropped from 15% (in 2009) to 9% (in 2017) [14,34–36]. The presented results showed that employees of hospitality venues were significantly more often exposed to ETS at work than persons participating in the study conducted by Bogdanovica et al. [5]. Results of studies carried out in other European countries after the introduction of complete ban on smoking in hospitality venues show that the percentage of employees exposed to ETS significantly decreased [1,37–41].

In Ireland, new regulations helped reduce the incidence of ETS in restaurants from 85% to 3%, and in bars and pubs from 98% to 5%, whereas the cotinine level in the saliva of nonsmokers working in bars and restaurants fell by around 80% [3,19].

In Scotland, cotinine levels decreased in the saliva of nonsmoking bar and restaurant staff by 89% [42]. Studies in France also confirmed high effectiveness of new laws. There was a significant decrease in the incidence of ETS, i.e., in bars from 95.9% to 3.7% and in restaurants from 64.7% to 2.3%. The next stage of the study conducted in 2012 found out that the occurrence of ETS in bars was at the level of 6.6%, while in restaurants, it was at the level of 1.4% [19]. Finland is another positive example of the implementation of anti-tobacco laws. In this country, after the introduction of the restrictive anti-tobacco law, the number of restaurant employees who were not exposed to the ETS increased from 54% to 82%, while the number of unexposed employees of bars and pubs increased from 10% to as much as 70% [34]. The changes observed in selected countries by other authors differ significantly from the results obtained in this study. In the above-mentioned countries, there was a much greater decrease in the exposure of hospitality venues staff to ETS at work. It results from a comprehensive anti-tobacco legislation, introducing a total and absolute ban on smoking in public places, including hospitality venues. The degree of employee exposure to ETS is also influenced by the type of room in which they work [19,23,34,42].

In our survey conducted in 2010, over 31% of respondents said that they worked in a smoking room or in both a smoking and nonsmoking room, while the percentage of staff working only in the nonsmoking room was 37.6%. However, in the study conducted in 2015, there was a significant increase in the percentage of employees (over 73%) of hospitality venues who claimed to work only in a nonsmoking room. A multifactor logistic regression analysis carried out in 2015 showed that the odds ratio of exposure to ETS among employees of hospitality venues was increased the most when the establishment had both smokers and nonsmokers rooms. The results confirm partial implementation of the new tobacco laws. As a result of the ban on smoking in hospitality venues that was introduced at the end of 2010, there was a decrease in the number of bars, cafes, restaurants, and night and music clubs with rooms excluded from the smoking ban. However, in 2015, the percentage of staff who worked in smoking rooms was 4.7%, while 21.7% of staff worked both in smoking and nonsmoking rooms, which clearly indicates that the almost 5-year period of smoking ban did not completely eliminate the problem of smoking rooms. A significant impact on the incomplete implementation of the ban is a possibility left by the legislator to let owners or managers of venues with at least two rooms, i.e., closed, ventilated rooms, for customers to exclude from the general smoking ban; thus, smoking rooms may be created, while there is also lack of effective control measures for the smoking ban. As a result of legal possibility of creating separated smoking rooms, in 2015, there was a fourfold increase in the percentage of people who claimed that there was a room for smokers in their workplace. Finland is an example of a country where, similarly to Poland, it was possible to apply for a permission for smoking rooms; however, such solution was quickly recognized as insufficient protection against exposure to ETS, and regulations were introduced to strictly prohibit smoking in bars and restaurants [19,34,43]. In addition, research conducted in Finland, Australia, Canada, Germany, and Italy confirmed that only a total ban on smoking in hospitality venues protects against ETS exposure and leads to a decrease in the number of cigarettes smoked by employees [7,8,44–48].

The exposure to ETS of hospitality venues employees may be assessed with biomarker analysis (cotinine in saliva, urine, and plasma) or environmental monitoring (nicotine, PM2.5, and PM10), as well as an analysis of subjective feelings of employees (number of hours of ETS exposure) [49].

For the purposes of deepening the analysis, the authors of this study reviewed the literature and presented the analysis of publications on exposure biomarkers [49–52].

The biomarker for exposure to ETS used in all studies was cotinine, the main metabolite of nicotine. In the analysed studies, the reduction of the biomarker concentration in individuals exposed to ETS after the introduction of new legislative solutions compared to the concentration observed before the introduction amounted to 57–89% in establishments where the smoking ban was in force. For example, in studies conducted in Scotland, the concentration of cotinine in saliva before the introduction of new legislation was 2.9 ng/mL, while after the introduction, it decreased significantly to 0.4 ng/mL (by 89%, 95% CI: 85–92%). Fernandez et al. (2009) recorded a statistically significant decrease in the concentration of the exposure biomarker (from 1.6 to 0.5 ng/mL after the introduction of new legislative solutions; $p < 0.01$) (Table 7) only in hospitality venues where a total smoking ban was introduced [49–57].

Table 7. Changes in the level of ETS exposure after the implementation of legislative smoking bans—analysis based on biomarkers of exposure.

No.	Country	Characteristic			Reduction	References
1.	Ireland	Legislation (date of implementation and type of restrictions)		III 2004 ban of smoking in indoor workplaces	80% ($p < 0.001$)	Allwright, 2005 [49], Mulcahy, 2005 [50]
		Populations		111 bar staff		
		Biomarker of exposure		Cotinine level in saliva Md (IQR)		
		Biomarker level	Before smoking ban	29.0 nmol/L (18.2–43.2 nmol/L)		
			After smoking Ban	5.1 nmol/L (2.8–13.1 nmol/L)		
2.	Norway	Legislation (date of implementation and type of restrictions)		Ban on smoking in indoor workplaces	$p < 0.001$	Ellingsen, 2006 [51]
		Populations		25 employees in bars and restaurants		
		Biomarker of exposure		Cotinine level in urine GM (95% CI)		
		Biomarker level	Before smoking ban	Evening urine collection: 9.5 µg/g (6.5–13.7 µg/g) creatinine/Morning urine collection: 15.3 µg/g (10.3–22.7 µg/g) creatinine		
			After smoking Ban	Evening urine collection: 1.4 µg/g (0.8–2.5 µg/g) creatinine/Morning urine collection: 1.6 µg/g (0.9–3.0 µg/g) creatinine	$p < 0.001$	

Table 7. Changes in the level of ETS exposure after the implementation of legislative smoking bans—analysis based on biomarkers of exposure.

No.	Country	Characteristic			Reduction	References
3.	Italy	Legislation (date of implementation and type of restrictions)		I 2005 ban on smoking in indoor workplaces	$p < 0.0001$	Valente, 2007 [52]
		Populations		37 hospitality workers		
		Biomarker of exposure		Cotinine level in urine GM (95% CI)		
		Biomarker level	Before smoking ban	15.4 ng/mL (913–18.3 ng/mL)		
			After smoking ban	2.6 ng/mL (1.4–4.9 ng/mL)		
4.	Sweden	Legislation (date of implementation and type of restrictions)		VI 2005 ban on smoking in indoor workplaces	bd.	Larsson, 2008 [53]
		Populations		43 hospitality workers		
		Biomarker of exposure		Cotinine level in urine/percentage of people with cotinine level below the limit of detection		
		Biomarker level	Before smoking ban	37%		
			After smoking ban	67%		
5.	Scotland	Legislation (date of implementation and type of restrictions)		III 2006 ban on smoking in indoor workplaces	89% (85–92%)	Semple, 2007 [54], Menzies, 2006 [55]
		Populations		126 hospitality workers		
		Biomarker of exposure		Cotinine level in saliva GM (GSD)		
		Biomarker level	Before smoking ban	2.9 ng/mL (2.3 ng/mL)		
			After smoking ban	0.4 ng/mL (3.7 ng/mL)		
6.	England	Legislation (date of implementation and type of restrictions)		Ban on smoking in indoor workplaces	75% ($p < 0.001$)	Gotz, 2008 [56]
		Populations		75 workers		
		Biomarker of exposure		Cotinine level in saliva M (GM; SD)		
		Biomarker level	Before smoking ban	3.4 ng/mL (2.4 ng/mL; 2.5 ng/mL)		
			After smoking ban	0.8 ng/mL (0.4 ng/mL; 3.2 ng/mL)		

GSD—geometric standard deviation; Md—mediana; SD—standard deviation; IQR—interquartile range; bd.—no data. Źródło: Polańska, K.; Hanke, W.; Konieczko, K. Hospitality workers's exposure to ETS before and after implementation of smoking ban in public places: a review of epidemiological studies. Medycyna Pracy 2011, 62, 211–224 [57].

5. Conclusions

1. Exposure to ETS among hospitality venues staff decreased in 2010–2015; however, it remained high.
    1.1. Despite the introduction of more restrictive regulations that prohibited smoking in hospitality venues (amendment to the Act on the Protection of Health against the Consequences of Use of Tobacco and Tobacco Products from 2010), more than half of nonsmoking employees were exposed to ETS in the workplace.
2. Regulations prohibiting smoking in hospitality venues were often violated by employees and customers.

   2.1. There is a need for effective control of the implementation and compliance with anti-tobacco laws in hospitality venues.

3. Only a total ban on smoking in all enclosed work spaces can serve as an appropriate protection of employees' health (including employees of hospitality and commercial venues). Legislative solutions should be introduced to reduce exposure to ETS by placing a total ban on smoking.

Author Contributions: Conceptualization, E.K. and K.S.; methodology, E.K.; software, K.S.; validation, M.S., E.C. and J.K.; formal analysis, E.K.; investigation, K.S.; resources, M.S.; data curation, E.C.; writing—original draft preparation, K.S.; writing—review and editing, J.K.; visualization, M.S.; supervision, E.C.; project administration, K.S.; funding acquisition, E.K. All authors have read and agreed to the published version of the manuscript.

Funding: This research received no external funding.

Conflicts of Interest: The authors declare no conflict of interest.

References

1. Protano, C.; Vitali, M. The New Danger of Thirdhand Smoke: Why Passive Smoking Does Not Stop at Secondhand Smoke. *Environ Health Perspect.* **2011**, *119*, a422. [CrossRef] [PubMed]
2. Protano, C.; Manigrasso, M.; Avino, P.; Vitali, M. Second-hand smoke generated by combustion and electronic smoking devices used in real scenarios: Ultrafine particle pollution and agerelated dose assessment. *Environ. Int.* **2017**, *107*, 190–195. [CrossRef] [PubMed]
3. Bondy, S.J.; Zhang, B.; Kreiger, N.; Selby, P.; Benowitz, N.; Travis, H.; Florescu, A.; Greenspan, N.R.; Ferrence, R. Impact of an indoor smoking ban on bar workers' exposure to secondhand smoke. *J. Occup. Environ. Med.* **2009**, *51*, 612–619. [CrossRef] [PubMed]
4. Pechacek, T.; Kaufmann, R.; Trosclair, A.; Caraballo, R.; Caudill, S. Reduced secondhand smoke exposure after implementation of a comprehensive statewide smoking ban—New York, June 26, 2003—June 30, 2004. *MMWR Morb. Mortal. Wkly. Rep.* **2007**, *56*, 705–708. [CrossRef]
5. Farrelly, M.C.; Nonnemaker, J.M.; Chou, R.; Hyland, A.; Peterson, K.K.; Bauer, U.E. Changes in hospitality workers' exposure to secondhand smoke following the im-plementation of New York's smoke-free law. *Tob. Control* **2005**, *14*, 236–241. [CrossRef] [PubMed]
6. World Health Organization. *Protection from Exposure to Second-hand Tobacco Smoke. Policy Recommendations*; World Health Organization: Geneva, Switzerland, 2007; Available online: http://whqlibdoc.who.int/publications/2007/9789241563413_eng.pdf (accessed on 10 May 2020).
7. Heloma, A.; Jaakkola, M.S.; Kahkonen, E.; Reijula, K. The short-term impact of national smoke-free work place legislation on passive smoking and tobacco use. *Am. J. Public Health* **2001**, *91*, 1416–1418. [CrossRef]
8. Hyland, A.; Hassan, L.M.; Higbee, C.; Boudreau, C.; Fong, G.; Borland, R.; Cummings, K.M.; Yan, M.; Thompson, M.E.; Hastings, G. The impact of smoke free legislation in Scotland: Results from the Scottish ITC Scotland/UK longitudinal surveys. *Eur. J. Public Health* **2009**, *19*, 198–205. [CrossRef]
9. Verdonk-Kleinjan, W.M.I.; Knibbe, R.A.; Tan, F.E.S.; Willemsen, M.C.; de Groot, H.N.; de Vries, H. Does the workplace smoking ban eliminatek differences in risk for environmental tobacco smoke exposure at work. *Health Policy* **2009**, *92*, 197–202. [CrossRef]
10. Goodman, P.; Agnew, M.; McCaffrey, M.; Paul, G.; Clancy, L. Effect of the Irish smoking ban on respiratory health of bar workers and air quality in Dublin pubs. *Am. J. Respir. Critic. Care Med.* **2007**, *175*, 840–845. [CrossRef]
11. Nebot, M.; López, M.J.; Ariza, C.; Pérez-Ríos, M.; Fu, M.; Schiaffino, A.; Muñoz, G.; Saltó, E.; Fernández, E. Impact of the Spanish smoking law on exposure to secondhand smoke in offices and hospitality venues: Before-and-after study. *Environ. Health Perspect.* **2009**, *117*, 344–347. [CrossRef]
12. Twyman, L.; Bonevski, B.; Paul, C.; Bryant, J. Perceived barriers to smoking cessation in selected vulnerable groups: A systematic review of the qualitative and quantitative literature. *BMJ Open* **2014**, *4*, e006414. [CrossRef] [PubMed]
13. Cummings, K.M.; Hyland, A.; Giovino, G.A.; Hastrup, J.L.; Bauer, J.E.; Bansal, M.A. Are smokers adequately informel about the health risks of smoking and medicinal nicotine? *Nicotine Tob. Res.* **2004**, *6*, S333–S340. [CrossRef] [PubMed]

14. Drygas, W.; Niklas, A.; Piwońska, A.; Piotrowski, W.; Flotyńska, A.; Kwaśniewska, M.; Nadrowski, P.; Puch-Walczak, A.; Szafraniec, K.; Bielecki, W.; et al. Multi-centre National Population Health Examination Survey (WOBASZ II study): Assumptions, methods, and implementation. *Pol. Heart J.* **2016**, *74*, 681–690. [CrossRef] [PubMed]
15. World Health Organization. *Tobacco*; WHO: Geneva, Switzerland, 2019; Available online: https://www.who.int/news-room/fact-sheets/detail/tobacco (accessed on 26 October 2019).
16. Guillaumier, A.; Twyman, L.; Paul, C.; Siahpush, M.; Palazzi, K.; Bonevski, B. Financial Stress and Smoking withina Large Sample of Socially Disadvantaged Australians. *Int. J. Environ. Res. Public Health* **2017**, *14*, 231. [CrossRef] [PubMed]
17. Polakowska, M.; Kaleta, D.; Piotrowski, W.; Topór-Mądry, R.; Puch-Walczak, A.; Niklas, A.; Bielecki, W.; Kozakiewicz, K.; Pająk, A.; Tykarski, A.; et al. Tobacco smoking in Poland in the years from 2003 to 2014. Multi-centre National Population Health Examination Survey (WOBASZ). *Pol. Arch. Intern. Med.* **2017**, *127*, 91–99. [CrossRef]
18. Pursell, L.; Allwright, S.; O'Donovan, D.; Paul, G.; Kelly, A.; Mullally, B.J.; D'Eath, M. Befor and after study of bar workers'perceptions of the impast of smoke-free workplace legislation in the republic of Ireland. *BMC Public Health* **2007**, *7*, 131. [CrossRef]
19. World Health Organization. WHO Report on the Global Tobacco Epidemic, The MPOWER Package. 2008. Available online: https://www.who.int/tobacco/mpower/mpower_report_full_2008.pdf (accessed on 9 October 2019).
20. Kantar Public for Chief Sanitary Inspectorate. A Report from a Nationwide Survey on Attitudes towards Smoking. Available online: https://gis.gov.pl/wp-content/uploads/2018/04/Postawy-Polak%C3%B3wdopalenia-tytoniu-Raport-2017.pdf (accessed on 9 July 2019).
21. Pinkas, J.; Szymański, J.; Poznański, D.; Wierzba, W. Smoke-free policy in Poland on the example of the implementation into national law of the European Parliament and Council Directive 2014/40 /EU of 3 April 2014 on the approximation of laws, regulations and administrative provisions of the Member Statesconcerning the production, presentation and sale of tobacco and related products, and repealing Directive 2001/37/WE. *Ann. Agric. Environ. Med.* **2016**, *23*, 395–398.
22. Liu, X.; Lugo, A.; Spizzichino, L.; Tabuchi, T.; Pacifici, R.; Gallus, S. Heat-not-burnTobacco products: Concerns from the Italian experience. *Tob. Control* **2019**, *28*, 113–114. [CrossRef]
23. World Health Organization. Tobacco and Inequalities. In *Guidance for Addressing Inequities in Tobacco-RelatedHarm*; World Health Organization: Geneva, Switzerland, 2014.
24. Milcarz, M.; Polańska, K.; Bak-Romaniszyn, L.; Kaleta, D. How Social Care Beneficiaries in Poland Rate Relative Harmfulness of Various Tobacco and Nicotine-Containing Products. *Int. J. Environ. Res. Public Health* **2017**, *14*, 1029. [CrossRef]
25. Milcarz, K.; Bak-Romaniszyn, L.; Kaleta, D. Environmental Tobacco Smoke Exposure and Smoke-FreeRules in Homes among Socially-Disadvantaged Populations in Poland. *Int. J. Environ. Res. Public Health* **2017**, *14*, 447. [CrossRef]
26. Giovino, G.A.; Mirza, S.A.; Sarnet, J.M.; Gupta, P.C.; Jarvis, M.J.; Bhala, N.; Peto, R.; Zatonski, W.; Hsia, J.; Morton, J.; et al. Tobacco use in 3 billion individuals from 16 countries: An analysis of nationally representative cross-sectional household surveys. *Lancet* **2012**, *380*, 668–669. [CrossRef]
27. Milcarz, K.; Makowiec-Dąbrowska, T.; Bak-Romaniszyn, L.; Kaleta, D. Smoking Patterns and Smoking Cessation Willingness-A Study among Beneficiaries of Government Welfare Assistance in Poland. *Int. J. Environ. Res. Public Health* **2017**, *14*, 131. [CrossRef] [PubMed]
28. Brose, L.S.; Simonavicius, E.; Cheeseman, H. Awareness and Use of "Heat-not-burn" Tobacco Products in Great Britain. *Tob. Regul. Sci.* **2018**, *4*, 44–50. [CrossRef]
29. Smith, D.M.; Gawron, M.; Balwicki, L.; Sobczak, A.; Matynia, M.; Goniewicz, M.L. Exclusive versus dual use of tobacco and electronic cigarettes among adolescents in Poland, 2010–2016. *Addict. Behav.* **2019**, *90*, 341–348. [CrossRef]
30. Sansone, G.C.; Raute, L.J.; Fong, G.T.; Pednekar, M.S.; Quah, A.C.K.; Bansal-Travers, M.; Guptak, P.C.; Sinha, D.N. Knowledge of health effects and intentions to quitamong smokers in India: Findings from the Tobacco Control Policy (TCP) India Pilot Survey. *Int. J. Environ. Res. Public Health* **2012**, *9*, 564–578. [CrossRef]

31. Baig, M.; Bakarman, M.A.; Gazzaz, Z.J.; Khabaz, M.N.; Ahmed, T.J.; Qureshi, I.A.; Hussain, M.B.; Alzahrani, A.H.; AlShehri, A.A.; Basendwah, M.A.; et al. Reasons and Motivations for Cigarette Smoking and Barriersagainst Quitting Among a Sample of Young People in Jeddah, Saudi Arabia. *Asian Pac. J. Cancer Prev.* **2016**, *17*, 3483–3487.
32. Global Adult Tobacco Survey Poland 2009–2010. Available online: http://www.who.int/tobacco/surveillance/en_tfi_gats_poland_report_2010.pdf (accessed on 8 September 2018).
33. Oberg, M.; Jaakkola, M.S.; Woodward, A.; Peruga, A.; Prüss-Ustün, A. Worldwide burden of disease from exposure to second-hand smoke: A retrospective analysis of data from 192 countries. *Lancet* **2011**, *377*, 139–146. [CrossRef]
34. Hiscock, R.; Bauld, L.; Amos, A.; Platt, S. Smoking and socioeconomic status in England: The rise of the never smoker and the disadvantaged smoker. *J. Public Health.* **2012**, *34*, 390–396. [CrossRef]
35. Brownson, R.C.; Jackson-Thompson, J.; Wilkerson, J.C.; Davis, J.R.; Owens, N.W.; Fisher, E.B. Demographic and socioeconomic differences in beliefs about the health effects of smoking. *Am. J. Public Health* **1992**, *82*, 99–103. [CrossRef]
36. Sieminska, A.; Buczkowski, K.; Jassem, E.; Lewandowska, K.; Ucinska, R.; Chelminska, M. Patterns of motivations and ways of quitting smoking among Polishs mokers: Aquestionnaire study. *BMC Public Health* **2008**, *8*, 274. [CrossRef]
37. Goodchild, M.; Nargis, N.; Tursand'Espaignet, E. Global economiccost of smoking-attributable diseases. *Tob. Control* **2018**, *27*, 58–64. [CrossRef] [PubMed]
38. Demaio, A.R.; Nehme, J.; Otgontuya, D.; Meyrowitsch, D.W.; Enkhtuya, P. Tobacco smoking in Mongolia: Findings of a national knowledge, attitudes and practices study. *BMC Public Health* **2014**, *14*, 213. [CrossRef] [PubMed]
39. Gupta, B.; Kumar, N. A Cross-country comparison of knowledge, attitudes and practices about Tobacco use: Findings from the Global Adult Tobacco Survey. *Asian Pac. J. Cancer Prev.* **2014**, *15*, 5035–5042. [CrossRef] [PubMed]
40. Siahpush, M.; Yong, H.H.; Borland, R.; Reid, J.L.; Hammond, D. Smokers with financial stress are more likely to want to quit but less likely to try or succeed: Findings from the International Tobacco Control (ITC) four country survey. *Addiction* **2009**, *104*, 1382–1390. [CrossRef] [PubMed]
41. Raute, L.J.; Sansone, G.; Pednekar, M.S.; Fong, G.T.; Gupta, P.C.; Quah, A.C.K.; Bansal-Travers, M.; Sinha, D.N. Knowledge of health effects and intentions to quit among smokeless tobacco users in India: Findings from the International Tob Control Policy Evaluation (ITC) India Pilot Survey. *Asian Pac. J. Cancer Prev.* **2011**, *12*, 1233–1238. [PubMed]
42. Marcon, A.; Pesce, G.; Calciano, L.; Bellisario, V.; Dharmage, S.C.; Garcia-Aymerich, J.; Gislasson, T.; Heinrich, J.; Holm, M.; Janson, C.; et al. Trends in smoking initiation in Europe over 40 years: A retrospective cohort study. *PLoS ONE* **2018**, *13*, e0201881. [CrossRef]
43. Mitchell, D.; Moodie, C.; Critchlow, N.; Bauld, L. Adolescents' perceptions of standardised cigarette packaging design and Brand variant name post-implementation: A fokus group study in Scotland. *BMC Public Health* **2019**, *19*, 1227. [CrossRef]
44. Rosenthal, L.; Carroll-Scott, A.; Earnshaw, V.A.; Sackey, N.; O'Malley, S.S.; Santilli, A.; Ickovics, J.R. Targeting cessation: Understanding barriers and motivations to quitting among urban adult daily Tobacco smokers. *Addict. Behav.* **2013**, *38*, 1639–1642. [CrossRef]
45. Browning, K.K.; Ferketich, A.K.; Salsberry, P.J.; Wewers, M.E. Socioeconomic disparity in provider-deliveredas sistance to quit smoking. *Nicotine Tob. Res.* **2008**, *10*, 55–61. [CrossRef]
46. Cavelaars, A.E.; Kunst, A.E.; Geurts, J.J.; Crialesi, R.; Grötvedt, L.; Helmert, U.; Lahelma, E.; Lundberg, O.; Matheson, J.; Mielck, A.; et al. Educational differences in smoking: International comparison. *BMJ* **2000**, *22*, 1102–1107. [CrossRef]
47. Cohn, A.M.; Ganz, O.; Dennhardt, A.A.; Murphy, J.G.; Ehlke, S.; Cha, S.; Graham, A.L. Menthol cigarette smokingi associated with great er subjective reward, satisfaction, and "throathit", but not great er behavioral economic demand. *Addict. Behav.* **2019**, *101*, 106108. [CrossRef] [PubMed]
48. Hammett, P.J.; Fu, S.S.; Burgess, D.J.; Nelson, D.; Clothier, B.; Saul, J.E.; Nyman, J.A.; Widome, R.; Joseph, A.M. Treatment barriers among younger and older socioeconomically disadvantaged smokers. *Am. J. Manag. Care* **2017**, *23*, e295–e302. [PubMed]

49. Allwright, S.; Paul, G.; Greiner, B.; Mullally, B.J.; Pur-sell, L.; Kelly, A.; Bonner, B.; D'Eath, M.; McConnell, B.; McLaughlin, J.P.; et al. Legislation for smoke-free work-places and health of bar workers in Ireland: Before and after study. *Br. Med. J.* **2005**, *331*, 1117–1120. [CrossRef] [PubMed]
50. Mulcahy, M.; Evans, D.S.; Hammond, S.K.; Repace, J.L.; Byrne, M. Secondhand smoke exposure and risk following the Irish smoking ban: An assessment of salivary cotinine concentrations in hotel workers and air nicotine levels in bars. *Tob. Control* **2005**, *14*, 384–388. [CrossRef] [PubMed]
51. Ellingsen, D.G.; Fladseth, G.; Daae, H.L.; Gjolstad, M.; Kjaerheim, K.; Skogstad, M. Airborne exposure and biological monitoring of bar and restaurant wor-kers before and after the introduction of a smoking ban. *J. Environ. Monit.* **2006**, *8*, 362–368. [CrossRef] [PubMed]
52. Valente, P.; Forastiere, F.; Bacosi, A.; Cattani, A.; Di Carlo, S.; Ferri, M.; Figa'-Talamanca, R.; Marconi, A.; Paoletti, L.; Perucci, C.; et al. Exposure to fine and ultrafine particles from secondhand smoke in public places before and after the smoking ban, Italy 2005. *Tob. Control* **2007**, *16*, 312–317. [CrossRef]
53. Larsson, M.; Boëthius, G.; Axelsson, S.; Montgomery, S.M. Exposure to environmental tobacco smoke and health effects among hospitality workers in Sweden—before and after the implementation of a smoke-free law. *Scand. J. Work Environ. Health* **2008**, *34*, 267–277. [CrossRef]
54. Semple, S.; Maccalman, L.; Naji, A.A.; Dempsey, S.; Hilton, S.; Miller, B.G.; Ayres, J.G. Bar workers' exposure to secondhand smoke: The effects of Scottish smokefree legislation on occupational exposure. *Ann. Occup. Hyg.* **2007**, *51*, 571–580. [CrossRef]
55. Menzies, D.; Nair, A.; Williamson, P.; Schembri, S.; Al-Khairalla, M.; Barnes, M.; Fardon, T.C.; McFarlane, L.; Magee, G.J.; Lipwotrh, P.J. Respiratory symptoms, pulmonary function, and markers of inflammation among bar workers before and after a legislative ban on smoking in public places. *JAMA* **2006**, *296*, 1742–1748. [CrossRef]
56. Gotz, N.K.; van Tongeren, M.; Wareing, H.; Wallace, L.M.; Semple, S.; MacCalman, L. Changes in air quality and second-hand smoke exposure in hospitality sector busi-nesses after introduction of the English smoke-free legi-slation. *J. Public Health* **2008**, *30*, 421–428. [CrossRef]
57. Polańska, K.; Hanke, W.; Konieczko, K. Hospitality workers's exposure to environmental tobacco smoke before and after implementation of smoking ban in public places: A review of epidemiological studies. *Med. Pr.* **2011**, *62*, 211–224. [PubMed]

© 2020 by the authors. Licensee MDPI, Basel, Switzerland. This article is an open access article distributed under the terms and conditions of the Creative Commons Attribution (CC BY) license (http://creativecommons.org/licenses/by/4.0/).

International Journal of
Environmental Research and Public Health

MDPI

Article

In-Vitro Growth Inhibition of Bacterial Pathogens by Probiotics and a Synbiotic: Product Composition Matters

Jacek Piatek [1,*], Hanna Krauss [1], Arleta Ciechelska-Rybarczyk [2], Malgorzata Bernatek [3], Paulina Wojtyla-Buciora [1] and Henning Sommermeyer [1]

1 Department of Medicine, The President Stanisław Wojciechowski State University of Applied Sciences in Kalisz, Nowy Świat 4, 62-800 Kalisz, Poland; hjk12@poczta.fm (H.K.); paulinawojtyla@gmail.com (P.W.-B.); h.sommermeyer@pwsz-kalisz.edu.pl (H.S.)
2 District Sanitary-Epidemiological Station in Jarocin, Waska 2, 63-200 Jarocin, Poland; mikrobiologia@psse-jarocin.pl
3 State Hospital Jarocin, Szpitalna 1, 63-200 Jarocin, Poland; drpiatek@o2.pl
* Correspondence: drpiatek@interia.eu; Tel.: +48-883389788

Received: 17 April 2020; Accepted: 8 May 2020; Published: 11 May 2020

Abstract: A variety of activities potentially contribute to the beneficial effects of probiotic bacteria observed in humans. Among these is a direct inhibition of the growth of pathogenic bacteria in the gut. The present study characterizes head-to-head the in-vitro pathogen growth inhibition of clinically relevant infectious bacterial strains by different types of probiotics and a synbiotic. In-vitro growth inhibition of *Escherichia (E.) coli EPEC, Shigella (Sh.) sonnei, Salmonella (S.) typhimurium, Klebsiella (K.) pneumoniae* and *Clostridioides (C.) difficile* were determined. Investigated products were a yeast mono strain probiotic containing *Saccharomyces (Sac.) boulardii*, bacterial mono strain probiotics containing either *Lactobacillus (L.) rhamnosus GG* or *L. reuteri DSM 17938*, a multi strain probiotic containing three *L. rhamnosus strains (E/N, Oxy, Pen)*, and a multi strain synbiotic containing nine different probiotic bacterial strains and the prebiotic fructooligosaccharides (FOS). Inhibition of pathogens was moderate by *Sac. boulardii* and *L. rhamnosus GG*, medium by *L. reuteri DSM 17938* and the *L. rhamnosus E/N, Oxy, Pen* mixture and strong by the multi strain synbiotic. Head-to-head in-vitro pathogen growth inhibition experiments can be used to differentiate products from different categories containing probiotic microorganisms and can support the selection process of products for further clinical evaluation.

Keywords: gut microbiome; antimicrobial activity; pathogen overgrowth

1. Introduction

The invention of antibiotic therapy some 100 years ago was a major step forward in medical practice, allowing medical practitioners to manage otherwise deadly infections caused by pathogenic bacteria. However, in the recent past, the rapid increase of bacterial antibiotic resistance has become a pressing problem of global healthcare [1]. More sensible and less widespread use of antibiotics is necessary to counteract the increasing loss of their efficacy. While it is clinically challenging to implement, antibiotics should only be used where their usage is appropriate, and best only in confirmed cases of bacterial infections. In addition to a careful usage of antibiotics, probiotics or synbiotics should be considered as prophylactic measures, at least for patients at risk of certain bacterial infections (e.g., the elderly admitted to an intensive care unit), as complementary therapy during and after antibiotic therapy, or even as standalone therapy under certain specific circumstances for bacterial infections.

Every orally taken antibiotic is leading to alterations of the gut microbiota, in some patients with symptoms (e.g., diarrhea), in others without symptoms [2]. While antibiotic-associated diarrhea (AAD)

is worrying, the disturbance of the gut microbiota by antibiotics can have more severe consequences. Among these severe side effects is, for example, pseudomembranous colitis [3], which is caused by a pathogen overgrowth of the gut, in this case by *C. difficile*. Pathogenic bacteria are omnipresent in the human gut. However, as long as the gut microbiota is well balanced and diverse, no pathogen overgrowth will take place [4]. Disturbance of the equilibrium of the gut microbiota can result in loss of this pathogenic overgrowth control. Such a can loss then lead to a strong proliferation of pathogens in the gut and finally result in disease manifestation. Supplementation with certain probiotics or synbiotics (mixtures of probiotic bacteria with a prebiotic component) during and after antibiotic therapy has been demonstrated to reduce occurrence of AAD [5].

Today, a huge variety of products containing probiotic microorganisms are available. These products can be categorized according to their composition: (i) yeast and bacterial products, (ii) mono strain and multi strain products or (iii) probiotic and synbiotic products. In the present study, the inhibitions of five bacterial pathogens, *E. coli EPEC, Sh. sonnei, S. typhimurium, K. pneumoniae* and *C. difficile* by representatives of different categories of products containing probiotic microorganisms are studied. The evaluated probiotics are the clinically well-established mono strain probiotics containing either *Sac. boulardii* [6], *L. rhamnosus GG* [7] or *L. reuteri DSM 17938* [8], and two newer products, one being a multi strain probiotic containing a mixture of three different *L. rhamnosus* strains *(E/N, Oxy, Pen)*, the other being a complex multi strain synbiotic containing six *lactobacilli* (*L. acidophilus, L. casei, L. paracasei, L. plantarum, L. rhamnosus GG, L. salivarius*), three *bifidobacteria* strains (*Bifidobacterium (B.) bifidum, B. longum, B. lactis*) and the prebiotic FOS.

Most published in-vitro growth inhibition studies focused on the characterization of only one pathogenic bacterium and one or a limited number of products containing probiotic microorganisms. The present study evaluated the antagonistic activity of a range of products in head-to-head in-vitro growth inhibition experiments with a number of different bacterial pathogens, which can support the selection of products for future, more in-depth investigations.

2. Materials and Methods

2.1. Probiotics and Synbiotic

The yeast *Sac. boulardii* probiotic Enterol® (Biocodex, Gentilly, France) contains in each capsule 4.5×10^9 colony forming units (CFU) of the *Sac. boulardii* strain CNCM I-745. Dicoflor® (Bayer Sp. z o.o., Warszawa, Poland) contains 6×10^9 CFU of *L. rhamnosus GG* ATCC© 53103 per capsule. BioGaia®, (InfectoPharm Arzneimittel und Consilium GmbH, Heppenheim, Germany) contains 10^8 CFU *L. reuteri DSM 17938* per 5 drops. Lakcid® (Biomed-Lublin S.A., Poland) contains a total of 2×10^9 CFUs as mixture of the *L. rhamnosus* strains *E/N* (40%), *Oxy* (20%) and *Pen* (40%), [9]. The complex multi strain synbiotic Multilac® Baby (Vivatrex GmbH, Aachen, Germany) is a freeze-dried powder. Each sachet contains a total of 10^9 CFUs as a mixture of equal CFU amounts of *L. acidophilus LA-14; L. casei R0215; L. paracasei Lpc-3; L. plantarum Lp-115; L. rhamnosus GG, L. salivarius Ls-33, B. lactis Bl-04, B. bifidum R0071, B. longum R0175* and 1.43 g of the prebiotic ingredient FOS.

2.2. In-Vitro Growth Inhibition of *E. coli EPEC, Sh. sonnei, S. typhimurium* and *K. pneumoniae*

For the in-vitro pathogen inhibition experiments with *E. coli EPEC* (clinical isolate, collection number 3410/19 *E. coli* 025), *Sh. sonnei* ATCC© 9290™ (ATCC, Manassas, VA, USA), *S. enterica subsp. enterica serovar typhimurium* ATCC© 14280™ (ATCC, Manassas, VA, USA) and *K. pneumoniae subsp. pneumoniae* ATCC© 700603™ (ATCC, Manassas, VA, USA) the respective pathogen was inoculated on Columbia agar with 5% sheep blood (CM0331, Fisher Scientific GmbH, Schwerte, Germany) and incubated at 37 °C under aerobic conditions for 24 h [10]. Suspensions (100 μL) of the five tested products (mono strain probiotics *Sac. boulardii, L. rhamnosus GG, L. reuteri DSM 17938*, the multi strain *L. rhamnosus E/N, Oxy, Pen* and the complex multi strain synbiotic) each containing 10^6 CFU were inoculated on MRS agar (CM0361, Fisher Scientific GmbH, Schwerte, Germany) and incubated for 48 h in the presence of 5% CO_2 [11]. After the incubation, 10 mm diameter bars were cut out and transferred to a Mueller–Hinton agar (CM0337, Fisher Scientific GmbH, Schwerte, Germany) previously inoculated with the respective pathogen strain with a density of 2 on the McFarland scale. The tested cultures were stored at 4 °C for 4 h, followed by incubation at 37 °C for 24 h under aerobic conditions.

2.3. *In-Vitro Growth Inhibition of C. difficile*

For the in-vitro pathogen inhibition studies with *C. difficile* ATCC© 9689™ (ATCC, Manassas, VA, USA), the pathogen was cultivated under anaerobic conditions at 35–37 °C for 24–48 h on Schaedler agar (CM0437, Fisher Scientific GmbH, Schwerte, Germany) [12]. Suspensions each containing 10^6 CFU of the five evaluated products were inoculated on MRS agar and incubated for 48 h in the presence of 5% CO_2. 10 mm diameter bars were transferred to a Mueller–Hinton agar with 5% horse blood and 20 mg/L NAD (PP0972, E&O Laboratories Ltd, Bonnybridge, UK) and incubated under anaerobic conditions for 24 h.

2.4. *FOS Control and Measurement of Growth Inhibition*

For testing a potential pathogen growth inhibitory effect of FOS, 100 µL of a solution containing 14.3 mg/mL FOS (F8052, Sigma Aldrich, St. Louis, Missouri, USA) was applied to a 10 mm filter disk that was then administered to respective pathogen testing plates. The multi strain synbiotic was tested on the same plates as a positive control.

At the end of the incubation, measurements of inhibition zones around the tested colonies were taken from the outer edge of the colonies to the outer edge of the clear zones. Each test was performed in triplicate and the arithmetic means of the radii measuring from the edges of the colonies to the edges of the clear zones were calculated as well as the standard deviations SD (Excel, Microsoft, Redmont, WA, USA). Independent T-test statistical analyses of datasets were conducted with GraphPad Prism software version 8.2 (GraphPad Software, San Diego, CA, USA), Datasets were considered as significantly different when a p-value < 0.01 was achieved.

3. Results

In-Vitro Pathogen Growth Inhibition

In-vitro growth of all five tested pathogenic bacteria were inhibited by all evaluated products. However, the extent of in-vitro growth inhibitions by the products varied significantly for all tested pathogens (Figure 1).

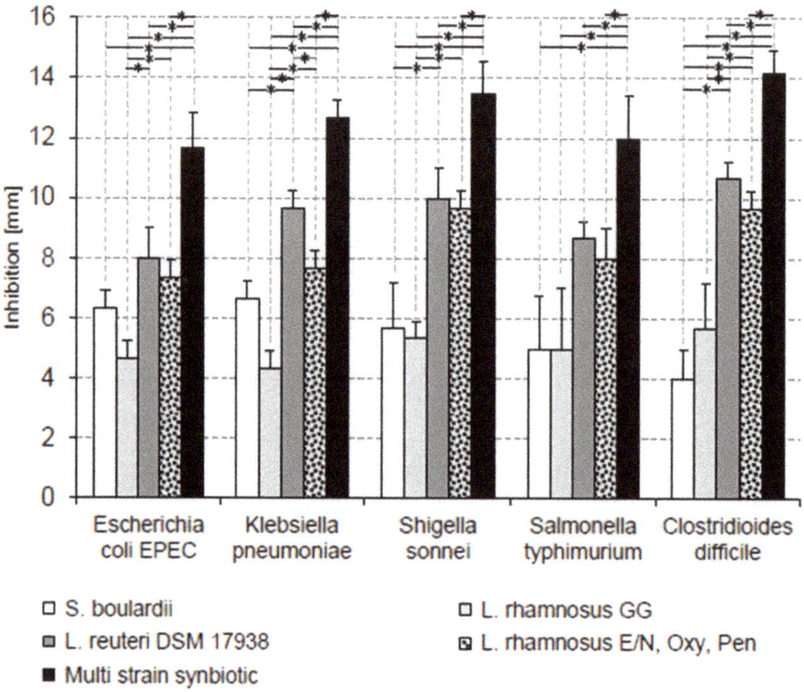

Figure 1. In-vitro growth inhibition of pathogens by different probiotics and a multi strain synbiotic. The *L. rhamnosus E/N, Oxy, Pen* mixture contains the three different probiotics in a CFU ratio of 40/20/40. The multi strain synbiotic contains a mixture of equal CFU amounts of *L. acidophilus LA-14; L. casei R0215; L. paracasei Lpc-3; L. plantarum Lp-115; L. rhamnosus GG, L. salivarius Ls-33, B. lactis Bl-04, B. bifidum R0071, B. longum R0175* and the prebiotic fructooligosaccharides (FOS). Significant differences (p-values < 0.01) between the inhibitions by different products are indicated by horizontal lines marked with an asterisk (*).

The inhibitory patterns found for *E. coli EPEC* and *K. pneumoniae* were similar. The antagonistic effects of the tested products ranked from weak to strong inhibition as follows: *L. rhamnosus GG* < *Sac. boulardii* < *L. reuteri DSM 17938* = multi strain probiotic *L. rhamnosus E/N, Oxy, Pen* < complex multi strain synbiotic.

For the inhibition of *Sh. sonnei* and *S. typhimurium* an inhibition ranking from weak to strong of *Sac. boulardii* = *L. rhamnosus GG* < *L. reuteri DSM 17938* = multi strain probiotic *L. rhamnosus E/N, Oxy, Pen* < complex multi strain synbiotic was found. While there was a significant difference (p-values < 0.01) between the inhibitions of *Sh. sonnei* by *Sac. boulardii* or *L. rhamnosus GG* and the *L. reuteri DSM 17938* probiotic or the multi strain probiotic *L. rhamnosus E/N, Oxy, Pen*, the difference of the inhibitions by these two groups of probiotics of *S. typhimurium* was not significant (p-values > 0.01).

The inhibition of the gram-positive *C. difficile* by the tested products resembled that of *Sh. sonnei* and *S. typhimurium*. The weakest inhibitions were found for *Sac. boulardii* and *L. rhamnosus GG*. Inhibitions by *L. reuteri DSM 17938* and the multi strain probiotic *L. rhamnosus E/N, Oxy, Pen* were intermediate and the best inhibitory effect was found for the complex multi strain synbiotic.

As shown in Figure 2, FOS alone had no inhibitory effect on *E. coli EPEC*. FOS also had no inhibitory effect on the growth of the other four tested pathogens (data not shown).

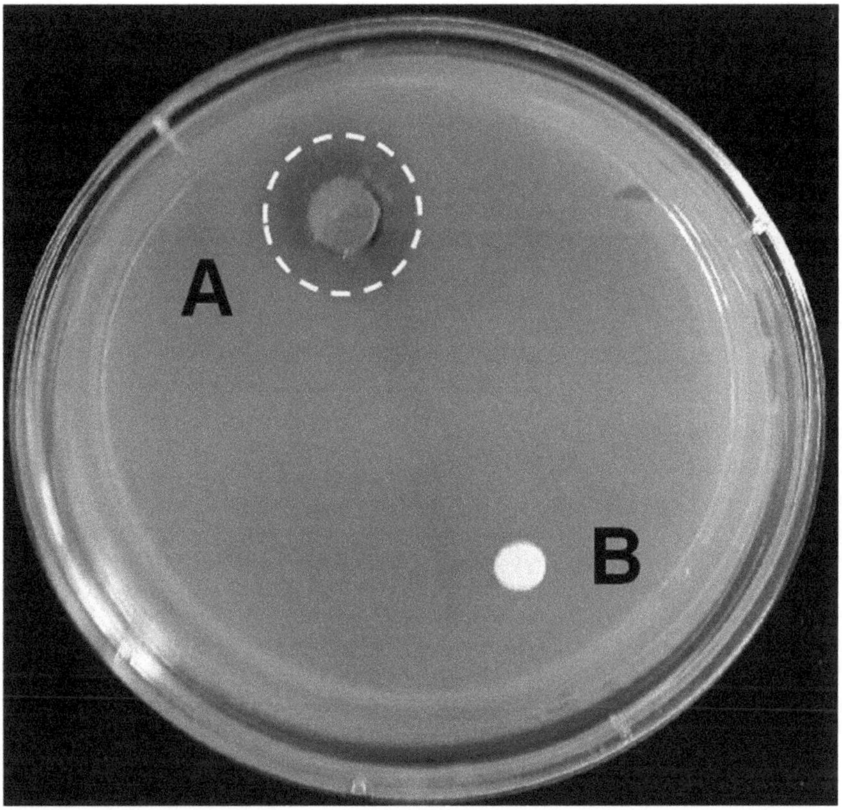

Figure 2. Example of the effects of the multi strain synbiotic (**A**) and FOS (**B**) on the in-vitro growth of *Enteropathogenic E. coli EPEC*.

4. Discussion

A large number of studies have been published which are characterizing the in-vitro pathogen growth effects of individual probiotic microorganisms. There are also studies which compare the growth inhibitory effects of individual strains in head-to-head experiments. Less data are available for the in-vitro growth inhibitory effect of individual multi-strain products. To our knowledge there is no study published which compares the in-vitro pathogen growth inhibitory effects of probiotic products from different categories (yeast, bacteria, mono-strain, multi-strain, probiotic, synbiotic) head-to-head. The present study compares the in-vitro pathogen growth inhibitory properties of products containing probiotic microorganisms commonly used by physicians in Poland. The in-vitro pathogen growth inhibition experiments of the present study are not aiming to identify the underlying mechanisms of growth inhibition or to identify the causes of potential synergistic effects of probiotic bacteria in the multi-strain products, but to focus primarily on establishing a ranking of efficacy in this one particular experimental set-up. Inhibition of pathogen growth is only one potential effect exhibited by probiotic bacteria in the gut and other activities (e.g., inhibition of the adherence of pathogens to the gut mucosa, stimulation of the host immune system) are also potential contributors to their efficacy. However, in-vitro pathogen growth inhibition testing is a helpful measure that can be used to support the selection of products for further investigation, e.g., in clinical trials.

For the in-vitro growth inhibition testing, five bacterial pathogens were selected which physicians encounter in their day-to-day practice. All five selected bacterial pathogens represent major health care concerns. As outlined in the following, infections with these pathogens are not always requiring antibiotic therapy. In addition, the increasing rate of developing antibiotic resistance in these bacteria has stimulated the interest in probiotics as prophylactic, alternative or adjuvant therapy for infections caused by them.

Enteropathogenic E. coli EPEC is a major cause of infant diarrhea [13]. As long as there is no evidence for a systemic infection, antibiotic therapy is rarely indicated and should be deferred until culture results are available. Due to this, and the emerging resistance of *E.coli* against antibiotics [14], probiotics are considered as additional options to manage *E. coli* infections [15]. In-vitro pathogen growth inhibition experiments found no clear *E. coli* growth antagonism by the yeast *Sac. boulardii* [16]. In contrast, in-vitro growth inhibitions of *E. coli* have been described for a number of mono strain bacterial probiotics, among them *L. rhamnosus GG* [17], *L. reuteri DSM 17938* [15] and multi strain probiotics [15,18,19].

Shigella infections are a major public health problem in areas of poor sanitation with especially high incidence, morbidity and mortality in children [20]. Shigellosis is spread by fecal-oral transmission and ingestion of a small number of *Shigella* bacteria can already cause clinical disease. Most patients recover from Shigellosis without antibiotic treatment within 5–7 days. Various antimicrobial agents are effective in the treatment of severe cases of Shigellosis, however, a globally emerging antibiotic resistance is observed [21]. Consequently, probiotics are considered as an alternative approach to manage Shigellosis. *Sac. boulardii* has been shown to interfere with *Shigella* pathogenesis [22], however, in-vitro growth inhibition of *Sh. sonnei* by *Sac. boulardii* has not been published. Inhibition of in-vitro growth of *Sh. sonnei* has been demonstrated for *L. rhamnosus GG* [17], for *L. reuteri* [23] and a number of other *lactobacilli* [24]. For the multi strain products investigated in this study, results from in-vitro growth inhibition of *Sh. sonnei* have not been demonstrated.

Most food-borne bacterial gastroenteritis is caused by *S. typhimurium*. Antibiotic therapy is not advised for most of the patients with *S. typhimurium* infection, as the infection is often self-limiting and antibiotic resistance of the bacterium is on the rise [25]. Therefore, especially for severe cases or patients with persistent infections [26], non-antibiotic treatment alternatives are needed [27]. There is a large amount of studies, including human clinical trials, indicating that *Sac. boulardii* has potential in the prevention and treatment of infections with *S. typhimurium* [28]. However, no study has been published comparing the in-vitro growth inhibition of *S. typhimurium* by *Sac. boulardii* with that of bacterial probiotics. In-vitro growth inhibition of *S. typhimurium* has been shown for a number of individual *lactobacilli* strains, [17,29,30]. The method has also been used for the search for new *lactobacilli* strains with growth inhibitory potential against *S. typhimurium* [19,31]. Our group has recently published that *S. typhimurium* was more strongly inhibited by a multi strain synbiotic (different from that investigated in the present study) when compared with the in-vitro growth inhibition caused by its individual bacterial strains [30].

K. pneumoniae is responsible for an alarming increase in hospital infections, especially in intensive care units. More and more antibiotics have lost their efficacy against *K. pneumoniae*, and strains of *K. pneumoniae* have emerged that are resistant against most of the presently available antibiotics [32,33]. Consequently, interest in probiotics/synbiotics as alternative options to manage infections with *K. pneumoniae* has emerged. Few studies have investigated the effect of probiotics on the in-vitro growth of *K. pneumoniae* [34–36], however, all with promising results.

For an initial *C. difficile* infection (CDI), the exposure to antibiotics is the most important risk factor [37]. In this context, it is important to note that certain antibiotics (e.g., clindamycin, quinolones, cephalosporins) are associated with a higher risk of causing CDI. The pathology of CDI is characterized by a disruption of the gut microbiota, resulting in an overgrowth by *C. difficile*, production of toxins and disease development. Characteristic symptoms are diarrhea and abdominal pain. In severe cases, the CDI can result in a life-threatening pseudomembranous colitis. Recurrence of CDI is a not uncommon observation. First step in the treatment of CDI is discontinuing the therapy with the inciting antibiotic as soon as possible. Depending on severity of the CDI, three antibiotics can be considered for therapy: metronidazole, vancomycin and fidaxomicin [38]. In addition, supporting the diversity of the gut microbiota is an important therapeutic objective in the management of CDI [39].

In-vitro growth inhibition of *C. difficile* by a number of mono strain probiotics, among them *L. rhamnosus*, and *bifidobacteria* has been described [40,41]. To our knowledge, the present study is the first which compares head-to-head the in vitro-growth inhibition of *C. difficile* by representatives from different categories of products containing probiotics.

The products evaluated in the present study can be categorized by the following criteria: (i) yeast/bacteria, (ii) mono strain/multi strain and (iii) probiotic/synbiotic (containing probiotic and prebiotic components). While the in-vitro pathogen growth inhibitory effects vary among the tested products, the inhibitory effects seem not to be pathogen-dependent. Weakest pathogen inhibition is observed for *Sac. boulardii* and *L. rhamnosus GG*, intermediate for *L. reuteri DSM 17938* and the *L. rhamnosus E/N, Oxy, Pen* mixture and strongest inhibition is found for the multi strain synbiotic. This finding is in line with the hypothesis that multi strain probiotics exhibit superior growth inhibitory effects towards pathogenic bacteria because they are combining a broader range of independent antibacterial activities, some of which might even act synergistically [42,43]. However, it has to be mentioned that in-vitro characterization of the potential synergistic effects in multi strain probiotics, especially those containing a larger number of different strains, can hardly be demonstrated due to the fact that thousands, if not millions of potential combinations would have to be investigated. Synbiotics might have an additional advantage over pure probiotics, as their prebiotic component provides a source of energy, potentially supporting the proliferation of their probiotic components. Based on our experiments, we can exclude that FOS itself has an inhibitory effect on one of the tested pathogenic bacteria.

An obvious limitation of the present study is that it is using only one experimental set-up to establish an in-vitro efficacy ranking among the evaluated products. No efforts have been undertaken to investigate the underlying mechanisms of the observed pathogen inhibitions or the underlying mechanisms of potential synergies in the multi strain products. Dedicated human clinical studies will be necessary to evaluate if these in-vitro findings will translate into clinical benefits for patients.

5. Conclusions

In-vitro growth inhibition of a variety of pathogens is helpful to differentiate products and product categories containing probiotic microorganisms. Based on the results of the present study, multi strain probiotics should be preferred in case a strong in-vitro growth inhibition of a broad range of pathogens is desired. Our study results can support the product selection for future clinical investigations.

Author Contributions: Conceptualization, H.S., J.P., H.K. and P.W.-B.; methodology, A.C.-R., and M.B.; validation, J.P. and H.K.; formal analysis, J.P., H.K.; data curation, A.C.-R. and M.B.; writing—original draft preparation, H.S. and J.P.; writing—review and editing, H.S., J.P. and H.K.; visualization, H.S.; supervision, J.P.; project administration, J.P. All authors have read and agreed to the published version of the manuscript.

Funding: This research received no external funding.

Acknowledgments: The authors would like to acknowledge the work of Sabine Hanna from Cambridge Assessment English for proof-reading, English style editing, and useful suggestions.

Conflicts of Interest: Henning Sommermeyer works as a consultant for Vivatrex GmbH, a company which is marketing Multilac® Baby in Germany. Vivatrex GmbH was not involved in the design, collection, analyses or interpretation of data, in writing the manuscript, or in the decision to publish the results. All other authors declare that they have no competing interests.

References

1. Aslam, B.; Arshad, M.I.; Khurshid, M.; Muzammil, S.; Rasool, M.H.; Nisar, M.A.; Alvi, R.F.; Aslam, M.A.; Qamar, M.U.; Salamat, M.K.F.; et al. Antibiotic resistance: A rundown of a global crisis. *Infect. Drug Resist.* **2018**, *11*, 1645–1658. [CrossRef] [PubMed]

2. McFarland, L. Antibiotic-associated diarrhea: Epidemiology, trends and treatment. *Future Microbiol.* **2008**, *3*, 563–578. [CrossRef] [PubMed]
3. Farooq, P.D.; Urrunaga, N.H.; Tang, D.M.; von Rosenvinge, E.C. Pseudomembranous Colitis. *Dis. Mon.* **2015**, *61*, 181–206. [CrossRef] [PubMed]
4. Bäumler, A.J.; Sperandio, V. Interactions between the microbiota and pathogenic bacteria in the gut. *Nature* **2016**, *535*, 85–93. [CrossRef] [PubMed]
5. Guo, Q.; Goldenberg, J.Z.; Humphrey, C.; El Dib, R.; Johnsto, B.C. Probiotics for the prevention of pediatric antibiotic-associated diarrhea. *Cochrane Database Syst. Rev.* **2019**, *4*, CD004827. [CrossRef]
6. Szajewska, H.; Kolodziej, M. Systematic review with meta-analysis: Saccharomyces boulardii in the prevention of antibiotic-associated diarrhoea. *Aliment. Pharmacol. Ther.* **2015**, *42*, 793–801. [CrossRef]
7. Szajewska, H.; Kolodziej, M. Systematic review with meta-analysis: Lactobacillus rhamnosusGG in the prevention of antibiotic-associated diarrhoea in children and adults. *Aliment. Pharmacol. Ther.* **2015**, *42*, 1149–1157. [CrossRef]
8. Urbanska, M.; Gieruszczak-Bialek, D.; Szajewska, H. Systematic review with meta-analysis: Lactobacillus reuteri DSM 17938 for diarrhoeal diseases in children. *Aliment. Pharmacol. Ther.* **2016**, *43*, 1025–1034. [CrossRef]
9. Jarocki, P.; Podlesny, M.; Wasko, A.; Siuda, A.; Targonski, Z. Differentiation of three Lactobacillus rhamnosus strains (E/N, Oxy and Pen) by SDS-PAGE and two-dimensional electrophoresis of surface-associated proteins. *J. Microbiol. Biotechnol.* **2010**, *20*, 558–562. [CrossRef]
10. Ellner, P.D.; Stoessel, C.J.; Drakeford, E.; Vasi, F. A new culture medium for medical bacteriology. *Tech. Bull. Regist. Med. Technol.* **1966**, *36*, 58–60. [CrossRef]
11. De Man, J.C.; Rogosa, M.; Sharpe, M.E. A medium for the cultivation of Lactobacilli. *Appl. Bact.* **1960**, *23*, 130–135. [CrossRef]
12. Schaedler, R.W.; Dubos, R.; Costello, R. The development of the bacterial flora in the gastrointestinal tract of mice. *J. Exp. Med.* **1965**, *122*, 59–66. [CrossRef] [PubMed]
13. Ochoa, T.J.; Contreras, C.A. Enteropathogenic, E. coli (EPEC) infection in children. *Curr. Opin. Infect. Dis.* **2011**, *24*, 478–483. [CrossRef] [PubMed]
14. Poirel, L.; Madec, J.Y.; Lupo, A.; Schink, A.K.; Kieffer, N.; Nordmann, P.; Schwarz, S. Antimicrobial Resistance in Escherichia coli. *Microbiol. Spectr.* **2018**, *6*. [CrossRef]
15. Fijan, S.; Sulc, D.; Steyer, A. Study of the In Vitro Antagonistic Activity of Various Mono-Strain and Multi-Strain Probiotics against Escherichia coli. *Int. J. Environ. Res. Public Health* **2018**, *15*, 1539. [CrossRef]
16. Rajkowska, K.; Kunicka-Styczynska, A.; Rygala, A. Probiotic Activity of Saccharomyces cerevisiae var. boulardii Against Human Pathogens. *Food Technol. Biotechnol.* **2012**, *50*, 230–236.
17. Hütt, P.; Shchepetova, J.; Lõivukene, K.; Kullisaar, T.; Mikelsaar, M. Antagonistic activity of probiotic lactobacilli and bifidobacteria against entero- and uropathogens. *J. Appl. Microbiol.* **2006**, *100*, 1324–1332. [CrossRef]
18. Shokryazdan, P.; Sieo, C.C.; Kalavathy, R.; Liang, J.B.; Alitheen, N.B.; Jahromi, M.F.; Ho, Y.W. Probiotic Potential of Lactobacillus Strains with Antimicrobial Activity against Some Human Pathogenic Strains. *BioMed Res. Int.* **2014**, *2014*, 927268. [CrossRef]
19. Jacobsen, C.N.; Rosenfeldt Nielsen, V.; Hayford, A.E.; Møller, L.; Michaelsen, K.F.; Pærregaard, A.; Sandström, B.; Tvede, M.; Jakobsen, M. Screening of Probiotic Activities of Forty-Seven Strains of *Lactobacillus* spp. by In Vitro Techniques and Evaluation of the Colonization Ability of Five Selected Strains in Humans. *Appl. Environ. Microbiol.* **1999**, *65*, 4949–4956. [CrossRef]
20. Niyogi, S.K. Shigellosis. *J. Microbiol.* **2005**, *43*, 133–143.
21. Puzari, M.; Sharma, M.; Chetia, P. Emergence of antibiotic resistant Shigella species: A matter of concern. *J. Infect. Public Health* **2018**, *11*, 451–454. [CrossRef] [PubMed]
22. Mumy, K.L.; Chen, X.; Kelly, C.P.; McCormick, B.A. Saccharomyces boulardii interferes with Shigella pathogenesis by postinvasion signaling events. *Am. J. Physiol. Gastrointest. Liver Physiol.* **2008**, *294*, G599–G609. [CrossRef]
23. Spinler, J.K.; Taweechotipatr, M.; Rognerud, C.L.; Ou, C.N.; Tumwasorn, S.; Versalovic, J. Human-derived probiotic Lactobacillus reuteri demonstrate antimicrobial activities targeting diverse enteric bacterial pathogens. *Anaerobe* **2008**, *14*, 166–171. [CrossRef] [PubMed]

24. Zhang, Y.C.; Zhang, L.W.; Ma, W.; Yi, H.X.; Yang, X.; Du, M.; Shan, Y.J.; Han, X.; Zhang, L.L. Screening of probiotic lactobacilli for inhibition of Shigella sonnei and the macromolecules involved in inhibition. *Anaerobe* **2012**, *18*, 498–503. [CrossRef] [PubMed]
25. Eng, S.K.; Pusparajah, P.; Ab Mutalib, N.S.; Ser, H.L.; Chan, K.G.; Lee, L.H. Salmonella: A review on pathogenesis, epidemiology and antibiotic resistance. *Front. Life Sci.* **2015**, *8*, 284–293. [CrossRef]
26. Marzel, A.; Desai, P.T.; Goren, A.; Schorr, Y.I.; Nissan, I.; Porwollik, S.; Valinsky, L.; McClelland, M.; Rahav, G.; Gal-Mor, O. Persistant Infections by Nontyphoidal Salmonella in Humans: Epidemilogy and Genetics. *Clin. Infect. Dis.* **2016**, *62*, 879–886. [CrossRef]
27. Liévin-Le Moal, V.; Servin, A.L. Anti-Infective Activities of Lactobacillus Strains in the Human Intestinal Microbiota: From Probiotics to Gastrointestinal Anti-Infectious Biotherapeutic Agents. *Clin. Microbiol.* **2014**, *27*, 167–199. [CrossRef]
28. Gut, A.M.; Vasiljevic, T.; Yeager, T.; Donkor, O.N. Salmonella infection—prevention and treatment by antibiotics and probiotic yeasts: A review. *Microbiology* **2018**, *164*, 1327–1344. [CrossRef]
29. Das, J.K.; Mishra, D.; Ray, P.; Tripathy, P.; Beuria, T.K.; Singh, N.; Suar, M. In vitro evaluation of anti-infective activity of a Lactobacillus plantarum strain against Salmonella enterica serovar Enteritidis. *Gut Pathog.* **2013**, *5*, 1. [CrossRef] [PubMed]
30. Piatek, J.; Sommermeyer, H.; Bernatek, M.; Ciechelska-Rybarczyk, A.; Oleskow, B.; Sommer Mikkelsen, L.; Bundvig Barken, K. Persistent infection by Salmonella enterica servovar Typhimurium: Are synbiotics a therapeutic option?—A case report. *Benef. Microbes* **2019**, *10*, 211–217. [CrossRef]
31. Dowarah, R.; Verma, A.K.; Agarwal, N.; Singh, P.; Singh, B.R. Selection and characterization of probiotic lactic acid bacteria and its impact on growth, nutrient digestibility, health and antioxidant status in weaned piglets. *PLoS ONE* **2018**, *13*, e0192978. [CrossRef]
32. Centers for Disease Control and Prevention (CDC). Vital Signs: Carbapenem-Resistant Enterobacteriaceae. *MMWR Morb. Mortal. Wkly. Rep.* **2013**, *62*, 165–170.
33. Xu, L.; Sun, X.; Ma, X. Systematic review and meta-analysis of mortality of patients infected with carbapenem-resistant Klebsiella pneumoniae. *Ann. Clin. Microbiol. Antimicrob.* **2017**, *16*, 18. [CrossRef] [PubMed]
34. Mogna, L.; Deidda, F.; Nicola, S.; Amoruso, A.; Del Piano, M.; Mogna, G. In Vitro Inhibition of Klebsiella pneumoniae by Lactobacillus delbrueckii Subsp. delbrueckii LDD01 (DSM 22106): An Innovative Strategy to Possibly Counteract Such Infections in Humans? *J. Clin. Gastroenterol.* **2016**, *50* (Suppl. 2). [CrossRef]
35. Savino, F.; Cordisco, L.; Tarasco, V.; Locatelli, E.; Di Gioia, D.; Oggero, R.; Matteuzzi, D. Antagonistic effect of Lactobacillus strains against gas-producing coliforms isolated from colicky infants. *BMC Microbiol.* **2011**, *11*, 157. [CrossRef]
36. Piatek, J.; Bernatek, M.; Ciechelska-Rybarczyk, A.; Oleskow, B.; Sommermeyer, H. Inhibition of Carbapenem-Resistant NDM-1 Klebsiella pneumoniae isolated from a Hospital Outbreak Patient by a Synbiotic: A Non antibiotic Treatment Option. *Int. J. Med. Res. Health Sci.* **2019**, *8*, 12–20.
37. Bignardi, G.E. Risk factors for Clostridium difficile infection. *J. Hosp. Infect.* **1998**, *40*, 1–15. [CrossRef]
38. McDonald, L.C.; Gerding, D.N.; Johnson, S.; Bakken, J.S.; Carroll, K.C.; Coffin, S.E.; Dubberke, E.R.; Garey, K.W.; Gould, C.V.; Kelly, C.; et al. Clinical Practice Guidelines for Clostridium difficile Infection in Adults and Children: 2017 Update by the Infectious Diseases Society of America (IDSA) and Society for Healthcare Epidemiology of America (SHEA). *Clin. Infect. Dis.* **2018**, *66*, e1–e48. [CrossRef]
39. Chang, J.Y.; Antonopoulos, D.; Kalra, A.; Tonelli, A.; Khalife, W.T.; Schmidt, T.M.; Young, V.B. Decreased Diversity of the Fecal Microbiome in Recurrent Clostridium difficile -Associated Diarrhea. *J. Infect. Dis.* **2008**, *197*, 435–438. [CrossRef]
40. Naaber, P.; Smidt, I.; Štšepetova, J.; Brilene, T.; Annuk, H.; Mikelsaar, M. Inhibition of Clostridium difficile strains by intestinal Lactobacillus species. *J. Med. Microbiol.* **2004**, *53*, 551–553. [CrossRef] [PubMed]
41. Schoster, A.; Kokotovic, B.; Permin, A.; Pedersen, P.D.; Dal Bello, F.; Guardabassi, L. In vitro inhibition of Clostridium difficile and Clostridium perfringens by commercial probiotic strains. *Anaerobe* **2013**, *20*, 36–41. [CrossRef] [PubMed]

42. Timmerman, H.M.; Koning, C.J.; Mulder, L.; Rombouts, F.M.; Beynen, A.C. Monostrain, multistrain and multispecies probiotics-A comparison of functionality and efficacy. *Int. J. Food Microbiol.* **2004**, *96*, 219–233. [CrossRef] [PubMed]
43. Ouwehand, A.C.; Invernici, M.M.; Furlaneto, F.A.C.; Messora, M.R. Effectiveness of Multistrain versus Single-Strain Probiotics: Current Status and Recommendations for the Future. *J. Clin. Gastroenterol.* **2018**, *52* (Suppl. 1), S35–S40. [CrossRef] [PubMed]

© 2020 by the authors. Licensee MDPI, Basel, Switzerland. This article is an open access article distributed under the terms and conditions of the Creative Commons Attribution (CC BY) license (http://creativecommons.org/licenses/by/4.0/).

International Journal of
Environmental Research and Public Health

MDPI

Article

Association between Living with Children and the Health and Health Behavior of Women and Men. Are There Differences by Age? Results of the "German Health Update" (GEDA) Study

Petra Rattay * and Elena von der Lippe

Department of Epidemiology and Health Monitoring, Robert Koch Institute, 12101 Berlin, Germany; E.vonderLippe@RKI.de

* Correspondence: P.Rattay@RKI.de; Tel.: +49-(0)30-18754-3323

Received: 6 April 2020; Accepted: 28 April 2020; Published: 2 May 2020

Abstract: Does the health of women and men living with and without minor children differ, and are age differences evident in the association? For self-rated general health, depression, back pain, overweight, smoking and sporting inactivity, the GEDA data 2009–2012 (18–54 years, n = 39,096) were used to calculate prevalence for women and men stratified by parental status (living with children: yes/no) and age. Moreover, we calculated odds ratios and predictive margins, performing logistic regressions with interaction terms of parental status and age. Women and men aged 45–54 living with children are healthier than those not living with children. Parents aged 18–24 smoke more frequently and do less sport; young mothers are also more likely to be overweight and suffer from back pain than women not living with children. In multivariable analysis, the interaction of living with children and age is significant for all outcomes (except depression and back pain in men). Living with children is an important social determinant of health, highly dependent on age. It is to be discussed whether the bio-psycho-social situation has an influence on becoming a parent, or whether parenthood in different phases of life strains or enhances health.

Keywords: family; living with children; parenthood; mother; father; self-rated health; health behavior; mental health; Germany

---

1. Introduction

Living together with children can contribute to good health through close social and emotional relationships as well as mechanisms of structuring everyday life, social control and meaningfulness of life. On the other hand, living with children is associated with a number of demands and obligations as well as conflicting role expectations that can lead to stress and poor health [1–3]. This is especially true when problems in the partnership or the sole responsibility for the upbringing of the children, financial worries or low social support occur. In addition to the effects of co-residence with children on the health of women and men ("causality" hypothesis), the health and health behavior of women and men can also have an influence on the probability of finding a partner or starting a family ("selectivity" hypothesis) [2,4].

1.1. State of Research

The current state of research on the health and health behavior of parents is inconsistent. Whereas some studies have reported better health and healthier behavior among parents than among women and men without children, other research reports poorer health among parents. In turn, some studies have found no association between parenthood and health.

Firstly, this variation is related to the large number of different health outcomes investigated. Inconsistent results have been found in particular in self-rated general health [5–13] and mental health [1,13–25]. For overweight or obesity, however, the majority of studies have reported higher prevalence or a higher BMI for mothers [13,26–31], while the findings for men are heterogeneous [13,32]. For physical activity and in particular for the exercise of sports, results are also found to the disadvantage of parents [27,33–38]. However, more household-related physical activity has been evident among women and men with children [39]. The diet of mothers follows the current recommendations to a greater extent than of nonparents [40]; nevertheless, mothers consume more calories [27,41]. With regard to tobacco and alcohol consumption, women and men with children behave more healthily than those without children [42–48].

Besides the dependence on health outcome, results on the association between parenthood and health also vary with the number and age of children [26,40,49–51] as well as the partner, employment and socioeconomic statuses [52,53]. For example, many studies have reported a higher psychosocial and health burden especially for single parents [5,15,16,54–56]. There are also differences by gender in the association between parental status and health, with significantly more studies analyzing the health of mothers (e.g., [7–10,12,17,23,55,57]) than of fathers [31,36,58–63]. Gender-comparative studies, in turn, do not produce consistent results (e.g., [54,64–69]), so that it is not possible to assess in general terms whether the health of parents is associated with gender. However, the social system and welfare state regulations also seem to influence the association between parental status and health. International comparisons have supported the thesis that parenthood in the USA is associated to a greater extent with poor health than in Western and especially Northern European countries, for which positive links between parenthood and health have often be found [44,64,70–74]. Hansen [75] assumes that these country-specific differences can be attributed to the fact that the Nordic welfare states offer families much more comprehensive public support (such as available and affordable daycare, flexible work schedules, job leave security, cash benefits, and paid parental leave) than the USA or Australia.

Besides this, it can be assumed that a large part of the inconsistent results can be attributed to their relationship to different age groups [76]. Although most results are adjusted for age, the moderating effect of age on the association between parenthood and health is rarely analyzed explicitly. When age-stratified results are reported, almost all studies have shown that very young parents in particular are severely affected by health problems [5,55,76–80]. This is supported by studies that include the age at first birth. The birth of the first child shortly after and especially before the mother's 20th birthday is associated with poorer health [12,20,25,52,78,81], less healthy behavior [45] and higher mortality [61,82].

Some studies have found that women and men who became parents late in life have a lower mortality rate and higher life satisfaction than childless women and men [61,73,82–84]. For example, in Margolis and Myrskylä [79] the association between parenthood and happiness at the age of 15 to 29 is negative, while it is not significant at the age of 30 to 39 and positive from the age of 40. Other studies have suggested that—especially in women—an age of around 30 years at the birth of the first child seems to be associated with good maternal health [20,85,86]. According to Graham [57] mothers show better physical and mental health and well-being compared to childless women, especially at the age of 34 to 44; while childless women perform better at a very young age and in old age. Carlson [85] has found that a deviation from the anticipated age at birth of the first child in younger and older mothers is accompanied by an increase in depressive symptoms.

With regard to Germany, Helbig et al. [18] have concluded that for 18- to 49-year-olds there is no moderating effect of age on the association between parental status and mental health. Stöbel-Richter et al. [23] have also found no age differences in the association of parenthood with depression or anxiety in 18- to 50-year-old women. In an analysis by a German health insurance company [78] on sick days of employed women and men with and without family-insured children, there are no differences in parental status in the middle working age (30–44 years), while young working parents have a slightly above average number of sick days and older men with children have a below average number of sick days. Becchetti et al. [87] have found higher happiness among parents compared to nonparents only from the age of 55 onwards. In a longitudinal study with data from the German Socio-Economic Panel, Myrskylä and Margolis [83] have shown that in women and men who became parents late in life (from the age of 35), happiness increased with the birth of the first and second child and remained then at a stable level for a longer period of time. On the other hand, happiness decreased in women and men who became parents between the ages of 18 and 22. Women and men who became parents at ages 23 to 34 have increasing happiness before a first birth and in the year of birth, but afterwards happiness decreases to the baseline level or below.

The short overview of the international and national research on the health of parents and on the moderating effect of age on the association between parenthood and health shows: (a) that the association between the parental status and different health outcome parameters varies, (b) that international findings on the health of parents cannot easily be transferred to Germany due to the varying social systems and family policies, and (c) that so far there are only a few gender-comparative studies on age differences regarding the association between parental and health status.

1.2. Aim of the Study

The aim of this analysis is to compare the health and health behavior of women and men living with and without children in Germany and to explore whether age differences in this association can be observed.

The research questions are:

- Are there differences in the health and health behavior of women and men according to parental status (living with children)?
- Does the association between health/health behavior and parental status in women and men vary with age?

2. Materials and Methods

2.1. Data

For the analyses, the pooled data of the study "German Health Update" (GEDA) from 2009, 2010 and 2012 were used, which were carried out by the Robert Koch Institute as part of health monitoring [88]. The GEDA study is a nationally representative telephone survey of German-speaking adults who live in private households and have a landline phone connection. Data were collected using computer-assisted telephone interviews. Random samples of telephone numbers from the German fixed-line network were generated, adapting the Gabler–Häder method [89]. The sample consists of 62,606 women and men aged 18 and over. The cooperation rate at respondent level (the proportion of interviews that were completed after initial contact with a potential participant), was 51.2% in 2009, 55.8% in 2010 and 76.7% in 2012. The response rate (the proportion of completed interviews to the number of neutral non-responses in the adjusted gross sample) amounted to 29.1% in 2009, 28.9% in 2010 and 22.1% in 2012 [88]. The absolute number of cases per survey year in the total sample were 21,262 in 2009, 22,050 in 2010 and 19,294 in 2012.

The GEDA surveys of the years 2009–2012 provide a large representative data set on health and disease, determinants of health, and the use of health services for use in national and European health reporting systems, health policy, and public health research [88]. Owing to the sample size of the pooled data, it is well suited for the analysis of subsamples.

Each GEDA round was approved by The Federal Commissioner for Data Protection and Freedom of Information, and verbal informed consent was obtained in advance from all participants.

For this analysis, the sample was limited to participants aged 18 to 54. The number of cases of women aged 55 years and above living with minor children is relatively small. After plausibility checks of the age data of the household members, 50 participants were excluded, so that the analyses were based on data from 39,096 women and men (weighted proportions: women: 49.1%, men: 50.9%). The sample description is given in Table 1.

Table 1. Description of the sample.

Variables	Women			Men		
	n Unweighted	% Weighted	Missing % Unweighted	n Unweighted	% Weighted	Missing % Unweighted
Total	21,379	100.0	0	17,717	100.0	0
Outcome variables						
Poor self-rated general health			0.1			0.1
Yes	3921	19.9		2663	17.5	
No	17,442	80.1		15,044	82.5	
Depression			0.3			0.1
Yes	1931	8.6		889	5.2	
No	19,382	91.4		16,804	94.8	
Back pain			0.1			0.1
Yes	3868	18.9		2115	13.3	
No	17,484	81.1		15,580	86.7	
Overweight			2.8			1.0
Yes	6875	35.0		9054	53.7	
No	13,898	65.0		8,486	46.3	
Smoking			<0.1			<0.1
Yes	6780	33.4		6613	40.0	
No	14,590	66.6		11,100	60.0	
Sporting inactivity			0.1			<0.1
Yes	5490	29.2		4451	29.0	
No	15,878	70.8		13,260	71.0	
Predictor and control variables						
Age groups			0			0
18–24 years	3221	15.9		3262	16.2	
25–34 years	4408	24.1		3650	24.0	
35–44 years	6771	27.1		5187	27.0	
45–54 years	6979	33.0		5618	32.8	
Parental status			0			0
Child(ren) in household	9731	55.3		6082	37.0	
No child in household	11,648	44.7		11,635	63.0	
Pre-school child in household (0–6 years)			0			0
Yes	4109	19.2		2688	15.2	
No	17,270	80.8		15,029	84.8	
Partner status			0.4			0.6
Partner	13,022	67.6		10,097	63.7	
No partner	8266	32.3		7521	36.3	
Socioeconomic status			0.2			0.2
High	6492	16.0		6120	17.0	
Middle	12,608	61.6		9680	59.2	
Low	2236	22.4		1889	23.8	
Employment status			0.5			0.4
Full-time employed	7919	34.2		13,581	76.4	
Part-time employed	8496	39.8		1692	9.1	
Not employed	4867	26.0		2379	14.5	
Residential region			0			0
West Germany	17,195	80.5		14,216	79.9	
East Germany	4184	19.5		3501	20.1	
Survey year			0			0
2009	7909	37.5		6058	34.2	
2010	8186	38.5		6269	34.6	
2012	5284	24.0		5390	31.2	

2.2. Variables

Outcome variables: The analysis included self-rated general health, depression, back pain, overweight, smoking, and sporting inactivity as outcome variables. Self-rated general health was surveyed using the question "How is your general state of health?". The response categories were dichotomized into "very good/good" versus "fair/poor/very poor". Depression was assessed by asking whether the respondent was suffering from depression or depressive mood in the last 12 months diagnosed by a doctor or psychotherapist (12-month prevalence). The 12-month prevalence of back pain is based on the question as to whether the respondent has had at least three months of persistent back pain in the last 12 months. Overweight data are based on respondents' height and weight data. According to the WHO definition, overweight occurs when the Body Mass Index (BMI) is greater than 25 kg/m^2. Information on smoking ("yes"/"no") was asked with the question "Do you smoke from time to time—even if only occasionally?" In the present analysis, the response categories "daily" and "occasional" were combined to "yes". The data on sporting inactivity ("yes"/"no") are based on the respondents' self-declaration that they have not practiced any sport in the last 3 months.

Predictor variable: Parental status is based on information provided by participants on all persons living in the household (relationship to respondent and age). We define women and men as parents when they live together with at least one own child under 18 years. We did not differentiate among the respondents' biological children, adopted children, or stepchildren (social parenthood).

Moderator variables: Age and sex are defined as moderator variables. The results of the descriptive analyses are stratified by age groups (18–24 years, 25–34 years, 35–44 years and 45–54 years) while in the multivariable analyses age (in completed life years) is included as a metric variable. All analyses are stratified by sex.

Control variables: Socioeconomic, employment and partner statuses as well as the age of the youngest child and the residential region are taken into account as control variables. The socioeconomic status was determined by means of an index which included information on school and vocational education, occupational status and net household income and which allows classification into "low", "medium" and "high" status groups [90]. For measurement of employment status (self-defined), we differentiate among "employed full-time", "employed part-time," and "not employed." The partner status takes into account living with a partner in the same household ("yes"/"no"), regardless of the marital status. With regard to the age of the youngest child, living with a pre-school aged child (0–6 years) ("yes"/"no") was included. Furthermore, we controlled the logistic models for the residential region ("West Germany"/"East Germany incl. Berlin").

2.3. Data Analysis

In the first step, prevalences were calculated for all outcome variables for women and men with and without children in the household (parental status) stratified by age groups and sex.

In the second step, we calculated multivariable logistic regressions over the entire age range for each outcome variable. In these models, age was used as a metric variable (completed life years, centered on the minimum age of 18 years). Age modelling (age vs. age squared) was performed by calculating the BIC (Schwarz's Bayesian information criteria) for selecting the (fully adjusted) model with the best "fit" (=lowest BIC value). For the analysis of the moderating effect of age on the association between parental status and health, the interaction between parental status and age or age squared was included in the models. Depending on the model with the best "fit", there were three variants of the interaction term: (a) parental status#age, (b) parental status#age and parental status#age squared, (c) parental status #age squared.

All models were adjusted for socioeconomic, employment and partner statuses, the age of the youngest child and the residential region. Odds ratios and predictive margins (adjusted at the mean) were calculated for all models [91]. In this paper, we present the predictive margins graphically. All calculations were carried out using a weighting factor that corrects for deviations within the sample from the population structure (as of 31 December, 2010) regarding age, sex, education and federal state. The analyses were conducted with the StataSE 15 statistical software (StataCorp, College Station, TX, USA) using the survey (svy) module. Statistical significance in the descriptive as well as in the multivariable analysis was determined using p-values ($p < 0.05$).

3. Results

Looking at the raw prevalence for the total group of 18- to 54-year-olds (Table 2, first column), it can be seen that women and men living with children are more often overweight and less frequently active in sports than women and men in childless households. Men with children also suffer more frequently from back pain. On the other hand, women and men with children smoke less often than those without children. Moreover, women living in households with underage children are less likely to report poor general health and suffer from depression than women in childless households.

However, a comparison of age-stratified prevalence for women and men with and without children shows that there are significant age differences (Table 2). In the youngest age group (18–24 years), the prevalences for all outcome variables are higher in women and men with children than in women and men without children—although not all differences reach statistical significance. In women, all prevalences (except for those for poor general health) are significantly higher in those living with children. Among men, significant differences according to parental status are only found in back pain, smoking and sporting inactivity.

In the group aged 45–54, the opposite association can be observed. In this age group, the prevalences of all health outcomes are higher in women and men from childless households than in women and men living with children. In women, for all health outcomes significant differences according to parental status were found, whereas in men the differences in prevalence for depression and overweight are not statistically significant.

The results of the multivariable models including the entire age range can be found in Tables 3 and 4 (odds ratios and p-values). In Figure 2, the predictive margins for each health outcome presenting the interaction of parental status and age are shown separately for women and men.

For women, the interactions between age and parental status are highly significant for all health outcomes ($p < 0.001$). Figure 2 shows that the association between parental status and health varies greatly with age. Young women (up to approx. 30 years of age) without children are healthier and behave healthier when comparing with women with children of the same age. Among women aged approximately 30–40 years, there are hardly any health differences according to the parental status. For women over the age of 40, those who live with children in the household are healthier and behave more healthily than women without children. From the age of 36, women with children are less likely than women without children to be depressed. It can also be observed that the predicted probabilities in women with children hardly change with age, while significantly higher predicted probabilities are found among women in childless households in the older age groups (except smoking).

The patterns of smoking and sporting inactivity in men are similar to that of women. For self-rated health and overweight, the differences between men with and without children are smaller than for women, but the interaction terms are still significant. For depression and back pain, however, there are no significant interactions between parental status and age in men.

Table 2. Health of women and men by parental status, stratified by age groups and sex (prevalence in %, 95% CI, p-values).

Women	Child in Household	Total			18–24 Years			25–34 Years			35–44 Years			45–54 Years		
		%	95% CI	p	%	95% CI	p	%	95% CI	p	%	95% CI	p	%	95% CI	p
Poor general health	No	21.3	20.4–22.2	***	11.6	10.5–12.9		13.5	11.8–15.3	*	21.6	19.3–24.1	**	31.4	29.7–33.1	***
	Yes	18.1	17.1–19.1		13.1	8.8–19.2		16.5	14.7–18.5		17.7	16.4–19.1		21.1	19.1–23.4	
Depression	No	9.7	9.1–10.4	***	4.3	3.6–5.2	*	8.9	7.6–10.3		11.7	9.9–13.6	***	13.1	12.0–14.3	***
	Yes	7.2	6.6–7.9		7.7	4.4–13.0		7.1	5.9–8.4		7.1	6.3–8.0		7.6	6.4-9.0	
Back pain	No	19.1	18.3–20.0		12.6	11.4–13.9	***	13.9	12.2–15.7	***	19.4	17.2–21.9		25.9	24.3–27.5	***
	Yes	18.6	17.7–19.6		27.1	20.5–34.8		18.7	16.8–20.6		18.1	16.8–19.4		19.2	17.2–21.3	
Overweight	No	33.4	32.4–34.5	***	15.7	14.3–17.2	***	26.7	24.5–29.0	***	35.9	33.1–38.8		47.6	45.8–49.4	***
	Yes	36.8	35.7–38.0		32.3	25.1–40.4		37.0	34.6–39.4		36.3	34.7–38.0		38.0	35.5–40.5	
Smoking	No	36.6	34.6–36.7	***	32.8	31.0–34.6	***	35.6	33.2–38.0		39.1	36.3–42.0	***	36.5	34.8–38.2	***
	Yes	30.6	29.5–31.8		49.0	40.9–57.1		33.7	31.4–36.1		30.7	29.2–32.3		25.1	23.0–27.4	
Sporting inactivity	No	25.9	24.9–26.9	***	14.4	13.1–15.9	***	21.2	19.1–23.4	***	31.1	28.4–34.0		34.2	32.5–35.9	***
	Yes	33.4	32.2–34.6		38.0	30.4–46.2		39.9	37.5–42.3		32.6	31.0–34.3		26.0	23.7–28.4	
Men	**Child in Household**	**Total**			**18–24 Years**			**25–34 Years**			**35–44 Years**			**45–54 Years**		
		%	95% CI	p	%	95% CI	p	%	95% CI	p	%	95% CI	p	%	95% CI	p
Poor general health	No	17.7	16.8–18.6		7.9	6.9–9.0		10.9	9.4–12.5	*	20.8	18.4–23.3	**	29.7	27.8–31.7	***
	Yes	17.1	15.9–18.4		8.8	3.0–22.8		14.1	11.3–17.4		15.8	14.2–17.7		20.6	18.3–23.0	
Depression	No	5.5	5.0–6.0		2.2	1.7–2.8		4.7	3.8–5.8	**	7.6	6.2–9.2	***	7.8	6.7–8.9	
	Yes	4.6	3.9–5.4		3.9	0.8–17.0		1.8	1.0–3.1		4.3	3.4–5.3		6.8	5.3–8.6	
Back pain	No	12.7	11.9–13.5	*	6.6	5.7–7.6	*	9.6	8.2–11.2		14.6	12.6–16.8		19.1	17.5–20.9	*
	Yes	14.3	13.2–15.4		16.4	7.7–31.7		11.3	9.2–13.8		14.2	12.8–15.9		15.9	13.9–18.1	
Overweight	No	48.8	47.6–50.0	***	25.0	23.3–26.7		42.0	39.7–44.3	***	58.5	55.9–61.2	*	68.3	66.4–70.2	
	Yes	62.2	60.7–63.6		29.7	17.5–45.6		54.8	50.9–58.7		62.8	60.7–64.8		65.9	63.4–68.3	
Smoking	No	41.4	40.3–42.6	***	37.3	35.5–39.1	***	44.8	42.5–47.2		45.6	42.9–48.4	***	39.8	37.8–41.9	***
	Yes	37.5	36.0–39.1		70.5	54.8–82.4		47.5	43.6–51.4		36.8	34.7–39.0		32.4	29.9–35.1	
Sporting inactivity	No	27.3	26.2–28.3	***	11.1	9.9–12.4	***	22.0	20.1–24.1	***	34.8	32.2–37.6		40.7	38.7–42.8	***
	Yes	32.0	30.5–33.5		39.4	25.2–55.6		33.4	29.8–37.2		33.1	31.0–35.3		29.6	27.1–32.2	

* $p < 0.05$; ** $p < 0.01$; *** $p < 0.001$.

Table 3. Results of the logistic regressions with interaction of parental status and age for women (odds ratios, p values; age (metric) centered at the minimum of 18 years).

Women	Poor Self-Rated General Health		Depression		Back Pain		Overweight		Smoking		Sporting Inactivity	
	OR	p	OR	p	OR	p	OR	p	OR	p	OR	p
Parental status: Child in household												
No child	Ref.		Ref.		Ref.		Ref.		Ref.		Ref.	
Child	2.43	<0.001	3.92	<0.001	2.19	<0.001	2.75	<0.001	6.15	<0.001	3.39	<0.001
Age	1.06	<0.001	1.16	<0.001	1.04	<0.001	1.10	<0.001	1.08	<0.001	1.13	<0.001
Age#Age			0.997	<0.001			0.999	<0.001	0.998	<0.001	0.998	<0.001
Child#Age												
No child#Age	Ref.		Ref.		Ref.		Ref.		Ref.		Ref.	
Child#Age	0.96	<0.001	0.87	<0.001	0.96	<0.001	0.92	<0.001	0.88	<0.001	0.91	<0.001
Child#Age#Age												
No child#Age#Age			Ref.				Ref.		Ref.		Ref.	
Child#Age#Age			1.002	0.005			1.001	0.013	1.002	<0.001	1.001	0.016
Pre-school child in household												
No	Ref.		Ref.		Ref.		Ref.		Ref.		Ref.	
Yes	0.61	<0.001	0.68	0.004	0.91	0.265	1.02	0.770	0.55	<0.001	1.42	<0.001
Partner status												
Partner	Ref.		Ref.		Ref.		Ref.		Ref.		Ref.	
No partner	1.15	0.007	1.85	<0.001	1.13	0.023	0.79	<0.001	1.45	<0.001	0.92	0.076
Socioeconomic status												
High	Ref.		Ref.		Ref.		Ref.		Ref.		Ref.	
Middle	1.78	<0.001	1.21	0.007	1.55	<0.001	1.75	<0.001	1.65	<0.001	2.13	<0.001
Low	2.92	<0.001	1.39	0.002	2.17	<0.001	2.51	<0.001	2.68	<0.001	4.56	<0.001
Employment status												
Full-time	Ref.		Ref.		Ref.		Ref.		Ref.		Ref.	
Part-time	1.06	0.278	1.49	<0.001	0.97	0.590	0.87	0.002	0.83	<0.001	0.78	<0.001
Not employed	2.07	<0.001	2.66	<0.001	1.35	<0.001	1.05	0.404	0.78	<0.001	1.10	0.071
Residential region												
West Germany	Ref.		Ref.		Ref.		Ref.		Ref.		Ref.	
East Germany	0.99	0.868	0.97	0.706	1.04	0.465	1.06	0.208	1.01	0.836	1.10	0.053

Table 4. Results of the logistic regressions with interaction of parental status and age for men (odds ratios, *p* values; age (metric) centered at the minimum of 18 years).

Men	Poor Self-Rated General Health		Depression		Back Pain		Overweight		Smoking		Sporting Inactivity	
	OR	***p***	**OR**	***p***	**OR**	***p***	**OR**	***p***	**OR**	***p***	**OR**	***p***
Parental status: Child in household												
No child	Ref.		Ref.		Ref.		Ref.		Ref.		Ref.	
Child	1.53	0.022	0.70	0.267	1.13	0.520	1.27	0.055	3.98	<0.001	4.15	<0.001
Age	1.12	<0.001	1.19	<0.001	1.09	<0.001	1.11	<0.001	1.09	<0.001	1.17	<0.001
Age#Age	0.999	0.007	0.997	<0.001	0.999	0.005	0.999	<0.001	0.998	<0.001	0.997	<0.001
Child#Age												
No child#Age									Ref.		Ref.	
Child#Age									0.886	<0.001	0.898	0.001
Child#Age#Age												
No child#Age#Age	Ref.		Ref.		Ref.		Ref.		Ref.		Ref.	
Child#Age#Age	0.999	<0.001	1.001	0.175	0.9997	0.164	0.9996	0.006	1.002	0.002	1.002	0.038
Pre-school child in household												
No	Ref.		Ref.		Ref.		Ref.		Ref.		Ref.	
Yes	0.75	0.010	0.59	0.007	0.90	0.356	0.93	0.332	0.81	0.015	1.04	0.651
Partner status												
Partner	Ref.		Ref.		Ref.		Ref.		Ref.		Ref.	
No partner	1.14	0.076	1.57	<0.001	0.84	0.028	0.82	<0.001	1.20	0.001	1.17	0.009
Socioeconomic status												
High	Ref.		Ref.		Ref.		Ref.		Ref.		Ref.	
Middle	2.48	<0.001	1.57	<0.001	1.77	<0.001	1.46	<0.001	1.77	<0.001	2.53	<0.001
Low	3.93	<0.001	2.04	<0.001	2.43	<0.001	2.00	<0.001	2.55	<0.001	4.53	<0.001
Employment status												
Full-time	Ref.		Ref.		Ref.		Ref.		Ref.		Ref.	
Part-time	1.61	<0.001	2.27	<0.001	1.28	0.024	0.68	<0.001	0.93	0.326	0.90	0.230
Not employed	3.47	<0.001	3.99	<0.001	2.04	<0.001	0.78	<0.001	0.87	0.035	0.98	0.744
Residential region												
West Germany	Ref.		Ref.		Ref.		Ref.		Ref.		Ref.	
East Germany	0.88	0.048	0.75	0.009	0.90	0.134	0.92	0.083	1.11	0.034	1.16	0.007

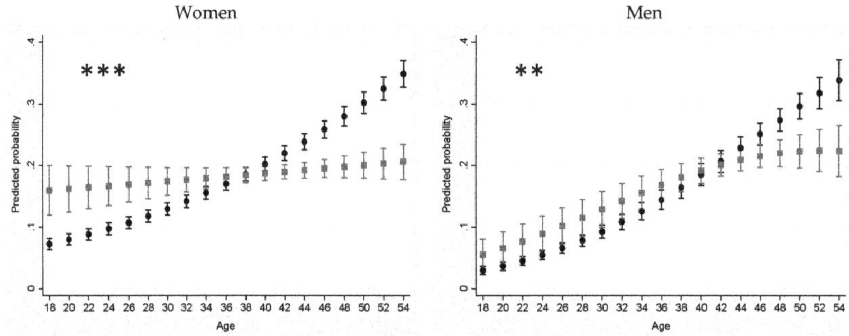

(**a**) Poor self-rated general health

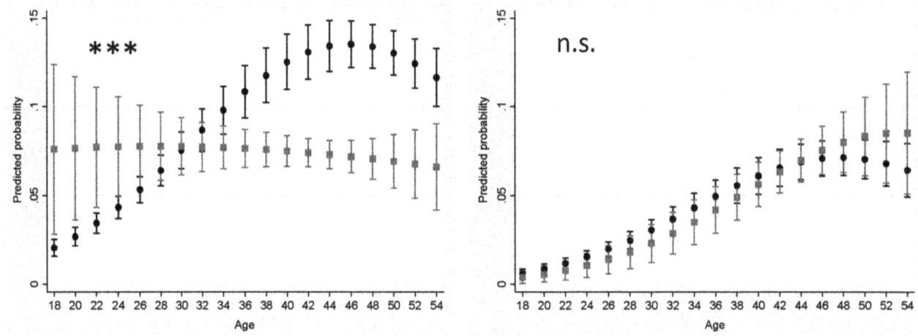

(**b**) Depression

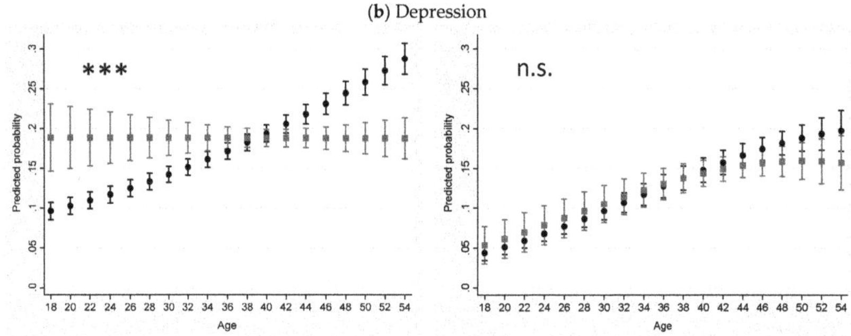

(**c**) Back pain

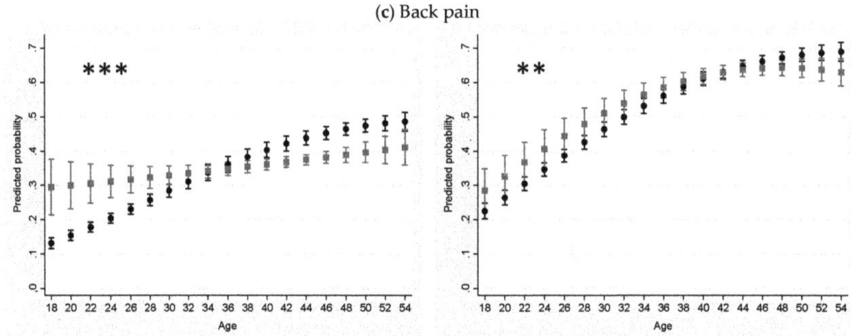

(**d**) Overweight

Figure 1. *Cont.*

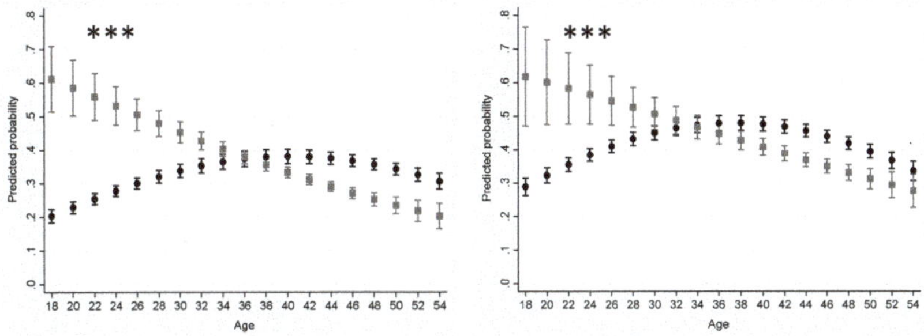

(e) Smoking

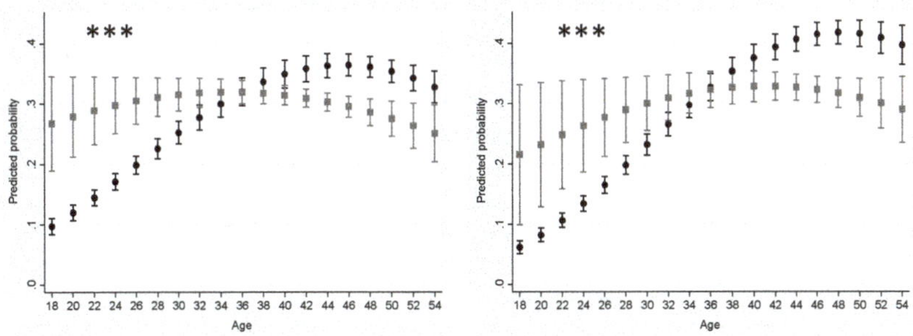

(f) Sporting inactivity

Figure 2. Health of women and men by parental status and age (in completed years) (predicted probabilities in %, 95% CI, adjusted for partner, socioeconomic and employment statuses, pre-school child in the household, region of residence). ■ Child in the household; • No child in the household; Interaction between parental status and age: * $p < 0.05$; ** $p < 0.01$; *** $p < 0.001$; n.s.: not significant.

4. Discussion

This paper examines whether there are associations between living with underage children and health as well as health behavior and whether these associations vary with age.

Without a differentiation according to age, a rather inconsistent picture emerges with regard to the association between parental status and health. For each health outcome, prevalence not stratified for age, nevertheless, largely coincides with the results of international research: women and men living with children are more often inactive in sports and more often overweight than women and men without children [26–28,31,33,37]. In contrast, women and men living with children smoke less often than childless women and men [45,48]. Moreover, mothers in particular often rated their general health better than women in childless households [5,6,9,10]. Mothers also suffer less from depression than women living without children, while men show no association between parental status and depression. This gender-differentiated pattern is also found for depression in Nomaguchi and Milkie [1] and Kalucza et al. [19], but not in Helbig et al. [18].

When age is taken into account, however, the present analysis reveals a rather clear pattern: the association between parental status and health or health behavior varies greatly with age. This is particularly true for women, where younger mothers are less healthy or behave less healthily and older mothers are healthier or behave more healthily than women of the same age without children. In the age between the end of 20 and the beginning of 40, however, there is only little or no difference in health by parental status. In men, this pattern is similarly pronounced only in health behavior (smoking and sporting inactivity) and slightly less pronounced in self-rated general health and overweight, but not in depression and back pain. This age gradient is consistent with some existing study results, which have also found poorer health for parents in the younger age group and better health or well-being for parents in the older age group than for nonparents of the same age [73,76,78,79,84]. The differences by sex described here are also found in the study by Margolis and Myrskylä [79] in the way that the association between parental status and happiness is greater among women aged 40 and over than among men. In contrast, evaluations of the insured-person data of the Techniker Krankenkasse [78] have shown the age effect particularly among men. However, this could be related to the fact that parenthood is represented in the insured-person data by the indicator of family-insured children, and this indicator identifies women with children worse than men with children. The result that the age differences in the association between co-residence with children and health in our analysis are more pronounced among women than among men could possibly be explained by the fact that the everyday life of women is influenced to a greater extent by living with children than that of men. This is reflected, for example, in greater responsibility for the care and upbringing of children as well as in greater problems in reconciling family and work life [79,92]. All in all, when comparing our findings with other study results, it must be taken into account that parenthood can be defined in different ways (fertility history versus living with children).

It is to be discussed whether the results reported here are to be interpreted as age effects or cohort effects. The observed age effect in the association between parental status and health could be explained on the one hand by the fact that the effect of parental caregiving changes with aging or varies between different phases of life. On the other hand, the described differences by age could also indicate cohort (or period) effects. Margolis and Myrskylä [79] can show, for the association between parenthood and happiness in a comparison of two cohorts, that the age gradient can be measured in both cohorts and thus cannot be attributed to cohort effects.

However, it can be assumed that it is less the age itself than the phase of life and the respective social context that have an influence on whether parental caregiving strains or enhances health. It can be presumed that, especially for young mothers and fathers, the close temporal link between education or entry into the labor market and the social role of parenthood can lead to reconciliation problems that can be accompanied by poorer health. Studies have shown, for example, that young mothers are often still in education at the time of their first child's birth, have not yet established themselves professionally and live in less stable partnerships, which often leads to social disadvantages in the further course of their lives [80,93]. Although Arnett [94] describes "emerging adulthood" as a "roleless role" in which young women and men have a wide range of opportunities and are much less restricted by role obligations than in all other phases of life, this does not apply equally to men and especially to women, who are already parents in this phase of life. In contrast to partnership, cohabitation, education or occupation, parenthood is not reversible. However, it is also quite conceivable that young women in particular, who have problems finding a satisfactory education and employment (possibly also due to poor health), consciously opt for the early formation of a family and a more family-oriented life plan [95]. Since studies also show that women who became mothers at an early age have experienced strong social disadvantages in childhood [93,96–98], it can be assumed that this is also reflected in poor health (social selection processes).

The analysis has some limitations. A main limitation is that the present analysis was carried out with cross-sectional data that do not allow statements regarding whether parental caregiving influences health or whether the health status influences the probability of starting a family (reverse causality or selection processes). In addition, the results can be distorted by unobserved heterogeneity in that—in addition to the control variables included here—other unobserved personality traits and social factors influence both the health status and the probability of parenthood, but in fact there is no causal relationship between parenthood and health [75,99]. In a review on parenthood and happiness, Hansen [75] concludes that findings from cross-sectional studies on the association between parental status and well-being can only be attributed to a small extent to selection processes and unobserved personality traits. Kalucza et al. [19] also came to the conclusion in longitudinal analyses regarding the association between parenthood and mental health that there is no evidence of selection processes in women, but there is in men. Myrskylä and Margolis [100] even assume that the effect of parenthood is underestimated rather than overestimated in cross-sectional studies, because they found stronger positive effects of parenthood on happiness in Germany in longitudinal analyses than in cross-sectional analyses. However, the results presented here do not clarify to what extent the bio-psycho-social situation has an influence on whether and at what age women and men have children (selectivity), or whether parenthood in different phases of life (also taking into account the number and age of children) strains or enhances health (causality) [4,101].

A further limitation is that the analysis presented here is based on data on persons in the common household; it can therefore only reflect social parenthood. Parents who do not live with their own children (e.g., after separation or divorce) cannot be identified. The same applies to parents whose children have already left their parents' home. Particularly in the group of older respondents who do not live with a child in the household, it is therefore not possible to differentiate between childless women and men and parents whose children had already left the parental household. Therefore, it cannot be ruled out that some of the reported results are confounded by the fact that women and men who had already become parents around the age of 20 and who may be less healthy are responsible for the poorer health status of women and men without children in the household over the age of 40.

Furthermore, no information is available on the age of the respondents at the birth of the first child. It can be assumed that—in addition to the phase of life in which the women and men were when taking part at the survey—the age of the first child's birth is also a decisive influencing factor, which is, moreover, associated with education and socio-economic situation [102]. Moreover, it could be important for health status whether parenthood was planned or childlessness was intended—especially since there are clear links with education [75]. Other important influencing factors can be assumed to be the quality of the parent-child relationship [22,103] or the parenting stage [104]. Simon and Caputo were able to show that parents whose children were 30 years and older reported better mental and physical health than parents of younger children [104].

There was also no analysis of whether the differences by age in the association between parental and health status varies with partner, employment or socioeconomic statuses, nor the region of residence (moderation effects). However, in a study by Margolis and Myrskylä [79] the moderating effect of income on age differences in the association between parenthood and happiness was small. It was only found in the group aged under-20 and in parents with three or more children. The authors have also shown that the age gradient in the association between parenthood and happiness is not moderated by partnership [79]. A moderating effect on age differences in the association between parental status and health could also be assumed with regard to employment status; however, there are no studies on this issue yet. For Germany in particular, it is also of interest for further analyses as to what extent there are differences between East and West Germany. Until the fall of communism in 1989, there were clear differences in reproductive biographies (e.g., a significantly younger age of mothers at the birth of their first child in East Germany). However, it can be assumed that this is more important with regard to the health status of older women and men, because since reunification there has been a convergence of reproductive biographies in East and West Germany [102]. In a longitudinal analysis using data from the German Socio-Economic Panel, Hank [105] was able to show that early motherhood in women aged 50 and over has been associated with poorer physical health in Western Germany, while late motherhood has been associated with lower psychological well-being in Eastern Germany. A comparison of Western and Eastern European countries has come to a similar finding. In both Western and Eastern Europe, early parenthood has been associated with poorer general health in later life, but the interrelationships have been much less pronounced in Eastern than in Western Europe [96].

Overall, a deeper understanding of the association between parenthood and health throughout the life course requires longitudinal analyses and a more precise distinction between fertility history and phases of living with children or social parenting.

5. Conclusions

The present analysis shows that the association between living with children and the health and health behavior of women and men varied greatly with age. This was particularly true for women, where younger mothers were less healthy or behaved in less healthy ways and older mothers were healthier or had healthier behaviors than women of the same age without children. Among men, this pattern was similar only for health behavior and it was slightly less pronounced for self-rated general health and overweight.

Thus, living together with children is an important social determinant of health strongly associated with age which is relevant for both research and health promotion.

With regard to health promotion for parents, a mixture of measures at the individual, the communal as well as the societal level seems promising—especially targeting young parental caregivers up to the age of 25. As Myrskylä and Margolis [100] can verify for Germany that changes in family policy (extension of parental leave benefits in 2007) were accompanied by a significant increase in the well-being of parents in the period around the birth of a child, measures at the societal level could play an important role in improving the health of parents [106]. Thus, Lee et al. [107] have shown that California's paid family leave policy had positive impacts on several parental health outcomes such as self-rated health, overweight, psychological distress, and alcohol consumption. As there are large differences between countries in family leave policies, the effects on parents' health would be an interesting topic for future studies [108,109]. It would also be of interest to analyze whether there are differences in the effects of the policy measures by age.

Nomaguchi and Milkie [3] emphasize the importance of parents' welfare for the parents themselves as well as for child development and the overall health of society.

Author Contributions: Both authors contributed to the study conception and design. Data analysis was performed by P.R. The first draft of the manuscript was written by P.R. and E.v.d.L. commented on previous versions of the manuscript. All authors have read and agreed to the published version of the manuscript.

Funding: The "German Health Update" (GEDA) is part of the Federal Health Monitoring (FHM) in Germany. The FHM is administered by the Robert Koch Institute (RKI) on behalf of the German Federal Ministry of Health. The ministry finances the RKI and gives substantial funds for the FHM. The funders had no role in study design, data collection and analysis, decision to publish, or preparation of the manuscript. All authors are employees of the RKI. No one of the authors received specific funding for this work.

Acknowledgments: We thank Analisa Avila, ELS, of Edanz Group (www.edanzediting.com/ac) for editing a draft of this manuscript.

Conflicts of Interest: The authors declare no conflict of interest.

References

1. Nomaguchi, K.M.; Milkie, M.A. Costs and rewards of children: The effects of becoming a parent on adults' lives. *J. Marriage Fam.* **2003**, *65*, 356–374. [CrossRef]
2. Hank, K.; Steinbach, A. Families and Health: A Review. In *A Demographic Perspective on Gender, Family and Health in Europe*; Doblhammer, G., Gumà, J., Eds.; Springer International Publishing: Cham, Switzerland, 2018; pp. 23–39.
3. Nomaguchi, K.; Milkie, M.A. Parenthood and well-being: A decade in review. *J. Marriage Fam.* **2020**, *82*, 198–223. [CrossRef]
4. McMunn, A.; Bartley, M.; Hardy, R.; Kuh, D. Life course social roles and women's health in mid-life: Causation or selection? *J. Epidemiol. Community Health* **2006**, *60*, 484–489. [CrossRef] [PubMed]
5. Fokkema, T. Combining a job and children: Contrasting the health of married and divorced women in the Netherlands? *Soc. Sci. Med.* **2002**, *54*, 741–752. [CrossRef]
6. Hewitt, B.; Baxter, J.; Western, M. Family, work and health: The impact of marriage, parenthood and employment on self-reported health of Australian men and women. *J. Sociol.* **2006**, *42*, 61–78. [CrossRef]
7. Khlat, M.; Sermet, C.; Le Pape, A. Women's health in relation with their family and work roles: France in the early 1990s. *Soc. Sci. Med.* **2000**, *50*, 1807–1825. [CrossRef]

8. Kostiainen, E.; Martelin, T.; Kestilä, L.; Martikainen, P.; Koskinen, S. Employee, partner, and mother: Woman's three roles and their implications for health. *J. Fam Issues* **2009**, *30*, 1122–1150. [CrossRef]
9. Lahelma, E.; Arber, S.; Kivelä, K.; Roos, E. Multiple roles and health among British and Finnish women: The influence of socioeconomic circumstances. *Soc. Sci. Med.* **2002**, *54*, 727–740. [CrossRef]
10. Roos, E.; Burström, B.; Saastamoinen, P.; Lahelma, E. A comparative study of the patterning of women's health by family status and employment status in Finland and Sweden. *Soc. Sci. Med.* **2005**, *60*, 2443–2451. [CrossRef]
11. von der Lippe, E.; Rattay, P. Association of partner, parental, and employment statuses with self-rated health among German women and men. *SSM Popul. Health* **2016**, *2*, 390–398. [CrossRef]
12. Floderus, B.; Hagman, M.; Aronsson, G.; Marklund, S.; Wikman, A. Self-reported health in mothers: The impact of age, and socioeconomic conditions. *Women Health* **2008**, *47*, 63–86. [CrossRef] [PubMed]
13. Namkung, E.H.; Mitra, M.; Nicholson, J. Do disability, parenthood, and gender matter for health disparities? A US population-based study. *Disabil. Health J.* **2019**, *12*, 594–601. [CrossRef] [PubMed]
14. Bird, C.E. Gender differences in the social and economic burdens of parenting and psychological distress. *J. Marriage Fam.* **1997**, *59*, 809–823. [CrossRef]
15. Bures, R.M.; Koropeckyj-Cox, T.; Loree, M. Childlessness, parenthood, and depressive symptoms among middle-aged and older adults. *J. Fam. Issues* **2009**, *30*, 670–687. [CrossRef]
16. Evenson, R.J.; Simon, R.W. Clarifying the relationship between parenthood and depression. *J. Health Soc. Behav.* **2005**, *46*, 341–358. [CrossRef] [PubMed]
17. Holton, S.; Fisher, J.; Rowe, H. Motherhood: Is it good for women's mental health? *J. Reprod. Infant Psychol.* **2010**, *28*, 223–239. [CrossRef]
18. Helbig, S.; Lampert, T.; Klose, M.; Jacobi, F. Is parenthood associated with mental health? Findings from an epidemiological community survey. *Soc. Psychiatry Psychiatr. Epidemiol.* **2006**, *41*, 889–896. [CrossRef]
19. Kalucza, S.; Hammarström, A.; Nilsson, K. Mental health and parenthood—A longitudinal study of the relationship between self-reported mental health and parenthood. *Health Sociol. Rev.* **2015**, *24*, 283–296. [CrossRef]
20. Mirowsky, J.; Ross, C.E. Depression, parenthood, and age at first birth. *Soc. Sci. Med.* **2002**, *54*, 1281–1298. [CrossRef]
21. Plaisier, I.; Beekman, A.T.F.; De Bruijn, J.G.M.; De Graaf, R.; Ten Have, M.; Smit, J.H.; van Dyck, R.; Penninx, B.W.J.H. The effect of social roles on mental health: A matter of quantity or quality? *J. Affect. Disord.* **2008**, *111*, 261–270. [CrossRef]
22. Pudrovska, T. Psychological implications of motherhood and fatherhood in midlife: Evidence from sibling models. *J. Marriage Fam.* **2008**, *70*, 168–181. [CrossRef]
23. Stöbel-Richter, Y.; Brähler, E.; Zenger, M. Lebenszufriedenheit und psychische Gesundheit von Müttern und Nichtmüttern im Vergleich. Repräsentative Ergebnisse [Life satisfaction and mental health of mothers and non-mothers compared. Representative results]. In *Was sagen die Mütter? Qualitative und quantitative Forschung rund um Schwangerschaft, Geburt und Wochenbett [What do the mothers say? Qualitative and quantitative research on pregnancy, childbirth and the puerperium]*; Makowsky, K., Schücking, B., Eds.; Beltz Juventa: Weinheim, Basel, Germany, 2013; pp. 305–324.
24. Giesselmann, M. Mutterschaft geht häufig mit verringertem mentalem Wohlbefinden einher [Motherhood is often associated with reduced mental well-being]. *DIW Wochenber. Nr. 35* **2018**. [CrossRef]
25. Holden, L.; Hockey, R.; Ware, R.S.; Lee, C. Mental health-related quality of life and the timing of motherhood: A 16-year longitudinal study of a national cohort of young Australian women. *Qual. Life Res.* **2018**, *27*, 923–935. [CrossRef] [PubMed]
26. Bastian, L.A.; West, N.A.; Corcoran, C.; Munger, R.G. Number of children and the risk of obesity in older women. *Prev. Med.* **2005**, *40*, 99–104. [CrossRef] [PubMed]
27. Berge, J.; Larson, N.; Bauer, K.W.; Neumark-Sztainer, D. Are parents of young children practicing healthy nutrition and physical activity behaviors? *Pediatrics* **2011**, *127*, 881–887. [CrossRef] [PubMed]
28. Brown, W.; Hockey, R.; Dobson, A. Effects of having a baby on weight gain. *Am. J. Prev. Med.* **2010**, *38*, 163–170. [CrossRef]
29. Brown, W.; Trost, S.G. Life transitions and changing physical activity patterns in young women. *Am. J. Prev. Med.* **2003**, *25*, 140–143. [CrossRef]

30. Burke, V. Changes in health-related behaviours and cardiovascular risk factors in young adults: Associations with living with a partner. *Prev. Med.* **2004**, *39*, 722–730. [CrossRef]
31. Garfield, C.F.; Duncan, G.; Gutina, A.; Rutsohn, J.; McDade, T.W.; Adam, E.K.; Coley, R.L.; Chase-Lansdale, P.L. Longitudinal study of body mass index in young males and the transition to fatherhood. *Am. J. Men's Health* **2015**. [CrossRef]
32. Syrda, J. The impact of marriage and parenthood on male body mass index: Static and dynamic effects. *Soc. Sci. Med.* **2017**, *186*, 148–155. [CrossRef]
33. Bellows-Riecken, K.H.; Rhodes, R.E. A birth of inactivity? A review of physical activity and parenthood. *Prev. Med.* **2008**, *46*, 99–110. [CrossRef] [PubMed]
34. Hull, E.E.; Rofey, D.L.; Robertson, R.J.; Nagle, E.F.; Otto, A.D.; Aaron, D.J. Influence of marriage and parenthood on physical activity: A 2-year prospective analysis. *J. Phys. Act. Health* **2010**, *7*, 577–583. [CrossRef] [PubMed]
35. Perales, F.; Del Pozo-Cruz, J.; Cruz, B.D.P. Long-term dynamics in physical activity behaviour across the transition to parenthood. *Int. J. Public Health* **2015**, *60*, 301–308. [CrossRef] [PubMed]
36. Pot, N.; Keizer, R. Physical activity and sport participation: A systematic review of the impact of fatherhood. *Prev. Med. Rep.* **2016**, *4*, 121–127. [CrossRef]
37. Laroche, H.; Wallace, R.B.; Snetselaar, L.; Hillis, S.; Cai, X.; Steffen, L.M. Weight gain among men and women who have a child enter their home. *J. Acad. Nutr. Diet.* **2013**, *113*, 1504–1510. [CrossRef]
38. Abell, L.P.; A Tanase, K.; Gilmore, M.L.; E Winnicki, A.; Holmes, V.L.; Hartos, J. Do physical activity levels differ by number of children at home in women aged 25–44 in the general population? *Women's Health* **2019**, *15*. [CrossRef]
39. Candelaria, J.I.; Sallis, J.F.; Conway, T.L.; Saelens, B.E.; Frank, L.D.; Slymen, D.J. Differences in physical activity among adults in households with and without children. *J. Phys. Act. Health* **2012**, *9*, 985–995. [CrossRef]
40. Roos, E.; Lahelma, E.; Virtanen, M.; Prättälä, R.; Pietinen, P. Gender, socioeconomic status and family status as determinants of food behaviour. *Soc. Sci. Med.* **1998**, *46*, 1519–1529. [CrossRef]
41. Laroche, H.H.; Hofer, T.P.; Davis, M.M. Adult fat intake associated with the presence of children in households: Findings from NHANES III. *J. Am. Board Fam. Med.* **2007**, *20*, 9–15. [CrossRef]
42. Balan, S.; Widner, G.; Chen, H.-J.; Hudson, D.; Gehlert, S.; Price, R.K. Motherhood, psychological risks, and Resources in relation to alcohol use disorder: Are there differences between black and white women? *ISRN Addict.* **2014**. [CrossRef]
43. Hajema, K.-J.; Knibbe, R.A. Changes in social roles as predictors of changes in drinking behaviour. *Addiction* **1998**, *93*, 1717–1727. [CrossRef] [PubMed]
44. Kuntsche, S.; Knibbe, R.A.; Kuntsche, E.; Gmel, G. Housewife or working mum-each to her own? The relevance of societal factors in the association between social roles and alcohol use among mothers in 16 industrialized countries. *Addiction* **2011**, *106*, 1925–1932. [CrossRef] [PubMed]
45. Oesterle, S.; Hawkins, J.D.; Hill, K.G. Men's and women's pathways to adulthood and associated substance misuse. *J. Stud. Alcohol Drugs* **2011**, *72*, 763–773. [CrossRef] [PubMed]
46. Paradis, C. Parenthood, drinking locations and heavy drinking. *Soc. Sci. Med.* **2011**, *72*, 1258–1265. [CrossRef] [PubMed]
47. Waldron, I.; Lye, D. Family roles and smoking. *Am. J. Prev. Med.* **1989**, *5*, 136–141. [CrossRef]
48. von der Lippe, E.; Rattay, P. Health-Risk Behaviour of Women and Men. Differences According to Partnership and Parenthood. Results of the German Health Update (GEDA) Survey 2009–2010. In *A Demographic Perspective on Gender. Family and Health in Europe*; Doblhammer, G., Gumà, J., Eds.; Springer: Cham, Switzerland, 2018; pp. 233–261.
49. Nomaguchi, K.M. Parenthood and psychological well-being: Clarifying the role of child age and parent-child relationship quality. *Soc. Sci. Res.* **2012**, *41*, 489–498. [CrossRef] [PubMed]
50. Hank, K.; Wagner, M. Parenthood, Marital Status, and Well-Being in Later Life: Evidence from SHARE. *Soc. Indic. Res.* **2013**, *114*, 639–653. [CrossRef]
51. Becchetti, L.; Ricca, E.G.; Pelloni, A. The Paradox of Children and Life Satisfaction. *Soc. Indic. Res.* **2013**, *111*, 725–751. [CrossRef]
52. Falci, C.; Mortimer, J.T.; Noel, H. Parental timing and depressive symptoms in early adulthood. *Adv. Life Course Res.* **2010**, *15*, 1–10. [CrossRef]

53. Kahn, R.S.; Certain, L.; Whitaker, R.C. A Reexamination of Smoking Before, During, and After Pregnancy. *Am. J. Public Health* **2002**, *92*, 1801–1808. [CrossRef]
54. Avison, W.R.; Davies, L. Family structure, gender, and health in the context of the life course. *J. Gerontol. B Psychol. Sci. Soc. Sci.* **2005**, *60*, S113–S116. [CrossRef] [PubMed]
55. Floderus, B.; Hagman, M.; Aronsson, G.; Marklund, S.; Wikman, A. Medically certified sickness absence with insurance benefits in women with and without children. *Eur. J. Public Health* **2012**, *22*, 85–92. [CrossRef] [PubMed]
56. Rattay, P.; von der Lippe, E.; Borgmann, L.-S.; Lampert, T. Gesundheit von alleinerziehenden Müttern und Vätern in Deutschland [Health of single mothers and fathers in Germany]. *JoHM* **2017**, 2. [CrossRef]
57. Graham, M. Is being childless detrimental to a woman's health and well-being across her life course? *Women's Health Issues* **2015**, *25*, 176–184. [CrossRef] [PubMed]
58. Bartlett, E.E. The effects of fatherhood on the health of men: A review of the literature. *J. Men's Health* **2004**, *1*, 159–169. [CrossRef]
59. Blackburn, C.; Bonas, S.; Spencer, N.; Dolan, A.; Coe, C.; Moy, R. Smoking behaviour change among fathers of new infants. *Soc. Sci. Med.* **2005**, *61*, 517–526. [CrossRef]
60. Condon, J.T.; Corkindale, C.J.; Boyce, P. The First-Time Fathers Study: A prospective study of the mental health and wellbeing of men during the transition to parenthood. *Aust. N. Z. J. Psychiatry* **2004**, *38*, 56–64. [CrossRef]
61. Einiö, E.; Nisén, J.; Martikainen, P. Is young fatherhood causally related to midlife mortality? A sibling fixed-effect study in Finland. *J. Epidemiol. Community Health* **2015**, *69*, 1077–1082. [CrossRef]
62. Eisenberg, M.L.; Park, Y.; Hollenbeck, A.R.; Lipshultz, L.I.; Schatzkin, A.; Pletcher, M.J. Fatherhood and the risk of cardiovascular mortality in the NIH-AARP Diet and Health Study. *Hum. Reprod.* **2011**, *26*, 3479–3485. [CrossRef]
63. Meadows, S.O. Family structure and fathers' well-being: Trajectories of mental health and self-rated health. *J. Health Soc. Behav.* **2009**, *50*, 115–131. [CrossRef]
64. Kuntsche, S.; Kuendig, H.; Bloomfield, K.; Gmel, G.; Knibbe, R.A.; Kramer, S.; Grittner, U. Gender and cultural differences in the association between family roles, social stratification, and alcohol use: A European cross-cultural analysis. *Alcohol Alcohol* **2006**, *41*, i37–i46. [CrossRef] [PubMed]
65. Kuntsche, S.; Knibbe, R.A.; Gmel, G. Parents' alcohol use: Gender differences in the impact of household and family chores. *Eur. J. Public Health* **2012**, *22*, 894–899. [CrossRef] [PubMed]
66. Mastekaasa, A. Parenthood, gender and sickness absence. *Soc. Sci. Med.* **2000**, *50*, 1827–1842. [CrossRef]
67. Nomaguchi, K.M.; Bianchi, S.M. Exercise Time: Gender Differences in the Effects of Marriage, Parenthood, and Employment. *J. Marriage Fam.* **2004**, *66*, 413–430. [CrossRef]
68. Shafer, K.; Pace, G.T. Gender differences in depression across parental roles. *Soc. Work* **2015**, *60*, 115–125. [CrossRef]
69. Huijts, T.; Kraaykamp, G.; Subramanian, S.V. Childlessness and Psychological Well-Being in Context: A Multilevel Study on 24 European Countries. *Eur. Sociol. Rev.* **2013**, *29*, 32–47. [CrossRef]
70. Burström, B.; Whitehead, M.; Clayton, S.; Fritzell, S.; Vannoni, F.; Costa, G. Health inequalities between lone and couple mothers and policy under different welfare regimes—The example of Italy, Sweden and Britain. *Soc. Sci. Med.* **2010**, *70*, 912–920. [CrossRef]
71. Hansen, T.; Slagsvold, B.; Moum, T. Childlessness and psychological well-being in midlife and old age: An examination of parental status effects across a range of outcomes. *Soc. Indic. Res.* **2009**, *94*, 343–362. [CrossRef]
72. Kendig, H.; Dykstra, P.A.; Van Gaalen, R.I.; Melkas, T. Health of aging parents and childless individuals. *J. Fam. Issues* **2007**, *28*, 1457–1486. [CrossRef]
73. Tanaka, K.; Johnson, N.E. Childlessness and mental well-being in a global context. *J. Fam. Issues* **2014**. [CrossRef]
74. Glass, J.; Simon, R.W.; Andersson, M.A. Parenthood and happiness: Effects of work-family reconciliation policies in 22 OECD countries. *AJS* **2016**, *122*, 886–929. [CrossRef] [PubMed]
75. Hansen, T. Parenthood and Happiness: A Review of Folk Theories Versus Empirical Evidence. *Soc. Indic. Res.* **2012**, *108*, 29–64. [CrossRef]

76. Nelson-Coffey, S.K.; Kushlev, K.; Lyubomirsky, S. The pains and pleasures of parenting: When, why, and how is parenthood associated with more or less well-being? *Psychol. Bull.* **2014**, *140*, 846–895. [CrossRef] [PubMed]
77. Sachs-Ericsson, N.; Ciarlo, J.A. Gender, Social Roles, and Mental Health: An Epidemiological Perspective. *Sex. Roles* **2000**, *43*, 605–628. [CrossRef]
78. Techniker, K. (Ed.) *Gesundheitsreport 2016. Gesundheit Zwischen Beruf und Familie [Health Report 2016. Health between Work and Family]*; Techniker Krankenkasse: Hamburg, Germany, 2016.
79. Margolis, R.; Myrskylä, M. A global perspective on happiness and fertility. *Popul. Dev. Rev.* **2011**, *37*, 29–56. [CrossRef]
80. Schlücker, F.U.; Blumenfelder, A.R. Effects of age at first birth on health of mothers aged 45 to 56. *J. Fam. Stud.* **2014**, *26*, 347–371. [CrossRef]
81. Henretta, J.C.; Grundy, E.M.; Okell, L.C.; Wadsworth, M.E. Early motherhood and mental health in midlife: A study of British and American cohorts. *Aging Ment. Health* **2008**, *12*, 605–614. [CrossRef]
82. Jaffe, D.H.; Kogan, L.; Manor, O.; Gielchinsky, Y.; Dior, U.; Laufer, N. Influence of late-age births on maternal longevity. *Ann. Epidemiol.* **2015**, *25*, 387–391. [CrossRef]
83. Myrskylä, M.; Margolis, R. Happiness: Before and after the kids. *Demography* **2014**, *51*, 1843–1866. [CrossRef]
84. Sabbath, E.L.; Mejía-Guevara, I.; Noelke, C.; Berkman, L.F. The long-term mortality impact of combined job strain and family circumstances: A life course analysis of working American mothers. *Soc. Sci. Med.* **2015**, *146*, 111–119. [CrossRef]
85. Carlson, D.L. Explaining the curvilinear relationship between age at first birth and depression among women. *Soc. Sci. Med.* **2011**, *72*, 494–503. [CrossRef] [PubMed]
86. Umberson, D.; Liu, H.; Mirowsky, J.; Reczek, C. Parenthood and trajectories of change in body weight over the life course. *Soc. Sci. Med.* **2011**, *73*, 1323–1331. [CrossRef] [PubMed]
87. Becchetti, L.; Ricca, E.G.; Pelloni, A. *Children, Happiness and Taxation*; SOEPpapers on Multidisciplinary Panel Data Research No. 230; German Institute for Economic Research (DIW): Berlin, Germany, 2009.
88. Lange, C.; Jentsch, F.; Allen, J.; Hoebel, J.; Kratz, A.L.; von der Lippe, E.; Müters, S.; Schmich, P.; Thelen, J.; Wetzstein, M.; et al. Data resource profile: German Health Update (GEDA)—The health interview survey for adults in Germany. *Int. J. Epidemiol.* **2015**, *44*, 442–450. [CrossRef] [PubMed]
89. Gabler, S.; Häder, S. Generierung von Telefonstichproben mit TelSuSa [Generating telephone samples using TelSuSa]. *Zuma-Nachr.* **1999**, *44*, 138–143.
90. Lampert, T.; Kroll, L.; Müters, S.; Stolzenberg, H. Messung des sozioökonomischen Status in der Studie „Gesundheit in Deutschland aktuell "(GEDA) [Measurement of socio-economic status in the „German Health Update" (GEDA)]. *Bundesgesundheitsblatt Gesundh. Gesundh.* **2013**, *56*, 131–143. [CrossRef]
91. Jann, B. Predictive Margins and Marginal Effects in Stata. In Proceedings of the 11th German Stata Users Group Meeting, Potsdam, Germany, 7 June 2013.
92. Ruppanner, L.; Perales, F.; Baxter, J. Harried and Unhealthy? Parenthood, Time Pressure, and Mental Health. *J. Marriage Fam.* **2019**, *81*, 308–326. [CrossRef]
93. Boden, J.M.; Fergusson, D.M.; Horwood, L.J. Early motherhood and subsequent life outcomes. *J. Child. Psychol. Psychiatry* **2008**, *49*, 151–160. [CrossRef]
94. Arnett, J.J. Emerging adulthood. A theory of development from the late teens through the twenties. *Am. Psychol.* **2000**, *55*, 469–480. [CrossRef]
95. Schoon, I.; Chen, M.; Kneale, D.; Jager, J. Becoming adults in Britain: Lifestyles and wellbeing in times of social change. *Longitud. Life Course Stud.* **2012**, *3*, 173–189. [CrossRef]
96. Grundy, E.; Foverskov, E. Age at First Birth and Later Life Health in Western and Eastern Europe. *Popul. Dev. Rev.* **2016**, *42*, 245–269. [CrossRef]
97. Raymo, J.M.; Carlson, M.; VanOrman, A.; Lim, S.; Perelli-Harris, B.; Iwasawa, M. Educational differences in early childbearing: A cross-national comparative study. *Demogr. Res.* **2015**, *33*, 65–92. [CrossRef] [PubMed]
98. Lee, C.; Ryff, C.D. Early parenthood as a link between childhood disadvantage and adult heart problems: A gender-based approach. *Soc. Sci. Med.* **2016**, *171*, 58–66. [CrossRef]
99. Pollmann-Schult, M. Parenthood and life satisfaction in Germany. *CPoS* **2013**, *38*, 85–108. [CrossRef]
100. Myrskylä, M.; Margolis, R. *Parental Benefits Improve Parental Well-Being: Evidence from a 2007 Policy Change in Germany*; MPIDR Working Paper WP-2013-010; Max Planck Institute for Demographic Research: Rostock, Germany, 2013.

101. Barclay, K.J.; Keenan, K.; Grundy, E.M.; Kolk, M.; Myrskylä, M. Reproductive history and post-reproductive mortality: A sibling comparison analysis using Swedish register data. *Soc. Sci. Med.* **2016**, *155*, 82–92. [CrossRef] [PubMed]
102. Statistisches Bundesamt [Federal Statistical Office] (Ed.) *Geburtentrends und Familiensituation in Deutschland 2012 [Birth Trends and Family Situation in Germany 2012]*; Statistisches Bundesamt [Federal Statistical Office]: Wiesbaden, Germany, 2013.
103. Koropeckyj-Cox, T. Beyond parental status: Psychological well-being in middle and old age. *J. Marriage Fam.* **2002**, *64*, 957–971. [CrossRef]
104. Simon, R.W.; Caputo, J. The Costs and Benefits of Parenthood for Mental and Physical Health in the United States: The Importance of Parenting Stage. *Soc. Ment. Health* **2019**, *9*, 296–315. [CrossRef]
105. Hank, K. Childbearing history, later-life health, and mortality in Germany. *Popul. Stud.* **2010**, *64*, 275–291. [CrossRef]
106. Bullinger, L.R. The Effect of Paid Family Leave on Infant and Parental Health in the United States. *J. Health Econ.* **2019**, *66*, 101–116. [CrossRef]
107. Lee, B.C.; Modrek, S.; White, J.S.; Batra, A.; Collin, D.F.; Hamad, R. The effect of California's paid family leave policy on parent health: A quasi-experimental study. *Soc. Sci. Med.* **2020**, *251*, 112915. [CrossRef]
108. Nandi, A.; Jahagirdar, D.; Dimitris, M.C.; Labrecque, J.A.; Strumpf, E.C.; Kaufman, J.S.; Vincent, I.; Atabay, E.; Harper, S.; Earle, A.; et al. The Impact of Parental and Medical Leave Policies on Socioeconomic and Health Outcomes in OECD Countries: A Systematic Review of the Empirical Literature. *Milbank Q* **2018**, *96*, 434–471. [CrossRef]
109. Aitken, Z.; Garrett, C.C.; Hewitt, B.; Keogh, L.; Hocking, J.; Kavanagh, A.M. The maternal health outcomes of paid maternity leave: A systematic review. *Soc. Sci. Med.* **2015**, *130*, 32–41. [CrossRef] [PubMed]

© 2020 by the authors. Licensee MDPI, Basel, Switzerland. This article is an open access article distributed under the terms and conditions of the Creative Commons Attribution (CC BY) license (http://creativecommons.org/licenses/by/4.0/).

International Journal of
Environmental Research and Public Health

MDPI

Review

Food Systems Transformation for Child Health and Well-Being: The Essential Role of Dairy

Gregory D. Miller [1,2], Mitch Kanter [2], Laurence Rycken [3], Kevin B. Comerford [4,*], Nicholas M. Gardner [5] and Katie A. Brown [1]

1 National Dairy Council, Rosemont, IL 60018-5616, USA; gregory.miller@dairy.org (G.D.M.); katie.brown@dairy.org (K.A.B.)
2 Global Dairy Platform, Rosemont, IL 60018-5616, USA; Mitch.Kanter@GlobalDairyPlatform.com
3 International Dairy Federation, 1030 Brussels, Belgium; lrycken@fil-idf.org
4 California Dairy Research Foundation, Davis, CA 95618, USA
5 U.S. Dairy Export Council, Arlington, VA 22201, USA; ngardner@usdec.org
* Correspondence: kbcomerford@ucdavis.edu

Abstract: Malnutrition, in all its forms, during the critical stages of child growth and development can have lifelong impacts on health and well-being. While most forms of malnutrition can be prevented with simple dietary interventions, both undernutrition and overnutrition remain persistent and burdensome challenges for large portions of the global population, especially for young children who are dependent on others for nourishment. In addition to dietary factors, children's health also faces the growing challenges of climate change, environmental degradation, pollution, and infectious disease. Food production and consumption practices both sit at the nexus of these issues, and both must be significantly transformed if we are to achieve the 2030 Sustainable Development Goals. Food sources (i.e., animal-source foods vs. plant-source foods), food production practices, the effects of food processing, the impacts of a more globalized food system, and food loss and waste have all been receiving growing attention in health and sustainability research and policy discussions. Much of this work points to recommendations to reduce resource-intensive animal-source foods, heavily processed foods, and foods associated with excessive waste and pollution, while simultaneously increasing plant-source options. However, some of these recommendations require a little more nuance when considered in the context of issues such as global child health. All types of foods can play significant roles in providing essential nutrition for children across the globe, and for improving the well-being and livelihoods of their families and communities. Dairy foods provide a prime example of this need for nuance, as both dairy production practices and consumption patterns vary greatly throughout the world, as do their impacts on child health and food system sustainability. The objective of this narrative review is to highlight the role of dairy in supporting child health in the context of food system sustainability. When considering child health within this context it is recommended to take a holistic approach that considers all four domains of sustainability (health, economics, society, and the environment) to better weigh trade-offs, optimize outcomes, and avoid unintended consequences. To ensure that children have access to nutritious and safe foods within sustainable food systems, special consideration of their needs must be included within the broader food systems transformation narrative.

Keywords: child health; child nutrition; dairy; food security; food systems; nutrition; sustainability; sustainable food systems

Citation: Miller, G.D.; Kanter, M.; Rycken, L.; Comerford, K.B.; Gardner, N.M.; Brown, K.A. Food Systems Transformation for Child Health and Well-Being: The Essential Role of Dairy. *IJERPH* **2021**, *18*, 10535. https://doi.org/10.3390/ijerph181910535

Academic Editors: Wioletta Zukiewicz-Sobczak, Paulina Wojtyla-Buciora, Izabela Rącka and Andrzej Wojtyla

Received: 3 September 2021
Accepted: 5 October 2021
Published: 8 October 2021

Publisher's Note: MDPI stays neutral with regard to jurisdictional claims in published maps and institutional affiliations.

Copyright: © 2021 by the authors. Licensee MDPI, Basel, Switzerland. This article is an open access article distributed under the terms and conditions of the Creative Commons Attribution (CC BY) license (https://creativecommons.org/licenses/by/4.0/).

1. Introduction

Supporting the health and well-being of children today is foundational to building a successful future for generations to come. Proper nourishment throughout childhood and adolescence is critical to supporting growth, development, and learning, as well as productivity and overall health later in life [1,2]. Despite the abundance of scientific evidence and global efforts to improve early life eating patterns, malnutrition, in all its forms, remains one of the most significant challenges to child health around the world. Global rates of undernutrition and overweight and obesity continue to rise, and one out of three children under the age of 5 is failing to thrive due to poor dietary patterns [3]. The global population is projected to grow from 7.7 billion today to nearly 10 billion by 2050, with most of this growth likely to occur in parts of the world where malnutrition is highest and healthy diets are most unaffordable [4]. Without proper nutritional interventions, the future of child health will continue to worsen, resulting in future generations being less healthy than current ones. The problem of growing global rates of childhood malnutrition does not arise in isolation, nor can it be resolved through an approach that only focuses on improving dietary patterns. Childhood malnutrition is inextricably linked to several other global sustainability issues; and while inadequate food intake and poor diet quality are both leading causes of childhood malnutrition, these are just symptoms of deeper and more systemic issues such as poverty and food insecurity. At present, food insecurity affects roughly one-quarter of the global population and has been consistently increasing at the global level for years, affecting the diets and health status of hundreds of millions of children around the world [5].

In addition to malnutrition, poverty, and food insecurity, the present generation of children must also contend with global health issues such as climate change, water scarcity, pollution, and global pandemics. These issues all affect, and are all affected by, food systems and dietary patterns. For example, COVID-19 has worsened all forms of malnutrition across the globe, causing significantly more wasting and stunting in the world's children [6]. This pandemic has also forced more than 1.6 billion children out of school around the world, putting critical school meals out of reach for many who rely on them for consistent daily nourishment [7]. While the COVID-19 pandemic will eventually become less of a threat to children's well-being, climate change will likely be more of a threat, and the devastation from both underscore the imminent need for all food system actors to prepare for future global crises by transitioning food systems to be more resilient to future shocks.

In the face of climate change and COVID-19, a 'business-as-usual' approach to food systems will be disastrous for the planet and its populations [8,9]. Multiple aspects of food systems need to drastically change if we are to achieve the 2030 Sustainable Development Goals (SDGs). A growing body of literature on healthy and sustainable diets points to recommendations to reduce resource-intensive animal-source foods, heavily processed foods, and foods associated with excessive waste and pollution, while simultaneously increasing plant-source options ([10–14]). However, some of these recommendations require a little more nuance when considered in the context of issues such as global child nutrition and health. For example, there are many plant-source and animal-source foods that are resource intensive, heavily processed, rich in nutrients to limit (i.e., salt, sugar, unhealthy fats), associated with high levels of food loss or waste, and/or linked to inordinate pollution and greenhouse gas (GHG) emissions. Yet, there are also many plant-source and animal-source food options that are minimally processed, locally available, relatively low in associated sustainability impacts, and rich in the exact nutrients that children need for proper growth, development, and immune function. For both health and sustainability purposes, the latter options should be prioritized in food systems transitions, regardless of whether they are sourced from plants or animals.

Among animal-source foods, meat and dairy are often singled out as food groups to limit in the sustainable food systems literature. However, these food groups each contain a wide variety of dietary options that are both nutritionally and sustainably heterogenous [15]. In other words, these food groups are not equivalent to each other, nor are they uniform in their makeup. Due to their nutritional and health distinctions, meat and dairy are treated very differently in global food-based dietary guidelines (FBDGs), with dairy being more commonly recommended than meat (dairy is recommended in >70% of FBDGs, while meat in recommended in <55%), and also recommended in higher frequency (dairy is recommended an average of one to four servings/day in FBDGs, meat is recommended at approximately five servings per week) [16,17]. Regarding sustainability, the differences between environmental impacts such as carbon footprint can range up to 25-fold between dairy and meat products, with intensive milk production systems being associated with some of the lowest emissions intensities among animal-source foods [15]. The global dairy sector is also linked to nearly 1 billion livelihoods, which is significantly more than any other food sector [18], and leaders across the sector are currently financing a growing number of science-based efforts toward more sustainable food systems [19–22]. Taken together, these factors make dairy an intriguing case study for its role in the health and sustainability of food systems for present and future generations. Therefore, the objective of this review is to highlight the roles of dairy in supporting child health and well-being in the context of food systems transformation.

2. The Global State of Child Nutrition

The present state of our global food systems and dietary patterns are failing to help millions of children meet their nutritional needs. Globally, 45% of deaths in children under the age of 5 are associated with undernutrition [23]. 144 million children among this age group are stunted, 47 million suffer from wasting, and 38 million are overweight [24]. While the prevalence of stunting has been declining globally since 2000, its prevalence is still increasing in regions such as Africa. At present, wasting is highest in Asia and Oceania, where 1 in 10 children under the age of 5 has an increased risk of death due to wasting [24]. In all regions, the rising prevalence of overweight children under the age of 5 is concerning. Most of these issues are predicted to worsen due to the dual crises of COVID-19 and climate change [25].

Inadequate food intake and poor diet quality (imbalanced nutrient intake and/or imbalanced energy intake) are two of the major drivers of malnutrition. Globally, about half (53.1%) of children aged 6–23 months reach the minimum recommended frequency of meals, with only 29.3% meeting minimum dietary diversity and 18.9% meeting the minimums for an acceptable diet for healthy growth and development [26]. Across childhood and adolescence, dietary patterns tend to be lower than national recommendations in nutrient-rich foods including fruits, vegetables, animal-source foods (e.g., meat, dairy, poultry, eggs, and fish) and fortified foods [26]. In all geographic regions, inadequate fruit and vegetable intake is common. Among adolescent girls in low- and middle-income countries (LMICs), only 16% meet the daily serving recommendation for dairy and 46% meet dietary recommendations for meat, poultry, or fish [27]. Diets with low intakes of fruits, vegetables, and animal-source foods put children at risk for nutrition deficiencies, with many of the micronutrients that are missing in children's diets globally (iron, calcium, zinc, folate, vitamin B12, vitamin D) found mainly in animal-source foods [28].

Not only are poor diets the leading cause of many forms of childhood undernutrition; they are also a leading cause of childhood overweight and obesity. Globally, more than 38 million children under the age of 5 years are overweight or obese, and more than 340 million between the ages of 5–19 years are overweight or obese [29]. Once considered only an issue in high-income countries (HICs), overweight and obesity are increasing in LMICs, which now must face the double-burden of malnutrition, encompassing the impacts of undernutrition as well as obesity and overweight. For example, since 2000, the number of overweight children under the age of 5 in Africa has risen by nearly 24%, while in Asia, almost half of the children under the age of 5 are overweight or obese according to the World Health Organization (WHO) [29]. Unfortunately, large data gaps exist on overall dietary quality in children, limiting characterization of the effects of early life dietary quality over the course of life [30].

Childhood overweight and obesity are major risk factors for non-communicable disease (NCD)-related deaths from cardiovascular disease (CVD), type 2 diabetes (T2D), cancer, and musculoskeletal disorders (e.g., osteoarthritis, osteoporosis) [1,31]. These largely preventable diseases claim more than 40 million lives each year, or nearly three-quarters of all global deaths [32]. The pathophysiology and related symptoms of many diet- related NCDs begin in childhood, creating a variety of costs and offsets that often last for decades, including detriments to physical health, social and economic status, and overall quality of life. These costs and offsets also carry over to social institutions, affecting the time and resources of families, schools, work, government programs, and healthcare systems. It is estimated that the global cost of CVD alone will rise from approximately $863 billion in 2010 to over $1 trillion by 2030, further underscoring the value of focusing on child nutrition and healthy weight early in life [33]. Prioritizing the nutrition and health of the young, especially those who are poor and/or vulnerable, is not only a way to prevent unnecessary suffering and social costs, but also a way towards improving the health and well-being of all future populations as well as the economic stability of our future social systems [34].

In addition to the negative effects on physical health (i.e., body size, body weight, risk for NCDs), poor childhood nutrition is also associated with substandard neurological development, cognitive function, and psychosocial outcomes [2,35]. These issues manifest in various ways, such as impairments to visual processing, attention, memory, motor skills, and overall intelligence. While many of these early impairments can have lifelong effects [36], many others can be minimized or reversed with early life dietary improvements [2]. Taken together, these factors highlight the massive impacts that early childhood nutrition can have on physical and mental health and show that a global shift towards healthier diets could greatly reduce our population's current and future disease burden.

3. Global Efforts Pushing Towards Healthier Diets for Children and More Sustainable Food Systems

In 2016, The United Nations (UN) established a set of 17 SDGs to "promote prosperity while protecting the planet" [37]. SDG #2, "Zero Hunger" aims to end hunger, achieve food security, improve nutrition, and promote sustainable agriculture by 2030,unifying healthy diets and sustainable food systems in one important goal. However, at present, progress in the fight against global malnutrition is insufficient to meet SDG targets for 2030 [24], with the prevalence of multiple forms of malnutrition among children under the age of 5 remaining stubbornly high year-over-year [5]. Alongside persistent malnutrition, children face the additional challenges of current and future impacts of environmental degradation from their food systems and consumption patterns. And while climate change receives the bulk of attention, there are many more sustainability issues that rarely receive media headlines.

Therefore, the UN and its collection of specialized organizations that prioritize child health and sustainability (e.g., WHO, United Nations Children's Fund (UNICEF), and the Food and Agriculture Organization (FAO)), are making it clear that sustainable food systems are about more than protecting the environment and improving dietary patterns. There are also numerous social and economic factors to consider for current and future generations. For example, the UN's Committee on World Food Security High Level Panel of Experts on Food Security and Nutrition (CFS HLPE) defines sustainable food systems as the intention "to provide food security and nutrition today, in such a way that does not compromise the environmental, economic, and social bases that generate food security and nutrition for future generations" [38]. The recent UN Food Systems Summit, which took place in September of 2021, echoed this sentiment. The UN Secretary-General made it clear that in addition to nourishing people and respecting nature, sustainable and resilient foods systems must also emphasize equitable livelihoods, decent work and empowered communities [39]. Summit participants are aiming to put these recommendations into action by supporting two new multi-stakeholder and multi-sectoral coalitions, one with a focus on healthy diets from sustainable food systems and the other focused on providing healthy school meals to children [40]. Also, with an eye towards the future, leaders of the 2021 Food Systems Summit convened a Global Youth Summit Dialogue focused on the critical roles of youth in achieving the SDGs and laying the groundwork for sustainable food systems transformation [41]. This call to action to food system actors to be more inclusive of younger generations is also reflected in shifting policy recommendations, such as in global FBDGs that are placing more focus on childhood health [42], and a more holistic approach to dietary patterns that are inclusive of sustainability factors [43].

4. Factors Affecting Food Systems Transformation

There are numerous ways to assess the health and sustainability impacts of foods, with factors such as food source (i.e., animal-source foods vs. plant-source foods), protein quantity and quality, and carbon footprint receiving significant attention of late [11,44–47]. Although these are all critical issues for achieving healthier diets from more sustainable foods systems, they by no means comprise a comprehensive list. There are hundreds of compounds in foods that matter for health (e.g., vitamins, minerals, amino acids, fatty acids, fibers, probiotics, phenolics), hundreds of factors that matter for food system sustainability (e.g., carbon footprint, water footprint, land use, fertilizer use, cost of production, food lost or wasted, eutrophication, food accessibility and affordability) and many ways to measure and compare foods (per kg, per 100 kcal, per serving, per g of nutrient, etc.). Any combination of these variables may provide useful information for understanding the health and sustainability impacts of foods or diets.

Furthermore, these variables are constantly changing, making it difficult to predict how altering one aspect (e.g., pesticide us, fertilizer use, grazing schedule) may affect another (e.g., biodiversity, crop pollination, crop yield). While certain indicators may be more important than others for global health and sustainability, decision-making that only focuses on simple or singular indicators (e.g., carbon footprint/kg food) will likely lead to missed opportunities to mitigate tradeoffs or capitalize on food system synergies. For example, issues such as intensification, incentivization, and subsidization are potential game changers when it comes to improving the health and sustainability of food systems, but at the same time, these types of policies and practices can also have unintended consequences across all sustainability domains [48]. Similarly, research models show that reducing a major food system variable, such as the availability of animal-source foods in the food supply, may moderately reduce GHG emissions, but at the same time negatively impact hundreds of millions of agricultural livelihoods, as well as the availability of essential nutrients in the food supply, and the overall well-being of a large portion of the global population [49–51]. If we are to collectively achieve the SDGs by 2030, all major food system actors must unite behind "win-win" (i.e., synergistic) solutions that aim to solve multiple sustainability issues at once without causing new ones [52]. Accomplishing this task of optimizing food systems transformation will require a huge shift in industry and government support towards transdisciplinary approaches to research and policy decisions, with many of these new strategies dependent on the development and deployment of evidence-based innovations and technologies [53,54]. These "win-win" science-backed strategies, such as food fortification, regenerative agriculture, sustainable intensification, carbon capture/sequestration, blockchain technologies, upcycling of byproducts, waste valorization, improvements in biofuels, and the creation of safer agrochemicals [55–58], will all play key roles towards enabling safer, healthier, and more sustainable futures for the next generations.

5. Dairy's Essential Role in Child Health Within Sustainable Food Systems

5.1. Meeting the Nutrition and Health Needs of Children

To meet the nutrient needs necessary for proper growth and development, children require a balanced and varied diet of safe, nutrient-rich foods. In general, FBDGs from around the world recommend that these nutrient-rich foods come from both plant- and animal-source food groups on a daily basis [17]. The most commonly recommended food groups in FBDGs are fruits/vegetables and dairy foods-with approximately 70% of FBDGs recommending daily intake of milk and other dairy foods, and most of the other 30% of FBDGs consolidating dairy food recommendations into those for "protein foods" or "foods from animals" [17]. Among animal-source food groups, dairy foods, such as milk, cheese, and yogurt, contain an unmatched set of essential nutrients, which are rich in several of the nutrients of public health concern for underconsumption in both LMICs and HICs (e.g., calcium, iodine, potassium, zinc, vitamin A, vitamin D) [42,59]. Dairy foods are primarily recommended in FBDGs because they contribute nutrients critical for healthy growth and development, including calcium, vitamin D and high-quality protein [16]. Consuming adequate amounts of these nutrients is especially important to support skeletal health in childhood and adolescence [60]; and emerging research also suggests that the beneficial microbes found in fermented dairy foods may act in concert with calcium and vitamin D to benefit bone health later in life [61].

Although not often mentioned in FBDGs, dairy foods are also rich in magnesium, phosphorus, riboflavin, selenium, and vitamin B12 [42]. Many of the essential nutrients (and bioactive compounds in the case of probiotics) that dairy foods provide are underconsumed globally, and their adequate consumption has been shown to help improve growth in young children and to contribute to proper brain development and/or overall health in humans of all ages [3,42,62,63]. Additionally, the consumption of milk and other animal-source foods by undernourished children in LMICs has been linked to reduced risk of morbidity and mortality as well as improved cognitive function [64]. Obtaining similar quantity and quality of protein and micronutrients strictly from plant-source foods can be challenging for young children, given their small stomach size and the amount of plant-source foods required to meet requirements [65]. Research from the U.S. also shows that replacing dairy foods with nutrient-equivalent non-dairy options can lead to increases in both the cost and energy content of dietary patterns, primarily having to do with the difficulty in achieving nutrient adequacy for calcium and vitamin D from low/no dairy diets [66].

Higher consumption of foods that are sources of vitamin A, B, complex vitamins, vitamin C, vitamin D, zinc, iron, and/or selenium are recommended both for children and various vulnerable populations, including those who are malnourished, immune-compromised, and/or at a higher risk for contracting COVID-19 and other infections [32,67,68]. In the absence of dairy and other animal-source foods in the diet, it can be extremely difficult for young children to meet requirements for several of these micronutrients, unless they are consuming highly fortified options [65]. While fortified foods can provide a safe and cost-effective public health strategy for preventing nutrient deficiencies around the world, several of the main vehicles for fortification tend to be highly processed plant-based ingredients (salt, sugar, flour, oils) [69] which are commonly listed in global FBDGs as dietary components to limit [17]. Although fortified plant-source foods can possess a host of the essential nutrients required to prevent deficiencies, experts regard animal-source foods as the most effective first-line choice in the treatment of mild and moderately undernourished children [70]. The importance of animal-source foods is also recognized by Codex Alimentarius (which is an internationally recognized collection of guidelines and codes aimed at ensuring the safety, quality, and fairness of the international food trade), which requires that ready-to-eat therapeutic foods contain 50% dairy protein, which further demonstrates the importance of high-quality macronutrients for interventions within extremely vulnerable populations [71].

FBDGs inform national governments on food policy, with many of their dairy food recommendations resulting in the inclusion of daily dairy foods in school meals programs across the globe. In the U.S., dairy has been a mainstay in dietary guidelines since their inception, resulting in milk being a staple in U.S. school meal programs and serving as the main food source of nine essential nutrients (i.e., protein, calcium, phosphorus, magnesium, potassium, vitamins A, B12, D and riboflavin) in the diets of U.S. children and adolescents [72]. Healthy school meals that include dairy have been shown to close nutrient gaps and support improved academic performance and behavior [73]. And while not generally considered in FBDGs, many chronic diseases are diet-dependent and frequently begin in childhood, affecting quality of life and well-being throughout adolescence and adulthood. Adequate dairy intake in youth can beneficially impact health later in life [74], and adults who consume higher levels of dairy food, especially fermented dairy foods, may have a lower risk of several chronic diseases such as CVD, T2D, hypertension, and stroke [75].

The scientific evidence clearly shows that the inclusion of adequate dairy foods in children's diets can benefit their development and overall health in multiple ways (e.g., by preventing nutrient deficiencies, reducing the risk for later-onset chronic disease, providing beneficial probiotics through cultured products, for use as a vehicle for fortification, and as a base ingredient for therapeutic foods) [60–62,71,76,77]. The potential health benefits of consuming adequate dairy are well-documented, as are the reductions in healthcare costs and unnecessary suffering from deficiencies, infections, fractures, and NCDs [78–80]. In addition to these essential and potential benefits for human nutrition and health, the FAO and WHO also include milk and dairy products among the foods that fit within sustainable healthy diets for several other reasons as well (e.g., acceptability, accessibility, adaptability, and affordability), all of which may can impact child well-being [81].

5.2. Helping Families, Communities, and Economies Thrive

Approximately one billion people globally rely on the dairy sector to support their livelihoods and sustain their local communities [18]. Beyond the farm, milk collection, processing, distribution, and retail all generate direct and indirect employment and income. A 2018 systematic literature review on the dairy sector's impact on poverty reduction demonstrated dairy cow ownership and/or improvement of dairy cow production had a positive impact across several indicators, including gross household income, household nutrition, crop yields, and household expenditure [82]. This finding and its consistency across study types, countries, and indicators demonstrates that engagement in dairy production is a key factor in improved household welfare around the world. The review also provided strong evidence that dairy production can be part of the solution for improving sustainable development, as it can significantly contribute to poverty reduction at both the family and community levels, both of which are directly related to improving child health and well-being outcomes.

Dairy farms and businesses also provide substantial support to the economy in developed countries. In the U.S., dairy farming and production supports more than three million jobs that generate more than $150 billion in annual wages and over $600 billion in overall economic impact [83]. Dairy farmers also strengthen rural economies in the U.S. by promoting economic opportunity, maintaining and modernizing infrastructure, and establishing human services for their communities. For example, the National Milk Producers Federation has provided high-speed broadband internet and connectivity to rural areas through policy advocacy, funding and coordination, and enhanced mapping of coverage to encourage adequate development [84].

Dairy farming also empowers rural women around the world by increasing their income and influence over vocational and household expenditures. Globally, 37 million dairy farms are led by women, and 80 million women are engaged in dairy farming to some extent [85]. Income generated by dairy farming often enables families in LMICs to provide better educational opportunities to their children. For example, a 2006 study which evaluated the impact of a dairy cattle transfer program on Tanzanian families, showed that participating families that were considered highly income insecure prior to the program were able to send their children to secondary school and, in some cases, to more costly private schools, within three to four years of starting the program [82].

For rural, low-income families, household milk production is associated with positive community impacts. A 2016 study examining the impact of livestock distribution by Heifer International in Zambia showed that distribution of dairy cows to a subset of families led to a statistically significant increase in milk consumption among the community [86]. Importantly, even the households that did not receive an animal in the study demonstrated increased milk consumption due to the increase in the availability and affordability of milk within the community. This type of community "spill-over effect" from increased milk production does not only apply to the milk either, as the addition of dairy cows to certain low-income communities can also contribute to a more abundant, affordable, and accessible supply of draught power, fuel, clothing, bedding, and other animal-derived necessities for life [82,87]; all of which in turn can impact the health and well-being of the children living within those households and communities.

5.3. Supporting Environmental Sustainability with a Focus on Carbon and Climate

Climate change is predicted to worsen child health and well-being for both present and future generations [88]. The continued rise in GHG emissions will likely lead to an increase in natural disasters, infectious disease, food and water insecurity, mass migration, psychological stress, and respiratory disease, among other harms [89,90]. While all food production systems produce GHGs, the amount can vary considerably based on factors such as the production methods being used, the geographical location of production, the time of the year of production, and the type of distribution networks and end markets being used [91–93]. In the case of livestock production there are even more granular factors that can make a difference such as the species, breed, age, and diet of the animal. As a result, there may be up to a 50-fold difference in associated GHGs between similar food products [94].

Cow's milk provides an interesting case study in how agricultural practices can impact GHG emissions. The emission intensity for milk, which roughly equates to the amount of GHGs produced per unit of milk, is lowest in developed regions with large-scale milk production systems such as the U.S., ranging on average between 1.3 to 1.4 kg CO_2 eq. per kg fat-and-protein corrected milk, while developing dairy regions with smaller scale farms have higher emission intensities ranging on average between 4.1 to 6.7 kg CO_2 eq. per kg fat-and-protein corrected milk [95]. The differences in these numbers are largely dependent on the farming practices used, with more intensive production methods resulting in lower GHG emissions [96,97]. Importantly, emission intensities for milk across all regions has declined by almost 11% from 2005 to 2015, reflecting improvements in on-farm efficiency through improved animal productivity and better management [95]. These declines in emission intensities will likely continue into the future as further improvements are made to animal feeds, genetics, and manure management systems [98].

While GHGs get the most attention in sustainability discussions, it is important to note that a food's carbon footprint is only one sustainability measure out of many that matter. Although often underappreciated or ignored in the media, livestock production does play several vitally important roles in healthy and sustainable food systems. For one, ruminants utilize land for grazing that is not suitable for growing crops, and they also upcycle low-nutritional quality foods (including agricultural byproducts), that would otherwise rot or be burned and release GHGs. In this manner, ruminants are able to convert low-quality/inedible plants and byproducts into nutrient-rich foods for human consumption while at the same time minimizing the release of GHGs from unused byproducts [99,100]. Livestock can also improve the health and value of non-arable lands through ecosystem services such as manure fertilization, land aeration, improved biodiversity, improved ecosystem water productivity, and improved carbon sequestration [65].

There are still notable opportunities in all food sectors, including the dairy sector, to reduce environmental impacts. In contrast to many industries, the dairy sector has continuously invested in research and innovation over the last several decades to reduce its impacts and natural resource use by making significant advancements in areas such as crop production, water use, animal genetics, animal care, and food safety [101]. The dairy sector has further committed to advancing sustainable dairy production around the world through the development of a Dairy Sustainability Framework focused on tracking and improving key impacts related to GHGs, soil, water, waste, biodiversity, animal care, working conditions, and product safety [102]. The Innovation Center for U.S. Dairy's recently established 'Net Zero Initiative' aims to drive the industry to achieve carbon neutrality, optimized water usage, and improved water quality by 2050 [19]. Additionally, its Dairy Stewardship Commitment defines a rigorous set of standards that demonstrate positive social, economic, and environmental impact [101].

Global thought leaders increasingly recognize that dairy production and other livestock farming practices can and are on the path to being more sustainable [65,103]. Additional efforts, including research and advancements in innovation and technology, need to be made to ensure all food production systems, including those in LMICs, become more sustainable [65]. Mitigating climate change, preserving biodiversity, and preserving natural resources through sustainable farming practices will support the healthy environments needed for today's children and future generations to flourish.

6. Conclusions

Hunger and malnutrition continue to persist among children at devastatingly high levels. Despite global efforts, significantly greater support for child health and welfare is urgently required for achieving the SDGs. Scalable solutions are needed to address these issues and their underlying causes, such as poverty, poor sanitation, and food insecurity. These solutions should come from collective efforts by major food systems actors (i.e., academia, industry, governance, civil society), and aim to consider the potential trade-offs and synergies of food system transformation. But the science will only take us so far. These innovations and interventions will need to be designed and delivered with care, going beyond food system stakeholders to the socioeconomic support systems for children–namely caregivers, families, schools, and communities. Food Systems transformation must engage children and adolescents as agents of change, not just as dependents and consumers. While it is critical to provide adequate nutrition for growth and development, engaging children in nutrition education and sustainability actions also can help pave a path to a healthier and more sustainable future for all.

Due to their interconnectedness and everchanging nature, it may not be possible to simultaneously maximize the health and sustainability of complex food systems. However, it is imperative that we make significant and continuous improvements towards both. Although dairy production can be resource intensive and is associated with greater environmental impacts than many plant-source options, the dairy industry as a whole continues to transform towards more environmentally sustainable practices, while at the same time maintaining positive contributions to local and global economies, human health (especially child nutrition), and community well-being. Finding the right balance between these types of trade-offs among health, society, economics, and the environment will play a key role in determining the successes and failures of food systems transformation and, in time, the future well-being of our population and planet.

Author Contributions: Conceptualization, K.A.B., K.B.C. and G.D.M.; writing—original draft preparation, K.A.B., G.D.M., writing—review and editing, K.B.C., K.A.B., G.D.M., L.R., M.K., N.M.G. All authors have read and agreed to the published version of the manuscript.

Funding: This project received financial support from National Dairy Council.

Institutional Review Board Statement: Not applicable.

Informed Consent Statement: Not applicable.

Data Availability Statement: Not applicable.

Conflicts of Interest: G.M., K.B. work for National Dairy Council. M.K. works for Global Dairy Platform. L.R. works for International Dairy Federation, K.C. works for the California Dairy Research Foundation, and N.G. works for U.S. Dairy Export Council.

References

1. Akseer, N.; Al-Gashm, S.; Mehta, S.; Mokdad, A.; Bhutta, Z.A. Global and regional trends in the nutritional status of young people: A critical and neglected age group. *Ann. N. Y. Acad. Sci.* **2017**, *1393*, 3–20. [CrossRef]
2. Suryawan, A.; Jalaludin, M.; Poh, B.; Sanusi, R.; Tan, V.; Geurts, J.; Muhardi, L. Malnutrition in early life and its neurodevelopmental and cognitive consequences: A scoping review. *Nutr. Res. Rev.* **2021**, *10*, 1–14. [CrossRef] [PubMed]

3. UNICEF. *The State of The World's Children 2019. Children, Food and Nutrition: Growing Well in a Changing World*; UNICEF: New York, NY, USA, 2019. Available online: https://www.unicef.org/media/60806/file/SOWC-2019.pdf (accessed on 4 January 2021).
4. United Nations: Department of Economic and Social Affairs: Policy Brief #102: Population, Food Security, Nutrition and Sus-tainable Development. 2021. Available online: https://www.un.org/development/desa/dpad/publication/un-desa-policy-brief-102-population-food-security-nutrition-and-sustainable-development/ (accessed on 9 January 2021).
5. Food and Agriculture Organization of the United Nations (FAO); International Fund for Agricultural Development; UNICEF; World Food Programme and World Health Organization. *The State of Food Security and Nutrition in The World 2020. Trans-forming Food Systems for Affordable Healthy Diets*; FAO: Rome, Italy, 2020. Available online: https://www.unicef.org/media/72676/file/SOFI-2020-full-report.pdf (accessed on 9 January 2021).
6. Fore, H.H.; Dongyu, Q.; Beasley, D.M.; A Ghebreyesus, T. Child malnutrition and COVID-19: The time to act is now. *Lancet* **2020**, *396*, 517–518. [CrossRef]
7. Sova, C. *The Three Waves of Hunger: The Devastating Ripple Effects of COVID-19*; World Food Programme USA: Washington, DC, USA, 2020. Available online: https://www.wfpusa.org/articles/the-three-waves-of-hunger-the-devastating-ripple-effects-of-covid-19/ (accessed on 19 November 2020).
8. Hamilton, H.; Henry, R.; Rounsevell, M.; Moran, D.; Cossar, F.; Allen, K.; Boden, L.; Alexander, P. Exploring global food system shocks, scenarios and outcomes. *Futures* **2020**, *123*, 102601. [CrossRef] [PubMed]
9. Kahiluoto, H. Food systems for resilient futures. *Food Secur.* **2020**, *12*, 853–857. [CrossRef] [PubMed]
10. Wilson, N.; Cleghorn, C.L.; Cobiac, L.J.; Mizdrak, A.; Nghiem, N. Achieving Healthy and Sustainable Diets: A Review of the Results of Recent Mathematical Optimization Studies. *Adv. Nutr.* **2019**, *10*, S389–S403. [CrossRef]
11. Willett, W.; Rockström, J.; Loken, B.; Springmann, M.; Lang, T.; Vermeulen, S.; Garnett, T.; Tilman, D.; DeClerck, F.; Wood, A.; et al. Food in the Anthropocene: The EAT–Lancet Commission on healthy diets from sustainable food systems. *Lancet* **2019**, *393*, 447–492. [CrossRef]
12. Clark, M.A.; Springmann, M.; Hill, J.; Tilman, D. Multiple health and environmental impacts of foods. *Proc. Natl. Acad. Sci. USA* **2019**, *116*, 23357–23362. [CrossRef]
13. Goldstein, B.P.; Hauschild, M.Z.; Fernández, J.E.; Birkved, M. Contributions of Local Farming to Urban Sustainability in the Northeast United States. *Environ. Sci. Technol.* **2017**, *51*, 7340–7349. [CrossRef]
14. Nelson, M.E.; Hamm, M.W.; Hu, F.B.; Abrams, S.A.; Griffin, T.S. Alignment of Healthy Dietary Patterns and Environmental Sustainability: A Systematic Review. *Adv. Nutr.* **2016**, *7*, 1005–1025. [CrossRef]
15. Comerford, K.B.; Miller, G.D.; Kapsak, W.R.; Brown, K.A. The Complementary Roles for Plant-Source and Animal-Source Foods in Sustainable Healthy Diets. *Nutrients* **2021**, *13*, 3469. [CrossRef]
16. Comerford, K.B.; Miller, G.D.; Boileau, A.C.; Schuette, S.N.M.; Giddens, J.C.; Brown, K.A. Global Review of Dairy Recommendations in Food-Based Dietary Guidelines. *Front. Nutr.* **2021**, *8*, 671999. [CrossRef] [PubMed]
17. Herforth, A.; Arimond, M.; Álvarez-Sánchez, C.; Coates, J.; Christianson, K.; Muehlhoff, E. A Global Review of Food-Based Dietary Guidelines. *Adv. Nutr.* **2019**, *10*, 590–605. [CrossRef] [PubMed]
18. Hill, J.P.; Van Belzen, N. *Assessing the Overall Impact of the Dairy Sector*; Burleigh Dodds Science Publishing: Cambridge, UK, 2017; pp. 291–314.
19. Food Systems Summit 2021 Community: Commitments to Action—U.S. Dairy Net Zero Initiative. Available online: https://foodsystems.community/u-s-dairy-net-zero-initiative/ (accessed on 27 September 2021).
20. Fonterra. Fonterra Sustainability Report. Available online: https://indd.adobe.com/view/3451be70-971f-4965-9a74-645dde740377 (accessed on 22 December 2020).
21. Arla. Sustainable Dairy Production. Available online: https://www.arla.com/company/responsibility/sustainable-dairy-production/ (accessed on 22 December 2020).
22. Drewnowski, A. Measures and metrics of sustainable diets with a focus on milk, yogurt, and dairy products. *Nutr. Rev.* **2017**, *76*, 21–28. [CrossRef]
23. World Health Organization. Fact Sheet: Malnutrition. Available online: https://www.who.int/news-room/fact-sheets/detail/malnutrition (accessed on 4 January 2021).
24. UNICEF; World Health Organization (WHO); World Bank Group. *Levels and Trends in Child Malnutrition: Key Findings of the 2020 Edition of the Joint Child Malnutrition Estimates*; WHO: Geneva, Switzerland, 2020. Available online: https://www.who.int/publications/i/item/9789240025257 (accessed on 9 June 2021).
25. Strong, K.L.; Requejo, J.; Agweyu, A.; Billah, S.M.; Boschi-Pinto, C.; Horiuchi, S.; Jamaluddine, Z.; Lazzerini, M.; Maiga, A.; McKerrow, N.; et al. Revitalizing child health: Lessons from the past. *Glob. Health Action* **2021**, *14*, 1947565. [CrossRef]
26. Global Nutrition Report. *Global Nutrition Report: Action on Equity to End Malnutrition*; Global Nutrition Report: Bristol, UK, 2020. Available online: https://globalnutritionreport.org/reports/2020-global-nutrition-report/inequalities-global-burden-malnutrition/ (accessed on 6 January 2021).
27. Keats, E.C.; Rappaport, A.; Shah, S.; Oh, C.; Jain, R.; Bhutta, Z.A. The Dietary Intake and Practices of Adolescent Girls in Low-and Middle-Income Countries: A Systematic Review. *Nutrients* **2018**, *10*, 1978. [CrossRef]
28. Beal, T. Achieving dietary micronutrient adequacy in a finite world. *One Earth* **2021**, *4*, 1197–1200. [CrossRef]
29. WHO. Obesity and Overweight. Available online: https://www.who.int/news-room/fact-sheets/detail/obesity-and-overweight (accessed on 11 August 2021).

30. Kupka, R.; Siekmans, K.; Beal, T. The diets of children: Overview of available data for children and adolescents. *Glob. Food Secur.* **2020**, *27*, 100442. [CrossRef]
31. Drewnowski, A.; The Ecosystem Inception Team; Havelaar, A.; Sere, C.; De Fraiture, C.; Mitloehner, F.; Steinfeld, H.; Melgar-Quinonez, H.; Ingram, J.; Heller, M.; et al. The Chicago Consensus on Sustainable Food Systems Science. *Front. Nutr.* **2018**, *4*, 74. [CrossRef]
32. World Health Organization. COVID-19 Significantly Impacts Health Services for Noncommunicable Diseases. Available online: https://www.who.int/news/item/01-06-2020-covid-19-significantly-impacts-health-services-for-noncommunicable-diseases (accessed on 19 November 2020).
33. Allen, N.B.; Siddique, J.; Wilkins, J.T.; Shay, C.; Lewis, C.E.; Goff, D.C.; Jacobs, D.R.; Liu, K.; Lloyd-Jones, D. Blood Pressure Trajectories in Early Adulthood and Subclinical Atherosclerosis in Middle Age. *JAMA* **2014**, *311*, 490–497. [CrossRef]
34. Richter, L.M.; Daelmans, B.; Lombardi, J.; Heymann, J.; Boo, F.L.; Behrman, J.R.; Lu, C.; E Lucas, J.; Perez-Escamilla, R.; Dua, T.; et al. Investing in the foundation of sustainable development: Pathways to scale up for early childhood development. *Lancet* **2017**, *389*, 103–118. [CrossRef]
35. Pizzol, D.; Tudor, F.; Racalbuto, V.; Bertoldo, A.; Veronese, N.; Smith, L. Systematic review and meta-analysis found that malnutrition was associated with poor cognitive development. *Acta Paediatr.* **2021**, *110*, 2704–2710. [CrossRef] [PubMed]
36. Mwene-Batu, P.; Bisimwa, G.; Baguma, M.; Chabwine, J.; Bapolisi, A.; Chimanuka, C.; Molima, C.; Dramaix, M.; Kashama, N.; Macq, J.; et al. Long-term effects of severe acute malnutrition during childhood on adult cognitive, academic and behavioural development in African fragile countries: The Lwiro cohort study in Democratic Republic of the Congo. *PLoS ONE* **2020**, *15*, e0244486. [CrossRef]
37. United Nations. Sustainable Development Goals. Available online: https://www.un.org/sustainabledevelopment/ (accessed on 8 January 2021).
38. High Level Panel of Experts on Food Security and Nutrition, Committee on World Food Security (CFS). *Food Security and Nutrition: Building a Global Narrative Towards 2030*; CFS: Rome, Italy, 2020. Available online: http://www.fao.org/3/ca9731en/ca9731en.pdf (accessed on 6 June 2021).
39. UN Food Systems Summit 2021: Press Release—Food Systems Hold Power to 'Realise Vision of a Better World', Says UN Sec-retary-General at first Food Systems Summit. Available online: https://www.un.org/en/food-systems-summit/news/food-systems-hold-power-%E2%80%98realise-vision-better-world%E2%80%99-says-un-secretary-general (accessed on 23 September 2021).
40. World Health Organization. New Coalitions Announced at the UN Food Systems Summit to Increase Access to Healthy Diets from Sustainable Food Systems. Available online: https://www.who.int/news/item/23-09-2021-new-coalitions-announced-at-the-un-food-systems-summit-to-increase-access-to-healthy-diets-from-sustainable-food-systems (accessed on 24 September 2021).
41. UN Food Systems Summit: Global Youth Summit Dialogue Spotlights the Critical Role of Youth in Ensuring Good Food for All. Available online: https://www.un.org/en/food-systems-summit/news/global-youth-summit-dialogue-spotlight (accessed on 23 September 2021).
42. U.S. Department of Agriculture. Dietary Guidelines for Americans 2020–2025. Make Every Bite Count with the Dietary Guidelines. Available online: https://www.dietaryguidelines.gov/sites/default/files/2020-12/Dietary_Guidelines_for_Americans_2020-2025.pdf (accessed on 15 February 2021).
43. Food and Agriculture Organization of the United Nations (FAO). Food-Based Dietary Guidelines—Dietary Guidelines and Sustainability. Available online: http://www.fao.org/nutrition/education/food-dietary-guidelines/background/sustainable-dietary-guidelines/en/ (accessed on 4 January 2021).
44. Schiano, A.; Harwood, W.; Gerard, P.; Drake, M. Consumer perception of the sustainability of dairy products and plant-based dairy alternatives. *J. Dairy Sci.* **2020**, *103*, 11228–11243. [CrossRef]
45. Fresán, U.; Craig, W.J.; Martínez-González, M.A.; Bes-Rastrollo, M. Nutritional Quality and Health Effects of Low Environmental Impact Diets: The "Seguimiento Universidad de Navarra" (SUN) Cohort. *Nutrients* **2020**, *12*, 2385. [CrossRef] [PubMed]
46. Päivärinta, E.; Itkonen, S.; Pellinen, T.; Lehtovirta, M.; Erkkola, M.; Pajari, A.-M. Replacing Animal-Based Proteins with Plant-Based Proteins Changes the Composition of a Whole Nordic Diet—A Randomised Clinical Trial in Healthy Finnish Adults. *Nutrients* **2020**, *12*, 943. [CrossRef] [PubMed]
47. Fabek, H.; Sanchez-Hernandez, D.; Ahmed, M.; Marinangeli, C.P.F.; House, J.D.; Anderson, G.H. An examination of contri-butions of animal- and plant-based dietary patterns on the nutrient quality of diets of adult Canadians. *Appl. Physiol. Nutr. Metab.* **2021**, *46*. [CrossRef]
48. Clay, N.; Garnett, T.; Lorimer, J. Dairy intensification: Drivers, impacts and alternatives. *Ambio* **2020**, *49*, 35–48. [CrossRef]
49. White, R.R.; Hall, M.B. Nutritional and greenhouse gas impacts of removing animals from US agriculture. *Proc. Natl. Acad. Sci. USA* **2017**, *114*, E10301–E10308. [CrossRef]
50. De Neve, J.-E.; Sachs, J.D. The SDGs and human well-being: A global analysis of synergies, trade-offs, and regional differences. *Sci. Rep.* **2013**, *10*, 15113. [CrossRef]
51. Sun, H.; Weaver, C.M. Decreased Iron Intake Parallels Rising Iron Deficiency Anemia and Related Mortality Rates in the US Population. *J. Nutr.* **2021**, *151*, 1947–1955. [CrossRef] [PubMed]
52. Girardello, M.; Santangeli, A.; Mori, E.; Chapman, A.; Fattorini, S.; Naidoo, R.; Bertolino, S.; Svenning, J.-C. Global synergies and trade-offs between multiple dimensions of biodiversity and ecosystem services. *Sci. Rep.* **2019**, *9*, 5636. [CrossRef]

53. Boer, A.D.; Kok, K.; Gill, M.; Breda, J.; Cahill, J.; Callenius, C.; Caron, P.; Damianova, Z.; Gurinovic, M.; Lähteenmäki, L.; et al. Research and innovation as a catalyst for food system transformation. *Trends Food Sci. Technol.* **2021**, *107*, 150–156. [CrossRef]
54. Lynde, R. Innovation & entrepreneurship driving food system transformation. *Physiol. Behav.* **2020**, *220*, 112866. [CrossRef] [PubMed]
55. Nunes, P.A.D.A.; Laca, E.A.; Carvalho, P.C.D.F.; Li, M.; Filho, W.D.S.; Kunrath, T.R.; Martins, A.P.; Gaudin, A. Livestock integration into soybean systems improves long-term system stability and profits without compromising crop yields. *Sci. Rep.* **2021**, *11*, 1649. [CrossRef]
56. Wyckhuys, K.A.; Aebi, A.; van Lexmond, M.F.B.; Bojaca, C.R.; Bonmatin, J.-M.; Furlan, L.; Guerrero, J.A.; Mai, T.V.; Pham, H.V.; Sanchez-Bayo, F.; et al. Resolving the twin human and environmental health hazards of a plant-based diet. *Environ. Int.* **2020**, *144*, 106081. [CrossRef] [PubMed]
57. Balaine, L.; Dillon, E.J.; Läpple, D.; Lynch, J. Can technology help achieve sustainable intensification? Evidence from milk recording on Irish dairy farms. *Land Use Policy* **2020**, *92*, 104437. [CrossRef]
58. Ronaghi, M.H.; Mosakhani, M. The effects of blockchain technology adoption on business ethics and social sustainability: Evidence from the Middle East. *Environ. Dev. Sustain.* **2021**, 1–26. [CrossRef]
59. Bailey, R.L.; West, K.P., Jr.; Black, R.E. The Epidemiology of Global Micronutrient Deficiencies. *Ann. Nutr. Metab.* **2015**, *66* (Suppl. S2), 22–33. [CrossRef]
60. Golden, N.H.; Abrams, S.A. Optimizing Bone Health in Children and Adolescents. *Pediatrics* **2014**, *134*, e1229–e1243. [CrossRef]
61. Rizzoli, R.; Biver, E. Are Probiotics the New Calcium and Vitamin D for Bone Health? *Curr. Osteoporos. Rep.* **2020**, *18*, 273–284. [CrossRef] [PubMed]
62. Marco, M.L.; Hill, C.; Hutkins, R.; Slavin, J.; Tancredi, D.J.; Merenstein, D.; Sanders, M.E. Should There Be a Recommended Daily Intake of Microbes? *J. Nutr.* **2020**, *150*, 3061–3067. [CrossRef]
63. Cusick, S.; Georgieff, M.K. *The First 1000 Days of Life: The Brain's Window of Opportunity*; UNICEF: New York, NY, USA, 2013. Available online: https://www.unicef-irc.org/article/958-the-first-1000-days-of-life-the-brains-window-of-opportunity.html (accessed on 6 January 2021).
64. Dror, D.K.; Allen, L.H. The Importance of Milk and other Animal-Source Foods for Children in Low-Income Countries. *Food Nutr. Bull.* **2011**, *32*, 227–243. [CrossRef] [PubMed]
65. Global Alliance for Improved Nutrition (GAIN). *Animal-Source Foods for Human and Planetary Health: GAIN's Position*; Briefing Paper Series #2; GAIN: Geneva, Switzerland, 2020. Available online: https://www.gainhealth.org/sites/default/files/publications/documents/gain-briefing-paper-series-2-animal-source-foods-for-human-and-planetary-health.pdf (accessed on 6 June 2021).
66. Cifelli, C.J.; Auestad, N.; Fulgoni, V.L. Replacing the nutrients in dairy foods with non-dairy foods will increase cost, energy intake and require large amounts of food: National Health and Nutrition Examination Survey 2011–2014. *Public Health Nutr.* **2020**, *10*, 1–12. [CrossRef]
67. Petrelli, F.; Luciani, A.; Perego, G.; Dognini, G.; Colombelli, P.L.; Ghidini, A. Therapeutic and prognostic role of vitamin D for COVID-19 infection: A systematic review and meta-analysis of 43 observational studies. *J. Steroid Biochem. Mol. Biol.* **2021**, *211*, 105883. [CrossRef]
68. Akhtar, S.; Das, J.K.; Ismail, T.; Wahid, M.; Saeed, W.; Bhutta, Z.A. Nutritional perspectives for the prevention and mitigation of COVID-19. *Nutr. Rev.* **2020**, *79*, 289–300. [CrossRef]
69. Olson, R.; Gavin-Smith, B.; Ferraboschi, C.; Kraemer, K. Food Fortification: The Advantages, Disadvantages and Lessons from Sight and Life Programs. *Nutrients* **2021**, *13*, 1118. [CrossRef]
70. Food and Agriculture Organization of the United Nations (FAO); Global Dairy Platform (GDP) and IFCN Dairy Research Network (IFCN). *Dairy's Impact on Reducing Global Hunger*; FAO, GDP and IFCN: Chicago, IL, USA, 2020. Available online: http://www.fao.org/3/ca7500en/CA7500EN.pdf (accessed on 6 June 2021).
71. Electronic Working Group, Codex Committee on Nutrition and Foods for Special Dietary Uses, Joint FAO/WHO Food Standards Programme. *Proposed Draft Guideline for Ready-to-Use Therapeutic Foods*; Codex Alimentarius Commission: Rome, Italy, 2019. Available online: http://www.fao.org/fao-who-codexalimentarius/sh-proxy/en/%3Flnk%3D1%26url%3Dhttps%25253A%25252F%25252Fworkspace.fao.org%25252Fsites%25252Fcodex%25252FMeetings%25252FCX-720-41%25252FWD%25252Fnf41_06e.pdf (accessed on 6 June 2021).
72. Keast, D.R.; Fulgoni, V.L.; Nicklas, T.A.; O'Neil, C.E. Food Sources of Energy and Nutrients among Children in the United States: National Health and Nutrition Examination Survey 2003–2006. *Nutrients* **2013**, *5*, 283–301. [CrossRef] [PubMed]
73. Anderson-Wise, T. School Milk Programs Fuel Students, Support Healthy Bodies and Minds Around the World. International Dairy Federation. Available online: https://www.fil-idf.org/school-milk-programs-fuel-students-support-healthy-bodies-and-minds-around-the-world/ (accessed on 6 November 2020).
74. Givens, D.I. MILK Symposium review: The importance of milk and dairy foods in the diets of infants, adolescents, pregnant women, adults, and the elderly. *J. Dairy Sci.* **2020**, *103*, 9681–9699. [CrossRef]
75. Rice, B.H.; Quann, E.E.; Miller, G.D. Meeting and exceeding dairy recommendations: Effects of dairy consumption on nutrient intakes and risk of chronic disease. *Nutr. Rev.* **2013**, *71*, 209–223. [CrossRef]
76. Weir, R.R.; Johnston, M.; Lowis, C.; Fearon, A.M.; Stewart, S.; Strain, J.J.; Pourshahidi, L.K. Vitamin D3 content of cows' milk produced in Northern Ireland and its efficacy as a vehicle for vitamin D fortification: A UK model. *Int. J. Food Sci. Nutr.* **2021**, *72*, 447–455. [CrossRef] [PubMed]

77. Itkonen, S.T.; Erkkola, M.; Lamberg-Allardt, C.J.E. Vitamin D Fortification of Fluid Milk Products and Their Contribution to Vitamin D Intake and Vitamin D Status in Observational Studies—A Review. *Nutrients* **2018**, *10*, 1054. [CrossRef] [PubMed]
78. Javanbakht, M.; Jamshidi, A.R.; Baradaran, H.R.; Mohammadi, Z.; Mashayekhi, A.; Shokraneh, F.; Hamami, M.R.; Bakhsh, R.Y.; Shabaninejad, H.; Delavari, S.; et al. Estimation and Prediction of Avoidable Health Care Costs of Cardiovascular Diseases and Type 2 Diabetes Through Adequate Dairy Food Consumption: A Systematic Review and Micro Simulation Modeling Study. *Arch. Iran. Med.* **2018**, *21*, 213–222.
79. Scrafford, C.G.; Bi, X.; Multani, J.K.; Murphy, M.M.; Schmier, J.K.; Barraj, L.M. Health Care Costs and Savings Associated with Increased Dairy Consumption among Adults in the United States. *Nutrients* **2020**, *12*, 233. [CrossRef]
80. Hiligsmann, M.; Neuprez, A.; Buckinx, F.; Locquet, M.; Reginster, J.-Y. A scoping review of the public health impact of vitamin D-fortified dairy products for fracture prevention. *Arch. Osteoporos.* **2017**, *12*, 57. [CrossRef]
81. Food and Agriculture Organization of the United Nations (FAO); World Health Organization (WHO). *Sustainable Healthy Diets—Guiding Principles*; FAO; WHO: Rome, Italy, 2019. Available online: http://www.fao.org/3/ca6640en/ca6640en.pdf (accessed on 21 March 2021).
82. Food and Agriculture Organization of the United Nations (FAO); Global Dairy Platform (GDP); IFCN Dairy Research Network (IFCN). *Dairy Development's Impact on Poverty Reduction*; FAO; GDP; IFCN: Chicago, IL, USA, 2018. Available online: http://www.fao.org/3/ca0289en/CA0289EN.pdf (accessed on 6 June 2021).
83. International Dairy Foods Association. Dairy Delivers: The Economic Impact of Dairy Products. Available online: https://www.idfa.org/dairydelivers (accessed on 4 January 2021).
84. National Milk Producers Federation. Labor & Rural Policy. Available online: https://www.nmpf.org/policy_priorities/labor-rural-policy/ (accessed on 4 January 2021).
85. Slimko, M. How Dairy Farming Supports Women's Empowerment. Dairy Management Inc. Available online: https://www.usdairy.com/news-articles/how-dairy-farming-supports-womens-empowerment (accessed on 19 November 2020).
86. Jodlowski, M.; Winter-Nelson, A.; Baylis, K.; Goldsmith, P.D. Milk in the Data: Food Security Impacts from a Livestock Field Experiment in Zambia. *World Dev.* **2016**, *77*, 99–114. [CrossRef]
87. Food and Agriculture Organization of the United Nations (FAO). The Contributions of Livestock Species and Breeds to Eco-System Services. Available online: http://www.fao.org/3/a-i6482e.pdf (accessed on 7 June 2021).
88. Helldén, D.; Andersson, C.; Nilsson, M.; Ebi, K.L.; Friberg, P.; Alfvén, T. Climate change and child health: A scoping review and an expanded conceptual framework. *Lancet Planet. Health* **2021**, *5*, e164–e175. [CrossRef]
89. Watts, N.; Amann, M.; Arnell, N.; Ayeb-Karlsson, S.; Belesova, K.; Boykoff, M.; Byass, P.; Cai, W.; Campbell-Lendrum, D.; Capstick, S.; et al. The 2019 report of The Lancet Countdown on health and climate change: Ensuring that the health of a child born today is not defined by a changing climate. *Lancet* **2019**, *394*, 1836–1878. [CrossRef]
90. Ebi, K.L.; Hess, J.J. Health Risks Due to Climate Change: Inequity In Causes And Consequences. *Health Aff.* **2020**, *39*, 2056–2062. [CrossRef]
91. Rotz, C.A. Modeling greenhouse gas emissions from dairy farms. *J. Dairy Sci.* **2018**, *101*, 6675–6690. [CrossRef]
92. Laca, A.; Gómez, N.; Laca, A.; Díaz, M. Overview on GHG emissions of raw milk production and a comparison of milk and cheese carbon footprints of two different systems from northern Spain. *Environ. Sci. Pollut. Res.* **2020**, *27*, 1650–1666. [CrossRef]
93. Chen, W.; Jafarzadeh, S.; Thakur, M.; Ólafsdóttir, G.; Mehta, S.; Bogason, S.; Holden, N.M. Environmental impacts of animal-based food supply chains with market characteristics. *Sci. Total Environ.* **2021**, *783*, 147077. [CrossRef] [PubMed]
94. Poore, J.; Nemecek, T. Reducing food's environmental impacts through producers and consumers. *Science* **2018**, *360*, 987–992. [CrossRef] [PubMed]
95. Food and Agriculture Organization of the United Nations (FAO); Global Dairy Platform (GDP). *Climate Change and the Global Dairy Cattle Sector: The Role of the Dairy Sector in a Low-Carbon Future*; FAO: Rome, Italy, 2019. Available online: http://www.fao.org/3/CA2929EN/ca2929en.pdf (accessed on 7 December 2020).
96. Miller, G.D.; Wang, Y. Carbon and water footprint of U.S. milk, from farm to table—Special issue: Editorial. *Int. Dairy J.* **2013**, *31*, S1–S2. [CrossRef]
97. Thoma, G.; Popp, J.; Nutter, D.; Shonnard, D.R.; Ulrich, R.; Matlock, M.; Kim, D.; Neiderman, Z.; Kemper, N.; East, C.; et al. Data from: Greenhouse gas emissions from milk production and consumption in the United States: A cradle-to-grave life cycle assessment circa 2008. *Int. Dairy J.* **2019**, *31*, S3–S14. [CrossRef]
98. De Vries, M.; Al Zahra, W.; Wouters, A.P.; Van Middelaar, C.E.; Oosting, S.J.; Tiesnamurti, B.; Vellinga, T.V. Entry Points for Reduction of Greenhouse Gas Emissions in Small-Scale Dairy Farms: Looking Beyond Milk Yield Increase. *Front. Sustain. Food Syst.* **2019**, *3*, 3. [CrossRef]
99. Oltjen, J.W.; Beckett, J.L. Role of ruminant livestock in sustainable agricultural systems. *J. Anim. Sci.* **1996**, *74*, 1406–1409. [CrossRef] [PubMed]
100. Food and Agriculture Organization of the United Nations (FAO). *World livestock 2011: Livestock in Food Security*; FAO: Rome, Italy, 2011. Available online: http://www.fao.org/3/i2373e/i2373e.pdf (accessed on 7 December 2020).
101. Dairy Management Inc. The U.S. Dairy Stewardship Commitment. Available online: http://commitment.usdairy.com/ (accessed on 4 November 2020).

102. Dairy Sustainability Framework. Dairy Sustainability Framework Reporting Guidelines for Eleven High Level Indicators. Available online: https://dairysustainabilityframework.org/wp-content/uploads/2019/01/DSF-Reporting-Guidelines-for-11-high-level-indicators.pdf (accessed on 4 January 2021).
103. World Wildlife Fund. An Environmental and Economic Path Toward Net Zero Dairy Farm Emissions. Available online: https://www.worldwildlife.org/publications/an-environmental-and-economic-path-toward-net-zero-dairy-farm-emissions (accessed on 28 January 2021).

International Journal of
Environmental Research and Public Health

MDPI

Review

Adenomyosis and Infertility—Review of Medical and Surgical Approaches

Maria Szubert [1,*], Edward Koziróg [1], Olga Olszak [2], Klaudia Krygier-Kurz [1], Jakub Kazmierczak [1] and Jacek Wilczynski [1]

[1] Clinic of Surgical and Oncologic Gynecology, 1st Department of Gynecology and Obstetrics, Medical University of Lodz, 90-419 Lodz, Poland; edward.kozirog@gmail.com (E.K.); klaudia.krygier89@gmail.com (K.K.-K.); j.kazmierczak87@gmail.com (J.K.); jrwil@post.pl (J.W.)

[2] Clinic of Perinatology, 1st Department of Gynecology and Obstetrics, Medical University of Lodz, 90-419 Lodz, Poland; olgaapietrzak@gmail.com

* Correspondence: maria.szubert@umed.lodz.pl; Tel.: +48-426804722 or +48-604402786; Fax: +48-426860471

Abstract: The aim of this review is to clarify the relative association between adenomyosis and infertility and the possible treatment for an infertile patient. Although adenomyosis is detected more often in women of late reproductive age, its influence on pregnancy rates is important, especially considering the tendency to delay pregnancy among women in developed countries. In this article, we present a critical analysis of the literature data concerning the impact of adenomyosis on fertility. The possible effects of treatment on the pregnancy rate will also be discussed. We conducted a literature search; publications from Pubmed, Embase and Cochrane databases published from 1982 to 2019 were retrieved using terms 'adenomyosis and infertility' and 'adenomyosis and pregnancy outcomes', extensively studied in the aspects of diagnosis, pathogenesis of infertility and possible treatment methods. Molecular studies have given deep insight into the pathogenesis of adenomyosis in the recent few years, but there is a huge discrepancy between in vitro studies and praxis. Oral contraceptive pills, anti-prostaglandins, oral or parenteral progestins, danazol and gonadotrophin-releasing hormone (GnRH) analogues have all been used to control menstrual pain and menorrhagia in women with adenomyosis, but they temporarily suppress the menstrual cycle. Additionally, endometrial ablation and hysterectomy used to alleviate pain caused by adenomyosis exclude pregnancy planning. The development of imaging techniques—ultrasound and MRI—enables the diagnosis of adenomyosis with very high accuracy nowadays, but the methods of treatment mentioned above have not given satisfactory results in women planning pregnancy. For these patients, the high-intensity-focused ultrasound method (HIFU) and combined treatment before assisted reproductive techniques can prove beneficial in adenomyosis patients.

Keywords: adenomyosis; infertility; GnRH; in vitro fertilisation

Citation: Szubert, M.; Koziróg, E.; Olszak, O.; Krygier-Kurz, K.; Kazmierczak, J.; Wilczynski, J. Adenomyosis and Infertility—Review of Medical and Surgical Approaches. *IJERPH* **2021**, *18*, 1235. https://doi.org/10.3390/ijerph18031235

Academic Editors: Wioletta Zukiewicz-Sobczak, Paulina Wojtyla-Buciora, Izabela Rącka and Andrzej Wojtyla

Received: 19 December 2020
Accepted: 26 January 2021
Published: 30 January 2021

Publisher's Note: MDPI stays neutral with regard to jurisdictional claims in published maps and institutional affiliations.

Copyright: © 2021 by the authors. Licensee MDPI, Basel, Switzerland. This article is an open access article distributed under the terms and conditions of the Creative Commons Attribution (CC BY) license (https://creativecommons.org/licenses/by/4.0/).

1. Introduction

The aim of this article was to provide readers with the newest information useful in the management of infertility in adenomyosis patients. Adenomiosis, well-described at the end of the 19th century, still remains a mysterious disease with severe implications on fertility. We conducted a literature search—publications from Pubmed, Embase and Cochrane databases published from 1982 to 2019 were retrieved using terms 'adenomyosis and infertility' and 'adenomyosis and pregnancy outcomes'. Data of this search are provided in Figure 1. We focused on all aspects of infertility in adenomyosis: Symptoms, pathological background, diagnostics and possible management.

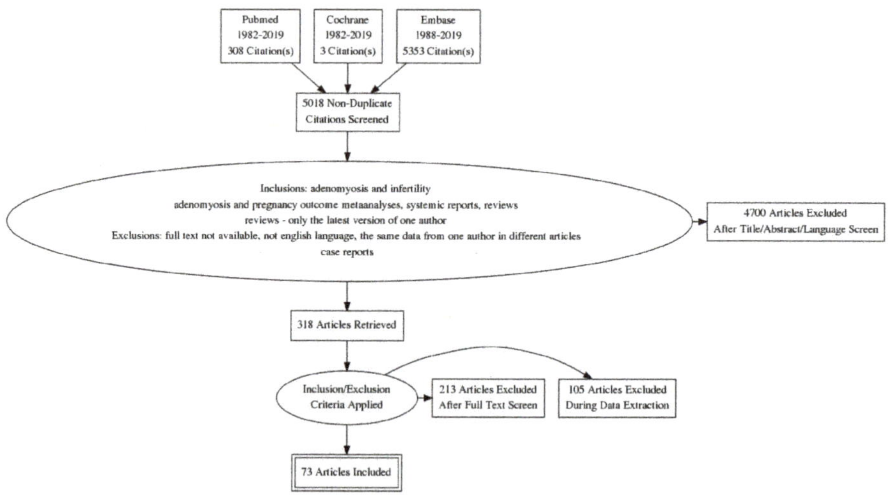

Figure 1. Process of data extraction.

2. Definition and Symptoms

Adenomyosis is defined as an invasion of the endometrium into the uterine myometrium, which results in an enlargement of the uterus, formation of adenomyotic tumours, profuse menstrual and inter-menstrual bleeding and recurrent pain. Microscopically ectopic nonneoplastic, endometrial glands and stroma surrounded by the hypertrophic and hyperplastic myometrium are noted.

The prevalence of adenomyosis fluctuates between 5 and 70% [1]. Before the age of 40 years, the disease affects 2 in 10 women, whereas between 40 and 50 years, the incidence increases to 8 in 10 women [2]. However, the incidence of adenomyosis is difficult to establish due to the lack of a unified definition and diagnostic criteria based on noninvasive diagnostic tests [3]. There are no pathognomonic clinical features for adenomyosis, nor laparoscopic criteria that could be implemented for the diagnosis [4].

In fact, adenomyosis was previously diagnosed in premenopausal women only on the basis of pathological examination after hysterectomy [5,6]. Nowadays, the diagnosis is based on imaging techniques such as transvaginal ultrasound scan (US) and magnetic resonance imaging (MRI) [7]. In one third of cases, adenomyosis is asymptomatic. The most common clinical symptoms are menorrhagia (up to 50% of patients), dysmenorrhea (30%) and metrorrhagia (20%), with other medical conditions such as enlarged uterus and infertility [2,6].

Adenomyosis may be accompanied by other mild oestrogen-dependent benign disorders such as endometriosis (70%), uterine fibroids (50%) and endometrial hyperplasia (35%). In the retrospective analysis of 945 patients who underwent hysterectomy, a significant positive correlation was found between the progression of adenomyosis and history of prior abortion, history of previous pregnancies and occurrence of leiomyoma. By contrast, there was no correlation with smoking, normal delivery, caesarean section, endometrial hyperplasia or ovarian endometriosis [8].

3. Pathogenesis

The pathogenesis of adenomyosis is still unclear. It may develop de novo from a metaplastic transformation of the embryological pluripotent mullerian remnants. The second theory, suggested by Bergeron et al., is an invasion of the basal endometrium into the myometrium through an altered or absent JZ (junctional zone—the area representing the internal myometrium) [9]. The invagination and intramyometrial spreading may be due to higher oestradiol receptor expression in the adenomyotic foci. Another hypothesis formulated by Leyendecker et al. concerns 'auto-traumatisation' of the uterus which leads to the TIAR (Tissue Injury And Repair) mechanism as the main cause of adenomyosis [10]. According to the authors, high intrauterine pressure, especially during menstruations, can cause a rupture of the archimyometrium mainly in the cornual region of the uterus. The TIAR mechanism causes a vicious cycle of hyper-estrogenic activity and expression of P450 aromatase. The TIAR mechanism can also occur after multiple D and C procedures (dilatation and curettage).

Many macromolecules such as hormones, cytokines and antigens might play a role in the pathogenesis of adenomyosis. The hyper-estrogenic environment promotes IL-10 expression. IL-10 may have an influence on the maintenance of host immunosuppression, augmenting the growth of adenomyotic foci. IL-1 and IL-6 mediate the inflammatory response by COX2- and PGE2-dependent pathways. An abnormal congestion of reactive oxygen species is also observed in adenomyosis, with the presence of ROS (reactive oxygen species), the nitric oxide derivative, being particularly abundant [11].

Moreover, the myometrium itself may be involved in disease development by influencing local biochemical factors like cytokines and oestrogens, which play a role in smooth muscle metaplasia and/or the trans-differentiation of fibroblasts into myofibroblasts [12].

A new direction of research is the expression of messenger RNA (mRNA) and long noncoding RNAs (lncRNAs) in adenomyosis. Lnc RNAs, earlier considered as transcriptional noise due to their low level of expression, are now being studied for a variety of biological functions, such as immune response and cell differentiation. Zhou et al. investigated the expression patterns of lncRNAs in human adenomyosis tissue. In microarray analysis, they showed that uc004dwe.2 lncRNA, which may affect the angio- and lymphangiogenetic function of NRP2 (neuropilin 2), is overexpressed in the ectopic endometrium compared to eutopic endometrium [13]. The limited understanding of lncRNA function in adenomyosis requires further research.

4. Diagnostics

Adenomyosis occurs in two different forms: Diffuse and focal, usually observed during transvaginal US examination. Focal forms of adenomyosis are presented as pseudo-widening, adenomyoma and haemorrhagic cysts [1]. So far, various systems have been proposed to describe adenomyosis [14].

4.1. Ultrasound

Ultrasound is the first-line imaging tool in the infertile patient. General diagnostic US features of the two forms of adenomyosis are presented in Table 1.

Table 1. Sonographic features of diffuse and focal adenomyosis.

Diffuse Adenomyosis	Focal Adenomyosis
globally enlarged uterus	focal disturbances in myometrium layer
asymmetric thickness anterior and posterior wall = pseudo-widening sign	sometimes focal form diagnosed as intramural myoma
cystic myometrium (cystic anechoic spaces)	anechoic cysts
junctional zone not clearly visible, thickening of the JZ	
heterogeneous echogenicity of the myometrium	

A characteristic symptom observed during ultrasound examination is the general enlargement of the uterus, which usually reaches 12 cm in length and cannot be explained by the presence of uterine fibroids, and is a characteristic finding in US examination. In addition, a sign called 'the question mark form of the uterus' (when the uterine corpus is flexed backwards, and the cervix is directed anteriorly towards the urinary bladder) is related to high sensitivity and specificity (92% and 75%, respectively) of US [1]. The transformation or the junctional zone is a layer that appears as a hypoechoic halo surrounding the endometrial layer [11]. Thickening of the junctional zone is a visible US sign of endometrial invasion into the myometrium. In a patient with suspected adenomyosis, a JZ thickness greater than 12 mm strongly suggests the presence of an affected myometrium, while a JZ thickness between 8 and to 12 mm may suggest solely adenomyosis, but other signs are required to confirm its presence [2].

Indeed, 2D—and 3D—TVS allows easy recognition of diagnostic signs (Figure 2). Eisnenberg et al. observed a very high overall prevalence (89.4%) of sonographic signs of adenomyosis in women undergoing laparoscopic surgery for endometriosis, much higher than in the control group [6]. In this study, a higher risk of infertility was correlated with a greater number of sonographic signs of adenomyosis. Additional examinations such as MRI are useful in identifying the variation in JZ and to exclude concomitant disorders [1].

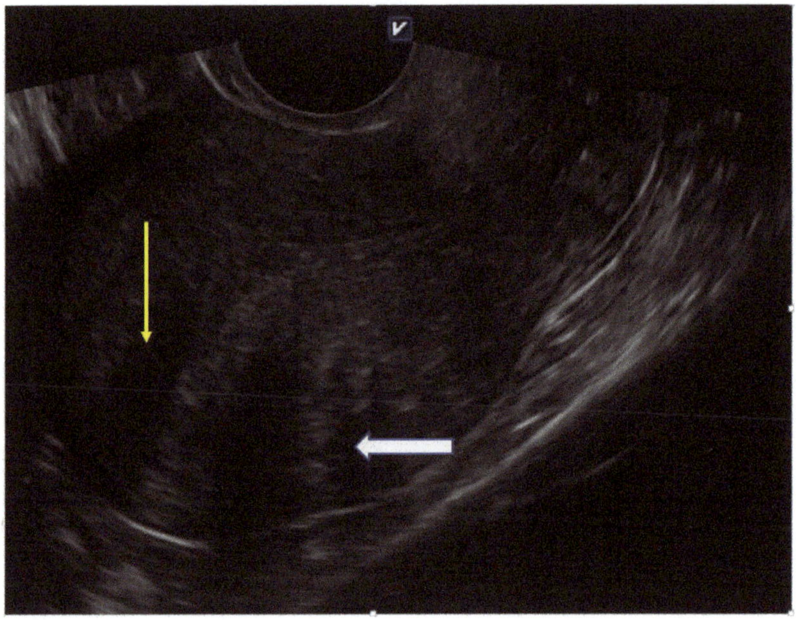

Figure 2. Globally enlarged uterus with heterogenic myometrium (yellow arrow) and asymmetric thickeness (white arrow); 41 year-old nulliparous woman with 5 years of infertility anamnesis (own material).

4.2. *Hysteroscopy*

In a specific group of patients (mainly those with an abnormal uterine bleeding pattern), hysteroscopy can be a valuable diagnostic technique which, on the one hand, enables direct visualisation of the uterine cavity, and, on the other hand, enables the collecting of material for histopathological examination. Although visual inspection does not allow a definite diagnosis, a number of features have been identified that may indicate the presence of adenomyosis: Pronounced hypervascularisation on the endometrial surface, an irregular endometrium with small openings, the so-called strawberry pattern of the endometrium and fibrous and haemorrhagic cystic lesions. More information can be obtained from the behaviour of the uterine muscle during the biopsy with a diathermy loop resectoscope. The presence of adenomyosis may be indicated by an irregular submedial myometrium, a distortion of the normal myometrial architecture noticeable during resection and the presence of intramural endometriomas [15].

4.3. MRI (Magnetic Resonance Imaging)

Pelvic MRI is the reference standard for the noninvasive detection of the adenomyosis in patients with infertility [16]. However, it requires expensive equipment and extensive knowledge and experience while assessing images. One should keep in mind that MRI better predicts adenomyosis while performed in the secretory phase of the menstrual cycle. The agreement is that a junctional zone of more than 12 mm in width is strongly associated with the disease. When the JZ measures 8–12 mm, the following key features should be assessed: More than 5 mm difference between the maximum and minimum thickness of the JZ, poor definition of borders of the JZ or high signal intensity foci in the myometrium [17].

4.4. Histological Evaluation

The key features in the diagnosis are endometrial glands within uterine muscles. The glands are circular in shape, filled with blood and surrounded by endometrial stroma. The epithelium of the glands could be pseudostratified, may have mitoses, and are densely packed spindle cells without nuclear atypia. Between epithelial cells, macrophages loaded with hemosiderin are observed. The surrounding myometrium is usually hyperplastic. There should be at least 2.5 mm distance between the endomyometrial junction and the adenomyotic gland to establish proper diagnosis of adenomyosis [9]. A histopathological report is obviously not required to treat qualified patients for infertility.

5. Biological Influence of Adenomyosis on Fertility—Possible Mechanisms

Recent studies show that adenomyosis negatively affects in vitro fertilisation, pregnancy and the live birth rate, as well as increases the risk of miscarriage. In addition, adenomyosis enhances the risk of obstetric complications, such as premature birth and preterm rupture of the amniotic membranes [18,19].

Fertility in adenomyotic patients could be disturbed by various mechanisms. Abnormal utero-tubal gamete and the embryo transport and disruption of endometrial function and receptivity have been described [2]. An enlarged uterus, anatomical distortion and intramural adenomyoma can all influence the shape of the uterine cavity. It may have a negative impact on sperm migration, embryo transfer and implantation potential [2,20]. The researchers suggested an association between spontaneous abortion and JZ function [21]. Chiang at al. observed a comparable dependence [22]. An average JZ greater than 7 mm was correlated with higher implantation failure [23].

Hyperactivity of the myometrium is also observed in adenomyosis. Changes in myocytes are also found on the cellular level—calcium circulation is distorted, which implies irregular muscle contractions, dysfunctional uterine hyperperistalsis with increased intrauterine pressure and the development of hyperplastic myometrial tissue. The thickening of the junctional zone is a visible sign of endometrial invasion into the myometrium. Alerted myometrial contractility may impair sperm progression towards the peritoneal opening of the tubes [2].

Distortion of the uterine cavity can be visualised in hysterosalpingography (HSG), and occurs in 78% of patients with diffuse adenomyosis and 54% cases of focal adenomyosis compared to 37% of women without adenomyosis. These findings may suggest the association between adenomyosis and the probability of abnormal utero-tubal transport [5].

Adenomyosis-associated changes may also worsen endometrial receptivity [24]. Endometrial receptivity is defined as physiological molecular and histological phenomena occurring during a restricted time of the menstrual cycle, making the uterus exclusively receptive to blastocyst attachment and implantation (so-called implantation window).

Evidence of reduced endometrial receptivity and impaired decidualisation in adenomyosis was found at the molecular level. Abnormal function of the implantation-associated molecules such as HOXA10, LIF, MMP, IL-6, IL -10, cytochrome P450 and RCAS1 has been described [25].

The decreased expression level of HOXA10 genes in the secretory phase endometrium appears to be involved in impaired implantation in women with adenomyosis. Similarly, a deregulation of leukaemia inhibitory factors (LIF) in uterine flushing fluid during the implantation window has been reported in adenomyosis [5,26,27]. Jiang et al. reported the down-regulation of the NR4A receptor and FOXO1A in adenomyotic tissue, which leads to incorrect decidualisation [28]. Whether these changes can be restored by the progestins given during the implantation window remains unknown due to the lack of conclusive data in humans [29,30].

Certain cell adhesion molecules, such as integrins, are also extensively studied in adenomyosis. Integrins are transmembrane receptors, which activate signalling pathways and mediate cellular signals such as regulation of the cell cycle. Integrins are responsible for endometrial receptivity. Abnormal expression of both integrin β-3 and OPN mRNA (osteopontin, responsible for the trophoblast-endometrium interaction) is found in adenomyosis patients, and it is suggested that this abnormal expression may be responsible for in vitro fertilisation (IVF) failure despite good embryo quality [31]. Integirn β3 together with osteopontin (OPN) are involved in cell–cell interactions, and their proper functioning is inevitably related to uterine receptivity. In the endometrium of adenomyotic patients, the levels of β3 and OPN were statistically lower compared to nonadenomyotic controls [25,27]. The influence of medical treatment (gonadotrophin-releasing hormone (GnRH) analogues, ovarian stimulation) on integrin expression and endometrial receptivity has so far been studied in animal models and could only partially be conclusive for human pathology [32,33].

It is a well-known fact that chronic inflammation has a negative impact on fertility [34–38]. In the case of a patient with adenomyosis, an increased expression of IL-1b and CRH (corticotrophin-releasing hormone) in the eutopic endometrium was observed [34]. In addition, the presented data showed differences in both cellular and humoral immunity in the eutopic endometrium of an adenomyotic uterus compared to the unaffected control [35]. Ishikawa et al. reported an increased inflammatory response in the endometrium due to the presence of a higher expression of pro- and anti-oxidative cytokines like Cu, Zn-SOD and Mn-SOD [36]. Other authors confirmed these findings by investigating the nitric oxide (NO) concentration in endometrium, macrophage activation, Il-6 and neurotrophins [25,37]. NO is involved in modulating uterine contractility during pregnancy and relaxing vascular smooth muscles. An abnormal high level of free radicals such as nitric oxide has a negative impact on sperm transport, implantation and decidualisation [38].

6. Impact of Adenomyosis Treatment on Fertility and Implication for Clinical Practice

As in the case of endometriosis, the management strategy of adenomyosis depends primarily on the presented symptoms. The pharmacological treatment of adenomyosis is similar to that of endometriosis, but the data concerning its influence on fertility remain inconclusive. Final treatment with a hysterectomy is the most effective way of achieving symptoms control and provides high satisfaction rates. However, for obvious reasons, it is unacceptable for women wishing to have children. Fertility-saving treatment has variable success rates for both pain and bleeding. Moreover, some of the available nonsurgical management methods severely interfere with fertility. The methods accepted for treatment of adenomyosis are listed in Table 2 and described in detail below:

Table 2. Adenomyosis treatment.

Pharmacological	Surgical
Anti-inflammatory drugs Oral contraceptives GnRH progestins	Endo-myometrial ablation High-intensity focused ultrasound Ablation Electrocoagulation of adenomyosis foci Resection of adenomyosis foci Hysterectomy

6.1. Surgical Methods

Endo-myometrial resection is effective and indicated in patients with the disease limited to the endo-myometrial junction and allows the reduction of heavy menstrual bleeding [39]. However, in patients who desire pregnancy, endo-myometrial resection is contraindicated [40]. Destruction of the endometrium together with JZ can cause serious complications in patients who managed to conceive, such as miscarriage, preterm labour and placentation complications. An unexpectedly high rate of pregnancy complications is reported in a systemic review by Kohn et al. [41]. Embolisation has also been described as an effective treatment of symptoms resulting from adenomyosis. This endovascular procedure causes the closing of vessels that supply the uterus. Premature ovarian insufficiency (POI) is mentioned as a consequence of embolisation and its rate should not be underestimated. It can affect both hormone production and ovarian reserve, leading to premature and iatrogenic amenorrhoea and infertility. Endometrial receptivity is also diminished after this procedure. Therefore, it should be contraindicated in women planning pregnancy but is useful in the post-reproductive age [42]. The high-intensity focused ultrasound method (HIFU) uses the thermal effect of the ultrasound beam, which causes coagulative necrosis within the targeted adenomyotic lesion. The lesion should be clearly visible in ultrasound or under MRI so that the beam could be precisely directed. This means that this method will be unsuitable for the diffuse form of adenomyosis. After the procedure, patients can attempt to conceive much earlier than after surgical treatment, but the exact time of delay in conception is unknown. Zhang et al. published data which indicate that the rate of uterine ruptures during pregnancy or delivery is lower than after classical surgical methods [43]. In a retrospective analysis of HIFU-treated patients, Zhou et al. found that 54 patients out of 68 conceived after HIFU, delivering 21 babies [44]. Although the miscarriages rate appears to be quite high, other severe complications like uterine rupture did not occur. It is suggested that in the HIFU method, myometrium around the adenomyotic lesion is intact, so there is no scar on the uterine wall. In classical surgical methods, the removal of significant amounts of myometrium with the adenomyotic lesion may result in a reduction in myometrial capacity of the uterus and in the formation of uterine scars. The first one can cause a lack of susceptibility of the uterus to grow during pregnancy, while the latter can cause a risk of uterine rupture. Otsubo at al. presented results of fertility-saving surgical excisions of diffuse adenomyotic foci and suggested that preservation of a 9 to 15 mm thickness of the uterine wall after excision with classical surgical methods is safe for future pregnancies [45].

Electrocoagulation has also been applied to focal or diffuse disease. However, the main disadvantage of electrocoagulation is that it may be less accurate than surgical excision, as well as poorly controlled during the procedure. Younes et al. concluded in the recent meta-analysis that conservative surgery in adenomyosis could improve fertility in some patients, but the rate of resulting successful pregnancies varied between surgical centres [46].

6.2. Pharmacological Methods

Nonsteroidal anti-inflammatory drugs (NSAIDs) are widely used in endometriosis-associated pain, but there are only a few randomised trials in endometriosis and none of them were performed in adenomyosis [47]. Their impact on fertility is negative; they can cause a delay in ovarian follicle rupture, but there is some evidence that NSAIDs can be used as a co-treatment in the IVF procedure [48]. OCs (oral contraceptives) are used in the treatment of adenomyosis to reduce the menstrual bleeding by decidualisation and subsequent endometrial atrophy. Patients with dysmenorrhoea and menorrhagia may benefit from the resulting amenorrhoea, which may alleviate symptoms. Medical therapy with OCs enables satisfactory, long-term pain control in two-thirds of women with symptomatic endometriosis or adenomyosis. However, there is no published information on the impact of OCs therapy on the subsequent fertility improvement [49]. GnRH analogues have been used to induce a constant hypoestrogenic state in women with histologically proven adenomyosis [40]. Although their use is accepted in patients with dysmenorrhoea due to adenomyosis, there are still very few data on their impact on future fertility. In one small case series published by Huang et al., authors did not show an improvement in fertility after the GnRH analogue management combined with conservative microsurgery [50]. Other results found in the literature are also vague—as presented in Table 3. An antiproliferative and anti-inflammatory effect of progestins was the basis of their use in adenomyosis treatment [51]. It has been found that they are at least partially effective in controlling pain symptoms associated with adenomyosis. They were able to reduce uterine volume and affect abnormal uterine bleeding, but their influence on fertility is poorly documented in either short case series or case reports [52–54].

7. IVF Outcome in Adenomyosis

The results of studies devoted to the efficacy of assisted reproductive techniques, such as in vitro fertilisation and embryo transfer (IVF-ET) and intracytoplasmic sperm injection (ICSI), on pregnancy rates in patients with adenomyosis showed conflicting results. Mijatovic et al. did not observe significant differences in clinical pregnancy rates in patients with adenomyosis in infertile women with proven endometriosis who were pretreated with long-term GnRH-agonist compared to controls [55].

Additionally, Thalluri and Tremellen noticed statistically significantly lower clinical pregnancy rates in patients subjected to IVF-ET with adenomyosis despite the GnRH stimulation protocol [56]. Contrarily, Costello et al. did not observe impaired fertility among patients with adenomyosis who received GnRH during IVF-ICSI [57].

Finally, Vercellini et al. published a meta-analysis confirming the negative effect of adenomyosis on the IVF-ICSI outcomes, impairing the rate of clinical pregnancy and implantation and increasing the risk of early pregnancy loss [58].

Adenomyosis was a risk factor reducing the rate of implantation and clinical pregnancy, as well as increasing the risk of early pregnancy loss. The presented data are generally inconclusive—see Table 3. The cohorts of patients were heterogeneous; patients were of different ages and differed in the clinical presentation of adenomyosis and concomitant endometriosis. The classification of patients to an appropriate group for studying purposes is difficult, which is connected to the lack of standards in the diagnosis of adenomyosis.

Table 3. Fertility outcomes and impact of different procedures on adenomyosis symptoms.

Author	Treatment	Patients		Results	
		N	Fertility	Bleeding	Pain
Kwack et al. 2018 [59]	conservative adenomyomectomy with TOUA (transient occlusion of uterine arteries)	116	5/116 conception by natura 15/11 conception by ART 7 live births	menorrhagia: 52/116 complete remission 53/116 partial remission	dysmenorrhea: 98/116 complete remission 18/116 partial remission
Al Jama et al. 2016 [60]	treatment with Gn-RH agonist	22	3/22 pregnancies 1/22 live birth	improvement in dysmenorrhea and menorrhagia was noted at the 6- and 12-month follow-up visits in both groups	
	combined conservative surgery and Gn-RHa therapy	18	8/18 pregnancies 6/18 live births		
Saremi et al. 2014 [61]	resection of adenomatosis lesions with a thin margin after sagittal incision in the uterine body	103	14/70 conception by ART 7/70 conception by natural 16/70 live births	decrease of 65% in the number of patients with a heavy bleeding pattern;	decrease of 41% in the number of patients with dysmenorrhoea symptoms;
Kishi et al. 2014 [62]	laparoscopic adenomyomectomy with laser	102	conception by natural: 16/75 (<40 y) and 0/27 (40 or more y) conception by ART: 15/75 (<40 y) and 1/27 (40 or more y) delivery: 26/75 (<40 y)	no data	no data
Chang et al. 2013 [63]	ultramini- or mini-laparotomy conservative surgery and Gn-RHa therapy	56	23/56 pregnancies 15/56 live births	no precise data	VNRS-6 (six-point verbal numeric rating scale) baseline of 3.96 ± 0.41 to 0.32 ± 0.46 1st year, 0.68 ± 0.78 2nd year 1.27 ± 1.22 3rd year,

Table 3. *Cont.*

Author	Treatment	Patients	Results		
		N	Fertility	Bleeding	Pain
Dai et al. 2012 [64]	local excision of adenomyoma at laparotomy	86	2/86 pregnancies	no data	alleviation of dysmenorrhea -12 months after treatment: >80% reduction in 77/79 (97.5%); 50–80% reduction in 2/79 (2.5%); -24 months after treatment: >80% reduction in 45/48 (93.8%); 50–80% reduction in 3/48 (6.2%);
Huang et al. 2012 [50]	excision of the adenomyosis tissue using a microsurgical technique and a six-month course of GnRHa therapy	9	6/18 conception by ART 3/18 conception by natural 2/18 live births	no data	pain score 4.7 ± 0.5 before treatment 0.33 ± 0.5 after 3 months 1.0 ± 0.9 after 12 months
Osada et al. 2011 [65]	adenomyomectomy with a triple-flap method, without overlapping suture lines	104	4/26 conception by natura 112/26 conception by ART 14 live births	VAS (visual analogue scale) hypermenorrhoea, 10 pre-surgically 3.27 ± 2.17 at 3 months, 2.89 ± 1.77 at 6 months, 2.63 ± 1.3 at 1 year, 2.87 ± 1.77 at 2 years post-surgery	The VAS findings (dysmenorrhoea, 10 pre-surgically) 1.61 ± 1.43 at 3 months, 1.54 ± 1.62 at 6 months, 1.44 ± 1.65 at 1 year, 1.67 ± 1.79 at 2 years post-surgery.
Takeuchi et al. 2010 [66]	laparoscopic enucleation of juvenile cystic adenomyoma	9	2/3 pregnancies	no data	dysmenorrhea 8–10 on the VAS before the surgery, decreased to 2 by 6 months after
Nishida et al. 2010 [67]	adenomyomectomy with unilateral salpingectomy	44	1/16 live births	reducing menstrual blood loss, no quantitative data	dramatic relief from dysmenorrhea, no quantitative data
Hadisaputra et al. 2006 [68]	laparoscopic resection +GnRH analogue after surgery	10	3/10 pregnancies	no change in the symptom of menorrhagia	75% reduction in dysmenorrhea after treatment
	myolysis +GnRH analogue after surgery	10	2/10 pregnancies	no change in the symptom of menorrhagia	58.31% reduction in dysmenorrhea after treatment
Rajuddin et al. 2006 [69]	laparotomic resection	32	3/32 pregnancies 2/32 live births	after intervention, 28/32 experienced disappearance of symptoms (dysmenorrhea, pelvic pain, menorrhagia, dyspareunia), while 4/32 had remaining symptoms;	
	treatment with aromatase inhibitor of anastrozole	23	2/23 pregnancies 1/23 live births	after therapy 14/23 experienced a disappearance of symptoms, while 9/23 had remaining symptoms;	
Takeuchi et al. 2006 [70]	laparoscopic adenomyomectomy and hysteroplasty	14	2/14 pregnancies	all 8 cases of polyhypermenorrhea improved, no precise data available	dysmenorrhea—VAS during menstruation decreased from 10 before operation to 2.5 after operation.
Fujishita et al. 2004 [71]	classical reduction surgery	5	0/5 pregnancies	2/5 relief of menorrhagia and dysmenorrhea	
	transverse H incision method and the reduction surgery	6	1/6 pregnancies	3/6 relief of symptoms	

8. Conclusions

There is no specific treatment for patients with adenomyosis who want to retain their uterus or wish to preserve fertility [72]. Sometimes, combined treatment can be proposed: Laparoscopy, GnRH treatment and in vitro fertilisation [73].

When comparing pharmacological and surgical treatment, the latter appears to be more effective but some details are unclear, i.e., how long pregnancy should be delayed after treatment and whether hormone treatment after surgery improves fertility outcome. Despite many studies on the pathogenesis of fertility failure in adenomyosis, their results are not correlated with treatment. Thus, it is of great importance to explore new, more effective, safe and less invasive managing strategies in women with infertility due to adenomyosis.

Author Contributions: M.S.—conceptualization, methodology, data curation, formal analysis, draft preparation, writing—review and editing; E.K.—resources, writing; O.O.—data analysis, methodology, writing—review and editing; K.K.-K.—data analysis, methodology, writing—review and editing; J.K.—data analysis, methodology, writing—review and editing, English corrections; J.W.—conceptualization, writing—review and editing, supervision. All authors have read and agreed to the published version of the manuscript.

Funding: This work was not supported by any Funding Agency. COPE guidelines were followed.

Institutional Review Board Statement: Not applicable.

Informed Consent Statement: Not applicable.

Conflicts of Interest: There is no potential conflict of interest to disclose.

References

1. Graziano, A.; Monte, G.L.; Piva, I.; Caserta, D.; Karner, M.; Engl, B.; Marci, R. Diagnostic findings in adenomyosis: A pictorial review on the major concerns. *Eur. Rev. Med. Pharmacol. Sci.* **2015**, *19*, 1146–1154. [PubMed]
2. Harada, T.; Khine, Y.M.; Kaponis, A.; Nikellis, T.; Decavalas, G.; Taniguchi, F. The Impact of Adenomyosis on Women's Fertility. *Obstet. Gynecol. Surv.* **2016**, *71*, 557–568. [CrossRef] [PubMed]
3. Li, J.J.; Chung, J.P.; Wang, S.; Li, T.C.; Duan, H. The Investigation and Management of Adenomyosis in Women Who Wish to Improve or Preserve Fertility. *BioMed Res. Int.* **2018**, *2018*, 6832685. [CrossRef] [PubMed]
4. Ecker, A.M.; Chamsy, D.; Austin, R.M.; Guido, R.S.; Lee, T.T.; Mansuria, S.M.; Rindos, N.B.; Donnellan, N.M. Use of Uterine Characteristics to Improve Fertility-Sparing Diagnosis of Adenomyosis. *J. Gynecol. Surg.* **2018**, *34*, 183–189. [CrossRef] [PubMed]
5. Senturk, L.M.; Imamoglu, M. Adenomyosis: What is new? *Womens Health* **2015**, *11*, 717–724. [CrossRef]
6. Eisenberg, V.H.; Arbib, N.; Schiff, E.; Goldenberg, M.; Seidman, D.S.; Soriano, D. Sonographic Signs of Adenomyosis Are Prevalent in Women Undergoing Surgery for Endometriosis and May Suggest a Higher Risk of Infertility. *BioMed Res. Int.* **2017**, *2017*, 8967803. [CrossRef]
7. Van den Bosch, T.; Van Schoubroeck, D. Ultrasound diagnosis of endometriosis and adenomyosis: State of the art. *Best Pract. Res. Clin. Obstet. Gynaecol.* **2018**, *51*, 16–24. [CrossRef]
8. Genc, M.; Genc, B.; Cengiz, H. Adenomyosis and accompanying gynecological pathologies. *Arch. Gynecol. Obstet.* **2015**, *291*, 877–881. [CrossRef]
9. Bergeron, C.; Amant, F.; Ferenczy, A. Pathology and physiopathology of adenomyosis. *Best Pract. Res. Clin. Obstet. Gynaecol.* **2006**, *20*, 511–521. [CrossRef]
10. Leyendecker, G.; Wildt, L.; Mall, G. The pathophysiology of endometriosis and adenomyosis: Tissue injury and repair. *Arch. Gynecol. Obstet.* **2009**, *280*, 529–538. [CrossRef]
11. Garavaglia, E.; Audrey, S.; Annalisa, I.; Stefano, F.; Iacopo, T.; Laura, C.; Massimo, C. Adenomyosis and its impact on women fertility. *Iran. J. Reprod. Med.* **2015**, *13*, 327–336. [PubMed]
12. Yen, C.F.; Huang, S.J.; Lee, C.L.; Wang, H.S.; Liao, S.K. Molecular Characteristics of the Endometrium in Uterine Adenomyosis and Its Biochemical Microenvironment. *Reprod. Sci.* **2017**, *24*, 1346–1361. [CrossRef] [PubMed]
13. Zhou, C.; Zhang, T.; Liu, F.; Ni, X.; Huo, R.; Shi, Z. The differential expression of mRNAs and long noncoding RNAs between ectopic and eutopic endometria provides new insights into adenomyosis. *Mol. Biosyst.* **2016**, *12*, 362–370. [CrossRef] [PubMed]
14. De Bruijn, A.M.; van den Bosch, T.; de Leeuw, R.A.; Dueholm, M.; Exacoustos, C.; Valentin, L.; Bourne, T.; Timmerman, D.; Huirne, J.A. A sonographic classification and reporting system for diagnosing adenomyosis. *Ultrasound Obstet. Gynecol.* **2019**, *53*, 151.
15. Di Spiezio Sardo, A.; Calagna, G.; Santangelo, F.; Zizolfi, B.; Tanos, V.; Perino, A.; De Wilde, R.L. The Role of Hysteroscopy in the Diagnosis and Treatment of Adenomyosis. *BioMed Res. Int.* **2017**, *2017*, 2518396. [CrossRef]
16. Khandeparkar, M.S.; Jalkote, S.; Panpalia, M.; Nellore, S.; Mehta, T.; Ganesan, K.; Parikh, F.R. High resolution magnetic resonance imaging in the detection of subtle nuances of uterine adenomyosis in infertility. *Glob. Reprod. Health* **2018**, *3*, e14. [CrossRef]
17. Kunz, G.; Beil, D.; Huppert, P.; Noe, M.; Kissler, S.; Leyendecker, G. Adenomyosis in endometriosis–prevalence and impact on fertility. Evidence from magnetic resonance imaging. *Hum. Reprod.* **2005**, *20*, 2309–2316. [CrossRef]
18. Vigano, P.; Corti, L.; Berlanda, N. Beyond infertility: Obstetrical and postpartum complications associated with endometriosis and adenomyosis. *Fertil. Steril.* **2015**, *104*, 802–812. [CrossRef]
19. Buggio, L.; Monti, E.; Gattei, U.; Dridi, D.; Vercellini, P. Adenomyosis: Fertility and obstetric outcome. A comprehensive literature review. *Minerva Ginecol.* **2018**, *70*, 295–302.
20. Vlahos, N.F.; Theodoridis, T.D.; Partsinevelos, G.A. Myomas and Adenomyosis: Impact on Reproductive Outcome. *BioMed Res. Int.* **2017**, *2017*, 5926470. [CrossRef]
21. Benagiano, G.; Brosens, I.; Habiba, M. Adenomyosis: A life-cycle approach. *Reprod. Biomed. Online* **2015**, *30*, 220–232. [CrossRef] [PubMed]

22. Chiang, C.H.; Chang, M.Y.; Shiau, C.S.; Hou, H.C.; Soong, Y.K. Effect of a sonographically diffusely enlarged uterus without distinct uterine masses on the outcome of in vitro fertilization-embryo transfer. *J. Assist. Reprod. Genet.* **1999**, *16*, 369–372. [CrossRef] [PubMed]
23. Maubon, A.; Faury, A.; Kapella, M.; Pouquet, M.; Piver, P. Uterine junctional zone at magnetic resonance imaging: A predictor of in vitro fertilization implantation failure. *J. Obstet. Gynaecol. Res.* **2010**, *36*, 611–618. [CrossRef] [PubMed]
24. Benagiano, G.; Brosens, I. The endometrium in adenomyosis. *Womens Health* **2012**, *8*, 301–312. [CrossRef]
25. Vannuccini, S.; Tosti, C.; Carmona, F.; Huang, S.J.; Chapron, C.; Guo, S.W.; Petraglia, F. Pathogenesis of adenomyosis: An update on molecular mechanisms. *Reprod. Biomed. Online* **2017**, *35*, 592–601. [CrossRef]
26. Fischer, C.P.; Kayisili, U.; Taylor, H.S. HOXA10 expression is decreased in endometrium of women with adenomyosis. *Fertil. Steril.* **2011**, *95*, 1133–1136. [CrossRef]
27. Xiao, Y.; Li, T.; Xia, E.; Yang, X.; Sun, X.; Zhou, Y. Expression of integrin β3 and osteopontin in the eutopic endometrium of adenomyosis during the implantation window. *Eur. J. Obstet. Gynecol. Reprod. Biol.* **2013**, *170*, 419–422. [CrossRef]
28. Jiang, Y.; Jiang, R.; Cheng, X.; Zhang, Q.; Hu, Y.; Zhang, H.; Cao, Y.; Zhang, M.; Wang, J.; Ding, L.; et al. Decreased expression of NR4A nuclear receptors in adenomyosis impairs endometrial decidualization. *Mol. Hum. Reprod.* **2016**, *22*, 655–668. [CrossRef]
29. Wu, D.; Li, G.; Lin, R.; Li, C.; Zhao, S. Progesterone Promotes HOXa-10 Expression in Mouse during Embryo Implantation Period. *J. Anim. Vet. Adv.* **2012**, *11*, 3076–3080.
30. Yao, M.W.; Lim, H.; Schust, D.J.; Choe, S.E.; Farago, A.; Ding, Y.; Michaud, S.; Church, G.M.; Maas, R.L. Gene expression profiling reveals progesterone-mediated cell cycle and immunoregulatory roles of Hoxa-10 in the preimplantation uterus. *Mol. Endocrinol.* **2003**, *17*, 610–627. [CrossRef]
31. Surrey, E.S.; Minjarez, D.A.; Schoolcraft, W.B. The incidence of aberrant endometrial alphavbeta(3) vitronectin expression in a high risk infertility population: Could prolonged GnRH agonist therapy play a role? *J. Assist. Reprod. Genet.* **2007**, *24*, 553–556. [CrossRef] [PubMed]
32. Garcıa, P.; Nieto, A.; Sanchez, M.A.; Pizarro, M.; Flores, J.M. Expression of alphav, alpha4, alpha5 and beta3 integrin subunits, fibronectin and vitronectin in goat peri-implantation. *Anim. Reprod. Sci.* **2004**, *80*, 91–100. [CrossRef]
33. Fayazi, M.; Boroujeni, M.B.; Salehnia, M.; Khansarinejad, B. Ovarian stimulation by exogenous gonadotropin decreases the implantation rate and expression of mouse blastocysts integrins. *Iran. Biomed. J.* **2014**, *18*, 8–15. [PubMed]
34. Carrarelli, P.; Yen, C.F.; Funghi, L.; Arcuri, F.; Tosti, C.; Bifulco, G.; Luddi, A.; Lee, C.L.; Petraglia, F. Expression of Inflammatory and Neurogenic Mediators in Adenomyosis. *Reprod. Sci.* **2017**, *24*, 369–375. [CrossRef]
35. Benagiano, G.; Brosens, I.; Habiba, M. Structural and molecular features of the endomyometrium in endometriosis and adenomyosis. *Hum. Reprod. Update* **2014**, *20*, 386–402. [CrossRef]
36. Ishikawa, M.; Nakata, T.; Yaginuma, Y.; Nishiwaki, K.; Goishi, K.; Saitoh, S. Expression of superoxide dismutase (SOD) in adenomyosis. *Am. J. Obstet. Gynecol.* **1993**, *169*, 730–734. [CrossRef]
37. Orazov, M.R.; Radzinsky, V.E.; Nosenko, E.N.; Khamoshina, M.B.; Dukhin, A.O.; Lebedeva, M.G. Immune-inflammatory predictors of the pelvic pain syndrome associated with adenomyosis. *Gynecol Endocrinol.* **2017**, *33* (Suppl. S1), 44–46. [CrossRef]
38. Ota, H.; Igarashi, S.; Sato, N.; Tanaka, H.; Tanaka, T. Involvement of catalase in the endometrium of patients with endometriosis and adenomyosis. *Fertil. Steril.* **2002**, *78*, 804–809. [CrossRef]
39. Busca, A.; Parra-Herran, C. The role of pathologic evaluation of endometrial ablation resections in predicting ablation failure and adenomyosis in hysterectomy. *Pathol. Res. Pract.* **2016**, *212*, 778–782. [CrossRef]
40. Wood, C. Adenomyosis: Difficult to diagnose, and difficult to treat. *Diagn. Ther. Endosc.* **2001**, *7*, 89–95. [CrossRef]
41. Kohn, J.R.; Shamshirsaz, A.A.; Popek, E.; Guan, X.; Belfort, M.A.; Fox, K.A. Pregnancy after endometrial ablation: A systematic review. *BJOG* **2018**, *125*, 43–53. [CrossRef] [PubMed]
42. De Bruijn, A.M.; Smink, M.; Lohle, P.N.; Huirne, J.A.; Twisk, J.W.; Wong, C.; Schoonmade, L.; Hehenkamp, W.J. Uterine Artery Embolization for the Treatment of Adenomyosis: A Systematic Review and Meta-Analysis. *J. Vasc. Interv. Radiol.* **2017**, *28*, 1629–1642. [CrossRef] [PubMed]
43. Zhang, L.; Rao, F.; Setzen, R. High intensity focused ultrasound for the treatment of adenomyosis: Selection criteria, efficacy, safety and fertility. *Acta Obstet. Gynecol. Scand.* **2017**, *96*, 707–714. [CrossRef] [PubMed]
44. Zhou, C.Y.; Xu, X.J.; He, J. [Pregnancy outcomes and symptom improvement of patients with adenomyosis treated with high intensity focused ultrasound ablation]. *Zhonghua Fu Chan Ke Za Zhi* **2016**, *51*, 845–849. (In Chinese) [PubMed]
45. Otsubo, Y.; Nishida, M.; Arai, Y.; Ichikawa, R.; Taneichi, A.; Sakanaka, M. Association of uterine wall thickness with pregnancy outcome following uterine-sparing surgery for diffuse uterine adenomyosis. *Aust. N. Z. J. Obstet. Gynaecol.* **2016**, *56*, 88–91. [CrossRef]
46. Younes, G.; Tulandi, T. Conservative Surgery for Adenomyosis and Results: A Systematic Review. *J. Minim. Invasive Gynecol.* **2018**, *25*, 265–276. [CrossRef]
47. Allen, C.; Hopewell, S.; Prentice, A.; Gregory, D. Nonsteroidal anti-inflammatory drugs for pain in women with endometriosis. *Cochrane Database of Syst. Rev.* **2017**, *1*, CD004753. [CrossRef]
48. Nyachieo, A.; Siristatidis, C.S.; Vaidakis, D. Nonsteroidal anti-inflammatory drugs for assisted reproductive technology. *Cochrane Database Syst. Rev.* **2019**, *10*, CD007618. [CrossRef]
49. Vercellini, P.; Viganò, P.; Somigliana, E.; Fedele, L. Endometriosis: Pathogenesis and treatment. *Nat. Rev. Endocrinol.* **2014**, *10*, 261–275. [CrossRef]

50. Huang, B.S.; Seow, K.M.; Tsui, K.H.; Huang, C.Y.; Lu, Y.F.; Wang, P.H. Fertility outcome of infertile women with adenomyosis treated with the combination of a conservative microsurgical technique and GnRH agonist: Long-term follow-up in a series of nine patients. *Taiwan. J. Obstet. Gynecol.* **2012**, *51*, 212–216. [CrossRef]
51. Vannuccini, S.; Luisi, S.; Tosti, C.; Sorbi, F.; Petraglia, F. Role of medical therapy in the management of uterine adenomyosis. *Fertil. Steril.* **2018**, *109*, 398–405. [CrossRef] [PubMed]
52. Matsushima, T.; Akira, S.; Fukami, T.; Yoneyama, K.; Takeshita, T. Efficacy of Hormonal Therapies for Decreasing Uterine Volume in Patients with Adenomyosis. *Gynecol. Minim. Invasive Ther.* **2018**, *7*, 119–123. [CrossRef] [PubMed]
53. Osuga, Y.; Fujimoto-Okabe, H.; Hagino, A. Evaluation of the efficacy and safety of dienogest in the treatment of painful symptoms in patients with adenomyosis: A randomized, double-blind, multicenter, placebo-controlled study. *Fertil. Steril.* **2017**, *108*, 673–678. [CrossRef] [PubMed]
54. Dasrilsyah, R.A.; Shan, L.P.; Kwang, N.B.; Shafiee, M.N.; Omar, M.H. Spontaneous conception following GnRHa and progestogen therapy in adenomyosis. *Horm. Mol. Biol. Clin. Investig.* **2016**, *27*, 77–79. [CrossRef] [PubMed]
55. Mijatovic, V.; Florijn, E.; Halim, N.; Schats, R.; Hompes, P. Adenomyosis has no adverse effects on IVF/ICSI outcomes in women with endometriosis treated with long-term pituitary down-regulation before IVF/ICSI. *Eur. J. Obstet. Gynecol. Reprod. Biol.* **2010**, *151*, 62–65. [CrossRef]
56. Thalluri, V.; Tremellen, K.P. Ultrasound diagnosed adenomyosis has a negative impact on successful implantation following GnRH antagonist IVF treatment. *Hum. Reprod.* **2012**, *27*, 3487–3492. [CrossRef] [PubMed]
57. Costello, M.F.; Lindsay, K.; McNally, G. The effect of adenomyosis on in vitro fertilisation and intra-cytoplasmic sperm injection treatment outcome. *Eur. J. Obstet. Gynecol. Reprod. Biol.* **2011**, *158*, 229–234. [CrossRef]
58. Vercellini, P.; Consonni, D.; Dridi, D.; Bracco, B.; Frattaruolo, M.P.; Somigliana, E. Uterine adenomyosis and in vitro fertilization outcome: A systematic review and meta-analysis. *Hum. Reprod.* **2014**, *29*, 964–977. [CrossRef]
59. Kwack, J.Y.; Kwon, Y.S. Conservative surgery of diffuse adenomyosis with TOUA: Single surgeon experience of one hundred sixteen cases and report of fertility outcomes. *Kaohsiung J. Med. Sci.* **2018**, *34*, 290–294. [CrossRef]
60. Al Jama, F.E. Management of adenomyosis in subfertile women and pregnancy outcome. *Oman Med. J.* **2011**, *26*, 178–181. [CrossRef]
61. Saremi, A.; Bahrami, H.; Salehian, P.; Hakak, N.; Pooladi, A. Treatment of adenomyomectomy in women with severe uterine adenomyosis using a novel technique. *Reprod. Biomed. Online* **2014**, *28*, 753–760. [CrossRef] [PubMed]
62. Kishi, Y.; Yabuta, M.; Taniguchi, F. Who will benefit from uterus-sparing surgery in adenomyosis-associated subfertility? *Fertil. Steril.* **2014**, *102*, 802–807. [CrossRef] [PubMed]
63. Chang, W.H.; Wang, K.C.; Lee, N.R.; Huang, N.; Su, W.H.; Chao, H.T.; Yen, M.S.; Fuh, J.L.; Wang, P.H. Reproductive performance of severely symptomatic women with uterine adenomyoma who wanted preservation of the uterus and underwent combined surgical-medical treatment. *Taiwan. J. Obstet. Gynecol.* **2013**, *52*, 39–45. [CrossRef] [PubMed]
64. Dai, Z.; Feng, X.; Gao, L.; Huang, M. Local excision of uterine adenomyomas: A report of 86 cases with follow-up analyses. *Eur. J. Obstet. Gynecol. Reprod. Biol.* **2012**, *161*, 84–87. [CrossRef] [PubMed]
65. Osada, H.; Silber, S.; Kakinuma, T.; Nagaishi, M.; Kato, K.; Kato, O. Surgical procedure to conserve the uterus for future pregnancy in patients suffering from massive adenomyosis. *Reprod. Biomed. Online* **2011**, *22*, 94–99. [CrossRef] [PubMed]
66. Takeuchi, H.; Kitade, M.; Kikuchi, I.; Kumakiri, J.; Kuroda, K.; Jinushi, M. Diagnosis, laparoscopic management, and histopathologic findings of juvenile cystic adenomyoma: A review of nine cases. *Fertil. Steril.* **2010**, *94*, 862–868. [CrossRef]
67. Nishida, M.; Takano, K.; Arai, Y.; Ozone, H.; Ichikawa, R. Conservative surgical management for diffuse uterine adenomyosis. *Fertil. Steril.* **2010**, *94*, 715–719. [CrossRef]
68. Hadisaputra, W.; Anggraeni, T.D. Laparoscopic resection versus myolysis in the management of symptomatic uterine adenomyosis: Alternatives to conventional treatment. *Med. J. Indones* **2006**, *15*, 9–17. [CrossRef]
69. Rajuddin, R.; Jacoeb, T.Z. Management of adenomyosis in infertile women: Comparison between laparotomic resection and administration of aromatase inhibitor (Experience in 55 cases). *Med. J. Indones* **2006**, *15*, 18–23. [CrossRef]
70. Takeuchi, H.; Kitade, M.; Kikuchi, I.; Shimanuki, H.; Kumakiri, J.; Kitano, T.; Kinoshita, K. Laparoscopic adenomyomectomy and hysteroplasty: A novel method. *J. Minim. Invasive Gynecol.* **2006**, *13*, 150–154. [CrossRef]
71. Fujishita, A.; Masuzaki, H.; Khan, K.N.; Kitajima, M.; Ishimaru, T. Modified reduction surgery for adenomyosis. A preliminary report of the transverse H incision technique. *Gynecol. Obstet. Investig.* **2004**, *57*, 132–138. [CrossRef] [PubMed]
72. Vannuccini, S.; Petraglia, F. Recent advances in understanding and managing adenomyosis. *F1000Research* **2019**, *8*. F1000 Faculty Rev-283. [CrossRef] [PubMed]
73. Sõritsa, D.; Saare, M.; Laisk-Podar, T.; Peters, M.; Sõritsa, A.; Matt, K.; Karro, H.; Salumets, A. Pregnancy rate in endometriosis patients according to the severity of the disease after using a combined approach of laparoscopy, GnRH agonist treatment and in vitro fertilization. *Gynecol. Obstet. Investig.* **2015**, *79*, 34–39. [CrossRef] [PubMed]

International Journal of
Environmental Research and Public Health

MDPI

Case Report

CHEK2 Mutation in Patient with Multiple Endocrine Glands Tumors. Case Report

Anna Szeliga [1], Aleksandra Pralat [2,†], Wiktoria Witczak [2,†], Agnieszka Podfigurna [1], Cezary Wojtyla [3,4], Anna Kostrzak [1] and Blazej Meczekalski [1,*]

1 Department of Gynecological Endocrinology, Poznan University of Medical Sciences, 61-701 Poznan, Poland; annamaria.szeliga@gmail.com (A.S.); agnieszkapodfigurna@gmail.com (A.P.); ankos30@op.pl (A.K.)
2 Students' Scientific Society of the Department of Gynecological Endocrinology, Poznan University of Medical Sciences, 61-701 Poznan, Poland; panikukulska@gmail.com (A.P.); wwitczak1@wp.pl (W.W.)
3 International Prevention Research Institute—Collaborating Centre, State University of Applied Sciences, 62-800 Kalisz, Poland; czwo@op.pl
4 Department of Oncological Gynecology and Obstetrics, Center of Postgraduate Medical Education, 00-416 Warsaw, Poland
* Correspondence: blazejmeczekalski@yahoo.com; Tel.: +48-61-65-99-366
† Contribution of authors is equal.

Received: 21 May 2020; Accepted: 17 June 2020; Published: 18 June 2020

Abstract: Background: Many studies show the occurrence of several multiple endocrine neoplasia syndromes caused by different mutations, for example, in MEN1 and RET genes. Nevertheless, there are less common mutations causing multiple endocrine glands tumors. Examples of such mutations are CHEK2 gene mutations, causing breast, kidney, gastric, colorectal, prostate, lung, ovarian, and thyroid cancers. Case description: In 2005, a 30-year-old woman was admitted to the hospital due to uncontrolled hypertension and obesity. Performed tests have shown ACTH (adrenocorticotropic hormone)—independent micronodular adrenal hyperplasia (AIMAH) as a cause. In 2010, the further diagnostic analysis revealed Cushing's disease caused by ACTH-secreting pituitary microadenoma. Additionally, in 2011, the patient underwent the strumectomy of multinodular struma. Papillary thyroid carcinoma was found in the excised tissue. In 2018, transvaginal ultrasonography revealed a tumor of the right ovary. After a performed hysterectomy with bilateral salpingo-oophorectomy, the histopathology result has shown female adnexal tumors of probable Wolffian origin (FATWO) located in the broad ligament of the uterus. Due to the history of multiglandular diseases, the patient was referred to genetic testing. We found a positive pathogenic mutation in CHEK2-suppressor gene involved in DNA repair, cell cycle arrest, and apoptosis in response to DNA damage. Conclusion: CHEK2 variants may predispose to a range of endocrine glands tumors, including those identified in our patient. Multiple endocrine glands tumors, as in the presented patient, are a serious problem of public health, due to numerous hospitalizations and necessary repeated surgical treatments. Moreover, the association between CHEK2 and ovarian cancer can be a serious problem with reproductive health.

Keywords: CHEK2; MEN; Cushing syndrome; hypertension; multiple endocrine glands tumors

1. Introduction

The term multiple endocrine neoplasia (MEN) encompasses several distinct syndromes featuring tumors of endocrine glands, each with its characteristic pattern. MEN type 1 describes the association of pituitary, parathyroid, and pancreatic islet cells [1]. In contrast, MEN type 2 (MEN 2) condition is associated with pheochromocytoma, medullary thyroid carcinoma, hyperparathyroidism, mucosal and digestive neurofibromatosis, and Marfanoid habitus [1]. Mutations in the PRKAR1A and KIT genes predispose to rare Carney's syndrome. Except for the triad of diseases—gastric stromal sarcoma, pulmonary chondroma, and paraganglioma, they can also cause micronodular adrenal hyperplasia, pituitary adenomas, and ovarian and thyroid tumors [2]. Tumor-associated mutations in the PTEN gene may cause Cowden's syndrome, in which multinodular struma, follicular thyroid cancer, uterine, and genitourinary tract tumors are observed [3]. CHEK2 gene encodes checkpoint kinase 2, which is a serine/threonine kinase activated upon DNA damage and is implicated in pathways that regulate DNA repair, cell cycle arrest (in gap 1 phase), or apoptosis in response to the initial damage. When activated, the encoded protein is known to inhibit CDC25C phosphatase, preventing entry into mitosis, and has been shown to stabilize the tumor suppressor protein p53 [4]. Loss of kinase function caused by mutations in the CHEK2 gene has been linked with Li–Fraumeni syndrome, sarcomas, breast, and colorectal cancer, brain, thyroid, lung, ovary, bladder, and prostate and kidney tumors [4]. In Poland, there are three polymorphic variants of CHEK2, which are present in 5.5% of the population [5]. The missense CHEK2 I157T allele was found in 4,8% of controls, and it was more common in cancer cases than in controls—the association was significant at the $p \leq 0.01$ level [5]. Studies reported no significant difference in the survival of the p.I157T carriers and non-carriers [6]. To summarize, CHEK2 is a multisite cancer gene that increases the risk of several cancers.

In this case report, we present a 44-year-old woman with multiple malignancies of endocrine glands without a pattern characteristic for described syndromes, caused by CHEK2 mutation. Multiple endocrine glands tumors, as in the presented patient, are a serious problem of public health due to numerous hospitalizations, necessary repeated surgical treatments, and quality of life [7].

2. Case Report

In 2005, a 30-year-old woman was admitted to the Department of Endocrinology, Poznan University of Medical Sciences, Poznan, Poland, with uncontrolled hypertension and obesity. On physical examination at admission, the patient was 163-cm tall and weighed 87 kg, with a body mass index (BMI) of 33 kg/m^2. The mean arterial pressure in daily measurements was 140/105 mmHg. In the clinical examination, Cushing syndrome was suspected. In the hormonal evaluation, elevated cortisol concentration with the normal circadian rhythm of cortisol secretion was found.

Nevertheless, in a dexamethasone suppression test, the patient presented an elevated cortisol concentration after administration of 1 mg of dexamethasone (Table 1). Serum and diurnal urine concentration of Na, K, Ca, and phosphates were normal. The serum glucose profile is presented in Table 2.

Table 1. Concentrations of chosen hormones in the described patient in 2005.

	Hormone	Result	Normal Range
Hormonal profile in 2005	ACTH	27.7 pg/ml	0–50 pg/ml
	ACTH after 1 mg of dexamethasone	4.4 pg/ml	
	Cortisol at 8 am	1354–1151–949.8 nmol/l	220–690 nmol/l
	Cortisol at 6 pm	967.9 nmol/l	50–165 nmol/l
	Cortisol at 8 pm	503 nmol/l	
	Cortisol after 1 mg dexamethasone	120.5 nmol/l	

ACTH, adrenocorticotropic hormone.

Table 2. Patient's blood glucose level before treatment in 2005.

Time of Measurement	8 AM	11 AM	5 PM	8 PM
Blood glucose concentration [mg/dl]	78	111	94	118

Among the complications of hypertension, the patient revealed a hypertensive retinopathy of stage I/II.

We also performed dual-energy X-ray absorptiometry: T-score at L1–L4 lumbar spine was −1.57, and the patient was diagnosed with osteopenia.

Due to adrenocorticotropic hormone (ACTH)-independent hypercortisolemia, we performed diagnostic imaging; an adrenal computed tomography (CT) scan revealed the presence of moderately enlarged left gland presenting homogeneous absorption with no focal changes, whereas the right adrenal gland appeared normal. Magnetic resonance imaging (MRI) did not confirm any lesions in the pituitary gland. The biological and radiological data were highly suggestive of ACTH-independent Cushing's syndrome. Surgical removal of the left adrenal gland was performed. Histopathology results showed micronodular adrenal hyperplasia. The patient was prescribed indapamide 1.5 mg once daily and perindopril 5 mg once daily. Moreover, sodium and potassium, calcium, and vitamin D3 supplementation were prescribed.

Despite surgical treatment, in 2010, the patient was admitted to the hospital due to recurrent uncontrolled hypertension. Her BMI was 35.8 kg/m^2, and mean arterial pressure in daily measurements was 155/120 mmHg. Skin hyperpigmentation was observed. The hormonal profile is presented in Table 3 and the patient's blood glucose level before treatment in 2010 is presented in Table 4.

Table 3. Concentrations of chosen hormones in the described patient in 2010.

<table>
<tr><th></th><th>Hormone</th><th>Result</th><th>Normal Range</th></tr>
<tr><td rowspan="5">Hormonal profile</td><td>ACTH at 8 am</td><td>47.6 pg/ml</td><td rowspan="2"><60 pg/ml</td></tr>
<tr><td>ACTH at 6 pm</td><td>47.35 pg/ml</td></tr>
<tr><td>Cortisol at 8 am</td><td>26.7 μg/dl</td><td>4.2–38.4 μg/dl</td></tr>
<tr><td>Cortisol at 6 pm</td><td>25.8 μg/dl</td><td>1.7–16.6 μg/dl</td></tr>
<tr><td>Cortisol in 24-h urine collection</td><td>359.304 μg/d</td><td>32–243 μg/d</td></tr>
</table>

ACTH, adrenocorticotropic hormone.

Table 4. Patient's blood glucose level before treatment in 2010.

Time of Measurement	8 AM	11 AM	5 PM	8 PM
Blood glucose concentration [mg/dl]	99.4	86.4	102.6	91.8

Adrenal angiography did not show any focal changes. MRI of the sella region was therefore performed, revealing a hypointense pituitary tumor of 5 mm in diameter located in the left anterior pituitary lobe. These MRI findings were highly suggestive of pituitary microadenoma.

Trans-sphenoidal total resection of the pituitary tumor was performed. Upon immunohistochemical examination, the resected pituitary microadenoma cells exhibited positive staining for the ACTH antibody. The patient was prescribed perindopril 5 mg once daily, losartan 50 mg once daily, spironolactone 50 mg once daily, bisoprolol 5 mg once daily, indapamide 1.5 mg once daily, acetylsalicylic acid 75 mg once daily, desmopressin 0.1 mg twice daily, and hydrocortisone 20 mg–10 mg–0 mg daily.

Subsequently, in 2011, the patient was admitted to the Department of Surgery for a scheduled strumectomy due to multinodular struma, which had been diagnosed in 2010. During the physical examination, an enlarged and fibroelastic thyroid was identified. The serum concentration of thyroid-stimulating hormone (TSH), free triiodothyronine (fT3), and free thyroxine (fT4) were normal, indicating euthyreosis. The hormonal profile is presented in Table 5.

Table 5. Concentrations of chosen hormones in the described patient.

Hormone	Result	Normal Range
TSH	1.97 μIU/ml	0.5–4.8 μIU/ml
fT3	3.26 pg/ml	2.3–4.2 pg/ml
fT4	1.22 ng/dl	0.8–2.3 ng/dl

TSH, thyroid-stimulating hormone; fT3, free triiodothyronine; fT4, free thyroxine.

Ultrasound imaging demonstrated multiple nodules in both thyroid lobes, with the biggest solid nodule with calcification and hypoechoic outline measuring 1.17 × 1.82 cm.

The patient underwent a total strumectomy. Histological sections from the left lobe showed features of papillary thyroid carcinoma. The lesion was 1 cm in diameter with hyaline-fibrous capsule and calcification inside. Histopathology also revealed a multinodular struma of the right lobe. The patient was prescribed levothyroxine 100 μg (1 × 1), and she was qualified for iodine treatment.

In 2018, the patient was admitted to the Department of Gynecological Endocrinology Poznan University of Medical Sciences, Poznan, Poland, due to irregular menstrual cycle.

In the transvaginal ultrasonography, multiple small uterine fibroids were revealed. The endometrial thickness was 7 mm. The right ovary was enlarged by a cystic-solid, unilocular tumor with a size of 97 mm × 69 mm with central blood flow. There were no abnormalities in the left ovary revealed in the ultrasound examination.

Additionally, a hormonal analysis was performed (Table 6).

Table 6. Hormonal analysis.

Hormone or Biochemical Marker	Result	Normal Range
FSH	6.05 mIU/ml	3.5–12.5 mIU/ml
LH	4.91 mIU/ml	1.4–9.6 mIU/ml
Estradiol	40.95 pg/ml	12.4–233 pg/ml
Prolactin	5.6 ng/ml	5–25 ng/ml
Testosterone	0.08 ng/ml	0.06–0.82 ng/ml
DHEA-S	1.53 μmol/l	0.96–6.95 μmol/l
HE4	72.21 pmol/l	<150 pmol/l
CA125	71.4 U/ml	<35 U/ml
ROMA	17.55% (high risk)	<11.4%

FSH, follicle-stimulating hormone; LH, luteinizing hormone; DHEA-S, dehydroepiandrosterone sulfate; HE4, human epididymis protein 4; CA125, cancer antigen 125; ROMA, Risk of Ovarian Malignancy Algorithm.

Due to ultrasound features and elevated Risk of Ovarian Malignancy Algorithm (ROMA) results, the patient was qualified for surgical treatment. Hysterectomy with bilateral salpingo-oophorectomy and retroperitoneal tumor excision were performed. Histopathology results have shown uterine glandular polyp (0.5 cm), serous superficial papillomas and superficial endometriosis in both ovaries, and female adnexal tumor of probable Wolffian origin (FATWO) located in the broad ligament of the uterus. The FATWO was 8 cm in diameter, contained a variable admixture of diffuse, solid, and sieve-like cystic areas filled with blood. Macroscopically, the tumor was soft and uncapsulated. Immunohistochemical examination revealed an interesting profile of this tumor: ER(+), PgR(+), AR(+), GATA3(+), calretinin(+), HBME(+), WT1(+), cytokeratins AE1/AE3(+), CK7(+), CKAE 1/3 (+), melan A(+), α-inhibin(+), CD99 (+), CEA mono(−), PAX8(−), TTF1(−), RCC(−), CK20(−), CD117(−), S100(−), EMA(−), HMB45(−), and Ki67 (5%).

Due to a history of multiglandular diseases: micronodular adrenal hyperplasia, pituitary adenoma, papillary thyroid carcinoma, and ovarian tumor, the patient was referred to genetic testing for mutations associated with multiple gland malignancy. We performed an analysis of 83 genes deletion/duplication (Invitae Multi Cancer Panel) in the whole blood sample. The Invitae Multi-Cancer Panel analyzes 83 genes associated with hereditary cancers across major organ systems, including: breast and gynecologic (breast, ovarian, uterine); gastrointestinal (colorectal, gastric, pancreatic); endocrine (thyroid, paraganglioma/pheochromocytoma, parathyroid, pituitary); genitourinary (renal/urinary tract, prostate); skin (melanoma, basal cell carcinoma); brain/nervous system sarcoma and hematologic (myelodysplastic syndrome/leukemia) cancers: AIP; ALK; APC; ATM; AXIN2; BAP1; BARD1; BLM; BMPR1A; BRCA1; BRCA2; BRIP1; CASR; CDC73; CDH1; CDK4; CDKN1B; CDKN1C; CDKN2A; CEBPA; CHEK2; CTNNA1; DICER1; DIS3L2; EGFR; EPCAM; FH; FLCN; GATA2; GPC3; GREM1; HOXB13; HRAS; KIT; MAX; MEN1; MET; MITF; MLH1; MSH2; MSH3; MSH6; MUTYH; NBN; NF1; NF2; NTHL1; PALB2; PDGFRA; PHOX2B; PMS2; POLD1; POLE; POT1; PRKAR1A; PTCH1; PTEN; RAD50; RAD51C; RAD51D; RB1; RECQL4; RET; RUNX1; SDHA; SDHAF2; SDHB; SDHC; SDHD; SMAD4; SMARCA4; SMARCB1; SMARCE1; STK11; SUFU; TERC; TERT; TMEM127; TP53; TSC1; TSC2; VHL; WRN; and WT1 gene mutations. We received positive results—pathogenic variant, exon 4. c.470T > C (p.Ile157Thr) heterozygous, with low penetrance identified in CHEK2, and a variant, exon 17. c.2484C > T (silent) heterozygous, of uncertain significance identified in KIT. Both mutations are not associated with MEN1 or MEN2 syndromes.

3. Discussion

The c.470T > C (p.Ile157Thr) variant in the CHEK2 gene is classified as a pathogenic variant with low penetrance. The isoleucine residue is moderately conserved, and there is a moderate physicochemical difference between isoleucine and threonine. Experimental studies find that this missense change reduces the binding of CHEK2 protein to Cdc25A, BRCA1, and p53 proteins in vitro and may have a dominant-negative effect on cells. However, it does not affect CHEK2 protein kinase activity [8]. The relationship between these experimental findings and cancer risk is unclear. Patients presenting CHEK2 mutations present a predisposition to certain cancers. The lifetime risk of breast cancer in females with a single pathogenic CHEK2 variant is 25–39% [9]. The risk was found to be more pronounced for lobular type breast tumors [10]. This variant has low penetrance because it does not confer the same level of cancer risk as other CHEK2 pathogenic variants. There is also a reported association between CHEK2 and other cancers, including male breast, kidney, gastric, colorectal, prostate, lung, ovarian, and thyroid cancers. Still, the lifetime risk of these other cancers is unknown [5,11–13]. Biological relatives have a chance of being at risk for CHEK2-related conditions and should consider testing if clinically appropriate.

A variant of uncertain significance, c.2484C > T (silent), was identified in KIT. This sequence change affects codon 828 of the KIT mRNA. It is a "silent" change, meaning that it does not change the encoded amino acid sequence of the KIT protein. This variant has not been reported in individuals with KIT-related diseases. The KIT gene is associated with autosomal dominant piebaldism [14], gastrointestinal stromal tumors (GISTs) [15], and familial mastocytosis [16]. The clinical significance of this result is uncertain. In summary, the available evidence is currently insufficient to determine the role of this variant in disease. Therefore, it has been classified as a Variant of Uncertain Signification.

Genetic testing was performed for our patient, and it was positive for the pathogenic variant c.470T > C (p.Ile157Thr) with low penetrance identified in CHEK2 and a variant c.2484C > T (silent) of uncertain significance identified in KIT. KIT mutation, as a silent one, is not suspected of being a cause of the patient's clinical condition. On the other hand, in the described patient, we identified CHEK2-associated typical tumors, like papillary thyroid cancer.

It was reported that the CHEK2 mutation was seen in 73 of 468 (15.6%) unselected patients with papillary thyroid cancer, compared to 28 of 460 (6.0%) age—and sex-matched controls (OR 3.3; $p < 0.0001$) [17]. On the other hand, it was a truncating mutation (IVS2 + 1G > A, 1100delC or del5395) that was associated with a higher risk of thyroid cancer (OR = 5.7; p = 0.006), than was observed in the missense mutation I157T (OR = 2.8; p = 0.0001) [15], which we identified in the patient. The patient was also diagnosed with micronodular adrenal hyperplasia. This disease's correlation with a mutation in CHEK2 is uncertain. It is known that CHEK2 is a TP53-associated gene [8]. Because of that, they are similarly linked to Li–Fraumeni syndrome and presumably hyperplasia and tumors of adrenal glands [18]. Moreover, the patient was suffering from pituitary microadenoma. CHEK2 mutations have also been found in some brain tumors, such as meningioma or glioma [19].

The patient was also diagnosed with a uterine glandular polyp and FATWO. Endometrial polyps carry a small risk of cancer. The risk of endometrial cancer in women with endometrial polyps is 1.3% [20]. While mutation in CHEK2 has been reported in patients with endometrial cancers [5], and because of that, it is worth considering if CHEK2 mutation affected our patient's uterine polyps. In the literature, we found information indicating that positive associations were seen with the CHEK2 I157T missense variant and ovarian cystadenomas (OR = 1.7; p = 0.005), borderline ovarian cancers (OR = 2.6; p = 0.002), and low-grade invasive cancers (OR = 2.1; p = 0.04) [21]. All these data indicate that CHEK2 variants may predispose to a range of ovarian tumor types of low malignant potential. It is difficult to refer to the association between FATWO and CHEK2-related cancers, because of its rarity.

FATWO is a very rare gynecologic neoplasm of low malignant potential, with fewer than 90 cases being reported worldwide [22]. It is presumed to be derived from the mesonephric (Wolffian) duct remnants in the upper female genital tract [23]. Moreover, some recurrent and metastatic cases have been reported in approximately 11% of cases [24]. Tumor-defining molecular alterations have yet to be discovered. FATWO originates from mesonephric (Wolffian) duct remnants. Tumors are present along the trajectory of the Wolffian duct: from the ovarian hilum, along the mesosalpinx, broad ligament, lateral aspects of the uterus and cervix, to the outer third of the vagina. There are no disease-specific mutations or genetic alterations identified [25]. Many patients remain asymptomatic, and the tumors are discovered incidentally [22]. Sometimes patients can present lower abdominal pain or distention, pelvic mass, abnormal vaginal bleeding, or ascites [23]. In most cases, tumors are not hormonally active. In rare cases, hormonally active tumors can cause endometrial hyperplasia [26]. Most tumors are unilateral, and rarely bilateral. On diagnostic imaging, FATWO are usually well defined, encapsulated, lobulated masses. As in our case, they may demonstrate solid and cystic areas. In an MRI, they present with low signal intensity rim around the tumor on T2 weighted MRI images [27,28]. Although most cases of FATWO are benign, a few aggressive cases occur [24].

Due to the found mutation, screening, and prevention of breast and colorectal cancers was implemented in the case of our patient, due to multiple neoplasia in the past. The patient is only 44 years old, and the lifetime risk of breast cancer in females with a single pathogenic CHEK2 variant is 20–44% [13]. Interestingly, mutations of CHEK2 have been associated with resistance to anthracycline-based chemotherapy in patients with breast cancer [29]. Significant associations of the CHEK2 I157T variant with colorectal cancer susceptibility were found (OR = 1.67, 95% CI = 1.24–2.26, p = 0.0008) [13]. There are limited guidelines for screening and prevention of CHEK2-related breast and colorectal cancers.

According to the correlation between CHEK2 and TP3 genes and the role of CHEK2 (suppressor gene that is involved in DNA repair, cell cycle arrest, or apoptosis in response to DNA damage)—mutations to the CHEK2 gene have been labeled as causes to a wide range of cancers. Thus, micronodular adrenal hyperplasia, pituitary microadenoma, papillary thyroid cancer, uterine, ovarian, and retroperitoneal (FATWO) lesions are possible after-effects of CHEK2 missense mutation identified in the patient.

4. Conclusions

The described case is undoubtedly significant and important for gynecologists, endocrinologists, geneticists, and other practitioners. It is not clearly established whether there is a significant link between a constellation of tumors presented in a described patient and the mutation found in the CHEK2 gene. Further evaluation of the neoplastic risk associated with CHEK2 gene mutation is necessary.

In conclusion, we want to underline that in cases of multiple endocrine neoplasias, thought should be directed at genetic testing and mutations. Additionally, possible predispositions and syndromes should be considered, because they may change how the patient is monitored clinically. Multiple endocrine glands tumors, as in the presented patient, are a serious problem of public health due to multiple hospitalizations and necessary repeated surgical treatments. Moreover, the association between CHEK2 and ovarian cancer can be a serious problem with reproductive health.

Author Contributions: Conceptualization, A.S., A.P. (Agnieszka Podfigurna), and B.M.; Methodology, A.S.; Software, A.P. (Aleksandra Pralat) and W.W.; Validation, A.S., A.P. (Agnieszka Podfigurna), and B.M.; Formal analysis, A.S.; Investigation, A.P. (Agnieszka Podfigurna) and A.S.; Resources, A.P. (Aleksandra Pralat) and W.W.; Data curation, A.P. (Aleksandra Pralat) and W.W.; Writing—original draft preparation, A.P. (Aleksandra Pralat), W.W., and A.S.; Writing—review and editing, A.S., A.P. (Agnieszka Podfigurna), C.W., A.K., and B.M.; Visualization, A.S.; Supervision, B.M.; Project administration, B.M.; Funding acquisition, Y.Y. All authors have read and agreed to the published version of the manuscript.

Funding: This research received no external funding.

Conflicts of Interest: The authors declare no conflict of interest.

References

1. Pizot, C.; Dragomir, M.; Macacu, A.; KoEchlin, A.; Bota, M.; Boyle, P. Global burden of pancreas cancer: Regional disparities in incidence, mortality and survival. *J. Health Inequal.* **2019**, *5*, 96–112. [CrossRef]
2. Carney, J.A. Carney triad: A syndrome featuring paraganglionic, adrenocortical, and possibly other endocrine tumors. *J. Clin. Endocrinol. Metab.* **2009**, *94*, 3656–3662. [CrossRef] [PubMed]
3. Milella, M.; Falcone, I.; Conciatori, F. PTEN: Multiple Functions in Human Malignant Tumors. *Front. Oncol.* **2015**, *5*, 24. [CrossRef] [PubMed]
4. Ciebiera, M.; Baran, A.; Słabuszewska-Jóźwiak, A.; Wojtyła, C.; Jakiel, G. Sertoli-Leydig cell tumour (SLCT)—The case of a 15 cm diameter ovarian tumour with negative markers and absent hormonal symptoms. *J. Health Inequal.* **2019**, *5*, 117–119. [CrossRef]
5. Cybulski, C.; Górski, B.; Huzarski, T.; Masojć, B.; Mierzejewski, M.; Debniak, T.; Teodorczyk, U.; Byrski, T.; Gronwald, J.; Matyjasik, J.; et al. CHEK2 is a multiorgan cancer susceptibility gene. *Am. J. Hum. Genet.* **2004**, *75*, 1131–1135. [CrossRef]
6. Huzarski, T.; Cybulski, C.; Wokolorczyk, D.; Jakubowska, A.; Byrski, T.; Gronwald, J. Survival from breast cancer in patients with $CHEK_2$ mutations. *Breast Cancer Res. Treat.* **2014**, *144*, 397–403. [CrossRef]
7. Berglund, G.; Lidén, A.; Hansson, M.G.; Oberg, K.; Sjöden, P.O.; Nordin, K. Quality of life in patients with multiple endocrine neoplasia type 1 (MEN 1). *Fam. Cancer* **2003**, *2*, 27–33. [CrossRef]
8. Falck, J.; Mailand, N.; Syljuåsen, R.G.; Bartek, J.; Lukas, J. The ATM-Chk2-Cdc25A checkpoint pathway guards against radioresistant DNA synthesis. *Nature* **2001**, *410*, 842–847. [CrossRef]
9. Cybulski, C.; Wokołorczyk, D.; Jakubowska, A.; Huzarski, T.; Byrski, T.; Gronwald, J.; Masojć, B.; Dębniak, T.; Górski, B.; Blecharz, P.; et al. Risk of breast cancer in women with a CHEK2 mutation with and without a family history of breast cancer. *J. Clin. Oncol.* **2011**, *29*, 3747–3752. [CrossRef]
10. Liu, C.; Wang, Y.; Wang, Q.S.; Wang, Y.J. The CHEK2 I157T variant and breast cancer susceptibility: A systematic review and meta-analysis. *J. Clin. Oncol.* **2011**, *29*, 3747–3752. [CrossRef]
11. Wasielewski, M.; den Bakker, M.A.; van den Ouweland, A.; Meijer-van Gelder, M.E.; Portengen, H.; Klijn, J.G.; Meijers-Heijboer, H.; Foekens, J.A.; Schutte, M. CHEK2 1100delC and male breast cancer in the Netherlands. *Breast Cancer Res. Treat.* **2009**, *116*, 397–400. [CrossRef] [PubMed]
12. Xiang, H.P.; Geng, X.P.; Ge, W.W.; Li, H. Meta-analysis of CHEK2 1100delC variant and colorectal cancer susceptibility. *Eur. J. Cancer.* **2011**, *47*, 2546–2551. [CrossRef] [PubMed]
13. Han, F.F.; Guo, C.L.; Liu, L.H. The effect of CHEK2 variant I157T on cancer susceptibility: Evidence from a meta-analysis. *DNA Cell Biol.* **2013**, *32*, 329–335. [CrossRef] [PubMed]
14. Jia, W.X.; Xiao, X.M.; Wu, J.B.; Ma, Y.P.; Ge, Y.P.; Li, Q.; Mao, Q.X.; Li, C.R. A novel missense KIT mutation causing piebaldism in one Chinese family associated with café-au-lait macules and intertriginous freckling. *Ther. Clin. Risk Manag.* **2015**, *11*, 635–638. [PubMed]

15. Ibrahim, A.; Chopra, S. Succinate Dehydrogenase-Deficient Gastrointestinal Stromal Tumors. *Arch. Pathol. Lab. Med.* **2020**, *144*, 655–660. [CrossRef] [PubMed]
16. Coltoff, A.; Mascarenhas, J. Relevant updates in systemic mastocytosis. *Leuk. Res.* **2019**, *81*, 10–18. [CrossRef]
17. Siołek, M.; Cybulski, C.; Gąsior-Perczak, D.; Kowalik, A.; Kozak-Klonowska, B.; Kowalska, A.; Chłopek, M.; Kluźniak, W.; Wokołorczyk, D.; Pałyga, I.; et al. CHEK2 mutations and the risk of papillary thyroid cancer. *Int. J. Cancer* **2015**, *137*, 548–552. [CrossRef]
18. Stratakis, C.A.; Boikos, S.A. Genetics of Adrenal Tumors Associated With Cushing's Syndrome: A New Classification for Bilateral Adrenocortical Hyperplasias. *Nat. Clin. Pract. Endocrinol. Metab.* **2007**, *3*, 748–757. [CrossRef]
19. Rajaraman, P.; Wang, S.S.; Rothman, N.; Brown, M.M.; Black, P.M.; Fine, H.A.; Loeffler, J.S.; Selker, R.G.; Shapiro, W.R.; Chanock, S.J.; et al. Polymorphisms in Apoptosis and Cell Cycle Control Genes and Risk of Brain Tumors in Adults. *Cancer Epidemiol. Biomark. Prev.* **2007**, *16*, 1655–1661. [CrossRef]
20. Wethington, S.L.; Herzog, T.J.; Burke, W.M.; Sun, X.; Lerner, J.P.; Lewin, S.N.; Wright, J.D. Risk and predictors of malignancy in women with endometrial polyps. *Ann. Surg. Oncol.* **2011**, *18*, 3819–3823. [CrossRef]
21. Szymanska-Pasternaka, J.; Szymanska, A.; Medreka, K.; Imyanitovb, E.N.; Cybulska, C.; Gorskia, B.; Magnowskic, P.; Dziubaa, I.; Gugalad, K.; Debniake, B.; et al. CHEK2 variants predispose to benign, borderline and low-grade invasive ovarian tumors. *Gynecol. Oncol.* **2006**, *102*, 429–431. [CrossRef] [PubMed]
22. Liu, L.; Fang, Q.; Xing, Y. Female adnexal tumor of probable wolffian origin arising from mesosalpinx: A case report and review. *J. Obstet. Gynaecol. Res.* **2018**, *44*, 1859–1863. [CrossRef] [PubMed]
23. Shalaby, A.; Shenoy, V. Female Adnexal Tumor of Probable Wolffian Origin: A Review. *Arch. Pathol. Lab. Med.* **2020**, *144*, 24–28. [CrossRef] [PubMed]
24. Syriac, S.; Durie, N.; Kesterson, J.; Lele, S.; Mhawech-Fauceglia, P. Female adnexal tumor of probable wolffian origin (FATWO) with recurrence 3 years postsurgery. *Int. J. Gynecol. Pathol.* **2011**, *30*, 231–235. [CrossRef]
25. Devouassoux-Shisheboran, M.; Silver, S.A.; Tavassoli, F.A. Wolffian adnexal tumor, so-called female adnexal tumor of probable Wolffian origin (FATWO): Immunohistochemical evidence in support of a Wolffian origin. *Hum. Pathol.* **1999**, *30*, 856–863. [CrossRef]
26. Inoue, H.; Kikuchi, Y.; Hori, T.; Nabuchi, K.; Kobayashi, M.; Nagata, I. An Ovarian Tumor of Probable Wolffian Origin With Hormonal Function. *Gynecol. Oncol.* **1995**, *59*, 304–308. [CrossRef]
27. Sakai, M.; Abiko, K.; Matsumura, N.; Kondoh, E.; Yamaguchi, K.; Minamiguchi, S.; Kurata, Y.; Kido, A.; Konishi, I. Two Cases of Wolffian Tumor With Novel Magnetic Resonance Imaging Findings Reflecting Characteristic Pathology. *J. Obstet. Gynaecol. Res.* **2016**, *42*, 1046–1051. [CrossRef]
28. Lucas, E.; Zheng, W.; Zynger, D. Female Adnexal Tumor of Probable Wolffian Origin (FATWO). Available online: http://www.pathologyoutlines.com/topic/ovarytumorfatwo.html (accessed on 9 June 2020).
29. Knappskog, S.; Chrisanthar, R.; Løkkevik, E.; Anker, G.; Østenstad, B.; Lundgren, S.; Risberg, T.; Mjaaland, I.; Leirvaag, B.; Miletic, H.; et al. Low expression levels of ATM may substitute for CHEK2/TP53 mutations predicting resistance towards anthracycline and mitomycin chemotherapy in breast cancer. *Breast Cancer Res.* **2012**, *14*, R47. [CrossRef]

© 2020 by the authors. Licensee MDPI, Basel, Switzerland. This article is an open access article distributed under the terms and conditions of the Creative Commons Attribution (CC BY) license (http://creativecommons.org/licenses/by/4.0/).

MDPI
St. Alban-Anlage 66
4052 Basel
Switzerland
Tel. +41 61 683 77 34
Fax +41 61 302 89 18
www.mdpi.com

International Journal of Environmental Research and Public Health Editorial Office
E-mail: ijerph@mdpi.com
www.mdpi.com/journal/ijerph